Handbook of Zoology

Gastrotricha, Cycloneuralia and Gnathifera

Volume 1:

Nematomorpha, Priapulida, Kinorhyncha, Loricifera

Handbook of Zoology

Founded by Willy Kükenthal
Continued by M. Beier, M. Fischer, J.-G. Helmcke, D. Starck, H. Wermuth

Gastrotricha, Cycloneuralia and Gnathifera

Edited by Andreas Schmidt-Rhaesa

DE GRUYTER

Gastrotricha, Cycloneuralia and Gnathifera

Volume 1: Nematomorpha, Priapulida, Kinorhyncha, Loricifera

Edited by Andreas Schmidt-Rhaesa

DE GRUYTER

Scientific Editor
Andreas Schmidt-Rhaesa
University Hamburg
Martin-Luther-King Platz 3
20146 Hamburg, Germany

ISBN 978-3-11-021938-8
e-ISBN 978-3-11-027253-6
ISSN 2193-4231

Library of Congress Cataloging-in-Publication Data
A CIP catalogue record for this book is available from the Library of Congress.

Bibliografic information published by the Deutsche Nationalbibliothek

The Deutsche Nationalbibliothek lists this publication in the Deutsche Nationalbibliografie; detailed bibliographic data are available in the Internet at http://dnb.d-nb.de.

Typesetting: Apex CoVantage
Printing and Binding: Hubert & Co. GmbH & Co. KG, Göttingen

Printed in Germany
www.degruyter.com

Editors' preface

Gastrotricha, Cycloneuralia and Gnathifera is the title for a series of chapters covering partly familiar animal taxa such as nematodes and rotifers, but to a large extent taxa that are not very familiar to most zoologists and which are rarely in the focus of actual research. At least for some of these taxa, summaries on their morphology, systematics and ecology are lacking or outdated. The chapters in the Handbook of Zoology aim to present a comprehensive overview, which integrates long grown knowledge and recent advances.

Not too long ago, the names mentioned in the title, Cycloneuralia and Gnathifera, were not available and the taxa dealt with would have been mentioned under names such as "Aschelminthes" or "Pseudocoelomates". Although most authors believed that such groups were paraphyletic, alternative hypotheses of relationship were developed particularly during the 1990s. They resulted in the recognition of two groups, Cycloneuralia and Gnathifera. Gastrotricha are often associated with Cycloneuralia, but their relationships are less clear and therefore they are treated here separately.

Gastrotricha, Cycloneuralia and Gnathifera are divided here into three print volumes. The first volume includes a historical and phylogenetic overview, a chapter on fossils and chapters on the cycloneuralian taxa Nematomorpha, Priapulida, Kinorhyncha and Loricifera. The second volume covers only Nematoda, the taxon with the highest species and ecological diversity. The third volume includes the taxa summarized as Gnathifera (Gnathostomulida, Micrognathozoa, Rotifera and Acanthocephala) and the Gastrotricha.

I thank all authors for the energy they put into their chapters. Many thanks also go to the publisher, De Gruyter, for starting the ambitious project of reviving the idea of a Handbook of Zoology, based on the excellent volumes from the 1920s and 1930s.

Andreas Schmidt-Rhaesa

Contents

Contributors

Iben Heiner Bang-Berthelsen
DTU, National Food Institute
Mørkhøj Bygade 19, Bldg. H
2860 Søborg, Denmark

Andreas Maas
University Ulm
Biosystematic Documentation
Helmholtzstraße 20
89081 Ulm, Germany

Reinhardt Møbjerg Kristensen
Natural History Museum of Denmark
University of Copenhagen Universitetsparken 15
2100 Copenhagen, Denmark

Birger Neuhaus
Museum für Naturkunde
Leibniz Institute for Research on Evolution and Biodiversity
Invalidenstraße 43
10115 Berlin, Germany

Andreas Schmidt-Rhaesa
Biozentrum Grindel und Zoologisches Museum
Martin-Luther-King Platz 3
20146 Hamburg

Andreas Schmidt-Rhaesa

1. Gastrotricha, Cycloneuralia and Gnathifera: General History and Phylogeny

Eleven taxa are treated in this volume: Gastrotricha, Nematoda, Nematomorpha, Priapulida, Kinorhyncha, Loricifera, Gnathostomulida, Micrognathozoa, Rotifera, Seisonida and Acanthocephala. The history of their discovery and especially the history of their phylogenetic relationships is a long, winding road.

1.1. The large worms

The large forms in this group, in particular nematodes and acanthocephalans, but also nematomorphs, were known for so long that the origin of their knowledge to humankind is hardly traceable. Endoparasitic nematodes and acanthocephalans are of particular medical interest, making their knowledge important for humans. Ascarid and strongylid nematodes (as endoparasites), the guinea worm and probably also nematomorphs, have already been mentioned in ancient times (Enigk 1986). Acanhocephalans were mentioned with certainty as parasites of the eel by Redi (1684) and Leeuwenhoek (1722), as mentioned by Lühe (1904). In Linnaeus´s 10th edition of his Systema Naturae (1758), we find among the "Vermes intestina" the genera *Ascaris*, *Trichocephalus*, *Filaria*, *Strongylus* and *Gordius*, which represent nematodes, acanthocephalans and nematomorphs. Under each of these taxa is listed a number of species, for example 78 species of *Ascaris*.

The intestinal worms did not only include nematodes, acanthocephalans and nematomorphs, but all other wormlike endoparasites such as trematodes and cestodes. Rudolphi (1808) introduced the name Entozoa for this group, although he later (Rudolphi 1819), as did other authors such as Von Baer (1826), stated that Entozoa is no natural unit and should not be treated as a class. In the following, two subgroups were distinguished, one including representatives with a more or less flat cross section and one with representatives having a more round cross section. Such a distinction has early roots (see Rauther 1928), but was formalized by Vogt (1851), who summarized trematodes, cestodes and nemerteans as Platyelmia, and nematodes (including nematomorphs) and acanthocephalans as Nematelmia. This last name was changed by Gegenbauer (1859) to Nemathelminthes.

The relationship between nematodes and nematomorphs is hard to clarify. Both correspond in several anatomical features, but differ in others. While nematodes were originally known only as endoparasites, nematomorphs share an endoparasitic and a free-living phase. Therefore authors meander between regarding nematomorphs as a member of nematodes or separating these taxa. This is complicated by the mermithid nematodes, which have a life cycle comparable to nematomorphs. Von Siebold (1843), for example, summarized both mermithids and nematomorphs as Gordiacea and regarded this taxon as an "order of helminths", separate from nematodes. A proper distinction between mermithids and nematomorphs and a clear separation of nematomorphs as a taxon of the same rank as nematodes was first made by Vejdovsky (1896) after the discovery of the marine nematomorph genus *Nectonema*.

The macroscopic priapulids, at least the species *Priapulus caudatus*, were also known for some time, although their roots in human knowledge are not easy to trace. Originally priapulids were considered to be holothurians. Linnaeus (1758) included *Priapulus caudatus* (as *Priapus caudatus*) with holothurian species under "Vermes Mollusca". It was De Lamarck (1816) who separated *Priapulus caudatus* from holothurians. Because the genus name *Priapus* was already in use for some sea anemones, he chose the new genus name, *Priapulus*. De Quatrefages (1847) created the taxon Gephyrea, a group uniting priapulids with sipunculids and echiurans.

1.2. The discovery of microscopically small animals

The invention of microscopes opened the window to the microscopically small fauna and it is likely that rotifers and gastrotrichs were among the first animals observed with early microscopes. The first name applied to all microscopically small animals was Infusoria. Essentially, these were protozoans, rotifers and gastrotrichs. Rotifers were treated as their own group, separate from other infusorians, by Ehrenberg (1838). The few freshwater gastrotrichs known at that time were considered as being rotifers. Ehrenberg first (1830) united gastrotrich species in a class among rotifers, but later (1838) included other rotifers in this class (see also Remane 1936). Gastrotrichs were subsequently separated from rotifers, for example by Mečnikow (1865), who introduced the name Gastrotricha. Mečnikow (1865) still agreed on a close relationship between rotifers and gastrotrichs, a relationship for which later Zelinka (1889) suggested the name Trochelminthes, because he derived them from a trochophore larva.

The exploration of the microscopic fauna from the marine environment started later. This may be due to the fact that many representatives of nemathelminth and gnathiferan taxa live within the sediment. Before techniques for the extraction of meiofauna were developed, this environment was considered to be poor in species diversity and abundance. A few representatives were known quite early, such as the marine gastrotrichs *Turbanella hyalina* (Schultze 1853) and *Hemidasys agaso* (Claparède 1867) or the kinorhynch *Echinoderes dujardini* (Dujardin 1851, Claparède 1863). Additionally, there were remarkable early investigations of free-living, marine nematodes, such as by Bastian (1865), Bütschli (1873, 1874) and De Man (1886).

A wave of discoveries began after Adolf Remane took a closer look at marine sediments. His original intention to characterize a species-poor environment by describing its entire abiotic and biotic components (see Remane 1952) failed, but failed fortunately, because Remane had opened the door to an extremely rich fauna. A diversity of marine meiofaunal species was discovered, including representatives of gastrotrichs (originally termed "aberrant" gastrotrichs, then later Macrodasyida, see Remane 1926), nematodes, rotifers and kinorhynchs (e.g. Higgins 1971). Among the animals observed in the early phase of meiofauna research was also a representative of gnathostomulids. Although being found by Remane in 1928, its formal description was delayed in consequence of the war, until finally described as *Gnathostomula paradoxa* by Ax (1956). First considered to be a new order of the "Turbellaria" (the paraphyletic free-living flatworms) (Ax 1956), it was later considered as a class separate from "Turbellaria" (Ax 1966) and finally as a phylum by Riedl (1969).

In marine sediments, it is essential to be able to temporally attach to hard substrates such as sand grains. Many meiofaunal organisms have developed adhesive glands. The ability to hold fast makes extraction difficult and this accounts especially for the taxon Loricifera. Only after the discovery of special extraction techniques (the "freshwater-shock"), were loriciferans first observed during the 70s and 80s and then formally described in 1983 (Kristensen 1983). Finally, a further microscopically small taxon awaited discovery. In 2000, Kristensen & Funch described *Limnognathia maerski* (Micrognathozoa) from a cold freshwater spring on Disko Island, Greenland.

1.3. Changing relationships

Any attempt to present the various relationships of the taxa treated in this volume in a complete way will end up in a confusing list, if it is in any way possible to make such a list complete. Almost for each taxon, a number of other taxa have been discussed as potentially being related. Therefore, the aim here is to outline the most important changes in hypotheses about relationships. Rauther (1928) and Hyman (1951) are sources for more detailed accounts up to their time period.

As has been mentioned above, the taxon name Nemathelminthes was introduced by Gegenbauer (1859), following older ideas to unite nematodes (including horsehair worms) and acanthocephalans. The name Nemathelminthes was nevertheless not used unequivocally. A comparison of Gegenbauer (1859) and Schneider (1866) may illustrate this. For Gegenbauer (1859), Nemathelminthes are one of four classes of the worms: Platyelminthes, Nemathelminthes, Oestelminthes (*Sagitta*, i.e. the chaetognaths) and Annulata (including besides the annelids the Gephyrea, among which priapulids are not mentioned). Nemathelminthes are further divided into the orders Acanthocephala and Nematoidea, which included nematodes ("Nematoden", not latinized) and Gordiacei (*Mermis* and *Gordius*). Schneider (1866) distinguishes among the Nemathelminthes two unnamed groups, one including the Gephyrea and Acanthocephala, the other including the Nematoidea, *Sagitta*, the Chaetopoda (= Annelida) and the imperfectly known *Rhamphogordius*.

Nevertheless, nematodes, nematomorphs and acanthocephalans remained the core taxa of the Nemathelminthes. In textbooks of the end of the 19th century, for example by Claus (1897), Nemathelminthes still include nematodes, acanthocephalans (and in this case chaetognaths), but this changes around the turn of the century, when newly discovered or established taxa were associated. In a new edition of the textbook, now edited by Claus & Grobben (1905), the name Nemathelminthes is replaced by Coelhelminthes, now uniting the taxa Rotifera, Gastrotricha, Kinorhyncha, Nematoda, Nematomorpha and Acanthocephala. The name Coelhelminthes proved to be an unfortunate one, because it had been used in another context before and was replaced by Aschelminthes (Grobben 1908). Both names, Nemathelminthes and Aschelminthes, have since then been used almost synonymously, although their exact content of taxa varies slightly among authors (see e.g. Tab. 1).

Whereas nematodes, nematomorphs and acanthocephalans have always been the core of Nemathelminthes/ Aschelminthes, the other taxa included by Claus & Grobben (1905) had more changing histories. Kinorhynchs, for example, had been associated and compared with acanthocephalans, nematodes, gastrotrichs, rotifers, nematomorphs, arthropods and annelids (see Reinhard 1887 and Schepotieff 1907 for overviews). None of these hypotheses was named, with the exception of Bütschli (1876), who introduced the name Nematorhynchia to express a closer relationship between gastrotrichs and kinorhynchs.

Gastrotrichs and rotifers, both discovered as "infusorians", remained in close association for some time. In contrast to nematodes, nematomorphs, acanthocephalans, kinorhynchs and priapulids, certain parts of their body remain ciliated. This was particularly important for hypotheses deriving gastrotrichs and rotifers from ciliated larvae such as the trochophore, which has for example led Zelinka (1889) to give the name Trochelminthes to a taxon including both groups. Nevertheless, several other taxa have been discussed as potential relatives of gastrotrichs, such as free-living flatworms ("turbellarians"), nematodes and kinorhynchs. For example, Hammarsten (1915) and Balzer (1930) suggested a relationship between

Tab. 1. Overview on taxon names and their contents according to particular authors. Nem = Nematoda, Nmo = Nematomorpha, Aca = Acanthocephala, Rot = Rotifera, Gas = Gastrotricha, Kin = Kinorhyncha, Pri = Priapulida, Gna = Gnathostomulida, Lor = Loricifera.

Nem	Nmo	Aca	Rot	Gas	Kin	Pri	Gna	Lor	Name	Author
■	■	■							**Vermes intestina / Entozoa**	Various authors
			■	■					**Infusoria** (+ protozoans)	Various authors
						■			**Gephyrea** (+ Sipunculida, Echiurida)	De Quatrefages 1847
■	■	■							**Nematelmia / Nemathelminthes**	Vogt 1851, Gegenbauer 1859
■	■	■				■			**Nemathelminthes** (+ Chaetognatha, Annelida)	Schneider 1866
				■	■				**Nematorhynchia**	Bütschli 1876
			■	■					**Trochelminthes**	Zelinka 1889
■	■	■	■	■	■				**Aschelminthes** (+ Chaetognatha)	Grobben 1908
■	■		■	■	■	■			**Aschelminthes**	Hyman 1951
■	■			■	■	■		■	**Nemathelminthes**	Ahlrichs 1995, Ehlers et al. 1996
■	■			■	■	■		■	**Cycloneuralia**	Nielsen 1995
■	■			■	■	■		■	**Nemathelminthes s.str.**	Neuhaus et al. 1996
■	■				■	■		■	**Cycloneuralia**	Ahlrichs 1995
■	■				■	■		■	**Introverta**	Nielsen 1995
■	■								**Nematoidea**	Ahlrichs 1995
■	■								**Nematoida**	Schmidt-Rhaesa 1996
■	■								**Nematozoa**	Zrzavý et al. 1998
	■				■	■			**Cephalorhyncha**	Malakhov 1980
	■				■	■		■	**Cephalorhyncha**	Malakhov & Adrianov 1995
					■	■		■	**Scalidphora**	Lemburg 1995
					■	■		■	**Cephalorhyncha**	Nielsen 1995
					■	■		■	**Scalidorhyncha**	Cavalier-Smith 1998
						■		■	**Priapozoa**	Cavalier-Smith 1998
						■		■	**Vinctiplicata**	Lemburg 1999
		■	■				■		**Gnathifera**	Ahlrichs 1995 ???
		■	■						**Syndermata**	Ahlrichs 1995 ???
		■	■						**Trochata**	Cavalier-Smith 1998
				■			■		**Monokonta**	Cavalier-Smith 1998
				■			■		**Neotrichozoa**	Zrzavý et al. 1998

rotifers, gastrotrichs, kinorhynchs and priapulids. From the present day perspective, an interesting fact concerning rotifer relationships are early comparisons between rotifers and acanthocephalans, e.g. by Von Haffner (1950).

As has been mentioned above, priapulids were long included into the Gephyrea. Although this grouping persisted in the literature for some time, doubts were raised every now and then as to whether this is a natural group. Convincing alternative concepts, however, were not forthcoming. Finally Hammarsten (1915) and Baltzer (1930) clearly expressed doubts that priapulids share any relationship with sipunculids and echiurans and brought priapulids into a closer relationship with nemathelminth/aschelminth taxa, in particular with rotifers and gastrotrichs.

Although Acanthocephalans almost always belonged to the "core Nemathelminthes", Hyman (1951) finds reasons to exclude them from her Aschelminthes (which then included Rotifera, Gastrotricha, Kinorhyncha, Priapulida, Nematoda and Nematomorpha).

Priapulids and kinorhynchs (as well as the later discovered loriciferans) are characterized by an eversible head/neck region, the introvert. Some authors, for example Bütschli 1876) pointed out that a similar region is also present in nematomorph larvae. Malakhov (1980) proposed the name Cephalorhyncha for a taxon including Nematomorpha, Priapulida and Kinorhyncha (see also Malakhov & Adrianov 1995, including Loricifera, and see Schmidt-Rhaesa 1998 for a critical review of this hypothesis).

The situation in the second half of the 20th century was as follows. Two names, Nemathelminthes and Aschelminthes were used more or less synonymously. The content of the group varied according to the author. Perhaps one may say that the usage of names has regional preferences. In the USA, the name Aschelminthes was commonly used, following Hyman´s influential series "The Invertebrates". In Germany, the name Nemathelminthes remained in use (see e.g. Gruner 1984 in a prominent German textbook). Besides preferring the name Aschelminthes, Hyman used a division of animals into acoelomate, pseudocoelomate and coelomate animals. This was not meant as a phylogenetic division, but "pseudocoelomates" was consequently used several times synonymously with Aschelminthes or Nemathelminthes.

1.4. From typological classification to phylogenetic systematics

Willi Hennig started with his publications on phylogenetic systematics (e.g. 1950, 1966); a new era in systematics, offering a powerful scientific tool for phylogenetic analyses. Although it took a while before his methodology was accepted, it formed the basis for an enormous expansion of systematic methodology, ranging from computer-aided cladistic analyses to the diversity of analytical methods available today. While the traditional classification searched for arguments to unite groups of animals into a common taxon of a certain hierarchical level, phylogenetic systematics systematically searched for sister taxa and the common ancestor they originated from.

The Nemathelminths/Aschelminthes have always been a major problem in systematics. Few analyses have suggested them as being probably monophyletic. Schram (1991) regarded Aschelminthes to be monophyletic, with the character "eutely" (constant number of somatic cells) as an autapomorphy, but later Schram & Ellis (1994) regarded eutely as not a good character to support monophyly of Aschelminthes. In fact, eutely has been shown only in some nematodes and rotifers and appears to be not present in a number of other "aschelminth" taxa. Lorenzen (1985), in the first phylogenetic analysis dedicated only to "pseudocoelomate" taxa, drew one common root and assigned some plesiomorphic characters to such a common ancestor. He nevertheless concluded that, as no autapomorphy for this common ancestor could be found, a monophyletic Pseudocoelomata could not be stated (see also Lorenzen 1996). Wallace et al. (1995, 1996) regarded characters such as the absence of statocysts and agametic reproduction, as well as the presence of special jaws, as potential characteristics of aschelminths, but each of these characters is highly debatable.

Several authors stated that Nemathelminthes/Aschelminthes/pseudocoelomates was no natural (monophyletic) taxon. Brusca & Brusca (1990), for example, state that "perhaps no other group of phyla is such a phylogenetic mystery as the pseudocoelomates" (p. 888). Ruppert (1991) concluded that any animal that does not clearly show an acoelomate or coelomate condition of body organization is regarded to be an aschelminth. Alternative concepts for the position of the taxa in question, however, were rarely made. Ax (1985, 1989), for example, excluded all such taxa except for gnathostomulids from his suggestion of basal bilaterian branching patterns. He regarded Gnathostomulida to be the sister group of Plathelminthes, together named Plathelminthomorpha. These were either the sister group of all remaining bilaterians (Ax 1985) or the sister group of Euspiralia (Ax 1989).

Lorenzen (1985) already indicated that there are two main lines, one leading to the Rotifera + Acanthocephala and one to the Nematoda + Nematomorpha + Gastrotricha + Priapulida + Kinorhyncha. This split became manifest in several following analyses (see, e.g., Nielsen 1995, 2001, Nielsen et al. 1996, Zrzavý et al. 1998, Sørensen et al. 2000).

The six taxa Gastrotricha, Nematoda, Nematomorpha, Priapulida, Kinorhyncha and Loricifera were united using the traditional name Nemathelminthes (Ahlrichs 1995, Ax 2003), Nemathelminthes s.str. (Neuhaus et al. 1996) or Cycloneuralia (Nielsen 1995). Several working groups came to very similar hypotheses about phylogenetic relationships within the Nemathelminthes. According to Ahlrichs (1995), Nielsen (1995, see also 2001), Ehlers et al. (1996) and Wallace et al. (1995, 1996), Gastrotricha is the first taxon to branch off. The remaining five taxa are named Cycloneuralia (Ahlrichs 1995) or Introverta (Nielsen 1995). Nematoda and Nematomorpha are hypothesized by all authors as sister taxa. The name Nematoidea used by Ahlrichs (1995) and Ehlers et al. (1996) was changed by Schmidt-Rhaesa (1996) to Nematoida. The remaining three taxa were named Scalidophora (Lemburg 1995) or Cephalorhyncha (Nielsen 1995; the same name had been used before, but including Nematomorpha, by Malakhov 1980). As discussed by Lemburg (1999), among the three taxa within Scalidophora there are possible arguments for each of the three possible sister-group relationships. A taxon Priapulida + Loricifera has been named Priapozoa (Cavalier-Smith 1998), but although this name has priority, the name Vinctiplicata (Lemburg 1999) is in more regular use. Some further names such as Nematozoa (Nematoda + Nematomorpha; Zrzavý et al. 1998) and Scalidorhyncha (Priapulida + Loricifera + Kinorhyncha; Cavalier-Smith 1998) ignore names introduced earlier and have to be regarded as nomina nuda. Different hypotheses of the internal relationships among Nemathelminthes were published by Lorenzen (1985): ((Priapulida + Kinorhyncha) (Gastrotricha (Nematoda + Nematomorpha))), Neuhaus et al. (1996): ((Gastrotricha + Nematoda) (unresolved Scalidophora); position of Nematomorpha unclear) and Lorenzen (1996): (unresolved trichotomy of (Gastrotricha + Nematoda), Nematomorpha and (Kinorhyncha (Priapulida + Loricifera)).

The second main line, a close relationship between rotifers and acanthocephalans, is supported, despite the vast morphological and ecological differences, by a unique structure of the syncytial epidermis and is found in most phylogenetic analyses (e.g. Wallace et al. 1995, 1996, Neuhaus et al. 1996, Nielsen et al. 1996). This common taxon was named Syndermata by Ahlrichs (1995). He also separated *Seison* (Seisonida) from the remaining rotifers (for which he kept the name Rotifera, other authors use the name Eurotifera) and assumed a sister-group relationship to Acanthocephala. Ahlrichs (1995) then hypothesized a sister-group relationship of Syndermata and Gnathostomulida, together named Gnathifera. The reason for this association was a comparable ultrastructure of the jaw-like hard elements in the anterior intestinal tract, a discovery that was published in parallel by Rieger & Tyler (1995). This phylogenetic hypothesis received corroboration by the discovery of a new taxon (preliminary mention by Kristensen 1995, formally described and named as Micrognathozoa (only species: *Limnognathia maerski*) by Kristensen & Funch 2000), which perfectly fitted into the earlier proposed system as a sister group of Syndermata (Ahlrichs 1997, Kristensen & Funch 2000, Sørensen et al. 2000).

There are several suggestions concerning the internal relationships of Syndermata, based on both morphological and molecular data (see Garey et al. 1998 for review). The four subtaxa Bdelloida, Monogononta, Seisonida and Acanthocephala each appear to be monophyletic. Traditionally, bdelloids, monogononts and seisonids have been named Rotifera (or Rotatoria, see Ricci 1983), but there are hypotheses for a sister group relationship of Seisonida + Acanthocephala (Ahlrichs 1995, 1997, Herlyn et al. 2003), Bdelloida + Acanthocephala (Winnepenninckx et al. 1995, Garey et al. 1996, Witek et al. 2008) and for (Seisonida (Acanthocephala (Monogononta + Bdelloida))) (Mark Welch 2000, 2005). Sørensen & Giribet (2006) receive support for a monophyletic (Seisonida + Bdelloida + Acanthocephala) as the sister group to Monogononta. As such contrasting relationships must influence the chapter division in this volume, we have decided to present Rotifera (as Bdelloida + Monogononta), Seisonida and Acanthocephala in separate chapters.

For the position of Nemathelminthes (or at least Cycloneuralia) and Gnathifera within the Bilateria, few suggestions have been made. Older views that the Nemathelminthes constitute miniaturized spiralians (e.g. Remane 1963) are no longer advocated. Nemathelminthes appear to belong to the Protostomia, but are not included in the Spiralia, because none of its subtaxa has a spiral cleavage. Therefore, Nemathelminthes are hypothesized as the sister group of Spiralia, together constituting the Protostomia (Ahlrichs 1995, Ehlers et al. 1996, Nielsen 1995, 2001). Gnathifera appear to belong into the Spiralia because the subtaxa show spiral cleavage (Gnathostomulida) or at least traces of it (Rotifera). As Ax (1985, 1989) regarded the Gnathostomulida to be the sister group of Plathelminthes, Ahlrichs (1995) extended this hypothesis to regard Gnathifera and Plathelminthes as sister taxa. However, Jenner (2004) concluded in an extensive review that the sister group of Plathelminthes cannot be unambiguously established. This remains the current status and Gnathifera have to be regarded as one subtaxon of Spiralia with an uncertain exact position.

1.5. The contribution of molecular studies

Since the late 1980s, phylogenetic analyses have been supplemented by molecular data coming from DNA comparisons; first from single (18S ribosomal DNA gene) and later from multiple genes. The initial study on metazoan relationships by Field et al. (1988) did not include any of the taxa treated in this volume. The inclusion of "strange" animals into the dataset was one important motivation, but it took until 1995 before a considerable selection of nemathelminth and gnathiferan taxa was included into molecular analyses. Besides Aboulheif et al. (1998; four nematodes, one priapulid, one acanthocephalan), Winnepenninckx et al. (1995; representatives of nematodes, nematomorphs, priapulids, gastrotrichs, rotifers and acanthocephalans) made a targeted approach to find the phylogenetic position of "aschelminth" taxa. The results were somewhat confusing, but become clearer from the recent perspective. The first nematode species, from which genes were sequenced, were more or less derived species and their genes showed evidence for a high number of substitutions. This caused analytical problems (the long-branch problems), often resulting in a basal position of nematodes (see Winnepenninckx et al. 1995). The artificial effects caused by fast evolving ("long branch") taxa were soon recognized and more emphasis on using slower evolving sequences helped to reduce such effects. When using sequences from less derived nematode species, nematodes do not come out in a basal position among Bilateria, but cluster with sequences from other cycloneuralian taxa (Aguinaldo et al. 1997).

In general, molecular analyses agree with morphological data in showing two clusters of taxa, which correspond to the Cycloneuralia and the Gnathifera. The most striking result of molecular analyses was the clustering of nematode, nematomorph, priapulid and kinorhynch sequences with those from arthropods, which led Aguinaldo et al. (1997) to propose that all moulting animals (Arthropoda + Cycloneuralia) are monophyletic and should be named Ecdysozoa. From the molecular perspective, Ecdysozoa is a relatively stable taxon through many analyses, using different single or multiple genes (e.g. Giribet et al. 2000, Garey 2001, 2003, Peterson & Eernisse 2001, Balavoine et al. 2002, Giribet 2003, Glenner et al. 2004, Mallatt et al. 2004, Petrov & Vladychenskaya 2005, Philippe et al. 2005, Mallatt & Giribet 2006, Dunn et al. 2008). Those few multigene analyses that do not support Ecdysozoa, but usually reveal a basal position of nematodes among Bilateria (e.g. Wolf et al. 2004) are likely to suffer from long-branch effects, because they again use *Caenorhabditis elegans* instead of slow evolving nematode species or do not correct for high rates of substitution (Dopazo & Dopazo 2005, Zheng et al. 2007).

In molecular analyses, the five cycloneuralian taxa appear not to form a monophyletic taxon. Instead, Nematoida (Nematoda + Nematomorpha) often appear as a sister taxon of Arthropoda (Aguinaldo et al. 1997, Garey 2001, 2003, Mallatt et al. 2004, Petrov & Vladychenskaya 2005, Mallatt & Giribet 2006), with Kinorhyncha + Priapulida being the sister group of Nematoida + Arthropoda. Telford et al. (2008) found priapulids as the sister taxon of arthropods. Sometimes, nematodes or nematodes + nematomorphs cluster with tardigrade sequences (Garey 2001, Roeding et al. 2007, Dunn et al. 2008, Sørensen et al. 2008, Hejnol et al. 2009). Loricifera were included into molecular analyses late; their inclusion did not yield stable results (Park et al. 2006) or offered an unconventional clustering as the sister group to Nematomorpha (Sørensen et al. 2008). Due to these contrasting results, the relationships of cycloneuralian taxa to arthropod taxa (including tardigrades and onychophorans) cannot be regarded as convincingly solved.

No molecular analysis has so far resulted in monophyletic Gnathifera, but sequences of rotifers, acanthocephalans, gnathostomulids and of *Limnognathia maerski* usually cluster with sequences from spiralian taxa (e.g. Giribet et al. 2004, Glenner et al. 2004, Dunn et al. 2008, Witek et al. 2008, Paps et al. 2009). The sequences from several rotifers, acanthocephalans and especially from gnathostomulids are fast evolving, thereby being probably responsible for unreliable phylogenetic positions, at least among spiralians (see e.g. Littlewood et al. 1998). The molecular analyses dealing with the internal relationships among Syndermata were included in the section above (From typological classification to phylogenetic systematics).

A considerable problem is posed by the gastrotrichs. Their sequences constantly cluster with spiralian sequences, in particular with those from flatworms (e.g. Winnepenninckx et al. 1995, Giribet et al. 2000, Dunn et al. 2008), although the support values are usually not very good (see also Schmidt-Rhaesa 2002). In molecular analyses, gastrotrichs are never associated with Ecdysozoa or Cycloneuralia. From a morphological perspective, there is a comparable structure of the cuticle (two layers in gastrotrichs, an additional third layer in cycloneuralians), of the muscular sucking pharynx with tridadiate lumen and in the cleavage (Teuchert 1968, Malakhov 1994, Schmidt-Rhaesa 2002), which led to the assumption of a sister-group relationship between Gastrotricha and Cycloneuralia. There are, however, differences, too. The nervous system of gastrotrichs shows several differences compared to cycloneuralians. The brain is composed of a dorsal commissure with only a delicate subpharyngeal commissure in gastrotrichs; neuronal cell somata are present lateral of the commissure. This architecture is in contrast to the cycloneuralian brain (Rothe & Schmidt-Rhaesa 2009, Rothe et al. 2011). Additionally, orthogonal patterns are present in the trunk nervous system of cycloneuralian taxa (e.g. Rothe & Schmidt-Rhaesa 2010 for priapulids), but they are absent in gastrotrichs. Due to this considerable conflict among data, the position of Gastrotricha has to be regarded as unresolved at present.

1.6. Outlook

Brusca & Brusca´s (1990) statement that the taxa presented in this volume present a phylogenetic mystery has fortunately changed, and today we have at least well-supported ideas. It turned out that a group (originally) named Nemathelminthes or Aschelminthes or pseudocoelomates is indeed a polyphyletic assemblage, consisting of at least two groups of animals, the Cycloneuralia and the Gnathifera. There are hypotheses for the internal relationships within these groups, but these are sometimes conflicting. In particular, the relationship of Cycloneuralia to the arthropods includes some open questions, as does the position of Gnathifera among Spiralia. A great challenge is the position of Gastrotricha, for which the greatest conflicts are present.

The discoveries of Loricifera and Micrognathozoa show that important discoveries can still be made. New technologies and methods, as well as analytical tools and the rapid advances in the analysis of DNA, especially the sequencing of entire genomes, will add much more data. These will certainly increase the number of conflicting hypotheses, but there is hope that careful evaluation of the available information will help to raise our understanding of nemathelminth and gnathiferan phylogeny and evolution to a higher level.

Literature

Abouheif, E., Zardoya, R. & Meyer, A. (1998): Limitations of metazoan 18S rRNA sequence data: implications for reconstructing a phylogeny of the animal kingdom and inferring the reality of the Cambrian explosion. *J. Mol. Evol.* 47: 394–405.

Aguinaldo, A. M. A., Turbeville, J. M., Linford, L. S., Rivera, M. C., Garey, J. R., Raff, R. A. & Lake, J.A. (1997): Evidence for a clade of nematodes, arthropods and other moulting animals. *Nature* 387: 489–493.

Ahlrichs, W. (1995): *Ultrastruktur und Phylogenie von* Seison nebaliae *(Grube 1859) und* Seison annulatus *(Claus 1876). Hypothesen zu phylogenetischen Verwandtschaftsverhältnissen innerhalb der Bilateria.* Cuvillier Verlag, Göttingen.

Ahlrichs, W. H. (1997): Epidermal ultrastructure of *Seison nebaliae* and *Seison annulatus*, and a comparison of epidermal structures within the Gnathifera. *Zoomorphology* 117: 41–48.

Ax, P. (1956): Die Gnathostomulida, eine rätselhafte Wurmgruppe aus dem Meeressand. – *Abh. Akad. Wiss. Lit. Math. Naturwiss. Kl.* 8: 535–562.

Ax, P. (1966): Eine neue Tierklasse aus dem Litoral des Meeres – Gnathostomulida. *Umschau in Wissenschaft und Technik* 1/1966: 17–23.

Ax, P. (1985): The position of the Gnathostomulida and Platyhelminthes in the phylogenetic system of the Bilateria. In: Conway Morris, S., George, J. D., Gibson, R. & Platt, H. M. (eds) *The Origin and Relationships of Lower Invertebrates*, pp. 168–180. Oxford University Press, Oxford.

Ax, P. (1989): Basic phylogenetic systematization of the Metazoa. In: Fernholm, B., Bremer, K. & Jörnvall, H. (eds) *The Hierarchy of Life*, pp. 229–245. Elsevier, Amsterdam.

Ax, P. (2003): *Multicellular Animals*, Vol. 3. Springer, Berlin.

Balavoine, G., de Rosa, R. & Adoutte, A. (2002): Hox clusters and bilaterian phylogeny. *Mol. Phyl. Evol.* 24: 366–373.

Baltzer, F. (1930): Priapulida. In: Krumbach, T. (ed.) *Handbuch der Zoologie, Vol.2, 2nd half: Vermes Polymera, Priapulida, Sipunculida, Echiurida*, pp. 1–14. Walter de Gruyter, Berlin.

Bastian, H. C. (1865): Monograph on the Anguillulidae, or free nematoids, marine, land, and freshwater; with descriptions of 100 new species. *Trans. Linn. Soc. Lond.* 25: 73–184.

Brusca, R. C. & Brusca, G.J. (1990): *Invertebrates*. Sinauer, Sunderland.

Bütschli, O. (1873): Beiträge zur Kenntnis der freilebenden Nematoden. *Nova Acta Acad. Nat. Curios.* 36: 1–124.

Bütschli, O. (1874): Zur Kenntnis der freilebenden Nematoden, insbesondere der des Kieler Hafens. *Abh. Senckenberg. Naturfosch. Ges.* 9: 236–292.

Bütschli, O. (1876): Untersuchungen über freilebende Nematoden und die Gattung *Chaetonotus*. *Z. Wiss. Zool.* 26: 363–413.

Cavalier-Smith, T. (1998): A revised six-kingdom system of life. *Biological Reviews* 73: 203–266.

Claparède, M.E. (1863): *Beobachtungen über Anatomie und Entwicklungsgeschichte wirbelloser Thiere an der Küste von Normandie angestellt.* Wilhelm Engelmann, Leipzig.

Claparède, M.E. (1867): Type d´un nouveau genre de gastrotriches. *Ann. Sci. Nat. Ser.* 5: 16–23.

Claus, C. (1897): *Lehrbuch der Zoologie, Sixth Edition*. N.G. Elwert´sche Verlagsbuchhandlung, Marburg.

Claus, C. & Grobben, K. (1905): *Lehrbuch der Zoologie. 7th, newly organized edition*. N.G. Elwert´sche Verlagsbuchhandlung, Marburg.

Cuvier, G. (1798): *Tableau Élémentire de l´histoire Naturelle des Animaux*. Paris.

De Lamarck, J.B. (1816): *Histoire naturelle des animaux sans vertebrès presentant les characteres generaux et particuliers des ces animaux, leur distribution, leurs classes, leurs familles, leurs genres,* et al. *citation des principales espeches qui s´y rapportent.* Verdiere, Paris.

De Man, J. G. (1886): *Anatomische Untersuchungen über freilebende Nordsee-Nematoden.* Paul Frohberg, Leipzig.

De Quatrefages, M.A. (1847): Mémoire sur l´Echiure de Gaertner (*Echiurus gaertnerii* Nob.). *Ann. Sci. Nat. Ser. 3, Zool.* 7: 307–343.

Dopazo, H. & Dopazo, J. (2005): Genome-scale evidence of the nematode-arthropod clade. *Genome Biol.* 6: R41.

Dujardin, F. (1851): Sur un petit animal marin, l´echinodère, formant un type intermédiare entre les crustacés et les vers. *Ann. Sci. Nat. Ser. Zool.* 15: 158–160.

Dunn, C. W., Hejnol, A., Matus, D. Q., Pang, K., Browne, W. E., Smith, S. A., Seaver, E., Rouse, G. W., Obst, M., Edgecombe. G. D., Sørensen, M. V., Haddock, S. H. D., Schmidt-Rhaesa, A., Okusu, A., Kristensen, R., Wheeler, W. C., Martindale, M. Q. & Giribet, G. (2008): Broad phylogenomic sampling improves resolution of the animal tree of life. *Nature* 452: 754–749.

Ehlers, U., Ahlrichs, W., Lemburg, C. & Schmidt-Rhaesa, A. (1996): Phylogenetic systematization of the Nemathelminthes (Aschelminthes). *Verh. Dtsch. Zool. Ges.* 89.1: 8.

Ehrenberg, C. J. (1830): *Organisation, Systematik und geographisches Verhältnis der Infusionsthierchen.* F. Dümmler, Berlin.

Ehrenberg, C. J. (1838): *Die Infusionsthierchen als vollkommene Organismen.* Leipzig.

Enigk, K. (1986): *Geschichte der Helminthologie im deutschaprachigen Raum.* Gustav Fischer Verlag, Stuttgart.

Field, K. G., Olsen, G. J., Lane, D. J., Giovannoni, J., Ghiselin, M. T., Raff, E. C., Pace, N. R. & Raff, R. A. (1988): Molecular phylogeny of the animal kingdom. *Science* 239: 748–753.

Garey, J. R. (2001): Ecdysozoa: the relationship between Cycloneuralia and Panarthropoda. *Zool. Anz.* 240: 321–330.

Garey, J. R. (2003): Ecdysozoa: the evidence for a close relationship between arthropods and nematodes. In: Legakis, A., Sfenthourakis, S., Polymeni, R. & Thessalou-Legakis, M. (eds) The new panorama of animal evolution, *Proc.18th Int. Congr. Zool.,* pp. 503–509. Pensoft Publishers, Sofia.

Garey, J. R., Near, T. J., Nonnemacher, M. R. & Nadler, S. A. (1996): Molecular evidence for Acanthocephala as a subtaxon of Rotifera. *J. Mol. Evol.* 43: 287–292.

Garey, J. R., Schmidt-Rhaesa, A., Near, T. J. & Nadler, S. A. (1998): The evolutionary relationships of rotifers and acanthocephalans. *Hydrobiologia* 387/388: 83–91.

Gegenbauer, C. (1859): *Grundzüge der vergleichenden Anatomie.* Verlag Wilhelm Engelmann, Leipzig [606 S.].

Giribet, G. (2003): Molecules, development and fossils in the study of metazoan evolution; Articulata versus Ecdysozoa revisited. *Zoology* 106: 303–326.

Giribet, G., Distel, D. L. D., Polz, M., Sterrer, W. & Wheeler, W. C. (2000): Triploblastic relationships with emphasis on the acoelomates and the position of Gnathostomulida, Cycliophora, Plathelminthes, and Chaetognatha: a combined approach of 18S rDNA sequences and morphology. *Syst. Biol.* 49: 539–562.

Giribet, G., Sørensen, M. V., Funch, P., Kristensen, R. M. & Sterrer, W. (2004): Investigations into the phylogenetic position of Micrognathozoa using four molecular loci. *Cladistics* 20: 1–13.

Glenner, H., Hansen, A. J., Sørensen, M. V., Ronquist, F., Huelsenbeck, J. P. & Willerslev, E. (2004): Bayesian inference of the metazoan phylogeny: a combined molecular and morphological approach. *Curr. Biol.* 14: 1644–1649.

Grobben, K. (1908): Die systematische Einteilung des Tierreichs. *Verh. Zool. Bot. Ges. Wien* 58: 491–511.

Gruner, H.-E. (1984): *Lehrbuch der Speziellen Zoologie. Vol. 1, part 2: Cnidaria, Ctenophora, Mesozoa, Plathelminthes, Nemertini, Entoprocta, Nemathelminthes, Priapulida. Fourth Edition.* Gustav Fischer Verlag, Stuttgart.

Hammarsten, O. (1915): Zur Entwicklungsgeschichte von *Halicryptus. Z. Wiss. Zool.* 112: 527–571.

Hejnol, A. et al. (2009): Assessing the root of bilaterian animals with scalable phylogenomic methods. *Proc. R. Soc. B.* 276: 4261–4270.

Hennig, W. (1950): *Grundzüge einer Theorie der Phylogenetischen Systematik.* Deutscher Zentralverlag, Berlin.

Hennig, W. (1966): *Phylogenetic Systematics.* University of Illinois Press, Urbana.

Herlyn, H., Pisurek, O., Schmitz, J., Ehlers, U. & Zischler, H. (2003): The syndermatan phylogeny and the evolution of acanthocephalan endoparasitism as inferred from 18S rDNA sequences. *Mol. Phyl. Evol.* 26: 155–164.

Higgins, R. P. (1971): A historical overview of kinorhynch research. *Smithsonian Contrib. Zool.* 76: 25–31.

Hyman, L. H. (1951): *The invertebrates. Vol. 3: Acanthocephala, Aschelminthes, and Entoprocta. The pseudocoelomate Bilateria.* McGraw-Hill, New York.

Jenner, R. A. (2004): Towards a phylogeny of the Metazoa: evaluating alternative phylogenetic positions of Platyhelminthes, Nemertea, and Gnathostomulida, with a critical rerappraisal of cladistic characters. *Contrib. Zool.* 73: 3–163.

Kristensen, R. M. (1983): Loricifera, a new phylum with Aschelminthes characters from the meiobenthos. *Z. Zool. Syst. Evolut.-forsch.* 21: 163–180.

Kristensen, R. M. (1995): Are Aschelminthes pseudocoelomate or acoelomate? In: Lanzavecchia, G., Valvassori, R. & Candia Carnevali, M. D. (eds) Body cavities: function and phylogeny. *Selected Symposia and Monographs U.Z.I., Mucchi, Modena* 8: 41–43.

Kristensen, R. M. & Funch, P. (2000): Micrognathozoa: A new class with complicated jaws like those of Rotifera and Gnathostomulida. *J. Morphol.* 246: 1–49.

Leeuwenhoek, A. (1722): *Arcana Naturae Detecta. Edition Novissima, Auctior et Correctior.* Lugduni Batavorum.

Lemburg, C. (1995): Ultrastructure of sense organs and receptor cells of the neck and lorica of the *Halicryptus spinulosus* larva (Priapulida). *Microfauna Marina* 10: 7–30.

Lemburg, C. (1999): *Ultrastrukturelle Untersuchungen an den Larven von Halicryptus spinulosus und Priapulus caudatus. Hypothesen zur Phylogenie der Priapulida und deren Bedeutung für die Evolution der Nemathelminthen.* Cuvillier Verlag, Göttingen.

Linnaeus, C. (1758): *Systema naturae per regna tria naturae, secundum classes, ordines, genera, species, cum characteribus, differentiis, synonymis, locis, Tenth Edition.* Holmiae, Laurentii Salvii.

Littlewood, D. T. J., Telford, M., Clough, K. A. & Rohde, K. (1998): Gnathostomulida – an enigmatic metazoan phylum from both morphological and molecular perspectives. *Mol. Phyl. Evol.* 9: 72–79.

Lorenzen, S. (1985): Phylogenetic aspects of pseudocoelomate evolution. In: Conway Morris, S., George, J. D., Gibson, D. I. & Platt, H. M. (eds) The origins and relationships of lower invertebrates. *Syst. Assoc. Spec.* 28: 210–223.

Lorenzen, S. (1996): Nemathelminthes (Aschelminthes). In: Westheide, W. & Rieger, R. (eds.) *Spezielle Zoologie*, pp. 682–684. Gustav Fischer Verlag, Stuttgart.

Lühe, M. (1904): Geschichte und Ergebnisse der Echinorhynchen-Forschung bis auf Westrumb (1821). *Zool. Ann. Z. Gesch. Zool.* 1: 139–250.

Malakhov, V. V. (1980): Cephalorhyncha, a new type of animal kingdom uniting Priapulida, Kinorhyncha, Gordiacea, and a new system of Aschelminthes worms. *Zool. Zh.* 59: 485–499. (in Russian with English summary)

Malakhov, V. V. (1994): *Nematodes. Structure, Development, Classification, and Phylogeny.* Smithsonian Institution Press, Washington.

Malakhov, V. V. & Adrianov, A. V. (1995): *Cephalorhyncha – a New Phylum of the Animal Kingdom.* KMK Scientific Press, Moscow.

Mallatt, J., Garey, J. R. & Shultz, J. W. (2004): Ecdysozoan phylogeny and Bayesian inference: first use of nearly complete 28S and 18S rRNA gene sequences to classify the arthropods and their kin. *Mol. Phyl. Evol.* 31: 178–191.

Mallatt, J. & Giribet, G. (2006): Further use of nearly complete 28S and 18S rRNA genes to classify Ecdysozoa: 37 more arthropods and a kinorhynch. *Mol. Phyl. Evol.* 40: 772–794.

Mark Welch, D. B. (2000): Evidence from a protein-coding gene that acanthocephalans are rotifers. *Invert. Biol.* 119: 17–26.

Mark Welch, D. B. (2005): Bayesian and maximum likelihood analyses of rotifer–acanthocephalan relationships. *Hydrobiologia* 546: 47–54.

Mečnikow, E. (1865): Ueber einige wenig bekannte niedere Thierformen. *Z. Wiss. Zool.* 15: 450–458.

Neuhaus, B., Kristensen, R. M. & Lemburg, C. (1996): Ultrastructure of the cuticle of the Nemathelminthes and electron microscopical localization of chitin. *Verh. Dtsch. Zool. Ges.* 89.1: 221.

Nielsen, C. (1995): *Animal Evolution. Interrelationships of the Living Phyla, First Edition.* Oxford University Press, Oxford.

Nielsen, C. (2001): *Animal Evolution. Interrelationships of the Living Phyla, Second Edition.* Oxford University Press, Oxford.

Nielsen, C., Scharff, N. & Eibye-Jacobsen, D. (1996): Cladistic analysis of the animal kingdom. *Biol. J. Linn. Soc.* 57: 385–410.

Paps, J., Baguñà, J. & Riutort, M. (2009): Lophortochozoa internal phylogeny: new insights from an up-to-date analysis of nuclear ribosomal genes. *Proc. R. Soc. B.* 276: 1245–1254.

Park, J.-K., Rho, H. S., Kristensen, R. M., Kim, W. & Giribet, G. (2006): First molecular data on the phylum Loricifera – an investigation into the phylogeny of Ecdysozoa with emphasis on the positions of Loricifera and Priapulida. *Zool. Sci.* 23: 943–954.

Peterson, K. J. & Eernisse, D. J. (2001): Animal phylogeny and the ancestry of bilaterians: inferences from morphology and 18S rDNA sequences. *Evol. Dev.* 3: 170–205.

Petrov, N. B. & Vladychenskaya, N. S. (2005): Phylogeny of molting protostomes (Ecdysozoa) as inferred from 18S and 28S rRNA gene sequences. *Mol. Biol.* 39: 503–513.

Philippe, H., Lartillot, N. & Brinkmann, H. (2005): Multigene analysis of bilaterian animals corroborate the monophyly of Ecdysozoa, Lophotrochozoa, and Protostomia. *Mol. Biol. Evol.* 22: 1246–1253.

Rauther, M. (1928): Nemathelminthes. Allgemeine Einleitung zur Naturgeschichte der Nemathelminthes. In: Krumbach, T. (ed.) *Handbuch der Zoologie, Vol.2, 2nd half*, pp. 1–7. Vermes Amera. Walter de Gruyter, Berlin.

Redi, F. (1684): *Osservazioni intorno agli animali viventi che si trovano negli animali viventi.* Firenze.

Reinhard, W. (1887): Kinorhynchen (*Echinoderes*), ihr anatomischer Bau und ihre Stellung im System. *Z. Wiss. Zool.* 45: 401–467.

Remane, A. (1926): Morphologie und Verwandtschaftsbeziehungen der aberranten Gastrotrichen. *I. Z. Morph. Ökol. Tiere* 5: 625–754.

Remane, A. (1936): Gastrotricha und Kinorhyncha. In: Bronn, H.G. (ed.) *Dr. H.G. Bronns Klassen und Ordnungen des Tierreichs. Vol. 4: Vermes. 2nd part: Askhelminthes, Trochhelminthes*, pp. 1–385. Akademische Verlagsgesellschaft, Leipzig.

Remane, A. (1952): Die Besiedlung des Sandbodens im Meere und die Bedeutung der Lebensformtypen für die Ökologie. *Verh. Dtsch. Zool. Ges.* 1951: 327–259.

Remane, A. (1963): The systematic position and phylogeny of the pseudocoelomates. In: Dougherty, E. C., Brown, Z. N., Hanson, E. D. & Hartman, W. D. (eds.) *The Lower Metazoa*, pp. 247–255. University of California Press, Berkeley.

Ricci, C. (1983): Rotifera or Rotatoria? *Hydrobiologia* 104: 1–2.

Riedl, R. J. (1969): Gnathostomulida from America. *Science* 163: 445–452.

Rieger, R. M. & Tyler, S. (1995): Sister-group relationship of Gnathostomulida and Rotifera-Acanthocephala. *Invert. Biol.* 114: 186–188.

Roeding, F., Hagner-Holler, S., Ruhberg, H., Ebersberger, I., von Haeseler, A., Kube, M., Reinhardt, R. & Burmester, T. (2007): EST sequencing of Onychophora and phylogenomic analysis of Metazoa. *Mol. Phyl. Evol.* 45: 942–951.

Rothe, B. H. & Schmidt-Rhaesa, A. (2009): Architecture of the nervous system in two *Dactylopodola* species (Gastrotricha, Macrodasyida). *Zoomorphology* 128: 227–246.

Rothe, B. H. & Schmidt-Rhaesa, A. (2010): Structure of the nervous system in *Tubiluchus troglodytes* (Priapulida). *Invertebrate Biology* 129: 39–58.

Rothe, B. H., Schmidt-Rhaesa, A. & Kieneke, A. (2011): The nervous system of *Neodasys chaetonotoideus* (Gastrotricha: *Neodasys*) revealed by combining confocal laserscanning and transmission electron microscopy – evolutionary comparison of neuroanatomy within the Gastrotricha and basal Protostomia. *Zoomorphology* 130: 51–84.

Rudolphi, C. A. (1808): *Entozoorum sive Vermium Intestinalium Historia Naturalis.* Vol. 1. Taberna Libraria et Artium, Amsterdam.

Rudolphi, C. A. (1819): *Entozoorum Synopsis cui Accedunt Mantissa Duplex et Indices Locupletissimi.* Berolini.

Ruppert, E. E. (1991): Introduction to the aschelminth phyla: a consideration of mesoderm, body cavities, and cuticle. In: Harrison, F.W. & Ruppert, E.E. (eds.) *Microscopic Anatomy of Invertebrates*, pp. 1–17. Wiley-Liss, New York.

Schepotieff, A. (1907): Die Echinoderiden. *Z. Wiss. Zool.* 88: 291–326.

Schmidt-Rhaesa, A. (1996): The nervous system of *Nectonema munidae* and *Gordius aquaticus*, with implications on the ground pattern of the Nematomorpha. *Zoomorphology* 116: 133–142.

Schmidt-Rhaesa, A. (1998): Phylogenetic relationships of the Nematomorpha – a discussion of current hypothesis. *Zool. Anz.* 236: 203–216.

Schmidt-Rhaesa, A. (2002): Two dimensions of biodiversity research exemplified by Nematomorpha and Gastrotricha. *Integr. Comp. Biol.* 42: 633–640.

Schneider, A. (1866): *Monographie der Nematoden*. Verlag Georg Reimer, Berlin.

Schram, F. R. (1991): Cladistic analysis of metazoan phyla and placement of fossil problematica. In: Simonetta, A. & Conway Morris, S. (eds.) *The early Evolution of Metazoa and the Significance of Problematic Taxa*, pp. 35–46. Cambridge University Press, Cambridge.

Schram, F. R. & Ellis, W. N. (1994): Metazoan relationships: a rebuttal. *Cladistics* 10, 331–337.

Schultze, M. (1853): Über *Chaetonotus* und *Ichthydium* Ehrenb. Und eine neue verwandte Gattung Turbanella. *Arch. Anat. Physiol.* 6: 241–256.

Sørensen, M. V., Funch, P., Willerslev, E., Hansen, A. J. & Olesen, J. (2000): On the phylogeny of the Metazoa in the light of Cycliophora and Micrognathozoa. *Zool. Anz.* 239: 297–318.

Sørensen, M. V. & Giribet, G. (2006): A modern approach to rotiferan phylogeny: combining morphological and molecular data. *Mol. Phyl. Evol.* 40: 585–608.

Sørensen, M. V., Hebsgaard, M. B., Heiner, I., Glenner, H., Willerslev, E. & Kristensen, R. M. (2008): New data from an enigmatic phylum: evidence from molecular sequence data supports a sister-group relationship between Loricifera and Nematomorpha. *J. Zool. Syst. Evol. Res.* 46: 231–239.

Telford, M. J., Bourlat, S. J., Economou, A., Papillon, D. & Rota-Stabelli, O. (2008): The evolution of the Ecdysozoa. *Phil. Trans. R. Soc. B.* 363: 1529–1537.

Teuchert, G. (1968): Zur Fortpflanzung und Entwicklung der Macrodasyoidea (Gastrotricha). *Z. Morphol. Tiere* 63: 343–418.

Vejdovsky, F. (1886): Zur Morphologie der Gordiiden. *Z. Wiss. Zool.* 43: 369–433.

Vogt, C. (1851): Zoologische Briefe. Naturgeschichte der lebenden und untergegangenen Thiere. *Literarische Anstalt, Frankfurt am Main*: 174–240.

Von Baer, K. E. (1826): Die Verwandtschaftsverhältnisse unter den niederen Thierformen. *Nova Acta C.L. Acad.* 13: 731–762.

Von Haffner, K. (1950): Organisation und systematische Stellung der Acanthocephalen. *Zool. Anz.* 145, Suppl.: Neue Ergebnisse und Probleme der Zoologie: 243–274.

Von Siebold, C. T. (1843): Bericht über die Leistungen im Gebiete der Helminthologie während des Jahres 1842. *Arch. Naturgesch.* 9: 302–310.

Wallace, R. L., Ricci, C. & Melone, G. (1995): Through Alice´´s looking glass: a cladistic analysis of pseudocoelomate anatomy. In: Lanzavecchia, G., Valvassori, R. & Candia Carnevali, M.D. (eds.) Selected Body cavities: function and phylogeny. *Symposia and Monographs U.Z.I., Mucchi, Modena* 8: 61–67.

Wallace, R. L., Ricci, C. & Melone, G. (1996): A cladistic analysis of pseudocoelomate (aschelminth) morphology. *Invert. Biol.* 115: 104–112.

Winnepenninckx, B., Backeljau, T., Mackey, L. M., Brooks, J. M., De Wachter, R., Kumar, S. & Garey, J. R. (1995): 18S rRNA data indicate that Aschelminthes are polyphyletic in origin and consist of at least three distinct clades. *Mol. Biol. Evol.* 12: 1132–1137.

Witek, A., Herlyn, H., Meyer, A., Boell, L., Bucher, G. & Hankeln, T. (2008): EST based phylogenomics of Syndermata questions monophyly of Eurotatoria. *BMC Evol. Biol.* 8: 345.

Wolf, Y. I., Rogozin, I. B. & Koonin, E. V. (2004): Coelomata and not Ecdysozoa: evidence from genome-wide phylogenetic analysis. *Genome Res.* 14: 29–36.

Zelinka, C. (1889): Die Gastrotrichen. Eine monographische Darstellung ihrer Anatomie, Biologie und Systematik. *Z. Wiss. Zool.* 49: 209–384.

Zheng, J., Rogozin, I. B., Koonin, E. V. & Przytycka, T. M. (2007): Support for the Coelomata clade of animals from a rigorous analysis of the pattern of intron conservation. *Mol. Biol. Evol.* 24: 2583–2592.

Zrzavý, J., Mihulka, S., Kepka, P., Bezděk, A. & Tietz, D. (1998): Phylogeny of the Metazoa based on morphological and 18S ribosomal DNA evidence. *Cladistics* 14: 249–285.

Andreas Maas

2. Gastrotricha, Cycloneuralia and Gnathifera: The Fossil Record

The preservation potential of individuals of gastrotrich, cycloneuralian and gnathiferan species is fairly low. Living species are mainly small and soft-bodied. Gastrotricha apparently lack a fossil record. Fossil gnathiferans are only known by a few specimens of the Rotifera found either in amber or in Pleistocene assemblages. Yet, fossil Cycloneuralia are well known from different sources and in a wide time frame from the Cambrian until Palaeogene deposits. A few specimens of nematodes are known from the Late Palaeozoic (Carboniferous) of North America. The youngest Cycloneuralia are represented by a number of nematode and nematomorph specimens preserved in Baltic amber. Strikingly, by far the highest number of fossil Cycloneuralia in terms of both species number and individual abundance is evident from Cambrian times (about 520 Million years old) and from various biota or preservational assemblages.

2.1. Gastrotricha

The fossilisation potential of the soft and tiny Gastrotricha is possibly among the lowest within Metazoa. Although cuticular remains are common in the fossil record, the cuticle of Gastrotricha seems to be too fragile to be able to become fossilized. Accordingly, no fossil specimen is known that could be assigned to Gastrotricha.

2.2. Gnathifera

The soft body of Gnathifera also has an extremely low fossilization potential. However, Gnathifera are characterized by specific hard parts within their body, i.e. their jaws forming their so-called jaw apparatus, which could be preserved. Although its small size makes it a challenge to identify it on a fossil specimen, the fossil record of equally small-sized material is indeed not rare (see below). However, no isolated jaws have been detected so far. Yet, the body covers, called thecae, of a bdelloid rotifer, *Habrotrocha angusticollis* (Murray 1906) have been reported from various places in the northern hemisphere, recently from 6000-year-old Pleistocene peat deposits in Ontario, Canada (see Warner & Chengalath 1988). Thecae of the same species have been reported from the Eocene (Poinar & Ricci 1992) and Miocene (Aquitanian; ~20–23 Ma) (Waggoner & Poinar 1993) in Dominican amber. Also the rotiferan taxon Monogononta is represented by finds from the Eocene (Lutetian; ~40–48 Ma) in South Australia (Southcott and Lange 1971; see also Labandeira 2002).

2.3. Cycloneuralia

Fossil Record of Cycloneuralia

Cycloneuralia is a taxon with a remarkably good fossil record, ranging from Cambrian to modern times (for the terminology of time periods and assigned absolute times, refer to the Geologic Time Scale of Ogg et al. 2008; see also Zhu et al. 2006). The reason for the good fossilisation potential of Cycloneuralia is obviously their cuticle. Also the fossil record of the likewise cuticle-bearing Arthropoda is excellent, although the potentially strongly sclerotized forms of the Arthropoda s. str. are unequally abundant in the fossil record from the Cambrian onward. The less sclerotized forms however, the lobopodians, are only mainly known from Cambrian times – strongly comparable to the fossil record of Cycloneuralia, which is almost restricted to Cambrian times; a clear bias obviously caused by the specific diagenetic and mineralization conditions in the phosphate-rich environment (e.g. Maas et al. 2006). Cuticular details, such as ornaments and fine annulations known from modern Cycloneuralia, are evident also from fossils. Again, specific in-group cycloneuralian characters such as the tooth-bearing pharynx, the scalid-bearing introvert or the loricae of Vinctiplicata are known from various fossils and help in identifying fossils as Cycloneuralia. Here we concentrate on Cycloneuralia from the Palaeozoic (see Chapter 3.7 for information on fossil nematodes and nematomorphs from Cretaceous and Palaeogene amber). Palaeozoic Cycloneuralian fossil species are mainly assignable only to the Cycloneuralia as a whole or to the Scalidophora, and within these to the Vinctiplicata, at most (for erroneous assignment to Priapulida see below).

By far the largest amount of fossil cycloneuralian material from the Palaeozoic (and fossil Cycloneuralia in general) is, again, between 520 and 495 million years old, i.e. is of Cambrian age. Complete, preserved fossils occur in different lagerstätten of highly distinct faunal composition, e.g. i) the Maotianshan-Shale Lagerstätten of southern China (~518–524 Ma; Maas et al. 2007a); ii) the Burgess-Shale Lagerstätten of British Columbia, Canada (~515 Ma; Conway Morris 1977); iii) the 'Orsten' fossil assemblages of Australia (~504–510 Ma; Maas et al. 2007b, 2009); iv) the Sirius Passet Lagerstätten of northern Greenland (~505–524 Ma; Conway Morris & Peel 2010; Peel & Ineson 2011). A number of additional lagerstätten has yielded fragmentary or compressed material of Cycloneuralia. Palaeoscolecidans are by far the most significant component of Cycloneuralia in the fossil record. They are worm-like forms characterized by an annulated trunk that bears cuticular thickenings, sclerites, of various sizes with a taxon-specific ornamentation with nodes and spines. Such sclerites may occur together with various other cuticular specialisations. The frontal body portion is spine-bearing and has been interpreted as the introvert. Palaeoscolecidans were initially described as compressed body fossils from Lower to Middle Ordovician shales (Ulrich 1878; Whittard 1953). Subsequently disarticulated, isolated phosphatic sclerites of palaeoscolecids were recovered from early Palaeozoic (especially Cambrian) fossil assemblages and treated as 'problematica' (e.g. Bengtson 1977; Gedik 1977; Wrona 1982, 1987, 2004; Peel & Larsen 1984; Bendix-Almgreen & Peel 1988; Märss 1988; Wrona & Hamdi 2000, 2001). These isolated phosphatic sclerites have been described under various generic names that can be regarded as form taxa. Examples are *Hadimopanella* Gedik, 1977 (= *Lenargyrion* Bengtson, 1977), *Kaimenella* Märss, 1988, *Milaculum* Müller, 1973, *Utahphospha* Müller & Miller, 1976. Recovery of articulated sclerite arrays arranged in annulated bands has provided evidence that the source of many of these sclerites can be attributed to palaeoscolecids, a recognition made independently by Kraft & Mergl (1989) and Boogaard (1989a, b; see also Hinz et al. 1990; Brock & Cooper 1993; Müller & Hinz-Schallreuter 1993; Harvey et al. 2010). Subsequent finds of exceptionally well preserved cuticular remains of palaeoscolecidans, e.g. Ordivician material of Bohemia (Hinz et al. 1990), Cambrian 'Orsten' fragmentary palaeoscolecidans from Australia (Müller & Hinz-Schallreuter 1993), Lower Cambrian material of China (Hou & Bergström 1994; Zhang & Pratt 1996) and Siberia (Bengtson 1977; Ivantsov & Wrona 2004) and restudy of older material of *Palaeoscolex* species by Conway Morris (1997) revealed a wide spectrum of morphological details to further understand the morphology, palaeobiology and taxonomy of Palaeoscolecida (see also Harvey et al. 2010).

However, such detailed information is not available for the frontal, i.e. introvert or pharynx region. Harvey et al. (2010) distinguished between palaeoscolecidan taxa, of which the characteristic cuticular substructure is evident (Palaeoscolecida s. str.; see Tab. 1) and those, of which the cuticle morphology has not yet been confirmed but which are slender worm-like introvert-bearing forms, i.e. have the same overall body shape. The latter group includes taxa such as *Louisella* from the Burgess Shale (Conway Morris 1977), *Maotianshania*, *Cricocosmia*, *Anningvermes*, *Tabelliscolex*, and *Tylotites* from the Maotianshan Shale (Hou & Bergström 1994; Huang et al. 2004a; Han et al. 2003a, 2007a,b).

Maotianshan-Shale Cycloneuralia

Cycloneuralia are extremely abundant in the Lower Cambrian Maotianshan Shales of southern China. These shales are a series of early Cambrian depositional layers in the Qiongzhusi formation (Brasier 1989), which contain numerous metazoan fossils of almost any taxon and in abundant individual numbers. The most famous fossil assemblage of the Maotianshan Shales is the Changjiang biota (name also used in the literature for these lagerstätten; Tab. 2; Hou & Sun 1988; Chen et al. 1991, 1996; Hou et al. 1991, 2004; Chen & Zhou 1997; Zhang et al. 2001; Hu 2005). Fossil Cycloneuralia include long worm-shaped forms, i.e. palaeoscolecidans, sac-like forms, loricate forms resembling larvae of the Vinctiplicata or adult Loricifera and species that lived in an elongate, obviously self-produced cone-shaped tube (Fig. 2.3.1 A–P; Maas et al. 2007a). One of the most abundant species among worm-like forms in the Chengjiang deposits is *Cricocosmia jinningensis* Hou & Sun, 1988 (Fig. 2.3.1 C; Tab. 2). New observations on this species revealed that trunk ornamentation is, in contrast to earlier reports, dorso-ventrally different (Han et al. 2007a). A similarly abundant species among the loricate forms is *Sicyophorus rara* Luo & Hu *in* Luo, Hu, Chen, Zhang & Tao, 1999 (Tab. 2). All other species occur in significantly less numbers.

Burgess-Shale Cycloneuralia

The Burgess-Shale lagerstätte of British Columbia, Canada, has yielded an enormous number of various metazoan taxa, with thousands of specimens. As with the Maotianshan-Shale deposits, the majority of the material has not been worked up yet. Probably the

Tab. 1. Palaeoscolecida sensu stricto (sensu Harvey et al. 2010). The taxa share a long and slender body ("worm-like") and characteristic cuticular ornamentation including sclerites, and other details such as putative flosculi and tubuli.

Taxon	Age/Occurrence	Appearance	References
Austroscolex Müller & Hinz-Schallreuter, 1993	Cambrian of Australia	cuticular fragments	Müller & Hinz-Schallreuter 1993
Bohemoscolex holubi Kraft & Mergl, 1989	Early Ordovician, Czech Republic	body fossils with cuticular details	Kraft & Mergl 1989
Chalazoscolex pharkus Conway Morris & Peel, 2010	Cambrian of Sirius Passet, Greenland	macrofossil	Conway Morris & Peel 2010
Corallioscolex gravus Müller & Hinz-Schallreuter, 1993	Cambrian of Australia	cuticular fragments	Müller & Hinz-Schallreuter 1993
Dispinoscolex decorus Duan, Dong & Donoghue, 2012	Cambrian of Hunan, China	cuticular fragments	Duan et al. 2012
Euryscolex paternarius Müller & Hinz-Schallreuter, 1993	Cambrian of Australia	cuticular fragments	Müller & Hinz-Schallreuter 1993
Gamascolex herodes Kraft & Mergl, 1989	Early Ordovician, Czech Republic	body fossils with cuticular details	Kraft & Mergl 1989; Harvey et al. 2010
Guandoscolex minor Hu, Luo & Fu, 2008	Cambrian, Kunming Region	macrofossil	Hu et al. 2008
Hadimopanella Gedik, 1977 form taxon (incl. *Lenargyrion* Bengtson, 1977)	Cambrian of various locations	disarticulated sclerites	e.g. Bengtson 1977; Gedik 1977; Wrona 1987; Hinz et al. 1990; Topper et al. 2010
Houscolex Zhang & Pratt, 1996	Cambrian of Shaanxi, China	disarticulated sclerites	Zhang & Pratt 1996
Kaimenella Märss, 1988 form taxon	Cambrian of various locations	disarticulated sclerites	Märss 1988; Hinz et al. 1990
Kaloscolex granulatus Müller & Hinz-Schallreuter, 1993	Cambrian of Australia	cuticular fragments	Müller & Hinz-Schallreuter 1993
Milaculum form taxon	Cambrian, Ordovician; Australia, Iran, Europe, USA, Canada	disarticulated sclerites	e.g. Müller 1973; Boogaard 1988; Hinz et al. 1990
Murrayscolex inaequalis Müller & Hinz-Schallreuter, 1993	Cambrian of Australia	cuticular fragments	Müller & Hinz Schallreuter 1993
Pantoioscolex oleschinskii Müller & Hinz-Schallreuter, 1993	Cambrian of Australia	cuticular fragments	Müller & Hinz-Schallreuter 1993
Plasmuscolex Kraft & Mergl 1989	Early Ordovician, Czech Republic	body fossils with cuticular details	Kraft & Mergl 1989
Palaeoscolex species	Cambrian to Ordivician, UK, Siberia, China, USA	macrofossil, cuticular fragments, disarticulated sclerites	e.g. Whittard 1953; Conway Morris & Robison 1986; Conway Morris 1997; Hou et al. 2004; Ivantsov & Wrona 2004
Protoscolex partim	various	macrofossil	e.g. Ulrich 1878; Bather 1920; Conway Morris et al. 1982
Rhomboscolex chaoticus Müller & Hinz-Schallreuter, 1993	Cambrian of Australia	cuticular fragments	Müller & Hinz-Schallreuter 1993
Sahascolex labyrinthus Ivantsov & Wrona, 2004	Cambrian of eastern Siberia	cuticular fragments, disarticulated sclerites	Ivantsov & Wrona 2004
Schistoscolex Müller & Hinz-Schallreuter, 1993	Cambrian of Australia	cuticular fragments	Müller & Hinz-Schallreuter 1993; Duan et al. 2012
Shergoldiscolex Müller & Hinz-Schallreuter, 1993	Cambrian of Australia	cuticular fragments	Müller & Hinz-Schallreuter 1993
Thoracoscolex armatus Müller & Hinz-Schallreuter, 1993	Cambrian of Australia	cuticular fragments	Müller & Hinz-Schallreuter 1993
Utahphospha Müller & Miller, 1976 form taxon	Cambrian of various locations	disarticulated sclerites	Müller & Miller 1976; Repetski 1981; Hinz et al. 1990

(Continued)

Tab. 1. *(Continued)*

Taxon	Age/Occurrence	Appearance	References
Wronascolex boreogyrus Ivantsov & Zhuravlev, 2005	Cambrian of Siberia	macrofossil	Ivantsov & Zhuravlev 2005
Xystoscolex boreogyrus Conway Morris & Peel, 2010	Cambrian of Sirius Passet, Greenland	macrofossil	Conway Morris & Peel 2010

Tab. 2. Palaeozoic cycloneuralian taxa of unclear affinities other than Palaeoscolecida s. str. of Tab. 1. Occurrence contains the most common localities, not all occurrences. For each species the reported maximum length and diameter and assignment to a particular type of body shape are noted. Remarks: *Selkirkia sinica* may be conspecific with *Selkirkia jinningensis* Hou, Bergström, Wang, Feng & Chen, 1999 (Hou et al. 2004). Another species that was interpreted as having cycloneuralian affinities is *Archotuba conoidalis* Hou, Bergström, Wang, Feng & Chen, 1999. It is almost only known from seemingly colonial elongated-conical tubes that have been compared with those of *Paraselkirkia sinica* (Hou et al. 1999, 2004). Yet this species might not even be a nemathelminth at all. Lately, it has been interpreted as a cnidarian (Huang et al. 2004a; Han et al. 2007c).

Species resp. taxa	Occurrence	Shape	Length [mm]	Max. Ø [mm]	References
Acosmia maotiania Chen & Zhou, 1997	Chengjiang County; Kunming region	sac	40–50	9	Fig. 2.3.1 A; Chen & Zhou 1997; Hou et al. 2004
Ancalogon minor (Walcott 1911)	Burgess Shale	sac	40–110	9	Fig. 2.3.2 A; Conway Morris 1977
Anningvermis multispinosus Huang, Vannier & Chen, 2004	Shankoucun village of Anning	sac	59	7	Huang et al. 2004a
Chalazoscolex pharkus Conway Morris & Peel, 2010	Sirius Passet	worm	90	9	Fig. 2.3.5 A; Conway Morris & Peel 2010
Corynetis brevis Luo & Hu *in* Luo, Hu, Chen, Zhang & Tao, 1999	Kunming region	sac	26	2	Fig. 2.3.1 B; Luo et al. 1999; Huang et al. 2004a
Cricocosmia jinningensis Hou & Sun, 1988	Kunming region	worm	50	4	Fig. 2.3.1 C; Luo et al. 1999; Hou et al. 2004
Fieldia lanceolata Walcott, 1912	Burgess Shale	worm	50	6	Fig. 2.3.2 B; Conway Morris 1977
Guandoscolex minor Hu, Luo & Fu, 2008	Guanshan Biota, Kunming region	worm	15	1.2	Fig. 2.3.1 D; Hu et al. 2008, 2010
Laojieella thecata Han, Zhang, Zhang & Shu, 2006	Kunming region	worm	120	2–3	Fig. 2.3.1 E; Han et al. 2006; Han et al. 2007c
Lecythioscopa simplex (Walcott 1931)	Burgess Shale	worm	4–7	3–4	Fig. 2.3.2 C; Conway Morris 1977
Louisella pedunculata Walcott, 1911	Burgess Shale	worm	150–200	12	Fig. 2.3.2 E, F; Conway Morris 1977
Maotianshania cylindrica Sun & Hou, 1987	Chengjiang area; Shankoucun of Anning	worm	40	2	Fig. 2.3.1 F; Sun & Hou 1987; Chen & Zhou 1997; Hou et al. 2004
Markuelia Valkov, 1983	various (see text)	worm	2	0, 5	Valkov 1983; Bengtson & Yue 1997; Dong et al. 2004; Haug et al. 2009a
Orstenoloricus shergoldi Maas, Waloszek, Haug & Müller, 2009	Cambrian of Duchess Embayment, Australia	loricate	0, 2	0, 1	Fig. 2.3.4; Maas et al. 2009
Ottoia prolifica Walcott, 1911	Burgess Shale	sac	30–150	10–20	Fig. 2.3.2 G; Conway Morris 1977
Palaeopriapulites parvus Hou, Bergström, Wang, Feng & Chen, 1999	Maotianshan, Chengjiang County	loricate	< 10	4	Fig. 2.3.1 G; Hou et al. 1999, 2004

(Continued)

Tab. 2. (*Continued*)

Species resp. taxa	Occurrence	Shape	Length [mm]	Max. Ø [mm]	References
Palaeoscolex sinensis Hou & Sun, 1988	Chengjiang County; Kunming region	worm	60–100	2–5	Fig. 2.3.1 H; Hou and Sun 1988; Hou et al. 2004
Paratubiluchus bicaudatus Han, Shu, Zhang & Liu, 2004	Jianshan of Kunming City	sac	8	3	Fig. 2.3.1 I; Han et al. 2004
Scolecofurca rara Conway Morris, 1977	Burgess Shale	sac	90	20	Conway Morris 1977
Selkirkia columbia Conway Morris, 1977	Burgess Shale	tubiform	20–60	8	Fig. 2.3.2 D; Conway Morris 1977
Selkirkia sinica (Luo & Hu in Luo, Hu, Chen, Zhang & Tao 1999)	Ercaicun and Shankoucun of Kunming City	tubiform	20	3	Fig. 2.3.1 J; Luo et al. 1999; Hou et al. 1999; Hou et al. 2004
Shergoldana australiensis Maas, Waloszek, Haug & Müller, 2007	Cambrian of Duchess Embayment, Australia	worm	0, 2	0, 1	Fig. 2.3.3; Maas et al. 2007b
Sicyophorus rara Luo & Hu *in* Luo, Hu, Chen, Zhang & Tao, 1999	Haikou and Anning of Kunming City	loricate	>10	4	Fig. 2.3.1 K; Luo et al. 1999; Hou et al. 1999, 2004
Sirilorica carlsbergi Peel, 2010	Sirius Passet	loricate	80	50	Fig. 2.3.5 C; Peel 2010
Tabelliscolex chengjiangensis Han, Liu, Zhang, Zhang & Shu, 2007	Ma'anshan Section, Changing County	worm	15	4	Fig. 2.3.1 M; Han et al. 2007a
Tabelliscolex hexagonus Han, Zhang & Shu, 2003	Shankoucun and Ercaicun of Kunming City	worm	15	4	Fig. 2.3.1 L; Han et al. 2003a
Tylotites petiolaris Luo & Hu, 1999	Ercaicun and Shankoucun of Kunming City	sac	15	4	Fig. 2.3.1 N; Luo et al. 1999; Han et al. 2003b
Xiaoheiqingella peculiaris Hu *in* Chen, Luo, Hu, Yin, Jiang, Wu, Li & Chen, 2002	Ercaicun of Kunming	sac	10	4	Fig. 2.3.1 O; Chen et al. 2002; Huang et al. 2004b, 2006
Xystoscolex boreogyrus Conway Morris & Peel, 2010	Sirius Passet	worm	120	10	Fig. 2.3.5 B; Conway Morris & Peel 2010
Yunnanpriapulus halteroformis Huang, Vannier & Chen, 2004	Ercaicun and Meishucun villages of Kunming	sac	15	4	Fig. 2.3.1 P; Huang et al. 2004b, 2006

best-known fossil Cycloneuralian from the Burgess Shale is *Ottoia prolifica* Walcott, 1911 (Fig. 2.3.2 G; cf. Tab. 2). Its morphology is known in considerable detail from the study of Conway Morris (1977), who accomplished a comprehensive review of the Burgess Shale Cycloneuralia. The high degree of overlap of the faunal composition of the Maotianshan Shale and the Burgess Shale, especially regarding arthropods, is also evident in the cycloneuralian taxa. One example are tube-building Cycloneuralia that have been referred to the taxon *Selkirkia* Walcott, 1911. The Burgess Shale counterpart of *S. sinica* (Luo & Hu in Luo, Hu, Chen, Zhang & Tao 1999) is the very similar *Selkirkia columbia* Conway Morris, 1977 (Fig. 2.3.2 D; Conway Morris 1977). Possible conspecifity has, however, never been critically worked up. Conway Morris (1977) interpreted these and other species he investigated (cf. Tab. 2) as Priapulida, based on the most likely structural homology of the tooth-bearing frontal apparatus and the likewise tooth-bearing pharynx of modern priapulids. Although there should not be any doubt about this homology, the structure itself is not an exclusive character of Priapulida but of a larger taxon Scalidophora or

Fig. 2.3.1. Examples of Cycloneuralia from the Maotianshan Shale of China. A, *Acosmia maotiania* Chen & Zhou, 1997; B, *Corynetis brevis* Luo & Hu *in* Luo, Hu, Chen, Zhang & Tao, 1999; C, *Cricocosmia jinningensis* Hou & Sun, 1988; D, *Guandoscolex minor* Hu, Luo & Fu, 2008; E, *Laojieella thecata* Han, Zhang, Zhang & Shu, 2006; F, *Maotianshania cylindrica* Sun & Hou, 1987; G, *Palaeopriapulites parvus* Hou, Bergström, Wang, Feng & Chen, 1999; H, *Palaeoscolex sinensis* Hou & Sun, 1988; I, *Paratubiluchus bicaudatus* Han, Shu, Zhang & Liu 2004; J, *Selkirkia sinica* Luo & Hu in Luo, Hu, Chen, Zhang & Tao, 1999; K, *Sicyophorus rara* Luo & Hu *in* Luo, Hu, Chen, Zhang & Tao, 1999; L, *Tabelliscolex hexagonus* Han, Zhang & Shu, 2003; M, *Tabelliscolex chengjiangensis* Han, Liu, Zhang, Zhang & Shu, 2007, detail of the trunk; N, *Tylotites petiolaris* Luo & Hu, 1999; O, *Xiaoheiqingella peculiaris* Hu *in* Chen, Luo, Hu, Yin, Jiang, Wu, Li & Chen 2002; P, *Yunnanpriapulus halteroformis* Huang, Vannier & Chen, 2004. A, C, F, J: courtesy of Hou Xianguang, Kunming; B, D, O: courtesy of Hu Shixue, Kunming; E, I, L–N: courtesy of Han Jian, Xi'an; G, K: courtesy of Chen Junyuan, Chengjiang. Scale bars: A 5 mm; B–L, N 2 mm; M 0.5 mm.

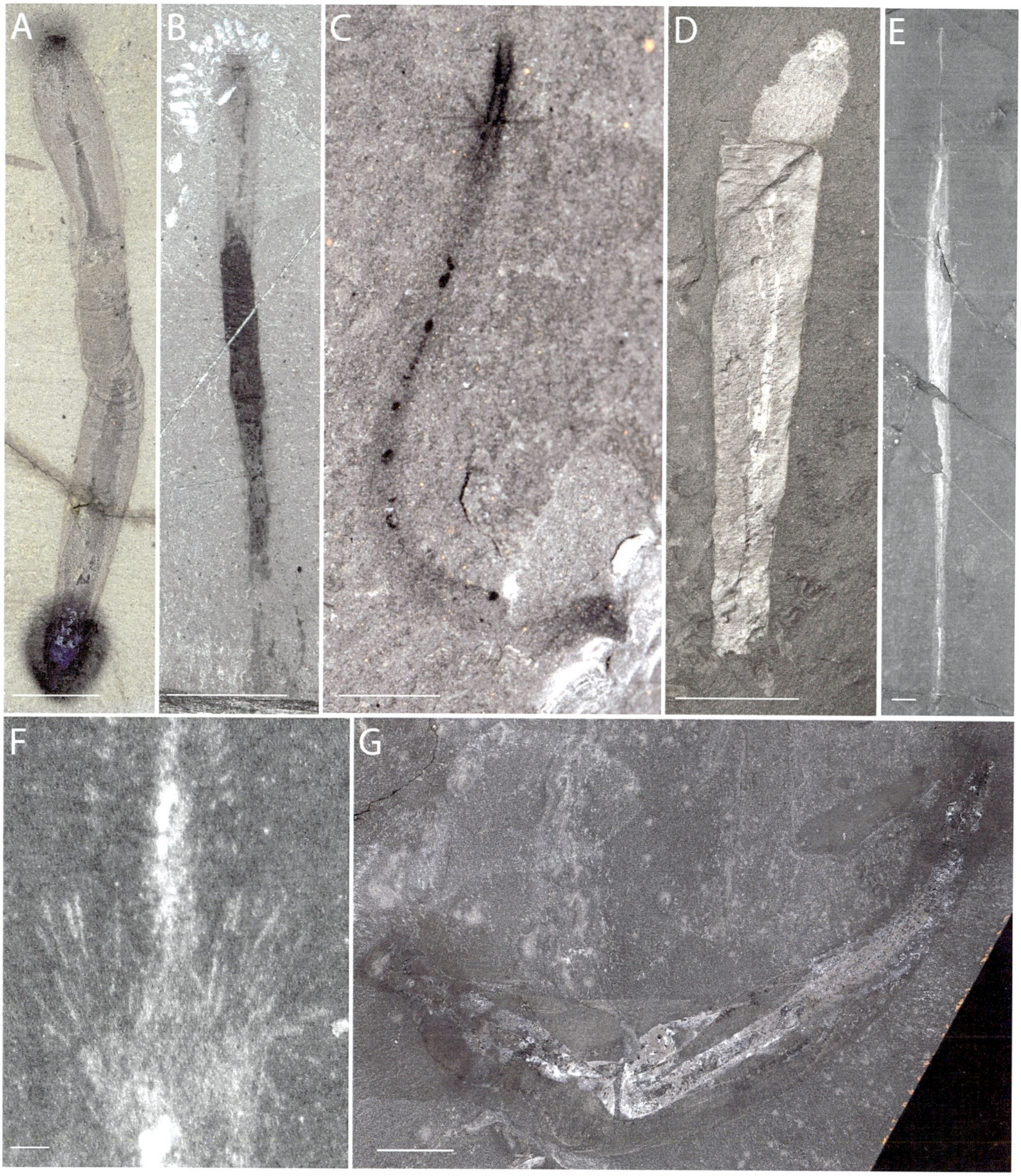

Fig. 2.3.2. Examples of Cycloneuralia from the Burgess Shale lagerstätte, British Columbia, Canada. USNM specimens are from the collection of the United States National Museum in Washington, D.C. Images of these were kindly taken by Jean-Bernard Caron, Toronto (A–D), Zhang Xingliang, Xi'an (F) and provided by Douglas Erwin, Washington, D.C. A. *Ancalogon minor* (Walcott, 1911) (Holotype USNM 57646; cf. Conway Morris 1977, pl. 24, fig. 7). B. *Fieldia lanceolata* Walcott, 1912 (USNM 198597; cf. Conway Morris 1977, pl. 27, fig. 1). C. *Lecythioscopa simplex* (Walcott, 1931) (USNM 83937b; cf. Conway Morris 1977, pl. 28, figs. 5, 6). D. *Selkirkia columbia* Conway Morris, 1977 (paralectotype USMN 83941a; cf. Walcott 1931, pl. 10, fig. 1; Conway Morris 1977, pl. 15, fig. 2). E, F. *Louisella pedunculata* Walcott, 1911 (paratype USNM 57616b). E. Complete specimen (cf. Conway Morris 1977, pl. 20, fig. 3). F. Close-up of transition between introvert and pharynx (cf. Conway Morris 1977, pl. 20, fig. 4). G. *Ottoia prolifica* Walcott, 1911 (IPB Waloszek 1 from the collection of the Steinmann Institute, University of Bonn; see also Chen et al. 2007, their fig. 4). The specimen is surrounded by numerous hyoliths, which are also present in the gut of the O. *prolifica* specimen. Scale bars: 1 cm except C, F (1 mm).

even the Cycloneuralia (see below). An assignment to Priapulida can therefore only be given if autapomorphies of Priapulida are evident in a fossil.

'Orsten' Cycloneuralia

The major difference between the 'Orsten' and the earlier mentioned lagerstätten is that the 'Orsten' is more of a type of preservation rather than a lagerstätte. Characteristic for this type of preservation is the impregnation of surface structures – largely restricted to the cuticle of cycloneuralian and arthropod fossils – by phosphate and the uncompressed preservation of fossils displaying very detailed structures such as setae, pores, spines and other surface ornaments still in place (e.g., Müller 1985; Müller & Walossek 1991; Maas et al. 2006). The fossils themselves are mainly in a size range of 100 to less than 1,000 µm, i.e. they are between ten and 100 times smaller than those of the Chengjiang and Burgess Shale deposits. This also explains the almost lacking faunal overlap of the 'Orsten' and the two other lagerstätten. The 'Orsten' became most famous for its extremely detailed arthropod fossils that are even preserved as a number of ontogenetic stages for a variety of crustacean species. Based on this it was, for instance, possible to reconstruct the ontogenesis for a number of Crustacea (a taxon completely missing from Burgess Shale and Chengjiang fossil assemblages) of different evolutionary levels (Müller & Walossek 1986, 1988; Walossek 1993; Maas et al. 2003; Haug, J. T. et al. 2009c), euarthropods (Müller & Walossek 1987 for *Agnostus pisiformis*), and the re-evaluation of the early evolution of Crustacea by recognition of specific heterochronic events during development (Haug et al. 2010a,b). Although 'Orsten' material has been found now on almost all continents and assigned to different time frames throughout the Palaeozoic and Mesozoic, the majority of the material is from the Cambrian of Sweden (Maas et al. 2006; Upper Series 3 and Furongian of the Cambrian, between 490 and 503 Ma old). However, this material lacks cycloneuralian fossils, whereas the slightly older material (Series 3 of the Cambrian, between 504 and 510 Ma old) from the 'Orsten' of Australia and China yielded presumably immature stages of three different types of Cycloneuralia. A 145 µm long, most likely larval and single specimen was described as representing the species *Shergoldana australiensis* Maas, Waloszek, Haug & Müller, 2007 (Fig. 2.3.3). Due to its fine preservation its morphology is now known in much detail (Maas et al. 2007b). The body of the only known stage of this species is tubular and almost circular in cross section. It is divided into

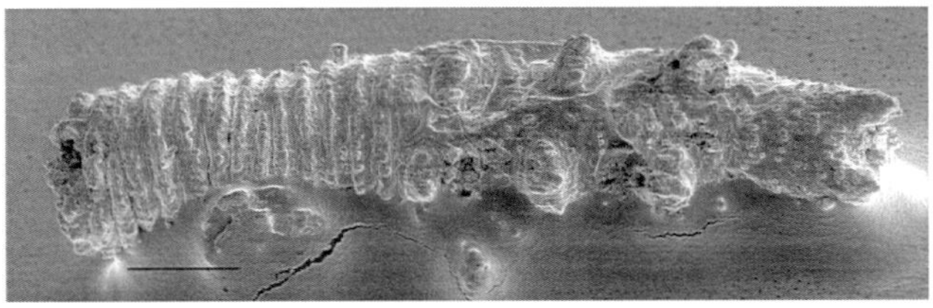

Fig. 2.3.3. *Shergoldana australiensis* Maas, Waloszek, Haug & Müller, 2007. Holotype and only known specimen CPC 23065 (Commonwealth Palaeontological Collection of Geoscience Australia, Canberra) in a supposed ventral view. Scale bar: 20 µm.

four distinct regions. The frontal region is restricted to two surrounding rings around the mouth. The inner ring comprises seven radial folds, each forming a triangular cushion enhousing a medially pointing spine. In the outer ring 18–19 humps form the frontal margin. This frontal region bears considerable similarity with that of larvae of nematomorphs. Here a ring of six (instead of seven) medially pointing spines is also situated on a cushion-like structure. The main difference is the significant backward extensions of these spines in nematomorph larvae. These spurs can be understood as a modification from a simple spine. Again, the ventral spine is drawn out into a pair of spurs, so the total number of spurs matches again the number of spines in *S. australiensis* (Maas et al. 2007b). The succeeding region of *S. australiensis*, making up almost one half of the body, is cylindrical and annulated, with 12–13 distinctly ridged helical folds that give this region an accordion-like appearance (Maas et al. 2007b). Again, this region is significantly similar to that following the mouth region of a nematomorph larva. A similarly long body region in *S. (= Shergoldana) australiensis* follows, in which the body bears 12 cuticular, pentagonal plates that are drawn out medially into a posteriorly curved spine that is flanked by secondary spines on either side. This body region does not show any correspondence to morphologies realized in extant Cycloneuralia. Although similar spine-bearing plates – scalids – occur among the cycloneuralian in-group Scalidophora, the scalids, constituting the introvert region directly behind the mouth resp. pharyngeal region (Lemburg 1999), the position of these structures in *S. australiensis* does not allow for determination of the respective region of *S. australiensis* as the introvert. The terminal region of *S. australiensis* makes up about 20% of the total body length. It ends bluntly and is drawn out into a pair of spines ventro-terminally. Both the plate-bearing and terminal region are covered with rows of microhairs (Maas et al. 2007b). A bifid body end is quite widespread among extant Cycloneuralia (paired paddles in loriciferan larvae; Kristensen 2002; paired cuticular outgrowths in kinorhynchs; Neuhaus & Higgins 2002) and also gastrotrichs (Schmidt-Rhaesa 2002). Presence of this structure in a Cambrian cycloneuralian indicates its plesiomorphic condition among Cycloneuralia. The evolutionary significance of morphological details of *S. australiensis* was discussed by Maas et al. (2007b). The mixture of features of this species shared with different cycloneuralian taxa indicates the cycloneuralian affinity of *S. australiensis*. Moreover it also underlines the finding gained from other Cambrian cycloneuralian species that Cycloneuralia comprised many morphologies in the Cambrian that are not realized anymore today.

Another 'Orsten' cycloneuralian species is *Orstenoloricus shergoldii* Maas, Waloszek, Haug & Müller, 2009 (Fig. 2.3.4). This species is known from 15, apparently larval, specimens (Maas et al. 2009). The known developmental stage of *O. (= Orstenoloricus) shergoldii* possessed an anterior body region, termed neck, that comprises ten annular folds, and a posterior body region comprising 20 axially arranged plates. Such morphology, i.e. cuticular plates arranged in a bag-like structure in the posterior body region, is also developed in the form of the lorica of modern priapulid larvae and larval and adult Loricifera (Lemburg 1999). Moreover, spines on the lorica of *O. shergoldii* are in the same positions as in extant species. Based on these characters Maas et al. (2009) discussed the affinities of *O. shergoldii* within the taxon *O. Orstenoloricus shergoldii*, Priapulida, Loricifera. The assignment of many other Cambrian cycloneuralian taxa to the Priapulida was founded merely on vinctiplicatan, scalidophoran or cycloneuralian characters, accordingly on symplesiomorphies rather than on autapomorphies of the in-group Priapulida (see above and Maas et al. 2007a).

The two aforementioned species from the Australian 'Orsten' material are not known from elsewhere. The same material also contains two presumably embryonic specimens of another species of Cycloneuralia, *Markuelia lauriei* Haug, Maas, Waloszek, Donoghue & Bengtson, 2009. This is the most recent description of a species of *Markuelia* Valkov, 1983, a taxon known from at least two earlier described species. *Markuelia secunda* Valkov, 1983 was first named by Valkov (1983) from the east Siberian Pestrotsvet Formation (Series 1 of the Cambrian, about 530 Ma old), near River Aldan, southern Yakutia (Bengtson & Yue 1997). *Markuelia hunanensis* Dong & Donoghue, 2004 is from the Bitiao-Formation of Wangcun (Series 1 of the Cambrian), Hunan Province, southern China (Donoghue et al. 2006a,b). Donoghue et al. (2006a, their fig. 1 E) depicted a single specimen of *Markuelia* sp. from the earliest Ordovician Vinini Formation (about 485 Ma old) of Battle Mountain, near Carlin, northern Nevada, USA. All specimens representing either species are an almost 2 mm long worm curled in a specific way into a ball and, occasionally, covered by an egg shell (Bengtson & Yue 1997; Donoghue et al. 2006b), i.e. they most likely represent a late embryonic stage shortly before hatching. Although Dong (2007) could document developmental changes during the time in the egg envelope for *M. hunanensis* nothing is known about a free-living phase of any of the species. It has so far not been possible to assign material to later stages of this species (Haug, J. T. et al. 2009a; see also Duan et al. 2012).

Sirius Passet Cycloneuralia

The Sirius Passet fossil Lagerstätte of Peary Land, Northern Greenland (Conway Morris et al. 1987) has yielded a

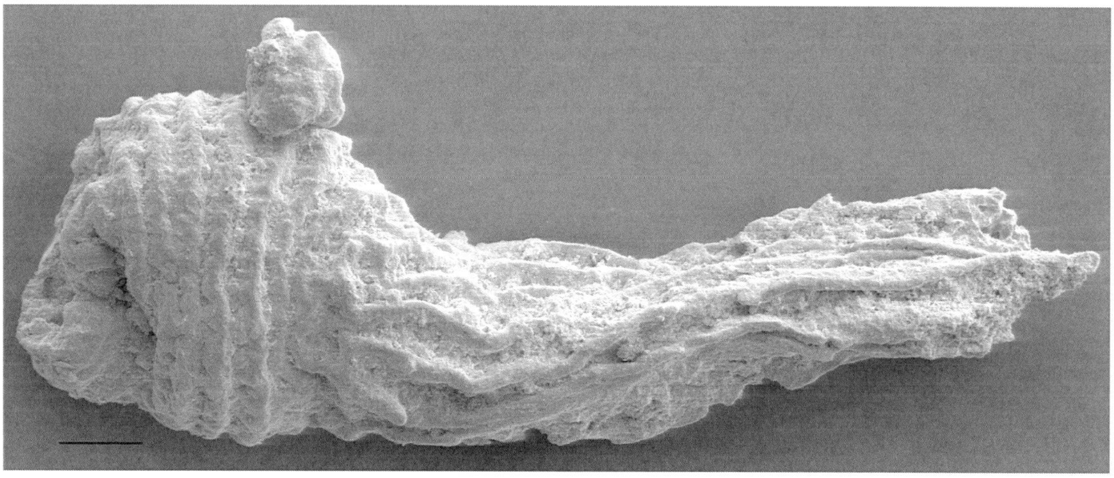

Fig. 2.3.4. *Orstenoloricus shergoldi* Maas, Waloszek, Haug & Müller, 2009. Lateral view of specimen CPC 39936 (see Maas et al. 2009, their fig. 4H). Scale bar: 50 µm.

number of mainly arthropod fossils from early Cambrian times (e.g., Budd 1995, 1998; Williams et al. 1996; Stein 2010). Recently, two species of soft-bodied Palaeoscolecida have been described, *Chalazoscolex pharkus* Conway Morris & Peel, 2010 (Fig. 2.3.5 A; Tab. 2) and *Xystoscolex boreogyrus* Conway Morris & Peel, 2010 (Fig. 2.3.5 B; Tab. 2; Conway Morris & Peel 2010). The material is assigned to Series 2, stage 3 of the Cambrian era, i.e. it is between 515 and 521 Ma old. The material of the two species reveals a scalid-bearing introvert, an ornamented trunk with cuticular sclerites and ridges, and a terminal end with circles of sclerites. Conway Morris & Peel (2010) interpreted the two species as being closely related, and the authors reconstructed the lifestyle of the two species as infaunal to epifaunal. Specimens of species of the arthropod taxon *Isoxys* Walcott, 1890 in the gut of the worms were interpreted as food (Conway Morris & Peel 2010). Another cycloneuralian species from the Sirius Passet material was described by Peel (2010) as *Sirilorica carlsbergi* Peel, 2010 (Fig. 2.3.5 C; Tab. 2). This species is known from various specimens possessing loricae with lengths up to 8 cm. These loricae consist of two circles of seven plates each, the plates of one circle being twice as long as those in the other circle. Peel (2010) suggested a close relationship between *Shergoldana carlsbergi* and the Loricifera, a systematic position that was also proposed for the Chengjiang species *Sicyophorus rara* (Maas et al. 2007a), a suggestion based on the presence of the lorica in rather large, i.e. interpreted as adult, specimens.

Cycloneuralia from other Cambrian Lagerstätten

More fossil Cycloneuralia are known from other areas around the world and other time periods (cf. Tab. 1, 2; see also Conway Morris 1977). This holds especially for the material of disarticulated palaeoscolecid sclerites, cuticular remains (Fig. 2.3.6 A–F) or compressed bodies. As stated above, palaeoscolecidans first became known in two different appearances: as complete worms with almost no structural details known (e.g. Ulrich 1878; Whittard 1953) and as disarticulated sclerites that have been given regular species names. Subsequently,

Fig. 2.3.5. Examples of Cycloneuralia from the Sirius Passet lagerstätte, Greenland. A. *Chalazoscolex pharkus* Conway Morris & Peel, 2010 (Specimen MGUH 29136; collection of the Geological Museum, Copenhagen, Denmark). B. *Xystoscolex boreogyrus* Conway Morris & Peel, 2010 (MGUH 29144). C. *Sirilorica carlsbergi* Peel, 2010 (MGUH 29156). A–C: courtesy of John Peel, Uppsala. Scale bars: A, B 10 mm; C 1 mm.

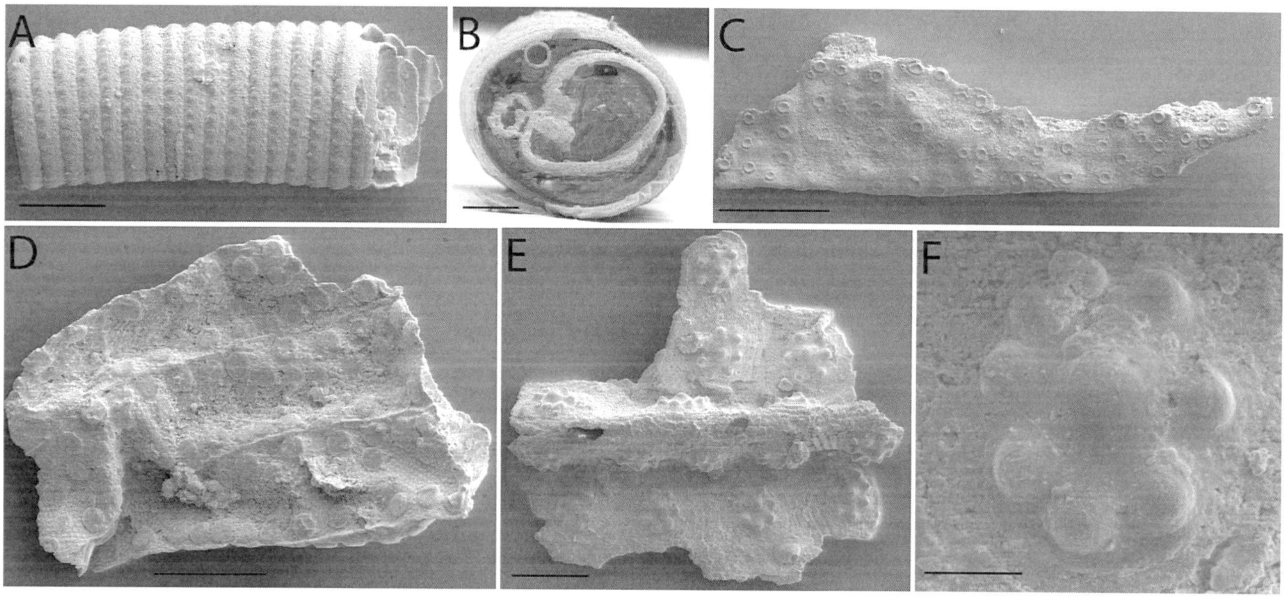

Fig. 2.3.6. Examples of remains of Palaeoscolecida based on microfossils from the Australian 'Orsten', extracted from Cambrian limestone rock from Mt. Murray, Duchess Embayment, Queensland (cf. Müller & Hinz-Schallreuter 1993; Walossek et al. 1993; Maas et al. 2007b, 2009). A, B, Fragment of *Austroscolex spatiolatus* Müller & Hinz-Schallreuter, 1993 (CPC 39944); A, Lateral view (see also Maas et al. 2009, their fig. 2A). Note the annulated surface with rows of button-shaped structures; B, Frontal view with internal layers of tissue preserved; C, Cuticular fragment with about 13 annuli preserved displaying their more or less regularly arranged button-shaped structures; D, Cuticular fragment with ornamented surface and distinctly corroded buttons; E, F, aff. *Hadimopanella oezguli* Gedik, 1977 (CPC 23019); E, Cuticle fragment with prominent button-shaped structures and surrounding ornamentation (see Maas et al. 2009, their fig. 2 B); F, Close-up of one of the button-shaped structures. Scale bars: A 200 µm; B 100 µm; C 500 µm; D 200 µm; E 100 µm; F 20 µm.

a connection between these two appearances could be drawn (Boogaard 1989b, Kraft & Mergl 1989). It is beyond the scope of this paper to present a comprehensive overview of the material found so far. Only a few examples are mentioned to emphasise the ecological significance of the Palaeoscolecida in the Cambrian. This is already evident by the high number of individuals found especially in the Maotianshan Shale of China, although most of them are not Palaeoscolecida s. str. (Harvey et al. 2010 and above).

The first palaeoscolecidan fossil found was the body fossil of *Protoscolex ornatus* Ulrich, 1878 from the Ordovician Eden Series (~455 Ma) of Covington, Kentucky, USA (Ulrich 1878). The animals are annulated and bear one or two rows of papillae per annulus (Conway Morris 1977). A single palaeoscolecidan specimen from the Cambrian Kinzers Formation (most likely Series 2, ~520 Ma) of Pennsylvania, USA was described as *Ottoia* sp. (Resser & Howell 1938; see also Conway Morris & Peel 2010 for additional compressed palaeoscolecidan material from this formation). *Palaeoscolex piscatorum* Whittard, 1953 has been described from the Tremadocian, i.e. the earliest Ordovician (~478–488 Ma), of Shropshire and Gloucestershire, England (Whittard 1953; Conway Morris 1997) and *Protoscolex latus* Bather, 1920 is from the Lower Ludlow beds in England (Silurian age, ~420 Ma; Bather 1920). Specimens with characteristic rows of papillae encircling the annuli of the body were described as *Protoscolex magnus* Miller & Faber, 1892 from the Ordovician Eden Series (~455 Ma) of Cincinnati, Ohio, USA (Miller & Faber 1892). Ruedemann (1925) described *P. batheri* Ruedemann, 1925 from the Silurian Lockport Limestone of Gasport, New York, USA. Kraft & Mergl (1989) reported material from several Ordovician localities from the Prague Basin, Czech Republic. Zhang & Pratt (1996) described disarticulated sclerites from the Cambrian of Shaanxi, China. Again, sclerites of fossil palaeoscolecids have been described as different taxa, such as *Hadimopanella* Gedik, 1977 (Lower to Upper Cambrian), *Kaimenella* Märss, 1988 (Cambrian-Ordovician boundary beds) and *Milaculum* Müller, 1973 (Lower Cambrian to Lower Silurian) (see above and Tab. 1). Some of these were eventually discovered in their cuticular context belonging to larger units rather than being just isolated pieces or even entire animals (Fig. 2.3.6 A–F; e.g., Hinz et al. 1990; Müller & Hinz-Schallreuter 1993; Hou & Bergström 1994). Such combination of isolated pieces and cuticular remains could also be reported by Ivantsov & Wrona (2004) and Ivantsov et al. (2005) from the Sinsk Formation at the Achchagyy Tuoydakh fossil Lagerstätte, eastern Siberia, Russia (~520 Ma). Yet, the palaeoscolecidan material including that from the 'Orsten' of Australia is rather fragmentarily preserved. Unfortunately no specimen reveals a well-preserved frontal body portion. Only the complete palaeoscolecidan specimens from the Maotianshan Shales of China give an indication of the frontal portion, which consisted of an introvert part with scalids and a spine-bearing pharynx in front of it, but the preservation of the material is too poor to exhibit this in more detail. Conway Morris (1977) suggested taxonomic revision of the *Protoscolex-Palaeoscolex* complex (see also Conway Morris et al. 1982; Müller & Hinz-Schallreuter 1993). However, nothing new has been done so far.

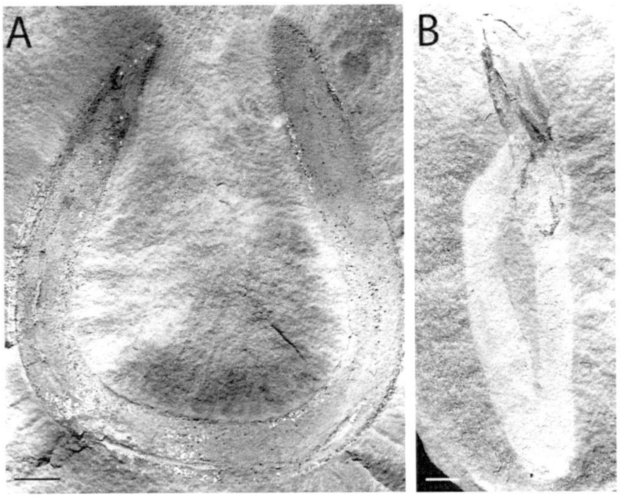

Fig. 2.3.7. *Nemavermes mackeei* Schram, 1973. Upper Pennsylvanian Mazon Creek Essex fauna of northern Illinois, USA. Images kindly provided by F. Schram, Langley, WA, USA. A, Holotype, slightly distorted specimen PE 21551 (collection of the Field Museum of Natural History, Chicago, Illinois); B, Complete specimen H124 (Private collection of Jerry Herdina, Berwyn, Illinois). Scale bars: 5 mm.

In-group cycloneuralian fossil taxa and trails assigned to Nematoda, Nematomorpha and Priapulida

Possibly the oldest fossil cycloneuralian that can be unambiguously assigned to the taxon Nematoda is *Palaeonema* phyticum Poinar, Kerp & Hass, 2008. It was reported from the Lower to Middle Devonian Rhynie Chert (~388–404 Ma old) and was probably associated with early land plants. Since the preservation of Rhynie Chert fossils is comparably good, the more than 500 specimens available revealed the finest morphological details, such as the buccal cavity, sensory structures and more (see Poinar et al. 2008 for more information). Additional fossils were reported that can be assigned to Nematoda (see Poinar et al. 1994 for a early Cretaceous (~120–135 Ma) nematode from Lebanese amber and Arduini et al. 1983 for *Eophasma jurasicum* Arduini, Pinna & Tereizzi, 1983 from the Jurassic (Sinemurian; ~192 Ma) of Italy). Schram (1973) reported material from the Upper Pennsylvanian Mazon Creek Essex marine fauna of northern Illinois, USA (~300 Ma; see Schellenberg 2002 for details on the Mazon Creek fossil beds) and described it as representing the nematode species *Nemavermes mackeei* Schram, 1973 (Fig. 2.3.7 A, B).

Subsequently Schram (1979) assigned nine specimens of a putative nematode from the Mississippian (~325 Ma) Bear Gulch limestone of central Montana, USA, based on gross similarity, to the same species. All specimens share a similar body shape and size comparable to that of a free-living nematode, such as *Caenorhabditis elegans* (Maupas, 1900). Schram (1973) pointed out that the fossils exhibit some structures, i.e. hair- or spine-like outgrowths present on the surface of the fossils, that make *N. mackeei* similar to modern species of the taxa Chromadorida and Monhysterida, which have a comparable cuticular ornamentation. Also some head structures observed on the fossil specimens may be interpreted as oral papillae or cirri (Schram 1973), but all this does not help to clarify the systematic position of *N. mackeei* any further because these characters are plesiomorphic characters for Nematoda. Re-investigation of the material using modern microscopical methods successfully applied to arthropod material from other fossil lagerstätten (Haug, C. et al. 2009; Haug, J. T. et al. 2009b, 2011) is needed and promising. Further reports on younger-dated fossil nematodes (and also nematomorphs) are associated, e.g., with finds of fossil mammals or arthropods, including amber fossils (see Poinar, in press and Schmidt-Rhaesa, this volume for details).

A further cycloneuralian species from the Mazon Creek fossil beds of Illinois, USA is *Priapulites konecniorum* Schram, 1973 assigned to the scalidophoran taxon Priapulida (Fig. 2.3.8 A, B; Schram 1973; Conway Morris 1977). Another, yet unconsidered specimen from the Mazon Creek may also be a representative of the same species (Fig. 2.3.6 C, D). The specimen is about 4 cm long. It comprises an ovoid introvert and a tubular trunk. The introvert bears rows of scalids, and the trunk is annulated. It ends in a slenderer trunk end piece. Details cannot be given until the specimen is investigated.

A number of fossil finds refer to nematodes and/or their traces in the sediment, i.e. sinusoidal trails known as *Cochlichnus anguineus* (Moussa 1970; Metz 1998). These characteristic traces are caused by the specific movement of nematodes in waves as their body wall comprises only longitudinal muscles but no ring musculature

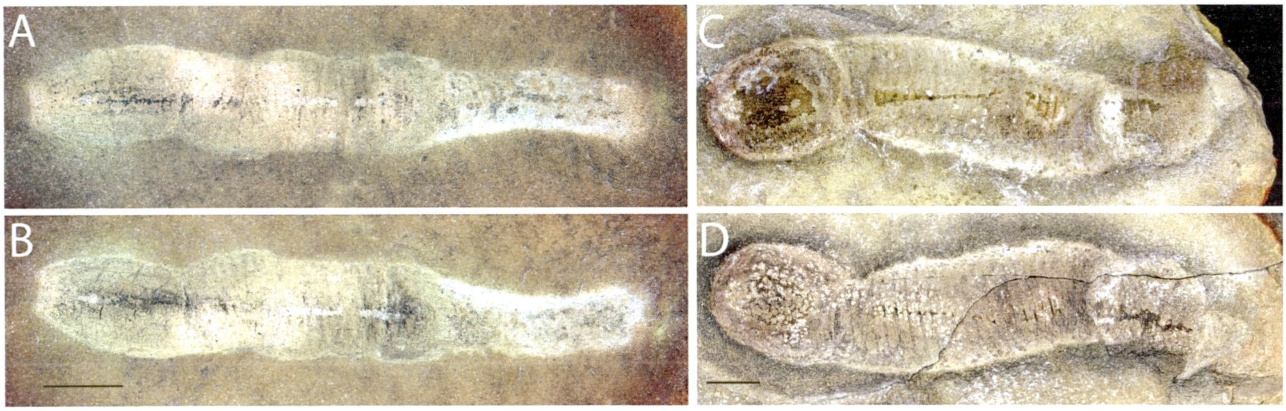

Fig. 2.3.8. Putative Priapulida from the Upper Pennsylvanian Mazon Creek Essex fauna of northern Illinois, USA. Images taken and kindly provided by Joachim & Carolin Haug, Greifswald. Scale bars: 5 mm. A, B, *Priapulites konecniorum* Schram, 1973 (see also Schram 1973; Fitzhugh & Sroka 1997). A, Part of specimen ROM 47975 (collection of the Royal Ontario Museum, Toronto, Canada); B, Counterpart of specimen ROM 47975; C, D, Yet undescribed Mazon Creek priapulid, possibly also representing *Priapulites konecniorum*; C, Part of specimen ROM 45522; D, Counterpart of specimen ROM 45522. Scale bars 5 mm.

(an autapomorphy of Nematoida). Although such traces may also be caused by other animals, some are associated with nematodes. Moussa (1970) reported nematode trails from the Green River Formation (Eocene, ~48 Ma) in the Uinta Basin, Utah, USA, and Metz (1998) documented nematode fossil trails from the Late Triassic (~210 Ma) of Pennsylvania, USA. Most recently, Knaust (2010) introduced a new fossil lagerstätte from the Middle Triassic (Upper Muschelkalk, ~237 Ma) close to Weimar, Germany including nematodes and their traces – among individuals and their traces of other taxa such as Foraminifera, Platyhelminthes, Nemertini, Annelida and Arthropoda.

Ecological significance of Palaeozoic Cycloneuralia

The number of species and especially the high abundance, in which some of these species are known, underline the ecological significance of Cycloneuralia already in the Cambrian. The specific morphology of the Cycloneuralia, with a scalid-bearing introvert mainly for locomotion and a spine-bearing pharynx, indicates that they were predators, at least originally, and hunted for prey that might even have reached a size range comparable to that of the cycloneuralian species themselves. However, little is known about their specific diet in the Cambrian. An exception is *Ottoia* prolifica, of which specimens were found with the gut full of hyoliths (Fig. 2.3.2; Conway Morris 1977; Chen et al. 2007). A lot of information has been accumulated about the possible life habits of Cambrian Cycloneuralia (Huang et al. 2004a; Han et al. 2007c; Maas et al. 2007a; Vannier et al. 2010). According to these studies the animals may have lived on or slightly within the soft sediment – the sea bottom of the original Chengjiang biota comprised a soft sandy sediment – where they could have searched for prey. Vice versa, the many various Cycloneuralia were, most likely, also prey organisms chased by hunters that lived close to the sea bottom, mainly arthropods (Liu et al. 2007; Maas et al. 2007a). Particularly remarkable is the morphological range of especially Cambrian Cycloneuralia. Long, worm-like animals with an introvert, such as the various palaeoscolecidans do not exist today anymore, but represented a major characteristic morphological design in the Cambrian. The same holds true for the about 1 cm large grape-shaped forms of the species *Sicyophorus rara* that made a main faunal component in the Maotianshan Shale fossil assemblage (Maas et al. 2007a). Today, the similarly grape-shaped loricate forms are either larvae, and hence about ten times smaller (priapulids and loriciferans), or they comprise the adult Loricifera, which are also at least ten times smaller than the Cambrian forms. This size difference is explained by a possible paedomorphic event in the early stem lineage towards Loricifera (Warwick 2000), a suggestion also supported by Kristensen & Brooke (2002) who pointed out that the high morphological complexity of the Loricifera can only be explained by its evolution from originally larger animals. A faunal overlap between the different Cambrian lagerstätten yielding Cycloneuralia is present, but it is almost insignificant. Large loricate species are only known from the Maotianshan Shale and the Sirius Passet, while one of the main components of the Burgess Shale, such as forms like *Ottoia prolifica*, is hardly present in the other lagerstätten; *Acosmia maotiania* may be regarded as a similar, but rather rare form from the Maotianshan Shale. An example of faunal overlap between Maotianshan Shale and Burgess Shale are two different species of the taxon *Selkirkia* (Tab. 2), introvert-bearing forms that lived in a self-made tube – a design that does not exist today. The clear separation of the taxa both temporally and spatially stresses the integration of such special morphologies in the palaeoecology of the Cambrian time. The 'Orsten' is specific in this respect in that it is a window into the microworld of the Cambrian – of Cycloneuralia. This exceptional view cannot be, technically (preservationally) and morphologically, provided by other lagerstätten. Yet, although such small forms should be expected elsewhere, also the Swedish 'Orsten' lacks cycloneuralian taxa completely, but their occurrence is, as known so far, restricted to the Australian localities (Maas et al. 2006).

Affinities of Palaeozoic Cycloneuralia with extant ones

Traditionally, Palaeozoic Cycloneuralia were treated, in the main, as priapulids. This assumption is based on the presence, in some of the forms, of a spiny introvert and a likewise spiny pharynx, features common to modern priapulids. However these features are not exclusive for Priapulida, but represent, at least, ground-pattern characters of the taxon Scalidophora, of which the Priapulida are an in-group (Lemburg 1999). Accordingly, Maas et al. (2007a) argued that the Cambrian species with a spine-bearing introvert and no other characters that would point to a closer relationship with any in-group taxon, are at most Scalidophora until more detailed information is available. If the introvert was present already in the ground pattern of Cycloneuralia (= Introverta sensu Nielsen 1995), as Nielsen (1995, 2001) suggested, the systematic assignment of Palaeozoic forms could be even more weakly substantiated (but see Nielsen 2012 where he followed Lemburg 1999 in the evolution of the introvert not before the scalidophoran ground pattern). Species such as *Maotianshania cylindrica* and *Cricocosmia jinningensis* were affiliated with nematomorphs. This assumption was based on spiralling of the animals, the introvert-like structure of the nematomorph larvae and the cuticle pattern (Hou & Bergström 1994; Hou et al. 2004). Again, such characters are merely plesiomorphies but not autapomorphies of Nematomorpha, therefore cannot hold for a systematic assignment. The best possible positioning is, at present, that the taxa in question (mainly all from Tab.s 1 and 2) are in-group Cycloneuralia, a conclusion also drawn by Budd (2001) and supported by Maas et al. (2007a). The more specific assignment can only be given for the lorica-bearing forms from the Cambrian, e.g. for *Sicyophorus rara, Orstenoloricus shergoldii* and *Sirilorica carlsbergi* (Tab. 2). These taxa must be components of the cycloneuralian in-group Vinctiplicata (Maas et al. 2007b, 2009) that are autapomorphically characterised by a lorica and include the Loricifera (a taxon of minute interstitial elongate animals) and Priapulida (interstitial and larger forms) (Lemburg 1999). A molecular study by Sørensen et al. (2008) resulted in the opinion of the authors of a close relationship between Loricifera and Nematomorpha. Accordingly, these authors concluded (a posteriori) that the lorica evolved twice independently, that the introvert was part of the ground pattern of Cycloneuralia (cf. Nielsen 2001) and that the cycloneuralian ground pattern included a larva. The latter two options were also proposed by Maas et al. (2007b). However, Maas et al. (2007b) based their conclusion on the presence of a larva on the material of the Cambrian *Shergoldana australiensis* (see above), a species that can at most be assigned to the Cycloneuralia and is only known by a larval/ immature stage. Again, the mixed character set of this species supports Nielsen's (1995, 2001) idea of an introvert already being present in the cycloneuralian ground pattern. A position of the Loricifera outside the Scalidophora would, in the scenario of Maas et al. (2007b), not be necessary. Peel (2010), in describing the Sirius Passet loricate form *Sirilorica carlsbergi* and discussing its putative relationships, supported the idea of a close relationship between Loricifera and Nematomorpha. He, applying the result of the molecular study by Sørensen et al. (2008), asked to consider the possibility that the lorica was already present in the ground pattern of Cycloneuralia. Although there are some morphological coincidences of the (putatively paedomorphic) Loricifera and the larvae of Nematomorpha (Sørensen et al. 2008) the assumption of an independent evolution of the lorica or, alternatively, its presence already in the ground pattern of Cycloneuralia is not supported by any palaeontological nor by any morphological data. Moreover, there is no evolutionary reason how and why a lorica should have become lost secondarily in Nematoda, Nematomorpha and Kinorhyncha. Again, descendance of the arthropods from introvert-bearing forms as supposed by Budd & Jensen (2000, 2003), Budd (2001, 2008) and Conway Morris & Peel (2010) is not supported by fossil Cycloneuralia or early fossil arthropods (Maas et al. 2007a,b). Accordingly the Palaeozoic Cycloneuralia are at least Cycloneuralia, either outside the crown group including

Nematoida and Scalidophora or as more closely related to one of these in-groups (see also Hou & Bergström 1994; Dong et al. 2004; Donoghue et al. 2006a; see also Harvey et al. 2010 for a detailed discussion of this matter). The possible outgroup Gastrotricha is, unfortunately, completely missing in the fossil record. Hence, they will not provide palaeontological data that could shed additional light on the question of the early cycloneuralian evolution.

Acknowledgments

I would like to thank my colleagues from the Workgroup Biosystematic Documentation, especially Dieter Waloszek, for inspiring discussions and the opportunity to work on 'Orsten' material. Again, Dieter Waloszek gave valuable suggestions to improve the manuscript. The depicted material is deposited as given in the figure captions. Jean-Bernard Caron, Toronto; Chen Junyuan, Chengjiang; Douglas Erwin, Washington, D.C., Carolin and Joachim Haug, Greifswald; Hou Xianguang, Kunming; Hu Shixue, Kunming; Han Jian, Xi'an; John Peel, Uppsala and Frederick Schram, Langley, WA, Zhang Xingliang, Xi'an kindly provided images as specified in the figure captions.

Literature

Arduini, P., Pinna, G. & Tereizzi, G. (1983): *Eophasma jurasicum* n. g., n. sp., a new fossil nematode of the Sinemucian of Osteno in Lombardy. *Atti Soc. Ital. Sci. Nat. Museo Civ. Storia Nat.* 124: 61–64.

Bather, F. A. (1920): *Protoscolex latus*, a new 'worm' from Lower Ludlow Beds. *Ann. Mag. Nat. Hist. Ser. Nine* 5: 125–132.

Bendix-Almgreen, S. E. & Peel, J. S. (1988): *Hadimopanella* from the Lower Cambrian of North Greenland: structure and affinities. *Bull. geol. Soc. Denmark* 37: 83–103.

Bengtson, S. (1977): Early Cambrian button-shaped phosphatic microfossils from the Siberian Platform. *Palaeont.* 20: 751–762.

Bengtson, S. & Yue Zhao (1997): Fossilized Metazoan Embryos from the Earliest Cambrian. *Science* 277: 1645–1648.

Boogaard, M. van den (1988): Some data on *Milaculum* Müller, 1973. *Scripta Geol.* 88: 1–25.

Boogaard, M. van den (1989a): A problematic microfossil, *Hadimopanella? coronata* sp. nov., from the Ordovician of Estonia. Rijksmus. *Geol. Min. Ser. B* 92: 179–190.

Boogaard, M. van den (1989b): Isolated tubercles of some Palaeoscolecida. *Scripta Geol.* 90: 1–12.

Brasier, M. D. (1989): China and the Palaeotethyan Belt (India, Pakistan, Iran, Kazakhstan, and Mongolia). In: Cowie, J. W. & Brasier, M. D. (eds) *The Precambrian-Cambrian Boundary*, pp. 40–74, Clarendon Press, Wotton-under-Edge.

Brock, G. A. & Cooper, B. J. (1993): Shelly fossils from the Early Cambrian (Toyonian) Wirrealpa, Aroona Creek, and Ramsay limestones of South Australia. *J. Paleontol.* 67: 758–787.

Budd, G. E. (1995): *Kleptothule rasmusseni* gen. et sp. nov.: an ?olenellinid-like trilobite from the Sirius Passet fauna (Buen Formation, Lower Cambrian, North Greenland). *Trans. Roy. Soc. Edinburgh, Earth Sci.* 86: 1–12.

Budd, G. E. (1998): Stem group arthropods from the Lower Cambrian Sirius Passet fauna of North Greenland. In: Fortey, R. A. & Thomas, R. H. (eds) *Arthropod Relationships*, pp. 125–138, Systematics Association Special Volume Series 55. Chapman & Hall, London.

Budd, G. E. (2001): Why are arthropods segmented? *Evol. Dev.* 3: 332–342.

Budd, G. E. (2008): The earliest fossil record of the animals and its significance. *Philos. Trans. R. Soc. B: Biol. Sci.* 363: 1425–1434.

Budd, G. E. & Jensen, S. (2000): A critical reappraisal of the fossil record of bilaterian phyla. *Biol. Rev.* 74: 253–295.

Budd, G. E. & Jensen, S. (2003): The limitations of the fossil record and the dating of the origin of the Bilateria. In: Donoghue, P. C. J. & Smith, M. P. (eds) Telling the Evolutionary *Time: Molecular Clocks and the Fossil Record*, pp. 166–189. Taylor & Francis, London.

Chen Junyuan & Zhou Guiqing (1997): Biology of the Chengjiang fauna. *Bull. Nat. Mus. Nat. Sci.* 10: 11–106.

Chen Junyuan, Bergström, J., Lindström, M. & Hou Xianguang (1991): Fossilized soft-bodied fauna. The Chengjiang fauna – oldest soft-bodied fauna on earth. *Nat. Geogr. Res. Expl.* 7(1), 8–19.

Chen Junyuan, Zhou Guiqing, Zhu Maoyan & Yeh K.Y. (1996): The Chengjiang Biota. A unique window of the Cambrian explosion. *Nat. Mus. Nat. Sci. Taichung, Taiwan*: 1–222.

Chen Junyuan, Waloszek, D., Maas, A. et al. (2007): Biodiversity, paleoecology, and the role of predation in marine shallow-water ecosystems of the Lower Cambrian Yangtze Plate – with emphasis on arthropods and nemathelminths. *Palaeogeogr. Palaeoclimatol. Palaeoecol.* 254: 250–272.

Chen Liangzhong, Luo Huilin, Hu Shixue, Yin Jiyun, Jiang Zhiwen, Wu Zhilian, Li Feng & Chen Ailin (2002): *Early Cambrian Chengjiang Fauna in Eastern Yunnan, China.* Yunnan Science and Technology Press, Kunming. (in Chinese)

Conway Morris, S. (1977): Fossil priapulid worms. *Spec. Papers Palaeont.* 20: 1–159, 30 pls.

Conway Morris, S. (1997): The cuticular structure of the 495-Myr-old type species of the fossil worm *Palaeoscolex, P. piscatorum* (? Priapulida). *Zool. J. Linn. Soc.* 119: 69–82.

Conway Morris, S. & Peel, J. S. (2010): New palaeoscolecidan worms from the Lower Cambrian: Sirius Passet, Latham Shale and Kinzers Shale. *Acta Palaeont. Pol.* 55: 141–156.

Conway Morris, S. & Robison, R. A., (1986): Middle Cambrian priapulids and other soft-bodied fossils from Utah and Spain. *Univ. Kansas Paleont. Contr.* 117: 1–22.

Conway Morris, S., Pickerill, R. K. & Harland, T. L. (1982): A possible annelid from the Trenton Limestone (Ordovician) of Quebec, with a review of fossil oligochaetes and other annulate worms. *Can. J. Earth Sci.* 19: 2150–2157.

Conway Morris, S., Peel, J. S., Higgins, A. K., Soper, N. J. & Davis, N. C. (1987): A Burgess shale-like fauna from the Lower Cambrian of North Greenland. *Nature* 326: 181–183.

Dong Xiping (2007): Developmental sequence of Cambrian embryo *Markuelia. Chin. Sci. Bull.* 52: 929–935.

Dong Xiping, Donoghue, P.C.J., Cheng, H. & Liu, J. (2004): Fossil embryos from the Middle and Late Cambrian of Hunan, South China. *Nature* 427: 237–240.

Donoghue, P. C. J., Kouchinsky, A., Bengtson, S., Cunningham, J., Dong Xiping, Repetski, J. E., Valkov, A. K. & Waloszek, D. (2006a): Fossilized embryos are widespread but the record is temporally and taxonomically biased. *Evol. Dev.* 8: 232–238.

Donoghue, P. C. J., Bengtson, S., Dong Xiping et al. (2006b): Synchrotron X-ray tomographic microscopy of fossil embryos. *Nature* 442: 680–683.

Gedik, I. (1977): Conodont stratigraphy in the Middle Taurus. *Bulletin of the Geological Society of Turkey* 20, 35–48. (in Turkish with English abstract)

Han Jian, Zhang Zhifei & Shu Degan (2003a): Discovery of the proboscis on *Tylotites petiolaris*. *Northwestern Geology* 36(1): 87–93.

Han Jian, Zhang Xingliang, Zhang Zhifei et al. (2003b): A new platy-armored Worm from the Early Cambrian Chengjiang Lagerstatte, *South China. Acta Geol. Sin.* 77(1): 1–6.

Han Jian, Shu Degan, Zhang Zhifei et al. (2004): The earliest-known ancestors of recent Priapulomorpha from the Early Cambrian Chengjiang Lagerstätte. *Chin. Sci. Bull.* 49: 1860–1868.

Han Jian, Zhang Xingliang, Zhang Zhifei et al. (2006): A new theca-bearing Early Cambrian worm from the Chengjiang fossil Lagerstätte, China. *Alcheringa* 30: 1–10.

Han Jian, Liu Jianni, Zhang Zhifei et al. (2007a): Trunk ornamentation on the palaeoscolecid worms *Cricocosmia* and *Tabelliscolex* from the early Cambrian deposits of China. *Acta Palaeont. Pol.* 52: 423–431.

Han Jian, Yang, Y., Zhang Zhifei et al. (2007b): New observations on the palaeoscolecid worm *Tylotites petiolaris* from the Cambrian Chengjiang Lagerstätte, south China. *Palaeont. Res.* 11: 59–69.

Han Jian, Zhang Zhifei, Liu Jianni et al. (2007c): Evidence of priapulid scavenging from the Early Cambrian deposits, Southern China. *Palaios* 22, 691–694.

Harvey, T. H. P., Dong Xiping & Donoghue, P. C. J. (2010): Are palaeoscolecids ancestral ecdysozoans? *Evol. Dev.* 12: 177–200.

Haug, C., Haug, J. T., Waloszek, D. et al. (2009): New methods to document fossils from lithographic limestones of southern Germany and Lebanon. *Palaeontol. Electron.* 12(3); 6T; 12p.

Haug, J. T., Maas, A., Waloszek, D. et al. (2009a): A new species of *Markuelia* from the Middle Cambrian of Australia. *Mem. Assoc. Austral. Palaeont.* 37: 303–313.

Haug, J. T., Haug, C., Maas, A. et al. (2009b): Simple 3D Images from Fossil and Recent Micromaterial Using Light Microscopy. *J. Micr.* 233: 93–101.

Haug, J. T., Maas, A. & Waloszek, D. (2009c): Ontogeny of two Cambrian stem crustaceans, †*Goticaris longispinosa* and †*Cambropachycope clarksoni*. *Palaeontographica A* 289: 1–43.

Haug, J. T., Waloszek, D., Haug, C. et al. (2010a): High-level phylogenetic analysis using developmental sequences: The Cambrian †*Martinssonia elongata*, †*Musacaris gerdgeyeri* gen. et sp. nov. and their position in early crustacean evolution. *Arthr. Str. Dev.* 39: 154–173.

Haug, J. T., Maas, A. & Waloszek, D. (2010b): †*Henningsmoenicaris scutula*, †*Sandtorpia vestrogothiensis* gen. et sp. nov. and heterochronic effects in early crustacean evolution. *Trans. Roy. Soc. Edinburgh* 100: 311–350.

Haug, J. T., Haug, C., Kutschera, V. et al. (2011): Autofluorescence Microscopy, Excellent Tool for Comparative Morphology. *J. Micr.* 244: 259–274.

Hinz, I., Kraft, P., Mergl, M. et al. (1990): The problematic *Hadimopanella, Kaimenella, Milaculum* and *Utahphospha* identified as sclerites of Palaeoscolecida. *Lethaia* 23: 217–221.

Hou Xianguang & Bergström, J. (1994): Palaeoscolecid worms may be nematomorphs rather than annelids. *Lethaia* 27: 11–27.

Hou Xianguang & Sun, W. (1988): Discovery of Chengjiang fauna at Meishucun, Jinning, Yunnan. *Acta Palaeont. Sin.* 27: 1–12, pls 1–4.

Hou Xianguang, Ramsköld, L. & Bergström, J. (1991): Composition and preservation of the Chengjiang fauna – a Lower Cambrian soft-bodied biota. *Zool. Scri.* 20: 395–411.

Hou Xianguang, Bergström, J., Wang Haifeng, Feng Xianghong & Chen Ailin (1999): *The Chengjiang Fauna. Exceptionally well-preserved animals from 530 million years ago.* Yunnan Science and Technology Press, Kunming. (in Chinese with English summary)

Hou Xianguang, Aldridge, R. J., Bergström, J., Siveter, Da. J., Siveter, De. J. & Feng Xianghong (2004): *The Cambrian fossils of Chengjiang, China – The flowering of early animal life.* Blackwell, Malden.

Hu Shixue (2005): Taphonomy and palaeoecology of the Early Cambrian Chengjiang Biota from Eastern Yunnan, China. *Berliner Paläobiol. Abh.* 7: 1–197.

Hu Shixue, Li Yong, Luo Huilin et al. (2008): New record of palaeoscolecids from the early Cambrian of Yunnan, China. *Acta Geol. Sin.* 82: 244–248.

Hu Shixue, Zhu Maoyan, Steiner, M. et al. (2010): Biodiversity and taphonomy of the Early Cambrian Guanshan biota, eastern Yunnan. *Sci. China Earth Sci.* 53: 1765–1773.

Huang Diying, Vannier, J. & Chen Junyuan (2004a): Anatomy and lifestyles of Early Cambrian priapulid worms exemplified by *Corynetis* and *Anningvermis* from the Maotianshan Shale (SW China). *Lethaia* 37: 21–33.

Huang Diying, Vannier, J. & Chen Junyuan (2004b): Recent Priapulidae and their Early Cambrian ancestors: comparison and evolutionary significance. *Geobios* 37: 217–228.

Huang Diying, Chen Junyuan & Vannier, J. (2006): Discussion on the systematic position of the Early Cambrian priapulomorph worms. *Chin. Sci. Bull.* 51: 243–249.

Ivantsov, A. Y. & Wrona, R. (2004): Articulated palaeoscolecid sclerite arrays from the Lower Cambrian of eastern Siberia. *Acta Geol. Pol.* 54: 1–22.

Ivantsov, A. Y. & Zhuravlev, A. Y. (2005): Paleontological Descriptions: Cephalorhynchs. In: Ponomarenko, A.G. (ed.) Unikal'nye sinskiye mestonakhozhdeniya rannekembriyskikh organizmov. *Trudy Paleontologicheskogo Instituta* 284: 61–72. (in Russian with English summary)

Ivantsov, A. Y., Zhuravlev, A. Y. & Leguta, A. V. (2005): Palaeoecology of the Early Cambrian Sinsk biota from the Siberian Platform. *Palaeogeogr. Palaeoclimatol. Palaeoecol.* 220: 69–88.

Knaust, D. (2010): Remarkably preserved benthic organisms and their traces from a Middle Triassic (Muschelkalk) mud flat. *Lethaia* 43: 344–356.

Kraft, P. & Mergl, M. (1989): Worm-like fossils (Palaeoscolecida; ?Chaetognatha) from the Lower Ordovician of Bohemia. *Sbornik Geologickych Vedeckych Rada P: Paleontologie* 30: 9–36.

Kristensen, R. M. (2002): An introduction to Loricifera, Cycliophora, and Micrognathozoa. *Int. Comp. Biol.* 2: 641–651.
Kristensen, R. M. & Brooke, S. (2002): Phylum Loricifera. In: Young, C. M. (ed.) *Atlas of Marine Invertebrate Larvae*, pp. 179–188. Academic Press, London and New York.
Labandeira, C. C. (2002): Paleobiology of predators, parasitoids, and parasites: death and accomodation in the fossil record of continental invertebrates. *Paleontol. Soc. Papers* 8: 211–249.
Lemburg, C. (1999): *Hypothesen zur Phylogenie der Priapulida und deren Bedeutung für die Evolution der Nemathelminthes.* Cuvillier, Göttingen.
Liu Yu, Hou Xianguang & Bergström, J. (2007): Chengjiang arthropod *Leanchoilia illecebrosa* (Hou, 1987) reconsidered. *GFF* 129: 263–272.
Luo Huilin, Hu Shixue, Chen Liangzhong et al. (1999): *Early Cambrian Chengjiang fauna from Kunming region, China*, pp. 1–127. Yunnan Science and Technology Press, Kunming. pls. 1–32. (in Chinese with English abstract)
Maas, A., Waloszek, D. & Müller, K. J. (2003): Morphology, Ontogeny and Phylogeny of the Phosphatocopina (Crustacea) from the Upper Cambrian 'Orsten' of Sweden. *Fossils Strata* 49: 1–238.
Maas, A., Braun, A., Dong Xiping et al. (2006): The 'Orsten' – more than a Cambrian Konservat-Lagerstätte yielding exceptional preservation. *Palaeoworld* 15: 266–282.
Maas, A., Huang Diying, Chen Junyuan et al. (2007a): Maotianshan-Shale nemathelminths – Morphology, biology, and the phylogeny of Nemathelminthes. *Palaeogeogr. Palaeoclimatol. Palaeoecol.* 254: 288–306.
Maas, A., Waloszek, D., Haug, J. T. et al. (2007b): A possible larval roundworm from the Cambrian 'Orsten' and its bearing on the phylogeny of Cycloneuralia. *Mem. Assoc. Austral. Palaeont.* 34: 499–519.
Maas, A., Waloszek, D., Haug, J. T. et al. (2009): Loricate larvae (Scalidophora) from the Middle Cambrian of Australia. *Mem. Assoc. Austral. Palaeont.* 37: 281–302.
Märss, T. (1988): Early Palaeozoic Hadimopanellids of Estonia and Kirgizia (USSR). *Proc. Acad. Sci. Esfonian SSR Geol.* 37: 10–17.
Metz, R. (1998): Nematode fossil trails from the Late Triassic of Pennsylvania. *Ichnos* 5: 303–308.
Miller, S. A. & Faber, C. (1892): Some new species and new structural parts of fossils. *J. Cincinn. Soc. nat. Hist.* 15: 79–87.
Moussa, M. T. (1970): Nematode trails from the Green River Formation (Eocene) in the Uinta Basin, Utah. *J. Paleont.* 44: 304–307.
Müller, K. J. (1973): *Milaculum* n. g., ein phosphatisches Mikrofossil aus dem Altpaläozoikum. *Paläontol. Z.* 47: 217–228.
Müller, K. J. (1985): Exceptional preservation in calcareous nodules. *Phil. Trans. Roy. Soc. Lond. B* 311: 67–73.
Müller, K. J. & Hinz-Schallreuter, I. (1993): Palaeoscolecid worms from the Middle Cambrian of Australia. *Palaeontology* 36: 549–592.
Müller, K. J. & Miller, J. F. (1976): The problematic microfossil *Utahphospha* from the Upper Cambrian of the western United States. *Lethaia* 9: 391–395.
Müller, K. J. & Walossek, D. (1986): *Martinssonia elongata* gen. et sp. n., a crustacean-like euarthropod from the Upper Cambrian of Sweden. *Zool. Scr.* 15(1): 73–92.
Müller, K. J. & Walossek, D. (1988): External morphology and larval development of the Upper Cambrian maxillopod *Bredocaris admirabilis. Fossils Strata* 23: 1–70.
Müller, K. J. & Walossek, D. (1991): "Orsten" arthropods – small in size but of great impact on biological and phylogenetic interpretations. *GFF Meeting Proc.* 113: 88–90, 4 Text figs.
Neuhaus, B. & Higgins, R. P. (2002): Ultrastructure, biology, and phylogenetic relationships of Kinorhyncha. *Integr. Comp. Biol.* 42: 619–632.
Nielsen, C. (1995): *Animal Evolution: Interrelationships of the Living Phyla, First Edition.* Oxford University Press, Oxford.
Nielsen, C. (2001): *Animal Evolution: Interrelationships of the Living Phyla, Second Edition.* Oxford University Press, Oxford.
Nielsen, C. (2012): *Animal Evolution: Interrelationships of the Living Phyla, Third Edition.* Oxford University Press, Oxford.
Ogg, J. G., Ogg, G. & Gradstein, F. M. (2008): *The Concise Geologic Time Scale (CGTS).* Cambridge University Press, Cambridge.
Peel, J. S. (2010): A Corset-Like Fossil from the Cambrian Sirius Passet Lagerstätte of North Greenland and Its Implications for Cycloneuralian Evolution. *J. Paleont.* 84: 332–340.
Peel, J. S. & Ineson, J. R. (2011): The extent of the Sirius Passet Lagerstätte (early Cambrian) of North Greenland. Bull. *Geosci.* 86(3): 535–543
Peel, J. S. & Larsen, N. H. (1984): *Hadimopanella apicata* from the Lower Cambrian of western North Greenland. *Rapport Grønlands Geol. Undersøgelse* 121: 89–96.
Poinar, G. O. (in press): Palaeontology of nematodes. In: Schmidt-Rhaesa, A. (ed.) *Handbook of Zoology, Gastrotricha, Cycloneuralia and Gnathifera, Vol. Nematoda.* Doi XXX.
Poinar, G. O. jr. & Ricci, C. (1992): Bdelloid rotifers in Dominican amber: evidence for parthenogenetic continuity. *Experientia* 48: 408–410.
Poinar, G. O. jr., Acra, A. & Acra, F. (1994): Earliest fossil nematode (Mermithidae) in Cretaceous Lebanese amber. *Fund. Appl. Nematol.* 17: 475–477.
Poinar, G. O. jr., Kerp, H. & Hass, H. (2008): *Palaeonema phyticum* gen. n., sp. n. (Nematoda: Palaeonematidae fam. nov.), a Devonian nematode associated with early land plants. *Nematol.* 10(1): 9–14.
Repetski, J. E. (1981): An Ordovician occurrence of *Utahphospha* Müller and Miller. *J. Paleont.* 55: 395–400.
Resser, C. E. & Howell, B. F. (1938): Lower Cambrian *Olenellus* zone of the Appalachians. *Bull. geol. Soc. Am.* 49: 195–248.
Ruedemann, R. (1925): Some Silurian (Ontarian) faunas of New York. *Bull. N. Y. St. Mus.* 265: 5–134.
Schellenberg, S. A. (2002): Mazon Creek: preservation in late Paleozoic deltaic and marginal marine environments. In: Etter, W., Hagadorn, J. W., Tang, C. M. & Bottjer, D. J. (eds) *Exceptional Fossil Preservation: A Unique View on the Evolution of Marine Life*, pp. 185–203. Columbia University Press, Columbia.
Schmidt-Rhaesa, A. (2002): Two dimensions of biodiversity research exemplified by Nematomorpha and Gastrotricha. *Integr. Comp. Biol.* 42: 633–640.
Schram, F. R. (1973): Pseudocoelomates and a nemertine from the Illinois Pennsylvanian. *J. Paleont.* 47: 985–989.
Schram, F. R. (1979): Worms of the Mississippian Bear Gulch Limestone of central Montana, USA. *Trans. San Diego Soc. Nat. Hist.* 19: 107–120.
Sørensen, M. V., Hebsgaard, M. B., Heiner, I. et al. (2008): New data from an enigmatic phylum: evidence from molecular sequence data supports a sister-group relationship between Loricifera and Nematomorpha. *J. Zool. Syst. Evol. Res.* 46: 231–239.

Southcott, R. V. & Lange, R. T. (1971): Acarine and other microfossils from the Maslin Eocene, South Australia. *Rec. South Australian Mus.* 16: 1–21.

Stein, M. (2010): A new arthropod from the Early Cambrian of North Greenland, with a 'great appendage'-like antennula. *Zool. J. Linn. Soc.* 158: 477–500.

Sun, W. & Hou Xianguang (1987): Early Cambrian worms from Chengjiang, Yunnan, China: *Maotianshania* gen. nov. *Acta Pal. Sin.* 26: 303–305, pls. 1–2.

Ulrich, E. O. (1878): Observations on fossil annelids, and descriptions of some new forms. *J. Cincinn. Soc. nat. Hist.* 1: 87–91.

Topper, T. P., Brock, G. A., Skovsted, C. B. et al. (2010): Palaeoscolecid scleritome fragments with *Hadimopanella* plates from the early Cambrian of South Australia. *Geol. Mag.* 147: 86–97.

Valkov, A. K. (1983): Rasprostranenie drevnejshikh skeletnykh organizmov i korrelyatsiya nizhnej granitsy kembriya v yugo-vostochnoj chasti Sibirskoj platformy. In: Khomentovsky, V. V., Yakshin, M. S. & Karlova, G. A. (eds) *Pozdnij dokembrij i rannij paleozoj Sibiri, Vendskie otlozheniya*, pp. 37–48. Instituta Geologii i Geofiziki, Novosibirsk.

Vannier, J., Calandra, I., Gaillard, C. & Żylińska, A. (2010): Priapulid worms: Pioneer horizontal burrowers at the Precambrian-Cambrian boundary. *Geology* 38: 711–714

Waggoner, B. M. & Poinar, G. O. Jr. (1993): Fossil habrotrochid rotifers in Dominican amber. *Experientia* 49(4): 354–357.

Walcott, C. D. (1931): Addenda to descriptions of Burgess Shale fossils (with explanatory notes by C. E. Resser). Smithson. misc. Collns. 85: 1–46.

Walossek, D. (1993): The Upper Cambrian *Rehbachiella kinnekullensis* Müller, 1983, and the phylogeny of Branchiopoda and Crustacea. *Fossils Strata* 32: 1–202.

Walossek, D., Hinz-Schallreuter, I., Shergold, J. H. & Müller, K. J. (1993): Three-dimensional preservation of arthropod integument from the Middle Cambrian of Australia. *Lethaia* 26: 7–15.

Warner, B. G. & Chengalath, R. (1988): Holocene fossil *Habrotrocha angusticollis* (Bdelloidea: Rotifera) in North America. *J. Paleolimnol.* 1(2): 141–147.

Warwick, R. M. (2000): Are loriciferans paedomorphic (progenetic) priapulids? *Vie et milieu – Life and environment* 50: 191–193.

Whittard, W. F. (1953): *Palaeoscolex piscatorum* gen. et sp. nov., a worm from the Tremadocian of Shropshire. *Quart. J. Geol. Soc.* 109: 125–135.

Williams, M., Siveter, Da. J. & Peel, J. S. (1996): *Isoxys* (Arthropoda) from the early Cambrian Sirius Passet Lagerstätte, North Greenland. *J. Paleont.* 70(6): 947–954.

Wrona, R. (1982): Early Cambrian phosphatic microfossils from southern Spitsbergen (Horsund region). *Palaeont. Polonica* 43: 9–16.

Wrona, R. (1987): Cambrian microfossil *Hadimopanella* GEDIK from glacial erratics in West Antarctica. *Palaeont. Polonica* 49: 37–48.

Wrona, R. (2004): Cambrian microfossils from glacial erratics of King George Island, Antarctica. *Acta Palaeont. Pol.* 49: 13–56.

Wrona, R. & Hamdi, B. (2000): Middle Cambrian *Hadimopanella* from Mila Formation in the Alborz Mountains, northern Iran. In: Acénolaza, G. F. & Peralta, S. (eds) Cambrian from the southern edge, Instituto Superior de Correlación Geológica (INSUGEO), *Miscelánea* 6: 143–146. San Miguel de Tucumán.

Wrona, R. & Hamdi, B. (2001): Palaeoscolecid sclerites from the Upper Cambrian Mila Formation of the Shahmirzad section, Alborz Mountains, northern Iran. *Acta Geol. Pol.* 51: 101–107.

Zhang Xiguang & Pratt B. R. (1996): Early Cambrian palaeoscolecid cuticles from Shaanxi, China. *J. Paleont.* 70: 275–279.

Zhang Xingliang, Shu Degan, Li Yong & Han Jian (2001): New sites of Chengjiang fossils: crucial windows on the Cambrian explosion. *J. Geol. Soc. Lond.* 158: 211–218.

Zhu Maoyan, Babcock, L. E. & Peng S.-C. (2006): Advances in Cambrian stratigraphy and paleontology: integrating correlation techniques, paleobiology, taphonomy and paleoenvironmental reconstruction. *Palaeoworld* 15: 217–222.

Andreas Schmidt-Rhaesa

3. Nematomorpha

Introduction

The taxon Nematomorpha, commonly known as horsehair worms, is a taxon including at present about 350 species of animals with an interesting life cycle including a parasitic and a free-living phase. Most species are found in fresh water as adults (Fig. 3.1.1) and larvae but almost all species develop in terrestrial hosts. The freshwater/terrestrial species are in the taxon Gordiida. Five species of the genus *Nectonema* live in the sea (Fig. 3.1.2); they are parasites of decapod crustaceans and swim freely in surface waters as adults.

Free-living horsehair worms have been known for over a century (see e.g. Meissner 1856, Girard 1882 and Camerano 1897a for a historical review). Several myths surround this intriguing group of parasites (see Anonymous 1831, Annandale 1904), for example they were thought to be poisonous ("if a cow swallows it she would be dead in a day or two", p. 193 in Baylis 1944). The name "horsehair worm" came from its long and slender form, but includes the belief that they may originate from horsehairs fallen into water in, for example, a watering trough. An experiment conducted by Leidy (1850) tested this idea by placing horsehairs in water. Leidy remarks: "I was at one time so silly to be led to try the experiment, with what success it is unnecessary for me to state." (Leidy 1850, p. 34).

The name *Gordius aquaticus*, which was, for example, used by Linné in his Systema Naturae, was used for free-living gordiids. Endoparasitic worms, probably including gordiids, were at that time summarized under the name *Filaria*. Although it was noted already by Lister (1672) that carabid beetles were host to hairworms, it took until the middle of the 19th century before it was established that horsehair worms have both a parasitic and a free-living phase in their life cycle. The characteristic larva, which is morphologically distinct from the adults, was described first by Grube (1849) and Leidy (1851). Until the end of the 19th century gordiids were often confused or even united with representatives of Mermithida, a nematode taxon with a comparable body size and life cycle (for some differences between both taxa see part 5 below). Long after gordiids were described from fresh water, the marine species *Nectonema agile* was described by Verrill (1879). Vejdovsky (1886) introduced the name Nematomorpha, clearly excluding mermithids.

With the increasing exploration of the world, the number of horsehair worms grew constantly, resulting in the description of increased numbers of species in about 20 genera. Nematomorphs are neither of medical nor of economical importance (although their potential as biological control agents still needs to be tested). This may explain why research on nematomorphs is still patchy, leaving several open questions concerning the life cycle, host–parasite relationships, inter- and intraspecific variability, early development and the exact structure of some organs.

3.1. Morphology

There are several extensive descriptions of the morphology based on histological sections, which still form the basis of the knowledge on nematomorph morphology. The most important ones are: Meissner (1856) on *Gordius aquaticus* and *Parachordodes* tolosanus (as *Gordius subbifurcus*); Grenacher (1868) on *Chordodes ornatus* and *Parachordodes tolosanus* (as *Gordius subbifurcus*); Vejdovsky (1886) on *Parachordodes tolosanus* (as *Gordius tolosanus*); Ward (1892a) on *Nectonama agile*; Vejdovsky (1894) on *Gordionus preslii* (as *Gordius preslii*), *Gordius aestivalis*, *Gordius vaeteri* and *Parachordodes pustulosus* (as *Gordius pustulosus*); Montgomery (1903) on *Paragordius varius*; Rauther (1904, 1905) on *Gordius aquaticus* and *Parachordodes tolosanus* (as *Gordius tolosanus*); Švábenik (1909, 1925) on *Parachordodes tolosanus* (as *Gordius tolosanus*), *Gordionus preslii* (as *Gordius preslii*), *Gordionus violaceus* (as *Gordius violaceus*) and *Gordius affinis*; and May (1919) on *Paragordius varius* and *Gordius robustus*.

Ultrastructural research was very sporadic until Bresciani's (1991) chapter in the Microscopic Anatomy of Invertebrates, which contained a lot of new information. Since then, several investigations provided detailed information on particular organs.

In the following, the morphology of juveniles and adults is described, the morphology of the larvae is described later in this section.

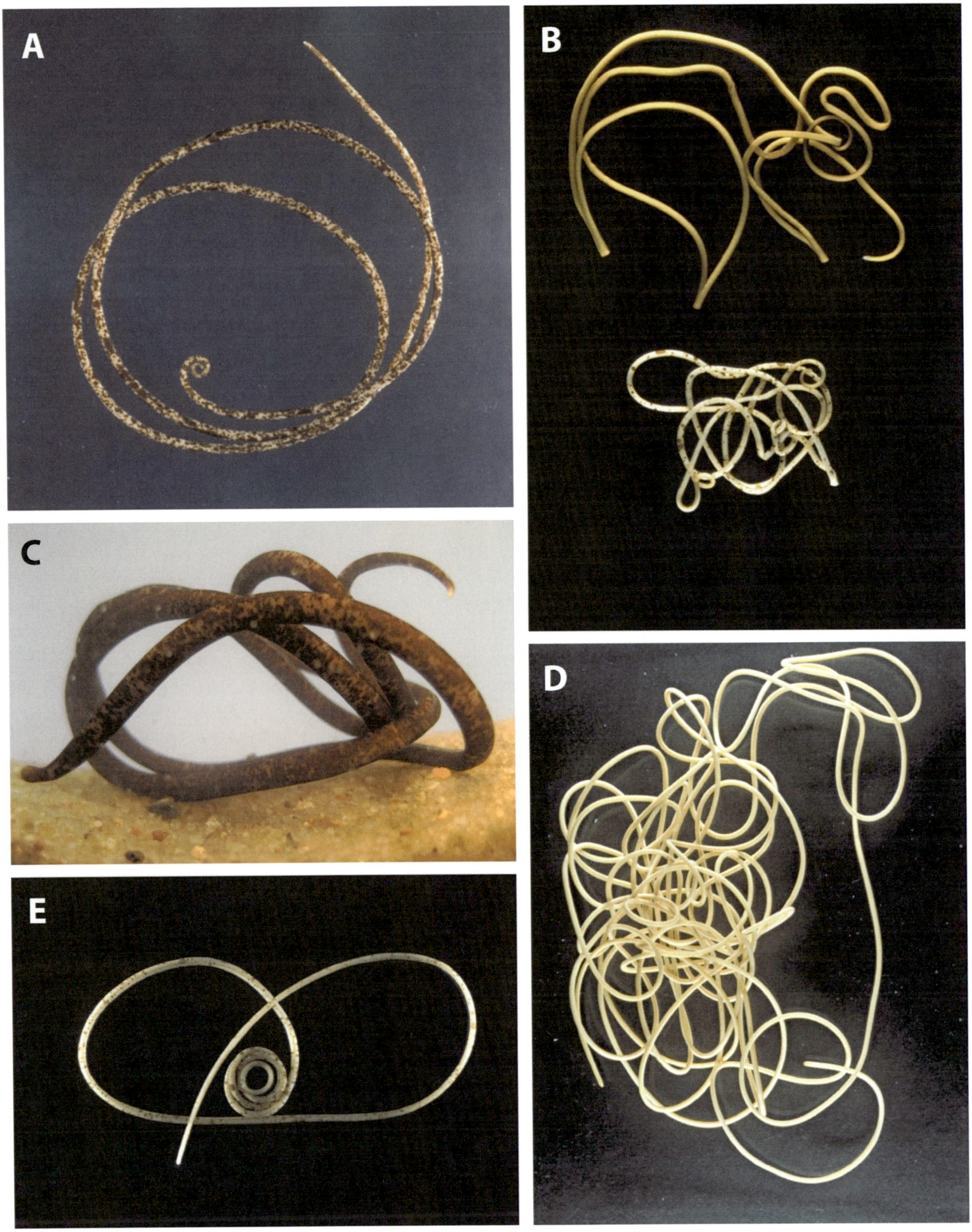

Fig. 3.1.1. A, Male of *Chordodes japonensis* (from Japan) with heavily spotted cuticle; B, Two males of *Spinochordodes tellinii* (from France), showing differences in size and body coloration; C, Female of *Chordodes japonensis* (from Japan); D, Female of *Gordius fulgur* from Fiji (specimen 1978.618–619 from Natural History Museum, London), with about 2200 mm the longest reported specimen of Nematomorpha; E, Male of *Spinochordodes tellinii* (from France).

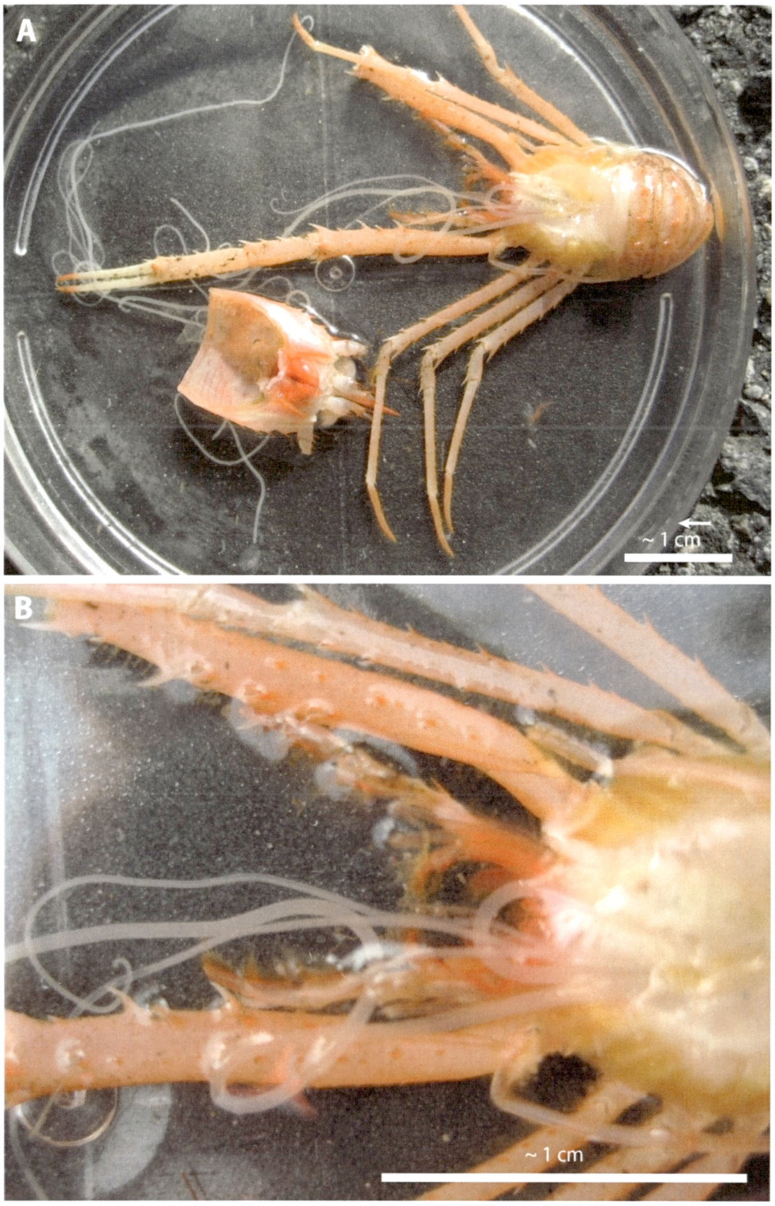

Fig. 3.1.2. A, B, Specimen of *Munida tenuimana* with dorsally opened carapax, infected by females (thicker specimens) and males (thinner specimens) of *Nectonema munidae* (from Korsfjorden near Bergen, Norway).

Morphology of juveniles and adults
General and external morphology

Gordiida. All juveniles and adults are wormlike, meaning that they have a thin and long body (Fig. 3.1.1). The diameter of adults is usually less than 1 mm, only females belonging to species in the genus *Chordodes* may have diameters up to almost 2 mm. The body length varies greatly, from a few centimeters to more than two meters (Fig. 3.1.1 D). Most worms, however, range from about 10 to 30 cm. Variation in length and diameter appears to be a usual pattern. As an example, Schmidt-Rhaesa et al. (2005) found more than twofold (*Paragordius tricuspidatus*) and threefold (*Spinochordodes tellinii*) length variation in a French population of both species.

The anterior end is either spherical (e.g. genus *Gordius*) or distinctly tapering (e.g. genus *Chordodes*). A mouth opening may be visible (but may also be closed) but no other structures are present in the anterior end. The posterior end differs distinctly between sexes. There is always one body opening, the cloacal opening. This is clearly ventral in all males (Fig. 3.1.3 A, B), but terminal or only slightly subterminal in all females (Fig. 3.1.3 C, D). The posterior end in males of several genera (including, e.g., *Gordius, Gordionus, Paragordius* and

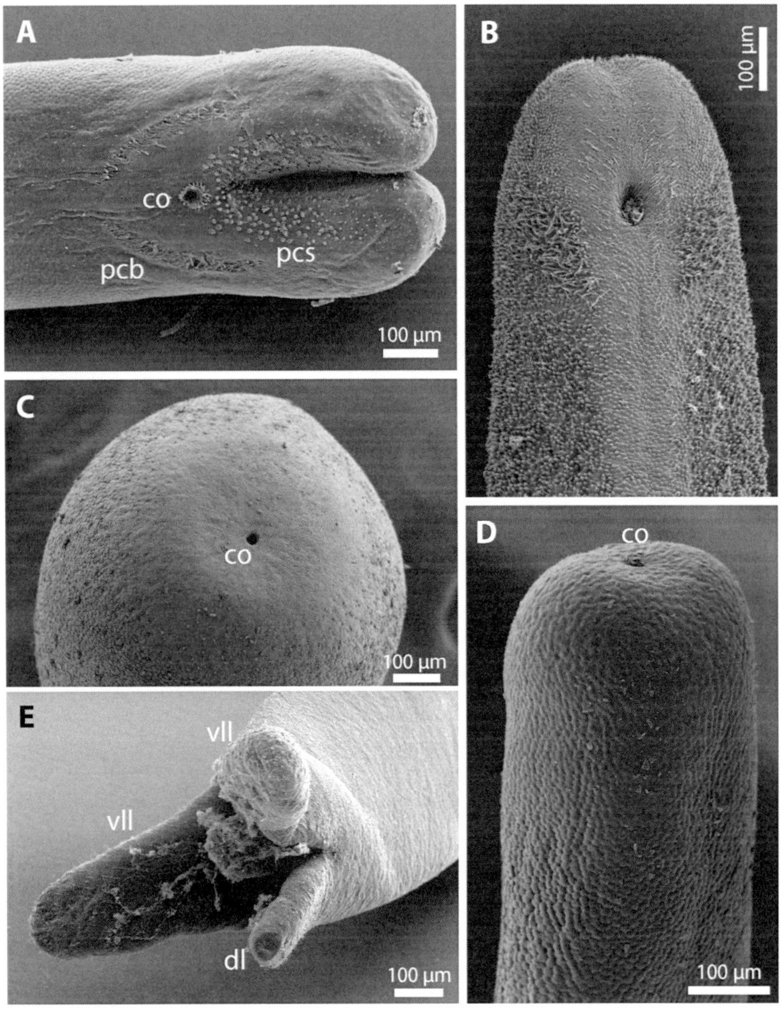

Fig. 3.1.3. Shapes of the posterior end. A, Ventral view on the bilobed posterior end of a male *Gordionus violaceus* (from Estonia), showing the cloacal opening (co), precloacal rows of bristles (pcb) and postcloacal spines (pcs); B, Ventral view on undivided posterior end of *Chordodes mizoramensis* (from India), showing the cloacal opening surrounded by several bristles, some arranged in denser anterolateral fields; C, Posterior end of female *Chordodes japonensis* (from Japan) with terminal cloacal opening (co); D, Posterior end of female *Gordionus violaceus* (from Austria); E, Posterior end of female *Paragordius obamai* (from Kenya) with two ventrolateral lobes (vll) and one dorsal lobe (dl). All images SEM. A from Schmidt-Rhaesa & Prous 2010, with kind permission from the Estonian Journal of Ecology. B from Schmidt-Rhaesa & Lalramliana 2011, with kind permission from Zookeys.

Parachordodes) is divided into two lobes diverging posterior of the cloacal opening (Fig. 3.1.3 A). The length of each tail lobe is usually around twice its diameter, in some species (e.g. from the genus *Paragordius*) they may be longer (see below). In the majority of cases, tail lobes are rounded on their posterior tip, only in the genus *Acutogordius* they are more or less pointed (see below). There are several cuticular structures on the male posterior end that are important for identification. Males of the genera *Gordius* and *Acutogordius* have a semicircular (sometimes parabolic or angled) cuticular fold, the postcloacal crescent (see below). Cuticular spines or bristles are present directly around the cloacal opening (as circumcloacal spines), in the region around the cloacal opening or on the inner side of the tail lobes in several genera. In several species of the genus *Gordionus* and one species of *Beatogordius* (*B. abbreviatus*, see Schmidt-Rhaesa & Bryant 2004), particular structures called adhesive warts are present (see below).

The female posterior end is in most cases rounded (Fig. 3.1.3 C, D). In some genera (e.g. *Chordodes*) it may be slightly enlarged compared to the remaining body (Fig. 3.1.3 D). Bristle-like cuticular structures have been reported only from species in the genus *Beatogordius* (Schmidt-Rhaesa & De Villalobos 2002, Schmidt-Rhaesa & Bryant 2004). Females in the genus *Paragordius* are peculiar in having a three-lobed posterior end, with usually one shorter dorsal and paired lateroventral lobes

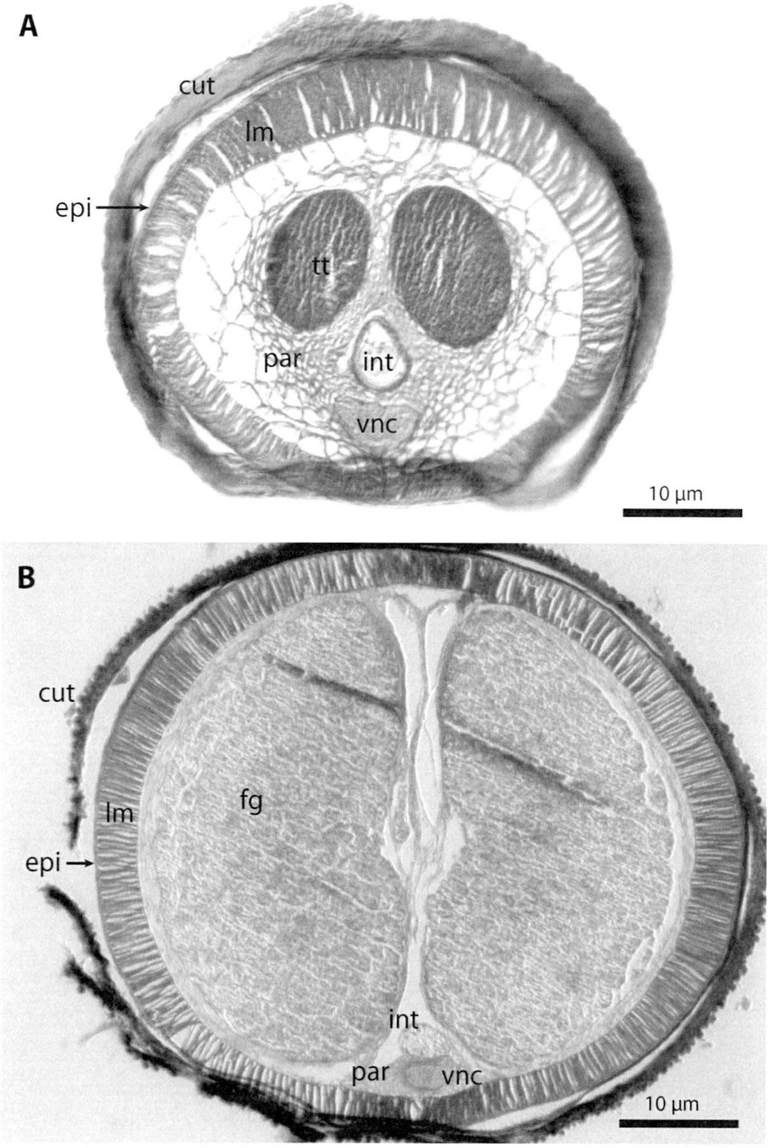

Fig. 3.1.4. A, Histological cross section in midbody region of a male *Pseudochordodes bedriagae* (freom Argentina), showing cuticle (cut), epidermis (epi), intestine (int), longitudinal musculature (lm), parenchyma (par), testes tubes (tt) and ventral nerve cord (vnc); B, Cross section through a female *Chordodes* sp. (from Argentina). Abbreviations as in A, in addition the female gonad (fg) is present.

(Fig. 3.1.3 E). The cloacal opening is central between these lobes (Fig. 3.1.3 C).

The cuticle of the entire body is either smooth or structured into so called areoles (see below under integument). The structure and arrangement patterns of these areoles are important for determination.

The color of gordiids ranges from white to dark brown, covering all shades of brown (Fig. 3.1.1). It is clear that the intraparasitic stages are white and it has been shown that coloration of the body starts within the host, probably in a postero-anterior manner (Schmidt-Rhaesa 2005). It is likely that coloration is linked to the replacement of the larval by the adult cuticle (see below) and it may represent some form of hardening of the cuticle. As white or lightly colored specimens can occasionally be found in the field, it appears that the degree of coloration is not essential for survival in the external environment. It is suspected that the color may change with time, but this has not yet been documented. However, specimens of one species usually display a range of coloration, showing that the color is not a reliable character for determination.

The internal anatomy of gordiids appears to correspond, as far as known data suggest, in general among all species (Fig. 3.1.4). A cuticle is secreted by a cellular epidermis, below which is a layer of radially flattened muscle cells representing only longitudinal musculature. There are intraepidermal nerves and a large ventral nerve

Tab. 1. Reported size measurements of *Nectonema* species.

Species	♂ Length in mm	♀ Length in mm	Author
N. agile	32–130 (n = 17)	34–40 (n = 3)	Ward 1892a
N. melanocephalum	10–47 (n = 9)	–	Nierstrasz 1907
N. svensksundi	–	190 (n = 1)	Bock 1913
N. munidae	90–155 (n = ?)[1]	170–845 (n = ?)[1]	Brinkmann 1930
N. munidae	100–138 (n = ?)[1]	180–960 (n = ?)[1]	Nielsen 1969
N. munidae	~100 (n = 2)	~400 (n = 15)	Schmidt-Rhaesa 1996a
N. zealandica	13–300 (sexes not differentiated)		Poinar & Brockerhoff 2001
Nectonema sp.	–	420 (n = 1)	Bakke 1975
Nectonema sp.	270 (n = ?)[2]	490 (n = ?)[2]	Oku et al. 1983
Nectonema sp.	30–320, females always larger than males		Skaing & MacKinnon 1988

[1] Many individuals, exact number not given.
[2] Estimated from fragments.

cord. A brain is present in the anterior region. Sensory structures are imperfectly known, but there are indications that at least integumental receptors are present. The intestine is present, but reduced and obviously no food uptake takes place in the free-living stages. The intestine joins the gonads in the posterior end, therefore forming a cloaca. Gonads are tube-like, with maturity they extend and carry a large quantity of gametes. The parenchyma, a tissue with assumed storing function, is present, but its extension appears to relate negatively to the extension of the gonads. Particular excretory organs are lacking.

Nectonema. The morphology of the five known *Nectonema* species is less well known compared to gordiids and is based mainly on the two species that have been found regularly, *N. agile* from the Northwestern Atlantic and *N. munidae* from fjords around Bergen (Norway). Like gordiids, *Nectonema* specimens are wormlike, long and very thin (Fig. 3.1.2). The males are always shorter than females. Reported lengths range from 10–270 mm in males and from 34–960 mm in females (see Tab. 1). The diameter is around 1 mm. The anterior and posterior ends are rounded in females; in males the posterior end is curved towards the ventral side and tapers distinctly (see Schmidt-Rhaesa 1999). In adults, the cuticle is smooth and does not contain any surface sculpturing comparable to the areoles in gordiids. The only cuticular structures are natatory bristles present along the dorsal and the ventral longitudinal midline of the animals (Fig. 3.1.5 A, B). Females of *N. munidae* are opaque, males appear to be almost transparent (Brinkmann 1930). Verrill (1879) reported *N. agile* to have a dark pigmentation on the ventral and the dorsal side, although this is not confirmed by Ward (1892a), who reported this species to be also opaque. The species *N. melanocephalum* and *N. svensksundi* are reported to have a darkly colored anterior end (Nierstrasz 1909, Bock 1913).

The internal anatomy of *Nectonema* species roughly corresponds to gordiids in respect of the cuticle, epidermis and musculature (Fig. 3.1.5 C). In addition to the ventral longitudinal nerve cord, a dorsal nerve cord is also present. Both are located in thickenings of the epidermis. The intestine is incomplete and ends blindly. Very different from gordiids is the structure of the anterior end. It contains a body cavity and conspicuous large cells of still unknown function, the giant cells (see below). Little data are known about the structure of the gonads and mature spermatozoa are completely unknown.

Integument

Gordiida. The integument consists of an epidermis (in older publications often termed hypodermis) and a cuticle (Fig. 3.1.6 A, 3.1.7 A, B). Both change during development, in relation to changing functions. During the parasitic phase, nutrients are absorbed through the integument. This has been shown by the demonstration of a differential presence of alkaline phosphatases (Inoue 1959a), the uptake of dyes (Kirjanova 1959) and the presence of voluminous cells rich in endoplasmic reticulum (Schmidt-Rhaesa 2005). While epidermal cells appear physiologically active in the parasitic phase, the cuticle is a thin layer that is a minimal barrier for nutrient uptake. During development, this cuticle (named larval cuticle) is replaced by a second cuticle, the adult cuticle, which is typical for the free-living phase. This is much thicker and consists of several layers. The epidermal

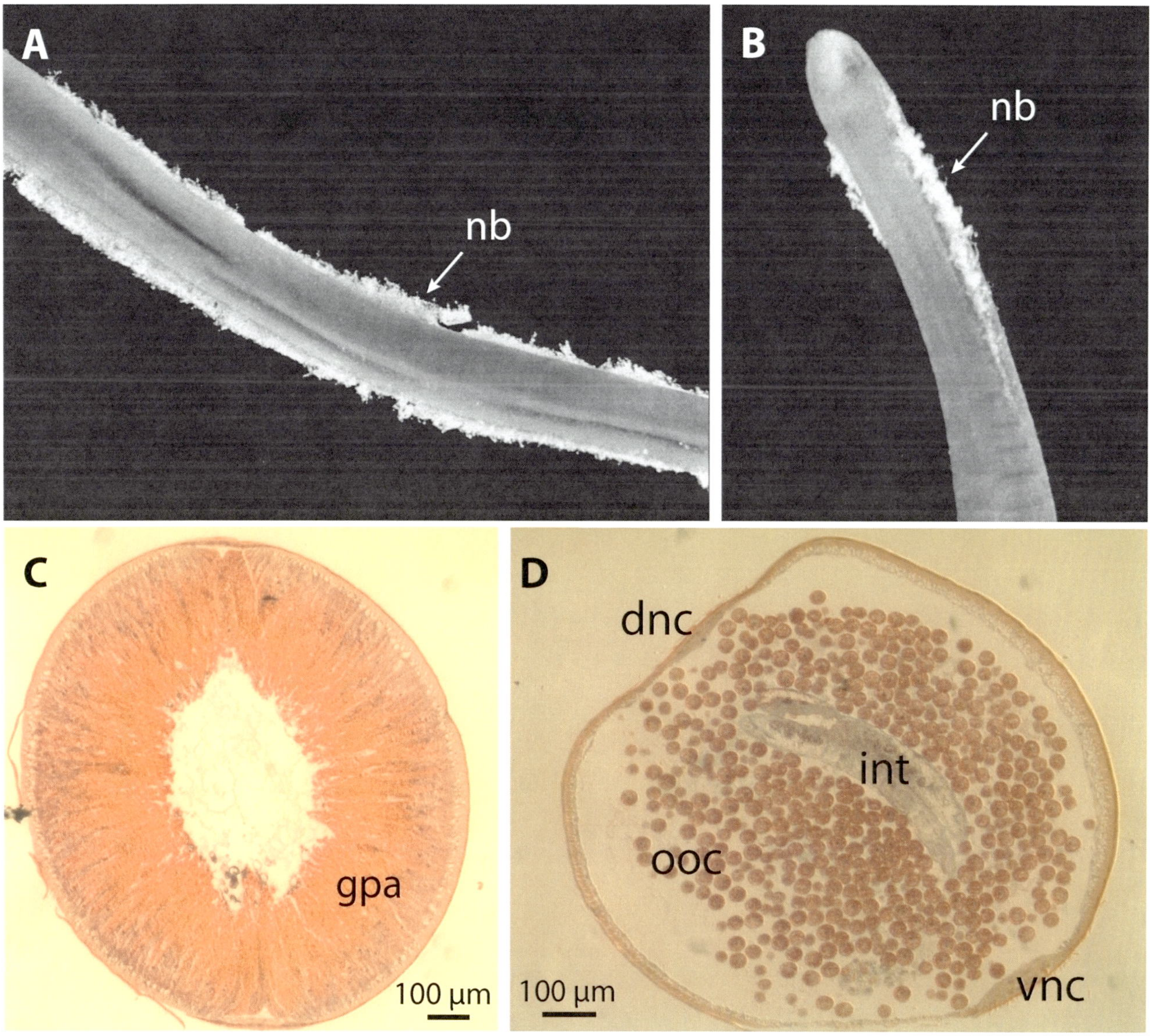

Fig. 3.1.5. Body organization in *Nectonema*. A, B, Midbody (A) and anterior end (B) of *N. agile* (from the Mediterranean Sea; Natural History Museum Berlin, accession number 5284); C, Cross section through an immature juvenile of *N. munidae* (from Norway) with gonoparenchyma (gpa); D, Cross section through mature female of *N. munidae* showing dorsal nerve cord (dnc), intestine (int), oocytes in the body cavity (ooc) and the ventral nerve cord (vnc).

cells shrink and are physiologically much less active compared to the parasitic stage (Schmidt-Rhaesa 2005). The cuticle appears to have a mainly protective function and nutrient uptake does not take place in the free living phase.

The available data suggest that the larval cuticle remains unchanged from the larva throughout the enormous growth in the juvenile and that it is replaced towards the end of the parasitic phase in one single moult by the adult cuticle (see below) (but see Chiu et al. 2011 for the hypothesis of further moults). The reason for this is that the characteristic hooks and stylets, which are differentiations from the larval cuticle (see e.g. Jochmann & Schmidt-Rhaesa 2007) from the anterior end of the larva remain, at least in some reports, as a cone on the anterior tip of juveniles (e.g. Von Linstow 1891a, Camerano 1892, Vejdovsky 1894, Švábenik 1925, Valvassori et al. 1988, Lanzavecchia et al. 1995). The larval cuticle therefore is capable of enormous growth, spanning from the larval size (about 100 µm) to the adult size of several to many centimeters. It should be noted, that what is called here the larval cuticle has been regarded to be a stocking formed by the host by Lanzavecchia et al. (1995). Based on the ultrastructure,

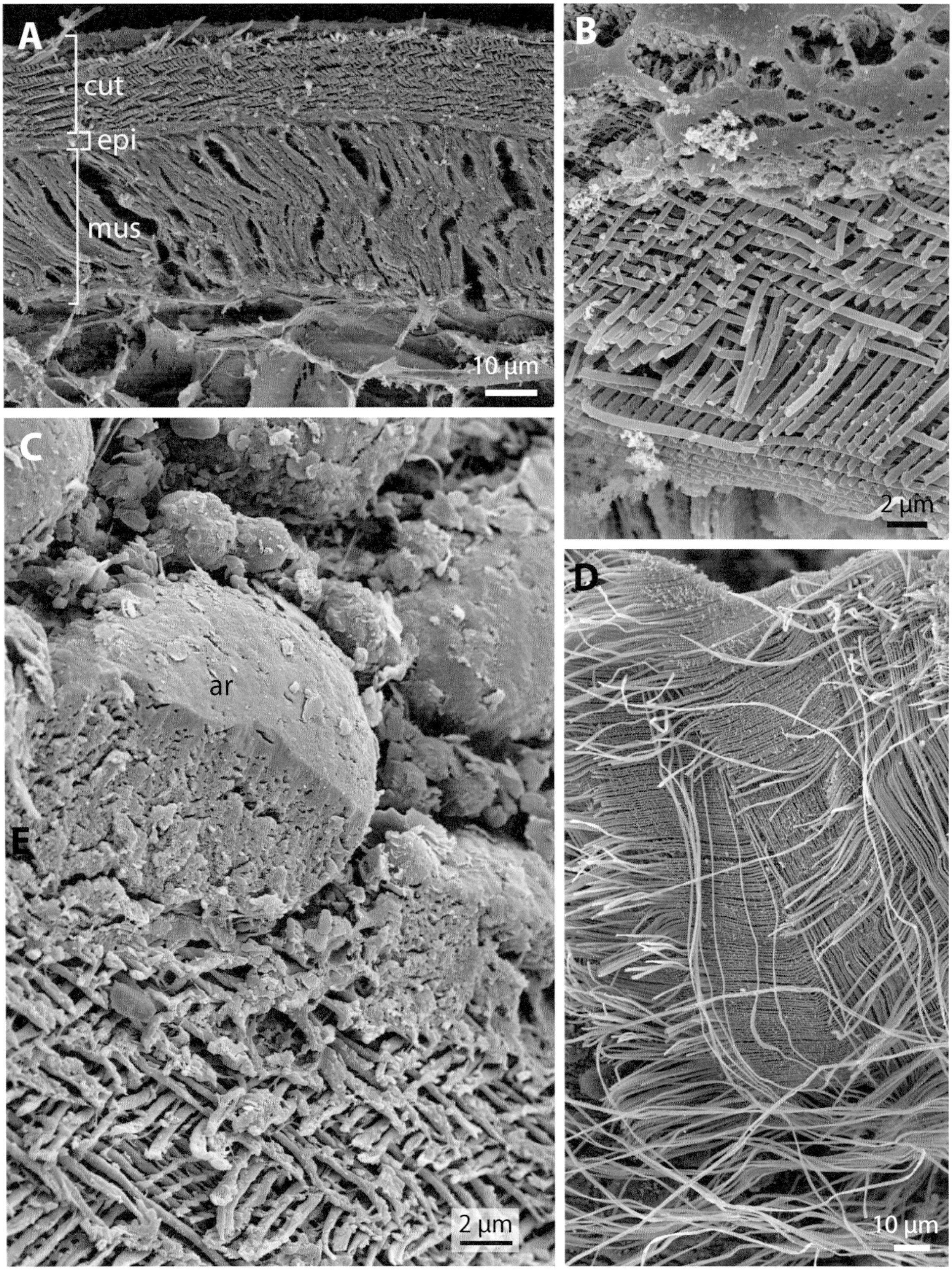

Fig. 3.1.6. Structure of the cuticle in gordiids. A, Cross section of a male specimen of *Gordius helveticus* (from Swizerland) showing size relations of cuticle (cut), epidermis (epi) and musculature (mus); B, Layers of fibres in the cuticle of an undetermined gordiid; C, Cross section through the fibre layer and an areole (ar) of *Chordodes* sp. (from Madagascar); D, Oblique section through the fibre layers in *Chordodes anthophorus* (from Georgia). All images SEM.

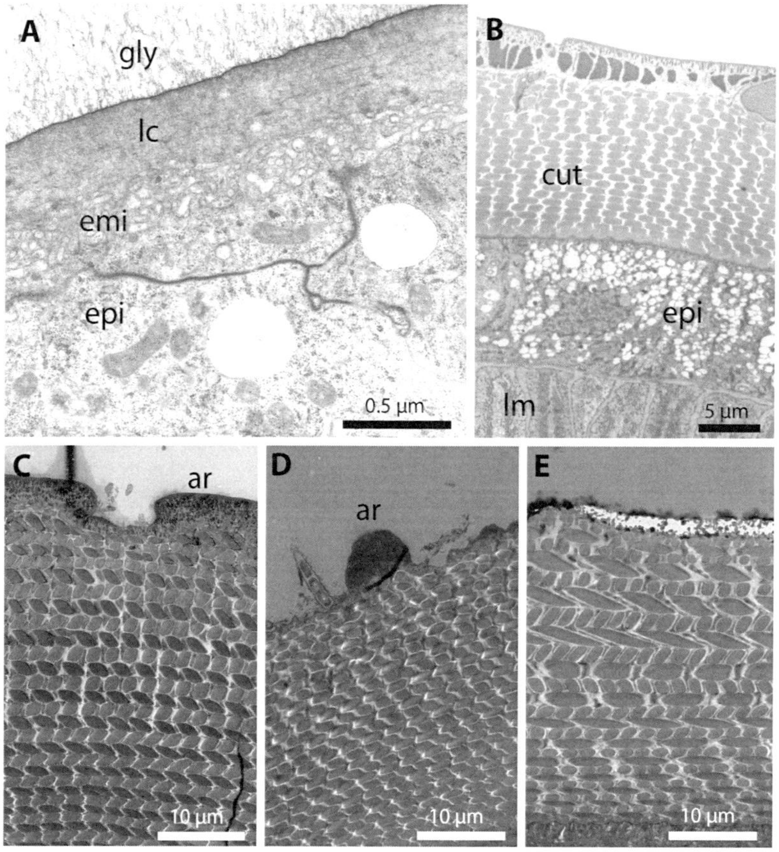

Fig. 3.1.7. Ultrastructure of the cuticle in gordiids. A, Larval cuticle (lc) and glycocalyx (gly) are formed by the epidermis (epi) and epidermal microvilli (emi) in a parasitic juvenile of *Paragordius varius*; B, Structure of cuticle (cut), epidermis (epi) and longitudinal musculature (lm) in an adult, free-living *P. varius*; C–E, Fibrillar layer and, in C and D, distal layer forming areoles (ar) in *Pseudochordodes bedriagae* (C), *Chordodes morgani* (D) and *Gordius difficilis* (E). Note that the white distal structure in the cuticle of *G. difficilis* is an artefact. All images TEM.

this seems to be a misinterpretation. Few ultrastructural data are available from the larval cuticle; these show a homogeneous, fine fibrillar layer and a thin apical darker layer (Lanzavecchia et al. 1995, Schmidt-Rhaesa 2005; Fig. 3.1.7 A). Externally, a glycocalyx is present (Schmidt-Rhaesa 2005). In *Paragordius varius*, the larval cuticle increases in size during development from 0.64 µm in the earliest investigated stage (10 days after artificial infection of a host with gordiid cysts) to 1.24 µm in the last stage where this cuticle was observed (Schmidt-Rhaesa 2005).

As the external structure of the adult cuticle is one of the most important characters for gordiid determination, many data concerning its surface are present (see Chapter 6). Much fewer data exist on the internal structure. Fine structural (ultrastructural) data are available for *Gordius aquaticus* (Bresciani 1970, 1991, Schmidt-Rhaesa & Gerke 2006), *G. villoti* (Brivio et al. 2000), *G. panighettensis* (Protasoni et al. 2003), *Pseudochordodes bedriagae* (De Villalobos & Restelli 2001), *Paragordius varius* (Zapotosky 1971, Schmidt-Rhaesa 2004a, 2005), *Chordodes nobilii* (Schmidt-Rhaesa & Gerke 2006) and two not exactly determined species (Eakin & Brandenburger 1974, Bresciani 1991) (see also Rajaram & Rajulu 1975 for additional data). Although the terminology used by different authors varies greatly, the adult cuticle of all species is composed of two main layers. The inner layer contains many sublayers of large fibers, the outer, distal one is very heterogeneous among genera. Schmidt-Rhaesa & Gerke (2006) suggested the terms "fibrillar layer" and "distal layer" for these two compartments.

The fibrillar layer consists of large fibers (between 0.2 and 0.35 µm in the small species *Paragordius varius* (Swanson 1974) and between 0.5 and 1 µm in other species (Eakin & Brandenburger 1974, Cham et al. 1983, Bresciani 1991)) that are orderly arranged in layers with parallel fibers (Fig. 3.1.6 A–E). The orientation of the fibers in each layer is at an angle of about 60° compared to the adjacent layers (Fig. 3.1.6 B, D, E). According to Protasoni et al. (2003) the angle ranges from 45–90° in different body regions. It is likely that the exact angle depends on the state in which the animal is stretched or curved (see also Seymour 1983). In most documented cases, the orientation of layers alters regularly, but in some cases two layers of the same orientation alter with one layer of a different orientation. This is present in the anterior end of *Gordius* sp., while in the remaining body the "usual" pattern of alternating orientations is present (Eakin & Brandenburger 1974). The number of layers reported by different authors varies. For example, for *Paragordius varius* 22 (Schmidt-Rhaesa 2004a, 2005), 24 (May 1919) or 45 (Swanson 1974) layers were reported. This may be explained on the one hand by a growing number of layers during development, although it is unclear whether layers can be added in the mature, free-living stage, and on the other hand by a varying number of layers in different body regions (see e.g. Protasoni et al. 2003). Ultrastructural investigations from *Gordius* species show the fibers to be surrounded by a net-like matrix with a honeycomb-like structure (Brivio et al. 2000, Protasoni et al. 2003, Schmidt-Rhaesa & Gerke 2006; Fig. 3.1.8 A). It is not clear whether this structure is typical for *Gordius* species or whether other species also show a comparable structure. With Scanning Electron Microscopy, a rectangular net-like structure was observed in *Chordodes* sp. between the fibers (Zoological Museum Hamburg, V 5778; unpublished observations; Fig. 3.1.9 A, B). Furthermore, radially oriented elements are found between the thick fibers; it is likely that this is a further matrix filling the spaces between the fibers. In addition to the honeycomb matrix in *Gordius*, radial components traversing the fibrillar layer are present (Fig. 3.1.8 A). The chemical nature of the cuticle was investigated by Brivio et al. (2000) and Protasoni et al. (2003), making it likely that it is not composed of collagen, but some other proteinaceous components (Protasoni et al. 2003) and that dithyrosine compounds are present as potential hardening elements (Brivio et al. 2000). Chitin was suspected to be absent from the gordiid cuticle already by Rauther (1905) and has not been detected since then.

The distal layer can be almost absent or it is more or less strongly developed (see Schmidt-Rhaesa & Gerke 2006 for overview; Fig. 3.1.7 E). This is obviously correlated with the presence or absence of structures called areoles. Areoles are elevations of various forms on the cuticular surface (Fig. 3.1.6 C). In most taxa, areoles of one or two distinguishable types are present and form a more or less regular pattern. Such regular areoles are reflected in cross section by a thick distal layer that is regularly subdivided by indentations going down to almost the fibrillar layer (Eakin & Brandenburger 1974, Bresciani 1991, De Villalobos & Restelli 2001; Fig. 3.1.7 C). Therefore, areoles are exclusively formed by the distal layer. Some species in the genus *Gordius*, e.g. *G. aquaticus*, have no areoles and in this case the distal layer is almost lacking (Schmidt-Rhaesa & Gerke 2006). More complex is the situation in *Paragordius*, where the homology of

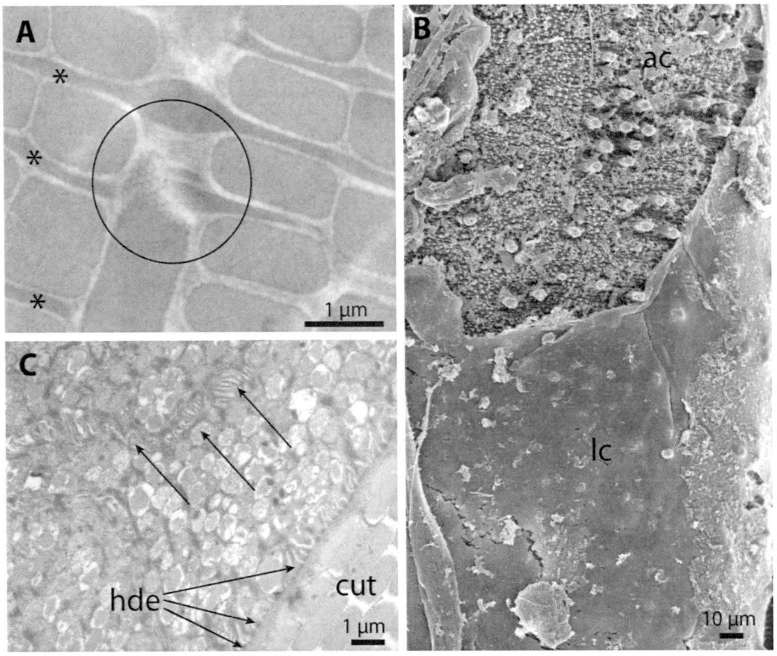

Fig. 3.1.8. A, The cuticle of *Gordius aquaticus* contains, between the fibres, radial elements (*) and a honeycomb-shaped matrix (in circle); B, Shedding of the smooth cuticle (lc) reveals the highly structured adult cuticle (ac) in *Chordodes morgani*; C, In the epidermis of adult gordiids (here *Paragordius varius*) cells are highly interdigitating (arrows). They attach to the cuticle (cut) by hemdesmosomes (hde). A,C by TEM, B by SEM.

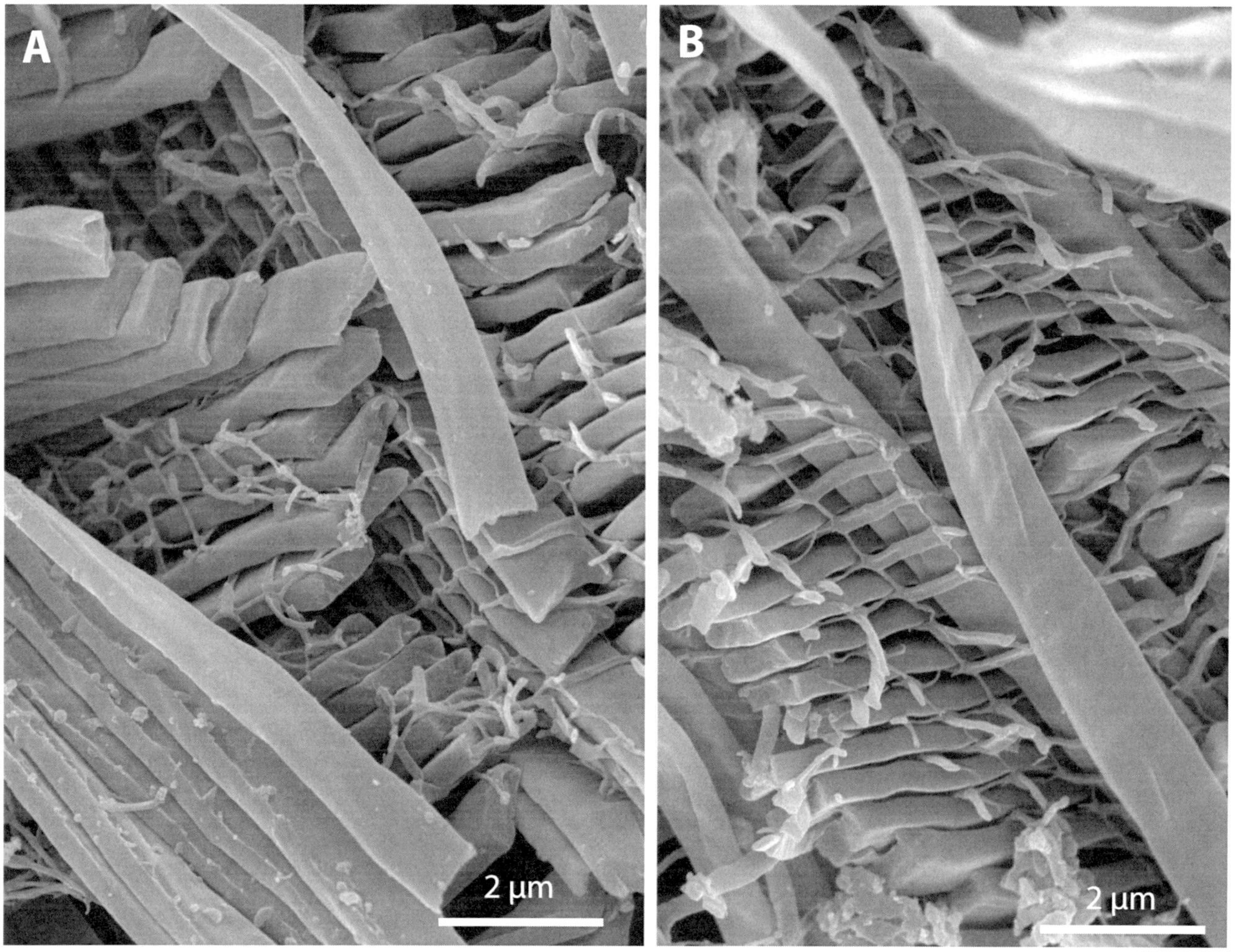

Fig. 3.1.9. A, B, Large fibers in the cuticle of *Chordodes* sp. (from Zoological Museum Hamburg, no V5778) with net-like structure surrounding fibres. All images SEM.

cuticular structures to areoles is not clear and in *Chordodes*, where areole diversity is greatest. In *Paragordius varius*, one large distal layer is present, which is not subdivided into areoles. The distal layer is composed of proximal material of moderate staining and distal material of dark staining; the distal material penetrates the proximal material with extensions or isolated drops (Fig. 3.1.7 B). This is comparable to the structure of other areoles. In addition, there are large, cushion-shaped structures in the distal layer (Fig. 3.1.7 B). These are, at least in some cases, connected to the epidermis by long extensions running straight through the fibrillar layer (Schmidt-Rhaesa 2004a, 2005; see below). The fine structure is still not completely clear, but it is likely that these structures represent some form of receptor (Schmidt-Rhaesa 2004a). Radial elements with a suspected secretory or receptive function have been reported by several authors; these are treated in more detail under "sensory structures".

More complex and even less well understood is the situation in the genus *Chordodes* (Fig. 3.1.7 D). The heterogeneity of areoles is highest compared with other gordiids, including distinctly elevated areolar types and the so-called crowned areoles, which are composed of a central stem and a crown of distal filaments (see Schmidt-Rhaesa et al. 2008 for overview and terminology). It is not clear what function this extended variety in areolar types has in *Chordodes* and whether particular areolar types may perform special functions such as receptive or secretive. The only available ultrastructural investigation within this genus (Schmidt-Rhaesa & Gerke 2006) gave first insights, but many open questions are remaining. Cushion-like structures, probably comparable to *P. varius*, are also present, including the straight connection to the epidermis. These cushions are present either at the base of areoles or within areoles and then appear to send extensions into the distal filaments.

The epidermis is cellular in all stages (in contrast to some earlier speculations that it might be syncytial, e.g. Bresciani 1970; the cellular character was reported already by Michel 1888), but undergoes strong changes during development, as reported already by Vejdovsky (1886). In young juvenile stages of *Paragordius varius*, the epidermis is 7–8 μm thick, the cells are densely packed with endoplasmic reticulum, mitochondria and glycogen. Irregular microvilli are present on the distal side of the cells, probably being involved in cuticle secretion (Schmidt-Rhaesa 2005; Fig. 3.1.7 A). The ventral nerve cord is completely integrated into the epidermis. In older juvenile stages and adults, the epidermis is only about 5 μm thick. The cells are almost devoid of recognizable organelles and contain the nucleus and either little cytoplasm or a large number of vesicles. The cell borders are highly interdigitating (see e.g. Eakin & Brandenburger 1974, Schmidt-Rhaesa 2005; Fig. 3.1.8 C), in contrast to the younger stages. Filament bundles traverse the epidermis from the basal to the distal side (Bresciani 1970, 1991). Adhaerens junctions between epidermal cells and hemidesmosomes connecting the basal sides of the cells to the basal lamina are present in all stages, but no other types of cell–cell connections have been observed. In older stages, where the adult cuticle is present, epidermal cells attach to the cuticle with abundant hemidesmosomes (Fig. 3.1.8 C). Filaments anchor at the hemidesmosomes and transverse the epidermal cells. It is therefore likely that the epidermis in early stages has a function in nutrient absorption, while in later stages it is modified to be a minimal barrier in the force-transferring system between cuticle and musculature.

Nectonema. As in gordiids, a cellular epidermis and a cuticle are present, with an adult cuticle following a larval cuticle. The adult cuticle is formed below the larval cuticle (Fig. 3.1.10 A, B), which is likely to be shed during a moult. The distinction between two cuticular layers has already been made by Ward (1892b); Nielsen (1969) named the larval cuticle and Bresciani (1991) the adult cuticle. The recognition of both layers as separate has not always been made; Skaling & MacKinnon (1988) described larval and adult cuticle together as "the" cuticle. The larval cuticle is between 1.4 and 3.0 μm thick (Schmidt-Rhaesa 1996b). Four layers have been described for *N. munidae* (Bresciani 1975, Schmidt-Rhaesa 1996b), these are (from internal to external) one up to 1.1 μm thick layer with fine fibers, one equally thick homogeneous layer, a very thin darkly stained layer and a thin external layer which might represent a glycocalyx. The first three layers are also described by Skaling & MacKinnon (1988) for *N. agile*. The larval cuticle includes chitin (Neuhaus et al. 1996). Comparable to gordiids, nutrient uptake through the larval cuticle is likely, as Skaling & MacKinnon (1988) found non-specific esterases, acid and alkaline phosphatases in the epidermis of parasitic stages that according to the figures only possess a larval cuticle.

In the anterior end the larval cuticle runs a short distance into the body at the mouth opening. In this region, rod- and tooth-like structures composed of darkly stained (sclerotized?) material are present (Schmidt-Rhaesa 1996b; see below). It is likely that they present remnants of the larval armature. About 15 μm from the mouth opening, the larval cuticle forms a sclerotized ring around the anterior part of the intestinal system. In the posterior end, the larval cuticle surrounds the animal like a sac (Schmidt-Rhaesa 1996a).

Below the larval cuticle the adult cuticle is formed in large parasitic stages. It can be observed in different thickness, which likely corresponds to different stages of formation (Skaling & MacKinnon 1988, Schmidt-Rhaesa 1996b). Both adult and larval cuticles are separated by a gap of varying thickness, in which for example the natatory bristles (see below) are present (Fig. 3.1.10 A, B). The thickness of the adult cuticle varies greatly between body regions and different individuals of *N. munidae* (Schmidt-Rhaesa 1996b); it is up to 8 μm thick. The adult cuticle is composed of two layers, a thick fiber layer and a small distal epicuticle (0.15–0.2 μm) (Fig. 3.1.10 C). The fiber layer includes up to 80 sublayers of fibers; the orientation of the fibers in the sublayers changes from sublayer to sublayer but corresponds in every other sublayer. The order of the sublayers is not always regular, in some regions there may be irregularities such as bulging of some layers against others. Additionally, the fiber layer may extend further into the internal tissues of the animal in the form of spherical structures (Schmidt-Rhaesa 1996a).

When both cuticle layers are present, the adult cuticle meets the larval cuticle at the mouth opening, but it does not line the oral cavity like the larval cuticle. In the posterior end, the adult cuticle lines part of the cloacal opening (Schmidt-Rhaesa 1996a).

Characteristic for *Nectonema* are the natatory bristles. These are long cuticular bristles attached in paired rows along the dorsal and the ventral longitudinal midline of the animals. Bürger (1891), Ward (1892a) and Rauther (1914) described the natatory bristles as hollow structures; Bock (1913) described them as filled with fine bristles which could be confirmed by ultrastructural studies (Bresciani 1991, Schmidt-Rhaesa 1996b). The bristles are

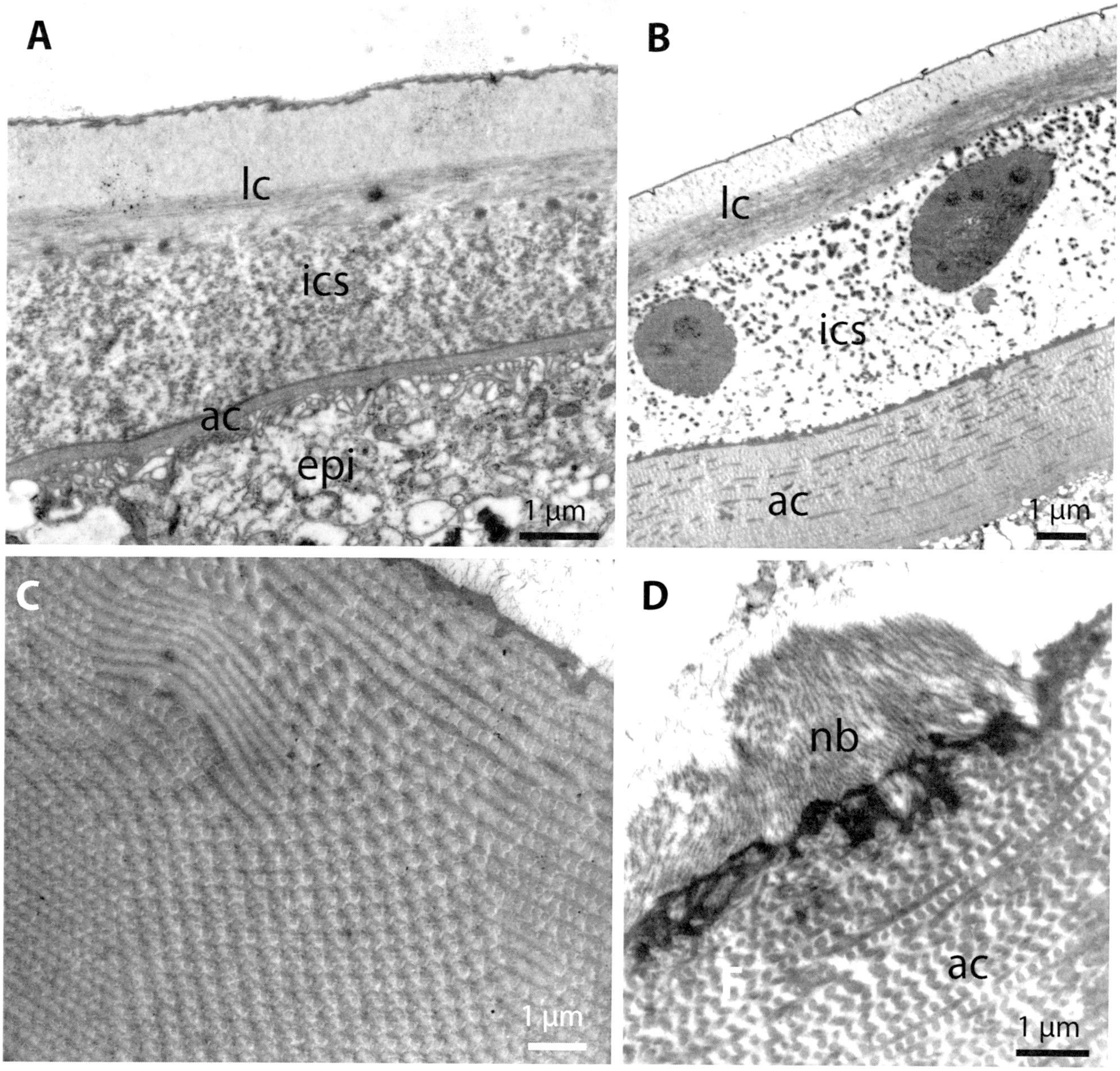

Fig. 3.1.10. Ultrastructure of the cuticle in *Nectonema munidae*. A, Structure of the larval cuticle (lc), which is separated by an intercuticular space (ics) from the adult cuticle (ac), which is starting to be secreted by the epidermis (epi); B, Older stage with thicker adult cuticle; C, Final adult cuticle, showing irregularities in the layering of fibres; D, Natatory bristles (nb) attach to the distal layer of the adult cuticle (ac). All images TEM.

composed of fine fibrillar material and attach directly to the distal layer (Fig. 3.1.10 D). The distal layer is irregular in this region, forming projections into the natatory bristle material.

The epidermis is a thin layer, which is thickened in the ventral and the dorsal regions to include nerve cords. It is either described as being cellular (Bock 1913) or syncytial (Bürger 1891, Nierstrasz 1907, Oku et al. 1983). In an ultrastructural investigation of *N. munidae* both states seem to be present. In the anterior body region the organization is cellular, in the midbody region it might be syncytial (Schmidt-Rhaesa 1996b). Epidermal cells contain rough ER and mitochondria; they are interconnected by desmosomes from which fibers run into the cells. Connection to the cuticle and to the basal lamina is with hemidesmosomes. In some regions irregularly distributed microvilli were observed between epidermis and cuticle (Schmidt-Rhaesa 1996a).

Musculature

Gordiida. In all gordiids, only longitudinal muscle cells are present in the body wall (Fig. 3.1.4 A, B). These are arranged in a layer below the epidermis. In cross section, many radially flattened muscle cells form the muscular layer; in longitudinal direction (i.e. in the anteroposterior axis of the animal) muscle cells are elongated. The musculature effectively forms a completely closed sheet surrounding all inner organs. On the ventral side this sheet is interrupted by a thin epidermal lamella connecting the ventral nerve cord to the epidermis, but as this lamella is very thin, the muscular layer functionally represents a closed sheet. The fibres are arranged in a helicoidal pattern around the animal (Lanzavecchia 1977).

In cross section, the flat muscle cells contain myofilaments along almost every margin, with a concentration along the long radial sides of the cell and less density on the proximal and the distal side (Schmidt-Rhaesa 1998a). On the proximal side, a region free of myofilaments can be present. The nucleus is flat end elongated; it is either in the central region or, when present, in the small proximal regions free of myofilaments. Mitochondria and glycogen are also present in the central, cytoplasmic part of the muscle cells. Z-rods and a sarcoplasmatic reticulum (Lanzavecchia et al. 1979), after Restelli et al. (2002) also T-tubules are present. In juvenile specimens, muscle cells are fewer and of the platymyarian type, consisting of a distal region with myofilaments and a proximal, cytoplasmic region (Vejdovsky 1894, Schmidt-Rhaesa 2005).

The myofilaments are characterized by very thick filaments (Fig. 3.1.11 D) containing paramyosin (Swanson 1971). These filaments are 98–150 nm thick in *Gordius robustus* and *Paragordius varius* (Swanson 1971) or 40–100 nm thick in *Gordius aquaticus* (Schmidt-Rhaesa 1998a). The paramyosin fibres are very long (up to 120 µm;

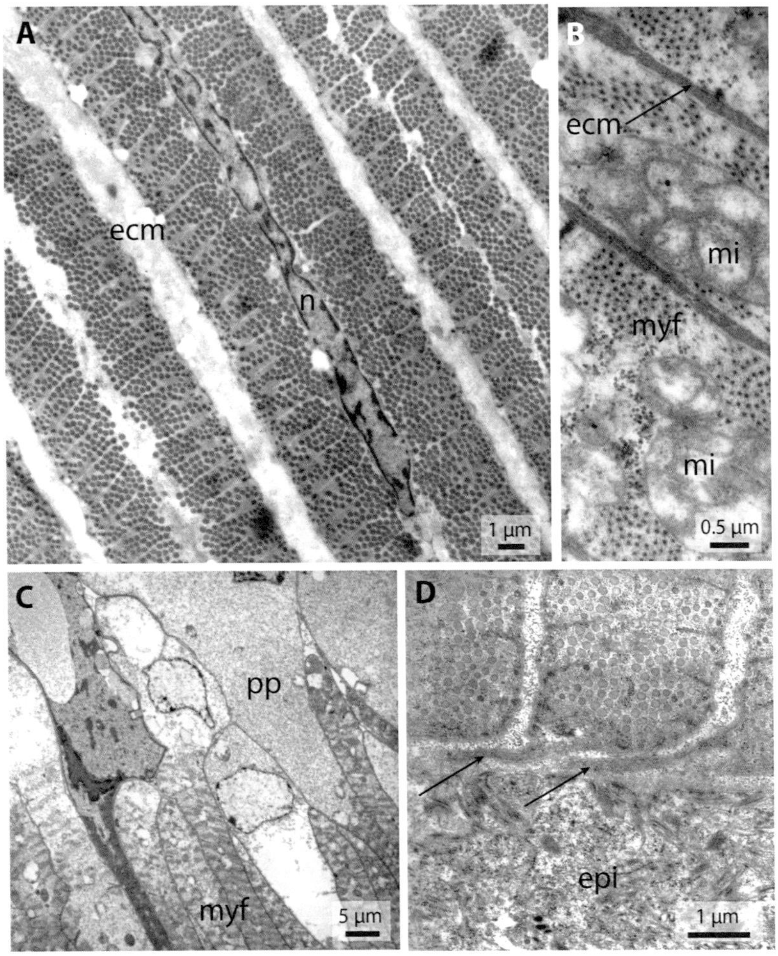

Fig. 3.1.11. Ultrastructure of the musculature. A, Flat, radial longitudinal muscle cells separated by ECM (ecm) in *Gordius aquaticus*. The elongate nucleus is between the myofilaments (thick dots = paramyosin); B, Fine structure of the muscle cells in *Nectonema munidae*. Cells are separated by ECM (ecm), myofilaments (myo) are peripheral, mitochondria (mi) are central; C, Transition between the part with myofilaments (myf) and the proximal, protoplasmic part (pp) in *N. munidae*; D, Extensions of the muscle cells (arrows) towards the epidermis (epi) in *Paragordius varius*. All images TEM.

Lanzavecchia et al. 1979), in longitudinal sections they show a characteristic asymmetric pattern of striation (see Lanzavecchia 1977, Lanzavecchia et al. 1977, 1979).

Each muscle cell is surrounded by extracellular matrix (ECM), which is composed of an intensive lamina fibroreticularis and a thin lamina densa. Muscle cells attach to the ECM by numerous hemidesmosomes (Schmidt-Rhaesa 1998a). Rarely, processes of the distal side of a muscle cell were observed that project into the ECM in the direction of the epidermis (Schmidt-Rhaesa 1998a, Restelli et al. 2002; Fig. 3.1.11 D). Such processes may play a role in innervation. A dense contact between muscle cells and epidermal neurites, as present in *Nectonema* (see below), could not be observed.

Nectonema. The musculature of *Nectonema* species corresponds with that in gordiids in the presence of numerous radially oriented, flattened longitudinal muscle cells (Fig. 3.1.11 A). Because *Nectonema* species have a dorsal and a ventral epidermal cord, the musculature is separated into paired lateral portions. Ultrastructural data have been given only for *N. munidae* by Bresciani (1991) and Schmidt-Rhaesa (1998a). The following description refers to these publications.

Each muscle cell contains a proximal region without myofibrils (Fig. 3.1.11 C), this is distinctly more voluminous in males compared to females. In some individuals, this protoplasmic region of the muscle cells extends so far that the muscle cells from both body sides meet in the center of the animal (Schmidt-Rhaesa 1996a). In younger female stages, the gono-parenchyma (see under reproductive system) is densely interdigitated with the small protoplasmatic part of the muscle cells. The nucleus is positioned either within the protoplasmatic part or at the transition between the protoplasmatic and fibrillar part.

In the fibrillar part, myofilaments are arranged along the sides of the cell, smaller clusters of myofilaments are separated by Z-elements (Fig. 3.1.11 A, B). Each cluster of myofilaments contains 35–45 thick filaments and two to three times as many thin filaments. The diameter of thick filaments is 45 nm (Schmidt-Rhaesa 1998a) up to 140 nm (Bresciani 1991). Flattened vesicles can be present close to the outer cell membrane, they probably represent sarcoplasmatic reticulum. The core of the fibrillar part contains cytoplasm with mitochondria and glycogen granules. Each muscle cell is surrounded by ECM, in which a thin lamina densa and a voluminous lamina fibroreticularis can be distinguished.

Additional muscles were described as some fibers within the sperm sac by Ward (1892a) in *N. agile* and as muscles in the posterior end of the males by Schmidt-Rhaesa (1998a) in *N. munidae*. The exact course of these muscles could not be followed and therefore their function remains unclear. However, as the curved posterior end of the males is used for copulation, this may be the functional context for these muscles. In cross section, myofilaments are distributed in the periphery of the muscles, either partially or around the entire periphery. The ultrastructure of the myofilaments corresponds to that of the body wall muscles. Each muscle is surrounded by ECM.

Nervous system

The nervous system of all nematomorphs consists of a brain in the head region, from which a dominant ventral longitudinal cord emerges. *Nectonema* species additionally possess a dorsal longitudinal cord. Basiepithelial neurites are present in the epidermis. All investigations of the nervous system were made by histological or ultrastructural investigations; the staining of neurogenic substances with immunohistochemical methods was so far not successful in nematomorphs.

Gordiida. The dominant element of the nervous system is the ventral nerve cord (Fig. 3.1.4 A, B), with a brain at the anterior end. The structure of the brain (also termed brain ganglion or cerebral ganglion) is described controversially in the literature. After Rauther (1904, 1905, 1930) and Švábenik (1909, 1925) the brain has a ring-like structure surrounding the anterior intestinal system or its remnants. It consists of a ring of neurites and associated cell bodies. In contrast, Vejdovsky (1886, 1894) and Montgomery (1903) described the brain to be mainly subpharyngeal, a dorsal (suprapharyngeal) commissure can be present (as in *Gordius preslii*, Vejdovsky 1886) or absent (as in *Parachordodes pustulosus*, Vejdovsky 1894; *Paragordius varius*, Montgomery 1903). In both cases the ventral nerve cord emerges directly from the ventral part of the brain.

Own investigations on serial sections of the anterior end of *Paragordius varius* (see Fig. 3.1.12 A–I) suggest the following structure. Close to the anterior tip of the animal, neurogenic tissue (probably already the brain) surrounds the cuticularized, ringlike esophagus (Fig. 3.1.12 A). Further towards the posterior, this tissue concentrates on the dorsal side of the esophagus and becomes quite voluminous (Fig. 3.1.12 B). The anterior end of a paired (!) ventral nerve cord occurs slightly more posterior (Fig. 3.1.12 C, D). The neurogenic tissue (brain) becomes smaller, the paired anterior ends of the ventral

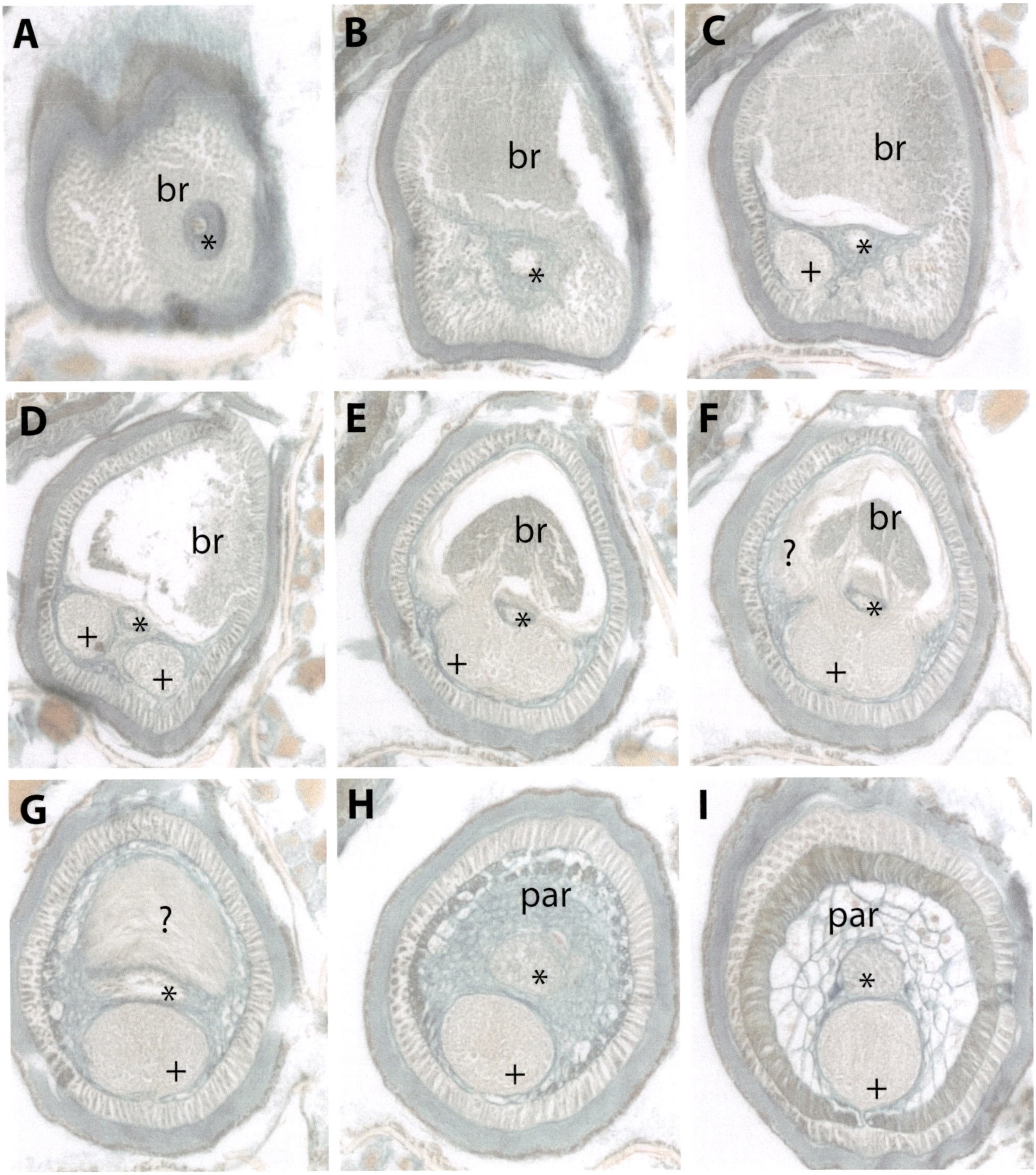

Fig. 3.1.12. Series of sections through the anterior end of a male *Paragordius varius* (from anterior to posterior). The structure interpreted here as the brain (br) starts as a ringlike structure (A), but the main part is dorsal (B–F). Posteriorly, it is replaced by an unknown structure (?). The esophagus is labeled by *, the ventral nerve cord by +. Note the paired origin of the ventral nerve cord in D. Parenchyma is indicated by blue color around the esophagus in anterior sections and becomes more developed further posterior (H, I). All images light microscopy from paraffin sections.

nerve cord fuse and this unpaired structure gains contact to the brain by extending fibres dorsally (Fig. 3.1.12 E, F). Soon after, the brain disappears and tissue of unknown nature extends for a very brief distance (Fig. 3.1.12 F, G), before the parenchyma starts (Fig. 3.1.12 H, I). The structure, here called brain, has a slightly different staining compared to the ventral nerve cord. It is probably the same structure that was interpreted by Montgomery

(1903) as an eye. The exact nature remains unknown, but at least it is in intense contact with the ventral nerve cord.

The ventral nerve cord originates as an unpaired anlage within the ventral epidermis (Vejdovský 1894, May 1919, Schmidt-Rhaesa 2005). During development, it is shifted from the intraepidermal position to a submuscular position, remaining connected to the epidermis by a thin epidermal lamella (Vejdovsky 1894, Lanzavecchia et al. 1995, Schmidt-Rhaesa 2005). The entire nerve cord is surrounded by ECM. Several ECM lamellae originate from this perineural ECM and run more or less radially into the nerve cord, separating particular regions or neurons (Schmidt-Rhaesa 1996c). Some authors (e.g. Montgomery 1903, Rauther 1930) describe the ventral nerve cord to be at first view tripartite in cross section, but on closer view such a tripartition is not evident any more. In the ultrastructural investigation of *Gordius aquaticus* by Schmidt-Rhaesa (1996c), the ECM strands divide the neural mass into several compartments, but the neural material within these compartments shows the same structure. Nuclei are found only in the basal region of the nerve cord. Some authors have described neurons of different size to be present in the ventral nerve cord, therefore the existence of giant fibers is probable (e.g. Rauther 1930).

Several authors describe a caudal ganglion in the form of a significant (females) or hardly significant (males) swelling of the ventral nerve cord (Villot 1874a, Montgomery 1903, Rauther 1905, May 1919). In females, the ventral nerve cord terminates with the caudal ganglion, while in males with a bifurcated posterior end it divides and a branch of the ventral nerve cord leads into each tail lobe (Fig. 3.1.13 A–D). In addition, some authors mention special nerves in the cloacal region or an intraepidermal longitudinal nerve below the lamella connecting the ventral nerve cord with the epidermis (Montgomery 1903 for *Paragordius varius*). These elements have to date been confirmed neither by ultrastructure nor by immunohistochemistry. Peripheral nerves are present in the epidermis according to e.g. Montgomery (1903) and Restelli et al. (2002), these may innervate the longitudinal musculature from its distal side, but this still has to be shown.

Nectonema. *Nectonema* species possess two longitudinal nerve cords, a dominant ventral one (about 50 μm in diameter) and a smaller dorsal one (10–15 mm in diameter). Both cords run within thickenings of the epidermis and are therefore completely intraepidermal (Schmidt-Rhaesa 1996c). In the ventral nerve cord, there is a central neuropil (axons and dendrites of nerve cells) and peripheral nerve cell somata. Histlological methods show these somata to be present scattered in the entire periphery of the nerve cord, with a concentration in the ventral side. Some nuclei are even present among the central neuropil, creating a tripartite impression of the ventral cord. Ultrastructural investigations could not confirm this tripartition and nuclei within the neuropil were not found (Schmidt-Rhaesa 1996c). The ventral nerve cord is in its dorsal and lateral region surrounded by ECM; part of this ECM runs into the cord, but this does not create a tripartition. In the ventral region, no ECM could be found between nerve and epidermal cells. The dorsal nerve cord is smaller, and neuropil and cell somata are not clearly separated from each other. Additionally, no ECM is present between nerve and epidermal cells.

In the anterior end, the ventral nerve cord extends in size to form the brain. The brain is generally described as being circumesophageal (Bürger 1891, Ward 1892a, Rauther 1914), but the dorsal part (the dorsal commissure) is quite weak. The ventral part of the brain is, in cross section, kidney shaped and lies below the esophagus. In *N. munidae*, a thin dorsal commissure was only found in juvenile stages, but in adults, this commissure is reduced (Schmidt-Rhaesa 1996b). A connection of the dorsal nerve cord to the brain could not be shown. The brain consists of a large neuropil and peripheral nuclei. The ventral nerve cord also has a posterior thickening (Ward 1892a).

Within the epidermis (other than the epidermal cords) of *N. munidae* some basiepidermal nerve fibers are present, forming a peripheral nerve system. Because of their basiepidermal position they are adjacent to the muscle cells, being separated from these only by the basiepidermal ECM. Vesicles (probably neurosecretory vesicles) have been shown to be present in such regions and the ECM between nerve and muscle cells may be distinctly smaller than in the remaining regions (Schmidt-Rhaesa 1996c). This indicates that the longitudinal muscle cells are probably innervated from these peripheral nerves.

One further nerve has been shown to be present associated with the intestine in *N. munidae* (Schmidt-Rhaesa 1996b).

Sensory structures

There are very few indications for the presence of sensory structures in nematomorphs in general. Some structures are candidates for a sensory function, such as different epidermal structures, but their exact structure remains unclear. The enigmatic giant cells in *Nectonema* might

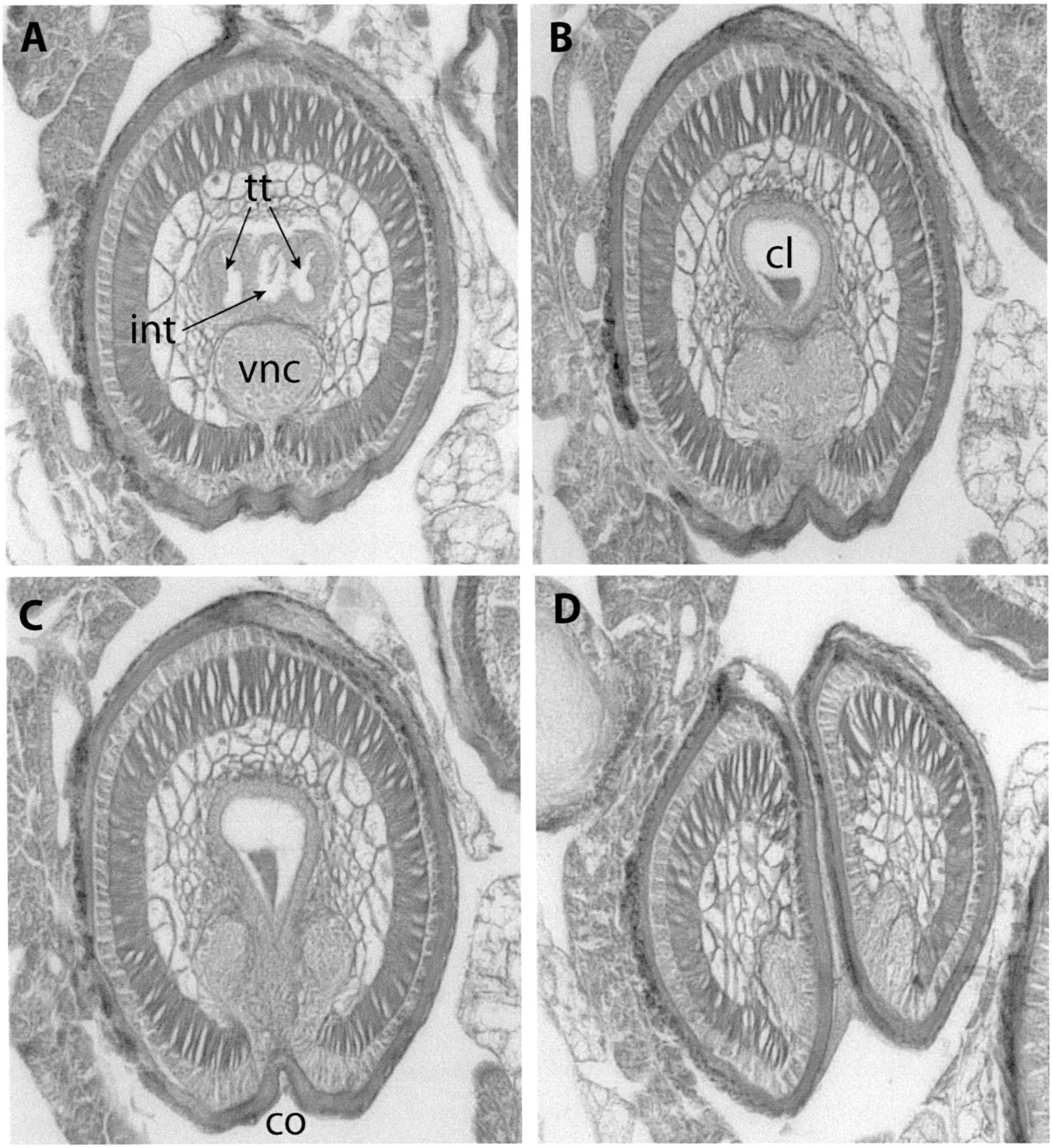

Fig. 3.1.13. Series of sections through the posterior end of a male *Paragordius varius* (from anterior to posterior). A, At the level where testes tubes (tt) and intestine (int) have not yet fused, the ventral nerve cord (vnc) is unpaired; B, At the level of the cloaca (cl), the cord starts to separate; C, Both paired arms of the cord are separated by the duct leading to the cloacal opening (co); D, Posterior of the cloacal opening the tail lobes are completely separated. All images light microscopy from paraffin sections.

have a sensory function. Photoreceptors have been stated to be present in few investigations, but convincing data for this are lacking.

Gordiida. The knowledge about sensory structures in gordiids is very rudimentary. Sensory structures receiving input from external stimuli should have a transcuticular character. Such transcuticular structures, originating in the epidermis and running towards the cuticle surface, are reported in several histological investigations (e.g. Villot 1874a, Rauther 1905), but the exact nature of these structures remains uncertain. Ultrastructural investigations confirm these observations, but offer little further resolution. Schmidt-Rhaesa (1996a) reported paired transcuticular structures in *Gordius aquaticus*, but as no details are recognizable, it remains uncertain whether these are cellular extensions at all. In *Paragordius varius*, the resolution of straight, transcuticular structures is better and it is probable that these include ciliary structures (Schmidt-Rhaesa 2004a; Fig. 3.1.14 A, B). The origin is a conspicuously darkly stained epidermal cell, probably a receptor cell (Fig. 3.1.14 B). The structures run straight through the cuticle (Fig. 3.1.14 A) and terminate in its distal layer, probably in the form of large cushion-like structures. There is no opening on the surface of the cuticle, so these structures could

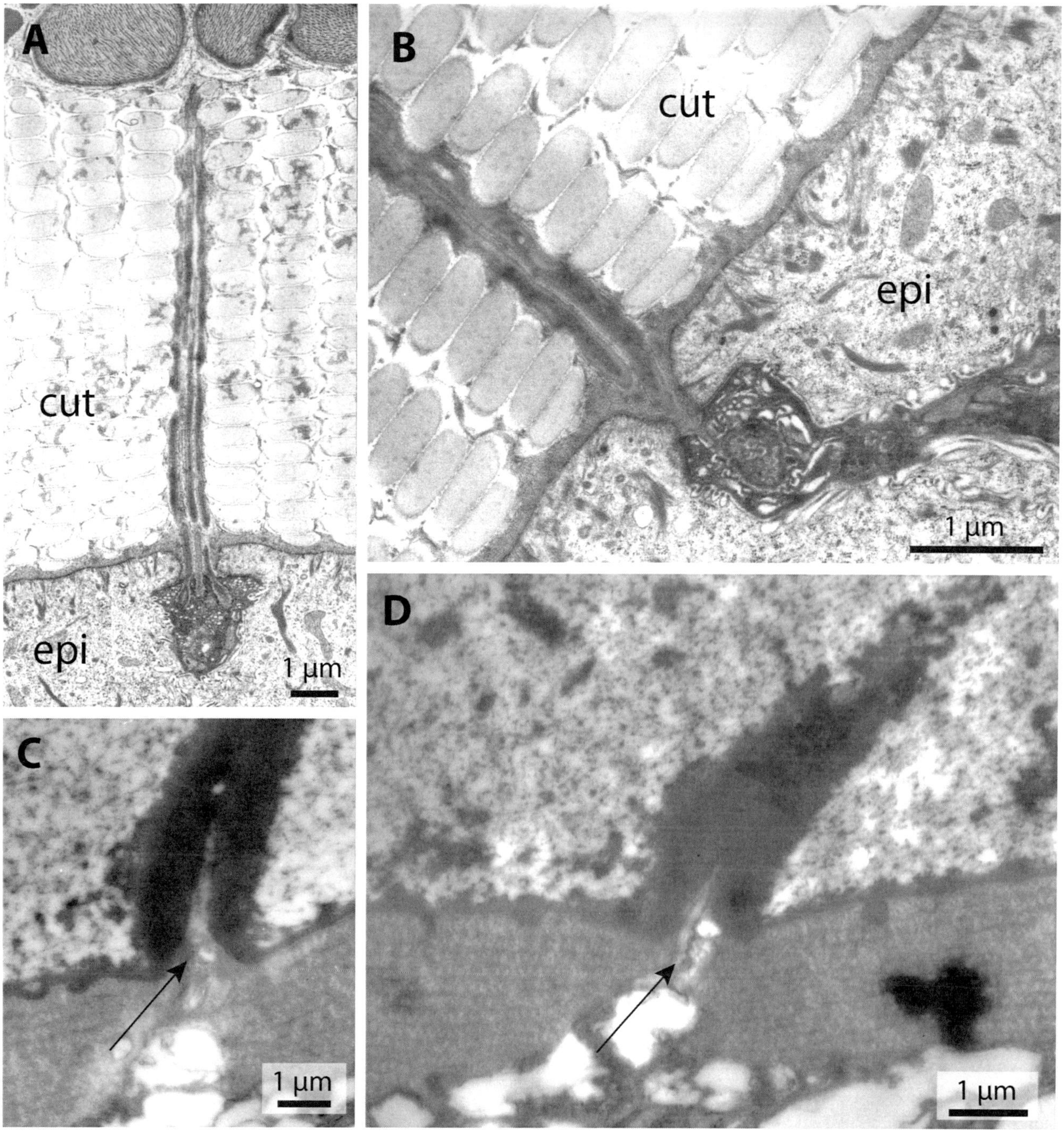

Fig. 3.1.14. Ultrastructure of sensory structures. A, B, Transcuticular structures in *Paragordius varius* traverse the fibrillar layer of the cuticle (cut) and originate in the epidermis (epi) in a darkly stained cell; C, D, Epicuticular bristles in *Nectonema munidae* have a cellular core originating in the epidermis, this is probably a sensory cilium (arrows). All images TEM.

represent mechanoreceptors sensing pressure or shearing of the cuticle. Comparable transcuticular structures ending in cushions in the distal cuticle layer have been found in *Chordodes nobilii* (Schmidt-Rhaesa & Gerke 2006), although the fine structural resolution is not equivalent to the one in *P. varius*.

The genus *Chordodes* remains an interesting taxon to search for sensory structures, because of the high diversity in cuticular structures. Transcuticular structures have been reported in several publications, but it remains unclear whether these are mechanoreceptive or represent pore channels for secretion. The only ultrastructural investigation of a species of this genus (*C. nobilii*, Schmidt-Rhaesa & Gerke 2006), provided initial details, but several aspects remained unresolved.

Among the several different cuticular structures present in gordiids, the conical spines present on the ventral side of the tail lobes in *Gordionus* species are also candidates for sensory structures, because some investigations show these spines to originate in a circular pit, a feature resembling mechanoreceptors in other taxa. All other cuticular structures are more or less unarticulated extensions of the outer cuticular layer.

All gordiids have an almost white anterior end, which either blends over to the normal body coloration (in e.g. *Chordodes*) or which is followed by a black ring-like coloration (the collar). Some authors have speculated that this light area might include photoreceptive structures (e.g. Von Linstow 1889, Rauther 1905), but this has not been confirmed by ultrastructural data (although few of these include the anterior body region). A large eye is reported to be present in *Paragordius varius* by Montgomery (1903). According to this investigation, a capsule is present, dorsal of the esophagus. The anterior wall is formed by long hypodermal cells; the remaining part of the capsule is formed by some kind of sheath. The capsule is filled with fluid, in its posterior part cells interpreted as retina cells are present. Pigment cells are lacking, but nerve fibers are reported to run into the capsule (see Fig. 3.1.12 for histological sections of the anterior end of *Paragordius varius*).

Nectonema. Very few fragmentary data indicate the presence of simple cuticular receptors in *Nectonema*. These consist of a fine epicuticular bristle, into which a structure, probably a cilium, runs (Fig. 3.1.14 C, D). This probable cilium traverses the cuticle. Below it, within the epidermis, is a cell which is, due to its richness in microtubuli, probably a nerve cell (Schmidt-Rhaesa 1996a).

Very characteristic structures of *Nectonema* species are the giant cells in the anterior end (Fig. 3.1.15 A). These have been described from *N. agile, N.munidae* and *N. zealanidcae*, but not from *N. melanocephalum* and *N. svensksundi* (Nierstrasz 1907, Bock 1913). A function in the sensory context is likely. When present, the giant cells lie within a body cavity that is separated from the cavity in the remaining body by a septum (Fig. 3.1.15 A). Such a septum is also described as an incomplete septum from *N. melanocephalum* (Nierstrasz 1907). In all species that possess giant cells, there are always four cells, with the exception of one individual of *Nectonema* sp. from Northern Norway, from which Bakke (1975) reports only three cells.

Giant cells are very large, spherical cells with a diameter of up to 400 μm. The following description is based on the only ultrastructural investigation by Schmidt-Rhaesa (1996a,b). The nucleus is located more or less centrally; the remaining volume of the cell is filled by granular material, no further organelles are visible. The cell membrane forms several tiny pockets; sometimes small neurons are present outside the giant cell. Directly below the cell membrane there is a layer of the granular material mentioned above. Towards the ventral nerve cord, the giant cells send out a funnel-shaped projection, which is filled with fine fibrous material. This material extends into the granular material described from within the cell and below the cell membrane, where it loses its fibrillar nature. The projection of the giant cells is densely integrated with neurons from the ventral nerve cord and vesicles indicate potential synapses between these structures.

With one side the giant cells are in contact with the epidermis (Fig. 3.1.15 A). The contact zone is about 100 μm broad. The epidermis is thickened in this region and consists of more than one layer. In the contact zone, the membrane of the giant cells forms a number of densely arranged microvilli.

Ward (1892a) already described the projection of the giant cells to the nervous system and assumed that they have a nervous function. He named the cells "dorsal cells" (see also Rauther 1914); Bresciani (1991) named them "giant nerve cells". Other authors did not recognize this nervous function, for example Fewkes (1883) labeled the giant cells as eggs and Bürger (1891) assumed that they are salivary glands. The probable synaptic connection of the giant cells to the nerve cord makes it very likely that the cells are some kind of sensory structure. The question remains, what kind of receptor? The large size and the probably fluid-filled interior may suggest that they are pressure receptors. As microvilli, which are present in the connection to the epidermis, are often present in photoreceptors, a photoreceptive function may also be possible. But these thoughts remain speculations.

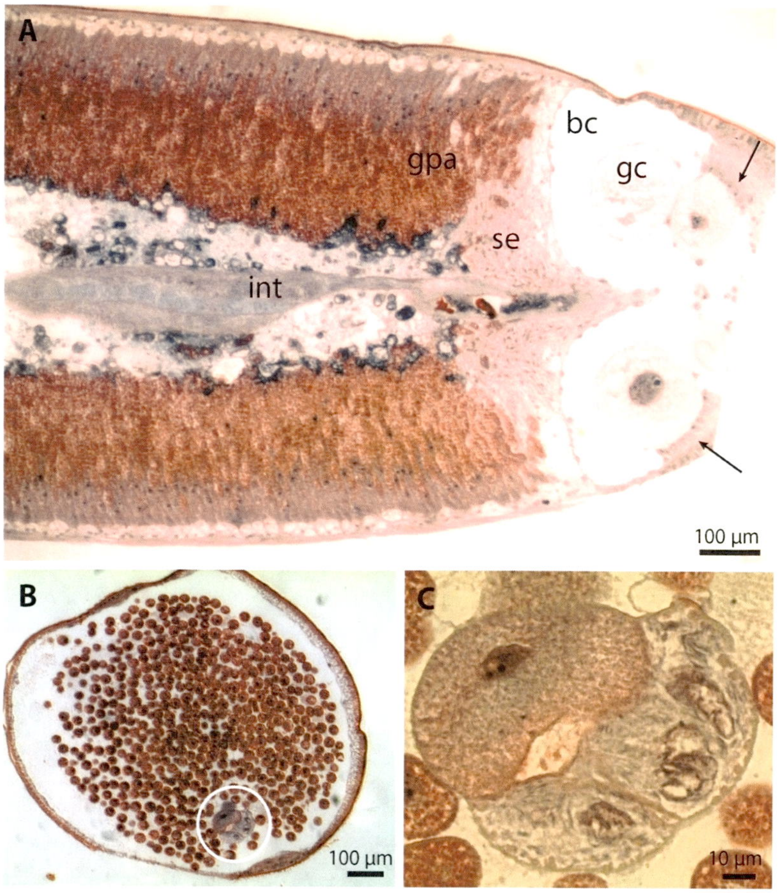

Fig. 3.1.15. Female *Nectonema munidae*. A, Longitudinal section through the anterior end (anterior is to the right, tip is cut off). Giant cells (gc) are present in the anterior body cavity (bc), they are in contact to the epidermis (arrows). A septum (se) divides the head from the remaining body with gono-parenchyma (gpa) and intestine (int); B, Cross section through the midbody region showing numerous oocytes and the intestine (in circle); C, Magnification of the intestine, which is composed of different cell types. All images light microscopy from paraffin sections.

The giant cells are absent in juvenile stages of *N. munidae*, which carry only the larval cuticle (Schmidt-Rhaesa 1996a,b). Therefore, they are formed quite late during development. It is tempting to assume that the absence of giant cells in *N. melanocephalum* and *N. svensksundi* (Nierstrasz 1907, Bock 1913) is due to the fact that the stages were too young, but as the few specimens of both species were free-swimming adults, it is more likely that giant cells are lacking in these species.

Intestinal system

The intestinal system of adult nematomorphs is reduced to a certain degree. This is probably linked to the short life span of adults. Adults do not feed and nutrient absorption is restricted to juveniles.

Gordiida. The main part of the intestinal system is the intestine, which runs through the entire body. Apart from some early misinterpretations (Berthold 1843 interpreted the intestine as testes, Meissner 1856 stated that an intestine is lacking and regarded the intestine as an excretory organ), most authors agree in their observations.

The intestine is positioned dorsal of the (ventral) nerve cord and is usually (but not always) surrounded by a small body cavity, the periintestinal cavity (see body cavities). In juveniles (at least of *Paragordius varius*, see Schmidt-Rhaesa 2005), the intestine is well-defined and consists of a layer of cuboidal cells surrounding a clearly visible lumen (Fig. 3.1.16 A). The intestinal cells have large, round nuclei and contain abundant endoplasmic reticulum (Fig. 3.1.16 A). Microvilli project towards the lumen (Fig. 3.1.16 A). This organization changed during development and in fully grown juveniles and adults the entire intestine appears collapsed in cross section (Schmidt-Rhaesa 2005). The circumference of the intestine is not roundish as in young stages, but irregularly folded. The lumen is very narrow and the intestinal cells contain, besides the irregular nuclei, few cytoplasm. The intestine is surrounded by a thick layer of

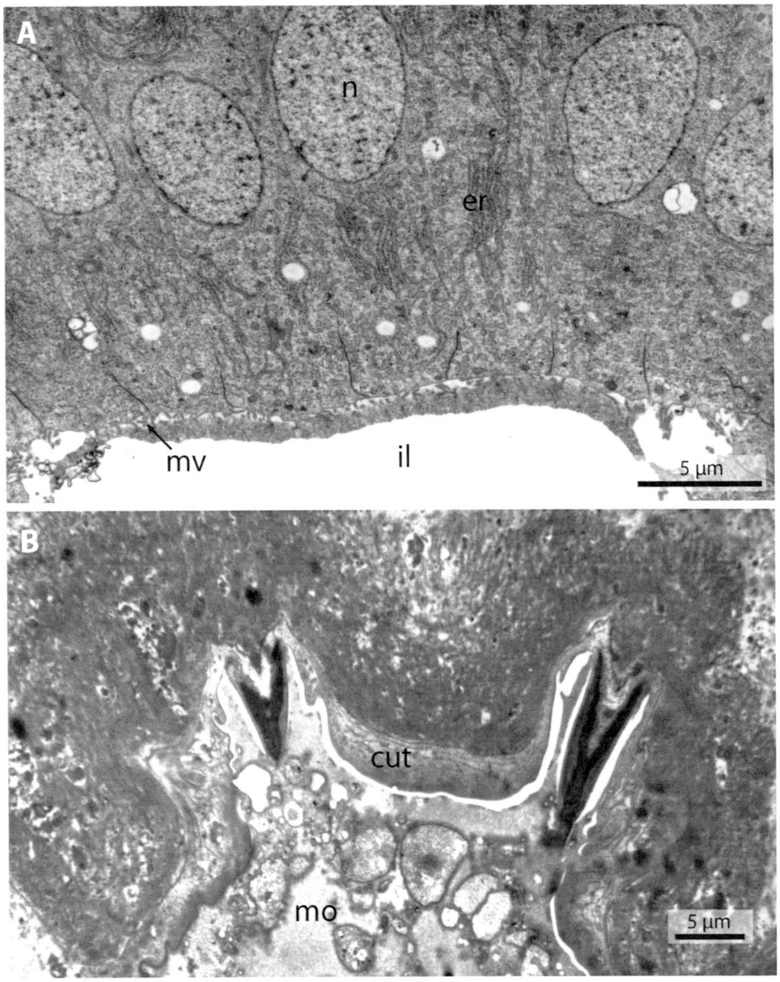

Fig. 3.1.16. Ultrastructure of the intestinal system. A, Intestinal cells in juvenile, parasitic *Paragordius varius* are rich in endoplasmatic reticulum (er) and have round nuclei (n). Few microvilli (mv) project towards the intestinal lumen (il); B, In the mouth region (mo) of *Nectonema munidae*, cuticular teeth-like structures are present. Cut = cuticle. All images TEM. B from Schmidt-Rhaesa 1996b, with kind permission of Blackwell Publishers.

extracellular matrix. The size of the entire organ decreases in diameter to more than half its original diameter (from about 50 µm in young stages to about 20 µm in adults), despite the increase of diameter of the entire animal. This all implies that the intestine is physiologically active in young stages and becomes functionless in adults.

The continuation of the intestine towards the anterior end appears to vary among species. In some species, a mouth opening and a thin, tube-like connection between mouth and intestine are reported (*Paragordius varius* Montgomery 1903; see Fig 3.1.12 A), but in others the lumen of these structures appears to be closed or the mouth opening is absent (e.g. *Parachordodes tolosanus* Vejdovsky 1886, *Gordius robustus* May 1919, *G. aquaticus* Rauther 1930, *G. villoti* Lanzavecchia et al. 1995). According to Inoue (1959b), *Chordodes japonensis* has a mouth opening, but the "midgut" is closed by cells. The mouth is, when present, terminal or slightly subterminal in position (Montgomery 1903). The connecting part is either called esophagus (= oesophagus) or pharynx, but because a special musculature is lacking, the latter term appears inappropriate. Montgomery (1903) reports the posterior part of the esophagus to be dilated, with radiating muscle fibers being attached. He calls this region the stomach. This region has not been reported in other investigations. Taking the structure of the intestine in *Paragordius varius* into consideration (see above) it appears probable that the mouth opens during young parasitic stages and then becomes closed in adults of some, but not all species. Vejdovsky (1894) reported a "brown gland" to be present in young stages of *Gordius vaeteri* and *Parachordodes pustulosus*, he assumed this gland to open into the esophagus. Such a structure has not been confirmed by later investigations.

Towards the posterior end, the intestine is joined by the reproductive system (see below). The presence of this cloaca was first recognized by Grenacher (1868).

Nectonema. The intestinal system of *Nectonema* starts with the mouth opening and consists of a cuticularized esophagus (sometimes named pharynx, e.g. Schmidt-Rhaesa 1996a) and a blindly ending intestine.

A region anterior of the esophagus is present in those young stages that possess a larval cuticle. In *N. munidae*, this region includes teeth and rods (Fig. 3.1.16 B) which are probably the remnants of the larval cuticular structures at the anterior end. Similar structures, at least the tooth-like ones, have also been found in *N. agile* (Feyel 1936) and *Nectonema* sp. (Oku et al. 1983). The adult cuticle attaches directly at the anterior margin of the esophagus.

The esophagus is a cuticularized tube which runs from the mouth opening through the body cavity in the head region and then passes the septum (Fig. 3.1.15). Behind the septum, the esophagus and the intestine, which starts directly at the septum, run in parallel for about 700 mm. Then the esophagus coils and the lumina of esophagus and intestine meet. The cuticular tube of the esophagus is surrounded by a small syncytium of cells (Schmidt-Rhaesa 1996a). A system of membranes is present on the outer side of this syncytium. Behind the septum, only one nucleus per section can be recognized and it is not clear whether the syncytial nature is still present. Besides the cuticular tube and the nucleus there is little further cytoplasm.

The intestine consists of few cells in cross section (Fig. 3.1.15 B, C). Actually, what appears to be cells in cross section may be syncytia in longitudinal direction, as no cell boundaries could be shown in this direction (Schmidt-Rhaesa 1996a). Per cross section, only 2–4 cells/syncytia are present; these consist of two types (Ward 1892a, Skaling & MacKinnon 1988, Schmidt-Rhaesa 1996b). One type is filled densely with darkly stained vesicles, the other does not contain such vesicles in the cytoplasm, only rough ER and mitochondria could be shown (Schmidt-Rhaesa 1996b). Towards the lumen, all cells/syncytia have few microvilli.

Body cavities and connective tissue

During development, several fluid-filled cavities can be observed in nematomorphs. The discussion on the nature of these cavities sometimes appears confusing (e.g. Švábenik 1908), due to different concepts being used for the definition of body cavities. The definitions followed here are based on the absence or presence of an epithelium immediately surrounding the cavity (see, e.g. Schmidt-Rhaesa 2007). A primary cavity is a (smaller or larger) cleft within the extracellular matrix (ECM) and is therefore completely surrounded by ECM. This is the case with all body cavities in nematomorphs. An epithelial lined cavity, a coelom, is not present. The reproductive organs represent epithelial lined cavities, but these are considered a particular organ (the reproductive organ) and not a body cavity.

Gordiida. A cross section through a gordiid can reveal very different states concerning the body cavity. Specimens may be found, in which there is no body cavity at all (Fig. 3.1.17 D), those which have one single large body cavity or several smaller cavities. The presence and size of the body cavities appears to be linked to the size of the reproductive organs and the parenchyma. In general, body cavities seem to be of greater extent in females compared to males.

Parenchyma is a tissue restricted to gordiids. The term summarizes those cells present between musculature and the inner organs such as nerve cord, intestine and reproductive organs (Fig. 3.1.17 A, B, D). Most parenchyma cells are very rich in vesicles and are surrounded by a thick ECM (Fig. 3.1.17 C). With histochemical methods, Reutter (1972) showed lipids and glycogen to be present in parenchyma cells; therefore it seems likely that they store nutrients. This makes sense considering a rapid nutrient uptake during the parasitic phase and a decrease in the extent of paranchyma with increasing gonad maturity. The relationships between body cavities and parenchyma may be best explained from a developmental perspective, although several open questions remain.

During development, young stages have only one large body cavity, in which the intestine and the developing reproductive organs are located (Lanzavecchia et al. 1995, Schmidt-Rhaesa 2005). This cavity is lined by developing muscle cells. As these are enclosed by ECM, the body cavity is a primary body cavity. Later, parenchyma cells start to appear. The further development is not completely clear. In adult males, the parenchyma is usually very abundant and fills either the entire animal or leaves just a small peri-intestinal body cavity (Schmidt-Rhaesa 1996a; Fig. 3.1.17 A, B). This is regardless of the developmental state of the testes tubes, be it narrow and hollow tubes or extensive, spermatozoa-filled tubes.

In adult females, several conditions have been documented and it is not clear how these conditions develop and what relation they have to each other. Some females

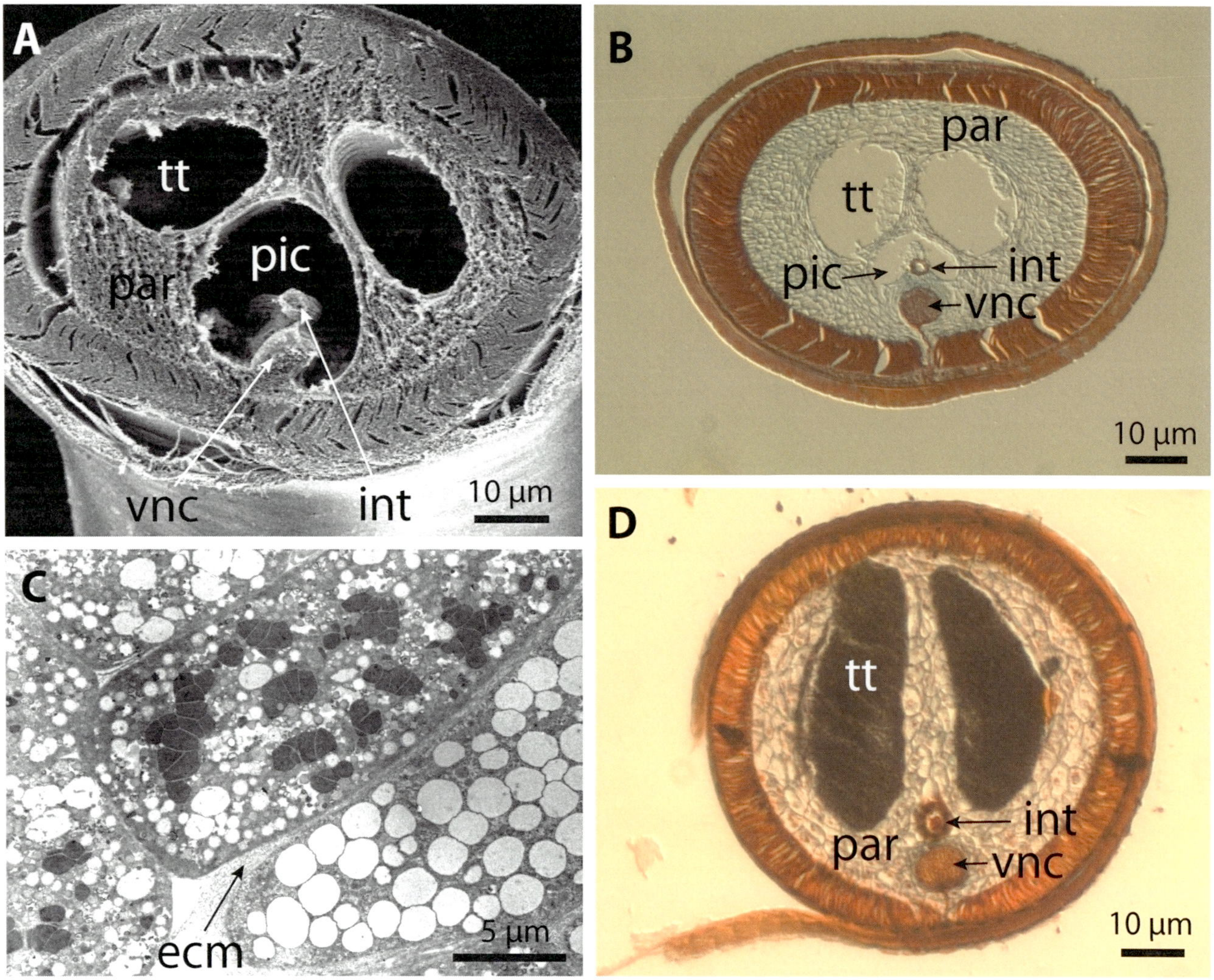

Fig. 3.1.17. Body cavities and parenchyma in gordiids. A, Cross section of *Acutogordius sulawensis* showing spatial relationships between testes tubes (tt), parenchyma (par), peri-intestinal body cavity (pic), ventral nerve cord (vnc) and intestine (int); B, Histological cross section of *Gordius aquaticus*, showing similar tissue distribution as in A; C, Ultrastructure of parenchyma cells of *Paragordius varius* showing different vesicles. Cells are surrounded by ECM (ecm); D, Cross section of *Gordius* sp. (from Iceland) showing absence of body cavities and extensive testes tubes. A with SEM, B, C light microscopy from paraffin sections, C with TEM.

have been reported with a total of three body cavities: unpaired around the intestine (as peri-intestinal cavity), as paired lateral structures (the lateral cavities) and as an unpaired dorsal cavity (e.g. Vejdovsky 1886, Montgomery 1903) (compare Fig. 3.1.4 B). Other reports show one large central cavity to be present. Oocytes can be present in the lateral cavities or, when present, in the large central cavity, their role is discussed in the chapter on the reproductive system.

Nectonema. A body cavity is generally present in *Nectonema* species; its extension appears to correspond with the developmental stage of the reproductive system. In females, the body cavity is smallest in immature animals and is restricted to a small central region within the "gono-parenchyma" (Schmidt-Rhaesa 1997a; Fig. 3.1.5 C). In mature females, the entire body proximal of the longitudinal musculature is a large body cavity (Schmidt-Rhaesa 1996a; Fig. 3.1.5 D). In males, the circumstances appear to be vice versa. The body cavity is largest in males, in which the sperm sac is smallest, whereas in species with a larger sperm sac, the proximal parts of the muscle cells extend so far that no body cavity is present any more and the animal is acoelomate. In addition to this body cavity, a second cavity is separated in the anterior end, at least in those species that have a complete septum (*N. agile, N. munidae, N. zealandica*) (Fig. 3.1.15). Ultrastructural investigations in *N. munidae* show that all body cavities are always completely surrounded by ECM and are therefore primary body cavity (according to e.g. Schmidt-Rhaesa 2007). In the anterior body cavity, isolated cells have been found within the cavity, i.e. on the proximal side of the ECM (Schmidt-Rhaesa 1996b). It is probable that such cells are "coelomocytes".

The septum is a solid structure separating the two body cavities at a distance of about 400 μm from the anterior tip. It attaches to the longitudinal musculature and is composed of cells with few mitochondria and some myofibrils (Schmidt-Rhaesa 1996b). In a young stage of *N. munidae*, a mass of cells with all characteristics of the septum cells is present directly behind the epidermis at the anterior tip of the animal (Schmidt-Rhaesa 1996b). It is likely that during development the septum is formed in this anterior-most region and then shifts backwards to form the anterior body cavity, in which the giant cells can develop.

Excretory system

There is no organ or structure with a clearly recognizable excretory function in any nematomorph. Some authors have speculated on the excretory function of particular structures, but such a function has never been proven. For example, Meissner (1856) regarded the intestine to be an excretory organ; Vejdovsky (1886) assumed the peri-intestinal body cavity to play a role in excretion; Villot (1889) assumed that the parenchyma cells may play a role in excretion; and Montgomery (1903) reported a "supra-intestinal organ", which he assumes to be excretory. This latter "organ" is described and figured as a lumen surrounded by obviously non-cellular material. As such a structure is not reported in other investigations, this may either be an artifact or a misinterpretation. Apart from these assumptions, none of the histological or ultrastructural investigations has clearly identified excretory structures. Therefore, it must be regarded as entirely unknown how nematomorphs excrete or perform osmoregulation. Probably, excretions are secreted in the same way as nutrients are absorbed, through the epidermis and the thin cuticle. During the adult stage, excretion may not be necessary because of the short life span and the absence of active food uptake. Osmoregulation during the free-living phase (especially in freshwater gordiids) may not be as necessary as in other freshwater animals, because the thick adult cuticle may prevent water exchange between the internal body and the water.

Reproductive organs

The reproductive organs are the organs in which Gordiida and *Nectonema* differ to the greatest extent from each other. In both taxa, the reproductive organs are a dominant system that produces enormous quantities of gametes, a fact that led Camerano (1888a) to state that free-living gordiids are more or less nothing more than enormous reproductive units.

Gordiida. Any section through a juvenile or adult gordiid (with the exception of the anterior-most end) will show some part of the reproductive system. In both sexes, gonads can be found in quite early stages. Gonads stretch throughout almost the entire body and produce large quantities of gametes. Past research has produced a number of described and figured cross sections, but it is not always clear how the sectioned regions of the reproductive system cohere along the entire individual or through development.

Sections through very young juveniles (May 1919, Lanzavecchia et al. 1995, Schmidt-Rhaesa 2005) show the gonads as a pair of dorsolateral strands of tissue surrounded by an ECM. The strands are composed of a more or less compact tissue of obviously undifferentiated cells. The cells contain a large nucleus and little cytoplasm. The strand appears to be compact in the earliest stages; later cells start to diverge from each other and clefts occur between the cells (Lanzavecchia et al. 1995, Schmidt-Rhaesa 2005). The outer cells form an epithelial layer (see also Vejdovsky 1894), while some cells remain in the lumen of the tube (Lanzavecchia et al. 1995, Schmidt-Rhaesa 2005).

In males, the strands observed in young stages form the reproductive organ throughout the life span. They are generally called testes or testes tubes. In mature individuals, these strands are very voluminous and densely filled with spermatozoa (e.g. Schmidt-Rhaesa 1997b, 1999, De Villalobos et al. 2005a; Fig. 3.1.4 A, 3.1.17 D), but they can also be present as empty, narrow tubes (e.g. Schmidt-Rhaesa 1997b, 1999). Such conditions could represent different regions of the testes tubes or different developmental stages of gamete production. Since gamete production appears to start early, one should assume gametes to be present in the testes throughout development. Because the parenchyma is well developed in stages with the narrow, hollow testes tubes, it seems unlikely that they represent a stage after releasing the gametes. Therefore, the testes tubes might be regionally differentiated into a gamete-producing part and a part comparable to a vas deferens. This, however, has not been investigated systematically.

An epithelial layer, which is evident in very young stages, is not reported in every investigation of older stages. An epithelial lining was observed for example by Meissner (1856), Montgomery (1903), Lanzavecchia et al. (1995), De Villalobos et al. (2005a) and Schmidt-Rhaesa (2005), but was absent in investigations by Rauther (1904, 1905) and May (1919). Vejdovsky (1894) suspected that an epithelium is only present in the anterior third of the testes tubes. Schmidt-Rhaesa (1997b) observed an epithelium to be present in some and absent in other specimens. Additionally, when present, the epithelium can be discontinuous, with not all cells touching their neighbors. These observations can best be interpreted as the presence of an epithelium in early stages, which then disintegrates and is finally lost. Additionally, there may be regional differences and the epithelium may be always absent from certain regions of the testes tubes. Considering the large quantities of gametes formed and the described conditions of the epithelium, it seems likely that male gametes develop from epidermal cells in the testes tubes.

In the posterior end, the testes tubes join the intestine and together form a cloaca. In *Gordius aquaticus*, the testes tubes appear to run from a dorsolateral position to a position lateroventral of the intestine, connected there by a horizontal duct, from which a connection runs to the intestine (Schmidt-Rhaesa 1997b). In *Paragordius varius*, the testes tubes run from dorsolateral to lateral of the intestine and then both tubes join the intestine separately (Fig. 3.1.13 A, 3.1.18 A). The cloaca opens on the ventral side by a cloacal opening (Fig. 3.1.18 B).

In females, understanding of the reproductive system is more difficult compared to males, because oocytes can be found in several compartments and their continuation is not always clear.

The youngest stages that can be identified as females are reported by Lanzavecchia et al. (1995). These have dorsolateral strands comparable as described above for early stages of testes, but on their ventral side, they proliferate into the spacious body cavity (which is surrounded by some peripheral parenchyma cells) to produce large spherical cells rich in vesicles (Fig. 3.1.18 A). These regions of proliferation are arranged serially along the longitudinal axis of the animal. A comparable state is reported by Vejdovsky (1894, see also 1886 and 1888), but here more parenchyma cells are present than in Lanzavecchia et al.'s stage. The parenchyma cells here form mesentery-like structures separating the body cavity into several compartments. Proliferation of presumed oocytes is into the large lateral cavities. This corresponds to figures of Montgomery (1903) for *Paragordius varius*

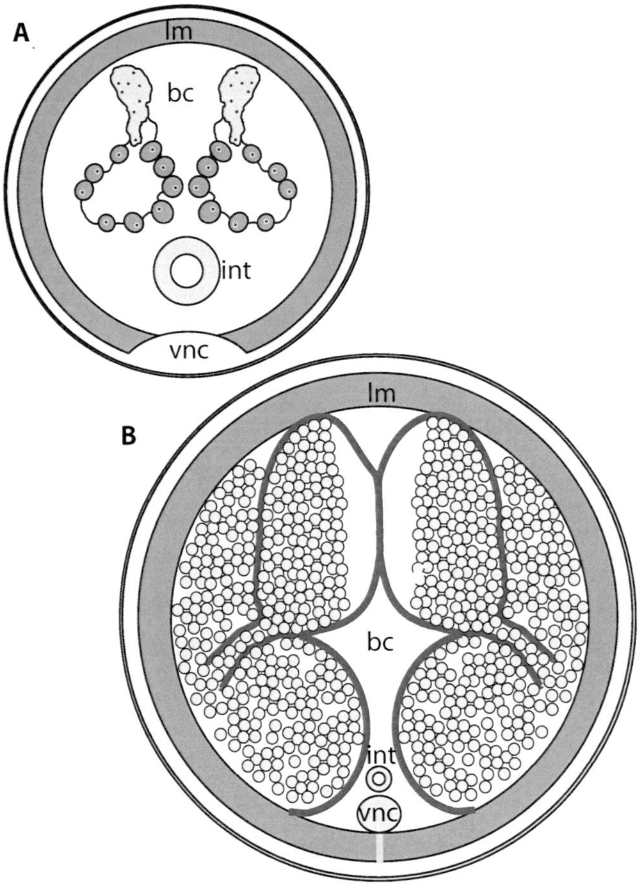

Fig. 3.1.18. Extension of the female reproductive system in gordiids. A, In young stages, dorsolateral strands (light grey) proliferate into the body cavity (bc). Int = intestine, lm = longitudinal musculature, vnc = ventral nerve cord; B, In older stages, membranes separate different compartmens, in which oocytes are found. Schematic drawings after Lanzavecchia et al. 1995 (A) and Montgomery 1903 (B).

and Rauther (1930) for *Parachordodes tolosanus*, here also oocytes appear to proliferate from dorsolateral strands into paired large lateral body cavities (Fig. 3.1.18 B). In very mature animals, oocytes occupy the majority of internal space, either in paired compartments as in *Paragordius varius* (Fig. 3.1.4 B) or even within one cavity, in which no remnants of former gonadal structures can be seen, as in *Gordius aquaticus* (Schmidt-Rhaesa 1996a).

In those female posterior ends that have been described (Montgomery 1903, May 1919, Rauther 1930, Inoue 1959b, Schmidt-Rhaesa 1996A; Fig. 3.1.19) the following elements are present: paired dorsolateral tubes containing oocytes, a voluminous ventral seminal receptacle, an atrium and the cloaca. The seminal receptacle is ventral of the intestine; it is connected by a narrower duct to a region called the atrium. The dorsolateral tubes join into this atrium in a lateral or ventrolateral position. Histological serial sections show the dosolateral tubes to

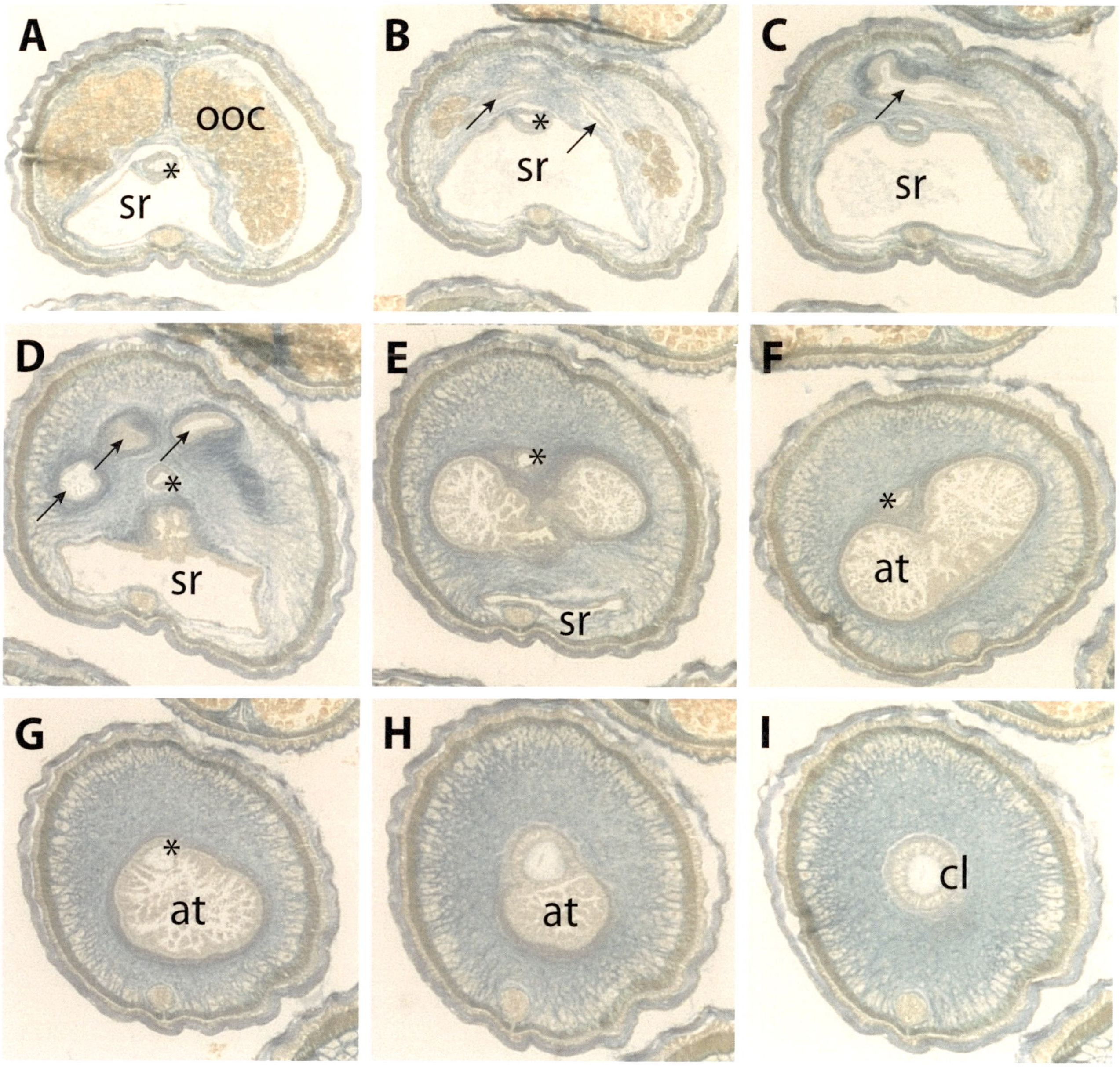

Fig. 3.1.19. Series of sections through the posterior end of a female *Paragordius varius* (from anterior to posterior). A, Section showing dorsolateral tubes containing oocytes (ooc), intestine (*) and anterior part of the seminal receptacle (sr). Dorsolateral tubes fuse (arrows in B) to form one structure (arrow in C), which divides into three cavities (arrows in D). The median structure terminates, the lateral ones enlarge (E) and fuse to form the atrium (at in F–H). The intestine finally joins the atrium to form the cloaca (I), which ends terminally between the three tail lobes. All images light microscopy from paraffin sections.

decrease in size towards the posterior (Fig. 3.1.19 A–C). They then merge by a duct (Fig. 3.1.19 B), which enlarges (Fig. 3.1.19 C) and from which very soon two lateral branches split off (Fig. 3.1.19 D). These branches run to a ventrolateral position, enlarge and form a strongly glandular epithelium characteristic of the atrium (Fig. 3.1.19 E). The exact joining of the seminal receptacle could not be observed, but is assumed to take place in this region (Fig. 3.1.19 E). The intestine "sinks" into the atrium (Fig. 3.1.19 G) and finally joins completely to form the cloaca (Fig. 3.1.19 I), which opens terminally between the three tail lobes.

All structures described for the female reproductive system have to be regarded as snapshots and the complete development, as well as the continuity of structures along the female body, are not satisfactorily clear. Do the dorsolateral tubes in the posterior end of *Paragordius varius*, for example, correspond to the body cavities, into which oocytes proliferate during development? Or are they continuous with the dorsolateral tubes producing oocytes in more anterior regions? How do sections showing one large cavity including oocytes in central regions of *Gordius aquaticus* relate to the posterior end, where tubes leading into an atrium are present?

Taking such questions into account, it is understandable that the terminology of the female reproductive system has confused authors on several occasions. The dorsolateral tubes are generally called the oviduct, the site of proliferation is often called the ovary. In some investigations the metameric arrangement of the ovaries caused interpretation problems, because an ovarial diverticula is connected to the oviduct in a small region but extends a bit further in the anterior and posterior direction. Depending on the position of a section, the tube and the proliferating region might be shown as continuous or as separate structures. Grenacher (1868), for example, concluded from this that the ovary and the oviduct originate as separate structures that later fuse (see also Von Linstow 1889). Vejdovsky (1886, 1888) called the body cavity, in which oocytes are found in later stages, "ovarial sacs". Rauther (1904, 1930) considered the dorsolateral tubes to be the site of origin for oogonia and assumed that their growth causes metameric diverticula that further communicate with adjacent body cavities. He therefore calls the entire structure an ovary, composed of ovarial longitudinal tubes and ovarial diverticula. He consequently assumes that oocytes are produced in the tubes, then develop in the body cavity and finally return into the tubes for release.

Nectonema. The reproductive system of *Nectonema* species is not very well understood and is not comparable in structure to that in gordiids or any other taxon. The male gametes are formed in a structure called the sperm sac, which is attached to the dorsal epidermal cord and either hangs into the body cavity or is completely bordered by the protoplasmic parts of the musculature (Schmidt-Rhaesa 1999). When the sperm sac is comparably small, its tissue is compact and cellular, with some muscle cells traversing the tissue. When the sperm sac is larger, many presumed early stages of gametogenesis are present within the sperm sac, mixed with non-gamete tissue-like cytoplasmic regions and nuclei. There is no epithelium lining the sperm sac, but the entire structure is surrounded by ECM. In the stages that have been investigated from *N. munidae*, the sperm sac terminates before it reaches the posterior end, therefore it is not clear how the gametes are released. In *N. melanocephalum*, the sperm sac divides in the anterior and in the posterior end into paired ducts, which are reported to lead into separate genital openings (Nierstrasz 1907).

Concerning the female reproductive system, two stages have been found. Either there is a large body cavity, which is completely filled with oocytes (Fig. 3.1.5 D) or there is a voluminous tissue of vesicle-filled cells that is named gono-parenchyma (Schmidt-Rhaesa 1997a; Fig. 3.1.5 C). The gono-parenchyma cells are densely interdigitated with the longitudinal muscle cells, which in this case do not possess a proximal protoplasmatic region. The cells contain rough ER, a nucleus and lots of darkly stained vesicles (Schmidt-Rhaesa 1997a). Towards the central lumen, the cells of the gono-parenchyma are rounded and, in some cases, appear to separate as spherical compartments (Fig. 3.1.20 A, B). This has been observed in *N. agile* (Feyel 1930, 1936) and *N. munidae* (Schmidt-Rhaesa 1997a) and very likely represents the mode of oocyte formation. An epithelial lining was stated to be present in *N. agile* by Bock (1913), but has not been found in other investigations. Between the putative oocytes there is some further tissue with nuclei and some vesicles (Fig. 3.1.20 C), this is probably some kind of nursing tissue. The vesicles in the "nurse cells" resemble those in the oocytes. Towards the posterior, the body cavity leads directly to the female genital opening and the oocytes are released through this opening without further structures such as oviducts.

Gametes

Gordiida. The spermatozoa of Gordiida are unique in form; comparisons to spermatozoa from other taxa are almost impossible. Spermatozoa were observed quite early (e.g. Villot 1874b), their fine structure was described by Lora Lamia Donin & Cotelli (1977), Schmidt-Rhaesa (1997b) and Valvassori et al. (1999).

The spermatozoa are aflagellate. Their form changes during sperm transfer (Mühldorf 1913, Schmidt-Rhaesa 1997b). Within the male reproductive system, spermatozoa are rod-shaped, they become more slender in their posterior part when they are within the external sperm drop on the female posterior end and are very slender and elongate within the female seminal receptacle.

The rod-shaped spermatozoa within the testes are composed of an elongate nucleus, which is present in the posterior part (Fig. 3.1.20 D). Anterior of it is a structure named acrosomal tube, surrounded by an acrosomal sheath. On the anterior tip of the acrosomal tube is a structure interpreted as the acrosome (Fig. 3.1.20 D). Filaments extend from the base of the acrosome into the acrosomal tube. The nucleus is surrounded by a perinuclear cisterna and two layers of the so-called multivesicular complex (Fig. 3.1.20 E, G). Outside the female posterior

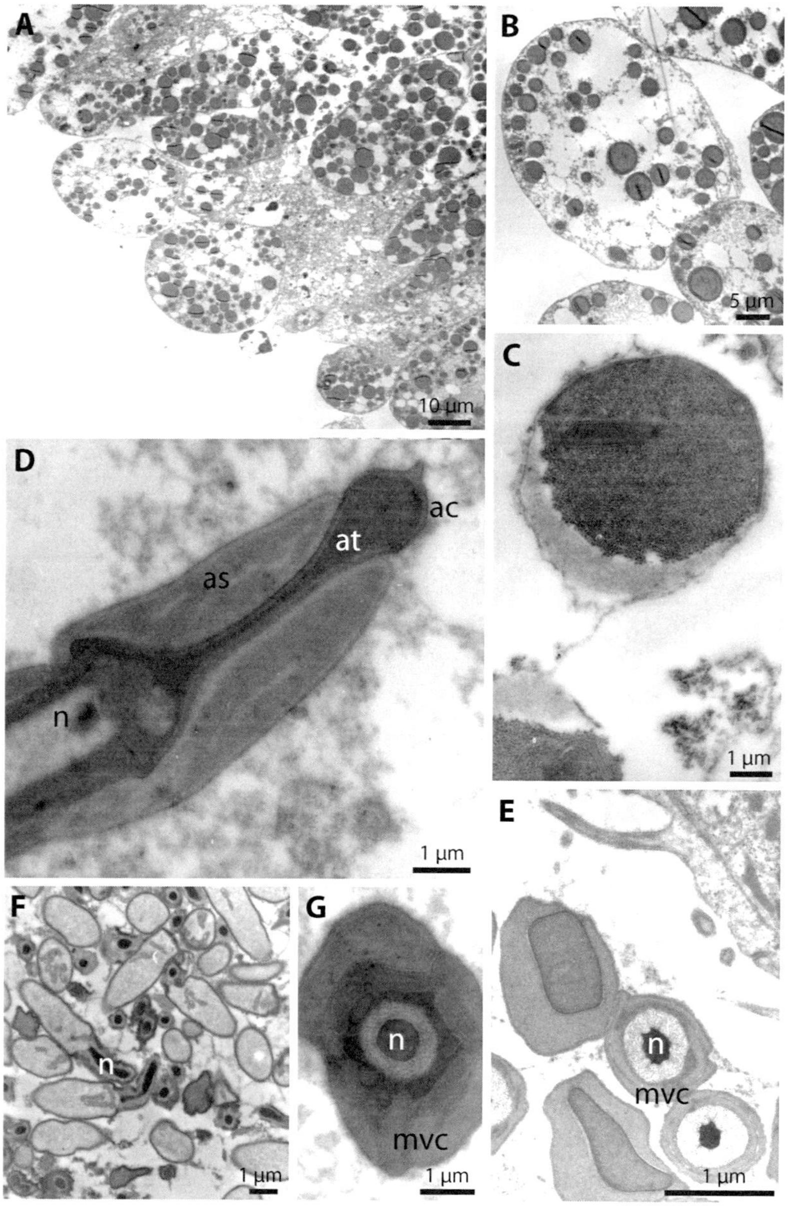

Fig. 3.1.20. A–C, *Nectonema mundae*; A, Proliferation of oocytes from the gono-parenchyma; B, Oocyte; C, Structure of one type of vesicle in the nurse cells, this type corresponds to the peripheral vesicles in the oocytes. D, F–G, Spermatozoa in *Gordius aquaticus*; D, Longitudinal section through the anterior part of a spermatozoon in the male testes showing nucleus (n), acrosomal tube (at), acrosomal sheath (as) and acrosome (ac); E, Cross section through spermatozoa from the testes in *Paragordius varius* showing nucleus (n) and multivesicular complex (mvc); F, Structure of the spermatozoa in the sperm drop deposited on the female posterior end; G, Cross section through spermatozoon with multivesicular complex (mvc). All images TEM. A from Schmidt-Rhaesa 1997a, E from Schmidt-Rhaesa 2005, both figures with kind permission of Springer Publishers.

end, the spermatozoa lose the outer layer of the multivesicular complex (Fig. 3.1.20 E) and anterior of the nucleus only one oval structure is present (Fig. 3.1.20 F), including the acrosome. Finally, within the seminal receptacle, this oval structure becomes filiform; the nucleus is surrounded by one thin layer which is not compartmentalized (Schmidt-Rhaesa 1997b).

Spermatogenesis starts with elliptic cells, which can have more than one nucleus and centrioles (Valvassori et al. 1999). During further development, an electron-dense, bell-shaped structure forms in the anterior region and the acrosomal vesicle forms there. The nucleus elongates and the structures characteristic of the mature spermatozoa develop subsequently.

Nectonema. Spermatozoa are documented very rarely. The only stage that has been found to date consists of more or less undifferentiated cells with a large nucleus and little cytoplasm. This probably represents an early stage of gametogenesis. These stages are present within the sperm sac or can be found within the body cavity (although it is not clear whether this is a preparation artifact). In several cases they were found in clusters attached to a central core with concentric membranes.

The oocytes of *N. munidae* contain two types of vesicles: homogeneously dark stained vesicles and dark vesicles with a lighter stained periphery. These latter vesicles are found in the periphery of the oocytes (Schmidt-Rhaesa 1997a). In the putative oocytes, which are freshly separated from the gono-parenchyma, quite similar vesicles are found (Fig. 3.1.20 B, C). There is a homogeneous one and another one with a dark staining and a crescent-shaped lighter structure at the margin (Fig. 3.1.20 C). The dark content of this latter type has a reglularly granulated substructure and contains some membranous compartments (Schmidt-Rhaesa 1997a). The oocytes of *N. agile* appear to be similar in structure (Ward 1892a, Bock 1913). According to Ward (1892a) and Huus (1932), the oocytes form spines when they are released into the water. Because one type of vesicles is found only in the periphery of the oocytes, it is likely that these vesicles are responsible for spine-formation.

Morphology of the larva

Gordiida. The early descriptions of the tiny larva of gordiids were not very detailed (Grube 1849, Leidy 1850, 1851, Villot 1872, 1875a, Tretjakow 1901), differing for example in the number of rings and the number of hooks within these rings. The first quite detailed descriptions were given by Montgomery (1904) and Schepotieff (1908); ultrastructural TEM data were provided by Zapotosky (1974, 1975) and Jochmann & Schmidt-Rhaesa (2007); SEM investigations were made by Valvassori et al. (1987), Bohall et al. (1997), Adrianov et al. (1998) and Bolek et al. (2010).

The larvae are very small, the size varies among species. Hanelt & Janovy (2002) measured larvae of *Gordius robustus* to be slightly more than 100 µm in length, larvae of *Paragordius varius* and *Chordodes morgani* are about half that size, with *Paragordius* being slightly longer than *Chordodes* larvae. Measurements in *Chordodes japonensis* (50 µm; Inoue 1958) and *C. janovyi* (48.5 µm; Bolek et al. 2010) make it likely that *Chordodes* larvae are in general very small. *Neochordodes occidentalis* larvae are also in this range (Poinar & Doelman 1974). Apart from the size, several parameters in cysts and larvae differ significantly among the three most abundant species in the USA, making these larvae distinguishable from each other (Hanelt & Janovy 2002). Dorier is the only one to note differences in larvae of certain European species. The most distinctive character is the structure of the posterior end. While in *Gordius aquaticus* larvae the posterior end terminates in one median-pointed tip (Fig. 3.1.21 A), *Gordionus violaceus* and *Parachordodes gemmatus* larvae have two small lateral tips, *Paragordius tricuspidatus* larvae have one pair of long lateral and one pair of more median small tips and *Gordionus alpestris* larvae have a total of six small tips (Dorier 1930, 1932, 1934). Dorier (1934) lists a number of differences enabling larvae of *Gordionus violaceus* and *Parachordodes gemmatus* to be distinguished, including different measurements and number of annuli on pre- and postseptum. Additionally, the terminal tips appear to be mobile in *P. gemmatus*, but fixed in *G. violaceus*.

The larval body is divided into two regions, which are divided by an internal septum and are therefore named the preseptum and the postseptum (terminology in early descriptions differs) (Fig. 3.1.21 A, 3.1.22 A). The preseptum itself is composed of a region that can be withdrawn and everted; this region is often called the proboscis (Fig. 3.1.21 C, D). The posterior part of the preseptum and the postseptum are superficially annulated by fine cuticular rings (Fig. 3.1.21 A, 3.1.22 A, C, F). When counted, some authors give a range (e.g. 22–24 rings in *Chordodes morgani* larvae; Bohall et al. 1997), only Inoue (1958) counted a fixed number of rings (24) in *Chordodes japonensis* larvae. The proboscis contains cuticular structures in the form of spines (sometimes named hooks) and stylets, which can be completely withdrawn into the remaining preseptum (Fig. 3.1.21 A–D, 3.1.22 A, C, D). The spines are pointed rod- or leaf-like structures, the tip points frontally when withdrawn, but when everted the tips point caudally. This obviously serves as an anchor in tissues such as the gut wall of the host. The spines are arranged in three rings. The numbering of rings varies in the literature (with the outermost ring being termed either the first or the third ring), therefore the rings are named here as the outermost (the one appearing first during eversion), median and innermost ring. The outermost ring includes six slender spines, which originate as long intracuticular rods within the cuticle of the proboscis and detach in their apical part (Jochmann & Schmidt-Rhaesa 2007; Fig. 3.1.22 A, C). The ventral spine is deeply furcated, leading to the external appearance of seven spines in this ring (Fig. 3.1.21 B, 3.1.22 A, C). The median ring is composed of six leaf-like spines (Fig. 3.1.21 B, D, 3.1.22 C, D). The rings of

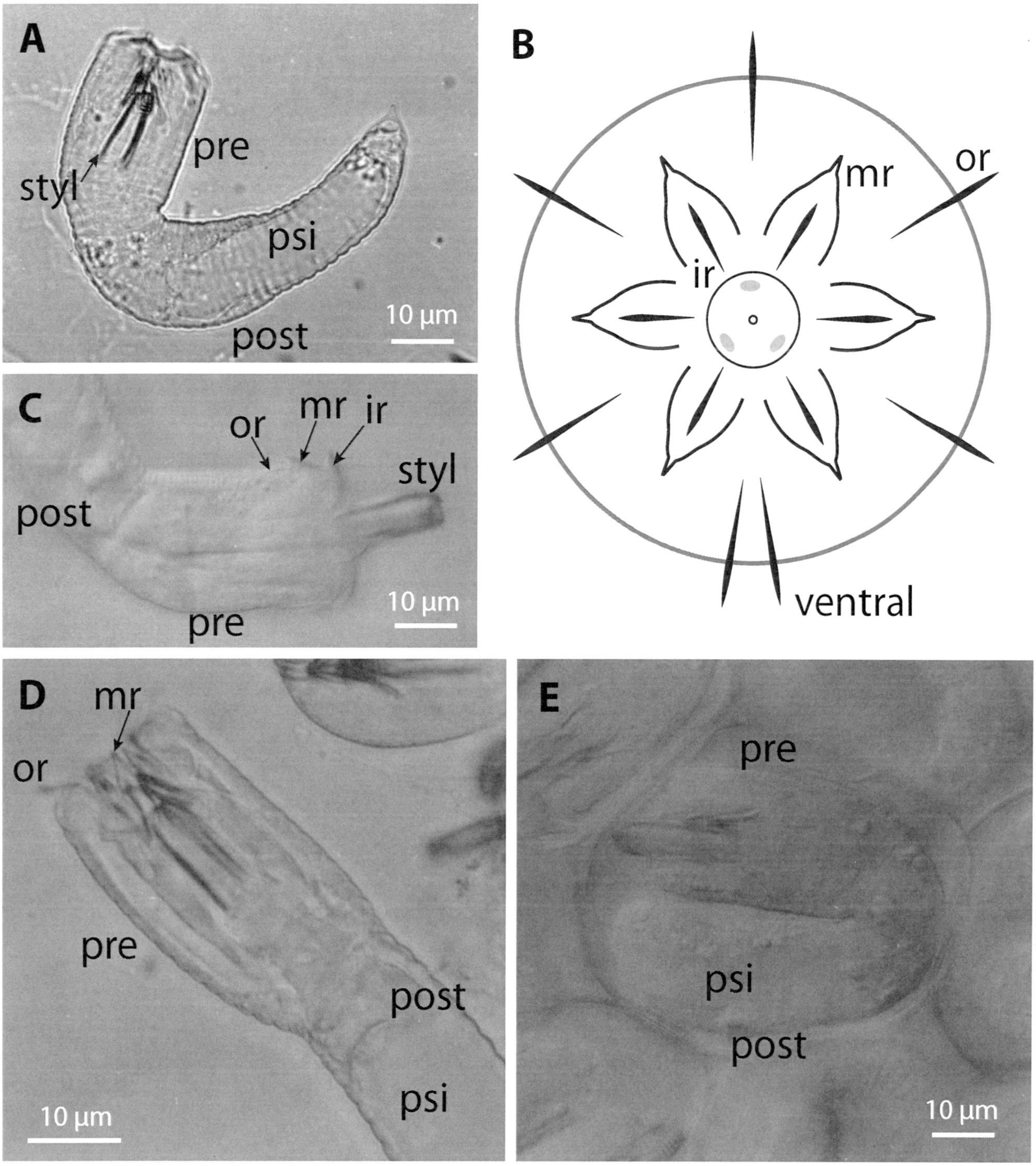

Fig. 3.1.21. Structure of the larva (all images are from *Gordius aquaticus*). A, D, Larva with inverted hooks and stylet (styl). Body is divided into preseptum (pre) and postseptum (post); the postseptum contains the pseudointestine (psi); B, Schematic drawing of arrangement of hooks in inner (ir), median (mr) and outer ring (or); C, Larva with everted hooks and stylet; E, Larva within the egg. For abbreviations see A and B. A, C–E with light microscope.

this and of the innermost ring are not in line with the outermost spines, but point in the region between two outermost spines. The innermost spines are also six in number (Fig. 3.1.21 B). All spines can be traced within the proboscis cuticle as dense rods; the median spines have a pair of rods each (Jochmann & Schmidt-Rhaesa 2007).

When the rings with hooks are everted, the central stylet is protruded. The distal part of this stylet is oval, at least in *Chordodes morgani* (Bohall et al. 1997) and

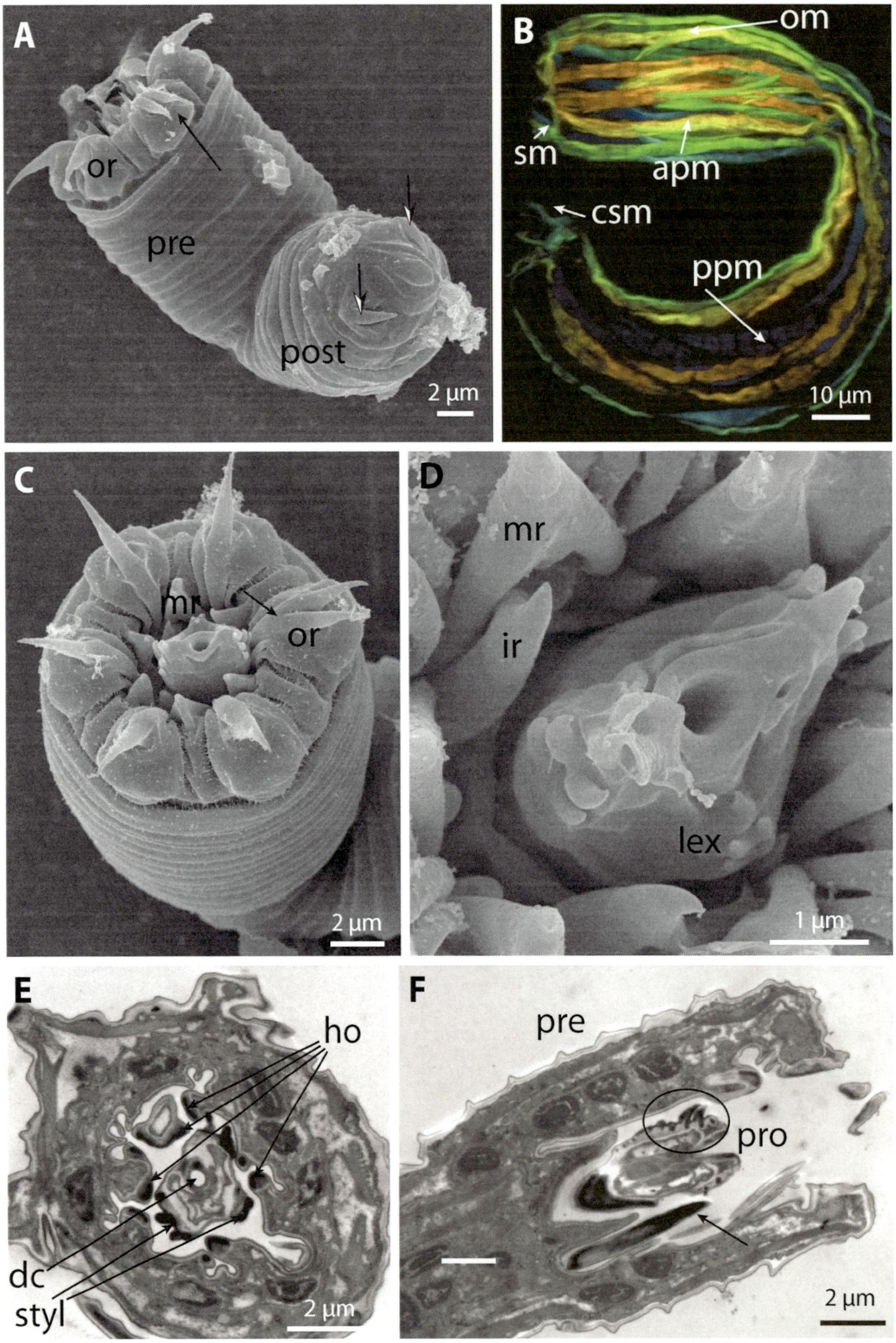

Fig. 3.1.22. A, C, D, Larva of *Paragordius varius* (from the USA). A, General organization of the larva with preseptum (pre) and partly everted proboscis, making outer ring of hooks (or) visible. Arrow points to the ventral double hook. Arrows in the postseptum (post) indicate posterior hooks; B, Staining of mucles in the larva of *Gordius aquaticus*, showing spine muscles (sm), oblique muscles (om), anterior parietal muscles (apm), posterior parietal muscles (ppm) and caudal spine muscles (csm); C, Frontal view on larva of *Chordodes janovyi*; D, Magnification of the stylet with lateral extension (lex), inner hooks (ir) and median hooks (mr); E, Cross section through preseptum of larva of *Paragordius varius* with central duct canal (dc). Stylet (styl) and hooks (ho) are recognizable as dark sclerotizations in the cuticle; F, Longitudinal section through preseptum (pre) of the *P. varius* larva. The proboscis (pro) with terminal teeth of one stylet (in circle) and one hook (arrow) are visible; A, C, D with SEM, B depth color-coded image from confocal laser scanning microscopy, E, F with TEM. B from Müller et al. 2004, with kind permission of Springer Publishers.

C. janovyi (Bolek et al. 2010; Fig. 3.1.22 C, D). There is a central opening and a series of stylet spines on each side (Fig. 3.1.22 D, F). A bit further down, there is a lateral extension (called the "lateral papilla" by Bolek et al. 2010), creating a more or less triangular cross section (Fig. 3.1.22 D, E). This is probably also the case for the *Paragordius varius* larva (Fig. 8 in Zapotosky 1974), but looks a bit different in *Gordius* sp. (Valvassori et al. 1987). In TEM sections it can be observed that the stylet is formed by three dense cuticular rods, which reach into the proboscis tissue (Zapotosky 1974, Jochmann & Schmidt-Rhaesa 2007; Fig. 3.1.22 F).

In the posterior end, there is one subterminal, slit-like opening (Jochmann & Schmidt-Rhaesa 2007, Bolek et al. 2010) and a varying number of terminal spines. Larvae of the genus *Gordius* have been described with only one terminal spine, which is more or less a pointed end of the body. Larvae of *Paragordius varius* have a pair of terminal spines (Zapotosky 1975), as do those of *Gordionus violaceus* (as *Parachordodes violaceus*; Dorier 1932) and *Parachordodes gemmatus* (Dorier 1934). *Chordodes* larvae appear to have four spines, as Bohall et al. (1997) report 3–4 spines and Bolek et al. (2010) found four spines. Six spines were described from *Gordionus alpestris* (as *Parachordodes alpestris*; Dorier 1932).

The internal morphology, in particular the structures of the postseptum, has been the source of some misinterpretations. The preseptum consists mainly of musculature. From the anterior opening on the stylet, a cuticularized canal runs through the preseptum (Fig. 3.1.22 E), penetrates the septum and terminates in a gland. Therefore, it is much more likely that the opening represents a gland opening instead of a mouth. Four sets of muscles can be distinguished in the preseptum (Müller et al. 2004). In the body wall are 12 longitudinal muscle bundles called the anterior parietal muscles (Fig. 3.1.22 B). Six oblique muscles function as retractors of the introvert (Fig. 3.1.22 B), a further six proboscoidal muscles function as proboscis retractors. Additionally, there are six small muscles associated with the spines of the outermost ring of spines (Fig. 3.1.22 B). Muscles are cross-striated. They attach by hemidesmosomes at the extracellular matrix (ECM) underlying the very thin epidermis on the one side and to the ECM surrounding the septal cells on the other side (Müller et al. 2004). It seems that the septal cells, which separate the two parts of the body, function mainly as attachment for the preseptal muscles.

In the postseptum are two structures with probable glandular function, the postseptal gland and the pseudointestine. Neither of these structures can be clearly homologized with the intestinal system because the further development of these structures remains unknown. The preseptal duct winds its way through some embracing cells and leads into the postseptal gland, which is composed of eight cells. Here, the duct wall is penetrated by pores; the surrounding gland cells form a system of microvilli-like processes towards the duct. The pseudointestine is a large structure; it is different in form and content among species (see Hanelt & Janovy 2002), but was investigated at the ultrastructural level only in *Paragordius varius* (Zapotosky 1975, Jochmann & Schmidt-Rhaesa 2007). Here, it is composed of four cells, forming an oval structure, in which there are anterior spherical or oval contents. A duct leads from the posterior part of the pseudointestine to the ventral side of the body. Dorier (1932, 1934) observed that the secretion from the pseudointestine is used for cyst formation. Jochmann & Schmidt-Rhaesa (2007) observed a further, syncytial tissue in the ventral, anterior region of the pseudointestine, but the function is not known. Several authors (Tretiakow 1901, Meyer 1913, Mühldorf 1913, May 1919, Malakhov & Spiridonov 1984) reported a ventral paired row of cells, which they assumed to represent the anlagen of the ventral nerve cord. No such structure was observed in the TEM investigations by Zapotosky (1975) and Jochmann & Schmidt-Rhaesa (2007). The musculature of the postseptum consists of a different number of peripheral parietal muscles and a few other muscles associated with the pseudointestinal duct and the terminal spines (Müller et al. 2004; Fig. 3.1.22 B).

The transition from larvae to juvenile worms and therefore the origin of most juvenile and adult tissues is unknown. It is likely that the preseptum, or at least the anterior part of it, is shed during a moult (see e.g. Von Linstow 1891a, Camerano 1892, Vejdovsky 1894, Švábenik 1925, Valvassori et al. 1988, Lanzavecchia et al. 1995 for reports of the tiny remnant of the preseptum on the juvenile body). In particular the origin of the intestine and the development of the nervous system remain completely unknown.

Nectonema. There is only one published observation of the larva of *Nectonema munidae* by Huus (1932). The youngest larva was found in the body cavity of *Munida tenuimana*. It is 350 µm long and may therefore already represent a stage in the transition between the larva and the juvenile. The anterior end was everted and withdrawn; on the everted region two rings of spines (number not reported) and terminally two cuticular structures named "jaws" are present. It is quite likely that the spines are homologous to the spines in gordiid larvae and the "jaws" may be homologous to the stylet. There is no separation of the body into different regions and there is no trace of a septum. An intestine is present, which traverses the entire body and terminates at the posterior end of the larva. Older stages observed by Huus (1932) still had the anterior armature and the capability to evert the anterior region, but here additionally a thin cuticularized esophagus was present.

3.2. Reproduction and development

All Nematomorpha are parasitic animals, which emerge from their host at maturity and reproduce in free water. All data on reproduction and life cycle must be considered as patchy and much more work is needed to get a more solid picture of the life cycle of horsehair worms.

Reproductive biology

Mating and pseudocopulation. After male and female gordiids have emerged from their host into water, they have to find each other for fertilization. It has been mentioned several times that males appear to be more agile than females, but it is not clear whether both sexes find each other by attraction and directed movement or by chance. Males tend to curl their posterior end ventrally and hold on to anything – vegetation and other worms. As a result, clusters of several or even large numbers of specimens can be formed (Fig. 3.2.1 A, C). Such a tangle of worms reminded people of the Gordian knot from Greek mythology and led to naming of the first genus *Gordius*. The formation of such tangles appears to occur at least in species of the genera *Gordius* and *Gordionus*; observations for other genera are few. In *Neochordodes occidentalis*, Poinar & Doelman (1974) observed (in the laboratory) that one male remained coiled around the female for 2 weeks (although it is not mentioned how many specimens were put together). In the genus *Chordodes*, knots appear to be absent. Gordian knots may consist of more than one species (De Miralles 1980).

Thomas et al. (1999) found a tendency with worms with gametes to be found in deeper and slower-flowing parts of a stream, while worms that had released their gametes were found in areas of higher velocity. Therefore it is suggested that the physical characteristics of a stream may influence mating.

When finding a female, males from those species with a bilobed posterior end glide along the female body using their spread tail lobes until both posterior ends are in close proximity (Dorier 1930). According to Hofmänner (1913), the female posterior end is even positioned between the male tail lobes. The male then deposits a volume of sperm over the entire posterior end of the female. It may be the case that males deposit their sperm drop also on other males (Müller 1926, Dorier 1930). Ultrastructural observation of spermatozoa from this external sperm "drop" showed radial elements between the spermatozoa; it is unclear whether these elements originate from the female cuticle or are formed by the sperm drop itself (Schmidt-Rhaesa 1997b). According to Müller (1926), the sperm drop disappears within 2 days. Sperm is also found in the female seminal receptacle and in the distal part of the female reproductive system (see e.g. Meissner 1856) and fertilization of the oocytes is internal. Due to these observations it appears that males do not inject sperm into the female, but deposit it externally. At least part of the sperm enters the female reproductive system and is stored in the seminal receptacle to be released into the atrium or cloaca for internal fertilization.

Mating may occur more or less directly after emergence from the host, for *Gordius difficilis* Bolek & Coggins (2002) report mating within 24–48 hours after emergence.

Rarely, copulatory organs have been reported from male gordiids (Kirjanova 1958a). SEM investigations document circumcloacal spines in several species (Fig. 3.2.2 A), but further structures are extremely rare (see e.g. Fig. 3.2.2 B). The few observations of sperm deposition never report any eversible structure or any direct contact with the female cloacal opening. Therefore, Kirjanova's (1958a) report may be due to a misinterpretation of structures in the posterior end. However, it must be noted that reproduction has been observed in only a few species and it is not clear how, for example, males from species with an undivided posterior end perform mating. Additionally, most gordiids have conspicuously diverse cuticular structures in the posterior end, including circumcloacal spines; the role of neither of them is known. To conclude, at present all observations support an external sperm deposition (pseudocopulation) with subsequent internal fertilization. It is neither known whether the entire external sperm is used by the females for fertilization, nor whether females copulate with several males.

Recently it was discovered that a newly described species, *Paragordius obamai*, reproduces by parthenogenesis (Hanelt et al. 2012). Specimens of this species were

Fig. 3.2.1. Mating and egg deposition. A, Gordian knot of several specimens of *Gordius aquaticus* (Hardegsen, Germany); B, Female *Paragordius varius* (brown) deposits a long string of eggs (white) (San Luis, Argentina); C, Several spcimens of *Pseudochordodes bedriagae* entangle with other specimens and vegetation (San Luis, Argentina); D, Curving eggstrings of *Chordode janovyi* are attached to vegetation; E, Female *Chordodes morgani* (black arrow) winds around a stick and deposits eggstrings (white arrow) (Nebraska, USA); F, Female *Gordius aquaticus* deposits her eggstrings (white) between rotting leaves in the vicinity of a pond (Bielefeld, Germany). A from Schmidt-Rhaesa 1996d, with kind permission from Elsevier Publishers; D photo by Matt Bolek, with kind permission; E from Schmidt-Rhaesa 2002c, with kind permission from Westarp Wissenschaften, Hohenwarsleben; F photo by Birgen H. Rothe, with kind permission.

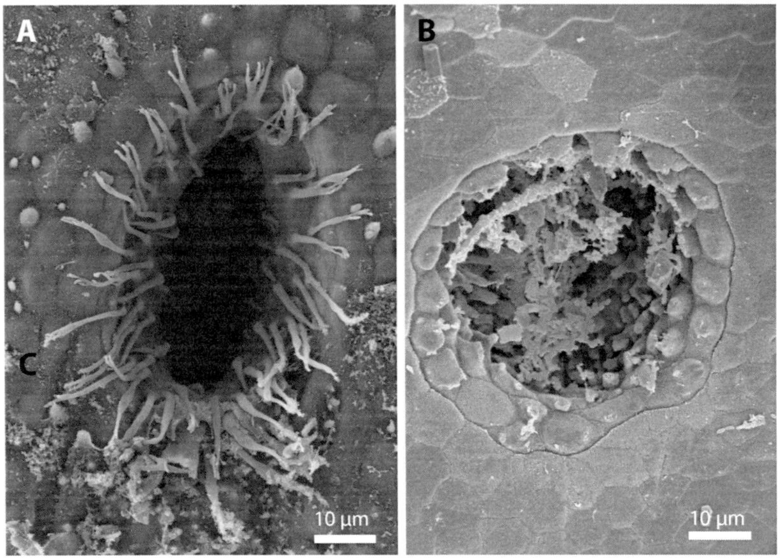

Fig. 3.2.2. Structure of the cloacal opening. A, Cloacal opening with circumcloacal spines in *Gordionus violaceus* (from Hessen, Germany); B, Unusual structure of the cloacal opening in an undescribed *Gordius* specimen from Atascadero, California, USA. Both images with SEM.

Tab. 2. Selection of sex ratios reported in the literature.

Species	Ratio ♂♂:♀♀	Author
Parachordodes (Gordius) tolosanus	7:3	Von Linstow 1891a
Gordius aquaticus	56:2	Mühldorf 1914
Parachordodes tolosanus	1:24	Mühldorf 1914
Parachordodes pustulosus	73:0	Mühldorf 1914
five species from 22 locations in Argentina	always male biased	De Miralles 1980
Gordius valnoxius	15:19	Degrange & Martinot 1995
Euchordodes nigromaculatus	61:0	Thomas et al. (1999)
Gordius difficilis	variable, but usually male biased	Bolek & Coggins 2002
Paragordius varius (lab raised)	258:215	Hanelt & Janovy 2004a
Gordius difficilis	strongly male biased	Cochran et al. 2004
Gordius albopunctatus	99:19	Schmidt-Rhaesa & Kristensen 2006
Gordius aquaticus, Gordionus violaceus	332: 93*	Weißbecker & Schmidt-Rhaesa 2010
Chordodes brasiliensis, Noteochordodes cymatium, N. talensis, Pseudochordodes dugesi	male biased	Salas et al. 2011

* This is the sum of all counts; data are broken down according to species, locality and year.

collected in Kenya and reared over several generations in the laboratory. Only females developed from larvae; these females laid eggs, from which new larvae developed.

In contrast to the pseudocopulation in gordiids, *Nectonema* species copulate. Males introduce their slender posterior end into the female genital opening (Huus 1932, see also Nielsen 1969, Schmidt-Rhaesa 1999). Nielsen (1969) showed in sections of females after copulation the presence of spermatozoa among oocytes. Copulation has been observed only a very few times and further details remain to be elucidated.

Sex ratio. Observations of the sex ratio in the field rarely show equal proportions of males and females; usually the collections are biased, mostly towards males (see Tab. 2).

Dorier (1928, see also 1930) found different sex ratios in *Gordius aquaticus* in France at different times of the year: 39♂♂:9♀♀ on July 28 (1928), 6♂♂:22♀♀ on August 13 (1927), 38♂♂:3♀♀ on August 24 (1928) and 1♂:45♀♀ on September 2 (1925). He concluded from these data that both sexes occur at different times of the year. The same appears to be valid for *Gordius dimorphus* in New Zealand, where Poulin (1996) found a female-biased sex ratio in spring and a male-biased sex ration in early summer.

Such observations can be explained in different ways. Males and females may occur sequentially, as has been suspected by Dorier (1929) and Poulin (1996). On the other hand, males may be detected and caught more easily due to their active behaviour.

A further explanation is based on unpublished observations on *Gordius aquaticus* in Zweischlingen, Bielefeld, Germany. While males can be caught within a shallow forest stream and pond, egg-laying females were detected under moist rotting leaves directly adjacent to the water (Fig. 3.2.1 F). Therefore, females maybe found in open areas of streams only for brief periods and seek hidden locations for egg deposition after the pseudocopulation. One further explanation is that adults die after they shed their gametes. As males shed only part of their gametes per pseudocopulation, it seems that they produce more sperm than they release. Females, in contrast, shed all their eggs and therefore may die sooner than males.

Egg laying. After mating (4–5 days after Dorier 1930, 2–4 days after Müller 1926; 15 days after emergence in *Gordius villoti*, see Valvassori et al. 1988), females lay enormous quantities of eggs. Estimations of the egg number range from 0.5 million (Dorier 1930) to 4 million (Valvassori et al. 1987) and 6 million (Leidy 1851). Hanelt (2008) counted between 200,000 and 5–8 million eggs, with egg numbers strongly correlating to female body length. This enormous quantity obviously compensates for relative low chances of an individual larva to complete the entire life cycle.

The egg-laying behaviour of only very few species is known, but this shows considerable differences. In *Gordius*, *Paragordius* and *Neochordodes occidentalis*, egg strings are given off from the females as long strings, which may break into longer (*Paragordius varius*, *N. occidentalis*) or smaller (*Gordius robustus*) pieces (Poinar & Doelman 1974, Hanelt & Janovy 2002) (Fig. 3.2.1 B). Valvassori et al. (1987) describe the female of *Gordius* sp. as depositing an egg string of the length of the entire female, which takes 48 hours. These strings are not attached anywhere in particular, but may cling more or less accidentally to vegetation or become trapped within a tangle of worms. This has been suspected by Wesenberg-Lund (1910) as some kind of brood care, which is not the case. In contrast to *Gordius* and *Paragordius*, *Chordodes* species (observations in *Chordodes japonensis* by Inoue 1958, *C. morgani* by Hanelt & Janovy 2002 and *C. janovyi* by Bolek et al. 2010) attach their eggs to vegetation in long winding lines (Fig. 3.2.1 D, E). Tangles ("Gordian knots") are not observed in these species.

Body sizes vary to some degree in gordiids. As the reproductive system is so dominant within the worm body, female size is a good indicator for fecundity, i.e. the quantity of shed eggs (Hanelt 2008). Shortly after shedding their gametes, the animals become quite flat and die (Dorier 1930, Inoue 1958).

Fresh egg strings are white, but they may also be found in shades of brown. Villot (1874b) suspected that the color of egg strings is characteristic of different species (white in *Gordius aquaticus*, brown in *Parachordodes* (as *Gordius) tolosanus*). Other reasons for color changes may be hardening (probably by inclusion of melanin; Meßner 1983) or contamination with detritus or other organisms, but the most likely interpretation is that the formation of hooks and stylets on the larval proboscis cause the darker appearance of the entire eggstring (Ben Hanelt, personal communication).

Under laboratory conditions and a temperature of about 23°C eggs of *Neochordodes occidentalis* develop in 30 days (Poinar & Doelman 1974) or 13 days for *Paragordius varius* (De Villalobos et al. 2003a). In the field, egg laying is reported in spring for *Chordodes japonensis* in Japan (Inoue 1958) and in July for *Gordionus violaceus* in France (Dorier 1932). Development of *Chordodes nobilii* is sensitive to extreme temperatures (Achiorno et al. 2008a); –3°C and 40.5°C inhibit egg development. Egg strings can be kept in the laboratory for some months (own investigations), after Müller (1926) larvae can emerge from the eggs even after 8 months.

Cleavage and early development

Cleavage and early development were investigated first by Villot (1874b), later by Camerano (1889a), Tretiakow (1901), Montgomery (1904), Meyer (1913), Mühldorf (1913), May (1919), Dorier (1930), Inoue (1958), Ochiai & Inoue (1970) and Malakhov & Spiridonov (1984).

The first cleavage is total and equal or almost equal, resulting in two blastomeres of almost equal size. From the beginning of the second cleavage, cell division is irregular, resulting in different patterns and making it impossible to trace a cell lineage (Montgomery 1904, Mühldorf 1913, Malakhov & Spiridonov 1984). The two blastomeres may not divide synchronously, resulting in intermediate three-cell-stages and later five-cell-stages (Meyer 1913). The four blastomeres can be arranged in very different constellations, including a row of cells (Meyer 1913) or a tetrahedral arrangement, in which two pairs of blastomeres are arranged perpendicular to each other. Malakhov & Spiridonov (1984) pointed out that such an arrangement also is present in the development of enoplid nematodes and of gastrotrichs. A coeloblastula forms approximately at the 16-cell-stage. All authors report a gastrulation by invagination. The origin of mesodermal tissue is described in different ways. Some cells already invade the coeloblastula by multipolar ingression according to Meyer (1913) and Inoue (1958). Most mesodermal cells, however, seem to originate during gastrulation (Montgomery 1904), after Meyer (1913) most mesodermal cells originate in the blastopore region, but can also come from other regions.

All authors correspond in describing the blastopore as becoming the posterior region of the larva, probably the opening of the pseudointestinal duct. The archenteron develops into the pseudointestine. The anterior region of the archenteron separates and becomes the postseptal gland (Montgomery 1904, Meyer 1913). The anterior end of the larva (the proboscis) forms by a second invagination on the opposite side, where the ectoderm has become thicker (Montgomery 1904). Montgomery (1904) and Meyer (1913) disagree regarding the origin of the preseptal duct, which connects the postseptal gland to the tip of the proboscis. Montgomery (1904) described the two invaginations as not communicating with each other, implying that the canal is a formation of the preseptum. Meyer (1913), in contrast, assumed that the developing intestinal tract grows anteriorly and joins the anterior invagination, therefore the duct would represent the anterior part of the intestinal system.

According to the above observations, the pseudointestine of the larva is directly derived from the archenteron. It is therefore of endodermal origin and should be homologous to the intestine. Although the transition of the pseudointestine to the intestine of juveniles and adults is not documented, this appears plausible. In that case, the later cloacal opening would go back to the blastopore, which is defined as deuterostomy, a rare phenomenon among protostome animals.

Times reported for larval development (i.e. from egg laying to hatching) vary; they may differ among species, but certainly depend on water temperature. Dorier (1932) reported a quite long development of 3 months at 13°C, other reported times are considerably shorter: 13 days in *Pseudochordodes bedriagae* (De Miralles 1980), 20 days in *Paragordius varius* (De Villalobos & Ronderos 2003) and 31 days in *Gordionus prismaticus* (De Miralles 1980). After Zanca et al. (2007) development of *Chordodes nobilii* larvae takes 20–25 days at 22°C, but 45–55 days at 5°C. Readily developed larvae stay in their eggshells for 6 months at 5°C.

Life cycle and host–parasite relations

One of the most fascinating aspects in gordiid biology is that their life cycle includes a free-living and a parasitic phase and that these phases span the aquatic as well as the terrestrial environment at least in most species. Nevertheless, several aspects of the life cycle remain unresolved. It is possible to set up a laboratory life cycle (Hanelt & Janovy 1999, 2004a), but this has been done for only a few species and it is not always clear how laboratory observations relate to field conditions. Recent overviews on the life cycle have been given by Schmidt-Rhaesa (2001a) and especially by Hanelt et al. (2005).

When we still don't have a clear picture of the gordiid life cycle, it is more than clear that early researchers had many more problems in making sense of the few available data and deriving hypotheses about the life cycle. The first one to observe that gordiid larvae form cysts was Meissner (1856). He also already speculated about the presence of intermediate hosts. Villot (1872, 1874b) reported that gordiid larvae penetrate the soft parts of the cuticle in chironomid larvae (Diptera). As chironomids are often eaten by fish, he concluded that the gordiid life cycle is completed in the fish intestine. The observed cases, in which terrestrial insects are infected by gordiids, were regarded by him as accidental hosts, lost for reproduction (Villot 1874b, 1875b). Later (Villot 1881) he changed his mind and accepted insects as being important hosts (like e.g. also Von Linstow 1883a). He assumed that there is no host specificity for gordiid larvae and that any animal can be infected by larvae.

Villot (1891) was also the first to perform some experiments, although these were not successful. Feeding of *Dytiscus marginalis* larvae with gordiid-infected leeches (*Erpobdella* (as *Nephelis) octoculata*), of *Dytiscus marginalis* with infected lymnaeid snails and of *Carabus monilis* with infected planorbid snails did not lead to infection. Additionally, *Carabus monilis* did not get infected by feeding on meat with free gordiid larvae and larvae of the fly *Musca vomitoria* even fed on meat by selectively leaving gordiid larvae behind. From the present day perspective, most of these experiments should have been successful.

Infection of paratenic hosts

Gordiid larvae are benthic and obviously not capable of directed movement. They survive free-living for up to 2 weeks (Hanelt et al. 2005; 3 days in *Neochordodes occidentalis* according to Poinar & Doelman 1974) and have to infect hosts during this time. There are four possibilities for how this can be managed: 1. larvae are consumed directly by the final host by drinking water containing larvae; 2. larvae encyst on vegetation and are taken up by the final host by eating vegetation containing cysts; 3. larvae actively penetrate the skin of intermediate hosts and encyst in their tissue; and 4. larvae are passively ingested and penetrate the epithelium of the digestive tract of intermediate hosts. In the following, I list the observations from experiments and sporadic observations supporting or contradicting these hypotheses.

Direct uptake of gordiid larvae appears to be possible, but experimental observations are few. Thorne (1940) assumed that orthopterans in Illinois (USA) become infected by the uptake of gordiid larvae while drinking water. In an experiment he fed 30 specimens with water containing gordiid larvae; two of them became parasitized. It should be noted that the orthopterans for this experiment came from a location where infection was unknown, but it cannot be excluded that they were already parasitized prior to the experiment. Hanelt & Janovy (1999) successfully infected *Tenebrio molitor* by drinking water with larvae of *Gordius robustus*; subsequent feeding of *Tenebrio* to *Gryllus firmus* leads to further development of the gordiids. In contrast, direct feeding of gordiid larvae to *Gryllus firmus* led to no infection (Hanelt & Janovy 1999). Direct infection of mantids, crickets, roaches, grasshoppers, wax moth larvae, tenebrionid beetles and tadpoles was not successful according to Poinar & Doelman (1974). The majority of final hosts are carnivorous or omnivorous, but there are also a few orthopteran species that are supposed to be vegetarian. If this is correct, at least infection of such hosts must exclude intermediate hosts.

The hypothesis that larvae encyst on vegetation was mainly brought forward by Dorier (1925, 1930). He observed that larvae of *Gordius aquaticus* can make cysts in water and survive for some time within these cysts, up to 7 months in water at 10–12°C. Larvae become active and cysts inflate and break when they are treated with macerated intestinal tracts of caddis fly larvae (*Stenophylax* sp.). Dorier (1930) tested intestinal extracts from a number of different potential hosts with the result that in most cases (caddis fly larvae, mayflies, chironomids, myriapods, *Gammarus*, *Asellus*, snails, fishes) the extracts activated the larvae. With extracts from some potential hosts (cockroaches, mantids and crickets), however, activation of larvae failed. Dorier (1930) tested the infection of a herbivore, the millipede *Glomeris*, by feeding cysts of *Gordius aquaticus* on leaves. This resulted in infections, even if they were few. Dorier´s observations of cysts on vegetation have not been confirmed since then. In contrast, Inoue (1960a) reports that larvae of *Chordodes japonensis* can't encyst in water or in air.

There are very few and quite old observations of gordiid larvae actively penetrating the integument of an intermediate host. Meissner (1856) did not observe such a penetration in insects, but he found gordiid larvae first occurring in the legs of insect hosts and therefore concluded that an integumental penetration makes most sense. Villot (1872, 1874b) is the only source for an observation of gordiid larvae penetrating the soft parts of the cuticle in chironomid larvae (Diptera). Švábenik (1925) reports that gordiid cysts are found in aquatic chironomid larvae and that larvae bore themselves into the body at the posterior end through the soft-skinned appendices. Both Meissner (1856) and Villot (1884) observed the active penetration of gordiid larvae into the foot of gastropods. There are no recent reports of integumental penetration, only some negative reports, where no affinity of gordiid larvae to the insect integument could be observed (Poinar & Doelman 1974, Schmidt-Rhaesa 1997c).

The uptake of gordiid larvae of the intermediate host by ingestion is supported by a number of infection experiments, the laboratory life cycle design and by observation. For example, *Neochordodes occidentalis* larvae were observed to penetrate the intestinal epithelium of culicids (Ochiai & Inoue 1970, Poinar & Doelman 1974) or were histologically documented during this process (Schmidt-Rhaesa 1997c; Fig. 3.2.3 A–D). After boring through the intestinal wall, gordiid larvae encyst in various tissues (Fig. 3.2.4 A, B). At least in insect hosts, they become distributed throughout the body cavity and can encyst quickly in any tissue. According to De Villalobos & Ronderos (2003) it is very likely that gordiid larvae (*Paragordius varius*) infect dipteran larvae (*Dasyhelea necrophila*) per os. One larva was also present in a pupa of *D. necrophila*, here infection probably took place through a spiracle, because pupae don't feed.

Inoue (1960b, 1962) conducted a series of infection experiments with larvae of *Chordodes japonensis*, using larvae, nymphs or pupae of *Cloeon* (Ephemeroptera), *Chironomus* or *Culex* (both Diptera) as intermediate hosts and *Tenodera sinensis* (Mantoptera) as final host. He showed that *Chordodes* larvae are unable to penetrate the integument of the intermediate hosts (Inoue 1960b) or the final host (Inoue 1962). Injection of gordiid larvae into the intermediate host was also unsuccessful. Intermediate hosts become infected only by oral ingestion of gordiid larvae. In a time series Inoue (1960b) could show that larvae were first found in, within or next to the intestine of the intermediate host and spread from

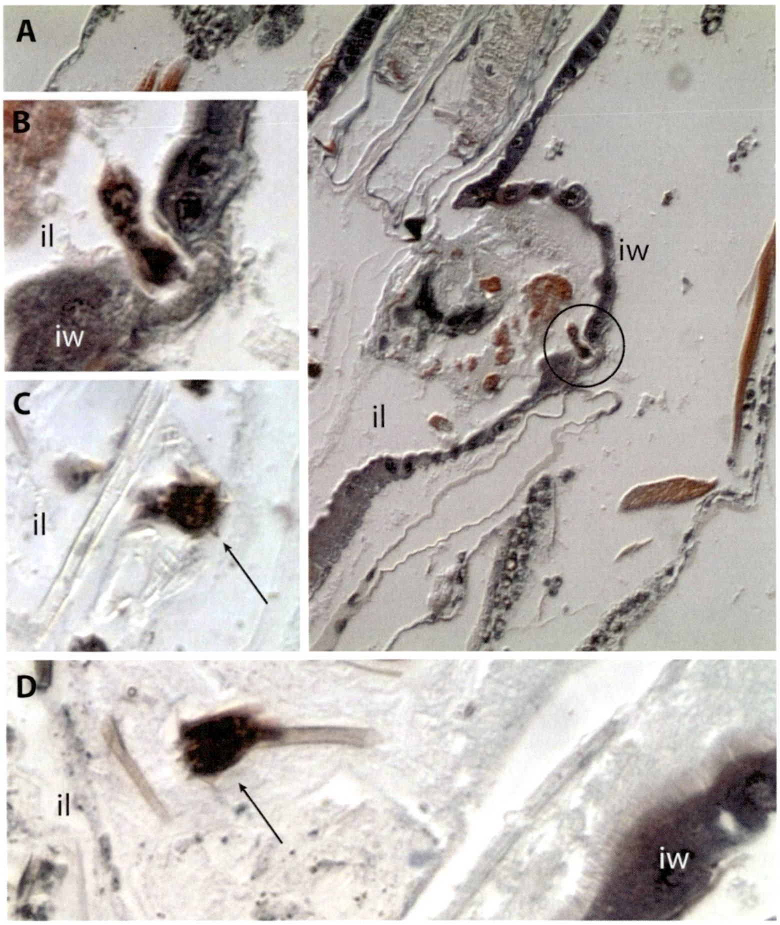

Fig. 3.2.3. Experimental infection of mosquito larvae (*Culex* sp.) with *Gordius aquaticus* larvae. Light microscopy from paraffine sections. A, Gordiid larva (in circle) penetrates the intestinal wall (iw) of the mosquito larva to migrate from the intestinal lumen (il) into the body cavity; B, Magnification from A; C, D, Gordiid larvae inside the intestinal lumen of the mosquito larva. Arrows indicate preseptum with hooks. Note: Images represent new photos from slides used in Schmidt-Rhaesa 1996d, 1997c.

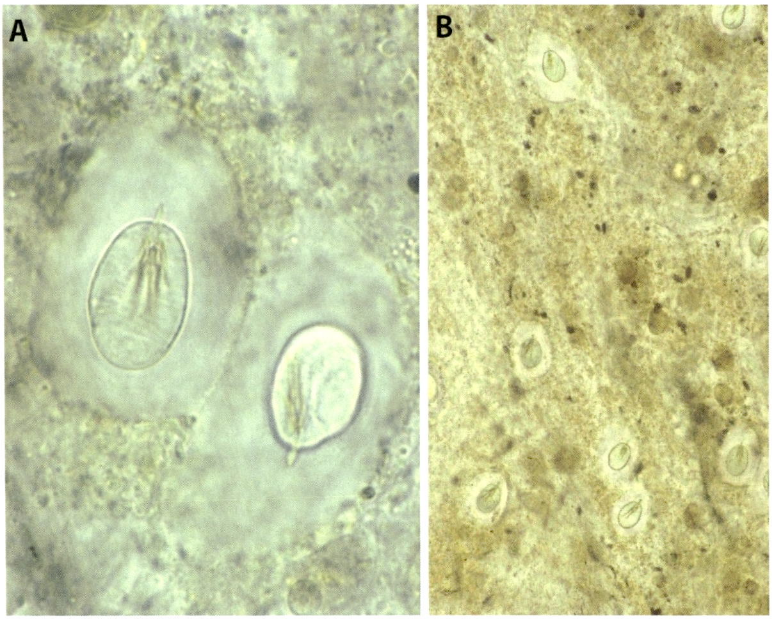

Fig. 3.2.4. Encystment of larvae. A, B, Cysts of *Paragordius varius* larvae in tissue of the snail *Physa gyrina* (from experimental infections, Nebraska, USA).

there into various tissues including the integument. Non-feeding nymphs or pupae are not infected. *Chordodes* larvae encyst within the intermediate host.

Other experiments were conducted by Hanelt & Janovy (2004b). They showed that three phylogenetically distinct species of gordiids (*Paragordius varius, Gordius robustus* and *Chordodes morgani*) indiscriminately entered most aquatic animals. They showed that cysts within insect larvae survived metamorphosis of the insect host and that crickets fed with *Paragordius varius* cysts from adult insects became successfully infected.

In summary, most life cycles appear to include the infection of aquatic intermediate (paratenic) hosts through passive ingestion, but it cannot be excluded that particular species are also capable of forming free cysts or infecting final hosts directly.

Cyst formation and structure

Dorier (1930) described the formation of free cysts (i.e. those cysts attached to vegetation) in *Gordius aquaticus*. The cyst material is secreted from the pseudointestine; the secretion turns into a gelatinous mass into which the larvae crawl. Secretion from the pseudointestine is also observed in larvae of other species (*Gordionus violaceus, Parachordodes gemmatus*; Dorier 1932, 1934, 1935), but here the secretion remains threadlike and does not form a cyst-like structure. Dorier (1935) suggests that this constitutes a difference between species, which means that only *Gordius aquaticus* larvae are capable of encysting on vegetation or substrate. Regardless of whether free cysts are possible or not, the cyst within the intermediate host also appears to be formed by secretions of the pseudointestine. For example, Poinar & Doelman (1974) report that jelly-like material appears first at the posterior end of the larva; this apparently comes from the pseudointestine, which empties during this procedure.

The cyst and the larva inside differ among species and at least the most common North American species can be recognized by larval and cyst morphology (Hanelt & Janovy 2002). Poinar (2009) found two types of cysts in the gut wall of *Stenoperla prasina* in New Zealand; this probably indicates that these are two different species. The presence of cysts in intermediate hosts can be used as a tool to record the presence of gordiids in an environment, because adults occur seasonally but larval cysts can be found in a much longer time range. For example, Hanelt et al. (2001) found cysts of gordiids in 70% of sites investigated in Nebraska, USA, but only found an adult gordiid at one site.

Cysts or even unencysted larvae are the target of defense reactions of the intermediate host, the most common reaction seeming to be the melanization of the cyst. According to Poinar & Doelman (1974), *Culex* larvae react by immediate melanization when larvae of *Neochordodes occidentalis* reach the body cavity. Melanization usually starts at the anterior end and proceeds over the entire larva. It seems that in second- and third-stage *Culex* larvae melanization had little effect on the cysts, but in fourth-stage larvae, almost all *Neochordodes* larvae were killed. Cappucci (1982) also reported melanization in fourth-stage larva of *Aedes aegypti*. Unencysted larvae are melanized to varying degrees in *Dasyhelea necrophila* (De Villalobos & Ronderos 2003). Most *Dasyhelea* larvae died, in particular those that were infected by >10 gordiid larvae (within 48 hours after exposure). Very few infected larvae pupated; adults died briefly after metamorphosis.

Inoue (1960b) observed two different types of cysts. In *Cloeon*, some cells, probably amoebocytes of the host, surround the gordiid cysts. In *Chironomus*, a darkly stained "cyst-wall" surrounds each cyst. This is interpreted by Inoue (1960b) as a chitinous wall, but, compared with other results, this may be a host defense reaction by melanization. In *Culex*, both types of cysts occur. Inoue (1962) successfully infected lab-raised mantids with infected intermediate hosts. The three intermediate hosts used showed different infection success. Almost all infection experiments with *Cloeon* led to infected mantids, while about 50% of the *Culex* experiments were successful and only few (three from 23) with *Chironomus*. In my opinion these results support the interpretation of the "cyst-wall" in the *Chironomus*-type cysts as the result of a defense reaction, showing that such a reaction is quite strong in *Chironomus*, less strong in *Culex* and absent in *Cloeon*.

Large numbers of infecting *Neochordodes* larvae (50+) had an effect on the *Culex* larvae, which were often killed, smaller numbers seemed to have no effect (Poinar & Doelman 1974). Larval mortality of *Aedes aegypti* larvae infected with *Paragordius varius* larvae was higher in the infected group (42.5%) than in the control (3%), especially in the first two instars (De Villalobos et al. 2006). In the 3rd instar (9.1%) and especially the 4th instar (68.4%) a defense reaction was observed.

A differential infection potential was reported from *Pedicia rivosa* (Diptera), where larvae are abundantly infected with gordiid cysts, but pupae never (De Villalobos et al. 2003a). Hanelt & Janovy (2004b) observed that in <5% of infected *Tanytarsus* sp. (Diptera, Chironomida) there was a host defense reaction by a layer of hemocytes around the cysts, but this did not affect viability of cysts.

In most cases, a rapid formation of a cyst is reported. However, in an experimental infection of the snail *Physella cubensis*, larvae did not encyst immediately, but rather over a period of 140 days. The percentage of encysted larvae continued to increase over time and after this period very few were not encysted (De Villalobos et al. 2003a). When fish (*Jenynsia multidentata*) were fed with crushed snail tissue, all got infected. Larvae were present in the gut epithelium and in the musculature, most larvae were encysted. Because the number of larvae observed in fish is higher than either the number of free larvae in snails or the number of encysted larvae, De Villalobos et al. (2003a) conclude that larvae must have been recruited from free as well as encysted ones in the snail.

When gordiid larvae encyst in a host, there is no indication that they develop any further. Therefore such hosts are intermediate or paratenic hosts (see, e.g., Sharpilo 2001).

Intermediate host taxa

Gordiid cysts have been reported from almost any aquatic taxon (see Tab. 3), including even hyperparasitism, in which gordiid larvae are found in trematode larvae, which are themselves infecting vertebrates (Tab. 4; Cort 1915, Fischthal 1942, Hanelt 2009).

This broad list indicates that almost any aquatic taxon can become infected by gordiid cysts. Hanelt & Janovy (2003) tested this experimentally and concluded that there is only low (intermediate) host specificity. From a variety of aquatic taxa, only free-living flatworms, water mites and mosquito larvae remained uninfected. Nevertheless, as infection appears to be a passive process, it remains questionable, which of the hosts is used to infect the final hosts (see below). It is assumed that many animals becoming infected with gordiid cysts (such as fish and snails) do not act as natural paratenic hosts and aquatic insect larvae in particular are likely to be responsible for spanning the gap between water and land (Hanelt & Janovy 2003).

Besides the potential to bridge the aquatic and terrestrial habitat, the defense potential of the intermediate host determines its capability to act as host, because the defense potential appears to be quite different among hosts (see above).

Infection of final hosts

The macroscopic, wormlike juveniles in the final hosts receive much more attention than the microscopically small larvae. A number of records exist. It is attempted here to report as many records as possible but it cannot be excluded that several further records remain "hidden" in the literature. Quite problematic are older lists of hosts, because these often include other taxa such as mermithids. For example, Von Siebold (1842) placed a request to entomologists to report intestinal worms (Fadenwürmer = Filariae) found in insects to receive more data to resolve the question about the identity of parasitic and free-living forms. He subsequently published such lists (Von Siebold 1842, 1843a, 1848, 1850, 1854, 1858), but species were rarely determined to species level. From the present day perspective, it seems, for example, likely that hymenopterans and butterfly caterpillars do not host nematomorphs; such reports probably refer to nematodes. Diesing (1851) describes 118 species of *Gordius* and 17 species of *Mermis*, all named after the particular host in which they occur, but he himself regards only four of these 135 species as sufficiently described. Other species lists which have to be approached with care are those from Diesing (1861) and Von Linstow (1878). In the following lists, most of these early records are excluded.

Hosts of gordiids. Representatives from several taxa have been reported as hosts of gordiids. The three major host groups are beetles, orthopterans and praying mantids.

Beetles are a very important host group for gordiids in Nearctic and Palaearctic regions (Tab. 5). Among beetles, carabids are the most important group, while other taxa, such as Silphidae, Tenebrionidae and a few others are parasitized to a lesser extent. It is also interesting to note that water beetles can be parasitized. Dytiscids played a major role in some reconstructions of the life cycle (e.g. Blunck 1922), but this seems to be an overestimation. The infection of aquatic beetles seems to be one possibility for gordiids, but the infection of terrestrial species is far more abundant. The gordiid species infecting beetles are from the genera *Gordius*, *Parachordodes* and *Gordionus*, with one additional report of *Spinochordodes bacescui* in *Calosoma maderae*. For a review of carabids as hosts see Poinar et al. (2004).

Another major host group is Orthoptera (Tab. 6; Fig. 3.2.5 D, E); they are used as hosts of gordiids worldwide. Most species belong to the Ensifera, representatives of which are quite often omnivorous. This is important for the consumption of intermediate hosts. However, purely vegetarian species have also been reported as hosts, which may require alternative ways of infection. The gordiid species using orthopterans as host come from a variety of genera.

Tab. 3. Hosts, in which gordiid cysts have been found.

	Host species	Geographic region	Author
Insecta			
Diplura	*Plusiocampa pucketti*	USA	Bareth 1974, 1975
	Plusiocampa bulgarica	Bulgaria	Bareth 1974, 1975
	Campodea paurocampa	Italy	Bareth 1974, 1975
Ephemeroptera	*Cloeon dipterum*	Germany	Von Linstow 1892
	Cloeon dipterum (larvae)	France (experiment)	Dorier 1930
	Cloeon sp.	Japan	Inoue 1960b
	Baetis sp.	USA	White 1966, 1969
	Leptophlebia sp.	USA	White 1969
	Ephemerella sp.	USA	White 1969
	different species	Poland	Arvy & Sowa 1976
	Callibaetis sp. larvae	USA	Hanelt & Janovy 2004b
Plecoptera	*Perla* sp. (larvae)	France	Dorier 1930
	Stenoperla prasina	New Zealand	Poinar 2009
Megaloptera	*Sialis lutaria*	Germany	Von Linstow 1891b
Heteroptera	*Sigara*	USA	White 1969
Diptera	*Tanypus* larvae	France	Villot 1872, 1874b
	Corethra larvae	France	Villot 1872, 1874b
	Chironomus larvae	France	Villot 1872, 1874b
	Chironomus tentans larvae	USA	Muttkowski 1918
	Pedicia rivosa	Germany	Müller 1920
	Dicranota sp.	Germany	Müller 1920
	chironomid larvae	Germany	Müller 1920
	Chironomidae (Tanytarssarinae, Chironominae, Orthocladiinae)	Czech Republic	Švábenik 1925
	Chironomus sp. (larvae)	France (experiment)	Dorier 1930
	Chironomus sp.	Japan	Inoue 1960b
	Culex sp.	Japan	Inoue 1960b
	Simulium	USA	White 1969
	larval tabanids		Andreeva 1978
	Dasyhelea necrophila	Argentina	De Villalobos & Ronderos 2003
	Aedes aegypti	Argentina	De Villalobos et al. 2006
Coleoptera	*Hydrophilus piceus* larvae	France	Villot 1881
Trichoptera	*Agapetus* sp. (larvae)	France	Dorier 1930
	Rhyacophila obliterata larvae	?	Russ 1912
	Brachycentrus	USA	White 1969
Annelida			
Enchytraeidae	*Enchytraeus vermicularis*[1]		Villot 1874b
	Fredericia sp.	Germany	Müller 1920
Naididae	*Chaetogaster* sp.	France (experiment)	Dorier 1930
Tubificidae	*Tubificides* sp.	France	Dorier 1930
Hirudinea	*Erpobdella (Nephelis) octoculata* (used in experiment)	France	Villot 1891
Mollusca			
Gastropoda[2]	*Planorbis* sp.	France	Villot 1874b, 1881
	Lymnaea vulgaris	Germany	Von Linstow 1877
	Lymnaeus ovatus	Germany	Von Linstow 1883a
	Planorbis carinatus	France (experimental)	Dorier 1930
	Physa fontinalis	France (experimental)	Dorier 1930
	Arianta (Helix) arbustorum	France	Dorier 1930
	Ancylus sp.	France	Dorier 1930

(*Continued*)

Tab. 3. (*Continued*)

	Host species	Geographic region	Author
	Physa gyrina	USA	Hanelt et al. 2001
	Physella cubensis	Argentina (experimental)	De Villalobos et al. 2003a
Vertebrata			
Cyclostmata	*Petromyzon planeri*	France	Villot 1881
	Petromyzon planeri	Germany	Von Linstow 1898
	Lampetra planeri	France	Dorier 1926
	Lampetra planeri and *L. fluviatilis*	Sweden	Malmqvist & Moravec 1978
Teleostei	*Phoxinus laevis*	France	Villot 1874b, 1881
	Cobitis barbatula	France	Villot 1874b, 1881
	Gobiomorphus breviceps and *Galaxias vulgaris*	New Zealand	Blair 1983
	Leucaspius delineatus	Russia	Sokolov & Spiridonov 1998
	Gobio gobio	Russia	Sokolov & Spiridonov 1998
	Rutilus rutilus	Russia	Sokolov & Spiridonov 1998
	Noemachilus barbatulus	Russia	Sokolov & Spiridonov 1998
	Harttiella crassicauda (Loricariidae)	Nassau Mountains, Suriname	new record[3]
Amphibia	mesenteries of frog	Germany (?)	Leydig 1853

[1] Villot refers to specimens reported by Claparède (1862), which were interpreted originally as scolices from cestodes.
[2] Chitwood & Chitwood (1937) report *Lymnaea vulgaris* and *L. ovata* as host of *Gordius aquaticus, L. ovata* for *G. villoti, L. ovata* for *Paragordius tricuspidatus, L.* sp. for *Parachordodes tolosanus,* without giving references or indicating whether these are larvae or adults. Mahler (1951) reports snails from terrestrial, but moist habitats such as *Vitrina* species, *Eucobresia diaphana* and *Phenacolimax pellucidus* as hosts for gordiids, although it is not specified if this refers to larval cysts or to juveniles.
[3] This new record was made available by Jan Mol, Naturalis Museum, Leiden, Netherlands.

Tab. 4. Cases of hyperparasitism, where gordiid cysts have been found in parasitic species.

Platyhelminthes

Host species	Host is parasite of	Geographic region	Author
Coitocalcum sp.		New Zealand	Blair 1983
Brachycoelium hospitale	*Diemictylus viridescens* (green newt)	USA	Cort 1915
Plagioporus sinitsini	*Hypentelium nigricans* (hog molly)	USA	Fischthal 1942
rediae of undetermined species of Notocotylidae	*Physa gyrina* (Gastropoda)	USA	Hanelt 2009

Praying mantids (Mantodea) (Tab. 7; Fig. 3.2.5 A) are the third major group of hosts; with few exceptions these are the dominant hosts in tropical and subtropical regions. Gordiid species come mainly from the genus *Chordodes*, but some species come also from other genera (see also Schmidt-Rhaesa & Ehrmann 2001).

There are several other insect taxa reported as host taxa. Among these, cockroaches appear to be quite important as a host group (Tab. 8; see Poinar 1999 for a review). Reliable reports of infection are also available from earwigs (Dermaptera; Tab. 9) and caddis flies (Trichoptera; Tab. 10). The latter are remarkable in the respect that the aquatic larvae serve as the final host, whereas aquatic larvae of other insect taxa are only intermediate hosts. Adult caddis flies seem not to be parasitized.

A further questionable record is: "*Filaria forficulae*" found in *Forficula auricularia* in England (Maund 1930) (see also Von Siebold 1850).

The other insect taxa have to be regarded as very irregular and unusual hosts or the parasites may have been misidentified. There are few reports of dragon flies as hosts; these may represent a rare host group (Tab. 11). The same may be the case for horse flies (Diptera, Tabanidae; Tab. 11). Reports of hymenopterans

Tab. 5. Reported cases of beetles (Coleoptera) as hosts for juvenile or adult gordiids.

Family	Host species	Gordiid species	Geographic region	Author
Carabidae	*Amara brunnea*	*Gordionus brunneus*	Sweden	Montén 1951
	Amara fusca	*Parachordodes tolosanus*		Camerano 1897a
		Parachordodes tolosanus (as *Gordius subbifurcus*)		Von Linstow 1878
	Amara similata	*Parachordodes* (as *Gordius*) *tolosanus*		Villot 1874c
		Parachordodes tolosanus (as *Gordius subbifurcus*)		Von Linstow 1878
		Parachordodes tolosanus		Camerano 1897a
	Calathus fuscipes	*Parachordodes pustulosus*	United Kingdom	Baylis 1944
	Calathus melanocephalus	*Gordionus linourgos*	Scotland	De Villalobos et al. 1999
	Calathus micropterus	*Gordionus violaceus*	Scotland	De Villalobos et al. 1999
	Calathus cisteloides	*Parachordodes* (as *Gordius*) *tolosanus*		Villot 1874c, Von Linstow 1889, 1898
		Parachordodes tolosanus (as *Gordius subbifurcus*)		Von Linstow 1878
		Gordionus (as *Parachordodes*) *violaceus*	United Kingdom	Camerano 1915a
		undetermined		Von Linstow 1889
		Gordius aquaticus		Von Linstow 1898
	Calathus sp.	*Gordius aquaticus*		Von Linstow 1878
	Calosoma maderae auropunctatum	*Spinochordodes* (as *Dacochordodes*) *bacescui*	Romania	Căpuşe 1971
	Carabus coriaceus	*Gordius aquaticus*		Von Linstow 1878
		Parachordodes tolosanus (as *Gordius subbifurcus*)		Von Linstow 1878
	Carabus coriaceus (presumed)	*Gordionus sulcatus*		Heinze 1937
	Carabus (as *Procrustes*) *coriaceus*	*Parachordodes* (as *Gordius*) *tolosanus*		Villot 1874c, Von Linstow 1889
		Gordius aquaticus		Von Linstow 1898
		Gordionus (as *Parachordodes*) *violaceus*	France	Dorier 1930
	Carabus gemmatus	*Gordius aquaticus*		Von Linstow 1878
	Carabus hortensis	*Parachordodes* (as *Gordius*) *tolosanus*		Villot 1874c
		Parachordodes tolosanus (as *Gordius subbifurcus*)		Von Linstow 1878
	Carabus monilis	*Gordionus* (as *Gordius*) *violaceus*		Villot 1891
	Carabus morbillosus	*Parachordodes* (as *Gordius*) *tolosanus*		Von Linstow 1898
	Carabus olyosidus	*Gordionus* (as *Parachordodes*) *violaceus*		Camerano 1915a
	Carabus smaragdinus	*Gordius aquaticus*	East Africa	Römer 1895a
	Carabus violaceus	*Gordius aquaticus*		Villot 1874c
		Gordionus (as *Gordius*) *violaceus*	United Kingdom	Baird 1853a, b

(*Continued*)

Tab. 5. (*Continued*)

Family	Host species	Gordiid species	Geographic region	Author
		Parachordodes (as *Gordius*) *tolosanus*		Von Linstow 1898
		Parachordodes tolosanus (as *Gordius subbifurcus*)		Von Linstow 1878
		Gordius villoti	Italy	Sciacchitano 1932
		Gordionus (as *Gordius*) *violaceus*		Von Linstow 1878
		Gordionus (as *Parachordodes*) *violaceus*		Camerano 1915a
	Chlaenius prasinus	*Gordius difficilis*	USA	Hanelt & Janovy 2000
	Chlaenius sericeus	*Gordius* sp.	USA	Leffler 1984
	Dolichus flavicornis	*Gordius aquaticus*		Assmuss 1858
	Harpalus aeneus	"*Dicranurus* sp."[1]	Probably Czech Republic	Presla 1836
	Harpalus affinis	*Gordionus harpali*	Germany	Heinze 1940
	Harpalus distinguendus	*Parachordodes speciosus*		Heinze 1937
	Harpalus fastiditus	*Gordius aquaticus*		Von Linstow 1878
	Harpalus hirtipes	*Parachordodes* (as *Gordius*) *tolosanus*		Von Linstow 1898
	Harpalus hottentotta	undetermined	Germany (?)	Gemminger 1849
		Parachordodes (as *Gordius*) *tolosanus*		Villot 1874c
	Harpalus latus	*Gordionus diligens*	Scotland	De Villalobos et al. 1999
	Harpalus pubescens	*Parachordodes* (as *Gordius*) *tolosanus*		Von Linstow 1898
	Harpalus (as *Calathus*) *ruficornis*	undetermined		Von Linstow 1889
	Harpalus sp.	*Parachordodes pustulosus* (as *Gordius pustulosus*)		Vejdovsky 1894
	Lesticus magnus	*Parachordodes lestici*	Japan	Heinze 1935a
	Loricera pilicornis	*Parachordodes tolosanus* (as *Gordius subbifurcus*)		Assmuss 1858
	Molops depressus	undetermined	Germany (?)	Gemminger 1849
	Molops elatus	*Parachordodes* (as *Gordius*) *tolosanus*		Villot 1874c
		Parachordodes tolosanus (as *Gordius subbifurcus*)		Von Linstow 1878
		juvenile specimen	Germany	Schmidt-Rhaesa 1996a
	Molops piceus	*Gordionus molopsis*		Heinze 1937
		Gordionus wolterstorffii		Heinze 1940
		juvenile specimen	Germany	Schmidt-Rhaesa 1996a
	Nebria brevicollis	*Gordionus violaceus*	Scotland	De Villalobos et al. 1999
		Gordionus linourgos	Scotland	De Villalobos et al. 1999
		undetermined	Scotland	De Villalobos 1999
	Nebria picicornis	undetermined	Germany	Müller 1920
		Parachordodes tolosanus		Müller 1926

(*Continued*)

Tab. 5. *(Continued)*

Family	Host species	Gordiid species	Geographic region	Author
	Ophones cephalotes	*Parachordodes tolosanus*		Heinze 1937
	Oregus aereus	undetermined	New Zealand	Townsend 1970
	Poecilus lepidus	*Gordius aquaticus*	Germany (?)	Gemminger 1849
		Parachordodes tolosanus (as *Gordius subbifurcus)*		Villot 1874c, Von Linstow 1878
	Pseudophanonus pubescens	undetermined		Von Linstow 1891b
		Parachordodes gemmatus		Heinze 1937
	Pterostichus amethystinus	*Gordionus tegonotus*	USA	Poinar et al. 2004
	Pterostichus castaneus	*Gordionus tegonotus*	USA	Poinar et al. 2004
	Pterostichus diligens	*Gordionus diligens*	Scotland	De Villalobos et al. 1999
	Pterostichus fastiditus	*Gordius aquaticus*		Villot 1874c
		Gordius robustus	USA	Leidy 1856
	Pterostichus inopinus	*Gordionus tegonotus*	USA	Poinar et al. 2004
	Pterostichus interstintus	*Gordionus dubiosus*	Romania	Căpuşe 1971
	Pterostichus (as *Steropus)* *madidus*	*Gordionus violaceus*	Spain	Dollfus 1943
		undetermined	France	Jolivet 1945
		Gordionus violaceus		Jolivet 1948
	Pterostichus madidus (as *Feronia madida)*	*Parachordodes pustulosus*	United Kingdom	Baylis 1944
		Gordionus violaceus and *G. wolterstorffii* (as *Parachordodes)*		Baylis 1944
	Pterostichus madidus	*Gordionus diligens*	Scotland	De Villalobos et al. 1999
	Pterostichus (as *Omaseus)* *melanarius*	*Gordius aquaticus*	Germany (?)	Gemminger 1849
	Pterostichus melanarius (as *Feronia melanaria)*	*Gordius aquaticus*	Germany	Von Siebold 1856, 1858, Villot 1874c
		Gordius aestivalis		Vejdovsky 1894
		Gordius preslii		Vejdovsky 1894
		Parachordodes (as *Gordius)* *tolosanus*		Villot 1874c
	Pterostichus melanarius	undetermined	Canada	Tomlin 1975
		Gordius difficilis	USA	Bolek & Coggins 2002
	Pterostichus (as *Omaseus)* *melas*	*Gordius aquaticus*		Villot 1874c
		Parachordodes (as *Gordius)* *tolosanus*		Villot 1874c
	Pterostichus melas	*Gordius aquaticus*		Von Linstow 1878
		Parachordodes (as *Gordius)* *tolosanus*		Von Linstow 1898
		Parachordodes tolosanus (as *Gordius subbifurcus)*		Von Linstow 1878
		Parachordodes tolosanus	Italy	Sciacchitano 1930
		Gordionus (as *Parachordodes)* *wolterstorffii*	Italy	Sciacchitano 1930
	Pterostichus metallicus (as *Feronia metallica)*	*Parachordodes* (as *Gordius)* *tolosanus*		Villot 1874c

(Continued)

Tab. 5. (*Continued*)

Family	Host species	Gordiid species	Geographic region	Author
	Pterostichus metallicus	*Parachordodes tolosanus* (as *Gordius subbifurcus)*		Von Linstow 1878
	Pterostichus metallicus (presumed)	*Paragordionus dispar*	Germany	Heinze 1937
	Pterostichus melas	*Parachordodes* (as *Gordius) pustulosus*	Italy	Camerano 1889b
	Pterostichus niger	*Parachordodes* (as *Gordius) tolosanus*	Germany	Von Linstow 1889, 1891b
		Parachordodes tolosanus		Müller 1926
		Gordionus scaber		Heinze 1937
		Gordionus diligens	Scotland	De Villalobos et al. 1999
		Gordionus linourgos	Scotland	De Villalobos et al. 1999
		Gordionus violaceus	Scotland	De Villalobos et al. 1999
		undetermined	Scotland	De Villalobos 1999
	Pterostichus niger (as *P. nigrita)*	*Parachordodes* (as *Gordius) tolosanus*		Villot 1874c
		Parachordodes tolosanus (as *Gordius subbifurcus)*		Von Linstow 1878
	Pterostichus strenuus	*Gordionus diligens*	Scotland	De Villalobos et al. 1999
	Pterostichus (as *Omaseus) vulgaris*	*Gordius aquaticus*		Von Linstow 1878
		Parachordodes tolosanus (as *Gordius subbifurcus)*		Von Linstow 1878
		Gordionus preslii		Heinze 1937
		Gordionus sulcatus		Heinze 1937
	Sphodrus leucophthalmus	*Parachordodes pustulosus* (as *Gordius pustulosus)*		Vejdovsky 1894
	undetermined Broscini	undetermined	New Zealand	Townsend 1970
	carabid beetle	*Parachordodes* (as *Gordius) tolosanus*	unknown	Römer 1895a
Chrysomelidae	*Timarcha tenebricosa*	*Gordionus violaceus*		Dorier 1930, Jolivet & Théodoridès 1950
Curculionidae	*Dinosius duhameli*	undetermined	Madagascar	Ferragu 1986
Dytiscidae	*Dytiscus marginalis* (larva and adult)	*Gordius aquaticus*		Von Linstow 1898
	Dytiscus (larvae)	*Gordius aquaticus*		Villot 1874c
		undetermined	Germany	Müller 1920
	Dytiscus marginalis	*Parachordodes* (as *Gordius) tolosanus*		Von Linstow 1898, Blunck 1915, 1922
		Gordius aquaticus	Russia	Spiridonov 1998
		undetermined	Germany	Schreitmüller 1911
		undetermined	France	Dollfus 1932
	Dytiscus semisulcatus	*Gordius aquaticus*		Heinze 1937
	Rhantus limbatus	*Gordius deltaensis*	Argentina	De Miralles & De Villalobos 1996a
Scarabaeidae	*Melolontha vulgaris*	*Gordius* sp.		Von Linstow 1878
	Oryctes crasicornis	*Gordius aquaticus*		Assmuss 1858
Silphidae	*Silpha atrata*	undetermined		Von Linstow 1889
		Gordionus silphae		Heinze 1937

(*Continued*)

Tab. 5. *(Continued)*

Family	Host species	Gordiid species	Geographic region	Author
	Silpha carinata	*Parachordodes* (as *Gordius*) *tolosanus*		Villot 1874c
		Parachordodes tolosanus (as *Gordius subbifurcus*)		Von Linstow 1878
	Silpha obscura	*Gordionus* (as *Parachordodes*) *violaceus*	France	Dorier 1930
	Silpha subrotundata	*Gordionus* (as *Parachordodes*) *violaceus*	Ireland	Camerano 1908a, Southern 1909
	Silpha tristis (presumed)	*Gordionus silphae*		Heinze 1937
	Silpha sp.	undetermined	unknown	Hofmänner 1913
Staphylinidae	*Staphylinus maxillosus*	*Gordius aquaticus*		Assmuss 1858
Tenebrionidae	*Blaps mortisaga*	*"Gordius aquaticus"*[2]		Gervais 1835
		Gordius pustulosus	France	Villot 1886
		Parachordodes (as *Gordius*) *pustulosus*	France	Camerano 1889b
	Blaps mucronata	*Parachordodes* (as *Gordius*) *pustulosus*	Italy	Camerano 1892
				Von Linstow 1878, Vejdovsky 1894
	Blaps obtusa	*Parachordodes* (as *Gordius*) *pustulosus*	United Kingdom	Baird 1853a, b, Camerano 1889b
			"Europe"	Örley 1881
		Gordius pustulosus	France	Villot 1886
Trogossitidae	*Thymalus* sp.	*Gordius germanicus*	unknown	Spiridonov 1998
?	*"Pelor blabtoides"*	*Parachordodes* (*Gordius*) *tolosanus*		Von Linstow 1898
?	undetermined	*Gordius aquaticus*	Russia	Römer 1895a

[1] The genus name *Dicranurus* was introduced by Presla (1836) for a worm parasitic in *Harpalus aeneus*. The worm appears to be a male gordiid because it has a bilobed posterior end, but no species name is given. The name *Dicranurus* has never been used again and must be regarded a nomen nudum.
[2] With threelobed posterior end, therefore probably *Paragordius*.

and lepidopterans as host taxa (Tab. 11) are quite ambiguous. In particular the hymenopteran records do not include determined gordiid species as parasites. All lepidopteran records are quite old and determinations (*Gordius aquaticus* or *G. subbifurcus*) are suspected to be "default determinations" by uncritically using the most common species names. Only in the report from Müller (1859), is a bilobed posterior end indicated, which would clearly point at a male gordiid. At one time, a stick insect (Phasmida), *Phybalosoma phythonis*, was reported as host for *Gordius aeneus* (Fiji-Islands; Römer 1895a). One aquatic backswimmer (*Notonecta undulata*, Hemiptera) was reported by Zalom (1977). Von Siebold (1850) reports one bug (Hemiptera), *Coriza striata* (= *Miris striatus*?) as host of a "*Gordius*-like" worm.

Finally, other arthropod taxa may also serve as hosts of gordiids. Diplopods and to a lesser extent chilopods can also be parasitized by gordiids (Tab. 12). There are some reports from parasitized spiders (Tab. 12), but few of them are very specific and further, more convincing records would be helpful. There are isolated further records from other arachnid taxa. These include *Beatogordius sankurensis* being reported to have emerged from a pedipalp of *Phrynus* (as *Phriniscus*) *bacillifer* (Sciacchitano 1958), an unconfirmed infection of an opilionid (Schmidt-Rhaesa 1997a) and a specimen of *Spinochordodes tellinii* found close to a dead scorpion (*Euscorpius flavicaudis*) (Schmidt-Rhaesa et al. 2005). Reports of crustaceans as hosts exist but are unreliable (e.g. *Monculus apus* in Von Siebold 1842; *Apus cancriformis*, *Caridina* sp. in Von Linstow 1878).

Sawyer (1971) reported two leeches (Annelida, Hirudinea), *Erpobdella punctata* and *Mooreobdella microstoma* as host for undetermined gordiids, but as the parasites are not described or figured, this report must be considered questionable. Quite unlikely also are reports of a bustard (*Otis*; Von Linstow 1883a) and of a turtle (*Clemmys* [as *Clemnys*] *caspica leprosa*) (Sciacchitano

Tab. 6. Reported cases of Orthoptera as hosts for juvenile or adult gordiids.

Host species	Gordiid species	Geographic region	Author
Acanthodis sp.	*Chordodes parasitus*	Brazil	Creplin 1847
	Acutogordius (Gordius) doriae	Burma	Camerano 1890a
Achaeta domestica	*Gordius* sp.	India	Joseph 1984
Acridium paranense	*Gordius acridiorum*	Argentina	Weyenbergh 1880, 1897
Anabrus purpurascens	*Gordius robustus*	USA (?)	Riley et al. 1877
Anabrus simplex	*Gordius robustus*	USA	Wakeland & Shull 1936, Thorne 1940
Anonconotus occidentalis	*Gordius* cf. *plicatissimus*	Italy	Schmidt-Rhaesa 2010
Antaxius pedestris	*Spinochordodes tellinii*	France	Schmidt-Rhaesa et al. 2005
Babitistes serricauda	*Spinochordodes tellinii*	France	Schmidt-Rhaesa et al. 2005
Caloptenus bivittatus	*Gordius seta*	USA (?)	Riley et al. 1877
Caloptenus spretus	*Gordius seta*	USA (?)	Riley et al. 1877
Ceuthophilus stygius	*Chordodes morgani*	USA	Studier et al. 1991
Decticus albifrons	*Gordius aquaticus*	France	Villot, 1874c, Dorier 1930
	Paragordius (as Gordius) tricuspidatus		Von Linstow 1878
Decticus pedestris	*Gordius aquaticus*		Von Linstow 1878
Decticus verrucivorus	*Gordius aquaticus*	France	Villot 1874c, Von Linstow 1878, 1898 Dorier 1928, 1930, Deronde 1936
	Gordius dectici		Heinze 1937
	Paragordius sp.	Austria	Ebner 1940
	Gordius villoti	United Kingdom	Baylis 1944
	Gordius sp.	Romania	Căpuşe 1966a
	Gordius valnoxius	France	Degrange & Martinot 1995
Decticus sp.	*Gordius villoti*	Italy	Camerano 1887a
Deinacrida parva	*Euchordodes nigromaculatus*	New Zealand	Poinar 1991
Diestrammena japonica	*Gordius* sp.	Japan	Ogawara 1980
Diestrammena sp.	undetermined	Japan	Sato et al. 2008
Eidmanacris sp.	*Paragordius esavianus*	Argentina	Zanca & De Villalobos 2006
Ephippiger cunii	*Spinochordodes tellinii*	France	Schmidt-Rhaesa et al. 2005
Ephippiger provincialis	Gordius aquaticus	France	Dorier 1930
Euanisous sp.	undetermined	Malaysia	Yang et al. 1990
Eupholidoptera chabrieri (?)	*Spinochordodes tellinii*	France	Schmidt-Rhaesa et al. 2005
Gomphocerus viridulus	*Gordius aquaticus*		Villot 1874c
Gryllus assimilis	*Paragordius flavescens*	Brazil	Carvalho 1944
Gryllus bordigalensis	*Paragordius* (as *Gordius*) *tricuspidatus*		Von Linstow 1878
Gryllus burdigalensis	*Gordius gratianopolensis*		Villot 1874c
Gryllus domesticus	*Gordius* sp.		Von Linstow 1878
Gryllus neglectus	*Gordius seta*	USA (?)	Riley et al. 1877
Gryllus sp.	*Gordius aquaticus*	France	Villot 1886
	Paragordius varius	USA	Poinar & Weissman 2004
	Gordius robustus	USA	Poinar & Weissman 2004
Hadenoecus subterraneus	*Chordodes morgani*	USA	Studier et al. 1991
Hemiandrus furcifer	*Gordius dimorphus*	New Zealand	Poinar 1991
	Parachordodes (as *Gordius*) *diblastus*	New Zealand	Poinar 1991

(*Continued*)

Tab. 6. *(Continued)*

Host species	Gordiid species	Geographic region	Author
Hemideina crassidens	*Parachordodes* (as *Gordius*) *diblastus*	New Zealand	Poinar 1991
Hemideina thoracica	*Parachordodes* (as *Gordius*) *diblastus*	New Zealand	Poinar 1991
Hieroglyphus nigropletus	*Gordius* sp.		Roonwal 1976
Leptophyes punctatissima	*Spinochordodes tellinii*	France	Schmidt-Rhaesa et al. 2005
Locusta hemitogia	*Gordius aquaticus*		Von Linstow 1878, 1898
Locusta sp.	*Gordius aquaticus*	Brazil	Von Linstow 1878
Madiga aberrans	*Chordodes pollonerae*	Democratic Republic of the Congo	Sciacchitano 1940
Meconema thalassinum	*Spinochordodes tellinii*	France	Schmidt-Rhaesa et al. 2005
Metridioptera brachyptera	*Gordius* sp.	Austria	Ebner 1940
Microcentrum sp.	*Chordodes lichyi*	Venezuela	Scorza 1952
Nemobius sylvestris	*Paragordius tricuspidatus*	France	Schmidt-Rhaesa et al. 2005, Thomas et al. 2002
Ontonotus onos	*Gordius annulatus*	Russia, Amur	Von Linstow 1906a
Orchelimum glaberrimum	*Gordius seta*	USA (?)	Riley et al. 1877
Orchelimum gladiator	undetermined	Canada	Morris et al. 1975
Orchelimum gracile	*Gordius robustus*	USA	Leidy (1870)
Pediodectes sp.	*Neochordodes occidentalis*	USA	Poinar & Weissman 2004
Phaneroptera sp.	undetermined	Austria (?)	Tappenbeck 2006
Phaulacridium marginale	undetermined	New Zealand	Harris 1990
Pholidoptera aptera	*Paragordius* sp.	Liechtenstein	Ebner 1940
Pholidoptera griseoaptera	*Spinochordodes tellinii*	France	Schmidt-Rhaesa et al. 2005
Phylloptera oblongifolia	*Gordius seta*	USA (?)	Riley et al. 1877
Pleioplectron simplex	*Gordius paranensis* undetermined	New Zealand New Zealand	Schmidt-Rhaesa et al. 2000 Thomas et al. 1998
Scapteriscus sp.	*Neochordodes puntatus* undetermined	Argentina Argentina	De Miralles & De Villalobos 1996b De Villalobos 1999
Sigaus australis (only females)	*Euchordodes nigromaculatus*	New Zealand	Poulin 1995
Sigaus obelisci (only females)	*Euchordodes nigromaculatus*	New Zealand	Poulin 1995
Stauroderus saulcyi	*Gordius aquaticus*	Spain	Codina 1925
Steiroxys sp.	*Gordius robustus*	USA	Poinar & Weissman 2004
Stenopelmatus fasciata	*Gordius robustus*	USA (?)	Riley et al. 1877
Stenopelmatus longispina	*Gordius robustus*	USA	Cappucci 1982
Stenopelmatus sp.	*Neochordodes occidentalis* *Gordius robustus* *Parachordodes platycephalus*	USA USA USA	Poinar & Weissman 2004 Poinar & Weissman 2004 Poinar & Weissman 2004
Stetophyma variegatum	*Gordius aquaticus*		Assmuss 1858
Tettigonia (as *Locusta*) *cantans*	*Gordius aquaticus* undetermined	 France	Von Linstow 1898 Dorier 1930
Tettigonia (as *Phasgonura*) *cantans*	*Gordius aquaticus*	France	Dorier 1930
Tettigonia (as *Phasgonura*) *viridissima*	*Gordius aquaticus* *Gordius albopunctatus* *Gordionus locustae* *Gordius seta*	France France unknown unknown	Jolviet 1946 Dorier 1946a Havlik 1954 Kollar 1857

(Continued)

Tab. 6. (*Continued*)

Host species	Gordiid species	Geographic region	Author
Tettigonia (as *Locusta*) *viridissima*	undetermined	France	Chavannes 1864, Villot 1874b
	Gordius aquaticus		Villot 1874c, Von Linstow 1878, 1898
Tettigonia viridissima	*Spinochordodes tellinii*[1]	France	Schmidt-Rhaesa et al. 2005
Thamnotrizon apterus	*Gordius aquaticus*		Von Linstow 1898
	Gordius hispidus	Austria	Von Linstow 1906a
Truxale sp.	undetermined	France	Villot 1874b
Uromenus rugosicollis	*Spinochordodes tellinii*	France	Schmidt-Rhaesa et al. 2005
Yersinella raymondi	*Spinochordodes tellinii*	France	Schmidt-Rhaesa et al. 2005
Zealandosandrus maculifrons	*Gordius dimorphus*	New Zealand	Poinar 1991
undetermined	*Gordius robustus*	USA	Leidy 1851
undetermined	*Gordius aquaticus*	Germany	Römer 1895a
"crickets"	undetermined	USA	Sanford 1853
large grasshopper	*Neochordodes* (as *Chordodes*) *occidentalis*	USA	Montgomery 1900
undetermined	undetermined	France	Marnay 1902
"in field cricket"	*Paragordius stylosus*	Italy	Sciacchitano 1932
undetermined cricket	*Paragordius varius*	Argentina	Zanca & De Villalobos 2006
undetermined cricket	*Chordodes bouvieri*	Ecuador	De Villalobos et al. 2009
undetermined cricket	*Gordius robustus*	Ecuador	De Villalobos et al. 2009

[1] Parasite found close to the host, parasitation is assumed but not proven.

Tab. 7. Reported cases of praying mantids (Mantodea) as hosts for juvenile or adult gordiids.

Host species	Gordiid species	Geographic region	Author
Acromantis australis	undetermined	New Guinea	Schmidt-Rhaesa & Ehrmann 2001
Ambivia popa	undetermined	Indonesia	Werner 1922
Coptopteryx gayi	*Gordius chilensis*	Chile	Lataste 1896
Deroplatys lobata	undetermined	Malaysia	Schmidt-Rhaesa & Ehrmann 2001
Deroplatys sp.	*Chordodes furnessi*	Indonesia	Montgomery 1898b
Hierodula dyaka	undetermined	Malaysia	Shelford 1903
	undetermined	Malaysia	Schmidt-Rhaesa & Ehrmann 2001
Hierodula majuscula	undetermined	Australia	Schmidt-Rhaesa & Ehrmann 2001
Hierodula membranacea	*Gordius verrucosus*	Sri Lanka	Bell 1885, 1886
Hierodula patellifera	*Chordodes japonensis*	Japan	Schmidt-Rhaesa 2004c
	Chordodes japonensis	Japan	Schmidt-Rhaesa & Urabe 2009
Hierodula perakana	undetermined	Malaysia	Schmidt-Rhaesa & Ehrmann 2001
	Chordodes compactus	Malaysia	Schmidt-Rhaesa & Brune 2008
Hierodula werneri	undetermined	Indonesia	Werner 1925
Hierodula sp.	*Chordodes furnessi*	Indonesia	Montgomery 1898b
	Chordodes cf. *furnessi*	India	Schmidt-Rhaesa & Yadav 2004
	Chordodes mizoramensis	India	Schmidt-Rhaesa & Lalramliana 2011
Idolomantis diabolica	*Gordius verrucosus*	Tanzania	Bell 1885, 1886
Kongobatha diademata	undetermined	Australia	Schmidt-Rhaesa & Ehrmann 2001
Mantis gayi	*Gordius chilensis*	Chile	Lataste 1896

(*Continued*)

Tab. 7. (*Continued*)

Host species	Gordiid species	Geographic region	Author
Mantis religiosa	*Gordius aquaticus*		Von Linstow 1878, 1898
	Gordius aquaticus	Brazil	Gliesch 1933
	Chordodes ibembensis	Democratic Republic of the Congo	Sciacchitano 1958
	Chordodes maculatus	Democratic Republic of the Congo	Sciacchitano 1958
	Chordodes madagascariensis	Democratic Republic of the Congo	Sciacchitano 1958
	Spinochordodes tellinii[1]	France	Schmidt-Rhaesa et al. 2005
Mantis sp.	*Chordodes shipleyi*	Malaysia	Camerano 1899
	Gordius sp.		Von Linstow 1878
Polypilota aeruginosa	*Chordodes ferox*	Tanzania	Schmidt-Rhaesa & Ehrmann 2001
Polyspilota pustulata	*Gordius* sp.	Madagascar	Von Linstow 1897
Popa spurca crassa	*Chordodes hawkeri*	Kenya	Camerano 1915a
Protohierodula ornatipennis	undetermined	Côte d´Ivoire	Roy 2003
Protohierodula picta	undetermined	Gabon	Roy 2003
Pseudocreobotra ocellata	undetermined	Ghana	Schmidt-Rhaesa & Ehrmann 2001
Rhombodera basalis	*Chordodes baramensis*	Malaysia	Römer 1895b, 1896
Rhomboidera basalis	undetermined	Malaysia	Shelford 1903
Rhomboderella scutata	*Chordodes capillatus*	Tanzania	Von Linstow 1901
	Chordodes madagascariensis	Democratic Republic of the Congo	De Beauchamp 1916
Rhombodera valida	undetermined	Malaysia	Schmidt-Rhaesa & Ehrmann 2001
	Chordodes villalobi	Malaysia	Schmidt-Rhaesa & Brune 2008
	Chordodes polycoronatus	Malaysia	Schmidt-Rhaesa & Brune 2008
	Chordodes compactus	Malaysia	Schmidt-Rhaesa & Brune 2008
Sphodromantis baccettii	undetermined	Kenya	Schmidt-Rhaesa & Ehrmann 2001
Sphodromantis gastrica	*Chordodes ferox*	South Africa	Schmidt-Rhaesa & Ehrmann 2001
Sphodromantis lineola lineola	*Chordodes madagascariensis*	Democratic Republic of the Congo	De Beauchamp 1916
Sphodromantis lineola	undetermined	Angola	Luna de Carvallho 1967
Sphodromantis tristis	*Chordodes queenslandi*	Australia	Schmidt-Rhaesa 2002b
Sphodromantis viridis	*Gordius verrucosus*	West-Africa	Bell 1885, 1886
Stagmatoptera hyaloptera	*Chordodes brasiliensis*	Argentina	Camerano 1897b, Carvalho 1946a
	Chordodes brasiliensis	Argentina	De Miralles 1976
	undetermined	Argentina	De Villalobos 1999
	Chordodes cornuta	Argentina	De Villalobos & Camino 1999
	Neochordodes semiluna	Argentina	De Villalobos & Camino 1999
	Chordodes brasiliensis	Argentina	Camerano 1897b
Stagmatoptera (as *Mantis*) *precaria*	*Gordius tenuis*	Argentina	Weyenbergh 1880, 1897
Stagmatoptera precaria	*Chordodes brasiliensis*	Brazil	Carvalho 1946b, Carvalho & Feio 1950
	Chordodes brasiliensis	Argentina	De Miralles 1976
Stagmatoptera sp.	*Chordodes brasiliensis*	Argantina	Carvalho 1946b, De Miralles 1976
Tenodera angustipennis	*Chordodes japonensis*	Japan	Inoue 1951
Tenodera aridifolia	*Chordodes tenoderae*	China	Kirjanova 1957
Tenodera sinensis	*Chordodes japonensis*	Japan	Inoue 1951
	Chordodes fukuii	Japan	Schmidt-Rhaesa 2004c

(*Continued*)

Tab. 7. *(Continued)*

Host species	Gordiid species	Geographic region	Author
Tenodera sp.	*Chordodes kolensis*	Democratic Republic of the Congo	Sciacchitano 1958
undetermined	*Chordodes aethiopicus*	Ethiopia	Inoue 1974
	Chordodes africanus	Angola	Sciacchitano 1961a
	Chordodes alticuspis	Ecuador	De Villalobos et al. 2009
	Chordodes amboinensis	Indonesia	Sciacchitano 1963
	Chordodes annandalei	Nepal	Camerano 1908b
	Chordodes brasiliensis	Brazil	Carvalho 1946b
	Chordodes caledoniensis	New Caledonia	Villot 1874a
		Australia	Camerano 1897a
	Chordodes capensis	Democratic Republic of the Congo	De Beauchamp 1923
		Tanzania	Baylis 1927
	Chordodes congolensis	Democratic Republic of the Congo	Sciacchitano 1933a
	Chordodes curvicillatus	Indonesia	Kirjanova & Spiridonov 1989
	Chordodes ferganensis	Uzbekistan	Kirjanova & Spiridonov 1989
	Chordodes ferox	Republic of the Congo	Camerano 1897a
		Democratic Republic of the Congo	Sciacchitano 1933a
	Chordodes hawkeri	South Africa	Camerano 1908c
		Sierra Leone	Camerano 1912a
	Chordodes ibembensis	Angola	Sciacchitano 1961a
	Chordodes kivuensis	Democratic Republic of the Congo	Sciacchitano 1958
	Chordodes kolensis	Democratic Republic of the Congo	Sciacchitano 1933a, 1958
		Angola	Sciacchitano 1961a
	Chordodes ligasiensis	Democratic Republic of the Congo	Sciacchitano 1933a
	Chordodes maculatus	Democratic Republic of the Congo	Sciacchitano 1958
	Chordodes (as *Gordius*) *madagascariensis*	Madagascar	Camerano 1893a, 1897a
	Chordodes madagascariensis	Democratic Republic of the Congo	De Beauchamp 1923, Sciacchitano 1933a, 1958
		Angola	Sciacchitano 1961a
	Chordodes mobensis	Democratic Republic of the Congo	Sciacchitano 1958
	Chordodes montgomeryi	Madagascar	Camerano 1901a
	Chordodes moutoni	Malaysia	Camerano 1903a
	Chordodes ornatus	Philippines	Grenacher 1868
	Chordodes ruginosus	Ecuador	De Villalobos et al. 2009
	Chordodes siamensis	India	Camerano 1908b
	Chordodes skorikovi	Sri Lanka	Camerano 1903b, 1904a,b
	Chordodes tuberculatus	Australia	Villot 1874a
	Chordodes undulatus	Australia	Von Linstow 1906b
	Chordodes (as *Gordius*) *verrucosus*	Republic of the Congo	Camerano 1893b
	Gordius kaschgaricus	Angola	Sciacchitano 1932
	undetermined	New Caledonia	Laboulbène 1864
		India	Green 1902
		South Africa	Von Pelser-Berensberg 1909
		France	Begon et al. 1990

[1] Parasite found close to the host, parasitation is assumed but not proven.

Tab. 8. Reported cases of cockroaches (Blattodea) as hosts for juvenile or adult gordiids.

Host species	Gordiid species	Geographic region	Author
Aethiomerus adelphus	*Chordodes* (as *Gordius*) *pardalis*	Madagascar	Camerano 1893a
Blabera gigantea	*Chordodes pilosus*	Venezuela	Möbius 1855
Blaptica dubia	undetermined	Argentina	De Villalobos 1999
Blaptica sp.	*Chordodes matensis*	Argentina	De Villalobos & De Miralles 1997
Blatta orientalis (?)	*Gordius aquaticus*	USA	Leidy 1878
Blatta orientalis	*Chordodes* (as *Neochordodes*) *moraisi* *Pseudochordodes bedriagae* *Beatogordius funis* *Neochordodes meridionalis*	Brazil Argentina Argentina Argentina	Carvalho 1942 De Miralles 1984a De Villalobos et al. 2003b De Villalobos et al. 2008a
Blatta sp.	*Beatogordius raphaelis* *Chordodes siamensis*	Democratic Republic of the Congo Democratic Republic of the Congo	Sciacchitano 1958 Sciacchitano 1958
"small *Blatta*"	*Beatogordius (Chordodiolus) echinatus*	Democratic Republic of the Congo	Sciacchitano 1958
Celatoblatta peninsularis	undetermined	New Zealand	Zervos 1989
Eurycotis floridana	undetermined	North America	Roth & Willis 1960
Ischnoptera brasiliensis	*Chordodes peraccae*	Brasil	Camerano 1901b
Kuchinga hemerobina	*Beatogordius raphaelis*	Republic of the Congo	Camerano 1893b
Methana marginalis	*Beatogordius lineatus*	Australia	Schmidt-Rhaesa & Bryant 2004
Minablatta sp.	*Chordodes moraisi*	Brazil	Carvalho 1944
Parahormetica bilobata	undetermined	Brazil	see Poinar 1999
Parellipsidion pachycerum	undetermined	New Zealand	Zervos 1989
Periplaneta americana	*Gordius?* Sp.	South Africa	see Poinar 1999
Periplaneta sp.	*Chordodes iturensis* *Chordodes iturensis*	Democratic Republic of the Congo Democratic Republic of the Congo	Sciacchitano 1958 Sciacchitano 1958
Phyllodromia hemerobina	*Beatogordius* (as *Gordius*) *raphaelis*	Republic of the Congo	Camerano 1893b
Phyllodromia parenthesis	*Beatogordius* (as *Gordius*) *raphaelis*	Republic of the Congo	Camerano 1893b
Pycnoscelus striatus	undetermined	Malaysia	Ng 1975
Supella sp. (?)	*Palaeochordodes protus*[1]	Dominican Republic (in amber)	Poinar 1999
Symploca parenthesis	*Beatogordius raphaelis*	Republic of the Congo	Camerano 1893b
undetermined	*Chordodes morgani* (as *C. puerilis*) *Chordodes morgani* undetermined undetermined *Neochordodes chordodides*	USA Australia Venezuela Costa Rica	Montgomery 1898a Montgomery 1907 Roth & Willis 1960 see Poinar 1999 Schmidt-Rhaesa & Menzel 2005

[1] Fossil specimen (15–45 million years).

1962) as hosts. In the latter case, a gordiid ("*Gordius aquaticus*" was found in the water in which the turtle had been placed.

The data above suggest that gordiids are not specific to one particular host species; this appears plausible at least for most species for which more than one host report is available. One of the few investigations targeting this question was made by Schmidt-Rhaesa et al. (2005) on a population of the two sympatric species *Paragordius tricuspidatus* and *Spinochordodes tellinii* in Southern France. While *P. tricuspidatus* was observed in only one host species, *Nemobius sylvestris*, *S. tellinii* was found in nine different orthopteran species. Among these, *Meconema thalassinum* was the most common, while other species occurred less often or just once.

The reported infection rates (Tab. 13) vary considerably, but usually do not reach high intensities. Some data indicate that males and females of one species can be parasitized to a different extent (Studier et al. 1991). As parasitized hosts aggregate near water, infection

Tab. 9. Reported cases of earwigs (Dermaptera) as hosts for juvenile or adult gordiids (see also Fig. 3.1.27 C).

Host species	Gordiid species	Geographic region	Author
Chelidura acanthopygia	*Parachordodes gemmatus*	France	Dorier 1930
	Parachordodes gemmatus	Germany	Schmidt-Rhaesa & Zaenker 2006
Forficula auricularia	undetermined	France (?)	Dufour 1828
	Gordionus (as *Parachordodes*) *violaceus*[1]	United Kingdom	Baylis 1944
	undetermined	Germany	Schmidt-Rhaesa 1997c, 2002c
undetermined	*Chordodes weberi*	Democratic Republic of the Congo	Sciacchitano 1958

[1] Parasite found close to crushed host.

Tab. 10. Reported cases of caddis flies (Trichoptera) as hosts for juvenile or adult gordiids (see also Fig. 3.2.5 B).

Host species	Gordiid species	Geographic region	Author
Allogamus auricollis	*Gordius villoti* (*Gordius* sp. in 1987)	Italy	Valvassori et al. 1987, 1988
Chaetopteryx vittatus (presumed)	*Gordius setiger*		Heinze 1937
Phrygaena sp.	undetermined	unknown	see Siltala 1906
Phrygaenidae larvae	undetermined	unknown	Von Siebold 1842
Pomatophylax cingulatus	*Gordius albopunctatus*	Denmark (Bornholm)	Schmidt-Rhaesa & Kristensen 2006
Stenophylax latipennis	*Parachordodes* (as *Gordius*) *tolosanus*		Von Linstow 1898
Stenophylax stellatus (presumed)	*Gordius setiger*		Heinze 1937
Stenophylax stellatus	*Gordius albopunctatus*		Heinze 1937
Stenophylax sp.	*Gordius aquaticus*	France	Dorier 1925, 1928
undetermined	undetermined[1]	Italy	Corallini Sorcetti & Moretti 1991

[1] It is not indicated whether cysts or juvenile worms were found.

rates may vary locally, depending on the vicinity of water. Thorne (1940) reported that up to an estimated 99% of Mormon crickets (*Anabrus simplex*) are infected in the direct vicinity of a lake, but that infection rates decrease with growing distance from this lake (Cut Bank, Montana, USA). In general it seems that gordiids do not have a large impact on host populations, a fact that makes them not well suited as biological control agents (Thorne 1940, Baker 1985).

In most cases, only one worm seems to be present within one parasite specimen, but multiple infections do also occur. For example, Kollar (1857) found seven specimens of "*Gordius seta*" in one cricket *Locusta viridissima*, Camerano (1892) found up to four specimens of *Parachordodes pustulosus* in *Blabs mucronata*, Valvassori et al. (1988) found usually one, but a maximum of seven specimens of *Gordius villoti* in the caddisfly *Allogamus auricollis* and Schmidt-Rhaesa et al. (2005) observed usually one, but up to four specimens of *Paragordius tricuspidatus* in *Nemobius sylvestris*. Space within the host is likely to be a limiting factor. As final hosts may receive numerous cysts with gordiid larvae it can be suspected that some competition among developing juveniles exists, but there are no observations on this. Higher numbers of parasites have been reported in laboratory infections. Hanelt (2008) counted up to 30 worms per host. This may be related to the laboratory host, *Gryllus firmus*, which probably is not the natural host of the laboratory worm, *Paragordius varius*.

Gordiid parasites may have different kinds of effects on their hosts. There might be a phenotypic change, an influence on the internal organs or a change of behaviour. All three kinds of effects seem to exist, but to varying degrees. According to Ebner (1940), infection by gordiids (and mermithids) may lead to shorter wings or an altered form of the cerci in *Metridioptera brachyptera* (Orthoptera). Roy (2003) reported a male of *Protohierodula ornatipennis* and one male of *P. picta* parasitized

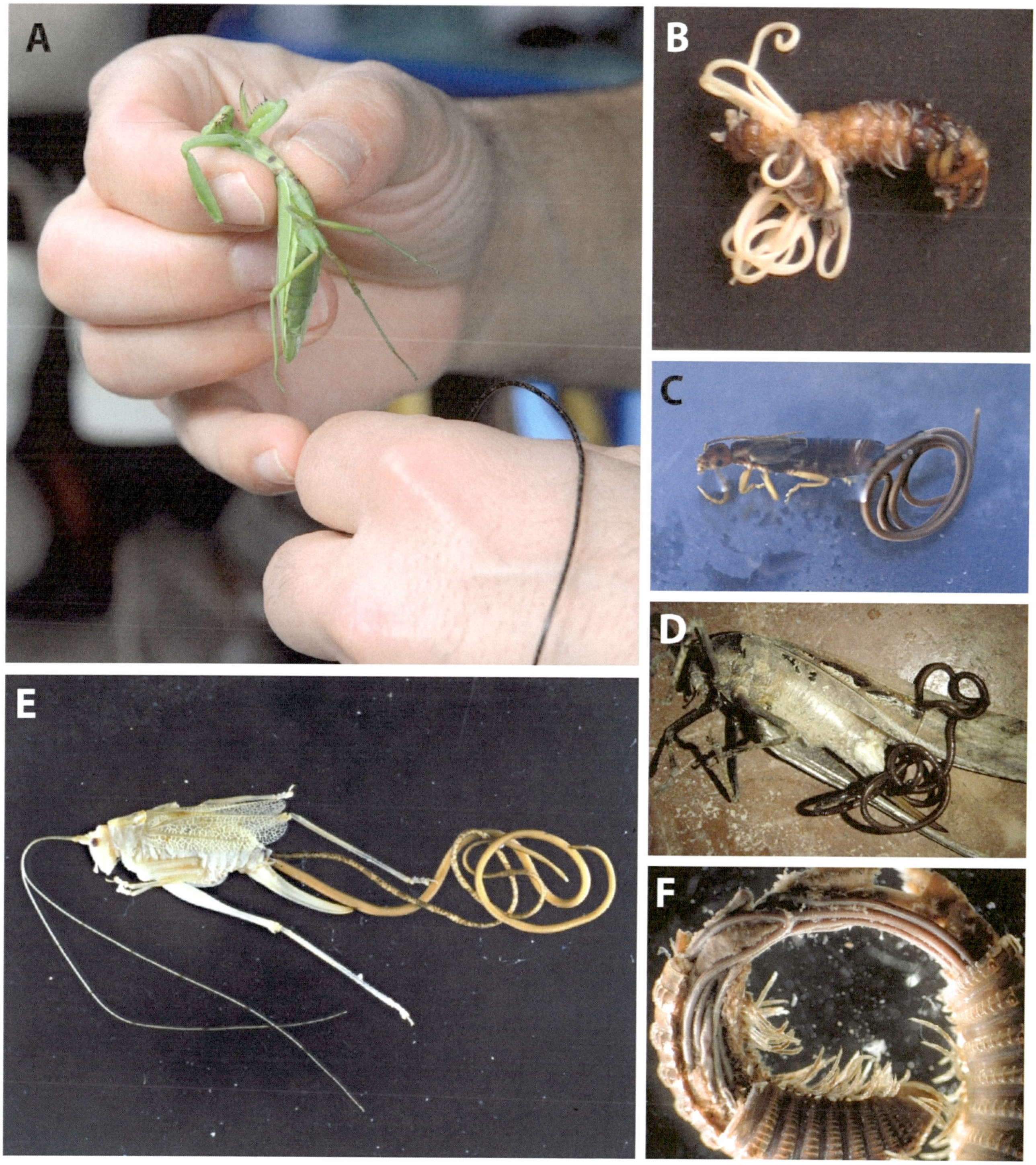

Fig. 3.2.5. Hosts of gordiids. A, Specimen of *Chordodes japonensis* escaped from a praying mantid (*Hierodula patellifera*) immediately after catching (Shiga, Japan); B, Several specimens of *Gordius albopunctatus* in a caddis-fly larva (*Potamophylax cingulatus*) (Bornholm, Denmark); C, Undetermined gordiid emerging from an earwig (Germany); D, *Chordodes* sp. emerging from an undetermined orthopteran (Panay, Philippines); E, Male (thin specimen) and female (thicker specimen) of *Spinochordodes tellinii* emerging synchronously from *Meconema thalassinum* (Avene les Bains, France); F, *Gordionus lineatus* inside the diplopod *Cambala annulata* (Ohio, USA). F from Schmidt-Rhaesa et al. 2009, with kind permission from the Northeastern Naturalist. D made by M. Gaulke and A. Demegillo, with kind permission.

Tab. 11. Reported cases of Odonata, Diptera, Lepidoptera and Hymenoptera as hosts for juvenile or adult gordiids.

Host species	Gordiid species	Geographic region	Author
Odonata			
Cordulia metallica	*Gordius aquaticus*		Assmuss 1858
Libellula cancellata	*Gordius subbifurcus*		Assmuss 1858
Libellula pedemontana	*Gordius aquaticus*		Assmuss 1858
libellulid nymphs	*Euchordodes libellulovivens*	Germany	Heinze 1937
Diptera			
Tabanus autumnalis	*Gordius borisphaenicus*	Russia	Spiridonov 1984, 1998
Tabanus bromius	*Gordius aquaticus*		Assmuss 1858
Lepidoptera[1]			
Abrostola urticae	*Gordius subbifurcus*		Assmuss 1858
Calpe libatrix	*Gordius aquaticus*		Assmuss 1858
caterpillar of *Cucullia scrophulariae*	undetermined ("Fadenwurm")	Germany	Fiedler 1909
Erebia euryale	undetermined	Austria	Chapman 1902
Hepiolus humuli	*Gordius subbifurcus*		Assmuss 1858
Plusia gamma	*Gordius aquaticus*		Assmuss 1858
Saturnia sp.	*Gordius longissimus*	"South Sea"	Römer 1895a
caterpillar of *Saturnia artemisiae*	*Gordius aquaticus*	Russia	Römer 1895a
caterpillar of *Vanessa vanessa album*	undetermined	probably Czech Republik	Müller 1859
Zerene grossularia	*Gordius aquaticus*		Assmuss 1858
Zygaena filipendulae	*Gordius aquaticus*		Assmuss 1858
Hymenoptera			
Paravespula germanica[2]	undetermined		Gauss 1970
Vespa crabro	*Filaria* ("Fadenwurm")	Sardinia	Von Siebold 1850
Vespa germanica[3]	undetermined	Germany	Kloft 1951
ants	*Filaria* ("*Gordius*-artiger Fadenwurm")		Von Siebold 1843a
"bees"	undetermined	Brazil	Cury 1946

[1] Further more unspecific records are filariae ("Fadenwürmer") from butterflies and caterpillars see Von Siebold (1848, 1850) and Von Linstow 1878, different species of butterflies from Kurland (Lithuania) (Slevogt 1909) and a not very specific report from *Danais chrysippus* caterpillars in South Africa (Von Pelser-Berensberg 1909).
[2] According to the figures, this parasite is a mermithid nematode (Heinze 1954).
[3] Judging from the pictures the parasite is a mermithid nematode.

Tab. 12. Reported cases of Diplopoda, Chilopoda (Myriapoda) and Araneae (Chelicerata) as hosts for juvenile or adult gordiids (see also Fig. 3.2.5 F).

Host species	Gordiid species	Geographic region	Author
Diplopoda			
Cambala annulata	*Gordionus lineatus*	USA	Schmidt-Rhaesa et al. 2009a
Cylindroiulus teutonicus	undetermined	France	Sahli 1972
Glomeris humbertiana	*Gordius* sp.	France	Dorier 1930
Glomeris marginata	*Gordius aquaticus*	France	Dorier 1930
Iulus sp.	*Gordionus* (as *Parachordodes)* *alpestris*	France	Dorier 1929, 1930
Narceus sp.	undetermined	USA	Cooper & Storck 1973
Ommatoiulus moreletii	undetermined	Portugal	Baker 1985

(*Continued*)

Tab. 12. (*Continued*)

Host species	Gordiid species	Geographic region	Author
Polydesmus complanatus	*Gordionus alpestris*	Romania	Căpuşe 1971
Schizophyllum sabulosum	undetermined	Germany	Sahli 1972
Tachypodioiulus albipes	undetermined	France	Sahli 1972
Chilopoda			
Lithobius sp.	*Gordius parisii*	Italy	Sciacchitano 1932
Lithobius forficatus	*Gordius aquaticus*	France	Villot 1886
	Gordius villoti		Camerano 1887a
Scolopendra cingulata	*Gordius aquaticus*	Marocco	Dorier 1929
Scolopendra cormocephalina	*Gordius robustus*	Uruguay	Carvalho 1946a
Scolopendra sp.	*Gordius aquaticus*	France	Camerano 1897a
Araneae[1]			
Araneus (as Epeia) diademata	undetermined	France	Villot 1881
Aranea sp.	undetermined	France	Villot 1881
Arctosa alpigena	*Gordionus violaceus*	Scotland	De Villalobos et al. 1999
Ctenus bryrrbus	*Neochordodes* sp.	Costa Rica	Acholonu 1968
Drassus fuscus	*Parachordodes* (as *Gordius*) *tolosanus*	France	Villot 1874c, 1881
Drassus lucifugus	undetermined	France	Villot 1881
Epeira ceropegia	undetermined	France	Villot 1881
"large spider"	undetermined	Canada	Clementi 1869
"from a spider"	*Gordius aquaticus*	Germany	Römer 1895a
from spider	*Parachordodes ciferrii*	Dominican Republic	Sciacchitano 1932

[1] See also: Filariae from spiders: Von Siebold 1842, 1843a, 1848.

Tab. 13. Infection rates reported for gordiids.

Infection rate	Parasite	Host	Region	Author
7–28%	undetermined	*Ommatoiulus moreletii* (Diplopoda)	Australia	Baker 1985
5.3–20.3%	*Gordius villoti*	*Allogamus auricollis* (Trichoptera)	Italy	Valvassori et al. 1988
0.26%	undetermined	*Celatoblatta peninsularis* (Blattodea)	New Zealand	Zervos 1989
0.5%	undetermined	*Parellipsidion pachycerum* (Blattodea)	New Zealand	Zervos 1989
2.9%	undetermined	male *Ceuthophilus stygius* (Orthoptera)	USA	Studier et al. 1991
16.9%	undetermined	female *Ceuthophilus stygius* (Orthoptera)	USA	Studier et al. 1991
0.8%	undetermined	male *Hadenoecus subterraneus* (Orthoptera)	USA	Studier et al. 1991
0.9%	undetermined	female *Hadenoecus subterraneus* (Orthoptera)	USA	Studier et al. 1991
3%	*Euchordodes nigromaculatus*	female *Sigaus australis* (Orthoptera)	New Zealand	Poulin 1995
13%	*Euchordodes nigromaculatus*	female *Sigaus obelisci* (Orthoptera)	New Zealand	Poulin 1995
4.6–8.4%	*Gordius albopunctatus*	larvae of *Pomatophylax cingulatus* (Trichoptera)	Denmark	Schmidt-Rhaesa & Kristensen 2006

by a gordiid with simplified genitalia, resembling an intersexuality, which might be caused by the parasites. Lombardo & Umbriaco (2011) also assume that gordiid infection caused intersex morphology in a specimen of *Parastagmatoptera abnormis*, a species described on the basis of only one infected male. On the other hand, Poulin (1995) found no size differences between parasitized and not parasitized short-horned grasshoppers in New Zealand and Thomas et al. (1998) found no effect of the parasite on cave wetas (*Pleioplectron simplex*) in New Zealand concerning coloration patterns or symmetrical patterns, a parasitological phenomenon called fluctuating asymmetry.

Internally, the growing gordiids appear to compete for space with some internal organs, especially the gonads and the fat body (Von Linstow 1891a, Ebner 1940, Sahli 1972, Studier et al. 1991), which might lead to castration. The long-term effects of infection on the hosts have not been tested, but when fat bodies are reduced, a reduced fitness appears plausible. Changes of behaviour range from reports of hosts becoming less active (Valvassori et al. 1988) or responding less to male song and failure to mate (Morris et al. 1975), to a directed water-seeking behaviour (see Emergence from final hosts below). Some strange reports come from infections of fish in the New York Aquarium (Nigrelli 1941). After infection, worms are reported to occur in the body cavity or in the sinus venosus of the circulatory system and some semi-tropical fish had distended bellies. As this report is only a brief communication, without any pictures, it must be regarded as questionable whether the parasites in this case are really nematomorphs.

Nectonema. All hosts of *Nectonema* species reported so far are decapod crustaceans (see Fig. 3.1.2 A, B; Tab. 14), ranging from swimming forms to benthic ones.

In some cases, further crustacean species were tested for *Nectonema* infection. Born (1967) dissected more than 2000 specimens of *Crangon septemspinosa* and additionally *Hippolyte zostericola* and *Pagurus longicarpus* without finding a *Nectonema* specimen. No *Nectonema* specimen was found in about 3000 specimens of *Eupagurus cuanensis* in Roscoff (Pérez 1934). In *Palaemonetes vulgaris*, *Nectonema* was only found in juvenile crabs (Born 1967).

Usually one specimen of *Nectonema* is found within one host, but more specimens also occur regularly. Brinkmann (1930) reported that from 61 infected specimens of *Munida tenuimana*, 51 had one specimen of *N. munidae*, eight had two, one three and one six specimens. Born (1967) found two specimens of *N. agile* in only one of 37 infected *Palaemonetes vulgaris*. Feyel (1936) found in 11 specimens of *Anapagurus hyndmanni* – three with two *N. agile* specimens and in another sample of five there was one with two *Nectonema* specimens. Leslie et al. (1981) found most crabs to have one parasite, but they found up to nine specimens per host.

Few records report a parallel infection of hosts by *Nectonema* and another parasite. Pérez (1934) found few cases in which *Nectonema* and *Fecampia crythrocephala* (Plathelminthes, Fecampiidae) were found together in *Anapagurus hyndmanni*. In *Macropodia rostrata*, there were few infections with *Nectonema* and *Sacculina fraissei* (Crustacea, Rhizocephala). Nielsen (1969) found simultaneous infection of *Munida* specimens with *N. munidae* and *Pseudione crenulata* (Crustacea, Isopoda).

Leslie et al. (1981) found that the size of worms was correlated with host size, with larger hosts (e.g. *Cancer irroratus*) having larger parasites and smaller hosts (e.g. *Pandalus montagui*) having smaller parasites. Nielsen (1969) found the contrary, according to him, host and parasite size are not correlated. According to Mouchet (1931) and Poinar & Brockerhoff (2001), both sexes of the host were infected to different rates. Whereas in *Anapagurus hyndmanni* more males are infected (50% of males, 10% of females; Mouchet 1931), in *Hemigrapsus edwardsi* more females are infected (Poinar & Brockerhoff 2001). Nielsen (1969) found no differences in the infection between host males and females; he only found that juvenile *Munida tenuimana* specimens are more heavily infected (32%) than adults (6.8%). Reported infection rates vary strongly from <1% to maximally 21.4% (Tab. 15). According to Poinar & Brockerhoff (2001) infection rates vary monthly in *N. zealandica*.

There are contrasting reports concerning the question of whether *Nectonema* specimens cause internal damage to their hosts. Brinkmann (1930), Nielsen (1969) and Nouvel & Nouvel (1934) observed no tissue alteration in parasitized host specimens, but Mouchet (1931) stated that males of *Eupagurus bernhardus* and *Anapagurus hyndmanni* become castrated by *Nectonema agile*. According to Leslie et al. (1981), *Nectonema agile* causes only minor degeneration in the reproductive organs of *Cancer irroratus*.

Emergence from final hosts

It is quite paradoxical that gordiids as aquatic animals infect terrestrial hosts. Although the extremely large number of eggs and larvae suggests the loss of a considerable

Tab. 14. Hosts of juvenile and adult *Nectonema*. Actual host names after Poinar & Brockerhoff (2001).

Host species	*Nectonema* species	Number of infected host specimens	Geographic region	Author
Anapagurus hyndmanni	*Nectonema* sp.	~30	vicinity of Roscoff, France	Pérez 1927a,b
	N. agile	?		Mouchet 1931
	N. agile	103		Pérez 1934
	N. agile	16	vicinity of Morlaix, France	Feyel 1936
Anapagurus laevis	*N. agile*	1	vicinity of Roscoff, France	Pérez 1934
Cancer borealis	*Nectonema* sp.		Bay of Fundy, Canada	Leslie et al. 1981
Cancer irroratus	*Nectonema* sp.		Bay of Fundy, Canada	Leslie et al. 1981
		7	Bay of Fundy (St. Andrews, Grand Manan), Canada	Brattey et al. 1985
Diogenes pugilator	*N. agile*	2	vicinity of Roscoff, France	Pérez 1934
Erimacrus isenbeckii	*Nectonema* sp.	4	Japan	Oku et al. 1983
Hemigrapsus edwardsi	*N. zealandica*		New Zealand, South Island	Poinar & Brockerhoff 2001
Homarus americanus	*N. agile*	1	Bay of Fundy (St. Andrews, Grand Manan), Canada	Schmidt-Rhaesa et al. 2012
Lebbeus (as *Spirontocaris*) *polaris*	*Nectonema* sp.	1	Greenland	Nouvel & Nouvel 1938
Macropipus (as *Portunus*) *arcuatus*	*Nectonema* sp.	4	vicinity of Dinard, France	Gallien 1949
Macropipus (as *Portunus*) *pusillus*	*Nectonema* sp.	1	vicinity of Roscoff, France	Pérez 1927b
Macropodia rostrata	*N. agile*	49	vicinity of Roscoff, France	Pérez 1934
		4	vicinity of Morlaix, France	Feyel 1936
Munida sarsi	*N. munidae*	10	vicinity of Bergen, Norway	Nielsen 1969
Munida tenuimana	*N. munidae*	61	vicinity of Bergen, Norway	Brinkmann 1930
		447		Nielsen 1969
		17		Schmidt-Rhaesa 1996a
Pagurus acadianus	*Nectonema* sp.		Bay of Fundy, Canada	Leslie et al. 1981
Pagurus (as *Eupagurus*) *bernhardus*	*N. agile*	2	vicinity of Roscoff, France	Mouchet 1931
		20		Pérez 1934
Pagurus cuanensis	*N. munidae*	1	vicinity of Bergen, Norway	Nielsen 1969
Pagurus (as *Eupagurus*) *pubescens*	*N. munidae*	1	vicinity of Bergen, Norway	Brinkmann 1930
Pagurus pubescens	*N. munidae*	2	vicinity of Bergen, Norway	Nielsen 1969
Palaemon elegans (as *Leander squilla)*	*Nectonema* sp.	1	vicinity of Roscoff, France	Nouvel & Nouvel 1938
		1	vicinity of Dinard, France	Gallien 1949
Palaemon (as *Leander*) *serratus*	*Nectonema* sp.	2	vicinity of Roscoff, France	Nouvel & Nouvel 1934
Palaemonetes vulgaris	*N. agile*	37	Woods Hole, Massachusetts, USA	Born 1967
Palaemonetes sp.	*N. agile*	1	?	Ward 1892b
Pandalus borealis	*Nectonema* sp.		Bay of Fundy, Canada	Leslie et al. 1981
Pandalus montagui	*Nectonema* sp.		Bay of Fundy, Canada	Leslie et al. 1981, Sakaling & MacKinnon 1988
Pontophilus norvegicus	*N. munidae*	1	vicinity of Bergen, Norway	Brinkmann 1930
Sergestes arcticus	*N. munidae*	1	Herdla near Bergen, Norway	Greve 1972

Tab. 15. Infection rates of *Nectonema*.

Host species	*Nectonema* species	Number host spp.: number *Nectonema* spp.	Rate	Author
Anapagurus hyndmanni	*N.* sp.	491:~20	~4.1	Pérez 1927a
	N. sp.	308:~12	~3.9	Pérez 1927b
	N. sp.	1359:103	7.6	Pérez 1934
	N. agile	110:5	4.6	Feyel 1936
Cancer borealis	*N.* sp.	133:2	1.5	Leslie et al. 1981
Cancer irroratus	*N.* sp.	22:4	18.2	Brattey et al. 1985
	N. sp.	66:3	4.5	Brattey et al. 1985
	N. sp.	538:115	21.4	Leslie et al. 1981
Eupagurus bernhardus	*N.* sp.	1024:10	1.0	Pérez 1934
	N. sp.	1300:10	0.8	Pérez 1934
	N. agile	152:2	1.3	Mouchet 1931
Hemigrapsus edwardsi	*N. zealandica*	636:81	12.7	Poinar & Brockerhoff 2001
Leander squilla	*N.* sp.	>250:1	>0.4	Gallien 1949
	N. sp.	600:4	0.7	Arvy 1963
	N. sp.	400:2	0.5	Arvy 1963
Macropodia rostrata	*N.* sp.	813:49	6.0	Pérez 1934
Munida sarsi	*N. munidae*	140:10	7.1	Nielsen 1969
Munida tenuimana	*N. munidae*	776:61	7.9	Brinkmann 1930
	N. munidae	6535:447	6.8	Nielsen 1969
	N. munidae	951:17	1.79	Schmidt-Rhaesa 1996a
Pagurus acadianus	*N.* sp.	591:11	1.9	Leslie et al. 1981
Pagurus cuanensis	*N. munidae*	500:1	0.2	Nielsen 1969
Pagurus pubescens	*N. munidae*	58:2	1.2	Nielsen 1969
Palaemonetes vulgaris	*N. agile*	>5000: 37	<0.74	Born 1967
Pandalus borealis	*N.* sp.	15:3	20	Leslie et al. 1981
Pandalus montagui	*N.* sp.	458:96	21	Leslie et al. 1981

number of these along the life cycle, mechanisms to encourage the final host to find water would optimize the life cycle. Indeed, it has been known for quite some time that hosts search the proximity of water or even enter the water. For example, McCook (1884) observed that crickets hold their abdomen into the water to let gordiids emerge; Müller (1920) and Jolivet (1945, 1948) reported that carabid beetles search for water and enter this to release their parasites. These observations were tested by Thomas et al. (2002) in field and laboratory experiments; they could clearly demonstrate that infected crickets (*Nemobius sylvestris*) actively jump into water to release their parasites.

The mechanism of this parasite-induced alteration in behaviour remains unclear. Traditionally, it has been assumed that the hosts become thirsty and therefore approach water (e.g. Blunck 1922, Jolivet 1948). This might indeed be the case, because growing gordiids likely take up a large quantity of nutrients and fluid from the body cavity. But thirst does not explain why hosts seek water only when the gordiids are mature and why they enter water instead of remaining at the water's edge. Thomas et al. (2003) compared the brains of infected and uninfected crickets and found differences in the concentrations of neurotransmitters and neuromodulators as well as in the mitotic index in the mushroom body. Biron et al. (2005a, 2006) investigated cricket brains before, during and after parasite manipulation as well as the parasite itself by two-dimensional gel electrophoresis. They found several proteins, which are differently expressed in infected versus uninfected crickets; among these are proteins linked to neurogenesis, neurotransmitter activity and circadian rhythm (Biron et al. 2006). One possible model is that the gordiid may directly influence the host´s nervous system by secretion of "effective molecules", for example proteins of the Wnt family.

Sanchez et al. (2008) showed that the manipulation of the host by the parasite consists of two steps. Juvenile gordiids can cause an erratic behaviour, such as leaving the natural habitat, without necessarily seeking water. This second step is induced by mature worms.

Biron et al. (2005b) approached the question of parasite manipulation from a different angle and asked whether the host has advantages to collaborate with the parasite by "bringing" it to water. Infected cooperative specimens (placed in a box with water) were compared to uncooperative specimens (placed in a box without water) and to an uninfected control group. The results showed a strong influence on infected specimens in general (reduced survival, castration in males), but only some differences between the cooperative and the uncooperative group (uncooperative females never had ovaries; in cooperative females some produced eggs but had problems during copulation and oviposition).

Emerging worms penetrate the soft intersegmental integument close to the posterior end and emerge head first (e.g. Blunck 1922, Schmidt-Rhaesa et al. 2005). Emergence follows soon after entrance of the host into water and takes about 1 minute. The size of the worm inside its host may make it possible that the integument is punctured by pure force. Especially during the last phase, emergence is aided by undulating movements.

Human infections?

There are a number of cases in which some kind of association between humans and gordiids is reported. A summary lists 67 such cases (see Tab. 16); other summaries (including more doubtful cases) were published for example by Stiles (1907), Herzog (1913), Riddell (1928) and Cappucci (1976). The question behind these cases is whether humans can be considered as a host for gordiids or whether the association is accidental, a "pseudoparasitism".

Additional reports, which could not be checked, are: Gao et al. (1989; in Chinese), Weng (1991; in Chinese), Cui (1995; digestive tract, in Chinese) and He (1995; in Chinese). More than ten cases of Gordian worms were passed per urethram in the Dominican Republic (Bundy & Steele 1984); they cite as source the Proceedings of the 5th International Congress of Parasitology in Toronto 1982, but a search within this volume did not reveal a paper which includes this topic.

Very early reports are difficult to interpret, because they seem to confuse gordiids with other parasitic worms, probably mainly nematodes. For example, Vieweg (1811) and Anonymous (1840) report gordiids to cause blisters (which was considered to be not very likely by Von Bolschwing 1842). Reports from other vertebrates, such as gordiids found in the lungs of sheep (Crisp 1873), are also very unlikely. However, gordiids were blamed for some kinds of illness (De Bacounin 1790). Remarkable is that De Bacounin (1790), based on these reports, conducted a self-test by swallowing gordiids without any effects.

The reported cases can be divided into three groups: 1. association with the human intestinal tract, 2. assumed infection of the human excretory system and 3. miscellaneous cases.

The reports from the intestinal tract are by far the most numerous (43 in Tab. 16). Gordiids are either vomited or found in the stool. Similar reports exist from domestic animals like cats and dogs (Horton 1986, Saito et al. 1987). It can almost be excluded that the presence of a gordiid within the human intestinal tract follows a proper infection or any kind of development of gordiids within the human body. It is far more likely that humans swallow either gordiids or infected hosts. This can occur where natural water is consumed, but gordiids have regularly been found even in water supplies (Baylis 1944, Reeker 1894, Foresi & Caroli 1968, Cappucci 1982). Von Linstow (1889) reported a case where a gordiid was found in a drinking bottle of water belonging to a soldier who might not have been able to notice the worm at night. The movements of gordiids in the upper intestinal system may cause nausea, but gordiids seem also to survive the passage through the entire intestinal system and can be passed *per anum*.

More difficult to explain are the cases in which gordiids are associated with the human excretory system. These are 11 cases in Tab. 16. In most of these cases the worms were found in the toilet (or equivalent containers) after urinating. None of the reports explicitly states that a gordiid was observed emerging through the urethra and it therefore remains questionable if the worms had really been present within the excretory system. An infection of the excretory system appears quite unlikely. Either adults or larvae would have to infect the system by actively moving into the urethra, for example during bathing. Judging from the more or less unspecific movements of both stages this is hard to imagine. More convincing is to assume that those gordiids observed in toilets, etc. were not expelled from the human body, but emerged there from their insect hosts. This has been assumed for example by Herter & Nesse (1989), who reported that a mother found gordiids after her girl had used the toilet, who said that she used to throw insects into the toilet bowl. Either by this habit or by the active search of parasitized insects for water in the domestic environment, gordiids can emerge and then be found in toilets. Although some reports state that the bowl had been clean before urinating (Schmidt-Rhaesa et al. 2003b) proof for the exceptional presence of gordiids within the human excretory system by a clear observation still remains to be given.

Tab. 16. Associations between humans and gordiids, ordered after year of report. Sex of gordiid is given in the first column – sign only means one specimen, "?" means undetermined sex. "°" indicates that the record was copied from the literature, but could not be checked personally.

Gordiid species and sex (according to source)	Location	Infested organ in human	Comments	Author
undetermined [1]	France	intestinal system	vomited by an 8-year-old child	Degland 1823
undetermined	France	intestinal system	passed *per anum* by a 15- or 16-year-old girl after colic attacks	Degland 1823
[Parachordodes tolosanus][2] (♂)	France	intestinal system	vomited by a farmer	Cloquet 1822a,b
undetermined	?	excretory system	gordiids were found in the urine of women on two occasions	Clark 1854°
undetermined	?	?	a gordiid was coughed up	Clark 1854°
Gordius aquaticus (♀)	Schliersee, Germany	intestinal system	vomited by a 22-year-old woman; a number of symptoms ceased after this	Von Siebold 1854
Gordius varius (♂)	Cleveland, Ohio, USA	intestinal system	passed *per anum* by a girl	Leidy 1856
Gordius trifurcatus (♂)[3]	USA	intestinal system	(probably) passed by a human after use of aloes	White 1859
Gordius aquaticus (♂)	Texas, USA	intestinal system	passed by a boy	Pepper 1872°
Gordius aquaticus	probably Austria	intestinal system	in the stool of an 8-year-old boy	Patruban 1875
undetermined (♂)	Italy	intestinal system	passed *per anum* by a 42-year-old woman	Fiori 1881°
Gordius aquaticus	Italy	intestinal system	vomited by a 7-year-old boy after severe gastrointestinal pains, which ceased after expelling the worm	Cerrutti & Camerano 1888°
undetermined	Lithuania ("Kurland")	?	only notice: "from a woman"	Von Linstow 1898
undetermined	France	intestinal system	a 15-year-old boy found a gordiid in his throat after drinking a glass of wine	Blanchard 1897
Gordionus (as *Gordius*) *violaceus* (♂)	Fougéres, France	intestinal system	vomited by a 34-year-old man	Topsent 1900
Parachordodes pustulosus (♀)	Genua, Italy	intestinal system	passed *per anum* by a 45-year-old woman after complaining of something moving in the rectum	Parona 1901
Gordionus alpestris[4]	Chabanais, France	intestinal system	vomited by a 34-year-old man after 5 days of intestinal problems	Guéguen 1905a,b
Paragordius cinctus (?)	Lydenburg, South Africa	not stated	only notice: "in *Homo sapiens*"	Von Linstow 1906a
Paragordius varius	Saco, Maine, USA	not stated	only notice: "host *Homo*"; case is probably identical with the one of White (1861) (see Cappucci 1976)	Stiles 1907
Paragordius varius	London, Ontario, Canada	intestinal system	a 13-year-old girl passed a "large quantity" of gordiids twice within 1 week. As the worms were stiff, Stiles (1907) suspected that they might have been in the toilet before this was used by the girl	Stiles 1907
Paragordius varius (♂+♀)	Chicago, Illinois, USA	intestinal system	vomited by a patient in hospital	Stiles 1907

(Continued)

Tab. 16. *(Continued)*

Gordiid species and sex (according to source)	Location	Infested organ in human	Comments	Author
undetermined	China	intestinal system	found in the stool of a 20-month-old child	Maxwell 1908
Gordius aquaticus (♂)	Dorentingen, Swizerland	intestinal system	passed *per anum* by a 2.5-year-old boy together with several *Oxyuris* (Nematoda); patient had nervous problems, which ceased after release of the parasites	Hall 1912, Zschokke 1912a,b, Herzog 1913
Gordius aquaticus (2♂♂)	Romania	intestinal system	passed *per anum* by a 14-year-old child with mild gastrointestinal problems	Léon 1914
undetermined	Geneve, Switzerland	intestinal system	passed *per os* after 2 weeks of gastrointestinal problems by a young girl	André 1927
Paragordius areolatus (2♀♀)	Ghana (as "Gold Coast")	unknown	"passed by a native girl"	Baylis 1927
Chordodes capensis (♀)	British East Africa	unknown	"passed by a native"	Baylis 1927
Gordius aquaticus (♂)	Hartford, Connecticut, USA	[intestinal system]	worm fragments found in the toilet bowl, the occurrence was tentatively linked to intestinal problems of a child	Reardon 1928
Gordius aquaticus	Belfast, Northern Ireland	intestinal system	found in a stool within a diaper of a 1-year-old child	Riddell 1928
[*Gordius* sp.]	Florida, USA	[in tumor-like swelling]	a histological section of a tumor-like swelling close to the eye of a 37-year-old man was interpreted as a female *Gordius* sp., but later determined as *Dirofilaria* sp. (Beaver & Orihel 1965)	Sayad et al. 1936
Gordius gesneri (♂)	Tegernsee, Germany	intestinal system	vomited by a woman	Heinze 1937, 1941
Gordius perronciti (♀)	Tegernsee, Germany	intestinal system	vomited by a woman	Heinze 1937, 1941
Gordius robustus	Provo, Utah, USA	excretory system	found in the urine of a female	Tanner 1939
Paragordius varius (♀)	Lincoln, Nebraska, USA	intestinal system	found in the stool of a child	Carvalho 1942
Paragordius esavianus (♀)	Espirito Santo, Brazil	excretory system	"expelled alive from the urethra" of a child	Carvalho 1942
Gordionus (as *Parachordodes*) *wolterstorffii*	North of England	excretory system	"passed *per vaginam*" from a woman	Baylis 1944
Gordius sp. (♀)	Missouri, USA	intestinal system	found in the stool of 4-year-old child	Craig & Faust 1951
undetermined	South Carolina, USA	excretory system	many immature gordiids found in the urine of a 45-year-old man	Craig & Faust 1951
undetermined	Malaysia	intestinal system	passed *per anum* from a hospital patient	Sandosham 1953°
Pseudogordius tanganyikae (♀)	Mwanza, Tanzania	excretory system	found in the urine of a child with abdominal pains, which ceased after release of the worm	Yeh & Jordan 1957
Beatogordius (Gordius) inesae (♂)	Santiago del Estero, Argentina	intestinal system	vomited from an adult man after gastrointestinal problems	Cavalieri 1961
Neochordodes columbianus (♂+♀)	Colombia	outer ear canal	removed from the outer ear canal of a "poorly nourished country girl"	Faust & Botero Ramos 1960, Faust et al. 1970

(Continued)

Tab. 16. (*Continued*)

Gordiid species and sex (according to source)	Location	Infested organ in human	Comments	Author
Chordodes sp. (♀+?)	Johore, Malaysia	intestinal system	a 3-year-old girl passed two gordiids *per anum*	Fernando & Fernando 1961
Chordodes skorikovi (♂)[5]	Colombo, Sri Lanka	[intestinal system]	found on the floor in a spot where a 3-year-old girl had vomited the previous night	Fernando & Dissanaike 1962
Gordius sp. (♀)	Gifu prefecture, Japan	intestinal system	vomited by a 5-year-old child	Kagei et al. 1966
Gordius reddyi (♂+3?)	India	tumor in the orbital region	four specimens, the largest measuring 266 mm, are reported to have been found in a tumor in the orbit. No further information, such as size of the tumor, are given	Singh & Rao 1966
Gordius sp. (♀)	Iwaki and Shirahama, Japan	intestinal system	vomited by 2-year-old girl following 15 days of nausea	Ichihara et al. 1967
Gordius sp.	Italy	intestinal system	found in the stools of six soldiers after having colic	Foresi & Caroli 1968
Paragordius esavianus (♀)	Texas, USA	excretory system	found after urination of a 23-year-old woman who suffered excretory problems for 3 years, the problems ceased after passing of the worm	Burger 1972
Paragordius sp. (♂+♀)	Receife, Brazil	excretory system	a 60-year-old man suffered from lower abdominal pains for 9 months; he passed within 1 month three gordiids (the last one in hospital, it was found in a chamber-pot")	De Lucena & Campelo 1975
Gordionus sp.	Vancouver area, Canada	intestinal system	four specimens were vomited from small children (number of children and number of worms per child are not given)	Cappucci 1976
undetermined	Puerto Rico	[intestinal system]	found in a toilet bowl after use by a child	Cappucci 1976
Paragordius varius (♂)	Quebec, Canada	intestinal system	found in the stool of a 58-year-old woman after diarrhea; the worm was alive and active	Ali-Khan & Ali-Khan 1977
undetermined	Shizuoka, Japan	[intestinal system]	reported as passed "*per anus* (?)"	Kagei 1977
Gordius sp. (♀)	Ishikawa, Japan	intestinal system	found in the stool of a 38-year-old woman following 3 days of stomach problems, which ceased after release of the worm	Yoshimura et al. 1977
Parachordodes sp.	Hubei province, China	excretory system	was expelled alive during urinating by a 26-year-old woman following 4 years of abdominal pains, which became severe before and ceased after releasing the worm	Wei & Yang 1981
Gordius sp. (♂)	Kagoshima, Japan	intestinal system	found in the stool of a 1-year-old girl	Uchikawa et al. 1987
Gordius sp.	Livorno (?, as Liverno), Italy	intestinal system	found in fecal samples	Cerini & Ristori 1988
Gordius sp.	Italy	intestinal system	several worms vomited by a 3-year-old boy	Cerini & Ristori 1988
Paragordius varius (♀+ several specimens)	Indianapolis, USA	intestinal system	a tangle with several specimens, one of which was determined, was found in the stool of a 30-year-old man	Smith et al. 1990
Neochordodes sp.	China	intestinal system	reported by Mao from a Chinese publication (not seen by the author) without giving further details	Mao 1991

(*Continued*)

Tab. 16. *(Continued)*

Gordiid species and sex (according to source)	Location	Infested organ in human	Comments	Author
Parachordodes sp.	China	excretory system		Mao 1991
Spinochordodes (as *Dacochordodes*) *bacescui* (♂+♀)	Doboj, Bosnia and Herzegovina	intestinal system	vomited by a 50-year-old man (who had vomited "a dozen" worms 20 days earlier)	Spiridonov et al. 1992, Pikula et al. 1996
Spinochordodes (as *Dacochordodes*) *bacescui* (♀)	Doboj, Bosnia and Herzegovina	intestinal system	found in the stool of a 35-year-old woman	Spiridonov et al. 1992, Pikula et al. 1996
undetermined (♂)	Biei, Japan	nasal cavity	gordiid occurred from a man after sneezing	Miyamoto & Uenishi 2002
Beatogordius raphaelis (♂)	South Africa	excretory system	passed by a child *per urethram*	Schmidt-Rhaesa & De Villalobos 2002[6]
Gordius sp. (♂+?)	Gwangju, Kyunggi-do, South Korea	intestinal system	vomited by a 3-year-old girl 15 minutes after having eaten a cricket-like insect	Lee et al. 2003
Parachordodes megareolatus (♂)	Gyeongnam prov., South Korea	excretory system	found in the bed-pan of a 77-year-old woman after urinating (the bed-pan was reported to be clean prior to urinating)	Schmidt-Rhaesa et al. 2003b
Paragordius varius (♀)	Nebraska, USA	?	"found alive in the underclothing of a woman"	Hanelt et al. 2005

[1] *Gordius tolosanus* (= *Parachordodes tolosanus*) according to Villot (1884).
[2] Worms named *Ophiostoma pouterii* (Cloquet 1822a) or *O. pontierii* (Cloquet 1822b) by Cloquet; later Blanchard (1890) and Stiles (1907) used the name *Parachordodes tolosanus* for the same specimens. Riddell (1928) and Cappucci (1976) note that no proper description of the specimens is given and that therefore both the original name and the synonymy to *P. tolosanus* are uncertain.
[3] Very likely a synonym of *Paragordius varius* (see Schmidt-Rhaesa et al. 2003a).
[4] According to the figures in Guéguen (1905), the specimen belongs to *Spinochordodes* (probably *S. tellinii*).
[5] Spelling "*skorikowi*" by Fernando & Dissanaike (1962) is wrong.
[6] This case seems to have been reported in the literature before, see Yeh & Jordan (1957). Schmidt-Rhaesa & De Villalobos (2002) report the case from notes in the files of the Natural History Museum.

Finally, there are some cases in which gordiids have been reported from other parts of the body. Two cases (Faust & Ramos 1960, Miyamoto & Uenishi 2002) report gordiids from external cavities in the ear and the nose. Gordiids probably reached these parts of the body by escaping from their hosts in close vicinity. More problematic are reports from gordiids that have been reported from blisters or nodules. There are two such cases. The first case reported by Sayad et al. (1936) turned out to be a misidentification; the worms were in fact nematodes (*Dirofilaria*; Beaver & Orihel 1965). The other case was reported by Singh & Rao (1966). Although specimens found in a tumor were determined as a new species, *Gordius reddyi*, several questions remain. Four specimens were found within the same tumor, two long (266 and 211 mm) and two short ones. The large specimens were figured; they show a flat and twisted appearance, which is somewhat unusual for gordiids. Further details are given only for one specimen. The cuticular pattern figured by Singh & Rao (1966) does not contain clearly typical gordiid structures such as areoles. The posterior end appears to be bilobed, but a cloacal opening was not observed. In their legend of figure 1 Singh & Rao (1966) describe a postcloacal crescent, which is, nevertheless, not clearly evident from the figure. Therefore, I regard this report as questionable.

A gordiid from a subcutaneous nodule has also been reported from a cat (Moisan et al. 1996), but the sections figured are not convincing and this probably also is not a gordiid. Another gordiid, *Chordodes guineensis*, was found in an inner abscess of a dog (Spiridonov 2001).

3.3. Physiology

There are almost no data on the physiology of nematomorphs. The nutrient uptake is explained above (Chapter 1). Czeczuga (1981) investigated *Gordius aquaticus* from deep wells in Poland and reported the presence of different carotenoids, even if the worms were very pale.

Kovalchuk (1988) measured respiration of *Gordius* sp. and found differences in oxygen consumption compared to nematodes and rotifers.

3.4. Phylogeny

A number of different phylogenetic associations have been discussed in the early literature, but only three hypotheses should be discussed here: the relationship between Nematomorpha and Nematoda, the relationship between Nematomorpha and Scalidophora and the relationship between Nematomorpha and Loricifera.

Many characters are shared by the taxa Nematoda and Nematomorpha: the long and thin form of the body, the presence of longitudinal dorsal and ventral epidermal cords including unpaired nerve cords, the absence of ring musculature, the absence of protonephridia and the cloaca in males. Both taxa have a number of individual autapomorphies, making it unlikely that Nematomorpha is a subtaxon of Nematoda, therefore both taxa should be considered as sister taxa, being united in a taxon Nematoida (Schmidt-Rhaesa 1998b). On first view, nematomorphs can be confused with some large parasitic nematode species, especially with representatives of the taxon Mermithida. Mermithids also abundantly use arthropods as hosts. In the past, gordiids and mermithids were often summarized as “Fadenwürmer”. Just as an example Meissner (1856) believed there to be strong parallels in the body organization of gordiids and mermithids, for example by interpreting the intestinal system as being quite similar. Grenacher (1868) was one of the first to state that both groups are so different that they should not be treated together. In practice, juvenile gordiids in their parasitic phase and mermithids can be hard to tell apart, but adult worms are quite easy to distinguish. Where mermithids are often pale in color or even transparent, adult nematomorphs are usually darker. Mermithids have pointed body ends, whereas gordiids have more rounded ends. The bilobed male posterior end is, when present, unique for gordiids. In mature mermithid females, comparably large eggs may be visible, whereas the numerous tiny eggs in gordiids cannot be recognized.

Whereas the relationships between nematodes and nematomorphs are mainly based on the resemblance of adults, the nematomorph larva has no equivalent in nematodes, but reminded early authors of kinorhynchs and priapulids. This resemblance is based on the presence of an eversible anterior region, which includes spine-like structures. Malakhov (1980) formalized this relationship as assuming a monophyletic taxon Cephalorhyncha including Nematomorpha, Kinorhyncha and Priapulida (and later also Loricifera; see also Malakhov & Adrianov 1995). Schmidt-Rhaesa (1998b) discussed this relationship and pointed out some differences. The eversible region in scalidophoran taxa (Kinorhyncha, Priapulida, Loricifera) is pentaradially organized, in the nematomorph larvae it is hexaradial. The spines are solid in nematomorphs, but in scalidophorans they include epidermal and sensory cells, which communicate to the exterior through apical pores on the spines. Additionally, different retractor muscles are present. Such differences and the number of potential synapomorphies between nematomorphs and nematodes make a relationship between Nematomorpha and Scalidophora unlikely.

DNA sequence comparison has suggested that Nematomorpha may be closely related to Loricifera (Sørensen et al. 2008). From a morphological perspective, there is no support for this relationship, which requires quite complicated evolutionary scenarios.

3.5. Systematics

The systematics of Nematomorpha is based mainly on morphological characters and to a lesser extent on molecular analyses. Historically, Vejdovsky's introduction of the name “Nematomorpha” in 1886 was a starting point, because he finally abandoned concepts to unite mermithid nematodes and nematomorphs, as had for example been done by Von Siebold (1843b) under the name “Gordiacea”.

In early publications, authors often used taxon names modified by their native language, such as the Italian forms “Gordiidi” and “Nectonemidi” (Camerano 1897a) or the French form “Gordiens” (Villot 1881). Vejdovsky (1886, 1888, 1894) and also Rauther (1905) used the German version “Gordiiden”. Later, the taxon names “Gordioidea” and “Nectomematoidea” (e.g. Heinze 1941, Hyman 1951, Brusca & Brusca 2003) or “Gordioida” and “Nectomematoida” (e.g. Ruppert et al. 2004) were widely used. In reference to Vejdovsky´s “Gordiiden” the name “Gordiida” was used in many recent publications and is also used here. For the marine forms the genus name *Nectonema* is sufficient, all further names are redundant.

Today, we classify Nematomorpha in 19 genera. Three further genera had been introduced, but species belonging to these genera were transferred to other genera: *Chordodiolus echinatus* (Heinze 1935b) was transferred to *Beatogordius* (Schmidt-Rhaesa 2001b, Schmidt-Rhaesa & De Villalobos 2002), *Pantachordodes europaeus* (Heinze 1952, 1954) and *Dacochordodes bacescui* (Căpuşe

1965) were transferred to *Spinochordodes* (Zanca & Schmidt-Rhaesa 2006). Some genera can be regarded as monophyletic taxa on the basis of morphological characters that are likely to be autapomorphies (see Schmidt-Rhaesa 2002a), but this is not the case for all genera. For some genera recent reviews are available, while in other genera a review is still necessary.

There are few morphological characters on which species determination is based; these are the general shape and cuticular structures in the posterior end, particularly from males, and cuticular structures from the entire cuticle. A phylogenetic analysis of the 18S rDNA gene included 15 species from nine genera (Bleidorn et al. 2002). It showed a clear distinction between *Nectonema* and Gordiida. Within Gordiida, only from the genera *Chordodes*, *Gordius*, *Paragordius* and *Paragordionus* was more than one species analyzed; according to these data *Chordodes* was not found to be monophyletic, but the other three genera were monophyletic. There is ann indication for a basal split between *Gordius* and the remaining gordiids, reflecting the classical division into the families Gordiidae and Chordodidae (introduced by May 1919). Within Chordodidae, monophyly of a taxon including those gordiids with an undivided male posterior end, the subfamily Chordodinae (after Heinze 1935c), is also supported.

For morphological analyses, the potential variation of characters is a substantial problem. Whenever larger numbers of specimens have been investigated, it was obvious that the cuticular characters can vary within a certain range (see e.g. May 1919, Čanadjija 1956, Degrange & Martinot 1995 for *Gordius valnoxius*, De Villalobos & Zanca 2001 for *Chordodes festae*, Schmidt-Rhaesa 2010 for examples in the genus *Gordius*). An investigation of *Gordionus* specimens from the United Kingdom showed a broad range of cuticle variation, spanning the descriptions of *Gordionus violaceus* (Baird, 1853) to *G. wolterstorffii* (Camerano 1888) (Schmidt-Rhaesa 2001c). It was suspected that these specimens represent one highly variable species, which would mean that several other *Gordionus* species fall within this range of cuticle variation and have to be included into *G. violaceus*. This strong variation within *Gordionus* still represents an unresolved problem. Recently, molecular approaches ("barcoding") have been initiated; such methods may provide a valuable aid to resolving species boundaries within morphologically variable taxa.

For the determination of gordiid specimens, some standards have recently been established. For many years researchers used to cut more or less random cuticle samples from a worm and investigate these under the microscope in different media (water, alcohol, glycerol, dried), which led to differing images of the cuticle. Investigation and documentation with Scanning Electron Microscopy (SEM) overcomes these problems and documents structures more reliably (see e.g. Schmidt-Rhaesa 2001d). In species with a strongly structured cuticle, in particular in the genus *Chordodes*, structures may vary regionally. They may differ between anterior, middle and posterior body regions, but also in dorsal, ventral and lateral regions. Therefore, Schmidt-Rhaesa (2002b) proposed to investigate entire pieces of the worm from the midbody region to observe potential variation along the circumference.

In the following, a brief statement on the characteristics of each genus, an evaluation, whether the taxon is monophyletic and a species list with geographical data are given for each genus. At present, a total of 356 species is regarded as valid; a further 79 species are uncertain.

Gordius Linné, 1758

The genus *Gordius* was the first genus name to be introduced for horsehair worms, long before the name Nematomorpha was available. At first, this name was used rather broadly for long and thin free-living or parasitic worms. Several of the names introduced (in particular from Diesing 1851, 1861) are not supported by a description or figures, but are only linked to a particular host; these names have to be regarded as invalid. A recent review by Schmidt-Rhaesa (2010) showed that the descriptions of a further 40 species are not sufficient enough so these names must also be regarded as invalid.

The diagnostic character for the genus *Gordius* is the postcloacal crescent, a semicircular or parabolic cuticular fold posterior of the cloacal opening in male specimens (Fig. 3.5.1 B, D). This was explicitly stated for the first time by Camerano (1897a); before him, specimens with or without a postcloacal crescent had been assigned to the genus. The postcloacal crescent is not alone a characteristic for *Gordius*; it must be combined with rounded posterior tips of the tail lobes. This is because the genus *Acutogordius*, which was introduced later (Heinze 1952) also has a postcloacal crescent, but pointed tail lobe tips (see below). For diagnosis both genera are easily distinguishable, but in a phylogenetic sense, the position of *Acutogordius* relative to *Gordius* is not clear, making *Gordius* a potential paraphyletic taxon. *Acutogordius* could be the sister taxon of *Gordius* as well as a sister group to only a part of *Gordius*. Both taxa together

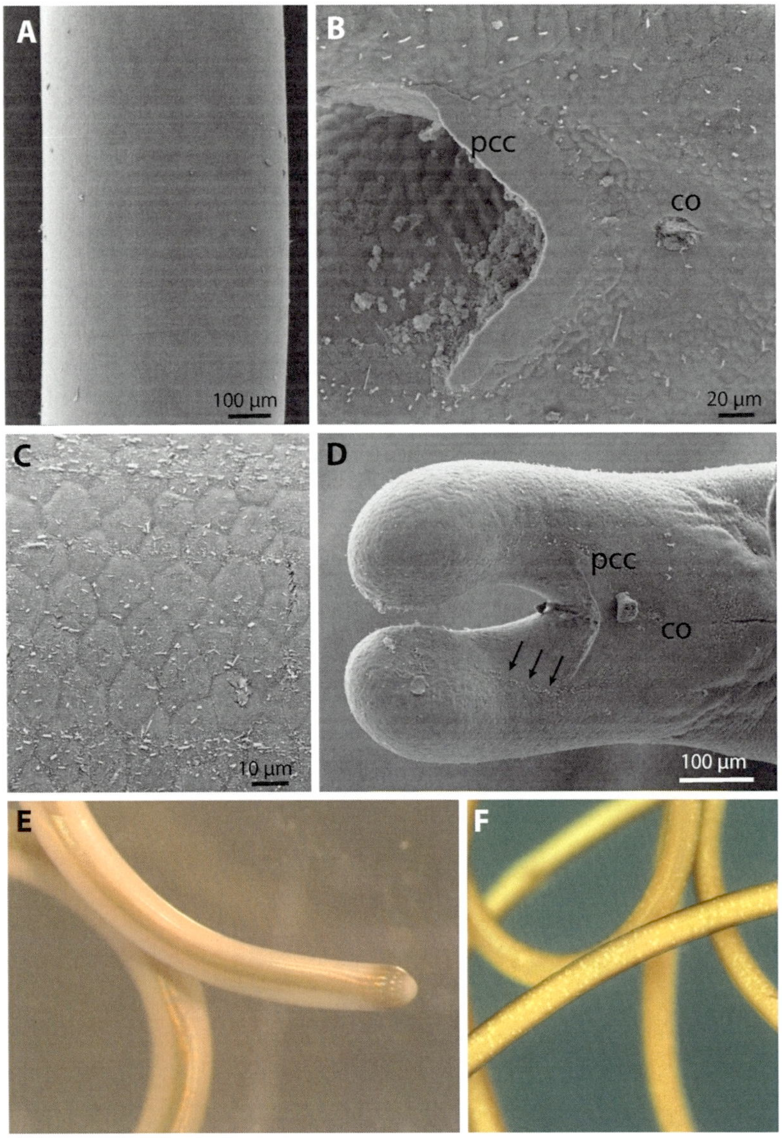

Fig. 3.5.1. Genus *Gordius*. A, Smooth cuticle of *Gordius aquaticus*; B, Postcloacal crescent (pcc) posterior of cloacal opening (co) in *Gordius* sp; C, Polygonal areoles on the cuticle of *G. spiridonovi* (from Estonia); D, Posterior end of *G. spiridonovi* (from Estonia) showing row of bristles on the tail lobes (arrows); E, Anterior end of *Gordius aquaticus* showing white tip, darkly pigmented ring in the head region and dark ventral line; F, Male of *Gordius albopunctatus* with white spots on the cuticle. A–D with SEM. B, D from Schmidt-Rhaesa & Prous 2010, with kind permission from the Estonian Journal of Ecology.

(*Gordius* + *Acutogordius*) are undoubtedly monophyletic with the postcloacal crescent as an autapomorphy (Schmidt-Rhaesa 2002a).

In the genus *Gordius*, the anterior end is not distinctly tapering and a white tip followed by a brown collar are usually present in both sexes (Fig. 3.5.1 E). Additionally, darker ventral and dorsal lines are also often present (Fig. 3.5.1 E). Color, length and diameter can be quite variable. White spots occur on the cuticle of males in several species (Fig. 3.5.1 F). The posterior end of the females is rounded, with a terminal cloacal opening; the male posterior end is bilobed (Fig. 3.5.1 D). Cuticular structures other than the postcloacal crescent are rare, except for fine cuticular bristles (see Schmidt-Rhaesa 2010; Fig. 3.5.1 B, D). The cuticle may be completely smooth (Fig. 3.5.1 A), contain minute bristles or areoles in the form of flat polygonal structures (Fig. 3.5.1 C) (see Schmidt-Rhaesa 2010 for further details).

At present, 89 species are recognized as valid species (Tab. 17). They are summarized in Schmidt-Rhaesa (2010), with further species descriptions in Schmidt-Rhaesa & Prous (2010) and new records in Schmidt-Rhaesa & Urabe (2009). A number of further species names (42) are invalid.

Tab. 17. List of described species in the genus *Gordius*. For references see Schmidt-Rhaesa (2010) and Schmidt-Rhaesa & Prous (2010).

Species	Ecozone	Distribution
Gordius albopunctatus Müller, 1926	Palearctic	Belgium, Denmark, Germany, Poland, Russia
Gordius alpinus Spiridonov, 1998	Palearctic	Italy
Gordius amoyensis Tang, 1934	Palearctic	China
Gordius angulatus Linstow, 1906	Palearctic	Madeira
Gordius aquaticus Linné, 1758	Palearctic	entire Europe, other records must be confirmed
Gordius attoni Redlich, 1980	Nearctic	USA, Argentina
Gordius austrinus De Villalobos, Zanca & Ibarra-Vidal, 2005	Neotropic	Chile
Gordius balcanicus Schmidt-Rhaesa, 2010	Palearctic	Serbia
Gordius balticus Schmidt-Rhaesa, 2010	Palearctic	Estonia
Gordius bivittatus Kirjanova, 1950	Palearctic	Tajikistan
Gordius borisphaenicus Spiridonov, 1984	Palearctic	Moldavia, Ukraine
Gordius cavernarum Inoue, 1972	Palearctic	Japan
Gordius cognettii Camerano, 1904	Indo-Malaya	Myanmar
Gordius corrugatus Camerano, 1895	Indo-Malaya	Indonesia (Sumatra)
Gordius dectici Heinze, 1937	Palearctic	Germany, Romania
Gordius deltaensis de Miralles & de Villalobos, 1996	Neotropic	Argentina
Gordius difficilis Smith, 1994	Nearctic, Neotropic	USA, Argentina
Gordius digitatus Schmidt-Rhaesa, 2010	Palearctic	Croatia
Gordius dimorphus Poinar, 1991	Oceania	New Zealand
Gordius elongiporus Schmidt-Rhaesa, 2000	Palearctic	Austria
Gordius emarginatus Villot, 1885	Palearctic	France
Gordius flavus Linstow, 1906	Australasia	Papua New Guinea
Gordius foochowensis Tang, 1934	Palearctic	China
Gordius fulgur Baird, 1861	Indo-Malaya, Palearctic	Indonesia, Japan, Myanmar, Nepal
Gordius georgensis Kirjanova, 1955	Palearctic	Georgia
Gordius germanicus Heinze, 1937	Palearctic	Germany, Poland, Romania
Gordius gesneri Heinze, 1937	Palearctic	Croatia, Germany
Gordius guatemalensis Linstow, 1902	Neotropic	Guatemala
Gordius heinzei Schuurmans Stekhoven, 1943	Palearctic	Belgium
Gordius helveticus Schmidt-Rhaesa, 2010	Palearctic	Swizerland
Gordius hispidus Linstow, 1906	Palearctic	Austria
Gordius horsti Camerano, 1895	Indo-Malaya	Indonesia (Borneo)
Gordius impressus Schneider, 1866	Palearctic	France, Germany
Gordius isohypsatus Čanadjija, 1956	Palearctic	Serbia
Gordius interjectus Heinze, 1940	Palearctic	Austria, Croatia
Gordius japonicus Inoue & Fukui, 1953	Palearctic	Japan
Gordius jarulini Zakarijev, 1967	Palearctic	Russia
Gordius karwendeli Schmidt-Rhaesa, 2010	Palearctic	Germany
Gordius kimmeriensis Spiridonov, 1998	Palearctic	Ukraine
Gordius labidus de Miralles & de Villalobos, 1996	Neotropic	Argentina
Gordius lapponicus Linstow, 1906	Palearctic	"Lapland"
Gordius lineatus Baek & Noh, 1992	Palearctic	South Korea

(*Continued*)

Tab. 17. (*Continued*)

Species	Ecozone	Distribution
Gordius locustae Havlik, 1954		unknown
Gordius longiformis Heinze, 1934	Oceania	Fiji
Gordius longissimus Römer, 1895		? ("South Sea")
Gordius lumpei Müller, 1926	Palearctic	Czech Republik
Gordius luteopunctatus Inoue, 1979	Palearctic	Japan
Gordius meruanus Camerano, 1910	Afrotropic	Tanzania, Democratic Republic of the Congo
Gordius mongolicus Spiridonov, 1998	Palearctic	Mongolia
Gordius mosaicus Čanadjija, 1956	Palearctic	Croatia
Gordius mülleri Heinze, 1933	Palearctic	Austria, Belgium, Germany
Gordius nonmaculatus Heinze, 1937	Palearctic	Austria, Germany, Macedonia, Romania
Gordius ogatai Inoue, 1979	Palearctic	Japan
Gordius omensis Wu & Tang, 1933	Palearctic	China
Gordius panighettensis Sciacchitano, 1955	Palearctic	Italy
Gordius paranensis Camerano, 1892	Neotropic, Oceania	Brazil, Chile, Paraguay, New Zealand
Gordius paronae Camerano, 1903	Indo-Malaya	India, Malaysia
Gordius parvaquaticus Heinze, 1952	Palearctic	Austria
Gordius pavlovskii Kirjanova, 1955	Palearctic	Russia
Gordius pesici Schmidt-Rhaesa, 2010	Palearctic	Montenegro
Gordius pioltii Camerano, 1887	Palearctic	Belgium, Italy, Kirgistan, Russia, Turkey
Gordius plicatissimus Heinze, 1952	Palearctic	Austria, Italy (as cf. *plicatissimus*)
Gordius plicatulus Heinze, 1937	Palearctic	Croatia, Germany, Spain
Gordius polychaetus Tang, 1934	Palearctic	China, Japan
Gordius robustus Leidy, 1851	Nearctic, Neotropic, Palearctic	USA, Canada, Argentina, Brazil, Chile, Mexico, Paraguay, South Korea, Trinidad, Uruguay, Venezuela
Gordius rosae Camerano, 1887	Palearctic	Italy
Gordius salvadorii Camerano, 1895	Indo-Malaya	Indonesia
Gordius samoensis Linstow, 1906	Oceania	Samoa
Gordius serratus Schmidt-Rhaesa, 2010	Australasia	Papua New Guinea
Gordius setiger Schneider, 1866	Palearctic	Denmark, Germany, Latvia, Netherlands
Gordius sinareolatus Havlik, 1949	Palearctic	Romania, Slovakia
Gordius solaris Kirjanova, 1950	Palearctic	Tadjikistan
Gordius spinosus De Miralles & De Villalobos, 1996	Neotropic	Argentina
Gordius spiridonovi Schmidt-Rhaesa, 2010	Palearctic	Estonia, Russia
Gordius stellatus Linstow, 1906	Palearctic	Germany, Italy
Gordius tatrensis Janda, 1893	Palearctic	"Tatra mountains", Croatia, Ukraine (as cf. *tatrensis*)
Gordius terminosetosus Schmidt-Rhaesa, 2010	Palearctic	Estonia
Gordius testaceus Kirjanova, 1950	Palearctic	Uzbekistan
Gordius tirolensis Heinze, 1937	Palearctic	Austria
Gordius toschii Sciacchitano, 1950	Palearctic	Libya
Gordius tuberculatus Čanadjija, 1956	Palearctic	Croatia
Gordius turcomanicus Spiridonov, 1998	Palearctic	Turkmenistan
Gordius undulatus Heinze, 1937	Palearctic	Croatia, Germany
Gordius valnoxius Degrange & Martinot, 1995	Palearctic	France

(*Continued*)

Tab. 17. (*Continued*)

Species	Ecozone	Distribution
Gordius villoti Rosa, 1882	Palearctic, Indo-Malaya	Bulgaria, England, Finland, France, Italy, Georgia, Germany, Ireland, Kirgistan, Russia, Uzbekistan, China (Tibet), Myanmar
Gordius willeyi Camerano, 1899	Australasia	Papua New Guinea
Gordius zammaranoi Sciacchitano, 1932	Afrotropic	Africa, Democratic Republic of the Congo
Gordius zavattarii Camerano, 1908	Indo-Malaya	India
Gordius zwicki Schmidt-Rhaesa, 2010	Palearctic	Russia

Invalid species names (compare Schmidt-Rhaesa 2010):

- *Gordius acridiorum* Weyenbergh, 1880 (Argentina)
- *Gordius aeneus* Villot, 1874 (Mexico, South Africa, Venezuela, "Viti-Islands (= Fiji?))
- *Gordius aestivalis* Vejdovsky, 1894 (Czech Republic)
- *Gordius affinis* Villot, 1885 (Italy)
- *Gordius agassizi* Montgomery, 1898 (USA)
- *Gordius alascensis* Montgomery, 1907 (USA)
- *Gordius ambiguus* De Beauchamp, 1913 (Kenya)
- *Gordius annulatus* Linstow, 1906 (Russia)
- *Gordius bonmartensis* Poisson, 1927 (Madagascar)
- *Gordius bouvieri* Sciacchitano, 1937 (Democratic Republic of the Congo) (is a mermithid nematode, see Sciacchitano 1958)
- *Gordius californicus* Camerano, 1915 (USA)
- *Gordius chilensis* Blanchard, 1849 (Chile)
- *Gordius chordodes* Diesing, 1851 (Brazil)
- *Gordius crispatus* Sciacchitano, 1933 (Democratic Republic of the Congo)
- *Gordius danielis* Camerano, 1894 (Argentina)
- *Gordius fasciatus* Baird, 1853 (USA)
- *Gordius gialensis* Sciacchitano, 1932 (Libya)
- *Gordius gracilis* Villot, 1874 (is a mermithid nematode)
- *Gordius granulosus* Linstow, 1897 (Madagascar)
- *Gordius hawaiiensis*, Heinze, 1933 (USA)
- *Gordius lacutris* Leidy, 1871 (USA)
- *Gordius laevis* Villot, 1874 (is a mermithid nematode)
- *Gordius leidyi* Montgomery, 1898 (no collection data)
- *Gordius montenegrinus* Švábenik, 1909 (Montenegro)
- *Gordius pallidus* Linstow, 1906 (Romania)
- *Gordius palustris* Linstow, 1883 (Turkestan)
- *Gordius parisii* Sciacchitano, 1932 (Italy)
- *Gordius platyura* Baird, 1853 (Jamaica)
- *Gordius reddyi* Singh & Rao, 1966 (India)
- *Gordius rhomboidalis* Sciacchitano, 1933 (Democratic Republic of the Congo)
- *Gordius semilunaris* Camerano, 1915 ("Korea")
- *Gordius spaerura* Baird, 1853 (Burma, India)
- *Gordius subareolatus* Villot, 1874 (France)
- *Gordius subbifurcus* Siebold, 1848 (Germany)
- *Gordius subspiralis* Diesing, 1861 (USA)
- *Gordius sumatrensis* Villot, 1892 (Indonesia)
- *Gordius tenuis* Leidy, 1878 (probably a mermithid nematode)
- *Gordius tuberculatus* Villot, 1874 (Australia)
- *Gordius vaeteri* Vejdovsky, 1894 (Czech Republic)
- *Gordius valgannensis* Sciacchitano, 1932 (Italy)
- *Gordius verrucosus* Baird, 1853 (probably South Africa)

Gordius longareolatus Schuurmans Stekhoven, 1943 does not have a postcloacal crescent and probably belongs to the genus *Gordionus* (but not synonymous to *Gordionus longareolatus* Montgomery, 1898 (see Montgomery 1898c).

Chordodes Creplin, 1847

After recognizing the raising diversity of nematomorphs as product result of broader sampling worldwide, more genera had to be introduced. Creplin (1847) introduced the new genus name *Chordodes*, although it is not sure whether his specimen, *C. parasitus*, shows the diagnostic characters of *Chordodes* (Janda 1894a, Schmidt-Rhaesa et al. 2008). Janda (1894a) was the first to characterize the genus and to name the presence of elevated papillae on the cuticle as characteristic. The cuticle of *Chordodes*-species is very rich in different types of areoles and particularly one type, the crowned areoles (Fig. 3.5.2 A–C), can be regarded as a diagnostic structure for the genus. Crowned areoles consist of an elevated stem and a crown of apical filaments. These filaments are of different length; in some species they branch (e.g. *C. mizoramensis*; see Schmidt-Rhaesa & Lalramliana 2011). For a

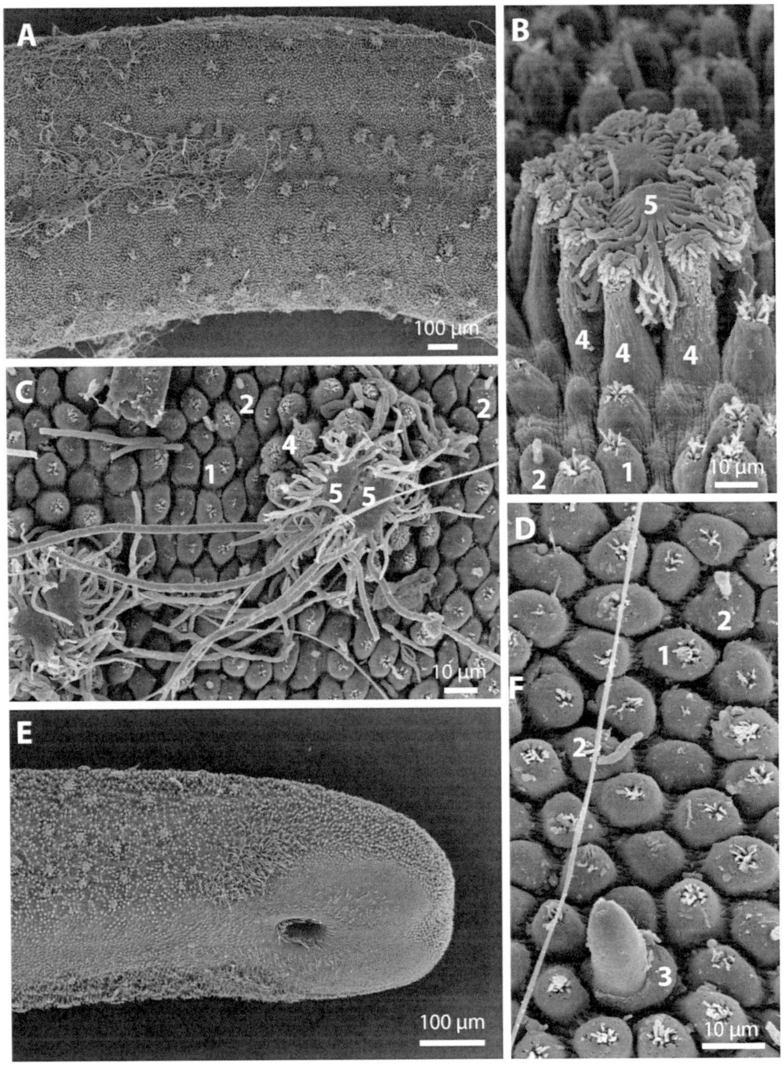

Fig. 3.5.2. Genus *Chordodes*. A, Ventral side of *Chordodes parabipilus* (from Georgia) showing clusters with crowned areoles as higher elevations and crowned areoles with very long apical filaments along the ventral midline; B, Clusters of paired crowned areoles (5) and circumcluster areoles (4) in *C. mizoramensis* (from India). Additionally, simple areoles (1) and tubercle areoles (2) are present; C, D, Areolar types labeled in B and additionally thorn areole (3 in D) in *C. parabipilus*; E, Ventral view on male posterior end in *C. mizoramensis* showing cloacal opening and distribution of areoles and bristles. All images by SEM. A, D from Kintsurashvili et al. 2011, with kind permission of Naturwissenschaftlicher Verein Hamburg. B, E from Schmidt-Rhaesa & Lalramliana 2011, with kind permission of Zookeys.

detailed terminology of areoles see Schmidt-Rhaesa et al. (2008) and Fig. 3.5.2 B–D. The crowned areoles can occur in two different types. If so, areoles along the ventral and sometimes also along the dorsal midline have very long apical filaments (Fig. 3.5.2 A, C). Sexual dimorphism in respect of the cuticular pattern is present probably in most species; this was documented e.g. by De Villalobos et al. (2004) for *C. balzani*.

The anterior end tapers distinctly towards the anterior. The apical tip is white or very light in color, blending into the usual body color soon after, without the presence of a darkly stained ring (compare *Gordius*). The color ranges from light to dark brown; quite often and especially in males, dark patches are present – this has been termed the leopard pattern (Fig. 3.1.1 A, B, E). In females, the posterior end is slightly swollen (Fig. 3.1.3 D). In males, the posterior end is undivided; a bilaterally symmetrical pattern can be indicated by a slight groove posterior of the cloacal opening (Fig. 3.1.3 B, 3.5.2 E). Bristles are abundant in the posterior end; they are present on the entire ventral side, around the cloacal opening (circumcloacal spines) and may be concentrated in two patches anterolateral of the cloacal opening (Fig. 3.1.3 B, 3.5.2 E).

Due to the presence of crowned areoles, *Chordodes* is certainly a monophyletic taxon (Schmidt-Rhaesa 2002a).

Only in a few species, such as *C. fukuii* and *C. morgani*, does the nature of the crowned areoles have to be reconsidered, because here crowned areoles resemble areoles present in species of the genus *Spinochordodes*.

Ninety-six species are currently recognized (Tab. 18), although some of them (35) have to be described in further detail (Schmidt-Rhaesa et al. 2008). Twenty-six further species names are invalid. Most species are distributed in the tropical regions, with a few species extending into more temperate regions. The main host group is Mantodea (see Schmidt-Rhaesa & Ehrmann 2001; Fig. 3.2.5 A).

Invalid species names (compare Schmidt-Rhaesa et al. 2008):

- *Chordodes annandalei* Camerano, 1908 (Nepal, India)
- *Chordodes annulatus* (von Linstow, 1906) (Australia)
- *Chordodes aquaeductus* Kirjanova, 1950 (Tajikistan)

Tab. 18. List of described species in the genus *Chordodes*. For references see Schmidt-Rhaesa et al. (2008).

Species	Ecozone	Distribution
Chordodes aelianus (Camerano, 1894)	Indo-Malaya	Indonesia
Chordodes aethiopicus Inoue, 1974	Afrotropic	Ethiopia
Chordodes africanus Sciacchitano, 1933	Afrotropic	Democratic Republic of the Congo, Angola
Chordodes albibarbatus Montgomery, 1898	Afrotropic	probably Gabon
Chordodes alticuspis De Villalobos, Zanca & Yanez, 2009	Neotropic	Ecuador
Chordodes amboinensis Sciacchitano, 1963	Indo-Malaya	Indonesia
Chordodes anthophorus Kirjanova, 1950	Palearctic	Georgia, Tajikistan, Uzbekistan
Chordodes auranthiacus von Linstow, 1906	Neotropic	Honduras
Chordodes balzani Camerano, 1896	Neotropic	Argentina, Brazil, Trinidad and Tobago, Venezuela
Chordodes baramensis Römer, 1895	Indo-Malaya	Malaysia
Chordodes betularius von Linstow, 1904	Palearctic	"Caspian Sea"
Chordodes bipilus Kirjanova, 1957	Palearctic	Iran
Chordodes boulengerii Camerano, 1912	Afrotropic	Cameroon, Democratic Republic of the Congo
Chordodes bouvieri (Villot, 1885)	Neotropic	Argentina, Colombia, Equador, Venezuela
Chordodes brasiliensis Janda, 1893	Neotropic	Argentina, Brazil, Uruguay
Chordodes brevipilus Schmidt-Rhaesa, 2002	Australasia	Australia
Chordodes bukavuensis Sciacchitano, 1958	Afrotropic	Democratic Republic of the Congo
Chordodes caledoniensis (Villot, 1874)	Australasia, Indo-Malaya	Australia, Indonesia, New Caledonia
Chordodes capenis Camerano, 1895	Afrotropic	Angola, Democratic Republic of the Congo, South Africa, Tanzania
Chordodes capillatus von Linstow, 1901	Afrotropic	Tanzania
Chordodes carmelitanus Carvalho & Feio, 1950	Neotropic	Brazil
Chordodes clavatus von Linstow, 1906	Afrotropic	Angola, Cameroon, Democratic Republic of the Congo
Chordodes compactus Schmidt-Rhaesa & Brune, 2008	Indo-Malaya	Malaysia
Chordodes compressus Römer, 1895	Indo-Malaya	Malaysia
Chordodes congolensis Sciacchitano, 1933	Afrotropic	Democratic Republic of the Congo
Chordodes corderoi Carvalho, 1946	Neotropic	Venezuela
Chordodes cornuta de Villalobos & Camino, 1999	Neotropic	Argentina
Chordodes curvicillatus Kirjanova & Spiridonov, 1989	Indo-Malaya	Indonesia
Chordodes defilippii (Rosa, 1881)	Palearctic	Georgia
Chordodes devius Kirjanova, 1950	Palearctic	Tajikistan

(*Continued*)

Tab. 18. (*Continued*)

Species	Ecozone	Distribution
Chordodes digitatus von Linstow, 1901	Afrotropic	Tanzania
Chordodes ecuadoriensis De Villalobos, Zanca & Yanez, 2009	Neotropic	Ecuador
Chordodes ferox Camerano, 1897	Afrotropic	Democratic Republic of the Congo, Republic of the Congo
Chordodes festae Camerano, 1897	Neotropic, Afrotropic	Equador, Venezuela, Democratic Republic of the Congo
Chordodes formosanus Chiu, Huang, Wu & Shiao, 2011	Palearctic	Taiwan, Japan
Chordodes fukuii Inoue, 1951	Palearctic	Japan
Chordodes furnessi Montgomery, 1898	Indo-Malaya	Indonesia, India (as cf. *furnessi*)
Chordodes gariazzi Camerano, 1902	Afrotropic	Democratic Republic of the Congo
Chordodes gestri Camerano, 1904	Neotropic	Guatemala
Chordodes guineensis Spiridonov, 2001	Afrotropic	Guinea
Chordodes hawkeri Camerano, 1902	Afrotropic	Democratic Republic of the Congo, Kenya, Sierra Leone, South Africa, Sudan
Chordodes heinzei Sciacchitano, 1937	Afrotropic	Democratic Republic of the Congo
Chordodes ikelensis Sciacchitano, 1961	Afrotropic	Democratic Republic of the Congo
Chordodes insidiator Camerano, 1899	Indo-Malaya	Malaysia
Chordodes iturensis Sciacchitano, 1958	Afrotropic	Democratic Republic of the Congo
Chordodes jandae Camerano, 1895	Indo-Malaya, Australasia	Indonesia, Australia
Chordodes janovyi Bolek, Schmidt-Rhaesa, Hanelt & Richardson, 2010	Afrotropic	Cameroon
Chordodes japonensis Inoue, 1951	Palearctic	Japan, South Korea
Chordodes joyeuxi Dorier, 1935	Indo-Malaya	Vietnam
Chordodes kakandensis Sciacchitano, 1958	Afrotropic	Democratic Republic of the Congo
Chordodes kallstenii Jägerskiöld, 1897	Afrotropic	Cameroon
Chordodes kivuensis Sciacchitano, 1958	Afrotropic	Democratic Republic of the Congo
Chordodes kolensis Sciacchitano, 1933	Afrotropic	Angola, Democratic Republic of the Congo, Ivory Coast, Kenia
Chordodes koreensis Baek, 1993	Palearctic	South Korea
Chordodes lefeburei (Sciacchitano, 1937)	Afrotropic	Democratic Republic of the Congo
Chordodes lenti Carvalho, 1944	Neotropic	Brazil
Chordodes ligasiensis Sciacchitano, 1933	Afrotropic	Democratic Republic of the Congo
Chordodes lotus De Miralles & De Villalobos, 1997	Neotropic	Argentina
Chordodes maculatus Sciacchitano, 1958	Afrotropic	Democratic Republic of the Congo
Chordodes madagascariensis (Camerano, 1893)	Afrotropic	Angola, Democratic Republic of the Congo, Guinea, Madagascar
Chordodes matensis De Villalobos & De Miralles, 1997	Neotropic	Argentina
Chordodes mizoramensis Schmidt-Rhaesa & Lalramliana, 2011	Indo-Malaya	India
Chordodes mobensis Sciacchitano, 1958	Afrotropic	Democratic Republic of the Congo
Chordodes modiglianii (Camerano, 1892)	Indo-Malaya, Australasia	Indonesia, West Papua
Chordodes montgomeryi Camerano, 1901	Afrotropic	Madagascar
Chordodes moraisi (Carvalho, 1942)	Neotropic	Brazil

(*Continued*)

Tab. 18. *(Continued)*

Species	Ecozone	Distribution
Chordodes morgani Montgomery, 1898	Nearctic	USA
Chordodes moutoni Camerano, 1895	Palearctic, Indo-Malaya	China, Malaysia
Chordodes muelleri Sciacchitano, 1937	Afrotropic	Ruanda
Chordodes nobilii Camerano, 1901	Neotropic	Argentina
Chordodes ornatus (Grenacher, 1868)	Indo-Malaya	Philippines
Chordodes parabipilus Kintsurashvili, Schmidt-Rhaesa & Gorgadze, 2011	Palearctic	Georgia
Chordodes penicillatus Camerano, 1895		Unknown[1]
Chordodes peraccae (Camerano, 1894)	Neotropic	Argentina, Bolivia, Brazil, Mexico (as *C. griffinii*)
Chordodes pilosus Möbius, 1855	Neotropic	Venezuela
Chordodes pollonerae Camerano, 1912	Indo-Malaya, Afrotropic	India, Somalia
Chordodes polycoronatus Schmidt-Rhaesa & Brune, 2008	Indo-Malaya	Malaysia
Chordodes polytuberculatus Schmidt-Rhaesa & Menzel, 2005	Neotropic	Costa Rica
Chordodes puncticulatus Camerano, 1895	Indo-Malaya	Indonesia, Malaysia
Chordodes queenslandi Schmidt-Rhaesa, 2002	Australasia	Australia
Chordodes rigatus Sciacchitano, 1937	Afrotropic	Democratic Republic of the Congo
Chordodes ruandensis Sciacchitano, 1937	Afrotropic	Ruanda
Chordodes ruginosus De Villalobos, Zanca & Yanez, 2009	Neotropic	Ecuador
Chordodes sajanensis Spiridonov, 2000	Palearctic	Russia
Chordodes sandoensis Sciacchitano, 1937	Afrotropic	Democratic Republic of the Congo
Chordodes schoutedeni Sciacchitano, 1933	Afrotropic	Democratic Republic of the Congo
Chordodes shipleyi Camerano, 1899	Indo-Malaya	Malaysia
Chordodes siamensis Camerano, 1903	Afrotropic, Indo-Malaya	Angola, India, Democratic Republic of the Congo, Birma, Thailand
Chordodes silvestri Camerano, 1895	Indo-Malaya, Palearctic	Indonesia, Japan
Chordodes skorikovi Camerano, 1903	Indo-Malaya	Sri Lanka, Thailand
Chordodes staviarskii Carvalho & Feio, 1950	Neotropic	Brazil
Chordodes tenoderae Kirjanova, 1957	Palearctic	China
Chordodes timorensis Camerano, 1895	Indo-Malaya	East Timor
Chordodes tuberculatus von Linstow, 1901	Afrotropic	Tanzania
Chordodes villalobi Schmidt-Rhaesa & Brune, 2008	Indo-Malaya	Malaysia
Chordodes wangi Wu & Tang, 1933	Palearctic	China

[1] The type specimen was labeled to be collected in "Cantraine: Italie", which was assumed by Camerano (1895) to be a wrong label.

Tab. 19. List of described species in the genus *Nectonema*.

Species	Distribution	Reference
Nectonema agile Verrill, 1879	Woods Hole, Newport (east coast USA), Naples (Italy, Mediterranean Sea), Roscoff and Dinard (France, North Atlantic, Channel Region), Black Sea	Verrill 1879, Fewkes 1883, Pintner 1899, Rauther 1914, Perez 1927a, b, 1934, Feyel 1936, Nouvel & Nouvel 1934, 1938, Gallien 1949, Arvy 1963, Born 1967, Kirjanova & Zakutsky 1967, Mouchet 1931
Nectonema melanocephalum Nierstrasz, 1907	Balabalagan-Islands (as Paternoster-Islands), Indonesia	Nierstrasz 1907
Nectonema munidae Brinkmann, 1930	Fjords off Bergen (Norway, North Sea)	Brinckmann 1930, Huus 1932, Nielsen 1969, Greve 1972, Schmidt-Rhaesa 1996a
Nectonema svensksundi Bock, 1913	Svalbard (North Atlantic)	Bock 1913
Nectonema zealandica Poinar & Brockerhoff, 2001	New Zealand	Poinar & Brockerhoff 2001

- *Chordodes butensis* Sciacchitano, 1937 (Democratic Republic of the Congo)
- *Chordodes cameranonis* Montgomery, 1900 (West coast of Central America)
- *Chordodes carioca* Carvalho & Feio, 1950 (Brazil)
- *Chordodes carvalhoi* de Miralles & Camino, 1983 (Argentina)
- *Chordodes cubanensis* Montgomery, 1898 (Cuba)
- *Chordodes delmae* De Villalobos, 1995 (Argentina)
- *Chordodes ferganensis* Kirjanova & Spiridonov, 1989 (Uzbekistan)
- *Chordodes gigliotosi*[1] Camerano, 1908 (India)
- *Chordodes hamatus* Römer, 1895 (Gaboon)
- *Chordodes ibembensis* Sciacchitano, 1958 (Angola, Democratic Republic of the Congo)
- *Chordodes lichyi* Scorza, 1952 (Venezuela)
- *Chordodes liguligerus* Römer, 1895 (India)
- *Chordodes longipilus* Kirjanova, 1949 (Turkmenistan)
- *Chordodes molukkanus* Römer, 1895 (Indonesia)
- *Chordodes oscillatus* Kirjanova, 1953 (Georgia)
- *Chordodes parasitus* Creplin, 1847 (Brazil)
- *Chordodes pardalis* (Camerano, 1893) (Madagascar)
- *Chordodes puerilis* Montgomery, 1898 (USA) (see Montgomery 1901)
- *Chordodes somaliensis* Sciacchitano, 1961 (Somalia)
- *Chordodes uncinatus* Sciacchitano, 1958 (Democratic Republic of the Congo)
- *Chordodes undulatus* von Linstow, 1906 (Australia)
- *Chordodes variopapillatus* Römer, 1895 ("Bahia")
- *Chordodes weberi* (Villot, 1892) (Democratic Republic of the Congo, Indonesia)

[1] Described as *Chordodes giglio-tosi* by Camerano (1908b); dashes are not allowed in species names according to article 27 of the international rules for zoological nomenclature.

Nectonema Verrill, 1879

Nectonema is the only marine genus of Nematomorpha and it shows several morphological differences to the freshwater/terrestrial forms (see sections on morphology). The first species, *N. agile*, was described by Verrill (1879), following a brief notice by Verrill (1873). Only five species have been described (Tab. 19), but there are further records not determined to species level. Only two species, *N. agile* and *N. munidae* have been collected on repeated occasions; all other species were single collections of few specimens.

Unconfirmed reports come from West-Greenland (Nouvel & Nouvel 1938), Northern Norway (Bakke 1975), Southeastern Canada (Leslie et al. 1981, Brattey et al. 1985, Skaling & MacKinnon 1988) and Japan (Oku et al. 1983). Reports from Brazil (Müller 1871) and Sierra Leone (Bate 1888) might also represent *Nectonema* specimens.

Paragordius Camerano, 1897

Paragordius is a conspicuous genus with easily recognizable males and females. Males have considerably long tail lobes (Fig. 3.5.3 B) and females have three lobes in their posterior end (Fig. 3.5.3 A). The cloacal opening is central between the three lobes (Fig. 3.1.19 I). The cuticle contains structures that may represent areoles, but the pattern is not as clear as in other genera (Fig. 3.5.3 D). Curiously, Camerano (1897a) and Montgomery (1898a) suggested the same genus name in parallel (see Montgomery 1898a, appendix). As Camerano´s publication occurred a bit earlier, the genus name includes his name.

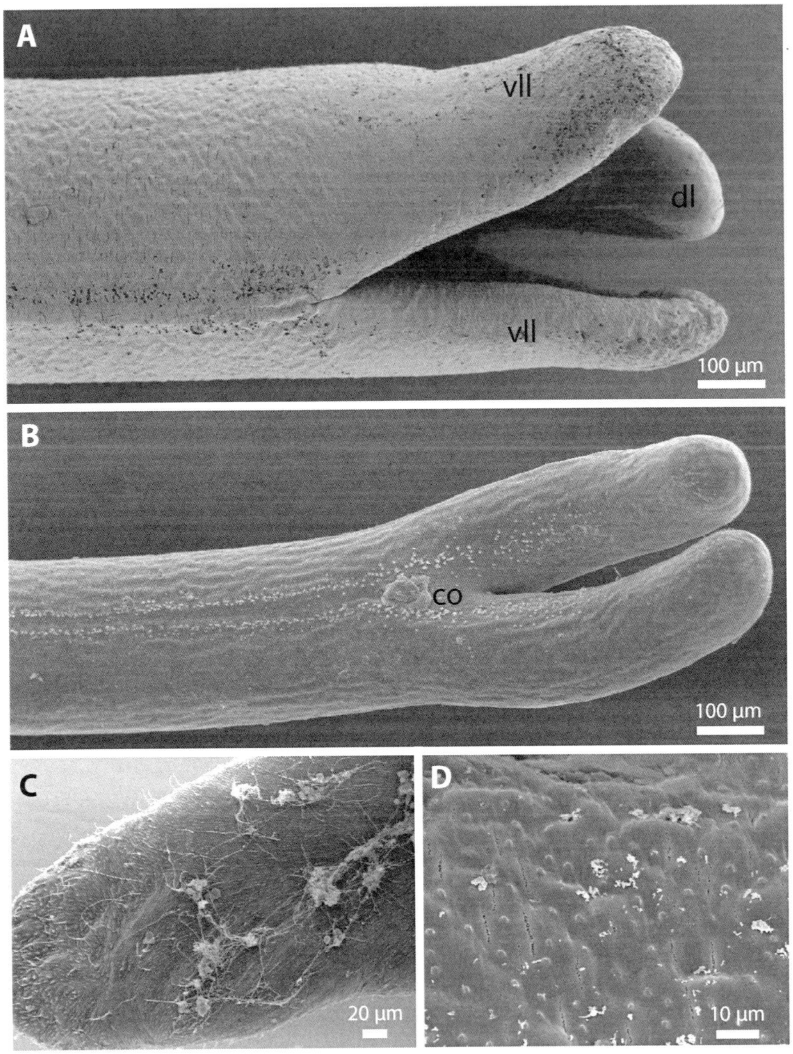

Fig. 3.5.3. Genus *Paragordius*. A, Female posterior end with three tail lobes: one dorsal (dl) and two ventrolateral (vll); B, Ventral view on male posterior end of *P. tricuspidatus* (from France) with cloacal opening (co) and spines; C, Inner side of tail lobe in *P. obamai* (from Kenya) with ridges and bristles (some dirt is attached to bristles); D, Cuticular pattern in *P. obamai*. All images with SEM.

The genus is likely to be monophyletic, but the status of the genera *Digordius*, *Progordius* and *Pseudogordius* in relation to *Paragordius* must be cleared. The males of *Digordius* and *Progordius* also have very long tail lobes (see description of these genera) and the female of *Pseudogordius* has three tail lobes with an additional fourth one (see description of this genus). All three genera have been described from very few specimens and are difficult to reinvestigate.

It is unknown why females differ, with their three-lobed posterior end, so much from the remaining gordiids. A similar egg string is produced (see Fig. 3.2.1 B). The lobes may be helpful in keeping the egg string in shape and to prevent the male sperm drop from being sloughed off. It is likely that the ridges and bristles present on the inner side of the tail lobes (Fig. 3.5.3 C; only very few species have been checked according to this character) play a role, too.

Today, 18 species are described, most of them from the Neotropic and the Afrotropic (Tab. 20). A review has not been carried out.

Parachordodes Camerano, 1897

When introduced, *Parachordodes* was a taxon for all nematomorphs that could not be assigned to the genera *Gordius*, *Chordodes* and *Paragordius*, the only other freshwater taxa available at that time (Camerano 1897a). Later, Heinze (1937) recognized very large areoles as diagnostic for *Parachordodes*. This type of areole was later named superareole (Schmidt-Rhaesa 2002a; Fig. 3.5.4 A–D).

Tab. 20. List of described species in the genus *Paragordius*.

Species	Ecozone	Distribution	References
Paragordius andreasii Zanca & De Villalobos, 2006	Neotropic	Argentina	Zanca & De Villalobos 2006
Paragordius areolatus Linstow, 1906	Afrotropic	probably South Africa (as "Botschabelo, South-East Africa")	Von Linstow 1906a
Paragordius cinctus Linstow, 1906	Afrotropic	Democratic Republic of the Congo, Somalia, South Africa, Tanzania	De Beauchamp 1923, Camerano 1915a, Sciacchitano 1932, 1940, 1958, Von Linstow 1906a
Paragordius dartevellei Sciacchitano, 1958	Afrotropic	Democratic Republic of the Congo	Sciacchitano 1958
Paragordius diversolobatus Heinze, 1935	Neotropic	Costa Rica	Heinze 1935b
Paragordius emeryi (Camerano, 1895)	Indo-Malaya	"Greater Sunda Islands"	Camerano 1895, 1897a
Paragordius esavianus Carvalho, 1942	Neotropic	Argentina, Brazil	Carvalho 1942, De Villalobos et al. 2000, Zanca & De Villalobos 2006
Paragordius flavescens Linstow, 1906	Neotropic, Afrotropic	Argentina, Brazil, Democratic Republic of the Congo	Heinze 1935b, Carvalho 1944, Sciacchitano 1958, Von Linstow 1906a
Paragordius laurae Sciacchitano, 1958	Afrotropic	Angola, Democratic Republic of the Congo	Sciacchitano 1958, 1961a
Paragordius marlieri Sciacchitano, 1958	Afrotropic	Democratic Republic of the Congo	Sciacchitano 1958
Paragordius minusculus Carvalho, 1944	Neotropic	Brazil	Carvalho 1944, Zanca & De Villalobos 2006
Paragordius mulungensis Sciacchitano, 1958	Afrotropic	Democratic Republic of the Congo	Sciacchitano 1958
Paragordius obamai Hanelt, Bolek & Schmidt-Rhaesa, 2012	Afrotropic	Kenya	Hanelt et al. 2012
Paragordius somaliensis Sciacchitano, 1963	Afrotropic	Somalia	Sciacchitano 1963
Paragordius stylosus (Linstow, 1883)	Palearctic, Indo-Malaya	Bulgaria, Croatia, Italy, Romania, Slovakia, Turkestan, India	Camerano 1897a, 1912b, Čanadjija 1956, Căpuşe 1966a, Havlik 1954, Heinze 1941, Konsuloff 1921, Sciacchitano 1929, 1932, 1963, Von Linstow 1883b
Paragordius tanganikensis Sciacchitano, 1958	Afrotropic	Democratic Republic of the Congo	Sciacchitano 1958
Paragordius tricuspidatus (Dufour, 1828)	Palearctic	Bulgaria, France, Italy, Romania, San Marino, Swizerland	Angelov 1959, Camerano 1887b,1888b, 1889b, 1897a, Căpuşe 1966a, De Villalobos et al. 2001, Heinze 1937, 1941, Schmidt-Rhaesa et al. 2005
Paragordius varius (Leidy, 1851)	Nearctic	Argentina, Bolivia, Brazil, Canada, Colombia, Costa Rica, Cuba, Mexico, Uruguay, USA, Venezuela	Caballero 1936, Camerano 1893b, 1895, 1897a,b, 1898, 1903c, Carvalho 1946a, Carvalho & Feio 1950, De Miralles 1975, 1988, De Villalobos et al. 2000, 2008a, Schmidt-Rhaesa et al. 2003a, Schmidt-Rhaesa & Menzel 2005, Scorza 1952, Vigueras 1934, Watermolen & Haen 1994, Zanca & De Villalobos 2006

Superareoles are probably the fusion product of several smaller areoles, because remnants of their borders can be preserved on the superareoles (see, e.g. Fig. 3.5.4 D). Superareoles carry a short tubercle in their central region. Superareoles are likely also to be an autapomorphy for a monophyletic taxon *Parachordodes*, similar areoles have only been described in *Pseudogordius tanganyikae* (Yeh & Jordan 1957; see below).

Specimens of *Parachordodes* usually show the typical coloration of a white anterior tip followed by a dark

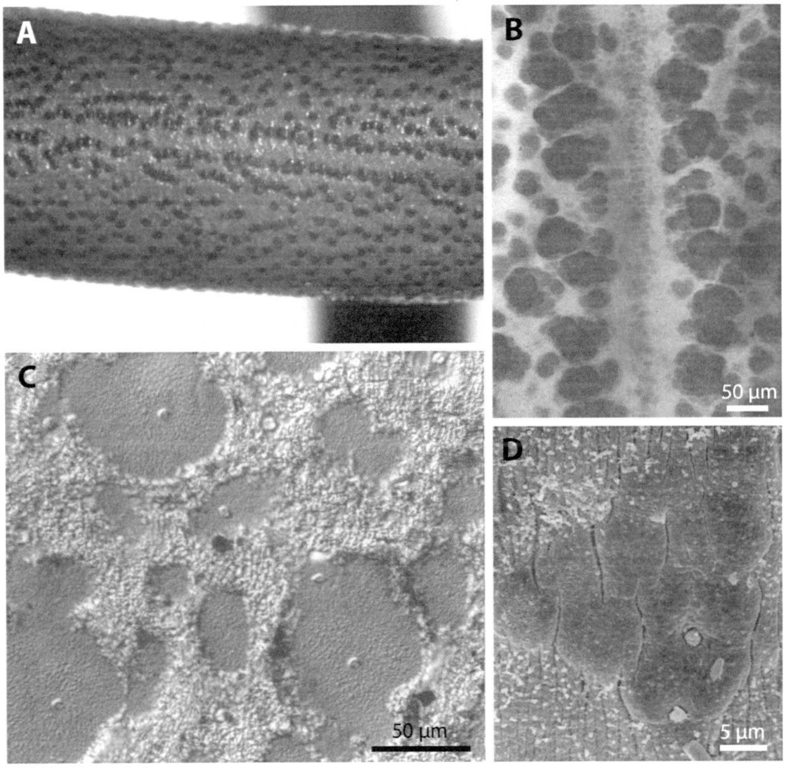

Fig. 3.5.4. Genus *Parachordodes* (all images *P. diblastus* from New Zealand). A, Superareoles are already visible at low magnifications, in *P. diblastus* they are darkly coloured; B, Linear arrangement of superareoles on both sides of the ventral midline (see also in A); C, D, Shape of superareoles. B, C with light microscopy, D with SEM. B–D from Schmidt-Rhaesa 2008, with kind permission from the Zoological Museum Hamburg.

collar. The male posterior end is bilobed, typical cuticular structures are circumcloacal spines, rows of bristles anterior to the cloacal opening and spines posterior to it. In females the posterior end shows, at least sometimes, a kind of longitudinal groove, creating a bilateral appearance (e.g. Heinze 1941 for *P. tolosanus*). The cloacal opening is in these cases slightly subterminal.

To date, 15 species have been described with superareoles (Tab. 21), a further nine species do not have superareoles but are still listed under the name *Parachordodes*. A further two species names are invalid. A review of this genus has not been carried out but appears necessary.

Species without superareoles, but described as members of *Parachordodes*:

- *Parachordodes capitosulcatus* Montgomery, 1898 (Cuba)
- *Parachordodes ciferrii* Sciacchitano, 1932 (Dominican Republic)
- *Parachordodes coreanus* Linstow, 1906 ("Korea")
- *Parachordodes densareolatus* Montgomery, 1898 (USA); see Schmidt-Rhaesa et al. (2003a)
- *Parachordodes lefeburi* Sciacchitano, 1937 (Democratic Republic of the Congo)
- *Parachordodes pleskei* (Camerano, 1896) (China) prob ably synonymous to *Gordionus wolterstorffii*, see Camerano 1904c)
- *Parachordodes roccatii* Camerano, 1912 (India)
- *Parachordodes sjöstedti* Camerano, 1910 (Tanzania)
- *Parachordodes tegonotus* Poinar, Rykken & LaBonte, 2004 (USA)

Invalid species names:

- *Parachordodes aequatorialis* Sciacchitano, 1937 and *Parachordodes bourguignoni* Sciacchitano, 1937 (both Democratic Republic of the Congo), are both mermithid nematodes (Sciacchitano 1958).

Gordionus Müller, 1926

When introducing the genus *Gordionus*, Müller (1926) introduced a new character that he had observed in

Tab. 21. List of described species in the genus *Parachordodes*.

Species	Ecozone	Distribution	References
Parachordodes arndti Heinze, 1935	Palearctic	Tadzhikistan, Turkestan	Heinze 1935c, Spiridonov 1989
Parachordodes diblastus (Örley, 1881)	Oceania	New Zealand	Örley1881, Poinar 1991, Schmidt-Rhaesa 2008
Parachordodes gemmatus (Villot, 1885)	Palearctic	Croatia, France, Germany, Italy, Montenegro, "Tatra Mountains" (Poland or Slovakia)	Camerano 1897a, 1905, 1915a, Čanadjija 1956, Dorier 1928, 1930, Heinze 1935c, 1937, 1941, Schmidt-Rhaesa & Zaenker 2006, Villot 1885, 1886
Parachordodes lestici Heinze, 1935	Palearctic	Japan	Heinze 1935a, Inoue 1955
Parachordodes maculatus (Linstow, 1883)	Palearctic	Turkestan	Camerano 1915a, Heinze 1935c, Von Linstow 1883b, 1905
Parachordodes magnus Spiridonov, 1989	Palearctic	Kazakhstan	Spiridonov 1989
Parachordodes megareolatus Schmidt-Rhaesa, Chung & Sohn, 2003	Palearctic	South Korea	Schmidt-Rhaesa et al. 2003b
Parachordodes modestus Spiridonov, 1989	Palearctic	Kazakhstan	Spiridonov 1989
Parachordodes okadai Inoue, 1955	Palearctic	Japan	Inoue 1955, 1988, Schmidt-Rhaesa et al. 2009b
Parachordodes orientalis Fukui & Inoue, 1940	Palearctic	Japan	Fukui & Inoue 1940
Parachordodes propareolatus Heinze, 1935	Palearctic	Kyrgyzstan, Turkestan	Heinze 1935c, Spiridonov 1989
Parachordodes pustulosus (Baird, 1853)	Palearctic	Belgium, Bulgaria, Czech Republic, England, China, Germany, India, Italy, Kyrgizstan, Romania, Russia, Uzebekistan, Tatra Mountains (Poland or Slovakia)	Baird 1853a, b, Camerano 1889b, 1890b, 1892, 1896, 1897a, 1903c, 1908b, d, 1912b, Căpuşe 1971, Dorier 1930, Havlik 1947, Heinze 1935c, 1937, 1941, Janda 1894a, Konsuloff 1921, Schuurmans Stekhoven 1943, Villot 1886
Parachordodes sciacchitanoi Heinze, 1935	Palearctic	China	Heinze 1935a
Parachordodes speciosus (Janda, 1893)	Palearctic	presumably Tatra Mountains (Poland or Slovakia), Germany	Janda 1893, 1894a, Römer 1896, Heinze 1935c, 1937, 1941
Parachordodes tolosanus (Dujardin, 1842)	Palearctic	Austria, Belgium, Czech Republic, Germany, England, France, Italy, Poland, Romania, Russia, San Marino	Baird 1853a, b, Camerano 1887c, 1888b, 1889b, 1890b, 1897a, Căpuşe 1966a, Dorier 1930, Dujardin 1842, Havlik 1947, Heinze 1935c, 1937, 1940, 1941, Hellriegl 1996, Janda 1894a, Levander 1908, Ritchie 1915, Römer 1895a, 1896, Rosa 1882, Schuurmans Stekhoven 1943, Sciacchitano 1930, 1932, 1963, Tappenbeck 2006, Vejdovsky 1886, Villot 1874c, 1886

the six species described by him, adhesive warts. These are special structures, probably modified areoles, which occur at some distance anterior of the cloacal opening along both sides of the ventral midline (Fig. 3.5.5 B, E). Most adhesive warts are elongated and have a keel-shaped appearance (Fig. 3.5.5 E), others have a round structure (e.g. *G. bageli*, Schmidt-Rhaesa & Gusich 2010). Nothing is known about the function of the adhesive warts, although their name implies this. Adhesive warts were mentioned as the diagnostic

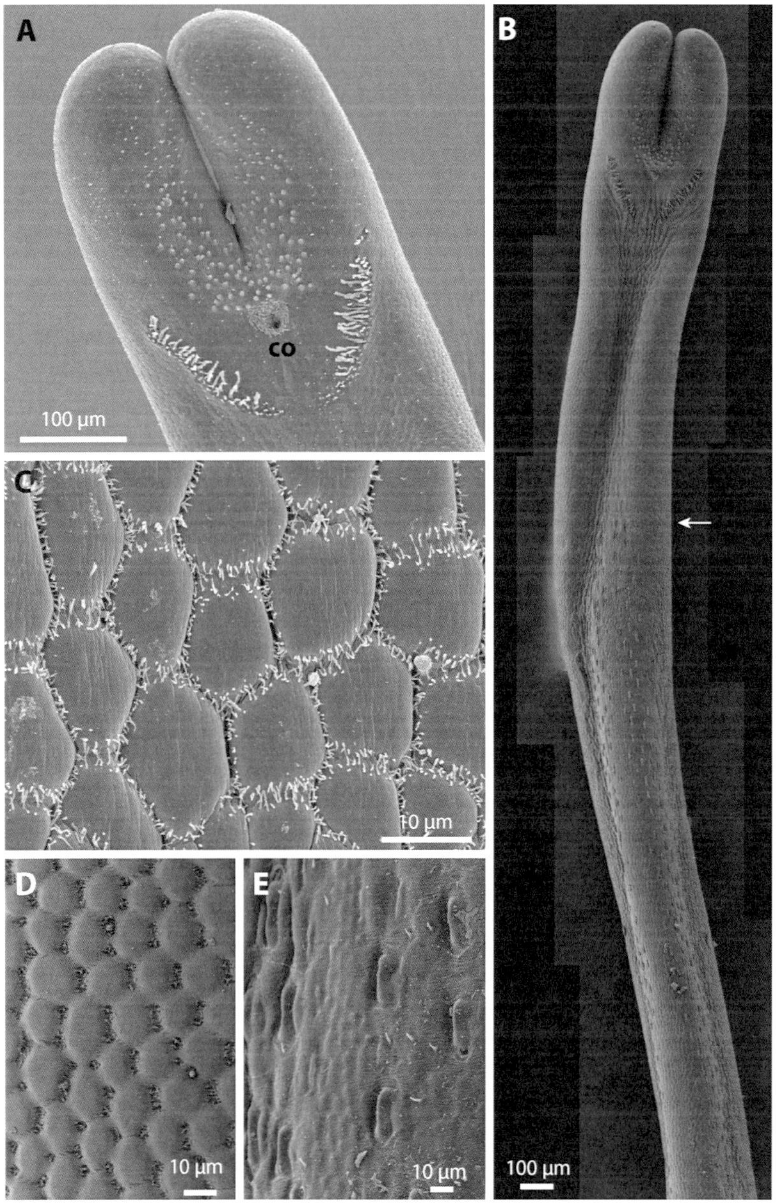

Fig. 3.5.5. Genus *Gordionus*. A, Ventral view on posterior end of *G. violaceus* (from Germany) with cloacal opening (co), precloacal rows of bristles and postcloacal spines; B, Overview showing the presence of adhesive warts on the ventral side of the posterior end of *G. violaceus* (from Estonia), starting some 100 µm posterior of the cloacal opening (arrow indicates beginning of adhesive warts); C, Areoles are completely surrounded by fine bristles; D, Areoles with reduced interareolar structures as typical for *G. wolterstorffii*, but transitions between patterns in C and D exist; E, Fine structure of keel-like adhesive warts (same specimen as B). All images with SEM. D from Schmidt-Rhaesa & Prous 2010, with kind permission from the Estonian Journal of Ecology.

character for the genus *Gordionus* by, e.g. Heinze (1937) and De Miralles (1976), but several descriptions lack the observation of these structures. This can be due to the fact that the posterior end is not always checked to a sufficient degree, but in some species they seem indeed to be lacking (see Schmidt-Rhaesa 2001c). Furthermore, adhesive warts have now also been found in *Beatogordius abbreviatus* (Schmidt-Rhaesa & Bryant 2004). Therefore, it is questionable whether adhesive warts are a characteristic feature of *Gordionus*.

In practice, *Gordionus* is recognized by the presence of one type of areole (Fig. 3.5.5 C), in combination with a bilobed male posterior end. This distinguishes *Gordionus* from *Parachordodes* and *Paragordionus*. *Beatogordius* is distinguished by its special arrangement of areoles (see below). When describing the number of different types of areoles, the arrangement patterns of areoles are usually neglected. There is, however, a conspicuous arrangement, in which two adjacent areoles surround a central tubercle (a small cuticular projection). In *Paragordionus*, such areoles are larger than the remaining ones (and are named megareoles), but in several *Gordionus* species, this arrangement pattern can be recognized with areoles indistinguishable from the remaining ones. Therefore, this pattern has been named the megareolar pattern (Schmidt-Rhaesa 2002a). The megareolar pattern creates a number of unresolved questions. It is not clear whether it is characteristic for certain species of *Gordionus* or is part of a cuticle variation (see e.g. Schmidt-Rhaesa 2001c for the occasional presence in specimens from the United Kingdom). In this case the distinction between megareoles and a megareolar pattern is not clearly worked out, as some *Paragordionus* species contain a megareolar pattern rather than megareoles (see below). Finally, an evolutionary transition from a megareolar pattern to megareoles and probably even superareoles is likely, but this may have the consequence that *Gordionus* is a paraphyletic group from which other genera such as *Parachordodes*, *Paragordionus* and probably also *Beatogordius* evolved.

Concerning the characters of the male posterior end, *Gordionus* corresponds to *Parachordodes* and *Paragordionus* (circumcloacal spines, precloacal rows of bristles and postcloacal spines; Fig. 3.5.5 A); the female posterior end is rounded.

At present, 56 species are described and only one further species is regarded as invalid (Schmidt-Rhaesa et al. 2003a; Tab. 22). A review has not been carried out, but appears necessary.

Tab. 22. List of described species in the genus *Gordionus*.

Species	Ecozone	Distribution	References
Gordionus alascensis (Montgomery, 1907)	Palearctic	USA (Alasca)	Montgomery 1907, Schmidt-Rhaesa et al. 2003a
Gordionus alpestris (Villot, 1885)	Palearctic	Belgium, Bulgaria, France, Germany, Romania, Slovakia, Sweden, Italy	Camerano 1887c, 1888b, 1897a, Căpuşe 1966a, 1971, Dorier 1928, 1930, Havlik 1949a,b, Heinze 1937, 1941, Konsuloff 1921, Montén 1951, Schuurmans Stekhoven 1943, Villot 1885, 1886
Gordionus annamensis Spiridonov, 1993	Indo-Malaya	Vietnam	Spiridonov 1993
Gordionus austriacus Heinze, 1952	Palearctic	Austria	Heinze 1952
Gordionus bageli Schmidt-Rhaesa & Gusich, 2010	Palearctic	Swizerland	Schmidt-Rhaesa & Gusich 2010
Gordionus barbatus Schmidt-Rhaesa & Cieslak, 2008	Palearctic	Spain	Schmidt-Rhaesa & Cieslak 2008
Gordionus bilinareolatus Schuurmans Stekhoven, 1943	Palearctic	Belgium, Croatia, Germany	Čanadjija 1956, Heinze 1941, Schuurmans Stekhoven 1943
Gordionus bilobatus Kirjanova, 1955	Palearctic	Russia	Kirjanova 1955
Gordionus blanchardi (Villot, 1874)	Palearctic	France	Villot 1874c, Dorier 1946b
Gordionus brunneus Montén, 1951	Palearctic	Sweden	Montén 1951
Gordionus chinensis (Villot, 1874)	Palearctic, Indo-Malaya	China, Indonesia, Japan	Camerano 1895, 1896, Dorier 1946b, Schmidt-Rhaesa & Sato 2009, Schmidt-Rhaesa et al. 2009b, Villot 1874c

(*Continued*)

Tab. 22. (*Continued*)

Species	Ecozone	Distribution	References
Gordionus conglomeratus Heinze, 1952	Palearctic	Austria	Heinze 1952
Gordionus cyrnensis Giudicelli & Nicoli, 1962	Palearctic	France	Giudicelli & Nicoli 1962
Gordionus diligens De Villalobos, Ribera & Downie, 1999	Palearctic	Scotland	De Villalobos et al. 1999
Gordionus dorieri Heinze, 1937	Palearctic	Germany	Heinze 1937, 1941
Gordionus divergens Schuurmans Stekhoven, 1943	Palearctic	Belgium	Schuurmans Stekhoven 1943
Gordionus dubiosus Heinze, 1937	Palearctic	Belgium, Croatia, Czech Republic, Germany, Poland, Romania	Čanadjija 1956, Căpuşe 1966a, 1971, Havlik 1949b, Heinze 1937, 1941, Schuurmans Stekhoven 1943
Gordionus enigmaticus De Villalobos, Zanca & Ibarra-Vidal, 2005	Nearctic	Chile	De Villalobos et al. 2005b
Gordionus formidatus Havlik, 1954	Palearctic	Slovakia	Havlik 1954
Gordionus fragmentarius Heinze, 1952	Palearctic	Austria	Heinze 1952
Gordionus harpali Heinze, 1940	Palearctic	Germany	Heinze 1940, 1941
Gordionus kaschgaricus Heinze, 1937	Palearctic	China and Turkestan (as "Kaschgaria – Turkestan chinese"), Angola, Eritrea, India, Indonesia	Camerano 1897a, 1903c, 1912b, 1915a, b, Sciacchitano 1932, 1940
Gordionus kii Schmidt-Rhaesa & Sato, 2009	Palearctic	Japan	Schmidt-Rhaesa & Sato 2009
Gordionus lapponicus Montén, 1951	Palearctic	Sweden	Montén 1951
Gordionus lenae Heinze, 1937	Palearctic	Germany	Heinze 1937, 1941
Gordionus lineatus Leidy, 1851	Nearctic	USA	Leidy 1851, Montgomery 1898a, Schmidt-Rhaesa et al. 2003a, 2009a, b, Smith 1991
Gordionus linourgos De Villalobos , Ribera & Downie, 1999	Palearctic	Scotland	De Villalobos et al. 1999
Gordionus longareolatus (Montgomery, 1898)	Nearctic, Palearctic	USA	Montgomery 1898c, Schmidt-Rhaesa et al. 2003a
Gordionus longistriatus Schmidt-Rhaesa, 2004	(Nearctic)	USA (Hawaii)	Schmidt-Rhaesa 2004b
Gordionus lunatus Müller, 1926	Palearctic	Germany, Romania	Căpuşe 1971, Heinze 1937, 1941, Müller 1926
Gordionus meissneri Müller, 1926	Palearctic	Germany	Heinze 1937, 1941, Müller 1926
Gordionus molopsis Heinze, 1937	Palearctic	Germany	Heinze 1937, 1941
Gordionus ondulatus De Miralles, 1984	Neotropic	Argentina	De Miralles 1984b, 1993, De Villalobos & Voglino 2000, De Villalobos et al. 2009

(*Continued*)

Tab. 22. (*Continued*)

Species	Ecozone	Distribution	References
Gordionus perronciti (Camerano, 1887)	Palearctic	Austria, Germany, Italy, Slovakia	Camerano 1887c , Franz 1954, Havlik 1954, Heinze 1937
Gordionus platycephalus (Montgomery, 1898)	Nearctic	USA	De Miralles et al. 1997, Montgomery 1898a, Schmidt-Rhaesa et al. 2003a
Gordionus porosus De Villalobos & Voglino, 2000	Neotropic	Argentina	De Villalobos & Voglino 2000
Gordionus preslii (Vejdovsky, 1886)	Palearctic	Czech Republic, Germany, Italy, probably Poland (as "Schlesien"), Slovakia	Camerano 1887c, 1888b, Havlik 1947, Heinze 1937, 1941, Janda 1894, Meßner 1983, Vejdovsky 1886, 1894
Gordionus prismaticus (Villot, 1874)	Neotropic	Argentina, Colombia	Camerano 1894, 1897a, 1915a, De Miralles 1967, 1969, 1976, Dorier 1946b, Villot 1874c
Gordionus punctulatus Heinze, 1937	Palearctic	Germany	Heinze 1937, 1941
Gordionus rautheri Heinze, 1937	Palearctic	Austria, Germany, Poland	De Miralles 1984b, Heinze 1937, 1940, 1941
Gordionus rugosus Čanadjija, 1956	Palearctic	Croatia	Čanadjija 1956
Gordionus scaber Müller, 1926	Palearctic	Austria, Croatia, Czech Republic, Denmark (Faroe Islands), Germany, Poland, Romania, Sweden	Čanadjija 1956, Căpuşe 1966a, Havlik 1947, Heinze 1937, 1940, 1941, Müller 1926, Montén 1951, Muus 1956
Gordionus scripturus Heinze, 1952	Palearctic	Austria	Heinze 1952
Gordionus semistriatus Heinze, 1937[1]	Palearctic	Germany	Heinze 1937, 1941
Gordionus senkovi Malakhov & Spiridonov, 1982	Palearctic		
Gordionus silphae Heinze, 1937	Palearctic	Germany	Heinze 1937, 1941
Gordionus sinepilosus Schmidt-Rhaesa, Hanelt & Reeves, 2003	Nearctic	Canada	Schmidt-Rhaesa et al. 2003a
Gordionus stenobothri Čanadjija, 1956	Palearctic	Croatia	Čanadjija 1956
Gordionus strigatus Müller, 1926	Palearctic	Germany, Poland	Heinze 1937, 1940, 1941, Müller 1926
Gordionus subalpinus Heinze, 1952	Palearctic	Austria	Heinze 1952
Gordionus sulcatus Müller, 1926	Palearctic	Germany	Heinze 1937, 1941, Müller 1926
Gordionus thienemanni Heinze, 1937	Palearctic	Germany	Heinze 1937, 1941
Gordionus thuringensis Müller, 1926	Palearctic	Germany	Heinze 1937, 1941, Müller 1926
Gordionus turkensis Schmidt-Rhaesa & Cieslak, 2008	Palearctic	Turkey	Schmidt-Rhaesa & Cieslak 2008

(*Continued*)

Tab. 22. (*Continued*)

Species	Ecozone	Distribution	References
Gordionus violaceus (Baird, 1853)	Palearctic, Nearctic	Armenia (as "province Eriwanensis"), Austria, Bulgaria, China (Tibet), Croatia, Czech Republic, Denmark, Finland, France, Germany, Ireland, Italy, Montenegro, Romania, Serbia, Spain, United Kingdom, USA	Baird 1853a, b, Camerano 1887c, 1888b, 1896, 1897a, 1903c, 1908a, 1915a, Čanadjija 1956, Căpuşe 1966a, De Villalobos et al. 1999, 2001, Dollfus 1943, Dorier 1928, 1930, Havlik 1949b, Heinze 1935c, 1937, 1941, Konsuloff 1921, Levander 1908, Montgomery 1898a, Muus 1956, Pešić & Ćurčić 2006, Pešić & Schmidt-Rhaesa 2008, Ritchie 1915, Römer 1895a, 1896, Schmidt-Rhaesa 2000, 2001c, 2009, Schmidt-Rhaesa & Kristensen 2006, Schmidt-Rhaesa & Zaenker 2006, Schmidt-Rhaesa et al. 2003a, Sciacchitano 1932, 1933b, 1963, Smith 1991, Southern 1907, 1909, Tappenbeck 2006, Villot 1886
Gordionus wolterstorffii (Camerano, 1888)	Palearctic	Austria, Belgium, China, Germany, Italy, Poland, United Kingdom	Baylis 1944, Camerano 1888b, 1904c, 1915a, De Villalobos et al. 2001, Heinze 1937, 1940, 1941, Schuurmans Stekhoven 1943, Sciacchitano 1930, 1932, 1933b, Schmidt-Rhaesa 2001c

[1] Described as *Gordionus semi-striatus* by Heinze (1937), corrected to *G. semistriatus* by Heinze (1941); dashes are not allowed in species names according to article 27 of the international rules for zoological nomenclature.

Invalid species names:

- *Gordionus reticulatus* Villot, 1874 (USA), see Schmidt-Rhaesa et al. (2003a)

Beatogordius Heinze, 1934

The species of the genus *Beatogordius* share a conspicuous arrangement of areoles in longitudinal rows parallel to the anteroposterior axis of the worms (Fig. 3.5.6 B). Males have a bilobed posterior end with rows of bristles anterior to the cloacal opening and spines posterior to it (Fig. 3.5.6 A). Females of some species also have a semicircular row of bristles on the ventral side of the posterior end, while the cloacal opening is, as in other females, almost terminal.

The presence of the row of bristles in the female posterior end has led to some confusion in the determination of sexes. In the most prominent case, a female specimen was described as the male of a new genus, *Chordodiolus*, by Heinze (1935b). Heinze had reinvestigated a specimen named *Chordodes echinatus* by Linstow (1901). A new reinvestigation showed that this specimen had oocytes and therefore was a female (Schmidt-Rhaesa 2001b). As the cuticular structure corresponds to the one in *Beatogordius*, the species was transferred to this genus (Schmidt-Rhaesa & De Villalobos 2002). Rows of bristles were also found in females of other *Beatogordius* species (De Villalobos et al. 2003b, Schmidt-Rhaesa & Bryant 2004).

The genus was reviewed (Schmidt-Rhaesa & De Villalobos 2002, De Villalobos et al. 2003b). At that time, only specimens from the Afrotropic and the Neotropic were known. The finding of species in Madagascar and Australia supported the reasoning that *Beatogordius* is distributed on the Gondwana continents (Schmidt-Rhaesa & Bryant 2004), but a new report from China extended the range beyond the Gondwana continents (Schmidt-Rhaesa 2011). To date, the genus includes 21 species (Tab. 23).

Some specimens were determined as *Beatogordius* cf. *echinatus* (from Burundi, the Democratic Republic of the Congo and Rwanda) or as *B.* cf. *raphaelis* (from the Democratic Republic of the Congo, South Africa, Tanzania and Zimbabwe), see Schmidt-Rhaesa & De Villalobos (2002).

Paragordionus Heinze, 1935

The genus *Paragordionus* includes species with two types of areoles. Besides the usual ones, there are larger areoles enclosing a tubercle between them (Fig. 3.5.6 D). These areoles are called megareoles. Megareoles can occur in larger clusters composed of several megareoles (Fig. 3.5.6 D). Two species without clear megareoles but with a megareolar pattern have been included into *Paragordionus* (*P. bohemicus* and *P. rautheri*), although they might

Tab. 23. List of described species in the genus *Beatogordius*. For references see Schmidt-Rhaesa & De Villalobos 2002, De Villalobos et al. 2003b, Schmidt-Rhaesa & Bryant 2004 and Schmidt-Rhaesa 2011.

Species	Ecozone	Distribution
Beatogordius abaiconus Carvalho, 1946	Neotropic	Brazil
Beatogordius abbreviatus (Villot, 1874)	Afrotropic	Reunion, Madagascar
Beatogordius alfredi (Camerano, 1894)	Neotropic	Argentina, Brazil
Beatogordius australiensis Schmidt-Rhaesa & Bryant, 2004	Australasia	Australia
Beatogordius brieni Sciacchitano, 1961[1]	Afrotropic	Democratic Republic of the Congo
Beatogordius chinensis Schmidt-Rhaesa, 2011	Palearctic	China
Beatogordius deshayesi (Villot, 1874)	Neotropic	Venezuela
Beatogordius echinatus (Linstow, 1901)	Afrotropic	Democratic Republic of the Congo, Tanzania
Beatogordius erythraeus (Camerano, 1915)	Afrotropic	Erithrea
Beatogordius funis De Villalobos, Schmidt-Rhaesa & Zanca, 2003	Neotropic	Argantina
Beatogordius inesae (Cavalieri, 1961)	Neotropic	Argentina
Beatogordius irregularis De Miralles, 1972	Neotropic	Argentina
Beatogordius latastei (Camerano, 1895)	Neotropic	Argentina, Chile
Beatogordius lineatus Schmidt-Rhaesa & Bryant, 2004	Australasia	Australia
Beatogordius palustre De Villalobos, Schmidt-Rhaesa & Zanca, 2003	Neotropic	Argentina
Beatogordius raphaelis (Camerano, 1893)	Afrotropic	Democratic Republic of the Congo, Republic of the Congo
Beatogordius regularis Heinze, 1934	Afrotropic	South Africa
Beatogordius sankurensis Sciacchitano, 1958	Afrotropic	Democratic Republic of the Congo
Beatogordius ugandensis Schmidt-Rhaesa & De Villalobos, 2002	Afrotropic	Uganda
Beatogordius variabilis De Miralles, 1981	Neotropic	Argentina
Beatogordius wilsoni Sciacchitano, 1958	Afrotropic	Democratic Republic of the Congo

[1]This specimen was described as a new species by Schmidt-Rhaesa & De Villalobos (2002) on the basis of a museum specimen labeled as a holotype, assuming that a descriptin had not been published. This was an error; the species was described by Sciacchitano (1961b).

better belong to *Gordionus* (see discussion of this genus; *P. rautheri* is here treated as *Gordionus rautheri*). In other respects, e.g. in the male posterior end (Fig. 3.5.6 C), *Paragordionus* closely resembles *Gordionus*. At present, five species are known – four from Europe and one from Japan (Tab. 24).

Euchordodes Heinze, 1937

Euchordodes is a small genus including four species (Tab. 25), which are geographically widely separated. A characteristic is the presence of two types of areoles, one of which are megareoles (see *Gordionus*). These are slightly more elevated or larger than the other "simple" areoles. The male posterior end is not bilobed. Bristles are present in rows or fields anterolateral of the cloacal opening (*E. libellulovivens* and *E. nigromaculatus*), or more or less on the entire ventral side of the posterior end (*E. malaysiensis*).

Remarkably, the majority of specimens found were males. In *E. malaysiensis* only 15 males are known, in *E. nigromaculatus* many males (Schmidt-Rhaesa et al. 1998) but only one female is known (Schmidt-Rhaesa 2008); in *E. libellulovivens* two females and one male are known.

The fourth species, *E. parvulus*, likely belongs to the genus *Spinochordodes* (Schmidt-Rhaesa 2002a), but has to be reinvestigated.

Neochordodes Carvalho, 1942

The genus *Neochordodes* includes Neotropic to Nearctic species with an undivided male posterior end and one type of areoles in the cuticle (Fig. 3.5.7 A). Some species have megareolar arrangement patterns. When the areoles in a megareolar pattern differ only slightly from the regular areoles (see Fig. 3.5.7 A), it is unclear whether these species belong to *Neochordodes* or to *Pseudochordodes*, where megareoles are present. Therefore, some species

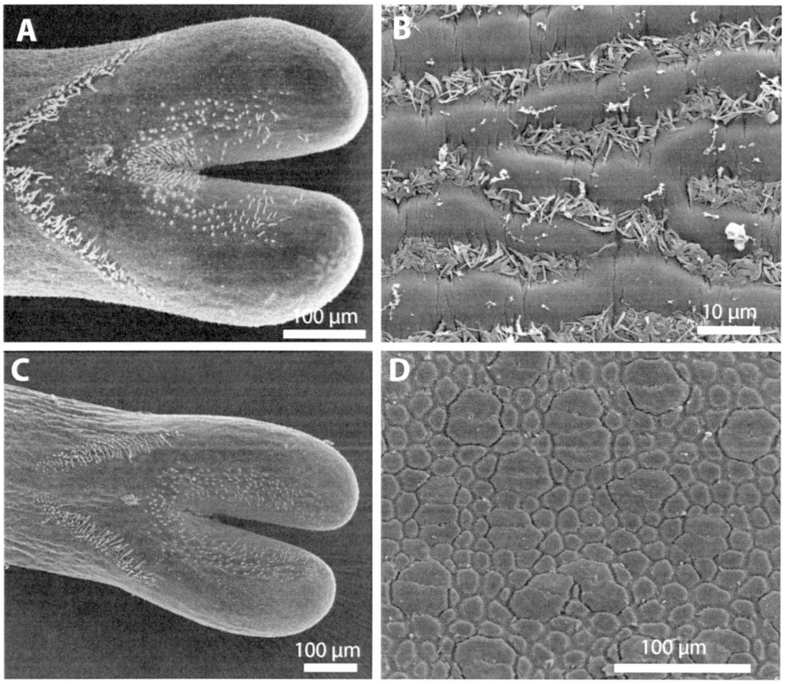

Fig. 3.5.6. A, B, Genus *Beatogordius*. A, Posterior end of male *B. abbreviatus* (from Madagascar); B, Areoles are fused to longitudinal lines (here *B. chinensis* from China). C, D, Genus *Paragordionus* (both *P. dispar* from Germany). C, Posterior end of the male; D, Cuticle with numerous megareoles among smaller areoles. All images with SEM. A from Schmidt-Rhaesa & Bryant 2004, with kind permission of Springer Publishers, Heidelberg; B from Schmidt-Rhaesa 2011, with kind permission of Naturwissenschaftlicher Verein Hamburg.

Tab. 24. List of described species in the genus *Paragordionus*.

Species	Ecozone	Distribution	References
Paragordionus bohemicus Havlik, 1947	Palearctic	Czech Republic	Havlik 1947
Paragordionus dispar (Müller, 1926)	Palearctic	Germany	Müller 1926, Heinze 1935c, 1937, 1941, Schmidt-Rhaesa & Zaenker 2006
Paragordionus ibericus Schmidt-Rhaesa & Cieslak, 2008	Palearctic	Spain	Schmidt-Rhaesa & Cieslak 2008
Paragordionus kawamurai Yamaguti, 1943	Palearctic	Japan	Yamaguti 1943, Inoue 1955, 1988, Schmidt-Rhaesa 2004c
Paragordionus vejdovskyi (Janda, 1894)	Palearctic	Germany, probably Czech Republic	Janda 1894b, Havlik 1947, Heinze 1935c, 1941

Tab. 25. List of described species in the genus *Euchordodes*.

Species	Ecozone	Distribution	References
Euchordodes libellulovivens Heinze, 1937	Palearctic	Germany	Heinze 1937, 1941, Zanca & Schmidt-Rhaesa 2006
Euchordodes malaysiensis Inoue, 1966	Indo-Malaya	Malaysia	Inoue 1966
Euchordodes nigromaculatus Poinar, 1991	Oceania	New Zealand	Poinar 1991, Schmidt-Rhaesa 2008, Schmidt-Rhaesa et al. 1998
Euchordodes parvulus Čanadjija, 1956	Palearctic	Croatia	Čanadjija 1956

have been shifted between these genera. The anterior end is tapering; a white tip blends into the brown body coloration with no separation by a darkly stained collar. To date, 12 species are known (Tab. 26); a revision has not been carried out. A clear distinction between the genera *Neochordodes* and *Pseudochordodes* is necessary.

Tab. 26. List of described species in the genus *Neochordodes*.

Species	Ecozone	Distribution	References
Neochordodes australis De Miralles & De Villalobos, 1996	Neotropic	Argentina	De Miralles & De Villalobos 1996c
Neochordodes bonaerensis De Miralles & De Villalobos, 1996	Neotropic	Argentina	De Miralles & De Villalobos 1996c
Neochordodes californensis De Miralles & De Villalobos, 1995	Nearctic	USA	De Miralles & De Villalobos 1995
Neochordodes chordodides Schmidt-Rhaesa & Menzel, 2005	Neotropic	Costa Rica	Schmidt-Rhaesa & Menzel 2005
Neochordodes colombianus Faust & Ramos, 1960	Neotropic	Colombia	Faust & Botero Ramos 1960
Neochordodes nietoi (Caballero, 1936)	Neotropic	Mexico	Caballero 1936, De Miralles & De Villalobos 1994
Neochordodes occidentalis (Montgomery, 1898)	Neotropic, Nearctic	Mexico, USA	Caballero 1936, De Miralles & De Villalobos 1993, Montgomery 1898a, Schmidt-Rhaesa et al. 2003a, Schmidt-Rhaesa & Menzel 2005, Watermolen & Chandler 1993
Neochordodes puntatus De Miralles & De Villalobos, 1996	Neotropic	Argentina	De Miralles & De Villalobos 1996b
Neochordodes semiluna De Villalobos & Camino, 1999	Neotropic	Argentina	De Villalobos & Camino 1999
Neochordodes serranensis De Miralles & De Villalobos, 1996	Neotropic	Argentina	De Miralles & De Villalobos 1996d
Neochordodes torrenticola De Miralles & De Villalobos, 1996	Neotropic	Argentina	De Miralles & De Villalobos 1996d
Neochordodes uniareolatus Carvalho, 1946	Neotropic	Argentina, Venezuela	Carvalho 1946a, Carvalho & Feio 1950, De Miralles 1969, 1984a, De Miralles & Camino 1983, De Miralles & De Villalobos 1996c

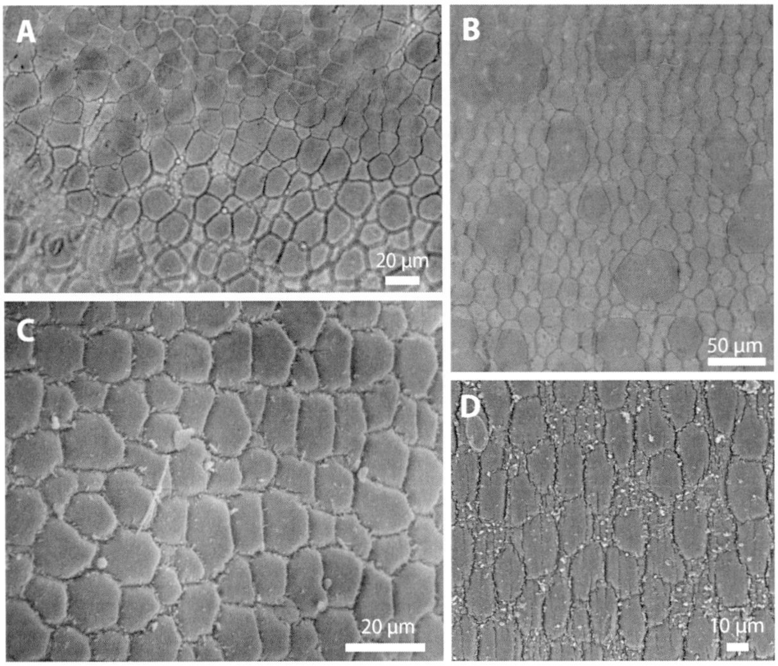

Fig. 3.5.7. A, Genus *Neochordodes*; Cuticular pattern of the *N. occidentalis* holotype (Museum of Comparative Zoology Harvard 1469). B–D, Genus *Pseudochordodes*; B, Clear distinction between small and megareoles in *P. manteri* (from Mexico); C, Areoles in *P. bedriagae* (from Argentina); D, Two types of areoles in *P. gordioides* (from the USA). A, B with light microscopy, C, D with SEM.

Pseudochordodes Carvalho, 1942

Species of the genus *Pseudochordodes* resemble those from *Neochordodes*, but have two types of areoles. One type are regular "simple" areoles, the other are elevated megareoles, quite similar to the ones in *Paragordionus*. This distinction between two areolar types is very clear in some species (e.g. *P. manteri* and *P. gordioides*, see Fig. 3.5.7 D, E), but can also be very insignificant (e.g. *P. bedriagae*, see Fig. 3.5.7 C). The males have an undivided posterior end (Fig. 3.5.7 B). At present, eight species are known, ranging from the Neotropic to the Southwestern Nearctic (Tab. 27). One further species name, *Pseudochordodes griffinii* (Camerano, 1898), is invalid (see De Villalobos & Zanca 2005).

Digordius Kirjanova, 1950

Kirjanova (1950a) described male specimens of *Digordius excavatus* from Tajikistan (Tab. 28). The anterior end terminates in three spherical compartments. The posterior end and the cuticle resemble those in *Paragordius* species. Kirjanova (1950a) cites established *Paragordius* species (e.g. *P. esavianus*) as belonging to *Digordius*, but it is not known to the author why this new genus name is introduced or whether it can be regarded as a synonym of *Paragordius*. A reinvestigation of the few specimens is needed.

Tab. 28. List of described species in the genus *Digordius*.

Species	Ecozone	Distribution
Digordius excavatus Kirjanova, 1950	Palearctic	Tajikistan
Digordius trinacris Kirjanova, 1954	?	?

Lanochordodes Kirjanova, 1950

The description of the single male specimen of *Lanochordodes zeravshanicus* is not very detailed (see Kirjanova 1950b; Tab. 29). The posterior end is undivided; the cuticle contains groups of elevated areoles, resembling the pattern in *Spinochordodes*. A reinvestigation is needed to decide whether *L. zeravshanicus* belongs to *Spinochordodes* or whether regarding it as a separate genus is justified.

Tab. 29. Distribution of the single species described in the genus *Lanochordodes*.

Species	Ecozone	Distribution
Lanochordodes zeravshanicus Kirjanova, 1950	Palearctic	Uzbekistan

Progordius Kirjanova, 1950

The one or few males of *Progordius maculosus* (Kirjanova 1950c; Tab. 30) resemble *Digordius excavatus* and it has

Tab. 27. List of described species in the genus *Pseudochordodes*.

Species	Ecozone	Distribution	References
Pseudochordodes bedriagae (Camerano, 1896)	Neotropic		Camerano 1896, De Miralles 1973, 1976, 1977, 1984a, De Miralles & De Villalobos 1993, 1995, De Villalobos et al. 2008a, Schmidt-Rhaesa et al. 2003a
Pseudochordodes bulbareolatus Schmidt-Rhaesa & Menzel, 2005	Neotropic	Guatemala, Mexico	Schmidt-Rhaesa & Menzel 2005
Pseudochordodes dugesi (Camerano, 1898)	Neotropic	Argentina, Mexico	Caballero 1936, Camerano 1898, De Miralles & Camino 1983, De Miralles & De Villalobos 2000, De Villalobos et al. 2008a
Pseudochordodes gordioides (Montgomery, 1898)	Nearctic	USA	Montgomery 1898a, De Miralles et al. 1997, Schmidt-Rhaesa et al. 2003a
Pseudochordodes guatemalensis De Miralles, De Villalobos & Rodriguez, 1997	Neotropic	Guatemala	De Miralles et al. 1997
Pseudochordodes manteri Carvalho, 1942	Neotropic, Nearctic	Mexico, USA	Carvalho 1942, Schmidt-Rhaesa et al. 2003a, Schmidt-Rhaesa & Menzel 2005
Pseudochordodes meridionalis Carvalho & Feio, 1950	Neotropic	Argentina, Chile, Mexico	Carvalho & Feio 1950, De Miralles 1977, 1984a, De Miralles & De Villalobos 1996c, De Villalobos et al. 2005b, 2008a, Schmidt-Rhaesa et al. 2003a, Schmidt-Rhaesa & Menzel 2005
Pseudochordodes texanus Schmidt-Rhaesa, Hanelt & Reeves, 2003	Nearctic	USA	Schmidt-Rhaesa et al. 2003a

Tab. 30. Distribution of the single species described in the genus *Progordius*.

Species	Ecozone	Distribution
Progordius maculosus Kirjanova, 1950	Palearctic	Tajikistan

been assumed that they also might belong to the genus *Paragordius* (Schmidt-Rhaesa 2002a).

Spinochordodes Kirjanova, 1950

Spinochordodes males have an undivided posterior end with circumcloacal spines and short bristles in the entire ventral region around the cloacal opening (Fig. 3.5.8 A). The cuticle contains two types of areoles, each are concentrated in patches (Fig. 3.5.8 C). One type is flat while the other type is elevated and ends apically with a serrated edge, resembling a crown (Fig. 3.5.8 B). In contrast to the "crown areoles" of *Chordodes*, the apical crown-like elements in *Spinochordodes* are shorter and not as slender as in *Chordodes* (see Spiridonv & Ivanova 2002, Zanca & Schmidt-Rhaesa 2006).

The characters typical for *Spinochordodes* are present in a few species that have been assigned to other genera. It has been mentioned that *Lanochordodes zeravshanicus* has a similar cuticular pattern. *Pantachordodes europaeus* Heinze, 1952 and *Dacochordodes bacescui* Căpuşe, 1965 have been transferred to *Spinochordodes* (Zanca & Schmidt-Rhaesa 2006), deleting the monospecific genera *Pantachordodes* and *Dacochordodes*.

At present, nine species have been described; a review has not been carried out (Tab. 31).

Acutogordius Heinze, 1952

Heinze (1952) created the genus *Acutogordius* for those species with a postcloacal crescent and pointed tail lobes (Fig. 3.5.8 D). In some species, the tail lobes are sharply pointed, while in others this is not so evident. *Acutogordius* is closely related to *Gordius*, see under description of this genus for possible relations between these two genera. The nine currently known species (Tab. 32) are distributed on the Southern continents (with the exception of *A. obesus* occurring in Morocco). A brief revision of the genus has been made by Schmidt-Rhaesa & Geraci (2006).

Semigordionus Heinze, 1952

Semigordionus circumannulatus (Tab. 33) resembles *Gordionus* species and differs from these only in the fact that the precloacal rows of bristles merge in the median line. It has been proposed, though not formally stated, to include *S. circumannulatus* into the genus *Gordionus* (Zanca & Schmidt-Rhaesa 2006, Schmidt-Rhaesa & Cieslak 2008). Another species with medially merging precloacal bristles has been described as a

Tab. 31. List of described species in the genus *Spinochordodes*.

Species	Ecozone	Distribution	References
Spinochordodes actiniphorus Kirjanova, 1950	Palearctic	Tajikistan	Kirjanova 1950b
Spinochordodes bacescui (Căpuşe, 1965)	Palearctic	Bosnia and Herzegovina, Montenegro, Romania	Căpuşe 1965, 1966a,b, 1971, Pešić & Ćurčić 2006, Pikula et al. 1996, Spiridonov et al. 1992, Zanca & Schmidt-Rhaesa 2006
Spinochordodes baeri (Camerano, 1896)	Palearctic	Georgia	Camerano 1896, Kirjanova 1950b
Spinochordodes cameranoi Kirjanova, 1950	Palearctic	Region of the Caspian Sea	Kirjanova 1950b
Spinochordodes europaeus (Heinze, 1952)	Palearctic	Austria	Heinze 1952, Zanca & Schmidt-Rhaesa 2006
Spinochordodes piliferus Kirjanova, 1950	Palearctic	Uzbekistan	Kirjanova 1950b
Spinochordodes skrjabini Kirjanova, 1958	Palearctic		Kirjanova 1958b
Spinochordodes tellinii (Camerano, 1888)	Palearctic	France, Italy, Macedonia	Camerano 1888b, 1897a, De Villalobos & Zanca 2002, Dorier 1931, 1942, Zanca & Schmidt-Rhaesa 2006
Spinochordodes vitiferus Kirjanova, 1950	Palearctic	Uzbekistan	Kirjanova 1950b

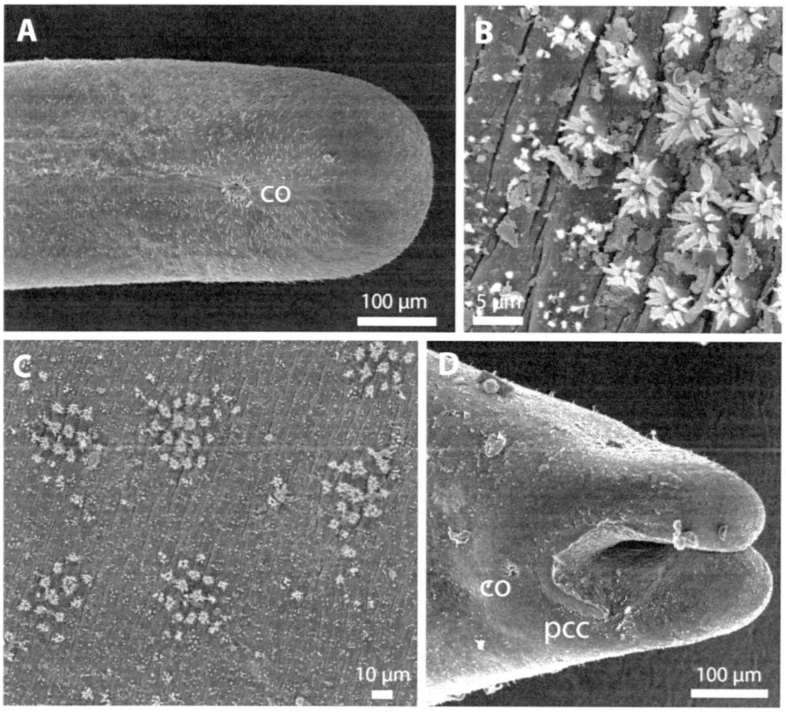

Fig. 3.5.8. A–C, Genus *Spinochordodes* (all *S. bacescui* from Montenegro). A, Ventral view on male posterior end (co = cloacal opening); B, Cuticular pattern showing elevated, flower-like areoles; C, Elevated areoles occur in patches among flatter regions. D, Genus *Acutogordius*, posterior end of male *A. sulawensis* from Indonesia showing cloacal opening (co) and postcloacal crescent (pcc). All images by SEM.

Tab. 32. List of described species in the genus *Acutogordius*.

Species	Ecozone	Distribution
Acutogordius acuminatus De Miralles & De Villalobos, 1998	Neotropic	Brazil
Acutogordius americanus De Miralles & De Villalobos, 1998	Neotropic	Costa Rica
Acutogordius australiensis Spiridonov, 1984	Australasia	Australia
Acutogordius doriae (Camerano, 1890)	Indo-Malaya	Burma, Malaysia, India
Acutogordius feae (Camerano, 1888)	Indo-Malaya	Myanmar, Sri Lanka
Acutogordius incertus (Villot, 1874)	Australasia	Australia
Acutogordius obesus (Camerano, 1895)	Neotropic, Palearctic	Equador, Morocco[1]
Acutogordius protectus Schmidt-Rhaesa & Geraci, 2006	Afrotropic	Madagascar
Acutogordius sulawensis Schmidt-Rhaesa & Geraci, 2006	Indo-Malaya	Indonesia

[1]In the original description, Camerano (1895) gave "Hollande" as the sampling locality; this appears unlikely and Camerano himself (1897a) regarded this as questionable.

Gordionus species (*G. barbatus* Schmidt-Rhaesa & Cieslak, 2008).

Pseudogordius Yeh & Jordan, 1957

Pseudogordius tanganyikae has been described by Yeh & Jordan (1957) from one single female specimen (Tab. 34). The description contains some unusual characters or character combinations, making a reinvestigation of this specimen desirable. However, the specimen could not be located in museums and seems to be lost (Schmidt-Rhaesa 2002a).

P. tanganyikae has three tail lobes, as in *Paragordius*, but an additional small fourth lobe. The cuticle contains structures resembling superareoles, which are otherwise

Tab. 33. Distribution of the single species described in the genus *Semigordionus*.

Species	Ecozone	Distribution
Semigordionus circumannulatus Heinze, 1952	Palearctic	Austria

Tab. 34. Distribution of the single species described in the genus *Pseudogordius*.

Species	Ecozone	Distribution
Pseudogordius tanganyikae Yeh & Jordan, 1957	Afrotropic	Tanzania

only present in *Parachordodes*. A reinvestigation should verify the presence and fine structure.

Noteochordodes De Miralles & De Villalobos, 2000

The youngest genus, *Noteochordodes*, was created for species with a particular type of areole (De Miralles & De Villalobos 2000, De Villalobos et al. 2008b). These are spherical in structure and have an apical depression. There is a second, variable type of areole; both types are scattered over the cuticle. The male posterior end is undivided. To date, five species have been described (Tab. 35).

3.6. Biogeography

With the exception of the Antarctic, gordiids have been reported from every continent. The species-rich genera are all distributed on several continents; some genera with fewer species are more restricted in their distribution. Distribution patterns that are connected to certain climate regions, as is the case for the genus *Chordodes*, may be linked to the occurrence of their main host group, praying mantids.

The genera *Neochordodes* and *Pseudochordodes* occur in South and Central America and in the Southwestern part of North America. Such a pattern is consistent with a spreading of these genera from South America northwards, probably after the connection of the two continents by the Panama land bridge.

The genus *Beatogordius* appeared to show a typical "Gondwana" distribution, occurring in South America, Africa and Australia (Schmidt-Rhaesa & Bryant 2004), but the recent finding in China (Schmidt-Rhaesa 2011) makes the explanation of the distribution pattern in *Beatogordius* more complex.

3.7. Paleontology

Only very few nematomorph specimens are known from the fossil record. The first record is *Gordius tenuifibrosus*, described by Voigt (1938) from Eocene brown coal (56–34 million years). Voigt described a wormlike organism, which appears to have a cuticle containing layers of fibers arranged at an angle of 52–55° to each other. This resembles the structure of the cuticle of nematomorphs, but as other taxa, such as mermithids or oligochaetes can also have layers of large fibers in their cuticle, the assignment to Nematomorpha is not as specific as Voigt implies (see also Poinar 1999). No structures such as areoles, which would be clearly indicative for Gordiida, are present.

Two clear reports of gordiids come from amber. *Palaeochordodes protus* has been described from a piece of Dominican amber (15–45 million years old) in association with its host, a cockroach closely resembling the genus *Supella* (Poinar 1999). There are two specimens – a female specimen is completely outside the host; the other specimen has its posterior end still inside the host. Specimens have clearly visible areoles, which appear uniform and are highly elevated.

Tab. 35. List of described species in the genus *Noteochordodes*.

Species	Ecozone	Distribution
Noteochordodes achosmosus De Villalobos, Zanca & Salas, 2008	Neotropic	Argentina
Noteochordodes cymatium De Villalobos, Zanca & Salas, 2008	Neotropic	Argentina
Noteochordodes desantini De Miralles & De Villalobos, 2000	Neotropic	Argentina
Noteochordodes saltae De Miralles & De Villalobos, 2000	Neotropic	Argentina
Noteochordodes talensis (Camerano, 1897)	Neotropic	Argentina, Bolivia, Brazil, Chile

Cretachordodes burmitis from Early Cretaceous Burmese amber (100–110 million years) is described by Poinar & Buckley (2006). The specimen is a male, with an undivided posterior end and a ventral cloacal opening. The cuticle contains a diversity of areoles, probably at least five different types. The worm is free, the host is unknown.

All other fossils that have been compared to nematomorphs are morphologically quite distinct, making comparison vague. For example, Hou & Bergström (1994) speculated that fossil worms of the taxon Palaeoscolecida may represent nematomorphs. Palaeoscolecids have an introvert covered with several to many spine-like structures. Some species (such as *Cricocosmia jinningensis*, see Hou & Bergström 1994) possess one terminal spine. The introvert, an annulated body and the terminal spine are compared by Hou & Bergström (1994) to nematomorph larvae. Apart from the fact that even small palaeoscolecids are larger than nematomorph larvae, there is no distinction of the body into pre- and postseptum; when one terminal hook is present in gordiid larvae it looks different and the introvert is distinctly different. According to Harvey et al. (2010), there is more support for the hypothesis that palaeoscolecids are stem-lineage priapulids. A further fossil organism, which bears some distant resemblance to nematomorph larvae, is *Shergoldana australiensis* from the Middle Cambrian of Queensland, Australia (Maas et al. 2007).

3.8. Ecology

The ecology of nematomorphs is strongly linked to their hosts, the parasitological aspects of which have been treated above. In the comparably short free-living phase, it is interesting to find out if there is a particular seasonality, what the preferred habitats are and if gordiids become important for other animals in their environment.

The question about seasonality in the occurrence of gordiids in water is likely to differ regionally and most data come from Central Europe. Here it seems that there is a peak in occurrence in late summer and fall, although gordiids have been found throughout the entire year (one example: Mühldorf 1913 collected *Gordius aquaticus* on February 9; see also Weißbecker & Schmidt-Rhaesa 2010). A late summer/fall peak seems to be the case in the entire Northern temperate regions, as Bolek & Coggins (2002) found *Gordius difficilis* in the USA from June to October and Inoue (1958) found *Chordodes japonensis* in Japan in September and October. In Argentina, Salas et al. (2011) report the highest abundance of the sympatric species *Chordodes brasiliensis, Noteochordodes cymatium, N. talensis* and *Pseudochordodes dugesi* in spring and fall.

Adult worms may be found within the sediment or among moist fallen leaves (Fig. 3.2.1 F), therefore some authors have assumed that adults may hibernate (Blunck 1915, 1922, Thorne 1940). The onset of rains towards the end of the summer may be a stimulus for the emergence in some cases, as massive emergence after rains has been observed at times (Godsall 1836, own observations). When species occur sympatrically, a succession in occurrence is possible, as has been shown by Schmidt-Rhaesa et al. (2005) in Avene les Bains, France, where *Paragordius tricuspidatus* occurs from June to August and *Spinochordodes tellinii* in August and September. Not so distinct is the temporal occurrence of *Gordionus violaceus* and *Gordius aquaticus* in Hessen, Germany. *Gordionus violaceus* appears first, with *Gordius aquaticus* very soon after. Both species then stay together in the water, until they disappear at the same time (Weißbecker & Schmidt-Rhaesa 2010).

Gordiids are found in a number of different aquatic habitats, ranging from the smallest water accumulations such as water troughs or water-filled tire tracts to large lakes (Fig. 3.8.1). It is very difficult to assign particular habitats to particular species, but if such a correlation exists, it appears likely that it is influenced by the habitat of the host. At least in temperate regions, small forest streams appear to be a common habitat, probably because of the abundance of carabids in forests. In this context it is interesting that in these regions the of smallest temporary water accumulations may be used to release gordiids (Milde 1868, Von Linstow 1891a, own observations). This may be caused by an unspecific release of gordiid worms by their hosts, although it is also possible that such habitats play a role in the life cycle, for example because snails, which also may occur in temporary waters, are more likely to be preyed upon by carabids when the water level decreases.

There is a considerable number of reports from gordiids in caves: from the USA (Arkansas: McDaniel & Smith 1976; Georgia: Holsinger & Peck 1971; Kentucky: Dearolf 1953, Conn 1981; North Carolina and Tennessee: Reeves 2000; Texas: Reddell 1965; Virginia: Goggin et al. 1989), Mexico (Schmidt-Rhaesa & Menzel 2005), Jamaica (Peck 1992), Belize and Guatemala (Reddell & Veni 1996), the Democratic Republic of the Congo (Sciacchitano 1955), Japan (Inoue 1972) and Italy (Sciacchitano 1952). In most of these cases the species is not determined

Fig. 3.8.1. Typical habitats in which gordiids have been collected. A, B, Two small mountain streams near Merlo, San Luis, Argentina; C, Small mountain stream near Otsu, Shiga, Japan; D, Elk Creek in Nebraska, USA; E, Ben Hanelt collecting *Chordodes morgani* from twigs hanging into Elk Creek; F, Small stream near St. Andreasberg, Harz Mountain, Germany. F with kind permission by Andreas Marten.

and to date there is no indication that particular "cave species" exist among gordiids. It appears more likely that single specimens are rinsed into the caves with water from outside the caves.

In the water, gordiids seem to be consumed regularly by fish (De Bacounin 1790, Kirjanova 1962; Fig. 3.8.2). Most records refer to trout (Lever 1963 in Ireland, Ponton et al. 2006 in France, Cochran et al. 1999 and Kinziger et al. 2002 in the USA, De Villalobos et al. 2008c in Argentina, McLennan & MacMillan 1984 in New Zealand). Other reports refer to the rock bass (*Ambloplites rupestris*) (USA, Cochran et al. 1999), longear sunfish (*Lepomis megalotis*) (USA, Kinziger et al. 2002), sterlet (*Acipenser ruthenus*) (Russia, Usynin 1978) and *Barbus barbus* (Danube river, Reichenbach-Klinke 1962) (see Cochran et al. 1999 for further records). In laboratory essays by

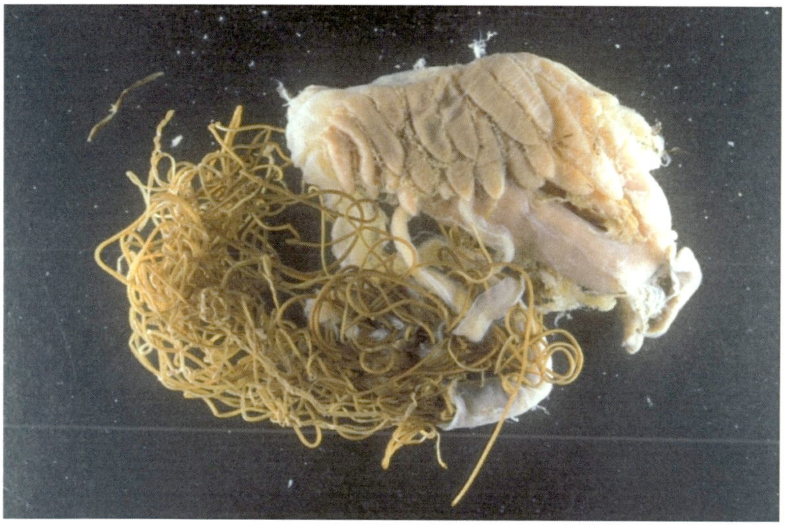

Fig. 3.8.2. Intestine of a trout containing numerous gordiids (Natural History Museum, London, 1947.12.17.7-12). From Schmidt-Rhaesa 2002c, with kind permission of Westarp Wissenschaften, Hohenwarsleben.

Cochran et al. (1999), most fish, except green sunfish (*Lepomis cyanellus*) rejected gordiids. This may be caused by fish feeding not directly on gordiids, but on infected insects (Reichenbach-Klinke 1962). Sato et al. (2008, 2011) assume that the behavioural alteration caused by the worms enhances the probability that infected insects are caught by fish. They found gordiids only in the stomach of fish that fed on camel crickets (*Diestrammena* sp.) in Japan. Particularly for the Japanese trout (*Salvelinus leucomaenis japonicus*) they could show that in the fall, infected orthopterans make up about 60% of the diet of trout (preferred over benthic invertebrates), which means that the energy input into the river ecosystem changes over the year, due to the parasitism caused by gordiids (Sato et al. 2011).

Ponton et al. (2006) observed in feeding experiments of infected crickets to predatory fish (trout *Oncorhynchus mykiss*, perch *Lepomis gibbosus*, bass *Micropterus salmoides*) and frogs (*Rana erythraea*) that a considerable number of worms (18–35%) escaped by winding out of the mouth, nose or gills of the predators. Whether this is due to active escape from predation or the effect of strong, undirected movements, may be left open. Escape from the host after this has been caught by a predator is also likely in a case where Fair et al. (2010) collected *Gordius robustus* from a number of nests of the Western Bluebird (*Sialia mexicana*) in New Mexico. It is quite likely that the birds caught infected insects and brought these to their nests. Gordiids emerged from the insect host before this was fed to nestlings.

A weird and certainly unusual case was reported by Notman & Yeates (1992) from New Zealand. A dead tomtit (*Petronica macrocephala*) was found with a gordiid coiled around its foot and a twig. It is likely that the tomtit was eating the hairworm or an infected insect and by the winding movements of the gordiid it tied the tomtit to the twig.

As parasites of arthropods, gordiids have been considered as potential candidates for biological control, e.g. in massive outbreaks of the Mormon Cricket (*Anabrus simplex*) (Wakeland & Shull (1936), see also Thorne (1940). Baker (1985) and Baker & Capinera (1997), however, concluded that nematomorphs are not suited as biological control agents because of their long development, large size, restricted number of potential hosts, infrequent occurrence in hosts and no *in vitro* production methods.

Very few studies have dealt with symbionts, commensals or parasites of nematomorphs. Epibionts of *Chordodes brasiliensis* in Argentina include diatoms, *Metrichia* sp. (Trichoptera) larvae and pupae and larvae of the beetles *Macrelmis* and *Phanocerus* (Elmidae) (De Villalobos et al. 2010). *Wolbachia* bacteria, which are commonly found in filarid nematodes and arthropods, are probably not present in nematomorphs (Duron & Gavotte 2007, Hudson & Floate 2009). This is especially interesting in the parthenogenetic species *Paragordius obamai*, because *Wolbachia* seem to be involved in parthenogenesis in other taxa (see Hanelt et al. 2012). Duron & Gavotte (2007) and Hudson & Floate (2009) additionally found no internal (symbiotic or not symbiotic) bacteria in gordiids. Parasites of nematomorphs are not known.

Gordiids have been subject to relatively few experiments concerning toxin resistance. Glyphosate, a weed-control herbicide does not affect larval development but affects the infective capacity of larvae of *Chordodes nobilii* and also has a 50% mortality in adult specimens (Achiorno et al. 2008b). Three standard toxicants, SDS (sodium dodecyl sulfate), cadmium chloride and potassium dichromate, were tested in larvae and juveniles of *Chordodes nobilii* by Achiorno et al. (2010). Egg development was not obviously altered, but larvae emerged from these eggs had a lower infection capacity. Potassium dichromate caused inhibition of larval development. Exposed larvae showed reduced survival for all three toxicants.

Acknowledgements

I thank my dear colleagues Ben Hanelt, Matt Bolek and Cristina de Villalobos for reading and improving the manuscript. Many thanks go to M. Bolek, A. Demegillo, M. Gaulke and A. Marten for sharing their photos with me. I am very thankful to numerous colleagues from all over the world, who shared their interest in horsehair worms with me. Over the years they have supplied material and discussions, without which the present review would not have been possible. Thank you all! Finally, many thanks to those of my students who were working with Nematomorpha; some of their data are included: Christoph Bleidorn, Sarah Gerke, Anne Görlitz, Ralf Jochmann, Anna-Sophie Rolof, André Schneider and Inga Wawrzynowicz.

Literature

This reference list is intended to represent all literature known to the author. For this list, only literature directly dealing with Nematomorpha has been considered. Chapters in textbooks are, if they are not cited in the text, omitted. Some popular records have also not been included. It is likely that there are a number of further records briefly mentioning nematomorphs.

In the following list, a regular citation indicates that the reference is cited in the text. An * indicates that the reference is mentioned only as part of a scientific taxon name (e.g. "*Gordius lineatus* Baek & Noh, 1992"). Parentheses [ ] indicate that the reference is not cited in the text. References are cited with their original titles; only references published in other writing systems, e.g. Cyrillic, Chinese or Japanese, are translated into English. In these cases, the original language is indicated (e.g. "Chinese, with English summary"). A° indicates that this record has been taken from the literature and could not be viewed by the author.

"Ä", "ö" and "ü" are treated like "ae", "oe" and "ue". References with more than two authors ("et al.") are ordered after the year and not alphabetically after the names of the second and consecutive authors.

Achiorno, C. L., Ferrari, L. & de Villalobos, C. (2008a): Effect of extreme temperature on egg development, larval and adult survival of *Chordodes nobilii* Camerano, 1901 (Gordiida, Nematomorpha). *Acta Parasitol.* 53: 392–396.

Achiorno, C. L., de Villalobos, C. & Ferrari, L. (2008b): Toxicity of the herbicide glyphosate to *Chordodes nobilii* (Gordiida, Nematomorpha). *Chemosphere* 71: 1816–1822.

Achiorno, C. L., de Villalobos, C. & Ferrari, L. (2010): Validation test with embryonic and larval stages of *Chordodes nobilii* (Gordiida, Nematomorpha): sensitivity to three reference toxicants. *Chemosphere* 81: 133–140.

Acholonu, A. D. (1968): *Neochordodes* sp. (Nematomorpha) as a parasite of the spider (*Ctenus bryrrbus*) in Costa Rica. *J. Parasitol.* 54: 1233–1234.

Adrianov, A. V., Malakhov, V. V. & Spiridonov, S. E. (1998): Fine structure of the larva of *Gordius* sp. (Nematomorpha). *Dokl. Akad. Nauk.* 361: 558–561. (in Russian)

Ali-Khan, F. E. A. & Ali-Khan, Z. (1977): *Paragordius varius* (Leidy) (Nematomorpha) infection in man: a case report from Quebec (Canada). *J. Parasitol.* 63: 174–176.

André, E. (1927): Sur la biologie des *Gordius*. *Bull. Soc. Vaudoise Sci. Nat.* 56: 285–286.

Andreeva, R. V. (1978): On parasitism of gordiids in larvae of *Tabanus autumalis* L. *Parazitologija* 12: 90–91. (in Russian with English summary)

Angelov, A. (1959): A new representative of the genus *Paragordius* (Camerano, 1898) from Bulgaria – *P. tricuspidatus* Dufour *minor* subsp. n. *God. Sof. Univ. Biol.* 51: 225–231. (in Russian with German summary)

Annandale, N. (1904): Note on the hair-worms in oriental and european folklore. *J. As. Soc. Bengal* 73: 75–77.

[Anonymus (1807): Note sur quelques habitudes observées chez des espéces d´un genre de ver nommé Dragonneaux (*Gordius*, Lin.). *Nouv. Bull. Sci. Soc. Philomant.* 1: 25–26.]

[Anonymus (1828): *Gordius aquaticus*. *Mag. Nat. Hist.* 1: 301.]

Anonymus (1831): no title. *Mag. Nat. Hist.* 4: 476–477.

Anonymus (1840): Ausziehung des *Gordius aquaticus* aus Wunden. *Z. Ges. Medizin* 14: 142–143.

[Anonymus (1870): Hair snakes. *Am. Ent. Bot.* 2: 180.]

Arvy, L. (1963): Données sur le parasitisme protélien de *Nectonema* (Nématomorphe), chez les Crustacés. *Ann. Parasitol.* 38: 887–892.

Arvy, L. & Sowa, R. (1976): Researches on internal parasites in Ephemeroptera around Krakow. *Ann. Parasitol. Hum. Comp.* 51: 111–120.

Assmuss, E. (1858): Verzeichniss einiger Insecten, in denen ich Gordiaceen antraf. *Wiener Entomol. Monatsschr.* 2: 177–181.

*Baek, K. M. (1993): Two species of genus *Chordodes* (Gordioidea, Nematomorpha) from Korea. *Korean J. Syst. Zool.* 9: 221–228.

*Baek, K. M. & Noh, Y. T. (1992): Two species of genus *Gordius* (Gordioidea, Nematomorpha) from Korea. *Korean J. Syst. Zool.* 8: 223–230.

Baird, W. (1853a): *Catalogue of the species of Entozoa or intestinal worms contained in the collection of the British Museum.* Presented by order of the trustees, London.

Baird, W. (1853b): Description of some new species of Entozoa from the collection of the British Museum. *Proc. Zool. Soc. Lond.* 18–25.

*Baird, W. (1861): Description of some new species of intestinal worms (Entozoa) in the collection of the British Museum. *Ann. Mag. Nat. Hist., 3rd. Ser.* 7: 228–230.

Baker, G. H. (1985): Parasites of the millipede *Ommatoiulus moreletii* (Lucas) (Diplopoda: Iulidae) in Portugal, and their potential as biological control agents in Australia. *Aust. J. Zool.* 33: 23–32.

Baker, G. L. & Capinera, J. L. (1997): Nematodes and nematomorphs as control agents of grasshoppers and locusts. *J. Entomol. Soc. Canada* 171: 157–211.

Bakke, A. (1975): En marin taggelmark *Nectonema* sp. (Nematomorpha) fra Nord-Norge. *Fauna (Oslo)* 28: 163–169.

Bareth, C. (1974): Présence de kystes et de larves de Gordiacés chez plusieurs espèces de Campodés cavernicoles (Diploures Campodéidés). *Ann. Spéléol.* 29: 657–662.

Bareth, C. (1975): Kystes et larves de Gordiacés chez 3 espèces de campodes cavernicoles. *Bull. Soc. Zool. France* 100: 247.

Bate, S. (1888): Report of scientific research: Crustacea Macrura. *Challenger Exp.* 24: 340–342.

Baylis, H. A. (1927): Notes on two Gordiids and a mermithid said to have been parasitic in man. *Trans. R. Soc. Trop. Med. Hyg.* 21: 203–206.

Baylis, H. A. (1944): Notes on the distribution of hairworms (Nematomorpha: Gordiidae) in the British Isles. *Proc. Zool. Soc. Lond.* 113: 193–197.

Beaver, P. C. & Orihel, T. C. (1965): Human infection with filariae of animals in the United States. *Am. J. Trop. Med. Hyg.* 14: 1010–1029.

Begon, M., Harper, J. L. & Townsend, C. R. (1990): *Ecology – Individuals, Populations, Communities. Second Edition.* Blackwell, Oxford.

Bell, F. J. (1885): Note on a nematoid worm (*Gordius verrrucosus*) obtained by Mr. H.H. Johnston on Kilima-njaro. *Proc. Zool. Soc. Lond.* 1885: 236.

Bell, F. J. (1886): Note on a nematoid worm (*Gordius verrucosus*) obtained by Mr. H.H. Johnston on Kilima-njaro. In: Johnston, H. H. (ed.) *The Kilima-Njaro expedition. A record of scientific exploration in eastern equatorial Africa*, pp. 364–365. Kegan Paul, Trench and Co., London.

Berthold, A. A. (1843): Ueber den Bau des Wasserkalbes (*Gordius aquaticus*). *Abh. Königl. Gesell. Wiss. Göttingen* 1: 1–18.

Biron, D. G., Marché, L., Ponton, F., Loxdale, H. D., Galéotti, N., Renault, L., Joly, C. & Thomas, F. (2005a): Behavioural manipulation in a grasshopper harbouring hairworms: a proteomics approach. *Proc. R. Soc.* B 272: 2117–2126.

Biron, D. G., Ponton, F., Joly, C., Menigoz, A., Hanelt, B. & Thomas, F. (2005b): Water-seeking behavior in insects harboring hairworms: should the host collaborate? *Behav. Ecol.* 16: 656–660.

[Biron, D. G., Joly, C., Marché, L., Galéotti, N., Calgano, V., Schmidt-Rhaesa, A., Renault, L. & Thomas, F. (2005): First analysis of the proteome in two nematomorph species, *Paragordius tricuspidatus* and *Spinochordodes tellinii* (Spinochordodidae). *Infect. Gen. Evol.* 5: 167–175.]

Biron, D. G., Ponton, F., Marché, L., Galeotti, N., Renault, L., Demey-Thomas, E., Poncet, J., Brown, S. P., Jouin, P. & Thomas, F. (2006): ´Suicide´ of crickets harbouring hairworms: a proteomics investigation. *Ins. Mol. Biol.* 15: 731–742.

Blair, D. (1983): Larval horsehair worms (Nematomorpha) from the tissues of native freshwater fish in New Zealand. *N.Z.J. Zool.* 10: 341–344.

Blanchard, M. R. (1897): Pseudo-parasitisme d´un *Gordius* chez l´homme. *Bull. Acad. Med. Paris* 37: 614–618.

Bleidorn, C., Schmidt-Rhaesa, A. & Garey, J. R. (2002): Systematic relationships of Nematomorpha based on molecular and morphological data. *Invertebrate Biol.* 121: 357–364.

Blunck, H. (1915): Ein kurzes Wort zur Kenntnis der Gordiidenbiologie. *Zool. Anz.* 45: 289–290.

Blunck, H. (1922): Zur Lebensgeschichte der im Gelbrand schmarotzenden Saitenwürmer. *Zool. Anz.* 54: 111–132 and 145–162.

[Blunck, H. (1924): Die Überwinterung von *Gordius aquaticus* L. in *Dytiscus marginalis* L. *Zool. Anz.* 61: 266–267].

Bock, S. (1913): Zur Kenntnis von *Nectonema* und dessen systematischer Stellung. *Zool. Bidr. Uppsala* 2: 1– 28.

[Bodenheimer, F. S. (1923): Die parasitären Beziehungen zwischen Würmern und Insekten. *Zentralbl. Bakteriol. Protozool. Parasitenkd. Infektionskr.* 58: 220–242.]

[Bogush, P. P. (1971): New data on *Chordodes longipilus* (Nemathelminthes: Nematomorpha). *Izvest. Akad. Nauk Turkmenskoi SSR Biol. Nauki* 5: 79. (in Russian)]

Bohall, P. J., Wells, M. R. & Chandler, C. M. (1997): External morphology of larvae of *Chordodes morgani* (Nematomorpha). *Invertebrate Biol.* 116: 26–29.

[Bohl, M. & Reichenbach-Klinke, H.-H. (1982): Saitenwürmer (Nematomorpha: Gordioidea) als gelegentliche Fisch- und Humanparasiten. *Fisch und Umwelt* 11: 31–35.]

[Bolek, M. G. (2000): Records of horsehair worms *Paragordius varius, Chordodes morgani* and *Gordius robustus* (Nematomorpha) from Indiana. *J. Freshwater Ecol.* 15: 421–423.]

Bolek, M. G. & Coggins, J. R. (2002): Seasonal occurrence, morphology, and observations of the lifecycle history of *Gordius difficilis* (Nematomorpha: Gordioidea) from Southeastern Wisconsin, United States. *J. Parasitol.* 88: 287–294.

Bolek, M. G., Schmidt-Rhaesa, A., Hanelt, B. & Richardson, D. J. (2010): Redescription of the African *Chordodes albibarbatus* Montgomery 1898, and description of *Chordodes janovyi* n. sp. (Gordiida, Nematomorpha) and its non-adult stages from Cameroon, Africa. *Zootaxa* 2631: 36–50.

Born, J. W. (1967): *Palaemonetes vulgaris* (Crustacea, Decapoda) as host for the juvenile stage of *Nectonema agile* (Nematomorpha). *J. Parasitol.* 53: 793–794.

Brattey, J., Elner, R. W., Uhazy, L. S. & Bagnall, A. E. (1985): Metazoan parasites and commensals of five crab (Brachyura) species from eastern Canada. *Can. J. Zool.* 63: 2224–2229.

Bresciani, J. (1970): The integument of *Gordius aquaticus* Duj. (Nematomorpha, Gordioida). *R. Vet. Agricult. Univ. Yearbook* 1970: 92–96.

Bresciani, J. (1975): Ultrastructure of the integument of the horsehair worm, *Nectonema munidae* Brinkmann 1930 (Nematomorpha). *2nd European Multicolloquium on Parasitology* p. 75.

Bresciani, J. (1991): Nematomorpha. In: Harrison, F. W. & Ruppert, E. E. (eds.) *Microscopic Anatomy of Invertebrates Vol. 4: Aschelminthes*, pp. 197–218. Wiley-Liss & Sons, New York.

Brinkmann, A. (1930): Über *Nectonema munidae* n.sp. *Bergens Museum Årbok* 9: 1–15.

Brivio, M. F., de Eguileor, M., Grimaldi, A., Vigetti, D., Valvassori, R. & Lanzavecchia, G. (2000): Structural and biochemical analysis of the parasite *Gordius villoti* (Nematomorpha, Gordiacea) cuticle. *Tiss. Cell* 32: 366–376.

Brusca, R. C. & Brusca, G. J. (2003): *Invertebrates, Second Edition*. Sinauer, Sunderland.

Bundy, D. A. P. & Steele, J. H. (1984): Parasitic zoonoses in the Caribbean region – a review. *Int. J. Zoonoses* 11: 1–38.

Burger, R. (1972): *Paragordius esavianus* passed per urethram. *J. Urology* 108: 469.

Bürger, O. (1891): Zur Kenntnis von *Nectonema agile Verr. Zool. Jb. Anat.* 4: 631–652.

Caballero, E. (1936): Contribucion al conocimiento de los *Gordius* de Mexico. *Ann. Inst. Biol. Univ. Nac. Mexico* 7: 477–488.

Camerano, L. (1887a): Ricerche intorno al parassitismo ed al polimorfismo dei Gordii. *Mem. R. Accad. Sci. Torino* 38: 395–413.

Camerano, L. (1887b): Del Gordius tricuspidatus (L. Dufour) in Italia. *Boll. Mus. Zool. Anat. Comp. R. Univ. Torino* 2, Nr. 28: 1–2.

Camerano, L. (1887c): Ricerche intorno alle specie italiane del genre *Gordius. Boll. Mus. Zool. Anat. Comp. R. Univ. Torino 2, Nr.* 20: 1–4.

*Camerano, L. (1887): Ricerche intorno alle specie italiane del genre *Gordius. Atti R. Accad. Sci. Torino* 22: 145–175.

[Camerano, L. (1887): Nuove osservazioni intorno ai caratteri diagnostici dei *Gordius. Zool. Anz.* 2: 602–604.]

[Camerano, L. (1887): Osservazioni sui charatteri diagnostici dei *Gordius* e sopra alcune specie di *Gordius* d´Europa. *Boll. Mus. Zool. Anat. Comp. R. Univ. Torino 2, Nr.* 24: 1–10.]

[Camerano, L. (1887): Nota intorno alla struttura della cuticula del *Gordius tricuspidatus* (L. Duf.). *Boll. Mus. Zool. Anat. Comp. R. Univ. Torino 2, Nr.* 25: 1–3.]

[Camerano, L. (1887): Observations sur les *Gordius. Arch. Ital. Biol.* 9: 59.]

Camerano, L. (1888a): Ricerche intorno alla anatomia ed istologia dei Gordii. *Boll. Mus. Zool. Anat. Comp. R. Univ. Torino 3, Nr.* 38: 1–6.

Camerano, L. (1888b): Ricerche sopra i Gordii d´Europa e descrizione di due nuove specie. *Boll. Mus. Zool. Anat. Comp. R. Univ. Torino 3, Nr.* 42: 1–10.

*Camerano, L. (1888): Descrizione di una nuova specie del genere *Gordius* raccolta in Birmania dal Sig. Leonardo Fea. *Ann. Mus. Civ. Storia Nat. Genova* 26: 168–170.

Camerano, L. (1889a): I primi momenti della evoluzione dei Gordii. *Mem. R. Accad. Sci. Torino* 40: 1–19.

Camerano, L. (1889b): Nuove osservazioni intorno ai Gordii Italiani. *Boll. Mus. Zool. Anat. Comp. R. Univ. Torino 4, Nr.* 66: 1–4.

[Camerano, L. (1889): Osservazioni intorno alla struttura dell´ integumento di alcuni Nematelminti. *Atti R. Accad. Sci. Torino* 24: 757–776.]

[Camerano, L. (1889): Sull´integumento dei Gordius. Boll. Mus. Zool. Anat. Comp. R. Univ. Torino 4, Nr. 54: 1–4.]

[Camerano, L. (1889): Osservazioni intorno alla struttura dell´ integumento di alcuni Nematelminti. *Atti R. Accad. Sci. Torino* 24: 3–23.]

Camerano, L. (1890a): Viaggio di Leornardo Fea in Birmania e regioni vicine. XXVII. Gordii. *Ann. Mus. Civ. Storia Nat. Genova* 10: 128–131.

Camerano, L. (1890b): Nuove osservazioni intorno ai Gordii italiani. III (1.). Gordii di Sardegna. *Boll. Mus. Zool. Anat. Comp. R. Univ. Torino 5, Nr.* 83: 1–2.

[Camerano, L. (1890): Intorno ad una specie di *Gordius* (*G. aeneus* Villot) raccolta dal Sig. G. B. Anselmo in Venezuela e intorno alle specie di questo genere fino ad ora descritte dell´America meridionale. *Ann. Mus. Civ. Storia Nat. Genova* 10: 123–127.]

Camerano, L. (1892): Ricerche intorno al parassitismo ed allo sviluppo del *Gordius pustulosus* Baird. *Atti R. Accad. Sci. Torino* 27: 598–607.

*Camerano, L. (1892): Descrizione di una nuova specie del genere *Gordius* raccolta nell´isola di Engano dal Dott. Elio Modigliani. *Ann. Mus. Civ. Storia Nat. Genova* 12: 539–541.

*Camerano, L. (1892): Descrizione di una nuova specie del genere *Gordius* di Palmeira (Paraná) raccolta dal Dott. G. Franco Grillo. *Ann. Mus. Civ. Storia Nat. Genova* 12: 965–966.

Camerano, L. (1893a): Descrizione di nuove specie di *Gordius* di Madagascar. *Boll. Mus. Zool. Anat. Comp. R. Univ. Torino 8, Nr.* 148: 1–2.

Camerano, L. (1893b): Sur quelques Gordiens nouveaux ou peu connus. *Bull. Soc. Zool. de France* 18: 213–216.

[Camerano, L. (1893): Ricerche intorno alla forza assoluta dei muscoli degli invertebrati. – I. Muscoli dei Gordii. *Atti R. Accad. Sci. Torino* 28: 221–233.]

*Camerano, L. (1894): Sopra due specie di *Gordius* raccolte in Sumatra dal Dott. Elio Modigliani. *Ann. Mus. Civ. Storia Nat. Genova* 14: 410–412.

Camerano, L. (1894): Viaggio del dottor Alfredo Borelli nella Repubblica Argentina e nel Paraguay. I. Gordii. *Boll. Mus. Zool. Anat. Comp. R. Univ. Torino 9, Nr.* 175: 1–6.

Camerano, L. (1895): Gordiens nouveaux ou peu connus du Musée d´ Histoire Naturelle de Leyde. *Notes Leyden Mus.* 17: 1–14.

*Camerano, L. (1895): Description d´une nouvelle espéce de Gordien de la Chine. *Bull. Soc. Zool. France* 20: 99–100.

*Camerano, L. (1895): Description d´un nouvelle espéce de *Gordius* du Chili. *Actes Soc. Sci. Chili* 5: 8–9.

Camerano, L. (1896): Gordiens nouveaux ou peu connus du Musée Zoologique de l´Academie Impériale des Sciences de St.-Pétersbourg. *Ann. Mus. Zool. Acad. Imp. St. Petersbourg* 1: 117–125.

*Camerano, L. (1896): Descrizione di una nuova specie di Gordio del Basso Beni (Bolivia) raccolta dal Prof. L. Balzan. *Ann. Mus. Civ. Storia Nat. Genova* 36: 9–10.

Camerano, L. (1897a): Monografia dei Gordii. *Mem R. Accad. Sci. Torino* 47: 339–415.

Camerano, L. (1897b): Viaggio del Dott. Alfredo Borelli nel Chaco boliviano e nella Repubblica Argentina. VIII Gordii. *Boll. Mus. Zool. Anat. Comp. R. Univ. Torino 12, Nr.* 294: 1–3.

*Camerano, L. (1897): Viaggio del Dr. Enrico Festa nella Repubblica dell´Equador e regioni vicine. III. Gordii. *Boll. Mus. Zool. Anat. Comp. R. Univ. Torino 12, Nr.* 293: 1–2.

[Camerano, L. (1897): Nuova classificazione dei Gordii. *Zool. Anz.* 20: 225–229.]

Camerano, L. (1898): Gordiens du Mexique. *Bull. Soc. Zool. France* 73–77.

Camerano, L. (1899): Gordii della Malesia e del Messico. *Atti R. Accad. Sci. Torino* 34: 460–469.

Camerano, L. (1901a): Gordii di Madagascar e delle isole Sandwich. *Boll. Mus. Zool. Anat. Comp. R. Univ. Torino 16, Nr.* 412: 1–2.

Camerano, L. (1901b): Viaggio del Dr. A. Borelli nel Matto Grosso e nel Paraguay. *Boll. Mus. Zool. Anat. Comp. R. Univ. Torino 16, Nr.* 411: 1–2.

*Camerano, L. (1901): Gordii raccolti dal Dottor Filippo Silvestri nella Repubblica Argentina e nel Paraguay. VI. Gordii. *Boll. Mus. Zool. Anat. Comp. R. Univ. Torino 16, Nr.* 410: 1–2.

[Camerano, L. (1901): Gordii raccolti della spedizione "Skeat" nella Penisola Malese 1899–1900. *Boll. Mus. Zool. Anat. Comp. R. Univ. Torino 16, Nr.* 408: 1–2.]

*Camerano, L. (1902): Nuova specie di *Chordodes* del Sudan. Boll. Mus. Zool. Anat. Comp. R. Univ. Torino 17, Nr. 416: 1–3.

*Camerano, L. (1902): Descrizione di una nuova specie di *Chordodes* del Congo. *Boll. Mus. Zool. Anat. Comp. R. Univ. Torino 17, Nr.* 426: 1–2.

Camerano, L. (1903a): Nematomorpha. In: Shipley, A. E.: On the ento-parasites, collected by the "Skeat Expedition" to Lower Siam and the Malay peninsula in the years 1899–1900. *Proc. Zool. Soc. Lond.* 2: 152–153.

Camerano, L. (1903b): Gordii di Ceylan. *Boll. Mus. Zool. Anat. Comp. R. Univ. Torino* 18, Nr. 438: 1–2.

Camerano, L. (1903c): Gordiens nouveaux ou peu connus du Musée Zoologique de l´Académie Impériale des Sciences de St. Pétersbourg. *Ann. Mus. Zool. Acad. Imp. St. Petersbourg* 8: 22–29.

*Camerano, L. (1903): Nuova specie di Gordii del Basso Siam. *Boll. Mus. Zool. Anat. Comp. R.* Univ. *Torino 18, Nr.* 437: 1–3.

Camerano, L. (1904a): Gordians of Ceylon. *Spolia Zeylanica* 1: 34–35.

Camerano, L. (1904b): Osservazioni intorno al *Chordodes skorikovi* Camer. *Boll. Mus. Zool. Anat. Comp. R. Univ. Torino* 19, Nr. 469: 1–3.

Camerano, L. (1904c): Della identitá specifica del *Parachordodes wolterstorffii* e del *Parachordodes pleskei. Boll. Mus. Zool. Anat. Comp. R. Univ. Torino* 19, Nr. 470: 1–4.

*Camerano, L. (1904): Nuova specie di *Chordodes* del Guatemala. *Ann. Mus. Civ. Storia Nat. Genova* 93–94.

*Camerano, L. (1904): Nuova specie di Gordio dell´Alta Birmania. *Boll. Mus. Zool. Anat. Comp. R. Univ. Torino 19, Nr.* 468: 1–3.

Camerano, L. (1905): Gordii dei Pirenei. *Boll. Mus. Zool. Anat. Comp. R. Univ. Torino* 20, Nr. 505: 1–2.

[Camerano, L. (1905): Osservazioni intorno al *Chordodes festae* Camer. *Boll. Mus. Zool. Anat. Comp. R. Univ. Torino* 20, Nr. 504: 1–2.]

[Camerano, L. (1905): Gordii dell´Isola d´Elba. *Monit. Zool.* 16: 249–250.]

Camerano, L. (1908a): Gordii d´Irlanda. *Boll. Mus. Zool. Anat. Comp. R. Univ. Torino* 23, Nr. 578: 1–2.

Camerano, L. (1908b): Gordiens du Musée Indien. *Rec. Indian Mus.* 2: 112–117.

Camerano, L. (1908c): Nota sul *Chordodes hawkeri*, Camer. *Boll. Mus. Zool. Anat. Comp. R. Univ. Torino* 23, Nr. 580: 1–2.

Camerano, L. (1908d): Sur les Gordiens recueillis par le Captaine F.H. Steward dans le Tibet. *Rec. Indian Mus.* 2: 315–316.

*Camerano, L. (1910): 4. Gordiidae. In: Sjöstedt, Y. (ed.) *Wissenschaftliche Ergebnisse der Schwedischen Zoologischen Expedition nach dem Kilimandjaro, dem Meru und den umgebenden Massaisteppen Deutsch-Ostafrikas 1905–1906.* 3. Vol., Palmquists Aktiebolag, Stockholm.

Camerano, L. (1912a): Osservazioni intorno al *Chordodes hawkeri* Camer. *Boll. Mus. Zool. Anat. Comp. R. Univ. Torino* 27, Nr. 646: 1–2.

Camerano, L. (1912b): Gordiens du Musée Indien. *Rec. Indian Mus.* 7: 215–216.

*Camerano, L. (1912): Nuova specie di *Chordodes* del Congo. *Boll. Mus. Zool. Anat. Comp. R.* Univ. *Torino 27, Nr.* 645: 1–2.

Camerano, L. (1915a): Revisione dei Gordii. *Mem. R. Accad. Sci. Torino* 66: 1–66.

Camerano, L. (1915b): Nota su due Gordii dell´Eritrea. *Boll. Lab. Zool. Gen. Agrar. R. Scuola Agricolt. Portici* 9: 335–336.

Čanadjija, S. (1956): Horsehair worms from the Zoological Museum in Zagreb. *Bull. Mus. Hist. Nat. Pays Serbe, Ser.* B 8: 1–38. (in Serbocroatian with German summary)

Cappucci, D. T. (1976): *The biology of* Gordius robustus *Leidy with a host list and summary of public health importance of the Gordioidea.* PhD thesis, University of California, San Francisco.

Camerano, L. (1982): Gordian worms (hairworms): *Biological and public health aspects*. 193–203.

Căpuşe, I. (1965): Un Gordiid nou pentru stiinta *Dacochordodes bacescui* gen.n., sp. n. (Ord. Chordodes, fam. Chordodidae). *Revista Muzeelor* 2: 134–136. (in Romanian with Russian and French summary)

Camerano, L. (1966a): Contributions a la connaissance de la répartition des Gordiens en Roumanie. *Trav. Mus. Hist. Nat. Grigore Antipa* 6: 9–16.

Camerano, L. (1966b): Sur le *Dacochordodes bacescui* Capuse (Ord. Chordodeia, Fam. Chordodidae). *Trav. Mus. Hist. Nat. Grigore Antipa* 6: 17–23.

Camerano, L. (1971): Nouvelles données sur la répartition et la biologie de quelques Gordiides de Roumanie. *Trav. Mus. Hist. Nat. Grigore Antipa* 10: 49–56.

Carvalho, J. C. M. (1942): Studies on some Gordiacea of North and South America. *J. Parasitol.* 28: 213–222.

Carvalho, J. C. M. (1944): Consideraçöes sôbre alguns Gordiáceos Brasileiros, com descricão de duas espécies novas (Gordiacea, Chordodidae). *Rev. Brasil. Biol.* 4: 485–491.

Carvalho, J. C. M. (1946a): Gordiáceos do Museu de História Natural de Montevideo. *Comun. Zool. Mus. Hist. Nat. Montevideo* 2: 1–7.

Carvalho, J. C. M. (1946b): Uma nova espécie de Gordiaceo do Brasil e consideraçöes sôbre *Chordodes brasiliensis* Janda (Nematomorpha, Gordoidea). *Bol. Mus. Nac. (N. S.)* 62: 1–8.

Carvalho, J. C. M. & Feio, J. L. A. (1950): Sôbre alguns Gordiáceos do Brasil e da Republica Argentina (Nematomorpha, Gordioidea). *Ann. Acad. Brasil Cienc.* 22: 193–216.

Cavalieri, F. (1961): Descripcion de una nueva especie de Gordiaceo *Gordius inesae* n.sp. (Gordioidea, Gordiidae). *Neotropica* 7: 3–6.

[Cazurro (1890): Nota sobre los Gordius. *An. Soc. Espanola Hist. Nat.* 19: 6–8.]

Cerini, G. & Ristori, G. F. (1988): Su alcuni casi di infestazione accidentale da *Gordius. Osped. Ital. Ped.* 23: 97–101.

°Cerutti, G. B. & Camerano, L. (1888): Di un nuovo caso di parasitissio di *Gordius. Giorn. R. Accad. Med. Torino 3*, 36: 264–269.

Cham, S. A., Seymour, M. K. & Hooper, D. J. (1983): Observations on a british hairworm, *Parachordodes wolterstorffii* (Nematomorpha: Gordiidae). *J. Zool.* 199: 275–285.

[Chandler, C. M. (1985): Horsehair worms (Nematomorpha, Gordioidea) from Tennessee, with a review of taxonomy and distribution in the United States. *J. Tennessee Acad. Sci.* 60: 59–62.]

[Chandler, C. M. & Wells, M. R. (1989): Cuticular features of *Chordodes morgani* (Nematomorpha) using Scanning Electron Micoscopy. *Trans. Am. Microsc. Soc.* 108: 152–158.]

Chapman, T. A. (1902): *Gordius* in a butterfly. *The Entomologist* 35: 42.

[Charvet, M. (1834): Observations sur deux espéces du genre Dragonneau, qui habitent dans quelques eaux courantes, aux environs de Grenoble. *Nouv. Ann. Mus. Hist. Nat. Paris* 3: 37–46.

Chavannes (1864): no title. *Bull. Soc. Vaudoise Sc. Nat.* 8: 183.

Chiu, M. C., Huang, C. -G., Wu, W. -J. & Shiao, S. -F. (2011): A new horsehair worm, *Chordodes formosanus* n. sp. (Nematomorpha: Gordiida) from *Hierodula* mantids of Taiwan and Japan with redescription of a closely related species, *Chordodes japonensis. Zookeys* 160: 1–22.

Chitwood, B. G. & Chitwood, M. B. (1937): Snails as hosts and carriers of nematodes and Nematomorpha. *The Nautilus* 50: 130–135.

Claparède, E. R. (1862): Recherches anatomiques sur les oligochètes. *Mém. Soc. Phys. Genève* 16 : 217–291.

°Clark (1854): Presentation of two specimens of very complete filiform animals said to have been voided from the bladder of two females. *Am. Med. Monthly* 2: 148.

Clarke, H. T. (1831): Gordii, beetles, and *Dytiscus* found in the intestine of a frog. *Mag. Nat. Hist.* 4: 95–96.

Clementi, V. (1869): Hair snakes. *Can. Entomol.* 1: 62.

Cloquet, H. (1822a): Sur une nouvelle espèce d´Entozoaire, du genre Ophiostome. *Bull. Soc. Philomath. Paris*, 3. Ser. 9: 32.

Cloquet, H. (1822b): Note sur un nouvelle espèce d´entozoaire. *J. Med. Chir. Pharm.* 13: 97–99.

[Cochran, P. A. (1999): Notes on the distribution and biology of the horsehair worm *Paragordius varius* (phylum Nematomorpha) in Wisconsin. *J. Freshwater Ecol.* 14: 137–139.]

[Cochran, P. A., Watermolen, D. J. & Haen, G. L. (1990): Horsehair worms (phylum Nematomorpha) in lake Michigan. *J. Great Lakes Res.* 16: 485–487.]

Cochran, P. A., Kinziger, A. P. & Poly, W. J. (1999): Predation on horsehair worms (phylum Nematomorpha). *J. Freshwater Ecol.* 14: 211–218.

Cochran, P. A., Newton, A. K. & Korte, C. (2004): Great gordian knots: sex ratio and sexual size dimorphism in aggregations of horsehair worms (*Gordius difficilis). Invert. Biol.* 123: 78–82.

Codina, A. (1925): Un cas d´un *Gordius* (Verm.: Nematoides) extret del cos d´un Stauroderus (Orth.: Acrydiinae). *Bull. Inst. Catalana Hist. Nat.* 5: 61–64.

Conn, D. B. (1981): *Cave life of Carter Caves State Park*, pp. 1–50. Appalachian Development Center, Morehead, Kentucky.

Cooper, C. L. & Storck, T. W. (1973): *Gordius* sp., a new host record. *Ohio J. Sci.* 73: 228.

Corallini Sorcetti, C. & Moretti, G. P. (1991): Alimentazione, parassiti ed epibionti delle larve di alcune specie di Tricotteri della Puglia. *Atti Congr. Naz. Ital. Entomol.* 16: 495–498.

Cort, W. W. (1915): *Gordius* larvae parasitic in a trematode. *J. Parasitol.* 1: 198–199.

[Cowen, W. W. & Williams, R. K. (1965): Horse hair worms or untying the Gordian knot. *Centr. Afr. J. Med.* 11: 359–360.]

Craig, C. F. & Faust, E. C. (1951): *Clinical Parasitology,Fifth Edition.* Lea & Febiger, Philadelphia.

Creplin, C. H. (1847): *Chordodes parasitus*, ein Schmarotzerwurm aus einer Heuschrecke. *Notizen Geb. Nat.– Heilkd.* 3: 161–166.

Crisp, E. (1873): *Gordius* in the lungs of the sheep. *Trans. Pathol. Soc. Lond.* 24: 276–279.

[Crivelli, G. B. (1843): Storia del genere *Gordius* e d´un nuovo elminto *Autoplectus protognostus* detto volgarmente gringo o filo dai contadini Lombardi. *Mem. R. Inst. Lombardo Sci. Lett. Art.* 2: 1–23.]

[Crivelli, G. B. (1851): Risposta all' analisi della memoria intitolata storia del genere *Gordius* e d'un nuovo elminto detto Gringo dai Contadini Lombardi. *Nuovi Ann. Sci. Nat.* 3(4): 73–76.]

Cury, R. (1946): Moléstias das abelhas. *O Biologico* 12: 241–254.

Cui, W. (1995): A case of Gordiacea infestation in human digestive tract. *Chin. J. Parasitic Dis. Contr.* 8: 298.

Czeczuga, B. (1981): Presence of carotenoids in *Gordius aquaticus* (Nematomorpha, Nemathelminthes) specimens from deep wells. *Hydrobiologia* 76: 13–15.

Dasgupta, R. & Khawas, P. (1986): A gordiid worm in Darjeeling. *J. Bengal Nat. Hist. Soc.* 5: 160–162.

De Bacounin, M. A. (1790): Mémoire sur les *Gordius* d´eau douce des environs de Turin. *Mem. Accad. R. Sci. Torino*: 23–42.

*De Beauchamp, P. (1913): *Turbellaries, Trématodes et Gordiacés del viaggio di Alluand e Jeannel nel l´Africa orientale.* Paris.

*De Beauchamp, P. [(1916): Sur quelques parasites provenant du Congo Belge. *Rev. Zool. Africaine* 4: 109–116.]

*De Beauchamp, P. (1923): Sur quelques Gordiens du Congo Belge. *Rev. Zool. Africaine* 11: 65–70.

[De Gravere, C. (1873): Mededeeling over eene soort van *Gordius* of *Mermis* gevonden te Batoer, residentie Banjoemas. *Natuurkd. Tijdschr. Nederlands-Indie* 33: 513.]

De Lucena, D. T. & Campelo, M. (1975): Accidental parasitism of man with two gordiids. *Rev. Inst. Med. Trop. São Paulo* 17: 390–393.

De Miralles, D. A. B. (1967): Nueva distribucion de *Gordionus prismaticus* (Villot) en Argentina. *Neotropica* 13: 195–196.

De Miralles, D. A. B. (1969): Lista preliminar de Gordiaceos de Argentina y datos biologicos sobre *Neochordodes uniareolatus* Carv. *Acta Zool. Lilloana* 24: 183–192.

*De Miralles, D. A. B. (1972): Sobre especies Sudamericanas del Genero *Beatogordius* (Nematomorpha Gordiidae). *Neotropica* 18: 121–125.

[De Miralles, D. A. B. (1972): Sobre algunos Gordiaceos del Museo de la Plata. *Commun. 3as Jorn. Argentinas Zool.* 262.]

De Miralles, D. A. B. (1973): Sobre la presencia de *Pseudochordodes bedrigae* (Camerano) en Argentina (Gordioidea Chordodidae). *Neotropica* 19: 101–103.

De Miralles, D. A. B. (1975): Nuevo aporte al conocimiento de la Gordiofauna Argentina. *Neotropica* 21: 99–103.

[De Miralles, D. B. (1976): Gordioidea. In : Ringuelet, R.A. (ed.). *Fauna de agua dulce de la Republica Argentina* Vol. 13, pp. 7–45. FECIC, Buenos Aires.]

De Miralles, D. A. B. (1976): Nota sobre Gordiaceos Argentinos. *Neotropica* 22: 77–80.

De Miralles, D. A. B. (1977): Nota sobre Gordiaceos Argentinos. *Neotropica* 23: 7–9.

[De Miralles, D. A. B. (1977): Gordiida. In : Hurlbert, S. H. (ed.) *Biota acuatica de Sudamerica Austral*, pp. 80–82. San Diego State University, San Diego.]

De Miralles, D. A. B. (1980): Observaciones bioecologicas sobre Gordiaceos de Ventania. *Neotropica* 26: 217–224.

*De Miralles, D. A. B. (1981): Nueva especie de Gordiaceo de Sierra de la Ventana, *Beatogordius variabilis* sp. nov. (Gordioidea Parachordodinae). *Neotropica* 27: 87–90.

[De Miralles, D. A. B. (1981): Gordioidea. In: Hurlbert, S. H., Rodriguez, G. & Dias Dos Santos, N. (eds) *Aquatic biota of tropical South America. Part 2: Anarthropoda*, pp. 151–154. San Diego State University, San Diego.]

De Miralles, D. A. B. (1984a): Nota sobre Gordiaceos Argentinos. III. *Neotropica* 30: 139–140.

De Miralles, D. A. B. (1984b): Estudio sobre Gordiaceos Argentinos. *Rev. Mus. La Plata* 13: 283–289.

[De Miralles, D. A. B. (1984): Nueva combinacion para Gordiaceos Argentinos. *Beatogordius inesae* (Cavalieri) conj. nov. y *B. raphaelis* (Camerano) conj. nov. *Neotropica* 30: 125–126.]

De Miralles, D. A. B. (1988): Nota sobre Gordiaceos Argentinos. IV. *Neotropica* 34: 62.

[De Miralles, D. A. B. (1989): Estructura cuticular de *Chordodes carmelitanus* Carvalho y Feio, 1950 nueva cita para la Argentina (Nematomorpha Chordodidae). *Neotropica* 35: 95–99.]

De Miralles, D. A. B. (1993): Contribución al estudio da la estructura cuticular de Gordiaceos argantinos. *Neotropica* 39: 101–102.

De Miralles, D. A. B. & Camino, N. B. (1983): Sobre la colleccion de Gordiaceos del Museo Argentino "Bernardino Rivadavia". *Neotropica* 29: 51–56.

[De Miralles, D. A. B. & De Villalobos, L. C. (1991): Consideraciones sobre la taxonomia y lista generica actualizada de la clase Gordiacea (Nematomorpha). *Biota* 7: 37–42.]

De Miralles, D. A. B. (1993): Redescripcion de *Gordius agasizzi* Montgomery, 1898, de *Pseudochordodes occidentalis* comb. n. y nuevo registro para *P. bedrigae* (Camerano, 1896) (Gordioidea, Nematomorpha). *Iheringia, Sér. Zool.* 75: 107–111.

[De Miralles, D. A. B. (1993): Distribucion geografica de los Gordiaceos en la Republica Argentina. In : De Castellanos, Z. A. (ed.) *Fauna de agua dulce de la Republica Argentina*, Vol. 13. pp. 5–32. PROFADU (CONICET), La Plata.]

[De Miralles, D. A. B. (1994): Redescripcion de *Chordodes jandae* Camerano, 1895 (Nematomorpha). *Comun. Mus. Cienc. Tecnol. PUCRS, Sér Zool.* 7: 195–203.]

De Miralles, D. A. B. (1994): Nueva combinación para gordiáceos mejicanos. *Neotropica* 40: 14.

De Miralles, D. A. B. (1995): A review of the Gordiacea (Nematomorpha) in the collection of the California Academy of Sciences with the description of a new species. *Graellsia* 51: 17–22.

De Miralles, D. A. B. (1996a): Nuevos *Gordius* para la Argentina (Gordiacea, Nematomorpha). *Acta Zool. Mex.* 68: 67–73.

De Miralles, D. A. B. (1996b): Nueva especie de *Neochordodes* (Gordiacea Nematomorpha) parásita de *Scapteriscus* sp. (Grillotalpidae). *Rev. Acad. Cienc. Zaragoza* 51: 237–246.

De Miralles, D. A. B. (1996c): Especies de *Neochordodes* de la Argentina (Gordiacea, Nematomorpha). *Iheringia, Sér. Zool.* 81: 145–150.

De Miralles, D. A. B. (1996d): Dos nuevas especies de Gordiáceos para Ventania, Argentina (Nematomorpha). *Comun. Mus. Cienc. Tecnol. PUCRS, Sér Zool.* 9: 3–9.

*De Miralles, D. A. B. (1996): Una nueva especie de *Gordius* (Linne, 1758) (Gordiaceae, Nematomorpha) de la Patagonia. *Nat. Patagónica Cienc. Biol.* 4: 41–45.

*De Miralles, D. A. B. (1997): Ultraestructura cuticular de una nueva especie de *Chordodes* Creplin, 1874, para la Argentina (Nematomorpha – Gordioidea – Chordodidae). *Estud. Biol.* 42: 41–48.

De Miralles, D. A. B. (1999): Consideraciones sobre los danos ocasionados en insectos por nematomorfos. *Rev. Soc. Cienc. Morfol.* 3: 11–16.

De Miralles, D. A. B., De Villalobos, L. C. & Rodriguez, A. (1997): Una nueva especie y una nueva combinación de gordiáceos (Nematomorpha). *Neotropica* 43: 53–56.

*De Miralles, D. B. & De Villalobos, L.C. (1998): Dos nuevas especies y una nueva combinación para el género *Acutogordius* (Nematomorpha: Gordiidae). *Neotropica* 44: 75–79.

De Miralles, D. A. B. (2000): *Noteochordodes*, un nuevo género para la República Argentina (Nematomorpha, Gordiaceo). *Revista del Museo de La Plata (N. S.)* 14: 279–283.

*De Villalobos, L. C. (1995): Una nueva especie del genero *Chordodes* Creplin, 1874, en la Argentina. *Neotropica* 41: 63–65.

De Villalobos, L. C. & Camino, N. B. (1999): Dos nuevas especies de Gordiaceos (Nematomorpha) parasitos de *Stagmatoptera hyaloptera* (Mantidae) en la Argentina. *Iheringia, Sér Zool.* 86: 71–76.

De Villalobos, L. C. & De Miralles, D. A. B. (1997): Una nueva especie de *Chordodes* (Gordiaceo, Nematomorpha) parasita de *Blaptica* sp. (Blaberidae, Blataria). *Comun. Mus. Cienc. Tecnol. PUCRS, Sér. Zool.* 10: 45–51.

De Villalobos, L. C. & Restelli, M. A. (2001): Ultrastructural study of the cuticle and epidermis in *Pseudochordodes bedriagae. Cell Tiss. Res.* 305: 129–134.

De Villalobos, C. & Ronderos, M. (2003): *Dasyhelea necrophila* Spinelli et Rodriguez, 1999 (Diptera, Ceratopogonidae) a new potential paratenic host of *Paragordius varius* (Leidy, 1851) (Gordiida, Nematomorpha). *Acta Parasitol.* 48: 218–221.

De Villalobos, L. C. & Voglino, D. (2000): Description of *Gordionus porosus* n.sp. and redescription of *Gordionus ondulatus* (Nematomorpha, Parachordodinae) with scanning electron microscopy. *J. Parasitol.* 86: 358–365.

De Villalobos, C. & Zanca, F. (2001): Scanning electron microscopy and intraspecific variation of *Chordodes festae* Camerano, 1897 and *C. peraccae* (Camerano, 1894) (Nematomorpha: Gordioidea). *Syst. Parasitol.* 50: 117–125.

De Villalobos, C. & Zanca, F. (2002): Scanning electron microscopy of *Spinochordodes tellinii* (Camerano, 1888), (Gordiacea, Nematomorpha). *Parasite* 9: 71–74.

De Villalobos, C. & Zanca, F. (2005): Ultrastructural redescription of *Chordodes moraisi* (Carvalho, 1942) and *Chordodes straviarskii* Carvalho and Feio 1950, and re-interpretation of *Chordodes gestri* Camerano 1904 and *Pseudochordodes griffinii* (Camerano 1898) (Gordiida, Nematomorpha). *J. Nat. Hist.* 39: 597–606.

De Villalobos, L. C., Ribera, I. & Downie, I. S. (1999): Hairworms found in Scottish agricultural land, with descriptions of two new species of *Gordionus* Müller (Nematomorpha: Gordiidae). *J. Nat. Hist.* 33: 1767–1780.

De Villalobos, C., Ramirez, G., Chavarria, S. & Zanca, F. (2000): Cuticular ultrastructure of *Paragordius esavianus* Carvalho and *P. varius* (Leidy) (Nematomorpha: Chordodidae: Paragordinae). *Proc. Acad. Nat. Sci. Philadelphia* 150: 135–144.

De Villalobos, L. C., Ribera, I. & Bilton, D. T. (2001): First data of Iberian Nematomorpha, with redescription of *Gordius aquaticus* Linnaeus, *G. plicatulus* Heinze, *Gordionus wolterstorffii* (Camerano) and *Paragordius tricuspidatus* (Dufour). *Contrib. Zool.* 70: 73–84.

De Villalobos, L. C., Schmidt-Rhaesa, A. & Zanca, F. (2003b): Revision of the genus *Beatogordius* (Gordiida, Nematomorpha). II. South American species with description of two new species. *Mem. Inst. Oswaldo Cruz* 98: 115–128.

De Villalobos, C., Rumi, A., Núñez, V., Schmidt-Rhaesa, A. & Zanca, F. (2003a): Paratenic hosts: larval survival strategy in *Paragordius varius* (Leidy, 1851) (Gordiida, Nematomorpha). *Acta Parasitol.* 48: 98–102.

De Villalobos, C., Hancock, E. G. & Zanca, F. (2004): Redescription and sexual dimorphism of *Chordodes balzani* Camerano, 1896 (Nematomorpha). *J. Nat. Hist.* 38: 2305–2313.

[De Villalobos, C., Zanca, F. & Schmidt-Rhaesa, A. (2004): New data on South American species of *Chordodes* (Nematomorpha). *Arq. Mus. Nac. Rio de Janeiro* 62: 375–386.]

De Villalobos, C., Restelli, M., Schmidt-Rhaesa, A. & Zanca, F. (2005a): Ultrastructural observations of the testicular epithelium and spermatozoa of *Pseudochordodes bedriagae* (Gordiida Nematomorpha). *Cell Tiss. Res.* 321: 251–255.

De Villalobos, C., Zanca, F. & Ibara-Vidal, H. (2005b): Redescription and new records of freshwater Nematomorpha (Gordiida) from Chile, with the description of two new species. *Rev. Chilena Hist. Nat.* 78: 673–686.

De Villalobos, C., Ronderos, M. M., Greco, N., Zanca, F., Diaz, F. & Achiorno, C. (2006): Susceptibility of *Aedes aegypti* larvae to parasitism by *Paragordius varius* under laboratory conditions. *Trans. Am. Entomol. Soc.* 132: 121–127.

[De Villalobos, C., Zanca. F. & Schmidt-Rhaesa, A. (2007): Revision of the genus *Chordodes* (Gordiida, Nematomorpha) from Africa. III. Ultrastructural redescription of *Chordodes capensis* Camerano, 1895, *C. clavatus* Linstow, 1906, *C.digitatus* Linstow, 1901, *C. tuberculatus* Linstow, 1901, and reinterpretation of *C ibembensis* Sciacchitano, 1958 and *C. uncinatus* Sciacchitano, 1958. *J. Nat. Hist.* 41: 2775–2788.]

[De Villalobos, C., Zanca, F. & Spiridonov, S. E. (2007): Fine morphology of the cuticle surface of *Chordodes anthophorus* and reinterpretation of *C. aquaeductus, C. ferganensis* and *C. oscillatus* (Gordiida Nematomorpha). *Zootaxa* 1397: 39–45.]

De Villalobos, C., Zanca, F. & Salas, L. (2008a): Especies de Gordiida (Nematomorpha) en la Provincia de Catamarca. *Rev. Cienc. (UNCA)* 3: 47–56.

De Villalobos, C., Zanca, F. & Salas, L. (2008b): Redescription of the genus *Noteochordodes*, reinterpretation of *Noteochordodes dugesi* (Camerano, 1898), a new combination of *Neochordodes talensis* (Camerano, 1897) and a description of a new species. *J. Nat. Hist.* 42: 187–202.

De Villalobos, C., Ortiz-Sandoval, J. J. & Habit, E. (2008c): Finding of *Gordius austrinus* de Villalobos, Zanca & Ibarra-Vidal, 2005 (Gordiida, Nematomorpha) in the stomach of *Salmo trutta* (Salmoniformes) in Patagonia. *Gayana* 72: 31–35.

De Villalobos, C., Zanca, F. & Yanez, A. B. (2009): Three new species of *Chordodes* and new data on hairworms (Nematomorpha) from Ecuador. *Zootaxa* 2205: 37–52.

[De Villalobos, C., Zanca, F. & Schmidt-Rhaesa, A. (2009): Revision of the genus *Chordodes* (Gordiida: Nematomorpha) from Africa IV. Ultrastructural redescription of *Chordodes congolensis* Sciacchitano, 1933, *Chordodes ferox* Camerano, 1897, *Chordodes madagascariensis* (Camerano, 1893), *Chordodes mobensis* Sciacchitano, 1958 and reinterpretation of *Chordodes maculatus* Sciacchitano, 1958 and *Chordodes kakandensis* Sciacchitano, 1958. *J. Nat. Hist.* 43: 2579–2595.]

De Villalobos, C., Salas, L., Zanca, F. & Atencio, A. (2010): The occurrence of epibionts of Gordiida (Nematomorpha) in Catamarca, Argentina. *Acta Parasitol.* 55: 281–285.

Dearolf, K. (1953): The invertebrates of 75 caves in the United States. *Proc. Pennsylvania Acad. Sci.* 27: 225–241.

Degland, C. D. (1823): Description d´un ver filiforme, rendu par le vomissement. *Rec. Trav. Soc. Amateurs Sci. Lille* 1819–1822: 166–169.

Degrange, C. & Martinot, J. -P. (1995): Gordiacés parasites d´orthoptères du Parc National de la Vanoise. *Trav. Sci. Parc Nat. Vanoise* 19: 161–182.

Deronde (1936): Le Dragonneaux (*Gordius aquaticus*) némathelminthe parasite des insectes et des myriapodes. *Bull. Soc. Hist. Nat. Doubs* 46: 43–48.

Diesing, K. M. (1851): *Systema Helminthum.* Vol. 2, pp. 83–112. Vindobonae, W. Braumüller.

Diesing, K. M. (1861): Revision der Nematoden. Sitzungsber. *Akad. Wiss. Wien* 42: 599–605.

Dollfus, R. P. (1932): Sortie de douze *Gordius* par les stigmates d´un *Dytiscus. Bull. Soc. Entomol. France* 17: 247–254.

Dollfus, R. P. (1943): *Steropus madidus* (Fabricius), nouvel hote du Gordien *Gordionus violaceus* (W. Baird). *Bull. Soc. Entomol. France* 48: 162–167.

Dorier, A. (1925): Sur la faculté d´enkystement dans l´eau de la larve du *Gordius aquaticus* L. *C. R. Acad. Sci. Paris* 181: 1098–1099.

Dorier, A. (1926): L'evolution regressive des kystes formes par les larves de Gordius dans la lamproie de planer, pp. 400–402. Assoc. Fr. Avanc. Sci. C. R. 49e Sess. Grenoble 1925.

Dorier, A. (1928): Quelques stations de Gordiacés en Dauphiné. *Trav. Lab. Hydrobiol. Piscicult. Univ. Grenoble* 20: 43–47.

Dorier, A. (1929): Sur les Gordiacés des Myriapods. *C. R. Acad. Sci. Paris* 188: 743–745.

Dorier, A. (1930): Recherches biologiques et systématiques sur les Gordiacés. *Trav. Lab. Hydrobiol. Piscicult. Univ. Grenoble* 22: 1–183.

Dorier, A. (1931): Sur la présence de *Chordodes tellinii* (Camerano) en Macédoine. *Bull. Soc. Zool. France* 56: 533–541.

Dorier, A. (1932): Observations sur deux larves de Gordiacés appartenant au genre *Parachordodes. Trav. Lab. Hydrobiol. Piscicult. Univ. Grenoble* 24: 47–61.

[Dorier, A. (1932): Sur la larve de *Parachordodes alpestris* (Villot). *C. R. Acad. Sci. Paris* 194: 2340–2342.]

[Dorier, A. (1933): Sur une petite collection de Gordiacés du Maroc rassemblée par Robert-Ph. Dollfus. *Bull. Soc. Sci. Nat. Maroc* 13: 109–111.]

Dorier, A. (1934): Étude biologique et morphologique de la larve de *Parachordodes gemmatus* (Villot). *Trav. Lab. Hydrobiol. Piscicult. Univ. Grenoble* 25: 147–161.

*Dorier, A. (1935): Un nouveau Gordius d´Indochine: *Chordodes joyeuxi* n.sp. *Bull. Soc. Zool. France* 60: 321–324.

Dorier, A. (1935): Sur le passage à la vie latente des larves de Gordiacés. *C. R. Acad. Sci. Paris* 200: 492–494.

[Dorier, A. (1936): Sur le développement des pontes de *Gordius aquaticus* Duj. en milieu humide. *C. R. Séances Soc. Biol.* 123: 955–956.]

Dorier, A. (1942): Sur la presence en France de *Chordodes tellinii* (Camerano) (Gordiacés). *Bull. Mus. Hist. Nat.* 2: 332–335.

Dorier, A. (1946a): *Phasgonura (Locusta) viridissima* (L.) hote de *Gordius albopunctatus* G.W. Müller. *Bull. Soc. Zool. de France* 71: 207–210.

Dorier, A. (1946b): Révision de quelques espéces de Gordiacés. *Bull. Mus. Hist. Nat.* 18: 480–494.

[Drummond (1835): no title. *Edinburgh New Phil. J.* 19: 400.]

Dufour, M. L. (1828): Notice sur la *Filaria forficulae*, espèce de ver trouvée dans l´abdomen du perce-oreille. *Ann. Sci. Nat. Zool.* 13: 66–68.
[Dufour, M. L. (1836): Recherches sur quelques entozoaires et larves parasites des insectes orthoptéres et hyménoptéres. C. R. Hebdom. *Séances Acad. Sci.* 3: 20–21.]
Dujardin, F. (1842): Mémoire sur la structure anatomique des *Gordius* et d´un autre Helminthe, le *Mermis*, qu´on a confondu avec eux. *Ann. Sci. Nat. Zool.* 18: 129–151.
Duron, O. & Gavotte, L. (2007): Absence of *Wolbachia* in nonfilarid worms parasitizing arthropods. *Curr. Microbiol.* 55: 193–197.
Eakin, R. M. & Brandenburger, J. L. (1974): Ultrastructural features of a gordian worm (Nematomorpha). *J. Ultrastruct. Res.* 46: 351–374.
Ebner, R. (1940): Veränderungen an Orthopteren durch parasitische Würmer. 6. *Congr. Int. Entomol. Madrid* 1: 341–347.
[Elmhirst, R. (1915): Faunistic notes: Record of *Parachordodes violaceus. Glasgow Naturalist* 7: 47.]
[Evans, W. (1909): Our present knowledge of the fauna of the Forth area. *Proc. R. Phys. Soc. Edinburgh* 17: 24–27.]
Fair, J. M., Hanelt, B. & Burnett, K. (2010): Horsehair worms (*Gordius robustus*) in nests of the Western Bluebird (*Sialia mexicana*): evidence for anti-predator avoidance? *J. Parasitol.* 96: 429–430.
Faust, E. C. & Botero Ramos, D. (1960): Extraordinario hallazgo de una nueva especie de *Neochordodes* (Gordiacea) en Colombia. Libro Homenaje al Doctor Eduardo Caballero y Caballero. *Inst. Politéc. Nac. Esc. Nac. Cienc. Biol. Mexico*: 523–527.
Faust, E. C., Russell, P. F. & Joung, R. C. (1970): Nematomorpha. Gordiacea, "hairworms" or "hair snakes". In: Faust, E.C. (ed.) *Craig & Faust´s clinical parasitology, Eighth Edition*, pp. 405–406. Lea & Febiger, Philadelphia.
Fernando, A. & Fernando, C. H. (1961): A second case of a hairworm (Gordiacea) "parasitic" in man in Malaya. *Singapore Med. J.* 2: 78–80.
Fernando, C. H. & Dissanaike, A. S. (1962): A hairworm (Gordiacea) "parasitic" in a child in Ceylon. *Ceylon J. Med. Sci.* 11: 47–49.
Ferragu, M. (1986): Insectes coléoptères Curculionidae Brachyderini. *Faune Madagascar* 66: 1–69.
Fewkes, J. W. (1883): On the development of certain worm larvae. *Bull. Mus. Comp. Zool. Harvard Coll.* 11: 167–208.
Feyel, T. (1930): Sur l´oogenèse du *Nectonema agile* Verr. *C. R. Soc. Biol. Paris* 82: 681–683.
[Feyel, T. (1930): Sur la couche musculaire du *Nectonema agile* Verr. *C. R. Soc. Biol. Paris* 82: 1174–1177.]
Feyel, T. (1936): Researches histologiques sur *Nectonema agile* Verr. Étude de la forme parasitaire. *Arch. Anat. Microsc.* 32: 197–234.
Fiedler, A. (1909): Fadenwurm aus einer Raupe von *Cucullia scrophularis. Z. Wiss. Insektenbiol.* 5: 135.
°Fiori, G. M. (1881): Un caso parasitissmo di *Gordius* adulto nell´ uomo. *Giorn. Accad. Med. Torino* 3, 29: 727–736.
Fischthal, J. H. (1942): A *Paragordius* larva (Gordiacea) in a trematode. *J. Parasitol.* 28: 167.
Foresi, C. & Caroli, G. (1968): Sulla presenza di elminti del genere *Gordius* in acque potabili condottate. *Riv. Ital. Igiene* 28: 580–588.
[Fowler, J. H. (1892): A curious parasite. *The Entomologist* 25: 247.]
Franz, H. (1954): Nematomorpha. In: Franz, H. (ed.) *Die Nordost-Alpen im Spielgel ihrer Landtierwelt*, pp. 191–193. Universitätsverlag Wagner, Innsbruck:.
[Fukui, T. & Inoue, I. (1939): Research on Japanese Nematomorpha. I. Research history. *Zool. Bot. Japan* 7: 44–46 (in Japanese).]
Fukui, T. & Inoue, I. (1940): A new hair-worm from Manchoukuo. In: Kawamura, T. (ed.) *Report of the limnobiological survey of Kwantung and Manchoukuo*, pp. 433–436. (in Japanese)
[Fukui, T. & Inoue, I. (1940): Research on gordian worms from Japan. *12th meeting of the Japanese Society of Parasitology* p. 37. (in Japanese)]
[Fukui, T. & Inoue, I. (1973): Nematomorpha. In: Uéno, M. (ed.) *Freshwater Biology of Japan*, pp. 269–271. Hokuryukan Publishing, Tokyo. (in Japanese)]
Gallien, L. (1949): Parasitisme protélien du genre *Nectonema* chez *Leander squilla* L. et *Portunus arcuatus* Leach. *Bull. Soc. Zool. France* 74: 179–184.
Gao, G. et al. (1989): A case report of human Gordiacea infection. *Chin. J. Parasitol. Parasitic Dis.* 7: 111.
Gauss, R. (1970): Beitrag zur Kenntnis von Parasitoiden bei aculeaten Hymenopteren. *Z. Angewandte Entomol.* 65: 239–244.
Gemminger (1849): no title. *Stettiner Entomol. Z.* 10: 63–64.
Gervais, M. (1835): No title. *Ann. Soc. Entomol. France* 4: LXX.
Girard, C. (1882): Historical sketch of Gordiaceae. *Proc. Acad. Nat. Sci. Philadelphia* 5: 279–284.
Giudicelli, J. & Nicoli, R. -M. (1962): Un *Gordionus* nouveau *Gordionus cyrnensis* n.sp. (Gordioidea). *Ann. Parasit. Humaine Comp.* 37: 616–622.
Gliesch, R. (1933): No title. *Egatea* 18: 218–219.
Godsall, W. (1836): Instances of the appearing of the *Gordius aquaticus* after plentiful rain has been produced. *Mag. Nat. Hist.* 9: 241–242.
Goggin, K., Hubbard, D. & Socky, D. (1989): Cave descriptions of Botecourt and Alleghany County. *Speleo Digest* 24: 209–210.
[Goodnight, C. S. (1940): A new technique for the study of Gordiaceans. *J. Tennessee Acad. Sci.* 16: 91.]
[Gorgadze, O. A. & Kitsurashvili, N. T. (2002): New data about hairworms (Gordiacea) of Georgia. *Proc. Inst. Zool.* 21: 68–70.]
[Gorgadze, O., Schmidt-Rhaesa, A. & Kintsurashvili, N. (2008): Hairworm *Spinochordodes* sp. (Nematomorpha, Gordiidae) from the fauna of Georgia. *Bull. Georgian Nat. Acad. Sci.* 2: 131–133.]
Green, E. E. (1902): Note on a species of *Gordius* parasitic in the body of a mantis. *J. Bombay Nat. Hist. Soc.* 14: 610.
Grenacher, H. (1868): Zur Anatomie der Gattung *Gordius* L. *Z. Wiss. Zool.* 18: 322–344.
[Grenacher, H. (1869): Ueber die Muskelelemente von *Gordius. Z. Wiss. Zool.* 19: 287–288.]
Greve, L. (1972): Reken, *Sergestes arcticus*, ny vert for den marine taggelorm, *Nectonema munidae. Fauna (Oslo)* 25: 219.
Grube, E. (1849): Über einige Anguillulen und die Entwicklung von *Gordius aquaticus. Arch. Naturgeschichte* 15: 359–375.
Guéguen, M. F. (1905): Nouveau cas de pseudo-parasitisme d´un *Gordius* dans le tube digestif de l´homme. *C. R. Soc. Biol. Paris* 59: 398–400.
Guéguen, M. F. (1905): Sur un nouveau cas de parasitisme occasionel, dans le tube digestif de l´homme, d´un nématode du genre *Gordius* Dujardin. *Bull. Sci. Pharmacol.* 12: 257–266.
[Gupta, P. D. & Ghosh, R. K. (1979): Catalogue of type specimens present in the Zoological Survey of India. Part 3. *Helminths. Zool. Sur. India* 74: 243–331.]
Hall, M. C. (1912): *Gordius aquaticus* L. as a parasite of man. *Science* 35: 636.

Hanelt, B. (2008): An anomaly against a current paradigm – extremely low rates of individual fecundity variability of the Gordian worm (Nematomorpha: Gordiida). *Parasitology* 135: 1–8.

Hanelt, B. (2009): Hyperparasitism by *Paragordius varius* (Nematomorpha: Gordiida) larva of monostome redia (Tramatoda: Digenea). *J. Parasitol.* 95: 242–243.

Hanelt, B. & Janovy, J. (1999): The lifecycle of a horsehair worm, *Gordius robustus* (Nematomorpha: Gordioidea). *J. Parasitol.* 85: 139–141.

Hanelt, B. & Janovy, J. (2000): New host and distribution record of *Gordius difficilis* (Nematomorpha: Gordioidea) from a vivid metallic ground beetle, *Chlaenius prasinus* (Coleoptera: Carabidae) from Western Nebrasca, U.S.A. *Comp. Parasitol.* 67: 107–108.

Hanelt, B. & Janovy, J. (2002): Morphometric analysis of nonadult characters of common species of American gordiids (Nematomorpha: Gordioidea). *J. Parasitol.* 88: 557–562.

Hanelt, B. & Janovy, J. (2003): Spanning the gap: experimental determination of paratenic host specificity of horsehair worms (Nematomorpha: Gordiida). *Invertebr. Biol.* 122: 12–18.

Hanelt, B. & Janovy, J. (2004a): Untying the gordian knot: the domestication and laboratory maintenance of a gordian worm, *Paragordius varius* (Nematomorpha: Gordiida). *J. Nat. Hist.* 38: 939–950.

Hanelt, B. & Janovy, J. (2004b): Lifecycle and paratenesis of american gordiids (Nematomorpha: Gordiida). *J. Parasitol.* 90: 240–244.

Hanelt, B., Grother, L. E. & Janovy, J. (2001): Physid snails as sentinels of freshwater nematomorphs. *J. Parasitol.* 87: 1049–1053.

Hanelt, B., Thomas, F. & Schmidt-Rhaesa, A. (2005): Biology of the phylum Nematomorpha. *Adv. Parasitol.* 59: 243–305.

Hanelt, B., Bolek, M. & Schmidt-Rhaesa, A. (2012): Going solo: discovery of the first parthenogenetic gordiid (Nematomorpha: Gordiida). PlosONE 7: e34472, doi: 10.1371/journal.pone.0034472.

Harris, A. C. (1990): *Phaulacridium marginale* (Walker) (Acrididae) parasitized by a gordian worm (Nematomorpha: Gordioida). *The Weta* 13: 27.

[Harris, E. A. (1982): Some observations on the Nematomorpha (Gordiacea) of the British Isles. *Parasitology* 84: IXIX–IXX.]

Harvey, T. H. P., Dong, X. & Donoghue, P. C. J. (2010): Are palaeoscolecids ancestral ecdysozoans? *Evol. Dev.* 12: 177–200.

[Hausman, L. A. (1923): The "horse-hair snake". An account of the extraordinary life history of one of our common worms. *Sci. Am.* 129: 402.

Havlik, O. (1947): The Gordioidea (Nematomorpha) of Bohemia. *Věstnik Čsl. Zool. Spol.* 11: 159–184. (in Czech with English summary)

Havlik, O. (1949a): Contribution a la répartition géographique de *Gordionus alpestris. Entomol. Listy (Folia Entomologica)* 12: 69–70. (in Czech with French summary)

Havlik, O. (1949b): Contribution to the description of some species of the genus *Gordionus* (Nematomorpha, Chordodidae) from Central Europe. *Věstnik Čsl. Zool. Spol.* 13: 101–110. (in Czech with English summary)

*Havlik, O. (1949): A new species of *Gordius* (Nematomorpha) from Czechoslovakia. *Věstnik Čsl. Zool. Spol.*13: 111–118. (in Czech with English summary)

Havlik, O. (1954): Nové druhy strunovcu z Československá (Gordioidea, Nematomorpha). *Českoslov. Parasitol.* 1: 85–95. (in Czech with German summary)

He, X. (1995): Report of a case passing *Gordius* via urethra. *Chin. J. Parasitol. Parasitic Dis.*13: 199.

[Heider, K. (1920): Über die Stellung der Gordiiden. *Sitzungsber. Preuss. Akad. Wiss.* 1920: 464–477.]

*Heinze, K. (1933): Revision von *Gordius flavus* Linstow 1906 und *Gordius flavus* G.W. Müller 1927. *Zool. Anz.* 105: 106–109.

*Heinze, K. (1934): Zur Systematik der Gordiiden. *Zool. Anz.* 106: 189–192.

*Heinze, K. (1935a): Weitere neue *Parachordodes*-Arten aus Asien. *Zool. Anz.* 112: 155–158.

*Heinze, K. (1935b): Über Gordiiden (Species inquirendae und Neubeschreibungen). *Zool. Anz.* 111: 23–32.

*Heinze, K. (1935c): Über das Genus *Parachordodes* Camerano 1897 nebst allgemeinen Angaben über die Familie Chordodidae. *Z. Parasitenkd.* 7: 657–678.

*Heinze, K. [(1936): Die bisher aus der Mark Brandenburg sicher bekannten Gordiiden. *Märkische Tierwelt* 2: 25–31.]

*Heinze, K. (1937): Die Saitenwürmer (Gordioidea) Deutschlands. Eine systematisch-faunistische Studie über Insektenparasiten aus der Gruppe der Nematomorpha. *Z. Parasitenkd.* 9: 263–344.

*Heinze, K. (1940): Neue deutsche Saitenwürmer aus den Gattungen *Gordionus* und *Gordius. Zool. Anz.* 132: 139–146.

*Heinze, K. (1941): Saitenwürmer oder Gordioidea (Nematomorpha). In: Dahl, F. (ed.) *Die Tierwelt Deutschlands und der angrenzenden Meeresteile.* Gustav Fischer Verlag, Jena.

*Heinze, K. (1952): Über Gordioidea, eine systematische Studie über Insektenparasiten aus der Gruppe der Nematomorpha. *Z. Parasitenkd.* 15: 183–202.

*Heinze, K. (1954): Bemerkungen zur Systematik der Gordioidea. *Z. Parasitenkd.* 16: 92.

[Hellier Baily, W. (1865): The hair-worm (*Gordius aquaticus). Science Gossip* 107.]

Hellrigl, K. (1996): *Die Tierwelt Südtirols.* Naturmuseum Südtirol, Bolzano.

Herter, C. D. & Nesse, R. E. (1989): Pseudoparasitism with *Gordius robustus. Am. Fam. Phys.,* March 1989: 139–142.

Herzog, H. (1913): *Gordius* als Parasit des Menschen. *Corr. Blätt. Schweizer Ärzte* 43: 1065–1068.

Hofmänner, B. (1913): Gordiiden und Mermithiden aus dem thurgauischen Naturhistorischen Museum zu Frauenfeld. Mitt. *Turgauischen Naturforsch. Ges.* 20: 282–286.

Holsinger, J. R. & Peck, S. B. (1971): The invertebrate cave fauna of Georgia. *Nat. Speleol. Soc. Bull.* 33: 23–44.

Horton, M. L. (1986): Nonparasitic worm vomited by a dog. *Mod. Vet. Pract.* 67: 905.

Hou, X. & Bergström, J. (1994): Palaeoscolecid worms may be nematomorphs rather than annelids. *Lethaia* 27: 11–17.

[Houdemer, E. (1932): A propos de Gordiacés, parasites d´insectes orthoptères du Tonkin. *Bull. Soc. Med.-Chir. Indochina* 10: 227–231]

[Houtman, G. (1977): Over *Gordius, Mermis* en andere wormen en hoe ze samenleven met insekten. *Natura (Hoogwoud)* 74: 29–34.]

Hudson, A. J. & Floate, K. D. (2009): Further evidence for the absence of bacteria in horsehair worms (Nematomorpha: Gordiidae). *J. Parasitology* 95: 1545–1547.

Huus, J. (1932): Über die Begattung bei *Nectonema munidae* Br. und über den Fund der Larve von dieser Art. *Zool. Anz.* 97: 33–37.
Hyman, L. H. (1951): *The invertebrates: Acanthocephala, Aschelminthes, and Entoprocta. The pseudocoelomate Bilateria.* Vol. 3. McGraw-Hill, New York.
Ichihara, A., Kamegai, S., Nonobe, H. & Machida, M. (1967): The second case of hair worm, *Gordius* sp. (Nematomorpha) vomited from human body in Japan. *Res. Bull. Meguro Parasit. Mus.* 1: 23–25.
[Ichikawa, A. (2007): Inventory. Nematomorpha in Japan. In: Enemies of Orthoptera in Japan. Parasite Series 1. *Battarigisu* 140: 12–22. (in Japanese)]
[Inoue, I. (1940): Studies on the biology of *Chordodes* spec. (a preliminary note). *Bot. Zool.* 8: 39–48. (in Japanese)]
Inoue, I. (1951): An additional new species of Gordiacean parasitizing praying mantids, and the domestic and overseas distribution of the genus *Chordodes. Zool. Mag.* (Tokyo) 60: 32. (in Japanese)
[Inoue, I. (1952): Classification of Japanese *Chordodes. Zool. Mag. (Tokyo)* 61: 27–32. (in Japanese with English summary)]
Inoue, I. (1955): Known species of Japanese Gordiacea, with a note on a new species of *Parachordodes. Bull. Biogeograph. Soc. Jap.* 16–19: 31–35.
Inoue, I. (1958): Studies on the life history of *Chordodes japonensis*, a species of Gordiacea. I. The development and structure of the larva. *Jap. J. Zool.* 12: 203–218.
Inoue, I. (1959a): Distribution of acid and alkaline phosphatases in the hairworm, *Chordodes japonensis. Annot. Zool. Jap.* 32: 209–213.
Inoue, I. (1959b): On the adult organization of *Chordodes japonensis. Zool. Mag. (Tokyo)* 68: 151–156.
Inoue, I. (1960a): Studies on the ecology of the larvae of the hairworm, *Chordodes japonensis. Bull. Tokyo Gakugei Univ.* 11: 21–25.
Inoue, I. (1960b): Studies on the life history of *Chordodes japonensis*, a species of Gordiacea. II. On the manner of entry into the aquatic insect-larvae of *Chordodes* larvae. *Annot. Zool. Jap.* 33: 132–141.
Inoue, I. (1962): Studies on the life history of *Chordodes japonensis*, a species of Gordiacea. III. The mode of infection. *Annot. Zool. Jap.* 35: 12–19.
Inoue, I. (1966): A new species of *Euchordodes* (Gordiacea) from Malaysia. *Annot. Zool. Jap.* 39: 60–62.
Inoue, I. (1972): The fauna of the lava caves around Mt. Fuji-san. VIII. Gordiacea. *Bull. Nat. Sci. Mus. Tokyo* 15: 167–169.
[Inoue, I. (1973): Life history of horsehair worms. *The Insectarium* 10: 76–80. (in Japanese)]
Inoue, I. (1974): A new species of the genus *Chordodes* (Gordiacea) from Ethiopia. *Annot. Zool. Jap.* 47: 107–109.
*Inoue, I. (1979): Two new species of Gordius (Gordioidea) from Japan. *Annot. Zool. Jap.* 52: 235–239.
Inoue, I. (1988): Nematomorpha. In: Okada, K., Uchida, S. & Uchida, T. (eds.) *New Illustrated Encyclopedia of the Fauna of Japan*, Vol. 1. pp. 465–466. Hokuryukan, Tokyo.
[Inoue, I. (1994): Gordioid specimens deposited at the Toyama Science Museum. *Bull. Toyama Sci. Mus.* 17: 101–103. (in Japanese)]
*Inoue, I. & Fukui, T. (1953): A note on a new species of *Gordius* from Japan. *Bull. Tokyo Gakugei Univ.* 4: 37–39. (in Japanese with English summary)
[James, J. (1865): More notes on the hairworm. *Science Gossip* 197–198.]
Janda, J. (1893): Prispevky K soustave Gordiidu. *Vestnik Kral. Česke Spol. Nauk.* 1893: 1–19. (in Czech)
Janda, J. (1894a): Beiträge zur Systematik der Gordiiden. I. Die Gordiiden Galliziens. – II. Ueber das Genus *Chordodes. Zool. Jahrb. Syst. Geogr. Biol. Tiere* 7: 595–612.
Janda, J. (1894b): Přispěvek k poznáni českých Gordiidů. *Sitzungsber. Königl. Böhm. Ges. Wiss. Math.-Naturwiss. Cl.* 1894: 1–4. (in Czech)
*Jägerskiöld, L. A. (1897): *Chordodes kallstenii,* eine neue Gordiide aus Kamerun. *Bihang K. Svensk. Vet. Akad. Handling.* 23: 3–9.
Jochmann, R. & Schmidt-Rhaesa, A. (2007): New ultrastructural data from the larva of *Paragordius varius* (Nematomorpha). *Acta Zool.* 88: 137–144.
[Johnston, G. (1836): Illustrations in British Zoology. *Mag. Nat. Hist.* 9: 353–357.]
Jolivet, P. (1945): De l´hydrotrophisme positif de *Steropus madidus*, Fabr. (Col., Pterostichidae). *Misc. Entomol.* 41: 102–106.
Jolivet, P. (1946): Nouvelle station française et nouvel hôte du ver *Gordius aquaticus. Misc. Entomol.* 42: 139–140.
Jolivet, P. (1948): Introduction a l´etude des Gordiacés, vers parasites d´insectes. *Misc. Entomol.* 45: 83–90.
[Jolivet, P. (1998): Manipulation du comportement chez les fourmis et les coléoptères sous l´influence de leurs parasites. *L´Entomologiste* 54: 211–222.]
[Jolivet, P. & Théodoridès, J. (1950): Les helminthes parasites de coléoptères Chrysomélides. *Ann. Parasitol.* 25: 340–349.]
Joseph, S.A. (1984): The occurrence of the horsehair worm (*Gordius* sp.) in the house cricket, *Achaeta domesticus. Cheiron* 13: 322–323.
Kagei, N., Oshima, T., Inoue, I. & Kumasaki, T. (1966): First human case on gordiid worm in Japan (Nematomorpha: Gordiidae). *Jap. J. Parasitol.* 15: 79–81. (in Japanese with English summary)
Kagei, N., Oshima, T., Inoue, I. & Kumasaki, T. (1977): Discussion on the human parasitism with "hair worms" (Gordioidea: Nematomorpha). *Bull. Inst. Publ. Health* 26: 14–18. (in Japanese with English summary)
[Kameoai, S. (1967): Gordiacea, Mesocestoides. *Igaku-no-ayumi* 61: 238–241. (in Japanese)]
[Kikuchi, H. (1956): Ecological survey of aquatic animals in the stream Yokoshibetsu, Sapporo. *J. Fac. Sci. Hokkaido Univ. Ser. VI Zool.* 12: 401–411.]
*Kintsurashvili, N., Schmidt-Rhaesa, A. & Gorgadze, O. (2011): *Chordodes parabipilus* (Nematomorpha: Gordiida), a new species of horsehair worms from Georgia. *Verh. Naturwiss. Ver. Hamburg* 46: 235–241.
Kinzinger, A. P., Cochran, P. A. & Cochran, J. A. (2002): Additional cases of predation on horsehair worms (phylum Nematomorpha), with a recent record for Missouri. *Trans. Missouri Acad. Sci.* 36: 11–13.
Kirjanova, E. S. (1950a): Nematomorpha *Digordius* (part. *Paragordius) excavatus* sp. nov. from the Lower river Piantsch (Tajikistan). *Dokl. Akad. Nauk. SSSR* 74: 157–158. (in Russian)
Kirjanova, E. S. (1950b): Nematomorpha from the basin of the river Zeravshan. *Trudy Zool. Inst. Akad. Nauka SSSR* 9: 255–280. (in Russian)
Kirjanova, E. S. (1950c): On a new species of hairworms (*Progordius* gen. nov.) from Tajikistan. *Dokl. Akad. Nauk. SSSR* 71: 1171–1173. (in Russian)

*Kirjanova, E. S. (1953): A new species of hairworms (*Chordodes oscillatus* sp. nov.) from Georgia. *Soobsc. Akad. Nauk. Gruz. SSSR* 14: 101–103. (in Russian)

[Kirjanova, E. S. (1954): Nematomorpha, their biology and economic significance. *Trudy Probl. Temat. Sov. Moskva* 4: 148–152. (in Russian)]

Kirjanova, E. S. (1955): New species of hairworms from the USSR. *Trudy Zool. Inst. Akad. Nauka SSSR* 21: 152–160. (in Russian)

*Kirjanova, E. S. (1957): Deux nouvelles espèces de Nematomorpha du genre *Chordodes* (Creplin, 1847) Möbius, 1855. *Zool. Zh.* 36: 1159–1166. (in Russian with French summary)

Kirjanova, E. S. (1958a): On the structure of the copulative organs of males of the freshwater hairworms (Nematomorpha, Gordioidea). *Zool. Zh.* 37: 359–372. (in Russian with English summary)

Kirjanova, E. S. (1958b): A new species, *Spinochordodes skrjabini*, nov.sp. from Armenia. In: *Papers in Helminthology to the celebration of the 8oth anniversary of K.I. Skrjabin*, pp. 152–157. Publication House of the USSR Academy of Sciences, Moscow. (in Russian)

Kirjanova, E. S. (1959): Permeability of the cuticle in freshwater hairworms (Nematomorpha, Gordioidea). *Zool. Zh.* 38: 509–519. (in Russian with English summary)

Kirjanova, E. S. (1962): Nematodes and hairworms (Nematomorpha) in the food of fishes. In: Fish of the Usa River and their food resources. *Acad. Nauk. SSSR, Moscow*: 212–218. (in Russian)

[Kirjanova, E. S. (1971): Nematomorpha – parasites of insects. *Zaščita Rastenij* 12: 32–34. (in Russian)]

Kirjanova, E. S. & Spiridonov, S. E. (1989): Two new species of the nematode genus *Chordodes* from mantids. *Parasitologiya (St. Petersburg)* 23: 358–362. (in Russian with English summary)

Kirjanova, E. S. & Zakutsky, U. P. (1967): *Nectonema agile* in the Black Sea. *Gidrobiol. Zh.* 3: 24–31. (in Russian with English summary)

[Klassen-Fischer, M. K., Abalos, F. A. & Neafie, R. C. (2000): Identification of gordian and mermithid worms submitted as human pathogens. *Lab. Invest.* 80: 170A.]

Kloft, W. (1951): Pathologische Untersuchungen an einem Wespenweibchen, infiziert durch einen Gordioiden (Nematomorpha). *Z. Parasitenkd.* 15: 134–147.

Kollar, V. (1857): Zur Naturgeschichte der Gattung *Gordius* und *Mermis*, Fadenwürmer der Insecten. *Verh. Dtsch. Zool. Bot. Ver. Wien* 7: 141.

Konsuloff, S. (1921): Notitzen über die Gordiiden Bulgariens. *Zool. Anz.* 53: 139–140.

Kovalchuk, A. A. (1988): Oxygen consumption by *Gordius* sp. (Gordiacea, Cephalorhyncha). *Nauchnye Dokl. Vysshei Shkoly Biol. Nauk.* 1988: 48–51. (in Russian with English summary)

[Krall, E. (1959): Hairworms and mermithids – parasites of insects. *Eesti Loodus* 4: 216–220.]

Laboulbène, A. (1864): Note sur un helminthe parasite du genre *Mermis* sorti du corps d'un orthoptère a la Nouvelle-Calédonie. *Ann. Soc. Entomol. France* 4: 678.

[Laboulbene, M. (1897): Sur la détermination d'un faux parasite du genre *Gordius*. *Bull. Acad. Nat. Med.* 37: 675.]

Lanzavecchia, G. (1977): Morphological modulations in helical muscles (Aschelminthes and Annelida). *Int. Rev. Cytology* 51: 133–186.

Lanzavecchia, G., Valvassori, R. & De Eguileor, M. (1977): Biopolarity in thick filaments of Nematomorpha. *J. Mol. Biol.* 111: 371–374.

[Lanzavecchia, G., De Eguileor, M. & Valvassori, R. (1978): La fibra muscolare dei Nematomorfi: riconstruzione tridimensionale del sistema contrattile. *Boll. Zool.* 45 (Suppl.): 27.]

Lanzavecchia, G., Valvassori, R., De Eguileor, M. & Lanzavecchia, P. (1979): Three-dimensional reconstruction of the contractile system of the Nematomorpha muscle fiber. *J. Ultrastruct. Res.* 66: 201–223.

Lanzavecchia, G., De Eguileor, M., Valvassori, R. & Scari, G. (1995): Body cavities of Nematomorpha. In: Lanzavecchia, G., Valvassori, R. & Candia Carnevali, M. D. (eds) *Body cavities: function and phylogeny*, pp. 45–60. Mucchi, Modena. Selected Symposia and Monographs U. Z. I. 8.

Lataste, F. (1896): *Gordius* et *Mantis*. *Actes Soc. Chile* 6: 71–73.

Lee, K. -J., Bae, Y. -T., Kim, D. -H., Deung, Y. -K., Ryang, Y. -S., Im, K. -I. & Yong, T. -S. (2003): A *Gordius* worm found in a three year old girl's vomitus. *Yonsei Med. J.* 44: 557–560.

Leffler, S. R. (1984): Record of a horsehair worm (Nematomorpha: Gordiidae) parasitizing *Chlaenius sericeus* Forster (Coleoptera: Carabidae) in Washington. *Coleopterists Bull.* 38: 130.

Leidy, J. (1850): Notes on the development of the *Gordius aquaticus*. *Proc. Acad. Nat. Sci. Philadelphia* 5: 98–100.

Leidy, J. (1851): On the Gordiaceae. *Proc. Acad. Nat. Sci. Philadelphia* 5: 262–263, 266 and 275.

Leidy, J. (1856): A synopsis of Entozoa and some of their ecto-congeners observed by the author. *Proc. Acad. Nat. Sci. Philadelphia* 8: 42–58.

[Leidy, J. (1857): No title. *Proc. Acad. Nat. Sci. Philadelphia* 9: 204.]

Leidy, J. (1870): The *Gordius*, or hair-worm. *Am. Entomol. Bot.* 2: 193–197.

*Leidy, J. (1871): Notice of some worms. *Proc. Acad. Nat. Sci. Philadelphia* 23: 305–307.

Leidy, J. (1878): Notices of *Gordius* in the cockroach and leech. *Proc. Acad. Nat. Sci. Philadelphia* 30: 383–384.

[Leidy, J. (1879): On *Gordius*, and on some parasites of the rat. *Proc. Acad. Nat. Sci. Philadelphia* 31: 10–11.]

Léon, N. (1914): Le *Gordius* parasite accidentel chez l'homme. *Proc. Congr. Int. Path. Comp.* 1: 182–183.

Leslie, H. A., Campbell, A. & Daborn, G. R. (1981): *Nectonema* (Nematomorpha: Nectonematidae) a parasite of decapod Crustacea in the Bay of Fundy. *Can. J. Zool.* 59: 1193–1196.

Levander, K. M. (1908): Zur Kenntnis der Verbreitung der Gordiiden in Finland. *Meddel. Soc. Fauna Flora Fennica* 34: 106–109.

Lever, R. A. (1963): Gordiid worms as trout food in Co. Wexford. *Irish Naturalist* 14: 155.

Leydig, F. (1853): Zoologische Notitz 2: Helminthologisches. *Z. Wiss. Zool.* 4: 377–387.

Lister (1672): Concerning animated horse-hairs, rectifying a vulgar error. *Philos. Trans. R. Soc. London* 1(VII): 717–718.

Lombardo, F. & Umbriaco, R. (2011): Taxonomic re-evaluation of *Parastagmatoptera abnormis* Beier, 1963 (Dictyoptera, Mantidae: Stagmatopterinae): an unusual case of "parasite-induced" synonymy. *Zootaxa* 2735: 31–34.

[Long, J. (1999): The parasitical relationships among a kind of Nematomorpha, a caddisfly and some mantises. *J. Changsga Univ. Electric Power (Nat. Sci.)* 14: 171–172.]

Lora Lamia Donin, C. & Cotelli, F. (1977): The rod-shaped sperm of Gordioidea (Aschelminthes, Nematomorpha). *J. Ultrastruct. Res.* 61: 193–200.

Luna de Carvalho, E. (1967): Terceira contribuição para dos estrepsipteros angolenses (Insecta – Strepsiptera). *Publ. Cult. Companh. Diam. Angola* 77: 13–56.

Maas, A., Waloszek, D., Haug, J. T. & Müller, K. J. (2007): A possible larval roundworm from the Cambrian "Orsten" and its bearing on the phylogeny of Cycloneuralia. *Mem. Assoc. Australasian Palaeontol.* 34: 499–519.

Mahler, F. (1951): Wie gelangt der *Gordius* in den Leib der Heuschrecken? *Arch. Hydrobiol.* 45: 393–394.

[Maison, E. (1902): *Gordius* et némertien. *Cosmos* 47: 547–548.]

Malakhov, V. V. (1980): Cephalorhyncha, a new type of animal kingdom uniting Priapulida, Kinorhyncha and Gordiacea, and a new system of Aschelminthes worms. *Zool. Zh.* 59: 485–499. (in Russian with English summary)

*Malakhov, V. V. & Spiridonov, S. E. (1982): *Gordionus senkovi* sp. n., a new species of hairworms (Nematomorpha, Gordiidae) and some stages of its embryonic development. *Zool. Zh.* 61: 810–816. (in Russian)

*Malakhov, V. V. & Spiridonov, S. E. (1984): The embryogenesis of *Gordius* sp. from Turkmenia, with special reference to the position of the Nematomorpha in the animal kingdom. *Zool. Zh.* 63: 1285–1296. (in Russian with English summary)

Malakhov, V. V. & Adrianov, A. V. (1995): *Cephalorhyncha – a New Phylum of the Animal Kingdom*. KMK Scientific Press, Moscow.

Malmqvist, B. & Moravec, F. (1978): *Gordius aquaticus* larvae found in the lampreys *Lampetra planeri* (Bloch) and *L. fluviatilis* (L.). *Folia Parasitol.* 25: 40.

[Manganelli, G. (1990): Gli invertebrati: Phylum Nematomorpha. *Carta della Natura* 1: 39–40.]

Mao, S. P. (1991): Protozoans and helminth parasites of humans in mainland China. *Int. J. Parasitol.* 21: 347–351.

Marnay, A. (1902): Les *Gordius* et les sorciers. *Arch. Parasitol.* 6: 297.

[Martin, W. E. (1973): Electron scan of a nematomorph cuticle. *Proc. Helminthol. Soc.* Washington 40: 173–177.]

Maund, B. (1930): Filaria forficulae. *Mag. Nat. Hist.* 3: 149.

Maxwell, J. P. (1908): A case of infection with one of the Gordiacea. *China Med. J.* 22: 98.

May, H. G. (1919): Contributions to the life histories of *Gordius robustus* Leidy and *Paragordius varius* (Leidy). *Illinois Biol. Monogr.* 5: 1–119.

McCook, H. C. (1884): Note on the intelligence of a cricket parasitised by a *Gordius*. *Proc. Acad. Nat. Sci. Philadelphia* 1884: 293–294.

McDaniel, V. R. & Smith, K. L. (1976): Cave fauna of Arkansas: selected invertebrate taxa. *Proc. Arkansas Acad. Sci.* 20: 57–60.

McLennan, J. A. & MacMillan, B. W. H. (1984): The food of rainbow trout (*Salmo gairdneri*) and brown trout (*Salmo trutta*) in the Mohaka and other rivers of Hawkes Bay, New Zealand. *New Zealand J. Mar. Freshwater Res.* 18: 143–158.

Meissner, G. (1856): Beiträge zur Anatomie und Physiologie der Gordiaceen. *Z. Wiss. Zool.* 7: 1–144.

Meßner, B. (1983): Das Eigelege von *Gordionus preslii* und anderer Saitenwürmer (Nematomorpha) wird gehärtet. *Entomol. Nachr. Ber.* 27: 279–280.

Meyer, N. T. (1913): Zur Entwicklung von *Gordius aquaticus* Villot. *Z. Wiss. Zool.* 105: 125–135.

Michel, A. (1888): De l´existence d´un véritable épiderme cellulaire chez les Nématodes, et spécialement les Gordiens. *C. R. Acad. Sci. Paris* 107: 1175–1177.

[Michel, A. (1889): Sur l´épiderme des Gordius. *Boll. Mus. Zool. Anat. Comp. R. Univ. Torino* 4, 59: 1–2.]

Milde, F. (1868): Mittheilung. *Verh. Dtsch. Zool. Bot. Ges. Wien* 18: 92.

Miyamoto, K. & Uenishi, H. (2002): A case of human parasitism by a hairworm in Hokkaido. *44th Joint Meeting of the Northern Japan Section of the Parasitological Society of Japan and the Japan Society of Medical Entomology and Zoology* 11. (in Japanese)

Möbius, K. (1855): *Chordodes pilosus*, ein Wurm aus der Familie der Gordiaceen. *Z. Wiss. Zool.* 68: 428–431.

Moisan, P. G., Mansfield, L. S., Murphy, A. J. & Hale, A. (1996): Incidental subcutaneous gordiid parasitism in a cat. *J. Vet. Diagn. Invest.* 8: 270–272.

Montén, E. (1951): Nematomorpha. In: Brinck, P. & Wingstrand, K. G. (eds) The Mountain Fauna of the Virihaure Area in Swedish Lapland. *Lunds Univ. Arskr.* 46: 162–170.

Montgomery, T. H. (1898a): The Gordiacea of certain American collections. *Bull. Mus. Comp. Zool. Harvard Coll.* 32: 23–59.

Montgomery, T. H. (1898b): Description of two new exotic species of the genus *Chordodes*. *Zool. Jahrb. Syst.* 11: 379–384.

Montgomery, T. H. (1898c): The Gordiacea of certain american collections, with particular reference to the North American fauna – II. *Proc. California Acad. Sci.* 1: 333–344.

[Montgomery, T. H. (1898): Description of the female of *Chordodes albibarbatus* Montg. *Zool. Jahrb. Syst.* 11: 493–496.]

[Montgomery, T. H. (1899): Synopses of North American invertebrates. II. Gordiacea (hair worms). *Am. Nat.* 33: 647–652.]

Montgomery, T. H. (1900): Gordiacea from the Cope collection. *Biol. Bull.* 1: 95–98.

Montgomery, T. H. (1901): The identity of the Gordiacean species, *Chordodes morgani* and *C. puerilis*. *Proc. Acad. Nat. Sci. Philadelphia* 53: 289–292.

Montgomery, T. H. (1903): The adult organisation of *Paragordius varius* (Leidy). *Zool. Jahrb. Ontog.* 18: 387–474.

Montgomery, T. H. (1904): The development and structure of the larva of *Paragordius*. *Proc. Acad. Nat. Sci. Philadelphia* 56: 738–755.

Montgomery, T. H. (1907): The distribution of the North American Gordiacea, with description of a new species. *Proc. Acad. Nat. Sci. Philadelphia* 59: 270–272.

Morris, G. K., Kerr, G. E. & Gwynne, D. T. (1975): Ontogeny of phototaxis in *Orchelimum gladiator* (Orthoptera: Tettigonidae: Conocephalidae). *Can. J. Zool.* 53: 1127–1130.

Mouchet, S. (1931): Spermatophores des Crustacés Décapodes Anomoures et Brachyoures et castration parasitaire chez quelques Pagures. *Ann. Stat. Océanogr. Salammbô* 6: 1–203.

Mühldorf, A. (1913): Studien über die Entwicklung der Nematomorphen (Vejd.). *Zool. Anz.* 42: 31–36.

Mühldorf, A. (1914): Beiträge zur Entwicklungsgeschichte und zu den phylogenetischen Beziehungen der Gordiuslarve. *Z. Wiss. Zool.* 111: 1–75.

Müller, F. (1871): Bruchstücke zur Naturgeschichte der Bopyriden. *Jenaische Z. Naturwiss.* 6: 53–73.

Müller, G.W. (1920): Beobachtungen an *Gordius*. *Zool. Anz.* 51: 225–229.

Müller, G.W. (1926): Über Gordiaceen. *Z. Morph. Ökol. Tiere* 7: 134–220.

[Müller, G.W. (1929): Die systematische Stellung der Gordiaceen. *Zool. Anz.* 84: 241–243.]

Müller, J. (1859): Ueber einen Fadenwurm aus einer Raupe von *Vanessa v. album* W.V. *Jahresheft der naturwissenschaftlichen Sektion der k. k. Mähr. Schles. Ges. Ackerbau Nat. Landeskd.* 1859: 109–110.

Müller, M. C. M., Jochmann, R. & Schmidt-Rhaesa, A. (2004): The musculature of horsehair worm larvae (*Gordius aquaticus, Paragordius varius*, Nematomorpha): F-actin staining and reconstruction by cLSM and TEM. *Zoomorphology* 123: 45–54.

Muttkowski, R. A. (1918): The fauna of Lake Mendota. *Wisconsin Acad. Sci. Art. Lett.* 19: 374–482.

Muus, B. J. (1956): A note on the Danish hairworms, Nematomorpha: Gordioidea. *Vidensk. Meddel. Dansk Naturhist. Forening Kobenhavn* 118: 33–39.

Neuhaus, B., Kristensen, R. M. & Lemburg, C. (1996): Ultrastructure of the cuticle of the Nemathelminthes and electron microscopical localization of chitin. *Verh. Dtsch. Zool. Ges.* 89.1: 221.

Ng, S. M. (1975): The biology of *Pycnoscelus striatus* Kirby, the cave roach, with special reference to nymphal development and instar differentiation. *Malay. Nat. J.* 28: 84–89.

Nielsen, S. -O. (1969): *Nectonema munidae* Brinckmann (Nematomorpha) parasitizing *Munida tenuimana* G.O. Sars (Crust. Dec.) with notes on host–parasite relations and new host species. *Sarsia* 38: 91–110.

Nierstrasz, F. (1907): Die Nematomorpha der Siboga-Expedition. *Siboga-Expeditie* 20: 1–21.

Nigrelli, R. F. (1941): Hairworms as parasites of fishes. *J. Parasitol.* 27: 38.

Notman, P. R. & Yeates, G. W. (1992): Tomtit killed by a worm. *Notornis* 39: 16.

Nouvel, H. & Nouvel, L. (1934): Sur deux crevettes parasitées par un *Nectonema*. Évolution des parasites. – Influence sur l´hote. *Bull. Soc. Zool. France* 59: 516–521.

Nouvel, H. & Nouvel, L. (1938): Sur deux hotes nouveaux de *Nectonema. Bull. Mus. Natn. Hist. Nat. Paris* 10: 507–508.

Ochiai, N. & Inoue, I. (1970): On the post-larval development of *Chordodes* (Gordiacea) – especially the formation of the intestine as well as its openings. *Bull. Tokyo Gakugei Univ.* 22: 78–87. (in Japanese with English summary)

Ogawara, T. (1980): A hairworm living in *Diestrammena japonica* Karny (Orthoptera, Stenopelmatidae). *New Entomol.* 29: 64. (in Japanese)

Oku, Y., Fukumoto, S. -I., Ohbayashi, M. & Koike, M. (1983): A marine horsehair worm, *Nectonema* sp., parasitizing a telecyclid crab, *Erimacrus isenbeckii*, from Hokkaido, Japan. *Jap. J. Vet. Res.* 31: 65–69.

Örley, L. (1881): On hair-worms in the collection of the British Museum. *Ann. Mag. Nat. Hist.* 8: 325–332.

Parona, C. (1901): Altro caso di pseudo-parassitismo di *Gordius* nell´uomo (*Parachordodes pustulosus* Baird). *Clinica Med.* 10: 3–8.

Patruban (1875): No title. *Anz. K. K. Ges. Aerzte Wien* 16: 69.

[Paulian, R. (1947): Les Gordiens, vers parasites d'Insectes. *L'Entomologiste* 3: 131–133.]

Peck, S. B. (1992): A synopsis of the invertebrate cave fauna of Jamaica. *NSS Bull.* 54: 37–60.

[Pellieux, M. (1825): Osservations sur le Dragonneaux d´eau douce. *Ann. Sci. Nat.* 6: 493–500.]

°Pepper, W. (1872): Gordius aquaticus. *Trans. Pathol. Soc. Philadelphia* 4: 239.

Pérez, C. (1927a): Sur le cycle évolutif du *Nectonema*. 10. *Congr. Int. Zool. Budapest*: 991–994.

Pérez, C. (1927b): Sur le parasitisme protèlien du *Nectonema. C. R. Acad. Sci. Paris* 185: 226–227.

Pérez, C. (1934): Non spécifité du parasitisme du *Nectonema agile* Verrill. *Bull. Soc. Zool.* France 59: 322–328.

[Pérez, C. (1935): Une ancienne observation relative au *Nectonema. Bull. Soc. Zool. France* 60: 467–468.]

Pešić, V.M. & Ćurčić, S.B. (2006): New records of hairworms (Nematomoorpha) from Montenegro (SE Europe). *Arch. Biol. Sci. (Belgrade)* 58: 5P–6P.

Pešić, V. & Schmidt-Rhaesa, A. (2008): First data on Serbian hairworms (Nematomorpha). In: Pavićević, D. & Perreau, M. (eds.) *Advances in the studies of the fauna of the Balkan Peninsula*, pp. 337–340. Institute for Nature Conservation of Serbia, Monograph No. 22, Belgrade.

Pikula, Z. P., Spiridonov, S. E. & Drljevic, E. T. (1996): *Dacochordodes bacescui* Capuse, 1966: report of the first two isolations from humans. *Can. J. Infect. Dis.* 7: 144–146.

Pintner, T. (1899): *Nectonema agile* Verrill in der Bai von Neapel. *Anz. K. Akad. Wiss. Wien* 36: 103–105.

[Pippet, J. R. & Fernando, C. H. (1961): Hairworms – a zoological curiosity. *Malay. Nat. J.* 15: 148–151.]

[Poinar, G. O. (2001): Nematoda and Nematomorpha. In: Thorp, J. H. & Covich, A. P. (eds) *Ecology and Classification of North American Freshwater Invertebrates, 2nd edition*, pp. 225–295. Academic Press, San Diego.]

Poinar, G. O. (1991): Hairworm (Nematomorpha: Gordioidea) parasites of New Zealand wetas (Orthoptera: Stenopelmatidae). *Can. J. Zool.* 69: 1592–1599.

Poinar, G. O. (1999): *Palaeochordodes protus* n.g., n.sp. (Nematomorpha, Chordodidae), parasites of a fossil cockroach, with a critical examination of other fossil hairworms and helminths of extant cockroaches (Insecta: Blattaria). *Invertebr. Biol.* 118: 109–115.

[Poinar, G. (2007): Nematomorpha. In: Yule, C. M. & Sen, Y. H. (eds) *Freshwater invertebrates of the Malaysian region*, pp. 157–161. Academy of Sciences, Malaysia.]

[Poinar, G. O. (2008): Global diversity of hairworms (Nematomorpha: Gordiaceae) in freshwater. *Hydrobiologia* 595: 79–83.]

Poinar, G. O. (2009): Phylum Nematomorpha, horsehair worms, gordian worms. In: Gordon, D. P. (ed.) *New Zealand inventory of biodiversity*, Vol. 1, pp. 491–497. Canterbury University Press, Christchurch.

Poinar, G. O. & Brockerhoff, A. M. (2001): *Nectonema zealandica* n. sp. (Nematomorpha: Nectonematoidea) parasitizing the purple rock crab *Hemigrapsus edwardsi* (Brachyura: Decapoda) in New Zealand, with notes on the prevalence of infection and host defense reactions. *Syst. Parasitol.* 50: 149–157.

Poinar, G. O. & Buckley, R. (2006): Nematode (Nematoda: Mermithidae) and hairworm (Nematomorpha: Chordodidae) parasites in early Cretaceous amber. *J. Invertebr. Pathol.* 93: 36–41.

[Poinar, G. & Chandler, C. M. (2004): Synopsis and identification of North American hairworms (Gordioidea: Nematomorpha). *J. Tennessee Acad. Sci.* 79: 1–7.]

Poinar, G. O. & Doelman, J. J. (1974): A reexamination of *Neochordodes occidentalis* (Montg.) comb. n. (Chordodidae: Gordioidea): larval penetration and defense reaction in *Culex pipiens* L. *J. Parasitol.* 60: 327–335.

Poinar, G., Rykken, J. & LaBonte, J. (2004): *Parachordodes tegonotus* n. sp. (Gordioidea: Nematomorpha), a hairworm parasite of ground beetles (Carabidae: Coleoptera), with a summary of gordiid parasites of carabids. *Syst. Parasitol.* 58: 139–148.

Poinar, G. & Weissman, D. B. (2004): Hairworm and nematode infections of North American Jerusalem crickets, field crickets, and katydids (Orthoptera; Stenopelmatidae, Gryllidae and Tettigonidae). *J. Orthoptera Res.* 13: 143–147.

*Poisson, M. H. (1927): Note sur un ver rond des eaux de Madagascar. *Bull. Acad. Malagache Tananarive N. S.* 9: 25–28.

[Polozhentsev, P. A. (1954): Helminths of insects (Nematoda and Gordiacea). A review of the literature. *Trudy Gelminthol. Lab. Akad. Nauk SSSR* 7: 282–303.]

Ponton, F., Lebarbenchon, C., Lefèvre, T., Biron, D. G., Duneau, D., Hughes, D. P. & Thomas, F. (2006): Parasite survives predation on its host. *Nature* 440: 756.

Poulin, R. (1995): Hairworms (Nematomorpha: Gordioidea) infecting New Zealand short-horned grasshoppers (Orthoptera: Acrididae). *J. Parasitol.* 81: 121–122.

Poulin, R. (1996): Observations on the free-living adult stage of *Gordius dimorphus* (Nematomorpha: Gordioidea). *J. Parasitol.* 82: 845–846.

[Presla, J. S. (1831): Hljsta w bruku, kraginjk zlatý (*Calosoma sycophanta*) zwaněm nalzená. *Krok* 2: 134–136.]

Poulin, R. (1836): Widlorep, *Dicranurus*, nowé pokolenj hlist. *Krok* 3: 160.

Protasioni, M., De Eguileor, M., Congiu, T., Grimaldi, A. & Reguzzoni, M. (2003): The extracellular matrix of the cuticle of *Gordius panigettensis* (Gordioiidae, Nematomorpha): observations by TEM, SEM and AFM. *Tiss. Cell* 35: 306–311.

Rajaram, B. & Rajulu, G. S. (1975): A study of the cuticle of *Gordius robustus*, a nematomorph worm. I. cuticle in the parasitic phase. *Acta Histochem.* 54: 257–271.

Rauther, M. (1904): Das Cerebralganglion und die Leibeshöhle der Gordiiden. *Zool. Anz.* 27: 606–614.

Rauther, M. (1905): Beiträge zur Kenntnis der Morphologie und der phylogenetischen Beziehungen der Gordiiden. *Z. Naturwiss.* 40: 1–94.

[Rauther, M. (1909): Morphologie und Verwandtschaftsbeziehungen der Nematoden und einiger ihnen nahe gestellter Vermalien. *Erg. Fortschr. Zool.* 1: 491–596.]

Rauther, M. (1914): Zur Kenntnis und Beurteilung von *Nectonema*. *Zool. Anz.* 43: 561–576.

Rauther, M. (1930): Nematomorpha = Saitenwürmer. In: Krumbach, T. (ed.) *Handbuch der Zoologie*, Vol. 2, pp. 403–448. Vermes Amera.

Reardon, L. (1928): No title. *J. Parasitol.* 15: 68.

Reddell, J. R. (1965): A checklist of the cave fauna of Texas. I. The invertebrata (exclusive of Insecta). *Texas J. Sci.* 17: 143–187.

Reddell, J. R. & Veni, G. (1996): Biology of the Chiquibul cave system, Belize and Guatemala. *J. Cave Karst Stud.* 58: 131–138.

*Redlich, A. (1980): Description of *Gordius attoni* sp.n. (Nematomorpha, Gordiidae) from Northern Canada. *Can. J. Zool.* 58: 382–385.

Reeker, H. (1894): No title. *Jahresber. Westfäl. Prov. Ver. Wiss. Kunst* 1894/95: 27.

[Rees, N. E. (1973): Arthropod and nematode parasites, parasitoids and predators of Acrididae in America north of Mexico. *US Dep. Agric. Tech. Bull.* 1460: 1–288.]

Reeves, W. K. (2000): Invertebrate cavernicoles of the Great Smokey Mountains National Park, USA. *J. Elisha Mitchell Sci. Soc.* 116: 334–343.

Reichenbach-Klinke, H. -H. (1962): Die Parasiten der Donaufische. *Arch. Hydrobiol. Suppl.* 27: 40–56.

Restelli, M. A., De Villalobos, C. & Zanca, F. (2002): Ultrastructural description of the musculature, the intraepidermal nervous system, and their interrelation in *Pseudochordodes bedriagae* (Nematomorpha). *Cell Tiss. Res.* 308: 299–306.

Reutter, K. (1972): *Gordius*, das Wasserkalb. *Mikrokosmos* 61: 198–204.

[Richter, R. (1930): Die Schwimmfährte des Wasserkalbs (*Gordius*). *Nat. Mus.* 60: 396–398.]

Riddell, W. (1928): Gordiidae as parasites of man. *Centralbl. Bacteriol. Parasitenkd. Infektionskr. Hyg.* 109: 331–338.

Riley, C. W., Packard, A. S. & Thomas, C. (1877): Hair-worms. *US Entomol. Comm. Ann. Rept.* (1877) 1: 326–334.

Ritchie, J. (1915): Scottish hairworms (Nematomorpha, Gordiidae), their occurrence, habits, and characteristics; with a key for the discrimination of the species recorded from Britain. *Scottish Nat.* 111–115, 136–142 and 255–262.

Römer, F. (1895a): Die Gordiiden des Naturhistorischen Museums in Hamburg. *Zool. Jahrb. Syst.* 8: 790–803.

Römer, F. (1895b): Drei neue Gordiiden von Borneo und Halmaheira. *Zool. Anz.* 18: 197–200.

Römer, F. (1896): Beitrag zur Systematik der Gordiiden. *Abh. Senckenberg. Naturforsch. Ges.* 23: 249–295.

Roonwal, M. L. (1976): Ecology and biology of the grasshopper *Hieroglyphus nigrorepletus* Bolivar (Acrididae). 2. Distribution, economic importance, life history, colour forms, and problems of control. *Z. Angew. Zool.* 63: 307–332.

*Rosa, D. (1881): Nota intorno ad una nuova specie del genere *Gordius*. *Atti R. Accad. Sci. Torino* 16: 572–574.

*Rosa, D. (1882): Nota intorno al *Gordius villoti* n.sp. ed al *G. tolosanus* Duj. *Atti R. Accad. Sci. Torino* 17: 333–341.

Roth, L. M. & Willis, E. R. (1960): The biotic associations of cockroaches. *Smiths. Misc. Coll.* 141: 1–470.

Roy, R. (2003): A propos de deux males parasités de *Prohierodula* Bolivar, 1908 (Dictyoptera, Mantidae). *Bull. Soc. Ent. France* 108: 181–183.

Ruppert, E. E., Fox, R. S. & Barnes, R. D. (2004): *Invertebrate Zoology, Seventh Edition*. Brooks/Cole, Belmont.

Russ, E. L. (1912): Contributions à l´étude des parasites des Trichoptères. *Ann. Sc. Univ. Jassay* 7: 255–259.

Sahli, M. F. (1972): Modifications des charactèrs sexuels secondairs mâles chez les Iulidae (Myriapoda, Diplopoda) sous l´influence de Gordiacés parasites. *C. R. Acad. Sci. Paris* 274: 900–903.

Saito, Y., Inoue, I., Hayashi, F. & Itagaki, H. (1987): A hairworm, *Gordius* sp., vomited by a domestic cat. *Jap. J. Vet. Sci.* 49: 1035–1037.

Salas, L., De Villalobos, C. & Zanca, F. (2011): Sexual size dimorphism, sex ratio and the relationship between seasonality and water quality in four species of Gordiida (Nematomorpha) from Catamarca, Argentina. *J. Helminthology* 85: 319–324.

Sanchez, M. I., Ponton, F., Schmidt-Rhasea, A., Hughes, D. P., Misse, D. & Thomas, F. (2008): Two steps to suicide in crickets harbouring hairworms. *Anim. Behav.* 76: 1621–1624.

°Sandosham, A. A. (1953): Malaysian parasites. XV. Seven new worms from miscellaneous hosts. *Stud. Inst. Med. Res. Malaya* 26: 212–226.

Sanford, S. N. (1853): On some points in the history of *Gordius*. *Proc. Am. Assoc. Sci.* 1853–1856: 250.

Sato, T., Arizono, M., Sone, R. & Harada, Y. (2008): Parasite-mediated allochtonous input: do hairworms enhance subsidized predation of stream salmonids on crickets? *Can. J. Zool.* 86: 1–5.

Sato, T., Watanabe, K. Kanaiwa, M., Niizuma, Y., Harada, Y. & Laferty, K. D. (2011): Nematomorph parasites drive energy flow through a riparian ecosystem. *Ecology* 92: 201–207.

Sawyer, R. T. (1971): Erpobdellid leeches as new hosts for the nematomorph, *Gordius* sp. *J. Parasitol.* 57: 285.

Sayad, W. Y., Johnson, V. M. & Faust, E. C. (1936): Human parsitization with *Gordius robustus*. *J. Am. Med. Ass.* 106: 461–462.

[Schenkel, G. (1959): Ein Erlebnis mit einer Gottesanbeterin. *Kosmos* 55: 456.]

Schepotieff, A. (1908): Über den feineren Bau der Gordiuslarven. *Z. Wiss. Zool.* 89: 230–241.

[Schmidt-Rhaesa, A. (1995): Rekonstruktion und Ultrastruktur des Vorderendes von *Nectonema munidae* (Nematomorpha). *Verh. Dtsch. Zool. Ges.* 88.1: 175.]

Schmidt-Rhaesa, A. (1996a): *Zur Morphologie, Biologie und Phylogenie der Nematomorpha – Untersuchungen an* Nectonema munidae *und* Gordius aquaticus. Cuvillier Verlag, Göttingen.

Schmidt-Rhaesa, A. (1996b): Ultrastructure of the anterior end in three ontogenetic stages of *Nectonema munidae* (Nematomorpha). *Acta Zool.* 77: 267–278.

Schmidt-Rhaesa, A. (1996c): The nervous system of *Nectonema munidae* and *Gordius aquaticus*, with implications on the ground pattern of the Nematomorpha. *Zoomorphology* 116: 133–142.

Schmidt-Rhaesa, A. (1996d): Die Biologie der Saitenwürmer (Nematomorpha) – Rätselhaftes Leben wenig bekannter Würmer. *Mikrokosmos* 85: 279–283.

[Schmidt-Rhaesa, A. (1996): Monophyly and systematic relationships of the Nematomorpha. *Verh. Dtsch. Zool. Ges.* 89.1: 23.]

Schmidt-Rhaesa, A. (1997a): Ultrastructural features of the female reproductive system and female gametes of *Nectonema munidae* Brinkmann 1930 (Nematomorpha). *Parasitol. Res.* 83: 77–81.

Schmidt-Rhaesa, A. (1997b): Ultrastructural observations of the male reproductive system and spermatozoa of *Gordius aquaticus* L. 1758 (Nematomorpha). *Invertebr. Reprod. Dev.* 32: 31–40.

Schmidt-Rhaesa, A. (1997c): Nematomorpha. In: Schwoerbel, J. & Zwick, P. (eds) *Süßwasserfauna Mitteleuropas*. Gustav Fischer Verlag, Stuttgart.

Schmidt-Rhaesa, A. (1998a): Muscular ultrastructure in *Nectonema munidae* and *Gordius aquaticus* (Nematomorpha). *Invertebr. Biol.* 117: 38–45.

Schmidt-Rhaesa, A. (1998b): Phylogenetic relationships of the Nematomorpha – a discussion of current hypothesis. *Zool. Anz.* 236: 203–216.

Schmidt-Rhaesa, A. (1999): Nematomorpha. In: Knobil, E. & Neill, J. D. (eds) *Encyclopedia of Reproduction*, Vol. 3, pp. 13–21. Academic Press, San Diego.

Schmidt-Rhaesa, A. (2000): Rasterelektronenmikroskopische Untersuchungen an *Gordionus violaceus* und *Gordius elongiporus* n. sp. (Nematomorpha) aus Niederösterreich. *Linzer Biol. Beitr.* 32: 455–461.

Schmidt-Rhaesa, A. (2001a): The lifecycle of horsehair worms (Nematomorpha). *Acta Parasitol.* 46: 151–158.

Schmidt-Rhaesa, A. (2001b): Reinvestigation and reinterpretation of *Chordodiolus echinatus* (Linstow, 1901), a peculiar nematomorph species. *Mitt. Mus. Naturkd. Berlin Zool. Reihe* 77: 91–94.

Schmidt-Rhaesa, A. (2001c): Variation of cuticular characters in the Nematomorpha: studies on *Gordionus violaceus* (Baird, 1853) and *G. wolterstorffii* (Camerano, 1888) from Britain and Ireland. *Syst. Parasitol.* 49: 41–57.

Schmidt-Rhaesa, A. (2001d): Problems and perspectives in nematomorph systematics. *Org. Div. Evol.* 1: 161–163.

Schmidt-Rhaesa, A. (2002a): Are the genera of Nematomorpha monophyletic taxa? *Zool. Scr.* 31: 185–200.

Schmidt-Rhaesa, A. (2002b): Australian species of *Chordodes* (Nematomorpha) with a description of two new species, remarks on the genus and its life history. *J. Nat. Hist.* 36: 1569–1588.

Schmidt-Rhaesa, A. (2002c): *Die Saitenwürmer. Neue Brehm-Bücherei*, Vol. 632. Westarp Wissenschaften, Hohenwarsleben.

[Schmidt-Rhaesa, A. (2002): Two dimensions of biodiversity research exemplified by Nematomorpha and Gastrotricha. *Integr. Comp. Biol.* 42: 633–640.]

Schmidt-Rhaesa, A. (2004a): Ultrastructure of an integumental organ with probable sensory function in *Paragordius varius* (Nematomorpha). *Acta Zool.* 85: 15–19.

Schmidt-Rhaesa, A. (2004b): *Gordionus longistriatus*, a new species of Nematomorpha from Hawai'i. *Records of the Hawaii Biological Survey for 2003, Bishop Museum Occasional Papers* 78: 53–55.

Schmidt-Rhaesa, A. (2004c): Japanese horsehair worms (Nematomorpha) from the Lake Biwa Museum. *Species Diversity* 9: 97–107.

Schmidt-Rhaesa, A. (2005): Morphogenesis of *Paragordius varius* (Nematomorpha) during the parasitic phase. *Zoomorphology* 124: 33–46.

Schmidt-Rhaesa, A. (2007): *The evolution of organ systems*. Oxford University Press, Oxford.

Schmidt-Rhaesa, A. (2008): New data on New Zealand horsehair worms (Nematomorpha, Gordiida). *Mitt. Hamburg. Zool. Mus. Inst.* 105: 5–12.

Schmidt-Rhaesa, A. (2009): Nematomorpha (Saitenwürmer). In: Rabitsch, W. & Essl, F. (eds) *Endemiten – Kostbarkeiten in Österreichs Tier- und Pflanzenwelt*, pp. 381–383. Naturwissenschaftlicher Verein für Kärnten und Umweltbundesamt GmbH, Klagenfurt.

Schmidt-Rhaesa, A. (2010): Considerations on the genus *Gordius* (Nematomorpha, horsehair worms), with the description of seven new species. *Zootaxa* 2533: 1–35.

Schmidt-Rhaesa, A. (2011): *Beatogordius chinensis* (Nematomorpha, Gordiida), a new species from China. Verh. Naturwiss. *Ver. Hamburg* 46: 243–248.

Schmidt-Rhaesa, A. & Brune, S. (2008): Description of one known and three new *Chordodes* species (Nematomorpha) parasitizing

praying mantids (Mantoptera) in Malaysia, with a discussion of sexual dimorphism in *Chordodes*. *Zoosystemat. Evol.* 84: 57–66.
Schmidt-Rhaesa, A. & Bryant, M. S. (2004): Filling the Gondwana gaps: new species and new reports of the genus *Beatogordius* (Nematomorpha) from Australia and Madagascar. *Syst. Parasitol.* 57: 173–181.
Schmidt-Rhaesa, A. & Cieslak, A. (2008): Three new species of *Paragordionus* and *Gordionus* (Nematomorpha, Gordiida) from Spain and Turkey, with comments on the taxon *Semigordionus*. *Mitt. Hamburg. Zool. Mus. Inst.* 105: 13–22.
Schmidt-Rhaesa, A. & De Villalobos, L. C. (2002): Revision of the genus *Beatogordius* (Nematomorpha). I. African species including *Chordodiolus echinatus*. *Ann. Mus. R. Afr. Centr. Zool.* 290: 1–25.
Schmidt-Rhaesa, A. & Ehrmann, R. (2001): Horsehair worms (Nematomorpha) as parasites of praying mantids. *Zool. Anz.* 240: 167–179.
Schmidt-Rhaesa, A. & Geraci, C.J. (2006): Two new species of *Acutogordius* (Nematomorpha), with a brief review of literature data of this genus. *Biodiv. Syst.* 4: 427–433.
Schmidt-Rhaesa, A. & Gerke, S. (2006): Cuticular ultrastructure of *Chordodes nobilii* Camernao, 1901, with a comparison of cuticular ultrastructures in horsehair worms (Nematomorpha). *Zool. Anz.* 245: 269–276.
Schmidt-Rhaesa, A. & Gusich, V. V. (2010): A new *Gordionus* Müller, 1927 from Swizerland (Nematomorpha, Gordiida). *Rev. Suisse Zool.* 117: 77–82.
Schmidt-Rhaesa, A. & Kristensen, P. (2006): Horsehair worms (Nematomorpha) from the Baltic Island Bornholm (Denmark), with notes on the biology of *Gordius albopunctatus*. *J. Nat. Hist.* 40: 495–502.
Schmidt-Rhaesa, A. & Lalramliana (2011): *Chordodes mizoramensis* (Nematomorpha: Gordiida), a new species of horsehair worm from Mizoram, North-East India. *ZooKeys* 75: 1–8.
Schmidt-Rhaesa, A. & Menzel, L. (2005): Central American and Caribbean species of horsehair worms (Nematomorpha), with the description of three new species. *J. Nat. Hist.* 39: 515–529.
Schmidt-Rhaesa, A. & Prous, M. (2010): Records of horsehair worms (Nematomorpha) in Estonia, with description of three new species from the genus *Gordius*. *Eston. J. Ecol.* 59: 39–51.
Schmidt-Rhaesa, A. & Sato, T. (2009): *Gordionus chinensis* (Villot, 1874) and *Gordionus kii* sp.n. (Nematomorpha, Gordiida), new reports of the genus *Gordionus* in Japan. *Spec. Div.* 14: 61–67.
Schmidt-Rhaesa, A. & Urabe, M. (2009): New records of *Gordius* specimens from Japan (Nematomorpha, Gordiida). *Mitt. Hamburg. Zool. Mus. Inst.* 106: 1–6.
Schmidt-Rhaesa, A. & Yadav, A. K. (2004): First report of *Chordodes* cf. *furnessi* (Nematomorpha) from a praying mantid in India with a note on Indian nematomorph species. *Curr. Sci.* 86: 1023–1027.
Schmidt-Rhaesa, A. & Zaenker, S. (2006): Saitenwürmer (Nematomorpha) aus hessischen Höhlen, Bergwerksstollen und Quellen, mit der Beschreibung einer Abnormalität bei einem Individuum. *Hess. Faun. Briefe* 24: 65–77.
Schmidt-Rhaesa, A., Pohle, G., Gaudette, J. & Burdett-Coutts, V. (2012): Lobster (*Homarus americanus*), a new host for marine horsehair worms (*Nectonema agile*, Nematomorpha). *J. Mar. Biol. Assoc. UK* (in press)
Schmidt-Rhaesa, A., Thomas, F. & Poulin, R. (1998): Scanning electron microscopy and intraspecific variation in *Euchordodes nigromaculatus* from New Zealand. *J. Helminthol.* 72: 65–70.
Schmidt-Rhaesa, A., Thomas, F. & Poulin, R. (2000): Redescription of *Gordius paranensis* Camerano, 1892 (Nematomorpha), a species new for New Zealand. *J. Nat. Hist.* 34: 333–340.
[Schmidt-Rhaesa, A., Kieneke, A., Muschiol, D. & Chung, P. -R. (2001): Tubular structures on the cuticle of two *Gordius* species (Nematomorpha, Gordioidea, Gordiidae) from Korea and New Zealand. *Kor. J. Syst. Zool.* 17: 165–169.]
Schmidt-Rhaesa, A., Hanelt, B. & Reeves, W. (2003a): Redescription and compiliation of Nearctic freshwater Nematomorpha (Gordiida), with the description of two new species. *Proc. Acad. Nat. Sci. Philadelphia* 153: 77–117.
Schmidt-Rhaesa, A., Chung, P. -R. & Sohn, W. -M. (2003b): *Parachordodes megareolatus*, a new species of horsehair worm (Nematomorpha) from Korea. *Kor. J. Syst. Zool.* 19: 161–166.
Schmidt-Rhaesa, A., Biron, D.G., Joly, C. & Thomas, F. (2005): Host-parasite relations and seasonal occurrence of *Paragordius tricuspidatus* and *Spinochordodes tellinii* (Nematomorpha) in Southern France. *Zool. Anz.* 244: 51–57.
Schmidt-Rhaesa, A., De Villalobos, L. C. & Zanca, F. (2008): Summary of *Chordodes* species (Nematomorpha, Gordiida), with a discussion of their diagnostic characters. *Verh. Naturwiss. Ver. Hamburg* 44: 37–114.
Schmidt-Rhaesa, A., Farfan, M. A. & Bernard, E. C. (2009a): First record of millipeds as host for horsehair worms (Nematomorpha) in North America. *Northeastern Nat.* 16: 125–130.
Schmidt-Rhaesa, A., Nishi, H., Tanabe, A. S. & Urabe, M. (2009b): New records of hairworms (Nematomorpha, Gordiida) from Japan. *Spec. Div.* 14: 131–135
*Schneider (1866): *Monographie der Nematoden* (chapter XXVIII. Gordius). G. Reimer, Berlin.
Schreitmüller, W. (1911): *Gordius aquaticus* L. (= Saitenwurm oder Wasserkalb). *Bl. Aquarien- und Terrarienkd.* 22: 495–497.
Schuurmans Stekhoven, J. H. (1943): Contribution a l´étude des Gordiides de la faune Belge. *Bull. Mus. R. Hist. Nat. Belgique* 19: 1–28.
Schuurmans Stekhoven, J. H. (1929): Note di Elmintologia Modenese. *Atti Soc. Nat. di Modena Ser.* 6(8): 147–150.
Schuurmans Stekhoven, J. H. (1930): Su alcuni Gordii nuovi per il Modenese e sulla distribuzione dei Gordii in Italia. *Boll. Zool.* 4: 143–148.
Sciacchitano, I. (1932): Su alcuni Gordii del Museo Civico di Milano. *Atti Soc. Ital. Sci. Nat.* 71: 241–259.
*Sciacchitano, I. (1932): Spedizione Scientifica all´oasi di Cufra. Gordii. *Ann. Mus. Civico Genova* 55: 305–306.
Sciacchitano, I. (1933a): Su alcuni Gordii del Congo Belga. *Rev. Zool. Bot. Afr.* 24: 50–59.
Sciacchitano, I. (1933b): Gordii di Romagna. *Boll. Zool.* 4: 187–189.
*Sciacchitano, I. (1937): Nuovi Gordii del Congo Belga. *Rev. Zool. Bot. Afr.* 30: 141–151.
Sciacchitano, I. (1940): Le attuali conoscenze sui Gordi dell'Africa Italiana. *Riv. Biol. Col.* 6: 441–444.
*Sciacchitano, I. (1950): Su un nuovo gordio della Cirenaica e sulla distribuzione del genere *Gordius* in Africa. *Act. Pontif. Acad. Sci.* 8: 213–216.
Sciacchitano, I. (1952): Irudinei e Gordii cavernicoli in Italia. *Arch. Zool. Ital.* 37: 439–443.

Sciacchitano, I. (1955): Su due Gordii cavernicoli del Congo Belga. *Rev. Zool. Bot. Afr.* 51: 65–67.

*Sciacchitano, I. (1955): *Gordius panighettensis* n. sp. di Gordioideo Italiano. *Monit. Zool. Ital.* 63: 56–59.

Sciacchitano, I. (1958): Gordioidea del Congo Belga. *Ann. Mus. R. Congo Belge* 67: 7–111.

Sciacchitano, I. (1961a): Gordioidea dell'Angola. *Publ. Cult. Compan. Diamantes Angola* 52: 119–126.

Sciacchitano, I. (1961b): Su alcuni gordioidei del Congo. *Rev. Zool. Bot. Afr.* 64: 351–356.

*Sciacchitano, I. (1961): Nuova specie di Gordioideo della Somalia. *Monit. Zool. Ital.* 68: 146–148.

Sciacchitano, I. (1962): Su alcuni Gordioidei della Tunisia. *Monit. Zool. Italia* 69: 103–105.

Sciacchitano, I. (1963): La collezione dei Gordioidei del Museo Civico di Genova. *Ann. Mus. Civ. Stor. Nat. Di Genova* 73: 272–281.

[Scortegagna (1851): Analisi della memoria intitolata – storia del genere *Gordius* e di un nuovo elminto – pubblicata nel II. volume delle Memoire dell'Instituto Lombardo. Milano, per Bernadoni 1845. *Nuovi Ann. Sci. Nat.* 3(3): 151–161.]

Scorza, J. V. (1952): Contribucion al estudio de los Gordiacea de Venezuela. *Acta Biol. Venezuelica* 1: 49–54.

Seymour, M. K. (1983): Some implications of helical fibres in worm cuticles. *J. Zool. Lond.* 199: 287–295.

Sharpilo, V. P. (2001): Paratenic parasitism in nematomorphs (Cephalorhyncha, Nematomorpha). *Vestn. Zool.* 35: 3–12. (in Russian with English summary)

Shelford, M. A. (1903): Binomical notes on some Bornean Mantidae. *The Zoologist* 7: 293–304.

[Shipley, A. E. (1903): On the ento-parasites collected by the "Skeat-Expedition" to Lower Siam and Malay Peninsula in the years 1899–1900. *Proc. Zool. Soc. Lond.* 145–156.]

Siltala, A. J. (1906): Zur Kenntnis der Parasiten der Trichopteren. 1. Beobachtungen über Parasiten der Trichopteren. *Z. Wiss. Insektenbiol.* 2: 382–385.

Singh, S. N. & Rao, V. G. (1966): On a case of human infection with a gordiid worm in the orbit. *Indian J. Helminthol.* 18 (Seminar Suppl.): 65–67.

Skaling, B. & MacKinnon, B. M. (1988): The absorptive surface of *Nectonema* sp. (Nematomorpha: Nectonematoidea) from *Pandalus montagui*: histology, ultrastructure, and absorptive capabilities of the body wall and intestine. *Can. J. Zool.* 66: 289–295.

Slevogt, B. (1909): Über Lepidopterenschmarotzer. *Z. Wiss. Insektenbiol.* 5: 197.

Smith, D. G. (1991): Observations on the morphology and taxonomy of two *Parachordodes* species (Nematomorpha, Gordioida, Chordodidae) in southern New England (USA). *J. Zool. Lond.* 225: 469–480.

[Smith, D. G. (1994): A reevaluation of *Gordius aquaticus difficilis* Montgomery, 1898 (Nematomorpha, Gordioidea, Gordiidae). *Proc. Acad. Nat. Sci. Philadelphia* 145: 29–34.

Smith, D. R., Watson, D. D., Clevenger, R. R. & McBride, R. C. (1990): *Paragordius varius*. *Arch. Pathol. Lab. Med.* 114: 981–983.

Sokolov, S. G. & Spiridonov, S. E. (1998): Gordiid larvae (Nematomorpha) from inner organs of fresh-water fishes in Tver region, Russia. *Vestn. Zool.* 32: 32.

Sørensen, M. V., Hebsgaard, M. B., Heiner, I., Glenner, H., Willerslev, E. & Kristensen, R. M. (2008): New data from an enigmatic phylum: evidence from molecular sequence data supports a sister-group relationship between Loricifera and Nematomorpha. *J. Zool. Syst. Evol. Res.* 46: 231–239.

Southern, R. (1907): Contributions to the Natural History of Lambay. Nematomorpha. *Irish Nat.* 84.

Southern, R. (1909): The Gordii of Ireland. *Irish Nat.* 115–117.

Spiridonov, S. E. (1984): Two new Nematomorpha species of the family Gordiidae. *Proc. Zool. Inst. USSR Acad. Sci.* 126: 97–101. (in Russian)

Spiridonov, S. E. (1989): Nematomorpha of the genus *Parachordodes* Camerano of Central Asia. *Proc. Zool. Inst. USSR Acad. Sci. Leningrad* 194: 76–82. (in Russian with English summary)

Spiridonov, S. E. (1993): *Gordionus annamensis* sp.n. (Nematomorpha: Parachordodidae) – a new species of gordian worm from central Vietnam. *Parasitologia* 5: 429–432. (in Russian with English summary)

Spiridonov, S. E. (1998): Horsehair worms of the genus *Gordius* Linne, 1776 (Gordiacea, Nematomorpha) in the collection of Scientific Zoological Museum, M.V. Lomonosov Moscow State University. *Bull. Mosk. Obshchestva Ispytatelei Prirodi* 103: 19–26. (in Russian)

*Spiridonov, S. E. (2000): New species of gordian worms of the genus *Chordodes* (Nematomorpha: Chordodidae) from the Minussinsk Valley. *Parazitologija* 34: 437–441. (in Russian with English summary)

Spiridonov, S. E. (2001): *Chordodes guineensis* sp.n., a new species of horsehair worms from West Africa (Nematomorpha: Chordodidae). *Zoosyst. Ross.* 10: 275–279.

Spiridonov, S. E. & Ivanova, E. S. (2002): Cuticle surface structure of *Spinochordodes* Kirjanova, 1950 (Nematomorpha, Chordodidae). *Proc. Inst. Parasitol. Russ. Acad. Sci.* 63: 246–252. (in Russian)

Spiridonov, S. E., Pikula, Z. P. & Drljevic, E. T. (1992): Redescription of *Dacochordodes bacescui* Capuse, 1966 (Nematomorpha: Chordodidae). *Helminthologia (Bratislava)* 29: 193–196.

[Standen, R. (1910): The hairworm (*Gordius aquaticus* Linn.): its folklore, and life history. *Lancashire Nat.* 2: 365–368.]

Stiles, C. W. (1907): Three new american cases of infection of man with horse-hair worms (species *Paragordius varius*), with summary of all cases reported to date. *US Publ. Health Serv. Hyg. Lab. Bull.* 34: 53–75.

[Stiles, C. W. & Hassall, A. (1920): Index catalogue of medical and veterinary zoology. *US Publ. Health Serv. Hyg. Lab. Bull.* 114: 316–318, 520–536, 608, 639–641.]

Studier, E. H., Lavoie, K. H. & Chandler, C. M. (1991): Biology of cave crickets, *Hadenoecus subterraneus*, and camel crickets, *Ceuthophilus stygius* (Insecta: Orthoptera): parasitism by hairworms (Nematomorpha). *J. Helminthol. Soc.* Washington 58: 248–250.

Švábeník, J. (1908): Studien an Nematomorphen. *Zool. Anz.* 33: 385–389.

Švábeník, J. (1909): Contributions to the anatomy and histology of Nematomorpha. *Sitzungsber. Ges. Wiss. Prag Math.-Naturwiss. Kl.* 1909: 1–64. (in Czech with German summary)

Švábeník, J. (1925): Parasitism and metamorphosis of the species *Gordius tolosanus* Duj. (Parasitismus a metamorfosa druhu *Gordius tolosanus* Duj.). *Publ. Fac. Sci. Univ. Masaryk* 58: 1–48. (in Czech with English summary)

Swanson, C. J. (1971): Occurrence of paramyosin among the Nematomorpha. *Nature New Biol.* 232: 122–123.

Swanson, C. J. (1974): Application of thin shell theory to helically-wound fibrous cuticles. *J. Theor. Biol.* 43: 293–304.

*Tang, S. T. (1934): Description of three new species of the Nematomorpha of Fukien. *Sinensia* 4: 201–208.

Tanner, V. M. (1939): Notes on the Gordiacea of Utah. *Great Basin Nat.* 1: 2.

Tappenbeck, L. (2006): Zum Vorkommen und zur Ökologie der Saitenwürmer (Nematomorpha) in Sachsen-Anhalt. *Abh. Ber. Mus. Heinearum* 7: 87–93.

[Templeton, R. (1836): A catalogue of the species of annulose animals, and of rayed ones, found in Ireland, as selected from the papers of the late J. Templeton, Esq., of Cranmore, with localities, descriptions, and illustrations. *Mag. Nat. Hist.* 9: 233–236.]

Thomas, F., Schmidt-Rhaesa, A. & Poulin, R. (1999): Microhabitat characteristics and reproductive status of male *Euchordodes nigromaculatus* (Nematomorpha). *J. Helminthology* 73: 91–93.

Thomas, F., Schmidt-Rhaesa, A., Martin, G., Manu, C., Durand, P. & Renaud, F. (2002): Do hairworms (Nematomorpha) manipulate the water-seeking behaviour of their terrestrial hosts? *J. Evol. Biol.* 15: 356–361.

Thomas, F., Ulitsky, P., Augier, R., Dusticier, N., Samuel, D., Strambi, C., Biron, D. G. & Cayre, M. (2003): Biochemical and histological changes in the brain of the cricket *Nemobius sylvestris* infected by the manipulative parasite *Paragordius tricuspidatus* (Nematomorpha). *Int. J. Parasitol.* 33: 435–443.

Thomas, F., Ward, D. F. & Poulin, R. (1998): Fluctuating asymmetry in an ornamental trait in the cave weta, *Pleioplectron simplex* Hutton (Orthoptera: Rhaphidophoridae): no role for parasites. *Can. J. Zool.* 76: 931–935.

Thorne, G. (1940): The hairworm, *Gordius robustus* Leidy, as a parasite of the Mormon cricket, *Anabrus simplex* Haldeman. *J. Washington Acad. Sci.* 30: 219–231.

Tomlin, A. D. (1975): Notes on the biology and rearing of two species of ground beetles, *Pterostichus melanarius* and *Harpalus pensylvanicus* (Coleoptera: Carabidae). *Can. Entomol.* 107: 67–74.

Topsent, E. (1900): Sur un cas de pseudoparasitisme chez l´homme de *Gordius violaceus* Baird. *Bull. Soc. Sci. Med. de l'Ouest* 9: 86–91.

Townsend, J. I. (1970): Records of gordian worms (Nematomorpha) from New Zealand Carabidae. *New Zealand Entomol.* 4: 98–99.

Tratiakow, D. (1901): Entwicklungsgeschichte von *Gordius aquaticus* Vill. *Trav. Soc. Imp. Nat. St. Petersbourg* 32: 24.

Uchikawa, R., Akune, K., Inoue, I., Kagei, N. & Sato, A. (1987): A human case of hairworm (*Gordius* sp.) infection in Kagoshima, Japan. *Jap. J. Parasitol.* 36: 358–360.

[Ulrey, A.B. (1897): Notes on the embryology of *Paragordius (Gordius) varius* (Leidy). *Proc. Indiana Acad. Sci.* 232–233.]

Usynin, V.F. (1978): Biology of the sterlet *Acipenser ruthenus* in the Chulyan River. *Vopr. Ikhtiol.* 18: 624–635. (in Russian)

Valvassori, R., Scari, G. & Lanzavecchia, G. (1987): Aspects of morphogenesis in nematomorphs. *Acta Embryol. Morphol. Exper.* 8: 453–456.

Valvassori, R., Scari, G., De Eguileor, M., Lernia, L. D., Magnetti, P. & Melone, G. (1988): *Gordius villoti* (Nematomorpha) lifecycle in relation with caddis fly larvae. *Boll. Zool.* 55: 269–278.

Valvassori, R., De Eguileor, M., Grimaldi, A. & Lanzavecchia, G. (1999): Nematomorpha. In: Adiyodi, K. G. & Adiyodi, R. (eds) *Reproductive biology of invertebrates*, Vol. 9, part A, pp. 213–227. John Wiley & Sons, Chichester.

Vejdovsky, F. (1886): Zur Morphologie der Gordiiden. *Z. Wiss. Zool.* 43: 369–433.

Vejdovsky, F. (1888): Studien über Gordiiden. *Z. Wiss. Zool.* 46: 188–216.

Vejdovsky, F. (1894): Organogenie der Gordiiden. *Z. Wiss. Zool.* 57: 642–703.

Vejdovsky, F. (1898): Bemerkungen zu den Gordiidenarbeiten von Linstow's. *Zool. Anz.* 21: 382–384.

Verrill, A. E. (1873): Report upon the invertebrate animals of Vineyard Sound and adjacent waters, with an account of the physical features of the region. *Rep. Comm. Fish Fisheries* 1871–1872: 295–778.

Verrill, A. E. (1879): Notice of recent additions to the marine invertebrates of the Northern coast of America. *Proc. US Nat. Mus.* 2: 165–204.

Vieweg (1811): Über den sogennanten Haarwurm. *Allg. Med. Ann. Altenburg* 423–424.

Vigueras, I. P. (1934): Nota sobre Gordidos de Cuba. *Mem. Soc. Cubana Hist. Nat.* 8: 353–354.

Villot, A. (1872): Sur la forme embryonnaire des Dragonneaux *(Gordius). C. R. Acad. Sci. Paris* 75: 363–365.

Villot, A. (1874a): Monographie des Dragonneaux (Genre *Gordius*, Dujardin), deuxième partie. Anatomie et physiologie. *Arch. Zool. Exp. Gen.* 3: 181–197.

Villot, A. (1874b): Monographie des Dragonneaux (Genre *Gordius*, Dujardin), troisième partie. Embryogénese et morphogénie + quatrième partie. Généralisation et affinités du genre *Gordius*. *Arch. Zool. Exp. Gen.* 3: 198–238.

Villot, A. (1874c): Monographie des Dragonneaux (Genre *Gordius*, Dujardin), introduction + première partie. Taxonomie et histoire naturelle. *Arch. Zool. Exp. Gen.* 3: 39–72.

Villot, A. (1875a): Sur la forme embryonnaire des Dragonneaux. *Bull. Soc. Stat. Dept. Isere*, 3me Ser. 4: 272–275.

Villot, A. (1875b): Sur la forme larvaire des Dragonneaux. *Bull. Soc. Stat. Dept. Isere*, 3me Ser. 4: 296–300.

[Villot, A. (1876): Sur la place des Dragonneaux dans la classification. *Bull. Soc. Stat. Dept. Isere*, 3me Ser. 5: 287.]

[Villot, A. (1876): Description de quartorze espéces de Dragonneaux du muséum de Paris. *Bull. Soc. Stat. Dept. Isere*, 3me Ser. 5: 267–268.]

[Villot, A. (1880): Sur l´organisation et le développement des gordiens. *C. R. Hebd. Séanc. Acad. Sci.* 90: 1569–1571.]

Villot, A. (1881): Nouvelles recherches sur l´organisation et le développement des gordiens. *Ann. Sci. Nat. Zool.* 11: 1–44.

Villot, A. (1884): Sur le parasitisme et la détermination spécifique des larves des gordiens. *Zool. Anz.* 7 : 84–88.

[Villot, A. (1885): No title. *Bull. Soc. Sci. Nat. Sud-Est* 3: 74–75.]

*Villot, A. (1885): Nouvelles espéces de gordiens. *Bull. Soc. Sci. Nat. Sud-Est* 3: 43–48.

Villot, A. (1886): Revision des gordiens. *Ann. Sci. Nat. Zool.* 1: 271–318.

[Villot, A. (1887): Sur le développenent et la détermination spécifique des gordiens vivant à l´état libre. *Zool. Anz.* 10: 505–509.]

[Villot, A. (1887): Sur l´anatomie des gordiens. *Ann. Sci. Nat. Zool.* 2: 189–212.]

[Villot, A. (1888): Encore un mot sur le développement et la détermination spécifique des Gordiens adultes. *Zool. Anz.* 11: 70–72.]

Villot, A. (1889): Sur l´ovogenése, la structure de l´ovaire et la régression du parenchyme des gordiens. *C. R. Acad. Sci. Paris* 109: 411–412.

[Villot, A. (1889): Sur la signification histologique, le mode de formation et l´usage de la cavité péri-intestinale des gordiens. *C. R. Acad. Sci. Paris* 108: 685–687.]

[Villot, A. (1889): Sur l´hypoderme et le systéme nerveux periphérique des gordiens. *C. R. Acad. Sci. Paris* 108: 304–306.]

Villot, A. (1891): L´évolution des gordiens. *Ann. Sci. Nat. Paris* 11: 329–401.

*Villot, A. (1892): Gordiens de Sumatra. Description de deux espèces nouvelles. In : Weber, M. (ed.) *Zoologische Ergebnisse einer Reise in Niederländisch Ost-Indien*, Vol. 2, pp. 136–138. Brill, Leiden.

[Villot, A. (1894): Le polymorphisme des gordiens. *C. R. Assoc. Franc. Avanc. Sci. (Congres de Caen)*: 179.]

[Villot, A. (1896): Réclamation de priorité sur l´embryogénie des gordiens et des nématoides. *Zool. Anz.* 19: 123–125.]

[Villot, A. (1896): Le polymorphisme du *Gordius violaceus*. *Assoc. Franc. Av. Sci. C. R. 24me Sess. Bordeaux*: 650–657.]

Voigt, E. (1938): Ein fossiler Saitenwurm (*Gordius tenuifibrosus* n.sp.) aus der eozänen Braunkohle des Geiseltales. *Nova Acta Leopoldina* 5: 351–360.

Von Bolschwing (1842): Ein Wort wider den *Gordius aquaticus* in Fussgeschwüren. *Med. ARGOS* 4: 339–341.

[Von Ebner, V. (1910): Über Fasern und Waben. Eine histologische Untersuchung der Haut der Gordiiden und der Knochen-grundsubstanz. *Sitzungsber. K. Akad. Wiss., Math.-Naturwiss. Klasse* 119: 285–325.]

Von Linstow, O. (1877): Parasitische Nematoden. *Arch. Naturgesch.* 43: 3–5.

Von Linstow, O. (1878): *Compendium der Helminthologie*. Hahn'sche Buchhandlung, Hannover.

Von Linstow, O. (1883a): Über die Zwischenwirthe des *Gordius aquaticus*. *Zool. Anz.* 6: 373–374.

Von Linstow, O. (1883b): Nematoden, Trematoden und Acanthocephalen, gesammelt von Prof. Fedtschenko in Turkestan. *Arch. Naturgesch.* 49: 274–314.

[Von Linstow, O. (1884): Helminthologisches. *Arch. Naturgesch.* 50: 137–138.]

Von Linstow, O. (1889): Ueber die Entwicklungsgeschichte und die Anatomie von *Gordius tolosanus* Duj. = *G. subbifurcus* v. Siebold. *Arch. Mikrosk. Anat.* 34: 248–268.

Von Linstow, O. (1891a): Weitere Beobachtungen an *Gordius tolosanus* und *Mermis*. *Arch. Mikrosk. Anat.* 37: 239–249.

Von Linstow, O. (1891b): Ueber die Entwicklungsgeschichte von *Gordius tolosanus* Duj. *Centralbl. Bacteriol. Parasitenkd.* 9: 760–762.

Von Linstow, O. (1892): Beobachtungen an Helminthenlarven. *Arch. Mikrosk. Anat.* 39: 325–343.

Von Linstow, O. (1897): Nemathelminthen grösstentheils in Madagascar gesammelt. *Arch. Naturgesch.* 63: 27–34.

Von Linstow, O. (1898): Zur Entwicklungsgeschichte von *Gordius aquaticus* Gmel. *Arch. Mikrosk. Anat.* 51: 747–763.

Von Linstow, O. (1901): Helminthen von den Ufern des Nyassa-Sees. *Jenaische Z. Naturwiss.* 35: 409–428.

*Von Linstow, O. (1902): Beobachtungen an neuen und unbekannten Helminthen. *Arch. Mikrosk. Anat.* 60: 217–232.

*Von Linstow, O. (1904): Neue Beobachtungen an Helminthen. *Arch. Mikrosk. Anat.* 64: 484–497.

Von Linstow, O. (1905): Neue Helminthen. *Arch. Naturgesch.* 71: 267–276.

Von Linstow, O. (1906a): Gordiiden und Mermithiden des Königlichen Zoologischen Museums in Berlin. *Mitt. Zool. Mus. Berlin* 3: 243–248.

Von Linstow, O. (1906b): Nematoden des Zoologischen Museums in Königsberg. *Arch. Naturgesch.* 72: 249–258.

*Von Linstow, O. (1906): On worms of the family Gordiidae from Corea. *Proc. Zool. Soc. Lond.* 556–557.

Von Pelser-Berensberg, H. (1909): Fadenwürmer in Insekten. *Z. Wiss. Insektenbiol.* 5: 284–285.

[Von Siebold, C. T. (1838): Helminthologische Beiträge. Vierter Beitrag: Ueber geschlechtslose Nematoideen. *Arch. Naturgesch.* 4: 302–314.]

Von Siebold, C. T. (1842): Ueber die Fadenwürmer der Insekten. *Stettiner Entomol. Z.* 3: 146–161.

Von Siebold, C. T. (1843a): Ueber die Fadenwürmer der Insekten (Erster Nachtrag). *Stettiner Entomol. Z.* 4: 78–84.

Von Siebold, C. T. (1843b): Bericht über die Leistungen im Gebiete der Helminthologie während des Jahres 1842. *Arch. Naturgesch.* 9: 302–310.

Von Siebold, C. T. (1848): Ueber die Fadenwürmer der Insekten (Zweiter Nachtrag). *Stettiner Entomol. Z.* 9: 290–301.

Von Siebold, C. T. (1850): Ueber die Fadenwürmer der Insekten (Dritter Nachtrag). *Stettiner Entomol. Z.* 11: 329–336.

Von Siebold, C. T. (1854): Ueber die Fadenwürmer der Insekten (Vierter Nachtrag). *Stettiner Entomol. Z.* 15: 103–121.

Von Siebold, C. T. (1856): Zusatz. *Z. Wiss. Zool.* 7: 141–144.

Von Siebold, C. T. (1858): Über die Fadenwürmer der Insekten (Fünfter Nachtrag). *Stettiner Entomol. Z.* 19: 325–332.

[Von zur Mühlen, M. & Schneider, G. (1920): Der See Wirzjerw in Livland – Biologie und Fischerei. *Arch. Naturkd. Estlands (Eesti Loodus. Arhiiv)* 14: 1–38.]

Wakeland, C. & Shull, W. E. (1936): The mormon cricket with suggestions for its control. *Univ. Idaho Agricult. Coll. Ext. Bull.* 100: 1–30.

[Wandsworth, W. W. (1829): A curious worm. *Mag. Nat. Hist.* 2: 103–104.]

Ward, H. B. (1892a): On *Nectonema agile*, Verrill. *Bull. Mus. Comp. Zool. Harvard Coll.* 23: 135–188.

Ward, H. B. (1892b): Preliminary communication on the host of *Nectonema agile*, Verr. *Proc. Am. Acad. Arts Sci.* 19: 260–261.

[Watermolen, D. J. & Chandler, C. M. (1992): A record of the horsehair worm *Chordodes morgani* (Nematomorpha) in Wisconsin. *J. Freshwater Ecol.* 8: 187–188.]

Watermolen, D. J. & Chandler, C. M. (1993): A record of the horsehair worm *Neochordodes occidentalis* (Nematomorpha) in the State of Washington. *J. Freshwater Ecol.* 8: 185–186.

Watermolen, D. J. & Haen, G. L. (1994): Horsehair worms (phylum Nematomorpha) in Wisconsin, with notes on their occurrence in the Great Lakes. *J. Freshwater Ecol.* 9: 7–11.

Wei, D. & Yang, W. (1981): *Parachordodes* sp. (Nematomorpha) human infestation of the lower urinary tract: the first case report in China. *Acta Acad. Med. Wuhan* 1: 40–45.

Weißbecker, M. & Schmidt-Rhaesa, A. (2010): Beobachtungen zur Lebensweise von Saitenwürmern mit Schwerpunkt in West- und

Mittelhessen im Zeitraum 2005 – 2008. *Hessische Faunistische Briefe* 29: 21–45.

Weng, F. -C. (1991): A case report of human Gordiacea infection. *Chin. J. Parasitol. Parasitic Dis.* 9: 158.

Werner, F. (1922): Zur Kenntnis der Mantodeenfauna der niederländischen Kolonien. *Zool. Med. Mus. Leiden* 7: 115–126.

Werner, F. (1925): Vierter Beitrag zur Mantodeenfauna von Niederländisch-Indien. *Treubia* 6: 476–486.

Wesenberg-Lund, C. (1910): Über eine eventuelle Brutpflege bei *Gordius aquaticus* L. *Intern. Rev. Ges. Hydrobiol.* 3: 122–127.

Weyenbergh, D. H. (1880): Descripciones de nuevos gusanos. *Per. Zool. Soc. Entomol.* 2–3: 106–111.

Weyenbergh, D. H. (1897): Descripciones de nuevos gusanos. *Bol. Acad. Cien. Argentina* 3: 213–219.

White (1859): No title. *Proc. Boston Nat. Hist.* 7: 175.

White, D. A. (1966): A new host record for *Paragordius varius* (Nematomorpha). *Trans. Am. Microsc. Soc.* 85: 579.

White, D. A. (1969): The infection of immature aquatic insects by larval *Paragordius* (Nematomorpha). *Great Basin Nat.* 29: 44.

[Whittaker, F.H. & Barker, R.L. (1983): Some observations on the egg string of a nematomorph worm, *Paragordius* sp. *Trans. Kentucky Acad. Sci.* 44: 76–77.]

*Wu, H. W. & Tang, S. F. (1933): Notes on the Nematomorpha of China. *Sinensia* 3: 173–178.

[Wülker, G. (1924): Parasitische Würmer bei Insekten. *Int. Entomol. Z.* 17: 138–144.]

Yamaguti, S. (1943): Saitenwürmer in Japan, I. *Paragordionus kawamurai* n.sp. *Bull. Biogeogr. Soc. Jap.* 13: 65–68.

Yang, C. M., Ng, P. K. L. & Lua, H. K. (1990): Record of a hairworm (phylum Nematomorpha) from a tettigonid grasshopper in Ulu Kinchin, Pahang, Malaysia. *Malay. Nat. J.* 43: 268–270.

Yeh, L. -S. & Jordan, P. (1957): On a new gordiid worm, *Pseudogordius tanganyikae* gen. et sp. nov., parasitic in man. *Ann. Trop. Med. Parasitol.* 51: 313–316.

[Yokoyama, T. & Inoue, I. (1971): On the adult organization of *Gordius* (Gordiacea). *Bull. Tokyo Gakugei Univ.* 23: 109–117. (in Japanese with English summary)]

Yoshimura, H., Kondo, K., Fukuda, K., Kagei, N. & Inoue, I. (1977): The third case of hair-worm, *Gordius* sp., discharged in feces of a woman in Japan. *Jap. J. Parasitol.* 26 (Suppl.): 185–186. (in Japanese)

*Zakarijev, A. J. (1967): Hairworms of Daghestan. *Zool. Zh.* 46: 943–946. (in Russian with English summary)

Zalom, F. G. (1977): A nematomorph parasitizing a back swimmer (Hemiptera: Notonectidae). *Am. Midl. Nat.* 97: 229–230.

[Zanca, F. & De Villalobos, C. (2005): Scanning electron microscopy of *Chordodes moutoni* Camerano, 1895 (Gordiida, Nematomorpha). *Zootaxa* 1082: 37–44.]

Zanca, F. & De Villalobos, C. (2006): South American species of *Paragordius* Camerano, 1897 (Gordiida, Nematomorpha), with redescription of *Paragordius minusculus* and a description of a new species. *Acta Parasitol.* 51: 217–222.

Zanca, F. & Schmidt-Rhaesa, A. (2006): Reinvestigation and new evaluation of representatives from the genera *Semigordionus, Euchordodes, Pantachordodes, Dacochordodes* and *Spinochordodes* (Nematomorpha). *Mitt. Mus. Naturkd. Berlin, Zool. Reihe* 82: 170–178.

[Zanca, F., De Villalobos, C. & Schmidt-Rhaesa, A. (2006): Revision of the genus *Chordodes* (Gordiida, Nematomorpha) from Africa. I. Ultrastructural redescription of *Chordodes gariazzi* Camerano 1902, *C. heinzei*, Sciacchitano 1937; *C. kolensis*, Sciacchitano 1933; *C. muelleri*, Sciacchitano 1937 and *C. ruandensis*, Sciacchitano 1937. *J. Nat. Hist.* 40:17–31.]

[Zanca, F., Schmidt-Rhaesa, A., De Villalobos, C. & Achiorno,C. (2006): Revision of the genus *Chordodes* (Gordiida, Nematomorpha) from Africa. II. Ultrastructural redescription of *C. africanus* Sciacchitano, 1933, *C. hawkeri* Camerano, 1902, *C. sandoensis* Sciacchitano, 1937, *C. schoutedeni* Sciacchitano 1937 and reinterpretation of *C. butensis* Sciacchitano, 1937. *Zootaxa* 1191: 49–59.]

Zanca, F., De Villalobos, C. & Rumi, A. (2007): Larval survival strategy during winter of *Chordodes nobilii* Camerano, 1901 (Gordiida, Nematomorpha). *Acta Parasitol.* 52: 382–385.

Zapotosky, J. E. (1971): The cuticular ultrastructure of *Paragordius varius* (Leidy, 1851) (Gordioidea: Chordodidae). *Proc. Helminthol. Soc. Washington* 38: 228–236.

Zapotosky, J. E. (1974): Fine structure of the larval stage of *Paragordius varius* (Leidy, 1851) (Gordioidea: Paragordidae). I. The preseptum. *Proc. Helminthol. Soc. Washington* 41: 209–221.

Zapotosky, J. E. (1975): Fine structure of the larval stage of *Paragordius varius* (Leidy, 1851) (Gordioidea: Paragordidae). II. The postseptum. *Proc. Helminthol. Soc. Washington* 42: 103–111.

Zervos, S. (1989): Stadial and seasonal occurrence of gregarines and nematomorphs in two New Zealand cockroaches. *New Zealand J. Zool.* 16: 143–146.

Zschokke, F. (1912a): *Gordius* als Parasit des Menschen. Zentralbl. *Bakteriol. Parasitenkd. Infektionskr.* 63: 64–67.

Zschokke, F. (1912b): *Gordius aquaticus* L. as a parasite of man. *Science* 35: 636.

Andreas Schmidt-Rhaesa

4. Priapulida

Introduction

With only 19 known extant species, Priapulida is a very small taxon of marine worms. Priapulid species nevertheless exhibit a fascinating range of differences, spanning from the microscopically small species to macroscopic ones.

The macroscopic priapulids have been known to humans for a very long time and the earliest description is creditied to Odhelius in 1754 (see Van der Land 1970). The species name used was *Priapus humanus*, which was also adopted by Linné (1758). Later Linné (1767) renamed it *Holothuria priapus* and assigned it to sea cucumbers. De Lamarck (1816) coined the still valid name *Priapulus caudatus*.

Priapulids were long considered to be related to either sea cucumbers or to sipunculids and echiurids (in a taxon Gephyrea, see De Quatrefages 1847). The general body organization, however, is joined by the taxa Kinorhyncha and Loricifera and a closer relationship of these three taxa in a monophyletic taxon Scalidophora is currently favoured (e.g. Nielsen 2012).

After the broadly distributed species *Priapulus caudatus*, a second species *Halicryptus spinulosus* was discovered (Von Siebold 1849). Macroscopic priapulids were consequently reported from various locations in cold and temperate waters and further species were described. Priapulids were originally found only on the continental shelf, but Menzies (1959) reported the first priapulid, *Priapulus abyssorum*, from the deep sea below 5000 m. Finally, with the discovery of microscopic priapulids in shallow tropical and subtropical waters (Van der Land 1968, 1970) a second size class of priapulids was discovered and priapulids became known also from warmer waters. Three genera of microscopic priapulids have been discovered, *Tubiluchus*, *Meiopriapulus* and *Maccabeus*. With ten known species they make up more than half of the recent priapulid diversity. Species are in the range of a few millimeters in size; therefore, although often considered to belong to the meiofaunal community, they are larger than the interstitial system between sand grains and therefore are small burrowing organisms rather than "real" interstitial organisms (see Paulay & Holthuis 1994). The largest species, *Halicryptus higginsi*, with a body length of almost 400 mm, was discovered quite late in the Beaufort Sea, Alaska and described by Shirley & Storch (1999).

Priapulids have a characteristic larval stage that can be recognized by the presence of a strong cuticular case, the lorica. Few data are known about the development before this loricate larva is formed. Priapulids grow by a series of moults. Loricate larvae moult into a postlarva that has lost the lorica, but some adult characters are still missing.

Most information on priapulids has been summarized by Van der Land (1970), Storch (1991), Adriano & Malakhov (1996a) and Lemburg (1999). If not otherwise indicated, information is taken from these publications.

4.1. Morphology

Most information on the morphology of priapulids is based on investigations of the species *Priapulus caudatus* and *Halicryptus spinulosus*. Important broad studies including histological sections were made by Ehlers (1862), Apel (1885) and Scharff (1885). Later publications added several data. Ultrastructural investigations were summarized by Storch (1991), Adrianov & Malakhov (1996a) and Lemburg (1999). Ultrastructural investigations are, besides data on *Priapulus caudatus* and *Halicryptus spinulosus*, available for *Tubiluchus* species (Alberti & Storch 1983, 1986, 1988, 1989, Storch & Alberti 1985a,b, Storch et al. 1985, Higgins & Storch 1989, Storch & Higgins 1989, Ferraguti & Garbelli 2006) and for *Meiopriapulus fijiensis* (Storch et al. 1989a,b). Confocal laser scanning microscopy was used to investigate the musculature (Rothe et al. 2006) and the nervous system (Rothe & Schmidt-Rhaesa 2010) of *Tubiluchus* species.

General and external morphology

The body of adult priapulids is divided into a trunk region and an introvert (Fig. 4.1.1 A, B). The introvert can be withdrawn completely into the trunk by the action of retractor muscles. Eversion of the introvert is created by applying pressure on the body cavity fluid by contraction of the trunk musculature. This process is very important during locomotion through muddy sediments; the sequence of movements is described by Friedrich & Langeloh (1936) and Hammond (1970a, 1980); see also Elder & Hunter (1980) and Hunter et al. (1983).

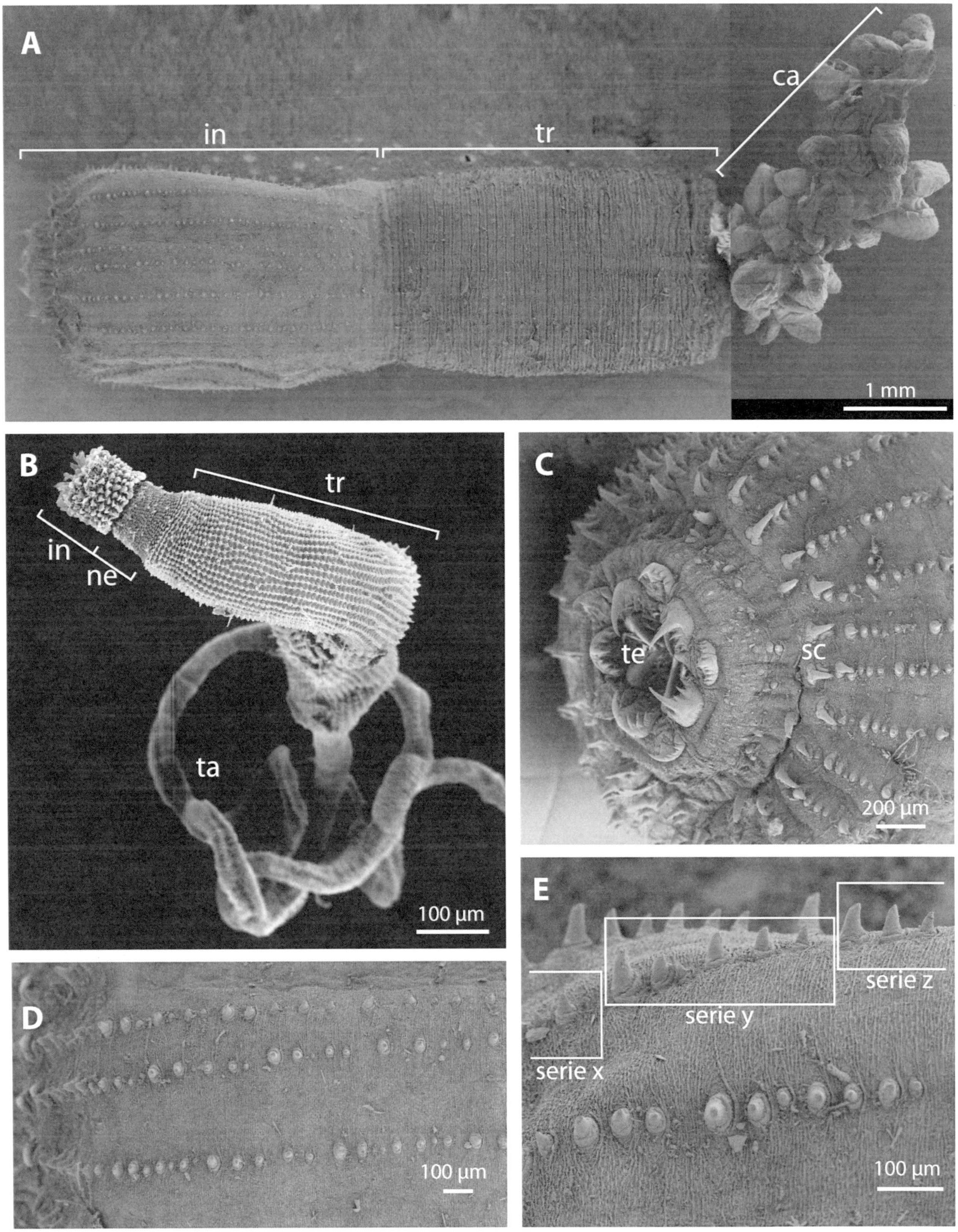

Fig. 4.1.1. External morphology of priapulids. A, Small specimen of *Priapulus caudatus*, showing introvert (in), trunk (tr) and caudal appendage (ca). Image composed of four single images; B, Body organization of *Tubiluchus* sp. (from Tenerife) with introvert, neck (ne), trunk and tail (ta); C–E, *Priapulus caudatus*; C, Anterior end of the introvert showing scalids (sc) and first pentagons of pharyngeal teeth (te); D, Scalids are arranged in longitudinal rows; E, Within rows, scalids are arranged into a series of decreasing scalid size (serie x–z mark three consecutive series within one row). All images SEM.

Fig. 4.1.2. A, *Priapulus caudatus*, small specimen of ~25 mm length; B, *Halicryptus spinulosus*, length of specimen ~20 mm; C, *Priapulus caudatus*, length about 80 mm; D, E, Larva of *Tubiluchus corallicola* from Lee Stocking Island (Bahamas), length of entire larva ~500 µm. an = anus, ca = caudal appendage, in = introvert, mg = midgut, ph = pharynx, pt = polythyridium, te = teeth, tr = trunk.

In the macroscopic species, the introvert and the trunk adjoin each other directly, without forming a transitory region (like the neck in larvae). A neck-like region, which differs in structure from trunk and introvert is present in *Tubiluchus* species (Fig. 4.1.1 B). In species of the genera *Halicryptus*, *Meiopriapulus* and *Maccabeus* the anterior trunk region differs from the remaining region and resembles a neck region. A terminal appendage, often called a tail or caudal appendage, is present in some species. This may be a long, further undivided tail as present in members of the genus *Tubiluchus* (Fig. 4.1.1 B), or one or two more or less branching appendages as in members of the genera *Priapulus*, *Priapulopsis* and *Acanthopriapulus* (Fig. 4.1.1 A, 4.1.2 A, C).

The entire introvert is very large compared to the trunk length in the genera *Priapulus*, *Priapulopsis* and *Acanthopriapulus* (Fig. 4.1.1 A, 4.1.2 A, C), but smaller in the microscopic species (e.g. Fig. 4.1.1 B) and in *Halicryptus* (Fig. 4.1.2 B). The first three genera are therefore summarized as Megintroverta (Lemburg 1999). The mouth opening is terminal on the introvert (Fig. 4.1.1 C). In the microscopic species, a mouth cone is present in the anterior region of the introvert. The anterior part of the pharynx or even a considerable part of the pharynx (e.g. in *Meiopriapulus fijiensis*, see Sørensen et al. 2012 and below) can be everted, exposing the pharyngeal teeth. The entire introvert is covered by spine-like cuticular structures, the scalids (Fig. 4.1.1 C–E). Scalids are arranged in a particular pattern that can be generalized as follows. The first ring (around the mouth opening) has eight scalids, the second ring has nine scalids. The following rings (3–5) have eight, eight and nine scalids. This pattern (8+8+9) is repeated in all following rings. The scalids of three consecutive rings are slightly displaced with regard to the previous ring and this arrangement creates 25 longitudinal rows of scalids on the introvert (Fig. 4.1.1 C, D). This pattern of scalid arrangement is found, for example in larvae and adults of *Priapulus caudatus* and *Halicryptus spinulosus* (e.g. Lemburg 1995a, Adrianov & Malakhov 2001a,b). On the large introverts of *Priapulus, Priapulopsis* and *Acanthopriapulus* the longitudinal rows of scalids are clearly recognizable; their organization into rings becomes less evident. Scalids within a longitudinal row are grouped into so-called "series", within which scalids decrease in size (Fig. 4.1.1 E). This means that in an antero-posterior direction, the smallest, last scalid in one series is followed by the largest scalid in the consecutive series and so on (see e.g. Théel 1911, Van der Land 1970).

Some species may deviate from the scalid pattern explained above; especially in the anterior region, rings may be modified or even missing. The most extreme modification is found in the *Maccabeus* species. Here the first ring of eight scalids is present, but this is followed by a ring of 16 strongly elongate scalids (see below). This arrangement can be interpreted as rings 2 and 3 forming one common ring with 16 scalids, while the fourth ring is reduced. More details on specific arrangement patterns and shapes of scalids are described below for each species.

Adrianov & Malakhov (2001a) analyzed the symmetrical proportions of the scalid arrangement in detail. When viewed from the oral perspective, the priapulid introvert appears to have a circular symmetry, but this is, in detail, not correct. The eight scalids of the first ring create an eight-point radial symmetry. No scalid of the other rings is positioned on the radii on which the eight first-ring scalids are; they all occur within the segments created by these radii. Three longitudinal rows of scalids occur in each segment, with the exception of the ventral radius, in which four rows are present. The two ventral rows are closer together than other rows. These ventral rows can also be modified by being shorter than the remaining rows or by slight displacement of single scalids in longitudinal direction (Adrianov & Malakhov 2001a).

In larvae, scalids are arranged in a similar pattern to those in adults, but their number is lower (e.g. Adrianov & Malakhov 1996b, 2001b; Fig. 4.1.3 D). Scalids are more or less uniform in appearance, even if adults have modified (e.g. elongate) scalids. Scalid size generally decreases in the posterior direction. In *Halicryptus spinulosus*, new scalids are generated in a region posterior of the second ring (Lemburg 1995a). In the genera *Priapulus, Priapulopsis* and *Acanthopriapulus*, which have series of scalids as adults, new scalids seem to arise within each series. This is supported by comparatively widely spaced scalid rings (with scalids of equal size) in postlarvae (e.g. Van der Land 1970) and the decreasing scalid size within each series.

The trunk is densely annulated in adults of the macroscopic priapulids (Fig. 4.1.1 A). This pattern is created by the spacing of circular muscle bundles of the main trunk musculature. Different cuticular structures can be present on the trunk, but this differs between species (see more detailed description below). Scalid-like spines occur in the anterior region of the trunk in *Meiopriapulus fijiensis* and *Maccabeus* species, or along the entire body in *Halicryptus spinulosus*. Structures with assumed sensory or secretory function, such as tubuli or flosculi, can occur along the trunk. In the posterior end, most species have rings of spine- or hook-like structures (Fig. 4.1.3 A, B).

Larvae are characterized by the presence of a lorica, a shelter formed by regionally thickened cuticle (Fig. 4.1.3 C, E). The introvert can be withdrawn completely into the lorica. A special neck region is present in larvae. The introvert is covered by scalids of uniform structure (see above).

In the genera *Priapulus, Priapulopsis, Halicryptus* and *Maccabeus* the lorica is composed of one large dorsal and one large ventral plate (Fig. 4.1.3 E). These are connected by three smaller lateral plates on each side. All plates are composed of strongly sclerotized cuticle and are connected by regions with a thinner, flexible cuticle. All described loricae are dorsoventrally flattened. Loricae differ from each other in shape and sculpture among species. In *Tubiluchus* species, the lorica is composed of 20 slender plates, which are arranged in a circular or almost circular pattern (e.g. Van der Land 1970, Kirsteuer 1976, Adrianov & Malakhov 1996a; Fig. 4.1.3 C). An early loricate larval stage of *Priapulus caudatus* also has a circular lorica composed of 8 plates (Wennberg et al. 2008), therefore this type of lorica is assumed to represent the ancestral structure (Higgins & Storch 1989, Lemburg 1999). The only species in which a lorica is missing is *Meiopriapulus fijiensis*; this is due to the direct development in this species (see below).

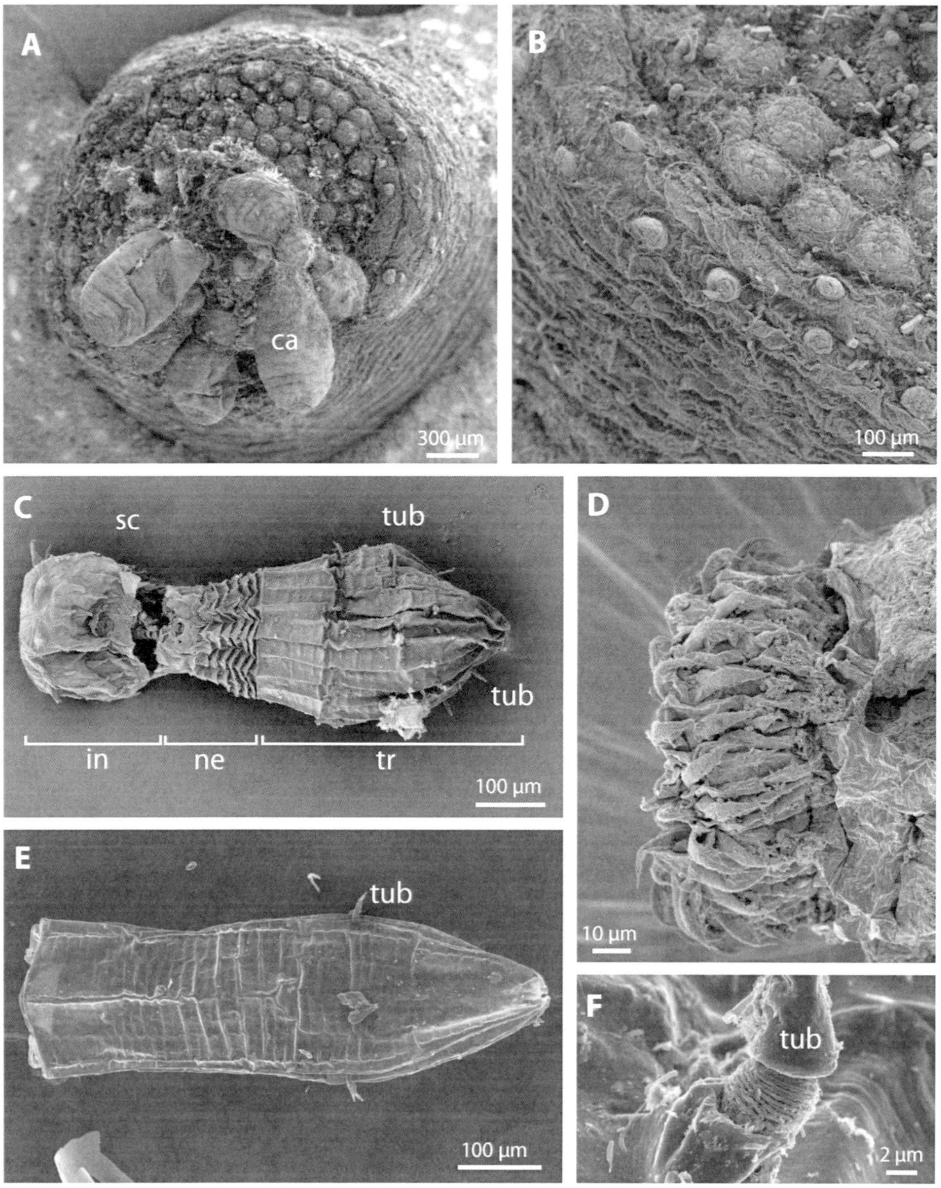

Fig. 4.1.3. A, B, Posterior end of *Priapulus caudatus*; A, Cuticular surface with warts and ring papillae, caudal appendage (ca) emerges from the centre; B, Magnification of A showing ring papillae; C, D, Larva of *Tubiluchus* sp. (from Tenerife); A, Overview showing introvert (in), neck (ne) and trunk (tr) as well as scalids (sc) and some lorica tubuli (tub); D, Introvert with scalids; E, Lorica of the larva of *Priapulus tuberculatospinosus* with lorica tubuli (tub); F, Ringed basal structure of the lorica tubuli in the *P. tuberculatospinosus* larva. All images SEM.

Very characteristic structures on the larval lorica are the tubuli. In *Tubiluchus* species, 15 (*T. vanatuensis*) or 18 (*T. corallicola*, *T. remanei*) lorica tubuli are present; these are arranged in three to five rings along the lorica (Fig. 4.1.3 C). In *Maccabeus*, a total of eight lorica tubuli is present, arranged in an anterior and a posterior ring with four tubuli in each ring. *Priapulus* and *Priapulopsis* larvae have four lorica tubuli in one ring, i.e. at one level; the position of the tubuli in relation to the lorica length differs among species (Fig. 4.1.3 E, F). In *Halicryptus spinulosus*, two rings with four lorica tubuli in each ring are present. They could be shown only in young larvae; in later larvae they seem to be missing (Storch & Higgins 1991, Lemburg 1999).

In most larvae the neck region is a flexible region with a thin cuticle that can be withdrawn with the introvert into the trunk. In species of the genus *Tubiluchus*, the neck region is formed by plates that are continuous with the 20 lorica plates and additionally transversally folded. This entire neck folds over the withdrawn introvert and forms a closing apparatus (e.g. Van der Land 1970; see below under *Tubiluchus*). In the loricate larvae of the other genera, one additional plate is present anterior of both the dorsal and the ventral lorica plate. When the introvert and neck region are withdrawn, these two plates fold over and close the lorica anteriorly.

In the neck region, special neck tubuli were found in *Priapulus caudatus* larva; larvae of *Halicryptus spinulosus* and *Tubiluchus* species have neck flosculi (for terminology and description see below under sensory structures) (Lemburg 1999).

Integument

The integument of priapulids is composed of an epidermis, an apical cuticle and the basal extracellular matrix (ECM, basal lamina). The epidermal cells are arranged in a single layer, as is usual for most invertebrates. Septate junctions and zonulae adhaerentes (desmosomes) are present between cells (Moritz 1972, Lemburg 1999). Neighboring cells interdigitate with each other; in these regions considerably long septate junctions are present (Moritz 1972, Higgins & Storch 1989, Storch et al. 1989a, Lemburg 1999). Many different lectin-binding sites were detected in *Halicryptus spinulosus*; lectin is important for cell–cell and cell–matrix interaction (Busch et al. 2001). In adult specimens of *Priapulus caudatus* and *Halicryptus spinulosus* it was found that the basal cell membrane forms folds into the epidermal cells, thereby creating a kind of lacunar system (e.g. Moritz & Storch 1970). In *Tubiluchus corallicola* larvae, epidermal cells are very flat and strongly interdigitating, and some of the cells are dominated by tonofilaments (=tendon cells) (Higgins & Storch 1989). This may be due to the rigid structure of the lorica.

The ultrastructure of the cuticle was first described by Shapeero (1962). Lemburg (1998) summarizes ultrastructural observations on the cuticle. Thereafter the cuticle is (at least in *Priapulus caudatus* and *Halicryptus spinulosus*) composed of three layers: a very thin epicuticle, a dense exocuticle and a fibrillar endocuticle. The endocuticle contains chitin. Chitin has been detected already by Carlisle (1959). After Saldarriaga et al. (1995) the entire cuticle contains 4–15% chitin and 9–85% proteins. They state that chitin in present in both layers of cuticle (both obviously means exo- and endocuticle), but Lemburg (1998) could detect chitin only in the endocuticle. According to Solovey (1983) the exocuticle consists of proteins, which are stabilized by phenolic polysaccarides and the endocuticle contains chitin, proteins and polysaccarides.

The cuticle is moulted periodically throughout the lifetime (Shapeero 1962, Van der Land 1970). Moulting was assumed to be present by Apel (1885) after histological studies. Hammarsten & Runnström (1920) found the first exuvia, Wesenberg-Lund (1929) observed the last phase of moulting and finally Lang (1948) observed the entire process (confirmed by Carlisle 1959, see also Lebsky 1974). Moulting starts with a local loosening of the cuticle; further separation of the "old" cuticle is aided by undulating movements of the animal (Lang 1948, 1953). Then the cuticle around the mouth opening splits open and the animal crawls through this gap, moulting the cuticle in one entire piece (Schulz 1931, Lang 1948, 1953). Moulting seems to be aided by the sediment, because it takes a long time in an aquarium without sediment (Lang 1948, 1953). In *Meiopriapulus fijiensis*, tubercles on the trunk are very small directly after moulting and grow thereafter (Paulay & Holthuis 1994). This is probably the reason for a colour change observed in this species.

Musculature

The musculature is composed of body wall musculature, some dominant retractor muscles, the musculature of the pharynx and some further muscles (Fig. 4.1.4, 4.1.5 A–C). In the trunk, a layer of outer circular and inner longitudinal musculature is present (e.g. Storch et al. 1989b, 1994, Storch 1991, Rothe et al. 2006; Fig. 4.1.6 D). Circular and longitudinal musculature is arranged in bundles; these form a regular grid (Fig. 4.1.6 B, D). In the introvert, Storch et al. (1990, 1994) describe the same structure as in the trunk, but Adrianov & Malakhov (1996a), Lemburg (1999) and Rothe et al. (2006) describe an additional layer of outer longitudinal musculature from the introvert.

Retractor muscles are important for withdrawl of the introvert. There are two sets, long and short retractors (see Mattisson et al. 1974, Storch et al. 1990, 1994; Fig. 4.1.4, 4.1.5 A–C, 4.1.6 A, C). Both retractors insert at the level of the brain, i.e. at the anterior margin of the introvert. There are always eight long retractors, which run through the body cavity deep into the trunk region. In larvae of *Tubiluchus*, there are only six such muscles (Van der Land 1968). The long retractors are only missing in adults of *Maccabeus tentaculatus* (Por & Bromley 1974); larvae of this species have seven or eight retractors (Von Salvini-Plawen 1974). According to the figure in Mattisson et al. (1974) the retractors attach to the body wall in the posterior part of the trunk in *Priapulus caudatus*, between trunk annulations 25 and 30 (Storch et al. 1990). In *P. tuberculatospinosus* attachment sites are even more posterior (Storch et al. 1990). In *Halicryptus spinulosus*, the attachment site is more anterior, at about the end of the anterior third of the trunk (Storch

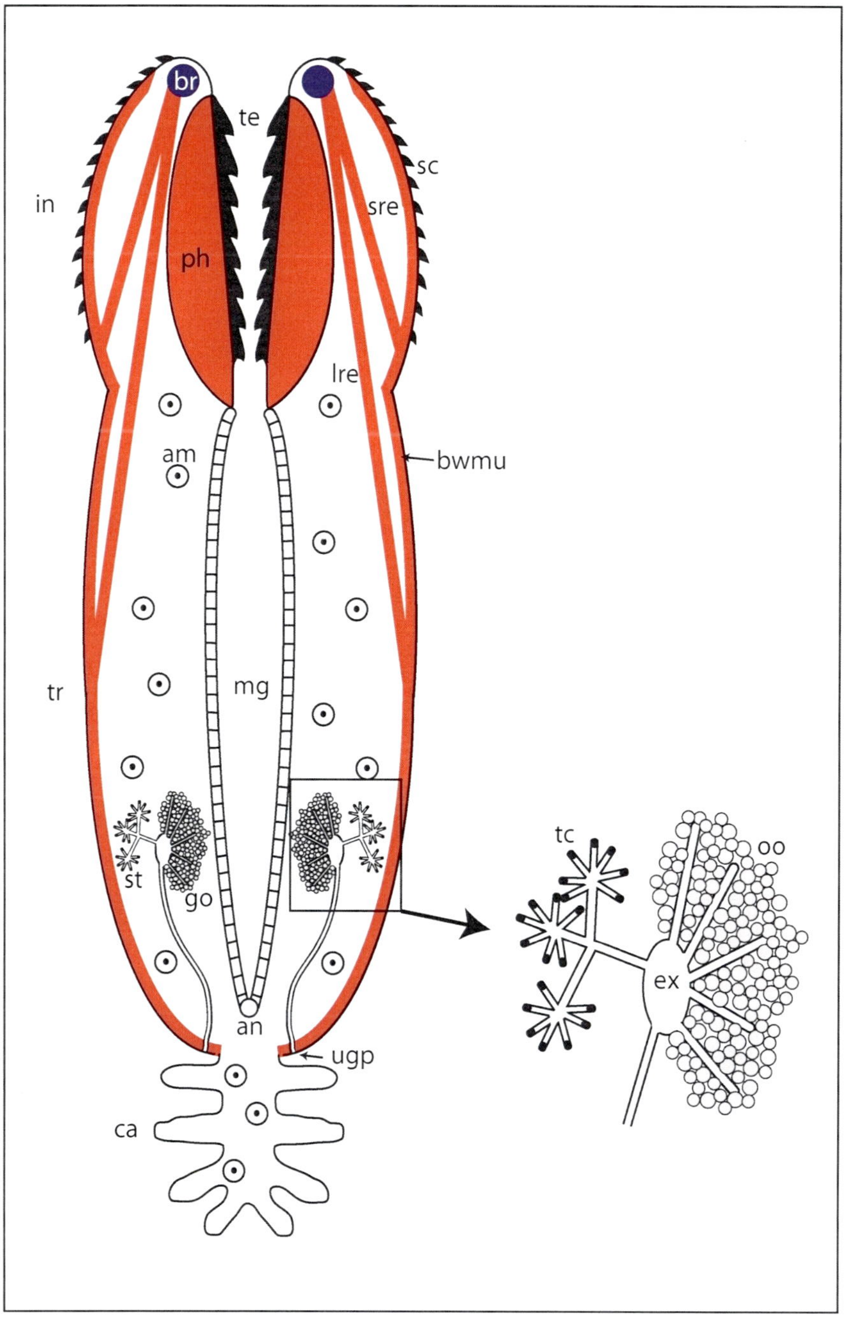

Fig. 4.1.4. Schematic representation of the body organization of priapulids, based on the genus *Priapulus*. Urogenital structures are magnified (to the right). Musculature in red, nervous system in blue. am = amoebocytes, an = anus, br = brain, bwmu = body wall musculature, ca = caudal appendage, ex = excretory sack, go = gonads, in = introvert, lre = long retractors, mg = midgut, oo = oocytes, ph = pharynx, sc = scalids, sre = short retractors, st = solenocyte tree, tc = terminal cells, te = pharyngeal teeth, tr = trunk, ugp = urogenital pore. Represenation of urogenital system based on Fig. 11.5 in Adrianov & Malakhov (1996a).

et al. 1994). The number of short retractors differs. After Hammond (1970b), 25 are present in *P. caudatus*, but later investigations found smaller numbers (10–25 after Mattisson et al. 1974). Storch et al. (1990, 1994) found eight short retractors in *H. spinulosus*, *P. caudatus* and *P. tuberculatospinosus*, but as these muscles can divide, a total of 16 posterior attachment sites can be present. In *Maccabeus tentaculatus* no short retractors are present (Por & Bromley 1974), but Por (1983) speculates that 25 radial muscles may be derived from the short retractors. These run from the region of the brain, where they insert between the long retractors, to the border between introvert and trunk. Because the introvert in *H. spinulosus* is very short, the short retractors are also considerably shorter than in *Priapulus* species.

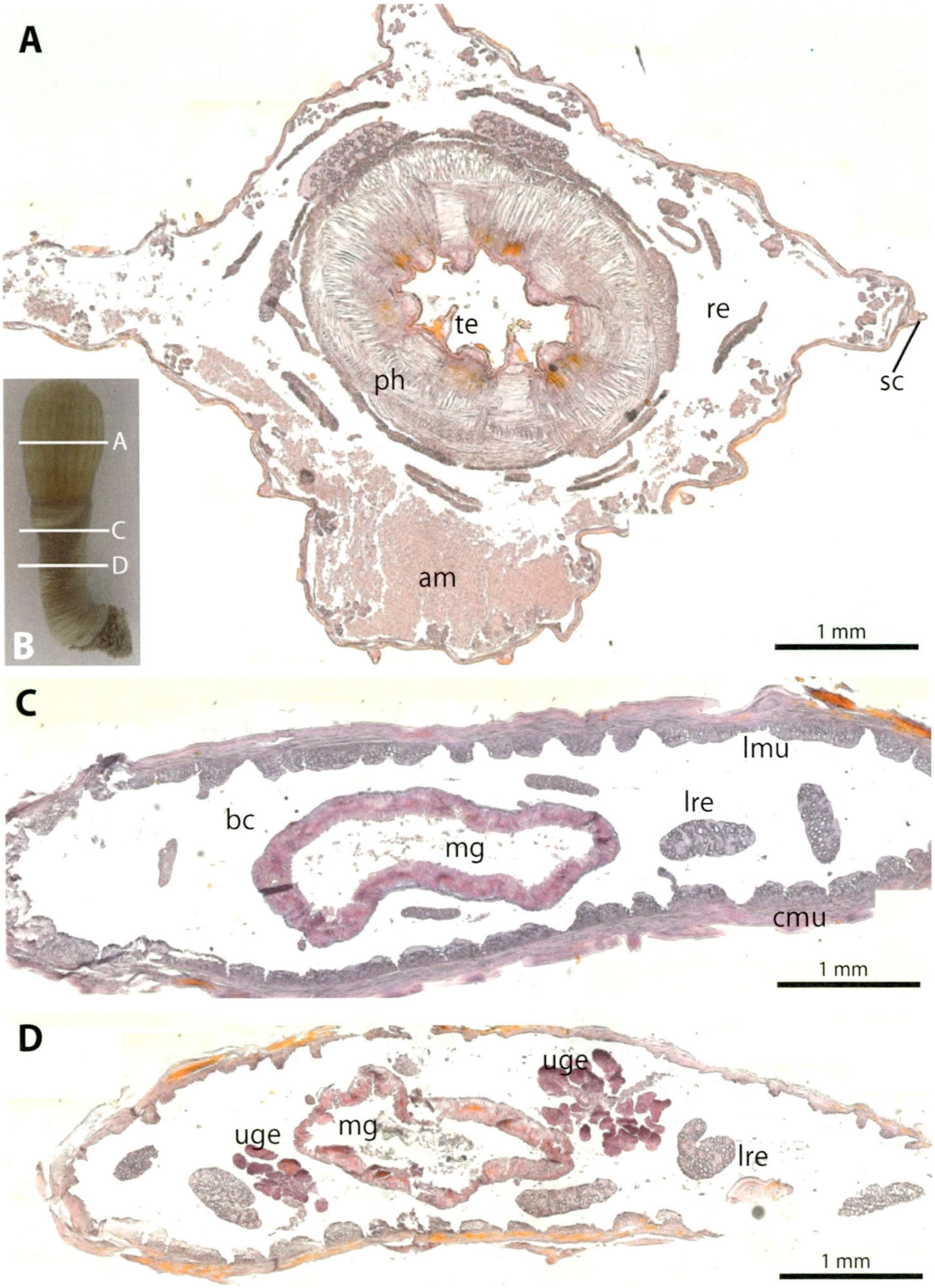

Fig. 4.1.5. Histological cross sections through different levels of body in *Priapulus caudatus* as indicated in B. All images composed of several single images. A, Section through the introvert showing some scalids (sc), the pharynx (ph) with teeth (te) and retractor muscles (re). Amoebocytes (am) are artificially concentrated on one side of the animal; C, Section through the anterior part of the trunk showing midgut (mg), primary body cavity (bc), the body wall with circular (cmu) and longitudinal (lmu) muscles and long retractor muscles (lre); D, Section through about the middle length of the trunk with anterior parts of the urogenital system (uge). Histological sections stained with Azan.

Other muscles have been documented less regularly. After Lemburg (1999; see further references therein), pharynx protractors, which run from the brain region to the anterior region of the pharynx are present, at least in *H. spinulosus* and *P. caudatus*. In larvae of these species, Lemburg (1999) additionally described six pharynx retractors running from the posterior end of the pharynx to the body wall and five short pharynx protractors. An additional muscle in adults has been described by Storch et al. (1994). This is a pharynx dilator muscle running from the anterior region of the pharynx to the body wall.

Descriptions on the musculature of the pharynx do not create a uniform picture. Radial, longitudinal and circular elements are present. After Rothe et al. (2006), the pharynx of *Tubiluchus troglodytes* is dominated in the anterior region by circular musculature and in the posterior part by longitudinal musculature. After Storch et al. (1994), radial musculature dominates the posterior part of the pharynx in *P. caudatus*. In *P. caudatus* and *P. tuberculatospinosus* six longitudinal musculature bands surround the pharynx over its entire length. In the anterior region, a series of overlying short longitudinal muscles makes a differential movement of the anterior pharyngeal teeth possible (Storch et al. 1994). A broad circular muscle band surrounds the anterior part of the pharynx in *P. tuberculatospinosus* (Storch et al. 1994). In the anterior region of the pharynx, strong circular musculature may have the function of a sphincter.

Musculature is also present around the intestine. After McLean (1985a) it is composed of outer circular and inner longitudinal musculature; the circular musculature is widely spaced.

Ultrastructural investigations show that muscles in adult priapulids are of the circomyarian type, in which the contractile filaments (actin and myosin) are arranged around the entire periphery of the muscle cells, while the nucleus, cytoplasm and further organelles are present in the center (Mattisson et al. 1974, Candia Carnevali & Ferraguti 1979, Storch 1991, Adrianov & Malakhov 1996a, Lemburg 1999; Fig. 4.1.6 C). Z-elements are spindle-shaped and most of them are not connected to the cell membrane (Lemburg 1999). The sarcoplasmatic reticulum and the T-system are quite extensive (Mattisson et al. 1974, Candia Carnevali & Ferraguti 1979, Lemburg 1999). The pattern of striation is not clear. While Janssen & Oeschger (1992) describe muscles of *Priapulus caudatus* as obliquely striated, Candia Carnevali & Ferraguti (1979) describe them as smooth. The longitudinal musculature of *Tubiluchus troglodytes* is cross striated (Rothe et al. 2006). Larvae of *Priapulus caudatus* and *Halicryptus spinulosus* differ in having platymyarian muscles in the body wall. In this muscle type, contractile filaments are arranged only in the apical periphery of the cell (the part directed towards the epidermis), while nucleus, cytoplasm and other organelles are concentrtated on the basal side of the cell (Lemburg 1999). This is also the case for *Tubiluchus corallicola* (Higgins & Storch 1989). Z-elements are connected to the cell membrane. The sarcoplasmatic reticulum and the T-system are much less developed in larvae compared to adults (Lemburg 1999). Adults of the microscopic species seem also to have muscles which are characteristic for larvae (Storch & Alberti 1985a, Storch et al. 1989a,b).

Nervous system

Information on the nervous system goes back to histological investigations (e.g. Schauinsland 1887), some ultrastructural data (in particular Rehkämper et al. 1989, but see also Higgins & Storch 1989, Storch 1991), silver nitrate staining (Punin & Zaitseva 2002) and immunohistochemical staining (Joffe & Kotikova 1988, Rothe & Schmidt-Rhaesa 2010).

The brain is ring-shaped and is located at the anterior margin of the introvert, at the transition into the buccal cavity (Fig. 4.1.6 F). It is composed of a ring-neuropil and a region of somata. The somata are present anterior of the ring neuropil in *Halicryptus spinulosus*, *Priapulus caudatus* and *Priapulopsis bicaudatus* (Horst 1881, Scharff 1885, Joffe & Kotikova 1988, Adrianov & Malakhov 1996a). In *Tubiluchus* species and in *Meiopriapulus* fijiensis, an additional ring with somata is present posterior of the neuropil (Rehkämper et al. 1989, Storch et al. 1989b, Rothe & Schmidt-Rhaesa 2010; Fig. 4.1.6 F). *Maccabeus tentaculatus* shows a completely different pattern; somata are here medial of the neuropil (Por & Bromley 1974). Adrianov & Malakhov (1996a) describe the brain as being locally thickened into eight ganglia, corresponding to the eight scalids in the first ring. Rothe & Schmidt-Rhaesa (2010) show that in *Tubiluchus troglodytes* the anterior somata (less evident in the posterior somata) form lobe-like clusters, but here 20–25 such clusters are present.

The ventral nerve cord is unpaired (Fig. 4.1.6 E, G). It originates in the ventral region of the brain neuropil and extends to the region anterior of the anus. Somata of the neurons in the ventral nerve cord are scattered along the cord. Their appearance sometimes appears regular, because somata are positioned between bundles of the circular musculature (therefore, for example, Schauinsland 1887 noted indications of segmentation in the priapulid nervous system). In *Tubiluchus troglodytes*, *Priapulus caudatus* and *Halicryptus spinulosus*, the anterior region of the ventral nerve cord is free of somata; these are present only in the trunk region. This might be explained functionally, because the absence of somata in the introvert makes the nerve cord more flexible during inversion and withdrawl. In larvae and postlarvae of *P. caudatus* and *H. spinulosus*, as well as in adults of *T. troglodytes*, some somata are concentrated in a structure commonly called neck ganglion (Adrianov & Malakhov 1996a, Lemburg 1999, Rothe & Schmidt-Rhaesa 2010). Another ganglion-like swelling is the so-called caudal ganglion in the posterior region of the ventral nerve cord. It seems to be present in all or most species. Whether the term "ganglion" for the neck ganglion and the caudal

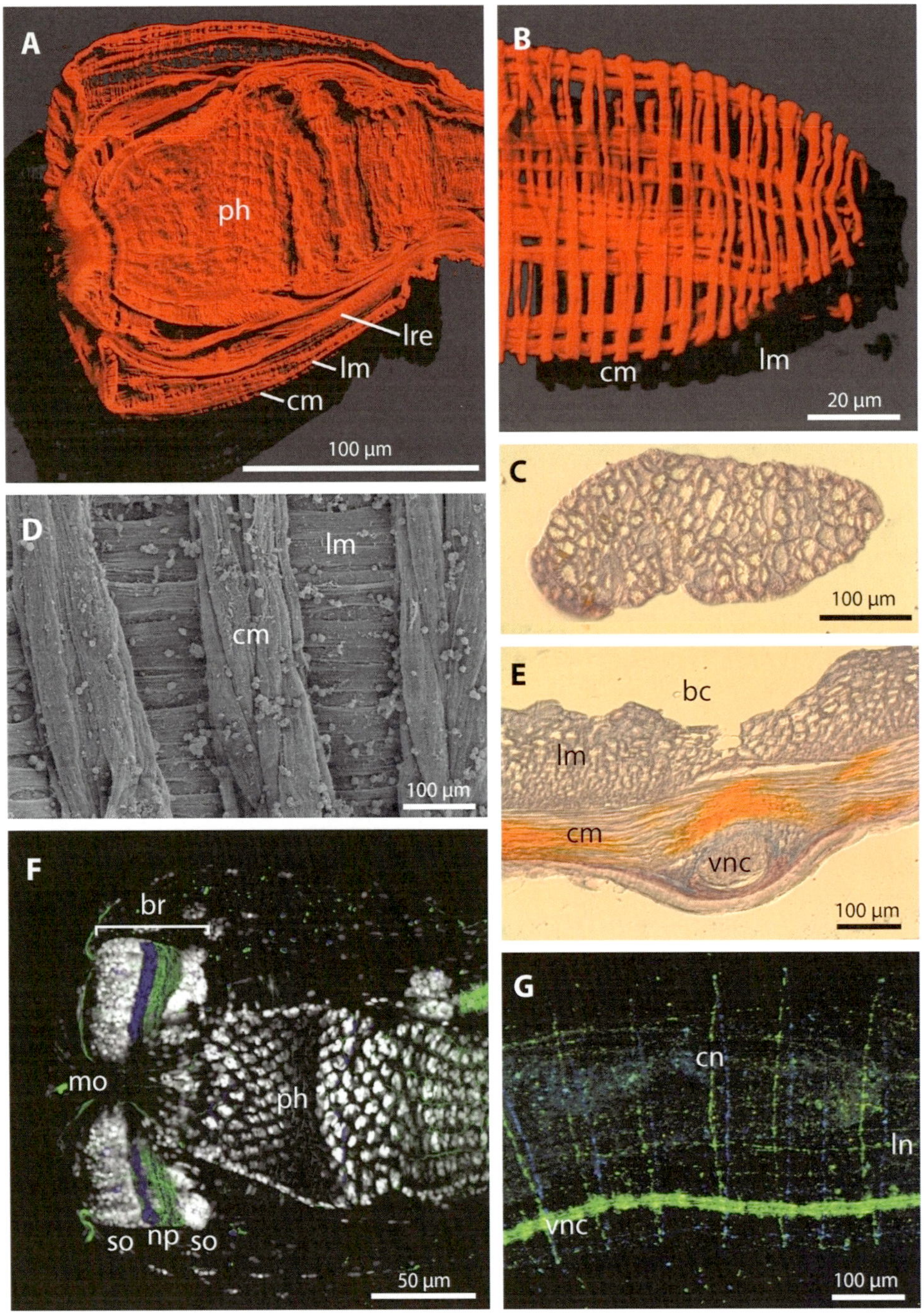

Fig. 4.1.6. Musculature and nervous system. A, B, Confocal images (SFP (= simulated fluorescence process algorithm) projection) of a phalloidin staining in *Tubiluchus troglodytes* (see Rothe et al. 2006 for details). A, Virtual longitudinal section from one half of the z-stacks showing pharynx (ph), long retractors (lre) and the circular (cm) and longitudinal (lm) musculature of the body wall. The additional external layer of longitudinal musculature is not visible in this image; B, Longitudinal and circular muscle bundles form a regular grid in the tail; C, E, Histological sections (Azan staining); C, long retractor muscles composed of circomyarian muscle cells; E, Body wall with unpaired ventral nerve cord (vnc); D, SEM image of body wall muscles viewed from internally; F, G, Confocal images from *Tubiluchus troglodytes*; F, Multiple stainings of nuclei with DAPI (white) and nervous structures with antibodies against α-tubulin (green) and RF-like amides (blue) (see Rothe & Schmidt-Rhaesa 2010 for details). The brain (br) forms a ring around the anterior part of the pharynx (ph) behind the mouth opening (mo). The neuropil (np; blue and green labelling of immunofluorescence) as well as anterior and posterior somata (so; white labelling) are clearly visible. In the pharynx, nuclei are concentrated in patches below the teeth; G, Immunofluorescence of β-tubulin and RF-like amides in the trunk showing the unpaired ventral nerve cord (vnc) and an orthogonal grid of circular (cn) and longitudinal (ln) neurites. A and B from Rothe et al. (2006), with kind permission of Pfeil Verlag, Munich.

ganglion is appropriate is under debate (see Richter et al. 2010).

In the introvert, intraepithelial longitudinal neurite bundles are present. After Joffe & Kotikova (1988), one neurite bundle runs below each longitudinal row of scalids in *Priapulus caudatus* and 24 additional bundles are present between them, making 49 bundles in total. In *Tubiluchus troglodytes*, Rothe & Schmidt-Rhaesa (2010) counted 47 bundles, but two bundles might have been overlooked. Longitudinal introvert neurite bundles were already seen by light microscopical investigations, but in smaller number. Apel (1885), for example, counted 26 bundles. Under each scalid, small accumulations of neurons and sensory cells are present. In the trunk, an orthogonal pattern of longitudinal and circular neurites is present (Joffe & Kotikova 1988, Lemburg 1995a, Rothe & Schmidt-Rhaesa 2010; Fig. 4.1.6 G); the more or less regular pattern becomes slightly irregular towards the posterior end.

The pharynx is innervated by fibers originating in the brain. Neurites form a regular network between the pharyngeal teeth (Lemburg 1999, Rothe & Schmidt-Rhaesa 2010). Concentrations of neurons are present below each tooth, equivalent to the scalids (Lemburg 1999, Rothe & Schmidt-Rhaesa 2010). Punin & Zaitseva (2002) describe two plexuses surrounding the intestine, one intraepithelial and one subepithelial plexus in the musculature.

Sensory structures

Priapulids have a variety of sensory structures on their body. Most such receptors seem to be mechano- or chemoreceptors. Photoreceptors and gravity receptors are unknown, although there is indication of a reaction to light and geotaxis (Langeloh 1936).

Apart from helping during locomotion, the scalids on the introvert are sensory structures and include receptor cells. The only known exceptions are the locomotory scalids (fourth and subsequent rings) in *Meiopriapulus fijiensis* (Storch et al. 1989b). In general, receptor cells are bipolar cells with an apical cilium. The cilium may be surrounded by seven microvilli ("stereocilia", "stereovilli"). Different kinds of receptor cells can be present in one scalid. For example, three different types of receptor cells are known from the scalids of adults in *Priapulus caudatus*, *P. tuberculatospinosus*, *Priapulopsis bicaudatus* and *P. australis* (Storch et al. 1994, 1995). One type includes an apical cilium that reaches the surface close to the pore in the scalid; the two other types have a cilium with surrounding microvili and end beneath the cuticle (Storch et al. 1995). Lemburg (1995a) described in detail different receptor types in scalids of the larva of *Halicryptus spinulosus*. Four receptor cell types are present. The first type is present in almost all scalids, vertical rootlet receptors only in the first ring, basal receptors in the second ring and coiled cilium receptors in the third and fourth ring. Receptor type 1 cells can be very numerous, with up to 50 receptor cells in each first ring scalid, but the number decreases in scalids of the other rings. The other receptor cell types occur singly or as a pair (Lemburg 1995a).

Scalids in adult macroscopic priapulids have several subapical tubules (Storch et al. 1990, Storch et al. 1994, 1995; Fig. 4.1.7 A–D). These contain the scalid receptors and cilia of the receptor cells reach the exterior of these tubules. The buccal papillae, which are present between scalids and teeth in *Halicryptus spinulosus*, also include receptor cells (Storch et al. 1990).

Conspicuous sensory structures are the flosculi (Fig. 4.1.7 E–G). These are flower-shaped structures that include a central cilium and seven surrounding microvilli (Moritz & Storch 1971, Storch 1991, Lemburg 1999). Flosculi are present in the neck region of larvae and on the trunk of adults and larvae.

There are further cuticular structures such as setae, anal hooks and tubuli, which also include receptor cells (Storch 1991, Lemburg 1999). Additionally, receptor cells can be present in the epidermis without having an elevated cuticular structure. For example, Lemburg (1995b) described groups of receptor cells close to epidermal secretory cells from the neck region of *Halicryptus spinulosus* larve.

Intestinal system

The intestinal tract is a straight, unbranched tube running from the terminal mouth opening to the almost terminal anus (Fig. 4.1.4). It can be divided into the ectodermal foregut, endodermal midgut and ectodermal hindgut. The ectodermal parts are covered by cuticle, which has led to the presence of several cuticular specializations in the foregut.

The dominant structure of the foregut is the pharynx (Fig. 4.1.4, 4.1.6 A). This is a strong muscular bulb (see above under musculature) with cuticular teeth (Fig. 4.1.5 A, 4.1.8 B). In species like *Priapulus caudatus*, the pharynx is the only component of the foregut (e.g. Storch et al. 1994). In *Tubiluchus* species and *Meiopriapulus fijiensis*, the forgut is more elaborate (see below). The entire pharynx can be withdrawn deeply into the trunk when the introvert is withdrawn. When the introvert is fully everted, the anterior region of the pharynx opens

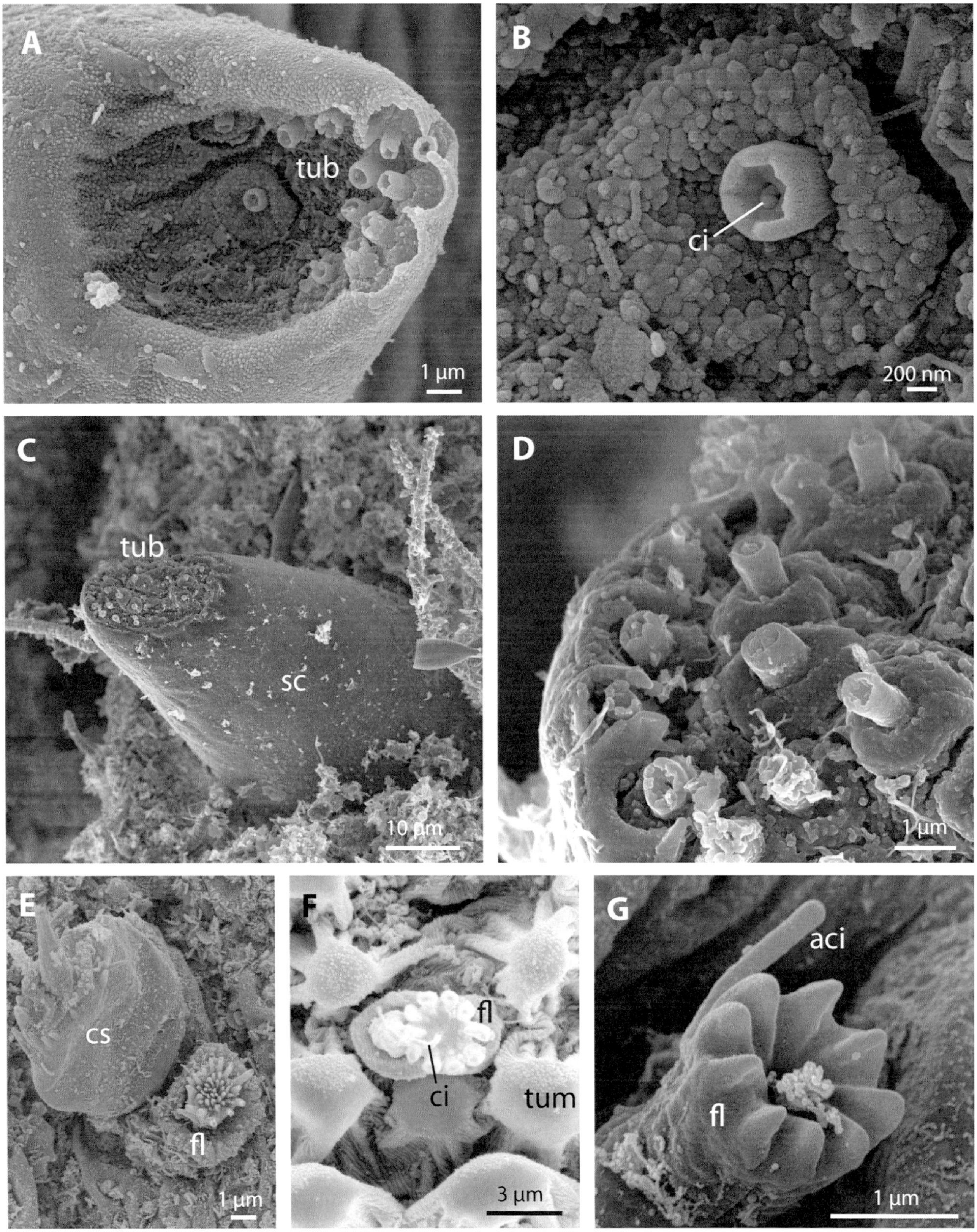

Fig. 4.1.7. Sensory structures. A–D, Scalid fine structure in *Priapulus caudatus*; A, B, overview on scalids from different specimens; B, D, magnification of the tubular structures. A central cilium (ci) is visible in B; E, Flosculus (fl) next to a "cone-shaped structure" (cs) in *Meiopriapulus fijiensis* (see Sørensen et al. 2012); F, Flosculus with central cilium next to tumuli (tum) on the trunk of *Tubiluchus* sp. from Tenerife; G, Flosculus with accessory cilium (aci) in *Tubiluchus corallicola*. E with kind permission from Martin V. Sørensen, Copenhagen. All images SEM.

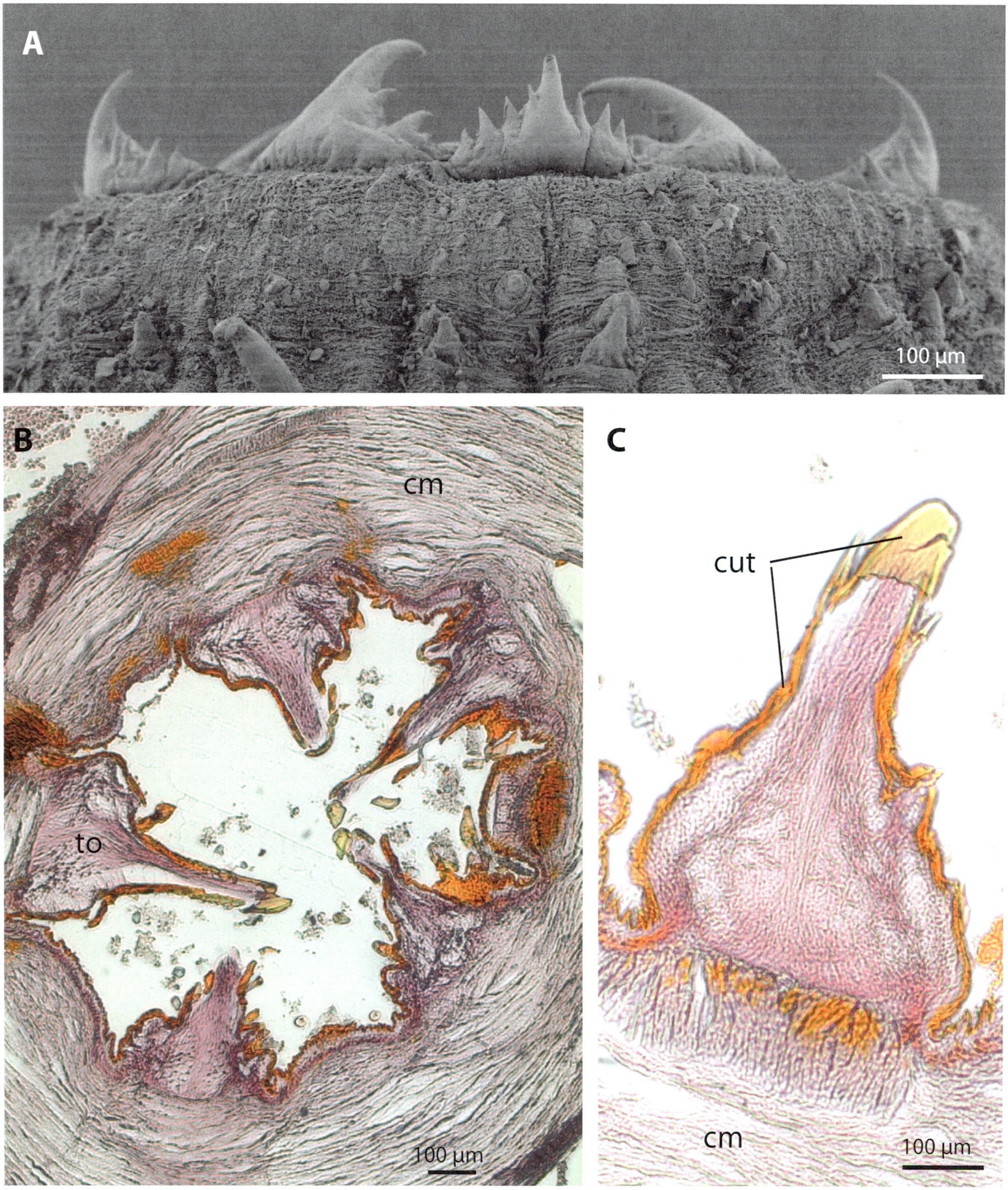

Fig. 4.1.8. Pharyngeal teeth in *Priapulus caudatus*. A, Lateral view on the first pentagon of teeth; B, Cross section through the pharynx (cm = circular musculature, to = tooth); C, Magnification of one tooth showing it as a structure covered by cuticle (cut) and filled with tissue. A with SEM, B, C histological section, Azan staining.

more or less and exposes the anterior teeth, with which prey can be grabbed (Fig. 4.1.1 C, 4.1.8 A).

Originally, priapulids were considered to be herbivorous, but Schulz (1931) observed the uptake of animal organisms. This was confirmed by Lang (1948, 1953), including cannibalism. Ankar & Sigvaldadottir (1981) analyzed the food in the intestine of *Halicryptus spinulosus* and confirmed it to be mainly carnivorous. Nevertheless,

Aarnio et al. (1998) found the stomach contents of *H. spinulosus* from the Northern Baltic Sea to consist of up to 68% detritus.

The pharyngeal teeth are of two types. In the macroscopic species, cuspidate teeth are present (Fig. 4.1.8 A–C, 4.1.9 A–D). These include one major spine and several smaller accessory spines. In *Tubiluchus* species and *Meiopriapulus fijiensis*, pectinate teeth are present; these have a comb of spines of more or less the same length (Fig. 4.1.9 E, F). Pectinate teeth are stalked; the stalk is called the manubrium (Van der Land 1970). In the posterior part of the pharynx in *Tubiluchus*, cuticular ridges with rows of fine bristles are present, these are called fimbrillae. Minute pharyngeal glands are present in the posterior part of the pharynx (Janssen 1994). For a comparative overview of teeth see Adrianov & Malakhov (1996b).

In adults of *Tubiluchus* and *Meiopriapulus fijiensis* a particular muscular structure is present between the pharynx and the midgut, the polythyridium (e.g. Van der Land 1970, Kirsteuer 1976, Morse 1981, Rothe et al. 2006; Fig. 4.1.2 D). The polythyridium is surrounded by a strong layer of circular musculature and includes cuticular structures called valvulae. Two kinds of valvulae are present, one with a fringe of fine bristles and one with a thickened apical region. The region between the pharynx and polythyridium is called the esophagus (= oesophagus).

In *Maccabeus tentaculatus*, the pharynx contains cuspidate teeth, but in the posterior region of the pharynx in particular, long cuticular structures called gastric spines are present (Por & Bromley 1974).

The midgut has no diverticula. In *Maccabeus tentaculatus*, the intestine is divided into an anterior "muscular stomach" and a posterior "glandular stomach" (Por & Bromley 1974), but such subdivision is not present in the other priapulid species. Microvilli border the cells towards the lumen (Higgins & Storch 1989, Storch 1991). Longitudinal and circular musculature is present around the midgut (Storch et al. 1990). The digestive fluid was shown to have strong extracellular proteolytic activity (Nilsson & Fänge 1967).

Quite regularly, the coccidian parasite *Alveocystis intestinalis* (Apicomplexa) was found in the intestine of priapulids (Lüling 1943, McLean 1984a, 1985b, Saldarriaga & Storch 1997).

Body cavities and connective tissue

Adult priapulids have a spacious body cavity (Fig. 4.1.4, 4.1.5 A–C). In larvae it is smaller, for example in the *Tubiluchus corallicola* larva (Higgins & Storch 1989). It extends into the caudal appendage of macroscopic priapulids (Fänge & Mattisson 1961) and into the tail of *Tubiluchus* species (Kirsteuer & van der Land 1970). The body cavity is sometimes stated to be a coelom (e.g. Shapeero 1961), but as it is bordered by extracellular matrix, it represents a primary body cavity (Lang 1953, McLean 1984b). Only in *Meiopriapulus fijiensis*, a small coelom is present surrounding the forgut (Storch et al. 1989b).

The body cavity contains a large number of free cells, the coelomocytes (e.g. Mattisson & Fänge 1973, Solovey & Chaga 1982). Two types are present; these are usually called amoebocytes (= leucocytes) and erythrocytes. Erythrocytes are flattened (Solovey 1984, Schreiber & Storch 1992), nucleated cells with vacuoles and occasional crystals, but with few organelles in the cytoplasm (Mattisson & Fänge 1973). Amoebocytes are irregularly shaped amoeboid cells with numerous organelles in the cytoplasm (Mattisson & Fänge 1973, Schreiber & Storch 1992). Erythrocytes outnumber amoebocytes (>90%). Amoebocytes are quite similar among species, but erythrocyte ultrastructure differs among genera (Schreiber & Storch 1992). Amoebocytes phagocytize bacteria and degenerating cells and tend to coagulate (Solovey 1987, Schreiber et al. 1991). They may also be involved in repair of injured regions of the cuticle (Solovey & Chaga 1982).

Erythrocytes contain hemerythrin (Fänge 1950, Fänge & Mattisson 1961). Hemerythrin is a respiratory pigment, even if in *Priapulus caudatus* its affinity to oxygen is comparably low (Weber & Fänge 1980). Weber & Fänge (1980) therefore speculate that it does not play an important role in respiration. This is connected to the question of whether the caudal appendage, which is present in species of *Priapulus*, *Priapulopsis* and *Acanthopriapulus*, is respiratory in function or not. It is a thin-walled structure and oxygen uptake through the thin epithelium into the body cavity appears plausible. Fänge & Mattisson (1961) observed that when specimens are kept in low oxygenated water and are then put into well-aerated sea water, the caudal appendage contracts repeatedly and pumps body cavity fluid into the trunk. This is contrasted by Lang´s (1953) observation that the caudal appendage is deep in the sediment, i.e. in layers with low oxygen content, and there is no change of position towards oxygenated water. Lang therefore concludes that the caudal appendage has no respiratory function.

Excretory system

The excretory system joins the reproductive system and therefore a urogenital system is formed (Fig. 4.1.4). The

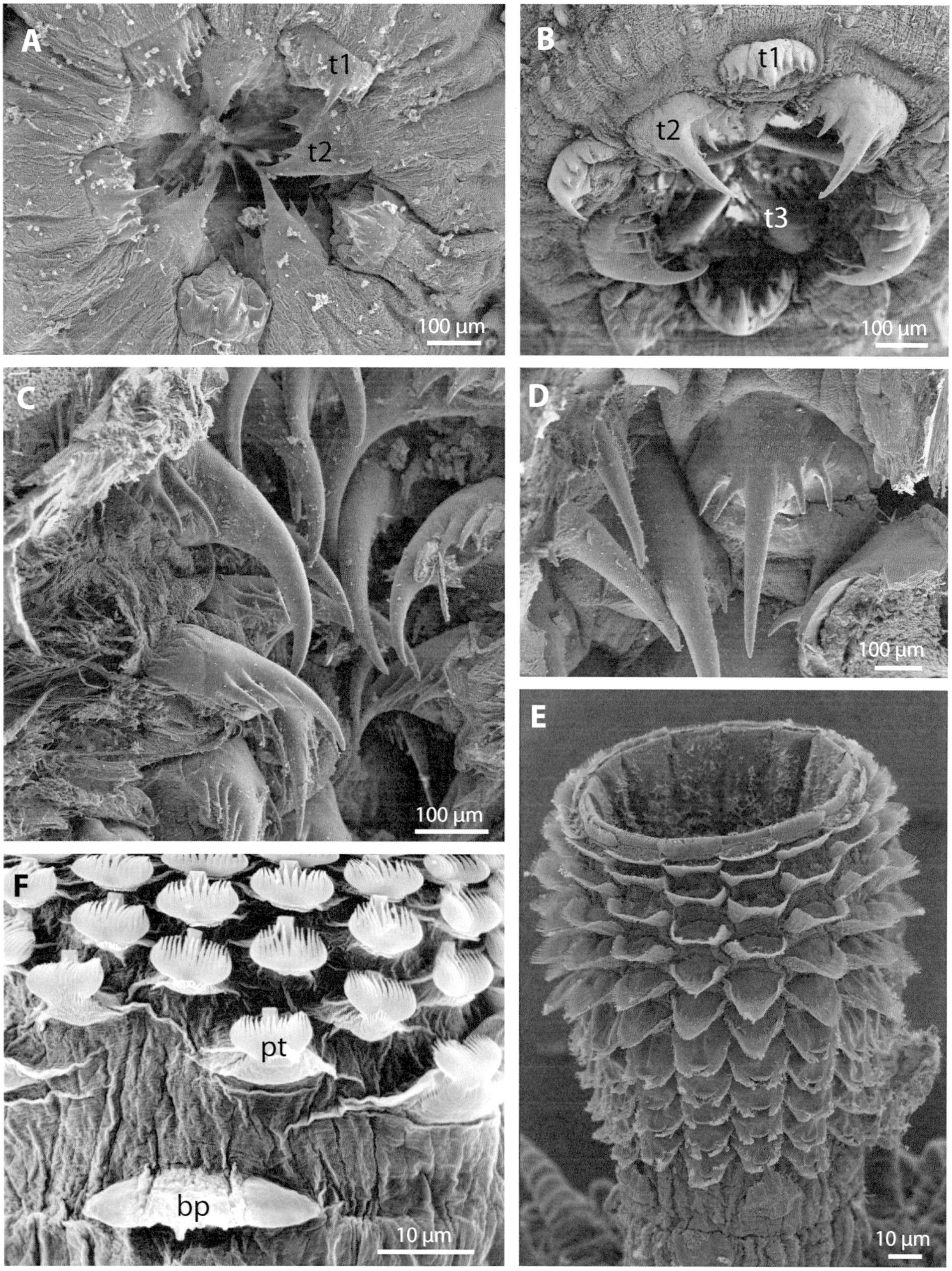

Fig. 4.1.9. Pharyngeal teeth. A–D. Cuspidate teeth in *Priapulus caudatus*. A, B, Teeth of first (t1), second (t2) or third (t3) pentagons; C, D, Teeth from about the middle of the pharynx (from longitudinal sections); E, Everted pharynx of *Meiopriapulus fijiensis* with regular arrangement of pectinate teeth; F, Buccal papilla (bp) and pectinate teeth (pt) on the everted pharynx of *Tubiluchus* sp. from Tenerife. E with kind permission from Martin V. Sørensen, Copenhagen. All images SEM.

excretory structures are protonephridia. Two (*Tubiluchus philippinensis*; Alberti & Storch 1986) or more terminal cells are extensively connected and form terminal organs (Kümmel 1964, Storch et al. 1989b, Lemburg 1999). Neighboring terminal cells interdigitate with microvilli-like extensions and form a weir, which acts as a filtration area (Lemburg 1999). Terminal cells in larvae of *Priapulus caudatus* and *Halicryptus spinulosus* have one cilium, in adults of these species they are multiciliary (Lemburg 1999). In adults of *Tubiluchus philippinensis* two–eight cilia are present in the terminal organs (Alberti & Storch 1986); larvae of *T. corallicola* have two cilia (Higgins & Storch 1989). Numerous microvilli are present. Several canal cells form the outlet tubes. The entire excretory system branches extensively and contains several to numerous such terminal organs. This entire structure has originally been termed the "solenocyte tree" (e.g. Nørrevang 1963).

Lüling (1940) described the development of the urogenital system in *Halicryptus spinulosus*. The earliest visible structure is a duct with few terminal cells. Therefore, the excretory system develops before the reproductive system. Later, reproductive organs develop and the number of terminal cells grows.

Reproductive organs

Priapulids are dioecious. Only in *Maccabeus tentaculatus* is parthenogenetic reproduction assumed, because only females have been found (Por & Bromley 1974, Von Salvini-Plawen 1974). Gonads are paired structures in the posterior part of the trunk; their ducts join with the excretory system (see above; Fig. 4.1.4). Only in *Meiopriapulus fijiensis* is one unpaired ovary present (Storch et al. 1989a). In the macroscopic species, mature gonads are voluminous and produce a large quantity of gametes for external fertilization. In microscopic species, gonads are smaller and produce fewer gametes. The entire urogenital system opens in paired pores lateral of the anus.

Gonads develop from the anterior part of the primary duct, which up to the formation of gonads has only an excretory function (Lüling 1940). Tissue proliferation leads to numerous branching tubes, which either develop into an ovary or a testis. Gametes originate in the epithelium of the gonads; they remain in contact with the epithelium until maturity. Some of the epithelial cells have a supporting function (nurse cells). Not all gametes mature at the same time and different developmental stages can be found. Gonads are attached to the body wall by mesenteries (ligaments). They are bordered towards the body cavity by extracellular matrix.

In *Tubiluchus philippinensis*, the ovary contains large oocytes rich in yolk. They are arranged in rows within tubular ovaries (Alberti & Storch 1989). Younger cells are surrounded by extensions of somatic cells in the ovarian tube; mature cells are only in part surrounded by cells. Nurse cells are not present.

Gametes

The shape of gametes differs considerably among species, depending on the mode of reproduction. Species with external fertilization have round-headed ("primitive") spermatozoa (Afzelius & Ferraguti 1978, Adrianov et al. 1992; Fig. 4.1.10 A), *Tubiluchus* species have modified, filiform spermatozoa (Alberti & Storch 1983, Ferraguti & Garbelli 2006; Fig. 4.1.10 B). An exception is *Meiopriapulus fijiensis* with round-headed spermatozoa, but assumed internal fertilization (Storch et al. 1989a; see below). See also Franzén (1983) and Lemburg & Schmidt-Rhaesa (1999) for summaries.

The round-headed spermatozoon of *Priapulus caudatus* (Afzelius & Ferraguti 1978), *Halicryptus spinulosus* (Adrianov et al. 1992) and *Meiopriapulus fijiensis* (Storch et al. 1989a) is composed of an oval main part and a cilium (= flagellum). The main part includes

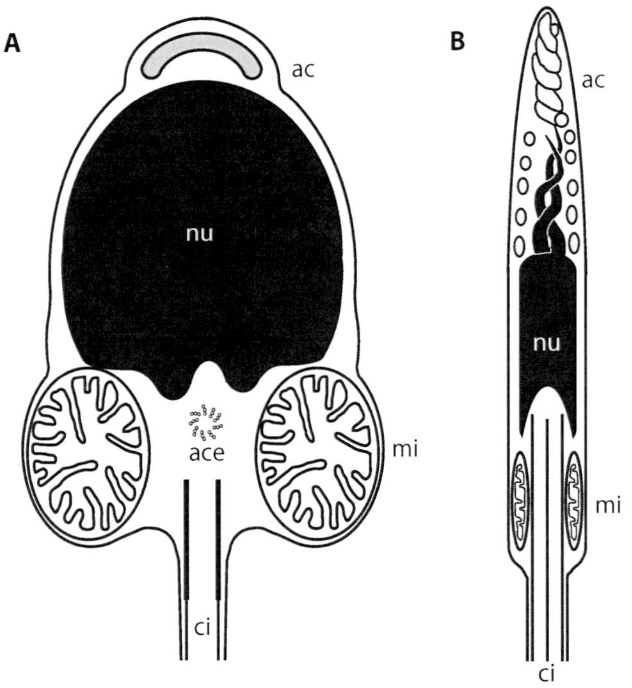

Fig. 4.1.10. Schematic representation of spermatozoa from *Priapulus caudatus* (A; after Afzelius & Ferraguti 1978) and *Tubiluchus philippinensis* (B; after Alberti & Storch 1983). ac = acrosome, ace = accessory centriole, ci = cilium, mi = mitochondria, nu = nucleus.

an anterior acrosome, followed by a large nucleus, four round mitochondria and an accessory centriole (an accessory centriole is not shown in *M. fijiensis*; Storch et al. 1989a). Adrianov et al. (1992) describe that in *H. spinulosus* numerous proacrosomal vesicles migrate to the anterior pole and fuse to form one acrosome.

The spermatozoa of *Tubiluchus* species (Fig. 4.1.10 B) have an elongate anterior part and a cilium. The nucleus is elongate and its anterior part is composed of two narrow extensions, which wind around each other. This anterior nuclear region is surrounded by the acrosome, which has two arms that wind in a corkscrew manner around each other in the anterior part and then around the anterior part of the nucleus (Alberti & Storch 1983, Ferraguti & Garbelli 2006). Small, elongate mitochondria are present and surround the anterior part of the cilium. Five mitochondria are reported for *Tubiluchus* sp. (Alberti & Storch 1983) and three for *T. troglodytes* (Ferraguti & Garbelli 2006). The cilium originates in a deep pit of the nucleus; an accessory centriole is not present. The cilium in the *T. troglodytes* spermatozoon has an unusual pattern of microtubules (Ferraguti & Garbelli 2006). In the anterior region, the "typical" pattern with two single central microtubules and nine peripheral duplets is accompanied by 27 accessory microtubules. Slightly more posterior, these 27 microtubules are arranged in obliquely radial rows distal of the duplets. Alberti & Storch (1983) also mention accessory microtubules for *Tubiluchus* sp., but the number is unknown. In the "free" part of the cilium of *T. troglodytes*, only one accessory microtubulus per duplet is present (Ferraguti & Garbelli 2006); in *Tubiluchus* sp. there are no accessory microtubules in this region (Alberti & Storch 1983). In the distal part of *T. troglodytes*, the pattern dissolves to a random distribution of microtubules.

Oocytes and oogenesis have been described in detail by Nørrevang (1965) for *Priapulus caudatus*. Oocytes originate from epithelial cells in the ovary, which lose their cilium and microvilli. Growing oocytes are held in place by neighboring epithelial cells and bulge in the direction of the body cavity. They remain surrounded by the extracellular matrix of the gonad. Oocytes contain numerous mitochondria, endoplasmic reticulum, yolk vesicles, lipid droplets and glycogen. Mature oocytes are between 50 and 60 μm in diameter in *Halicryptus spinulosus* (Janssen et al. 2009), 60 and 80 μm in *Priapulus caudatus* (Nørrevang 1965), 80–100 μm in *Maccabeus tentaculatus* (Von Salvini-Plawen 1974) and up to 250 μm in *Meiopriapulus fijiensis* (Higgins & Storch 1991).

4.2. Reproduction and development

Reproductive biology

The macroscopic species spawn their gametes into water, where fertilization takes place. It seems that the main season for spawning in *Priapulus caudatus* is winter. After Lang (1953), spawning takes place in the evening or before midnight. Males spawn first and spermatozoa remain without movement for about 15 minutes, then females release oocytes (Lang 1939, 1953).

In *Tubiluchus philippinensis*, Alberti & Storch (1988, 1989) found spermatozoa in the urogenital tract of the female, therefore internal fertilization is likely in this and other *Tubiluchus* species. This corresponds to the modified type of spermatozoa and the low number of gametes.

Internal fertilization is also likely in *Meiopriapulus fijiensis*, because Higgins & Storch (1991) observed that the embryo is brooded by the female. Only two–eight oocytes are present, with one oocyte maturing at a time. The size of oocytes is approximately 250 μm, about three times the size of oocytes in other species. A postembryonic stage without anal hooks and functional introvert is released from the unrogenital pore (Higgins & Storch 1991).

Cleavage and early development

Few studies describe the cleavage in priapulids. Lang (1939; further data in Lang 1953) and Žhinkin (1949) described cleavage in *Priapulus caudatus*. Žhinkin & Korsakova (1953) added the cleavage of *Halicryptus spinulosus* up to the gastrula stage. Finally, Wennberg et al. (2008) reinvestigated cleavage and development in *P. caudatus*. These descriptions correspond in part and disagree in other parts. Wennberg et al. (2008) discuss the reason for this and argue that Žhinkin (1949) and Žhinkin & Korsakova (1953) used too warm conditions for the culture (>10°C) and that such conditions lead to an artificially accelerated development and the formation of misshaped embryos. The following data are based on Wennberg et al. (2008), if not otherwise indicated.

The first cleavage steps are stereotyped; they follow a radial cleavage pattern. Cleavage is total. The first cleavage takes place about 15 hours after fertilization and results in two blastomeres of equal size. The second cleavage is slightly unequal and a cross furrow is formed between the four blastomeres. After the third cleavage a vegetal set of four larger blastomeres and an animal set of four smaller blastomeres are present. After the fourth cleavage four rings of cells are present between the vegetal and animal poles. A small blastocoel forms. Up to now all spindles are parallel, but during the fifth and sixth cleavage some spindles shift and new blastomeres are formed either towards the poles or within the rows of blastomeres. In the now formed 64-cell stage, gastrulation starts. After Wennberg et al.'s (2008) observations, gastrulation is by unipolar ingression of two enlarged cells into the blastocoel; a typical invagination gastrula as described by Lang (1953) is not present.

Wennberg et al. (2009) described the hatching and first stages of the *Priapulus caudatus* larva in detail. During development, a so-called hatching cap is formed on the eggshell. Larvae ready to emerge from the egg move their introvert strongly and finally tear off the breaking cap. Freshly emerged larvae are about 130 µm long and have a trunk, a neck and a functional introvert. No lorica is present. The trunk is circular in cross section. The neck is not very distinct; it carries a pair of neck tubuli. The introvert contains a ring of seven slender scalids and one to three additional scalids (second-ring scalids) can be present posterior to this ring. In a few cases the first-ring scalids were branched; the second-ring scalids are always apically branched. The mouth and anus are closed and therefore no pharyngeal teeth are visible. These observations for *P. caudatus* are roughly confirmed for *Halicryptus spinulosus* (Janssen et al. 2009). The hatching larva also does not have a lorica; it appears a bit more mobile than that of *P. caudatus*. Mouth and therefore also pharyngeal teeth are lacking. The first ring of scalids contains seven scalids, the second ring only one. Two neck flosculi were found in the hatching larvae (their number increases up to 15 in older larvae).

The first larval stage moults into the second larval stage with a lorica. In *Priapulus caudatus*, Wennberg et al. (2009) name this stage the first lorica larva. It has a trunk, neck and introvert region. The length is about 160 µm. The lorica is composed of eight cuticular plates, with the cross section of the trunk being still almost circular. It does not cover the extreme posterior end. Four lorica tubuli are present. A mouth field is recognizable in the anterior end, but the mouth itself is still not present. Four rows of scalids are present; in the first ring there are now eight scalids. Nine scalids are present in the second row; two of them are closer to each other than the other scalids. The third ring contains eight scalids. The fourth ring with four scalids is separated from the others by a gap.

The first lorica larva moults into a second lorica larva with a thicker lorica (Wennberg et al. 2009). It measures about 185 µm. Four lorica tubuli are present and the posterior end is still not completely covered by the lorica. A mouth opening is present now and four rings of pharyngeal teeth are present. The first, outermost ring contains three flat teeth covered with some denticles, the anlagen of two further teeth are visible. The teeth in the second ring are largest and have spines of varying length including a large median spine. Teeth in rings three and four have smaller spines and denticles. Four rings of scalids are present on the introvert, with 8–9–8–8 scalids being present per ring.

The results by Wennberg et al. (2009) show that the dorsoventrally flattened larvae that were found on most occasions (exception: *Tubiluchus* larvae with circular cross section) are later larval stages. They differ from each other in particular in the position and number of the lorica tubuli. Lorica tubuli are a rigid tube with a basal structure composed of ring-like elements (Fig. 4.1.3 F), which presumably cause some flexibility. *Halicryptus* larvae differ from the one of other taxa by the presence of a characteristic sculpture on the dorsal and ventral plates. It is difficult to say how many larval stages there are. Purasjoki (1944), for example, divided larval stages of *Halicryptus spinulosus* into four size classes, but it is unknown whether this is the entire series of larvae and whether the number of moults is fixed at all.

4.3. Physiology

Quite few data are known on the physiology of priapulids and most refer to the capacity of *Halicryptus spinulosus* to survive anoxia. This species is widely distributed in the Northern hemisphere and occurs regularly in the Baltic Sea. Here, anoxic conditions are often present, but *H. spinulosus* is able to deal with these conditions.

Anoxia causes changes in the metabolism. Metabolic rates drop down to finally about 2% (Oeschger et al. 1992). Anoxia changes the properties of enzymes, for example it was measured in *H. spinulosus* that the activity of phosphorylase drops one third under anoxia (Oeschger & Storey 1990). Under anoxic conditions, succinate, volatile fatty acids, acetate and propionate are produced (Oeschger 1990).

Halicryptus spinulosus can also tolerate high amounts of sulphide. Sulfide induces anaerobiosis, even in the presence of oxygen (Oeschger & Vetter 1992). Bacteria resembling *Thiobacillus* are present on the external surface; these epibacteria might be able to detoxify some part of the sulphide (Oeschger & Schmaljohann 1988, Juterzenka & Reichardt 1991, Oeschger & Janssen 1991), but worms must still rely on their own detoxification strategies.

Sulfide is immobilized by iron-sulfide formation, which is present on the exterior of the animal and darkens the surface and the blood (Oeschger & Janssen 1991, Oeschger & Vetter 1992). Thiosulfate is formed, depending on the oxygen availability (Oeschger & Vetter 1992). Janssen & Oeschger (1992) compared animals from oxygen-rich and hydrogen-sulfide-rich environments and found ultrastructural differences in epidermal cells. Mitochondria from the sulfide specimens are associated with dense granules of unknown function.

4.4. Phylogeny

Although being a diverse taxon, a number of autapomorphies support Priapulida as a monophyletic taxon (Lemburg 1999). The most prominent ones are:

- scalids in 25 longitudinal rows composed of rings with 8+8+9 scalids
- pharynx with pentagons of cuspidate teeth
- larvae with lorica tubuli
- urogenital system
- erythrocytes with hemocyanin.

There is little doubt that priapulids are related to the two other taxa with scalids on their introvert – Kinorhyncha and Loricifera (e.g. Lemburg 1999, Dunn et al. 2008, Nielsen 2012; for further data see Chapter 1).

With the traditional classification, genera were grouped to the families Priapulidae (*Priapulus, Priapulopsis, Acanthopriapulus*), Halicryptidae (*Halicryptus*), Tubiluchidae (*Tubiluchus*), Meiopriapulidae (*Meiopriapulus*) and Chaetostephanidae (*Maccabeus*) (e.g. Adrianov & Malakhov 1996c). Two proposed trees with phylogenetic relationships of priapulid subtaxa (Adrianov & Malakhov 1996c, Lemburg 1999; Fig. 4.4.1) differ from each other, in particular concerning the position of *Tubiluchus*. This genus is basal (together with *Meiopriapulus* as Tubiluchidae) in Lemburg (1999; Fig. 4.4.1 B), but sister taxon to Priapulidae in Adrianov & Malakhov (1996c; Fig. 4.4.1 A).

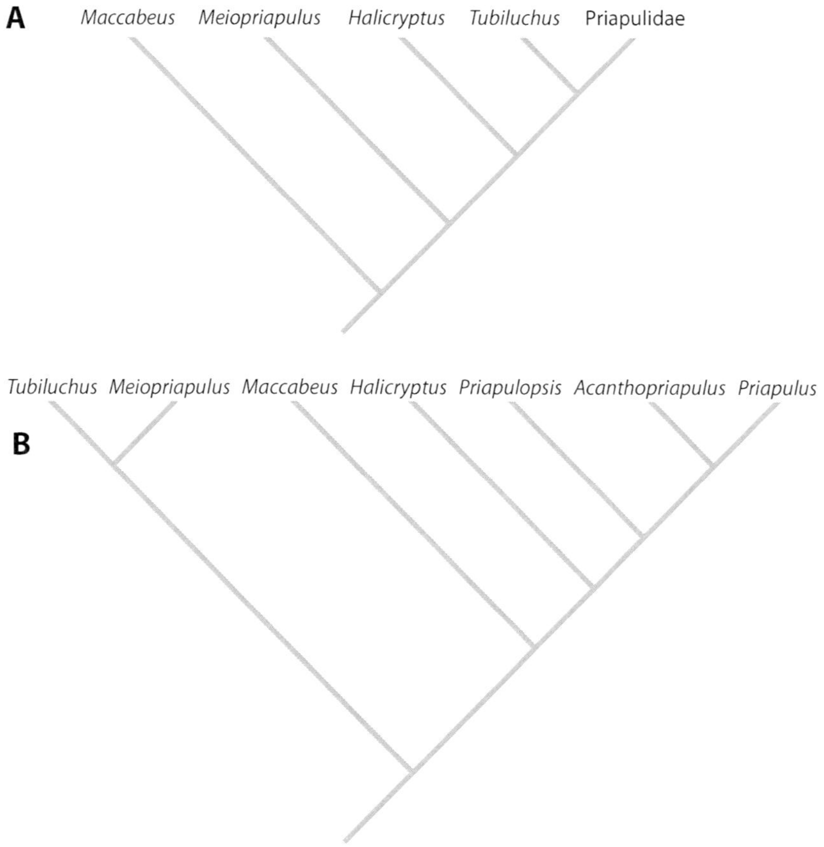

Fig. 4.4.1. Alternating phylogenetic hypothesis of internal priapulid relationships after A, Adrianov & Malakhov (1996b) and B, Lemburg (1999).

4.5. Systematics

In this section, the 19 currently known species are briefly described. It is not intended here to give extensive lists of synonyms and collection sites. Especially for *Priapulus caudatus*, it has been unclear for some time how exactly this species is characterized and therefore several synonyms are present (see Van der Land 1970, Adrianov & Malakhov 1996a,b). Some species names that have been introduced (e.g. in Ehlers 1862, Baird 1868, Sanders & Hessler 1962) are now considered as synonyms. If not otherwise indicated, information is taken from Van der Land (1970) and Adrianov & Malakhov (1996a,b). In the following, genera and species within these genera are ordered in the sequence of their description.

Genus *Priapulus* Lamarck, 1816. The genus name *Priapulus* was the first available name (apart from earlier names such as *Priapus*) and most species from the genera *Priapulopsis* and *Acanthopriapulus* have originally been described as *Priapulus* species. *Priapulus caudatus* is the most well known species and it must be stated that in some records no proper distinction between *P. caudatus* and other species within the genus was made. Today three species are regarded as valid in this genus. They are distributed worldwide, from intertidal regions to the deep sea. Specimens live in muddy sediments.

Species of the genus *Priapulus* have a comparably large introvert and one (unpaired) caudal appendage. Scalids are small conical structures, arranged in 25 clearly visible longitudinal rows. The trunk has 40–50 annuli, the surface is smooth with scattered papillae. Caudal rings of papillae are present. All teeth are cuspidate. The larva is dorsoventrally flattened; the lorica is composed of one larger dorsal and ventral plate and three lateral plates on each side. Four tubuli are present on the lorica. A closing apparatus of the larval lorica is composed of a smaller ventral and a dorsal plate (placids).

***Priapulus caudatus* De Lamarck, 1816.** This is the oldest and best known species and the one with most records. It occurs in the entire Northern hemisphere from Arctic waters to the Tropic of Cancer (see Van der Land 1970, Adrianov & Malakhov 1996b,d). There are several synonyms, including e.g. *P. profundus* (Sanders & Hessler 1962) (see Van der Land 1970, Adrianov & Malakhov 1996a,b for further synonyms). *Priapulus caudatus* has been reported from intertidal regions to about 2600 m depth, it has been found in sand, mud or silt and also in sulphidic biotopes (Adrianov & Malakhov 1996a,b). For specific characters see Tab. 1, for images see e.g. Fig. 4.1.1 A, 4.1.2 A, C.

***Priapulus tuberculatospinosus* Baird, 1868.** This species is distributed on the Southern hemisphere from Antarctic coasts northwards up to the Coast of Chile, Argentina and New Zealand (see Van der Land 1970, Adrianov & Malakhov 1996b). It occurs in muddy

Tab. 1. Comparison of characters in the three species of the genus *Priapulus*, after Adrianov & Malakhov (1996a,b).

Character	*P. abyssorum*	*P. caudatus*	*P. tuberculatospinosus*
maximal trunk length	30 mm	200 mm	200 mm
size of teeth in 1st pentagon in relation to teeth of the 2nd pentagon	2–3 times smaller	not much smaller	smaller
size of teeth in 4th pentagon	almost equal	almost equal	two dorsal teeth larger than others
number of scalids per series	9–10	5–6	4
size relation of 2nd scalid in a series in relation to 1st scalid	much smaller	not much smaller	not much smaller
number of posterior rings with papillae	4–7	1–3	1–2
warts in posterior end	absent	numerous	numerous
presence of enlarged annuli in the posterior end	absent	1–3 enlarged annuli present	absent
tail	unsegmented, with vesicles	slightly segmented in juveniles, unsegmented with densely arranged vesicles in adults	indistinctly segmented or smooth, with vesicles or tubercles
larval lorica tubuli	bristle-like	telescopic	telescopic
location of larval tubuli on lorica	1/5 of lorica length from posterior tip	1/5–1/4 of lorica length from posterior tip	2/5–3/5 of lorica length from posterior tip

sediments from 0 to 625 m depth (Adrianov & Malakhov 1996a,b,d). There is broad overall similarity with *P. caudatus* and obvious difficulties in distinguishing these species led for example to regarding specimens as a subspecies *P. caudatus antarcticus* (Michaelsen 1889), although Baird (1868) had already introduced the name *P. tuberculatospinosus* (as *tuberculato-spinosus*) for specimens from the Falkland Islands. Lang (1951) argued that *P. caudatus* and *P. tuberculatospinosus* form two different species. Adrianov & Malakhov (1996a,b) list a number of characters that indicate how these species can be distinguished (see Tab. 1).

***Priapulus abyssorum* Menzies, 1959.** This species was found in the Mid-America Trench off the West coast of Central America at 5680–5690 m depth (Menzies 1959). It was coined a "missing link" between *P. caudatus* from the Northern hemisphere and *P. tuberculatospinosus* from the Southern hemisphere. Several records are known worldwide (see Adrianov & Malakhov 1996b for a distribution map); species occur in the deep sea between 3000–8000 m, on silt or muddy sediments (Adrianov & Malakhov 1996a,b). For specific characters see Tab. 1.

Genus *Halicryptus* Von Siebold, 1849. Specimens described as *Halicryptus spinulosus* were described from the Baltic Sea, where they were washed upon the beach after a storm, by Von Siebold (1849, 1862). This species is broadly distributed in the Northern Hemisphere. A second species, *H. higginsi*, was described by Shirley & Storch in 1999. The discovery of this species followed a reinvestigation of large specimens from Alaska that were originally determined as *H. spinulosus* (see Shirley & Storch 1999). Observations on these specimens had already been published as *H. spinulosus* by Merriman (1981). Considerable size differences and further characters led to the recognition that these specimens represent a new species including the largest priapulids known.

Adults are characterized by a very short introvert compared to the trunk length (Fig. 4.1.2 B). Scalids are arranged in longitudinal rows, the two ventral rows being closer together than the other rows. Comparably few scalids are present per row (maximally 13, see Tab. 2); these are not arranged in series. Scalids are conical structures with a different ultrastructure in the two species (see Tab. 2). In *H. spinulosus*, scalids occur in two types; posterior scalids are apically bifurcated dentoscalids (Fig. 4.5.1 A, C). The trunk is annulated and covered by numerous pointed trunk spines (called setae by Van der Land 1970, trunk tubuli by Storch et al. 1990 and secretory tubuli by Shirley & Storch 1999) (Fig. 4.5.1 D, E). A caudal appendage is not present. A pair of anal setae is present on the posterior end. The pharynx has cuspidate teeth. Between pharynx and scalids are a varying number of so-called buccal papillae, which roughly resemble scalids, but have a different fine structure (Fig. 4.5.1 B). When the pharynx is retracted, the buccal papillae remain outside. The larva was described in detail for *H. spinulosus* (e.g. Hammarsten 1915, Van der Land 1970, Storch & Higgins 1991, Lemburg 1995a,b). The larva of *H. higginsi* is reported to be similar in size and structure to the one of *H. spinulosus*, but a description is not available (Shirley & Storch 1999).

***Halicryptus spinulosus* Von Siebold, 1849.** This is an abundant species, see Van der Land (1970) and Adrianov & Malakhov (1996b) for records. It is distributed on the entire Northern hemisphere north of about 45°N latidude. *Halicryptus spinulosus* is tolerant to low salinity (see e.g. Carter 1966, Aarnio et al. 1998), low concentrations of oxygen and high concentrations of sulphide (see above under Physiology). For characters in the adults see Tab. 2. Buccal teeth are cuspidate and have, apart from the central long spine, a varying number of lateral cusps – up to ten (Van der Land 1970).

The larva (see e.g. Hammarsten 1913, 1915, Purasjoki 1944, Van der Land 1970, Storch & Higgins 1991, Lemburg 1995a,b) is dorsoventrally flattened. The large lorica plates have a characteristic, "alveolate" sculpture. Lorica tubuli are minute (Storch & Higgins 1991, Lemburg 1999) and only present in young larvae. About five rows of scalids are present; these are of one type and decrease in size in the posterior direction. Behind the scalids, a ring of adhesive tubules is present. In the pharynx, six pentagons are present, followed by several rings with more than five teeth. Teeth decrease in size towards the posterior.

***Halicryptus higginsi* Shirley & Storch, 1999.** The species was described by Shirley & Storch (1999); additional data are in Merriman (1981), although under the name *H. spinulosus*. Specimens were found close to Barrow, Alaska. Larvae and juveniles appear to occur near the shore, while adults may live in slightly deeper waters (Shirley & Storch 1999). Large priapulids found in the stomach of walruses from the Bering Sea might represent, due to their size, also specimens of *H. higginsi* (Shirley & Storch 1999).

The trunk contains many annuli and many trunk spines; these are more dense in the anterior region than in the posterior region. Buccal papillae are present in different sizes; approximately 30 papillae are comparably large, with smaller and thinner buccal papillae present around the mouth. The pharyngeal teeth appear to have a fixed pattern of lateral cusps. Per side, six cusps are

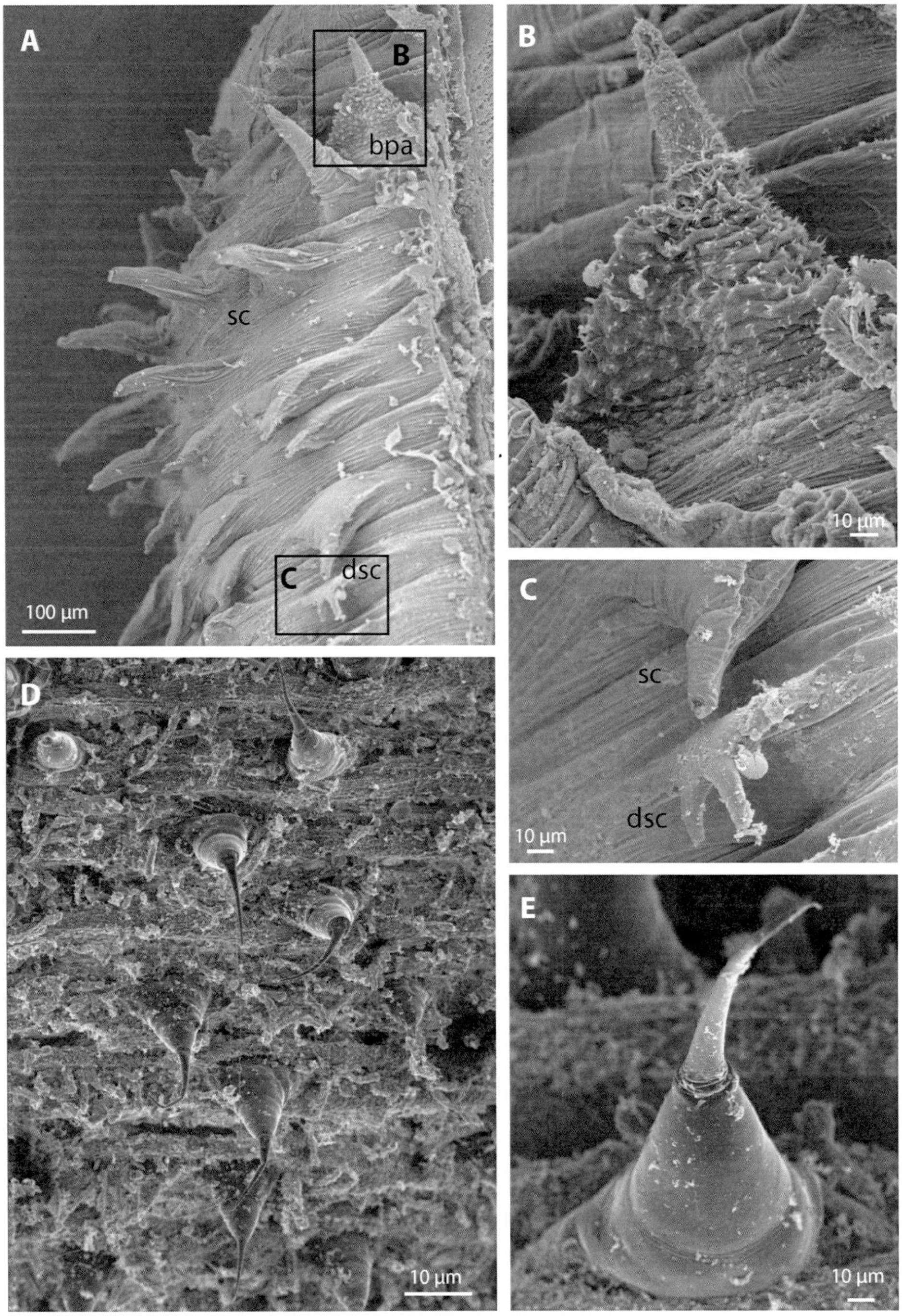

Fig. 4.5.1. *Halicryptus spinulosus.* A, Sequence of introvert structures with anterior buccal papilla (bpa), scalids (sc) and posterior dentoscalids (anterior is to the top, specimen is cut longitudinally, see right margin); B, Magnification of the buccal papilla; C, Magnification of dentoscalid; D, E, Trunk spines. All images SEM.

present in teeth of the first pentagon, one cusp in teeth of the second and third pentagon and no cusps in teeth of the fourth pentagon. In more posterior teeth, cusps can again be present and up to ten cusps can be present per side (Shirley & Storch 1999). For further characters see Tab. 2.

Genus *Priapulopsis* Koren & Danielsen, 1875. The genus was created by Koren & Danielsen (1875). It includes one broadly distributed species north of the Tropic of Cancer (*P. bicaudatus*), one species south of the Tropic of Capricorn with much fewer records (*P. australis*) and one questionable species (*P. cnidephorus*), which is nevertheless listed here.

Species resemble those in the genus *Priapulus*, the main difference being that a paired caudal appendage is present in *Priapulopsis* species. A large introvert is

Tab. 2. Comparison of characters in the two species of the genus *Halicryptus*, after Van der Land (1970), Storch et al. (1990) and Shirley & Storch (1999).

Character	*H. spinulosus*	*H. higginsi*
maximal trunk length	40 mm (usually below 25 mm)	385 mm
introvert length in comparison to entire body length	~10%	~2%
number of scalids per row	maximally 10	about 13
dentoscalids	present	absent
scalid ultrastructure	with large region of subapical tubuli	with scattered spots of few tubuli
small cuticular structures at introvert/trunk border	absent	present
trunk annuli	up to 100	up to 180
number of rings of pharyngeal teeth with five teeth (= pentagons)	6	5
largest teeth	in 2nd and 3rd pentagon	in 3rd and 4th pentagon
anal spines	large	small

Tab. 3. Comparison of characters in *Priapulopsis australis* and *P. bicaudatus*, after Van der Land (1970) and Adrianov & Malakhov (1996a,b).

Character	*P. australis*	*P. bicaudatus*
maximal trunk length	50 mm	100 mm
teeth in 1st pentagon	10 very small denticles	10 modified teeth
gap between anterior and posterior teeth	present	absent
number of scalids per (anterior) series	4–5	4–7
(anterior) series of scalids separated by a gap?	yes	no
scalids within one series	close together, basal parts can be fused	separated from each other
number of posterior rings with papillae	3–5	up to 5
tail branches (vesiculae)	with minute spines	smooth
closing apparatus in larva	made of dorsal and ventral placids	made of dorsal, ventral and lateral placids

present and scalids are small conical structures, arranged in 25 clearly visible longitudinal rows. A series of scalids is present. The trunk has 35–50 annuli, and is covered with small wart-like structures called tumuli. Numerous ring papillae are present in the posterior end of the trunk. The teeth of the first ring are more or less rudimentary; teeth of the third pentagon are the largest. A distinct gap is present between an anterior and a posterior group of teeth in *P. australis*. The larva is dorsoventrally flattened. The lorica is composed of one larger dorsal and ventral plate and three lateral plates on each side. Four tubuli are present on the lorica.

***Priapulopsis bicaudatus* (Danielssen, 1868).** There are numerous records from the entire Northern hemisphere, from Arctic waters to the Tropic of Cancer (see Van der Land 1970, Adrianov & Malakhov 1996b). There are some synonyms, including e.g. *P. atlantisi* (Sanders & Hessler 1962) (see Van der Land 1970, Adrianov & Malakhov 1996a,b for further synonyms). *Priapulopsis bicaudatus* has been reported from a few to 2000 m in silt, mud or sandy mud (Adrianov & Malakhov 1996a). For specific characters see Tab. 3.

***Priapulopsis australis* (De Guerne, 1886).** There are a few records from South America, New Zealand and South Africa (see Van der Land 1970, Adrianov & Malakhov 1996b). For synonyms see Van der Land (1970) and Adrianov & Malakhov (1996a). *Priapulopsis australis* has been reported from 90–400 m depth and from different sediments: mud, sand and coarse sand. For specific characters see Tab. 3.

***Priapulopsis cnidephorus* Von Salvini-Plawen, 1973.** Von Salvini-Plawen (1973) described a specimen from the Adria (Mediterranean) as "*Priapulopsis* (?) *cnidephorus*". It is obviously a juvenile specimen, but the paired caudal appendage and "warts" (= probably tumuli) on the cuticle make it likely that the specimen belongs to the genus *Priapulopsis*. Unexploded cnidocytes were found in the intestine and in the epidermis of the single specimen, making it likely that it feeds on cnidarians and uses unexploded cnidocytes for its own defense (cleptocnids). Neither of the characters listed in Tab. 3 is known from this species.

Genus *Acanthopriapulus* Van der Land, 1970. The single species in this genus, *A. horridus*, has been described as *Priapulus horridus* by Théel (1911) from 80 m depth off the coast of Uruguay. The new genus was established by Van der Land (1970). The particular morphology of the caudal appendage was the main argument for establishing a new genus.

***Acanthopriapulus horridus* (Théel, 1911).** This species is known from only a few specimens. Théel (1911) had for his original description only one specimen available. Amor (1975) reported one specimen, which was found in the intestine of a fish that was caught off Argentina. Adrianov & Malakhov (1996a) mention the species (with a single specimen) from New Zealand (Bay of Plenty, North Island; see Van der Land 2010). Reported depths are between 80 and 120 m.

The specimens are distinctly smaller than those in the genera *Priapulus* and *Priapulopsis*. The holotype measures 6 mm in total, including the tail (Van der Land 1970). Trunk and introvert resemble the specimens in *Priapulus* and *Priapulopsis*.

The introvert has scalids as small conical spines arranged in 25 longitudinal rows. Series are not clearly recognizable. The trunk has about 30 annuli; in the posterior end two rings of papillae are present. The tail is unpaired, club-shaped and segmented. It is covered by numerous hook-like spines and, in the anterior region, several extensions called "vesicles". The anterior pharyngeal teeth are not much smaller than the ones in the second pentagon, the teeth of the third pentagon are largest and teeth in the fourth pentagon are of equal size (Théel 1911, Van der Land 1970, Adrianov & Malakhov 1996a).

Genus *Tubiluchus* Van der Land, 1970. For long time, priapulids were known only as macroscopic specimens from cold and temperate waters. Remane & Schulz (1964) found in littoral regions in Al Ghardaqua (Egypt) a larva of what later turned out to be a new priapulid genus. Van der Land (1968) found microscopically small larvae and adults in Curaçao (Lesser Antilles in the Caribbean) and described these as *T. corallicola*. This was the first meiobenthic and also the first tropical priapulid species described. Originally, Van der Land (1968) described the new species only on the basis of the larva, but suspected that the adult specimens were conspecific. After further findings from Bermuda (Coull 1970, Van der Land 1970) it became clear that larvae and adults really belong to one species. Since their first discovery, several further records have been made and a total of seven species is now described (an eighth species will be described briefly; own results). This makes *Tubiluchus* the most diverse genus of priapulids. Species occur mainly in tropical and subtropical waters, as well as in the Mediterranean. Apart from this distribution, Parra & Benito (1994) report specimens not determined to species level from the coast of Northwest Spain and the species *T. arcticus* has the furthest northward distribution (see below). Species are mainly differentiated by the cuticular structures in the pregenital area of males; females are often indistinguishable, larvae may also differ.

The body size is about 2 mm, additionally a long tail (up to 4.5 mm) is present (Fig 4.1.1 B, 4.5.2 A). A neck region is present in adult specimens between introvert and trunk (Fig. 4.5.1 B). Scalids are arranged in 25 rows, but much denser compared to the macroscopic species. Scalids differ in shape; anterior scalids are slender and elongate, posterior scalids become conical (Fig. 4.5.2 C). Scalids are differently covered by fine bristles an have an apical seta on the distal part (e.g. Calloway 1975; Fig. 4.5.2 C, 4.5.3 A). The trunk is densely covered by cuticular structures called tumuli; these are arranged in rows (Fig. 4.1.7 F, 4.5.2 A, 4.5.3 B). Many tubuli and several flosculi (Fig. 4.1.7 F, G, 4.5.3 B) are present on the trunk. Sexual dimorphism is present and males have different cuticular structures in the genital region. For example, in *T. corallicola* these are one pair of clavulae (club-shaped structures), bullulae (hemispherical elevations), "normal" and perigenital setae (Van der Land 1970). The tail is densely covered with tubumuli and few flosculi. The anterior pharyngeal teeth are pectinate, and are equal in size (Fig. 4.1.9 F, 4.5.2 B). Fimbrillae are present in the posterior region (see e.g. Kirsteuer & Rützler 1973). The pharynx can be widely everted (Fig. 4.5.2 B). A pentamerous arrangement is present in the pharyngeal teeth, but teeth are not clearly arranged in circles. In the intestinal system, a polythyridium is present. The spermatozoon is filiform.

Larvae (Fig. 4.1.2 D, E, 4.1.3 C, 4.5.3 C, D) have a radial symmetrical lorica composed of ten longitudinal (primary) ridges, alternating with ten less distinct (secondary) ridges (Fig. 4.1.3 C). About 20 tubuli are present on the primary ridges. In the pharynx, cuspidate teeth are present (Van der Land 1970).

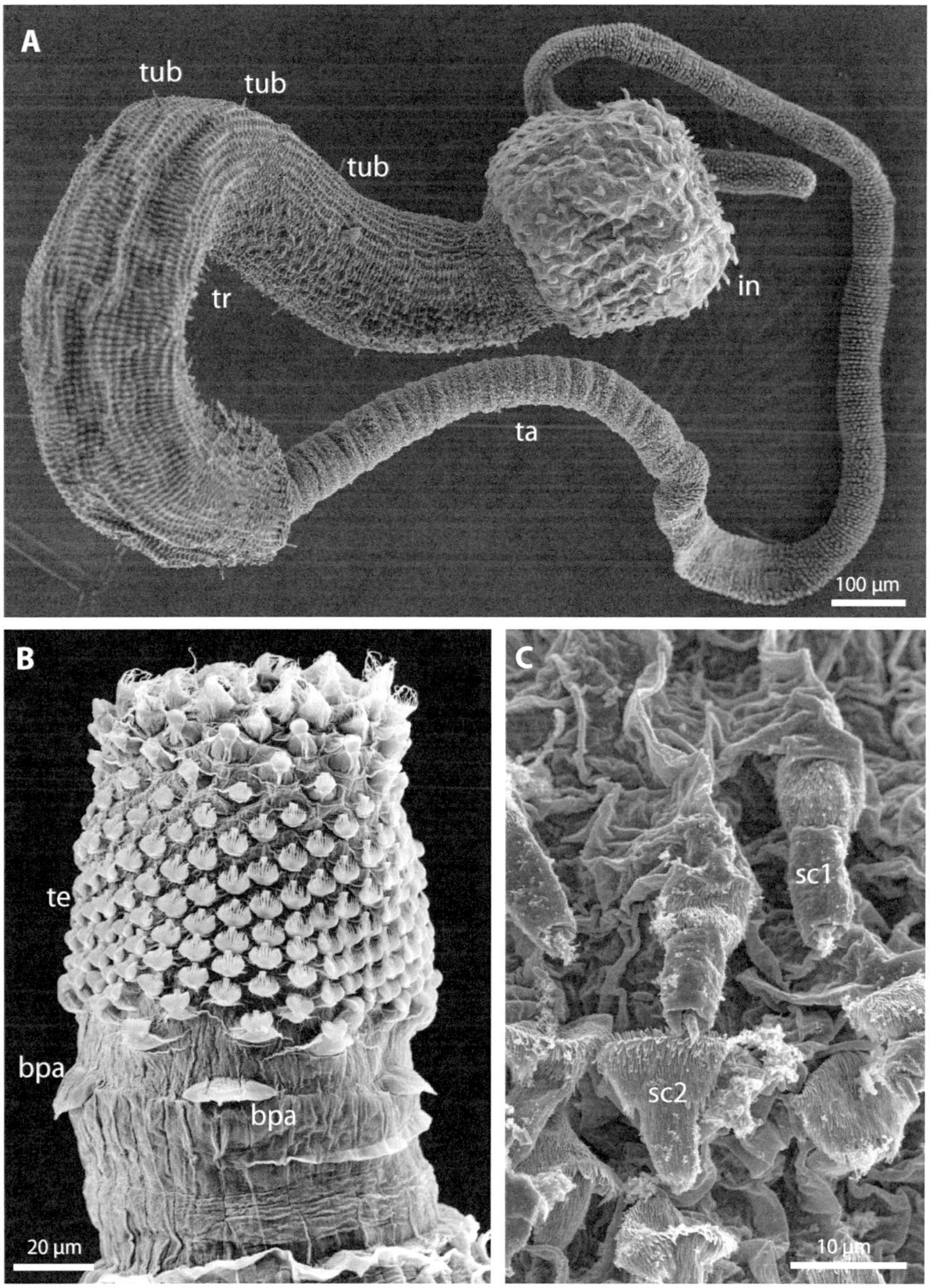

Fig. 4.5.2. Genus *Tubiluchus*. A, Body organization with introvert (in), trunk (tr) and tail (ta). Some trunk tubuli (tub) are labelled; B, Everted pharynx with buccal papillae (bpa) and pectinate teeth (te); C, Different morphology of scalids, anterior is to the top. Anterior scalids (sc1) are elongate, posterior scalids (sc2) are cone-shaped. A, B *Tubiluchus* sp. from Tenerife, C *Tubiluchus corallicola* from Lee Stocking Island (Bahamas). All images SEM.

***Tubiluchus corallicola* Van der Land, 1968.** Widely distributed in the Caribbean region (e.g. Bermuda (Coull 1970, Calloway 1975), Curaçao (Van der Land 1970), Barbados (Kirsteuer & Van der Land 1970), Bonaire (Van der Land 1970), Bahamas (own unpublished results)). Found usually in coarse sand. For a more detailled description see Van der Land (1968, 1970), Kirsteuer & Van der Land (1970), Kirsteuer & Rützler (1973), Calloway (1975) and Kirsteuer (1976).

***Tubiluchus remanei* Van der Land, 1982.** The species was described by Van der Land (1982) from coral sand in shallow waters near Hurgada, Red Sea (Egypt), close to the locality, where Remane & Schulz (1964) found the first *Tubiluchus* larva. For a more detailled description see Van der Land (1982).

***Tubiluchus australensis* Van der Land, 1985.** One male specimen was found on the Great Barrier Reef, Australia,

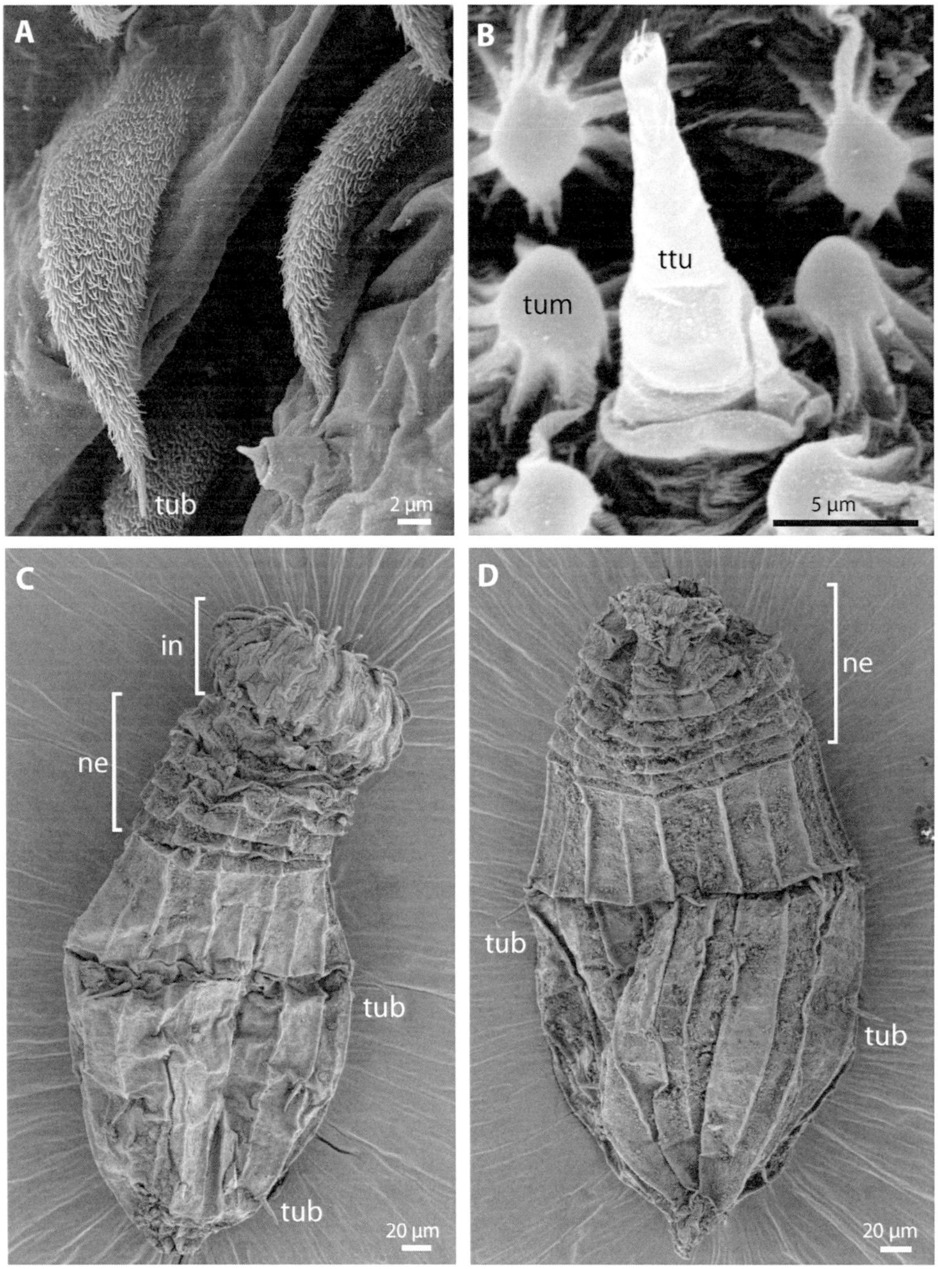

Fig. 4.5.3. Genus *Tubiluchus*. A, Fine structure of posterior scalids with covering of fine bristles and apical tubulus (tub); B, Trunk tubulus (ttu) and tumuli (tum) on the trunk; C, Larva with everted introvert (in) and neck region (ne). Some lorica tubuli (tub) are labelled; D, Larva with retracted introvert, neck region forms a closing apparatus. A *Tubiluchus corallicola* from Lee Stocking Island (Bahamas), B–D *Tubiluchus* sp. from Tenerife. All images SEM.

in a cave at 6 m depth in corase sand and gravel (Van der Land 1985). For a more detailled description see Van der Land (1985).

***Tubiluchus philippinensis* Van der Land, 1985.** Two males of this species were found on Cebu (Philippines) in shallow water (0.5 m at low tide) in silty coral sand (Van der Land 1985). Further specimens from the same location were used for ultrastructural investigations (Alberti & Storch 1983 [as *Tubiluchus* sp.], 1986, 1988, 1989, Storch & Alberti 1985a,b, Storch et al. 1985, Rehkämper et al. 1989). For a more detailled description see Van der Land (1985).

***Tubiluchus arcticus* Adrianov, Malakhov, Tchesunov & Tzetlin, 1989.** This is the only cold water species in the genus. It was found in the White Sea, Russia, between 5 and 100 m in coarse sand (Adrianov et al. 1989, see also Adrianov & Malakhov 1996a, 1999). For a more detailled description see Adrianov et al. (1989), Adrianov & Malakhov (1996a, 1999).

***Tubiluchus vanuatensis* Adrianov & Malakhov, 1991.** Adults and larvae of this species were found on the New Hebrides in shallow water (about 6 m) in coral sand. For a more detailled description see Adrianov & Malakhov (1991, 1996a, 1999).

***Tubiluchus troglodytes* Todaro & Shirley, 2003.** This species was found in a submarine cave in Apulia, Italy. Specimens were used for histochemical investigations by Rothe et al. (2006) and Rothe & Schmidt-Rhaesa (2010). For a more detailled description see Todaro & Shirley (2003).

Genus *Maccabeus* Por, 1973. Specimens belonging to the genus *Maccabeus* were discovered and described almost simultaneously from two locations in the Mediterranean Sea. Por (1973) found specimens, which he named *Maccabeus tentaculatus* near Cyprus; a detailed description followed a year later (Por & Bromley 1974). Von Salvini-Plawen (1974) described specimens from the Adriatic Sea as *Chaetostephanus praeposteriens*. This name had been mentioned in a footnote in an earlier publication (Von Salvini-Plawen 1973). The name *Maccabeus* became established as the genus name (see e.g. Calloway 1975). Additionally, both "species" seem to be identical and represent only one species, *M. tentaculatus*. A second species was described by Malakhov (1978).

Maccabeus belongs to the microscopic priapulids. It is unique among priapulids in having a hemisessile life style, living in tubes. Some of the scalids are strongly modified to form a kind of filtration apparatus. Reproduction is probably by parthenogenesis, because only females have been found.

***Maccabeus tentaculatus* Por, 1973.** (synonym *Chaetostephanus praeposteriens* Von Salvini-Plawen, 1973). Known from a few reports throughout the Mediterranean Sea, from the Alboran Sea (Murina 1984), the Gulf of Lyon (Von Salvini-Plawen 1977), the Adriatic Sea (Von Salvini-Plawen 1973, 1974, 1977) and near Cyprus (Por 1973, Por & Bromley 1974). Additionally, Por (1983) reports a specimen "very similar to *M. tentaculatus*" from the Moçambique Channel. Most specimens are reported from depths of 60–100 m, but they occur up to 550 m deep (Por & Bromley 1974, Von Salvini-Plawen 1974). They live on muddy sediments.

Adult specimens are 1.4–2.85 mm long and live in gelatinous tubes to which sediment is attached. It is likely that the tube material is secreted by glands close to certain scalids, which are therefore named glandular scalids.

Compared to other priapulids, some scalids on the introvert are strongly modified (Fig. 4.5.4). Around the terminal mouth opening there are eight short scalids called "trigger spines" by Por & Bromley (1974) or "hair setae (Haarborsten) by Von Salvini-Plawen (1974). The second ring of scalids includes 25 considerably long scalids, which are composed of two branches originating from a common base (double tentacles after Por 1974). Each of these branches bears further fine lateral bristles. These 25 scalids form a circular tentacle crown, which is used in prey capture. The crown is usually open, but closes on contact with potential food items such as copepods (Von Salvini-Plawen 1974). The entire crown can be withdrawn (Von Salvini-Plawen 1974). The tentacle scalids are surrounded by 16 sensory scalids (sensory spines after Por & Bromley 1974) associated with flosculi. The next circle of scalids is some distance away, leaving some part of the introvert "naked". Two rows with 25 trifid scalids (trifid spines after Por & Bromley 1974) in each row are followed by three rows with 25 glandular scalids (glandular spines after Por & Bromley 1974) in each row. A gland opens at the base of each of these spines.

At the posterior end of the trunk there is one ring with 40–65 hook-shaped spines. Pharyngeal teeth are cuspidate; they are arranged in six circlets with five teeth in each circlet. Posterior of these circlets are 15 longer spines, probably modified pharyngeal teeth, which are called gastral spines by Por & Bromley 1974. These spines seem to mechanically prevent food items from leaving the stomach. The stomach itself is divided into two regions, an anterior muscular and a posterior glandular stomach.

The larvae assigned to *M. tentaculatus* are 0.25–0.62 mm (Por & Bromley 1974) and about 0.7 mm (Von Salvini-Plawen 1974) long. The larvae described by Von Salvini-Plawen appear thick and swollen, but this is probably an artefact. The lorica resembles that known from macroscopic priapulids. It has a dorsal and a ventral plate and

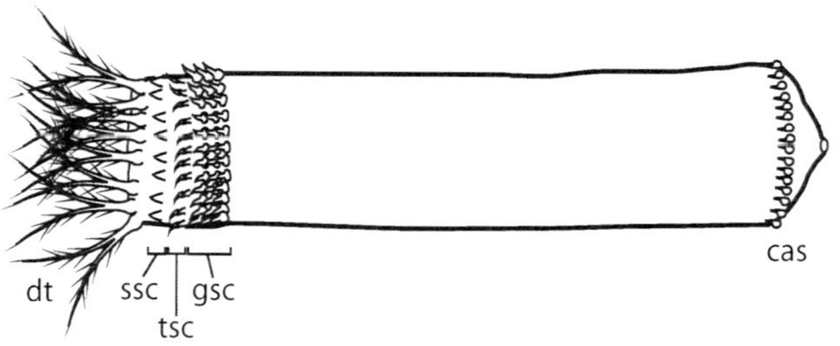

Fig. 4.5.4. Schematic drawing of *Maccabeus cirratus* after Por & Bromley (1974). cas = circumanal spines, dt = double tentacles, gsc = glandular scalids, ssc = sensory scalids, tsc = trifid scalids.

three smaller lateral plates on each side. Characteristic for *M. tentaculatus* is the presence of two sets of four tubuli, one at about two thirds of the lorica length and one at about one third of the lorica length. The larva has a small number of "simple" scalids. The postlarva is 0.5 mm long. The anterior scalids are present, but only one row of glandular scalids is present. The gastral spines are in the process of formation.

The tubes are gelationous secretions, very likely secreted from the glands at the base of the glandular spines. Sediment particles attach to the tubes. The tubes are open at both ends (Por & Bromley 1974), which makes sense because the anus is at the posterior tip of the animal and not, as in several other tube-living animals, close to the anterior end. The orientation of the tubes within the sediment, however, is not known. Von Salvini-Plawen (1974) observed from living animals that they are about two thirds of their length within the sediment.

***Maccabeus cirratus* (Malakhov, 1978).** This species was described as *Chaetostephanus cirratus* by Malakhov (1978) from about 2500 m depth from the Indian Ocean (see also Adrianov & Malakhov 1996a, 1999). It differes from *M. tentaculatus* especially by the structure of the scalids, in particular those from the third to fifth ring (Malakhov 1978, Adrianov & Malakhov 1996a, 1999).

Genus *Meiopriapulus* Morse, 1981. The genus *Meiopriapulus* includes only one species, *M. fijiensis* Morse, 1981. Specimens from Fiji were briefly mentioned by Morse (1980) and formally described by Morse (1981). Westheide (1990) reports specimens assigned to the same species from the Andaman islands, India. Paulay & Holthuis (1994) collected new specimens from the type locality and Sørensen et al. (2012) describe specimens from Jeju Island, South Korea. Gene sequences from further undetermined Korean *Meiopriapulus* specimens had been made available by Park et al. (2006) (further sequences by Sørensen et al. 2012) and the determination of Korean specimens as *M. fijiensis* by Sørensen et al. (2012) makes it likely that the specimens used by Park et al. (2006) also belong to this species.

Ultrastructural investigations of *M. fijiensis* were published by Storch et al. (1989a,b). Higgins & Storch (1991) gave information on reproduction and juvenile stages.

***Meiopriapulus fijiensis* Morse, 1981.** Adults are up to about 3.5 mm long (2.15 mm after Higgins & Storch 1991, 3 mm after Morse 1981).

The introvert has three rings of slightly elongate sensory scalids around the mouth opening (two rings after Morse 1981; Sørensen et al. 2012 discovered a third ring) and 19–22 rings of locomotory scalids on the introvert (Fig. 4.5.5 A, B). Additionally, 13–14 irregular rings of trunk scalids are present, as well as a single ring with 32–38 preanal hooks near the posterior end (Morse 1981, Sørensen et al. 2012).

The pharynx is widely eversible, presenting 16–21 rings of regular pharyngeal teeth (eight per row) (Fig. 4.1.9 E, 4.5.5 C). Additionally, two basal rings with modified teeth are present. Pharyngeal teeth are pectinate (Fig. 4.5.5 D). They differ in their fine structure along the pharynx (see Sørensen et al. 2012). Flosculi occur randomly distributed on the introvert and trunk (Fig. 4.1.7 E). The trunk is covered by minute tubercles. A polythyridium is present in the intestinal system. Protonephridia drain into a paired bladder, which also includes spermatozoa (Paulay & Holthuis 1994). It is unclear whether the testes are paired or unpaired. Morse (1981) reported paired testes, but has probably misinterpreted the bladder as testes (see Paulay & Holthuis 1994).

Sexes are separate; females have six to eight large, yolky eggs (Morse 1981, Higgins & Storch 1991). *Meiopriapulus fijiensis* is viviparous. One oocyte matures per time, reaching approximately 250 μm in size. A postembryonic stage without anal hooks and functional introvert is released from the unrogenital pore (Higgins & Storch 1991). Juveniles are more or less similar to adults, but have 13 anal hooks (instead of 25). Rings of pharyngeal teeth are added during development from a zone in the anterior of the everted pharynx and the number of locomotory scalids and trunk scalids grows during development (Sørensen et al. 2012). A loricate larval stage is missing.

Specimens were found from the intertidal (Morse 1981, Paulay & Holthuis 1994) to about 35 m depth (Sørensen et al. 2012). They have a patchy distribution and are probably actively gregarious (according to observations in a petri dish) (Paulay & Holthuis 1994).

4.6. Biogeography

The distribution of species within the genera *Priapulus* and *Priapulopsis* in cold waters on both hemispheres has been discussed under the topic "bipolar distribution" in the literature (see e.g. Baltzer 1934, Van der Land 1970, Adrianov & Malakhov 1996b). Schreiber et al. (1996) investigated allozymes in the sibling species in *Priapulus* (*caudatus/tuberculatospinosus*) and *Priapulopsis* (*bicaudatus*/*australis*). Species do not share identical allozyme alleles, but the variation within populations is low. This suggests that the cohesion within the populations, though broadly distributed, is efficient, even if there are no dispersal larvae.

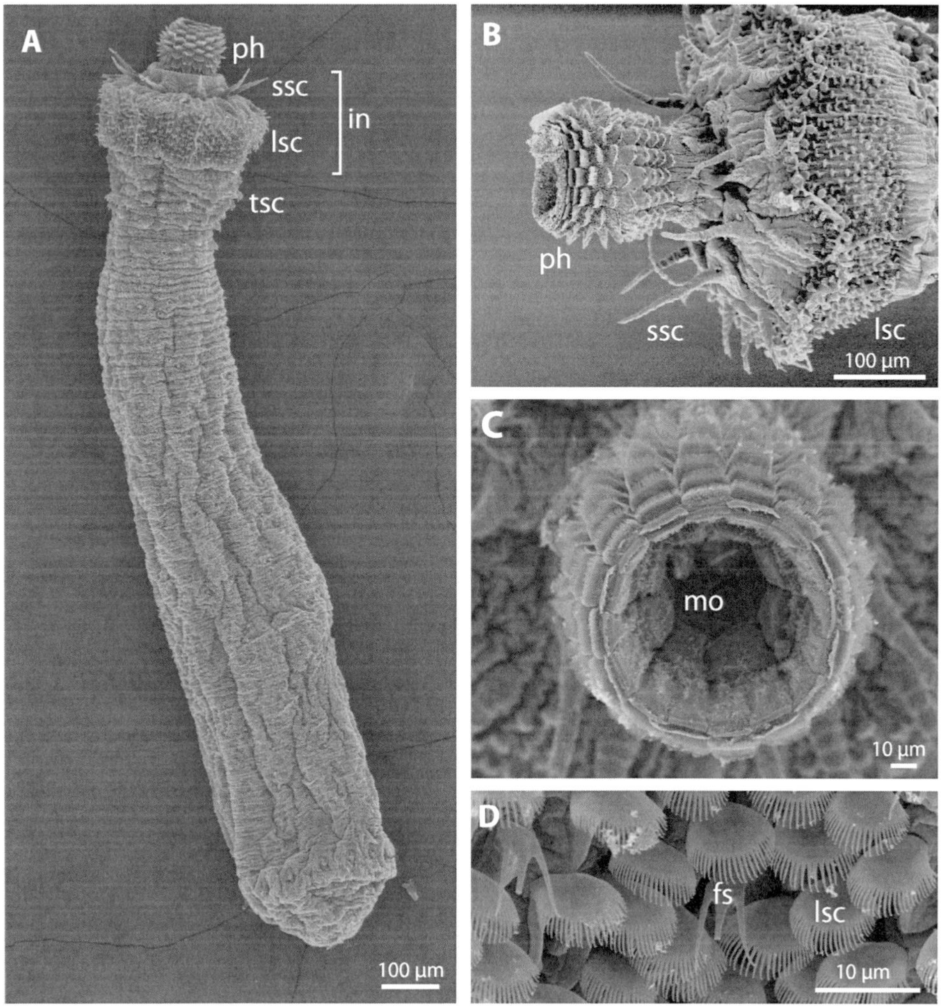

Fig. 4.5.5. *Meiopriapulus fijiensis*. A, Overview on entire specimen showing everted pharynx (ph), introvert (in) with sensory scalids (ssc) and locomotory scalids (lsc). Trunk scalids (tsc) are present in the anterior trunk region; B, Anterior end with pharynx and scalids (labelled as in A); C, Frontal view on mouth opening on the everted pharynx; D, Detail of locomotory scalids (lsc). Between the scalids, furcated (fs) structures are present. All figures with kind permission from Martin V. Sørensen, Copenhagen. All images SEM.

4.7. Palaeontology

The fossil record of priapulid-like worms is better than for any other cycloneuralian taxon. Trace fossils from the Precambrian/Cambrian border are assigned to priapulids (Vannier et al. 2010). A number of species has been found in Cambrian deposits such as the Burgess Shale (e.g. Conway Morris 1977) and the Chengjang biota (e.g. Huang et al. 2004a,b). Most such worms regarded as priapulomorph (often regarded as "archaeopriapulids") have a worm-like trunk and an introvert with scalid-like structures (Conway Morris 1977, Wills 1998). In some species, an additional "proboscis" is present with structures resembling teeth.

Analyses and opinions to assign fossil priapulomorph worms in the phylogenetic tree have been and are still controversial. Some fossils may belong within Priapulida, others may be stem lineage representatives, still others may belong to the stem lineage of Scalidophora, Cycloneuralia or even Ecdysozoa (see, e.g. Wills 1998, Dong et al. 2004). Budd & Jensen (2000) point out that a general resemblance to extant priapulids, for example concerning large body size and the possession of an introvert, are not sufficient to assume a priapulid relationship, because such characters are also expressed by other taxa such as palaeoscolecids and may therefore be plesiomorphic within Scalidophora, Cycloneuralia or even Ecdysozoa (see also Chapter 2 by Maas in this volume).

From the potential autapomorphies of priapulids named above, the particular scalid arrangement pattern and the presence of cuspidate teeth in pentagons of the pharynx may be the best characters to test a priapulid relationship of the fossil forms. A scalid arrangement into 25 longitudinal rows and probable pentagons of cuspidate teeth are present in some species in the Cambrian Chengjiang fauna: *Xiaoheiqingella peculiaris* Huang, Vannier & Chen, 2004, *Yunnanpriapulus halteroformis* Huang, Vannier & Chen, 2004 (see Han et al. 2004, Han & Hu 2006 for a probable synonymy of these species) and *Paratubiluchus bicaudatus* Han, Shu, Zhang & Liu, 2004 from the Maotianshan Shale (Chengjiang) (Han et al. 2004, Huang et al. 2004a). Scalids are arranged here in clear longitudinal rows and a number of 25 such rows is assumed. Cuspidate pharyngeal teeth appear to be present and an arrangement in a pentagon is also assumed (Huang et al. 2004a). Specimens of the two small species *Palaeopriapulites parvus* Hou, Bergström, Wang, Feng & Chen, 1999 and *Protopriapulites haikouensis* Hou, Bergström, Wang, Feng & Chen, 1999 appear to have longitudinal rows of spines on their introvert, too (Hou et al. 1999, 2004). The exact placement of these species and its impact on priapulid evolution is intensively and controversally discussed in Huang et al. (2006) and Han & Hu (2006).

In other Cambrian species from the Burgess Shale and the Chengjiang fauna, priapulid relationships are less clear. The "proboscis" is quite diverse among different taxa. In *Ottoia prolifica* Walcott, 1911 and *Selkirkia columbia* Conway Morris, 1977 the structures on the proboscis indeed resemble cuspidate teeth (Conway Morris 1977, Adrianov & Malakhov 1996b). Cuspidate teeth appear also to be present on the proboscis of *Anningvermis multispinosus* Huang, Vannier & Chen, 2004 (see Huang et al. 2004b). These species may therefore be stem lineage representatives of Priapulida.

A proboscis armed with some structures (but not clear teeth) is also present in a number of other species, which are assigned to priapulids, palaeoscolecids or nematomorphs, for example *Louisella pedunculata* Walcott, 1911, *Cricocosmia jinningensis* Hou & Sun, 1988, *Maotianshania cylindrica* Sun & Hou, 1987 and *Fieldia lanceolata* Walcott, 1912 (Hou et al. 2004, Harvey et al. 2010). Therefore, the proboscis itself may be a plesiomorphic character.

After the diversity of priapulids or priapulomorph species in the Cambrian, only one younger record is known, this is *Priapulites konecniorum* Schram, 1973 from the Carboniferous (Pennsylvanian) (Schram 1973).

4.8. Ecology

Priapulids may play an important role in their ecosystems as effective predators. Macroscopic species plough through muddy sediments in search of food. Microscopic species probably prey upon meiofaunal specimens and are therefore of importance for the local meiofauna community. Aarnio et al. (1998) investigated the impact of *Halicryptus spinulosus* in the Northern Baltic Sea in detail and observed a clear effect on the densities of meiofauna and macrofauna specimens, for example the recruitment of larvae in the bivalve *Macoma balthica*. The predation as well as the disturbance of the sediment by burrowing may be responsible for the effects of *H. spinulosus* on other animals.

Priapulids appear to have different population densities, but very few data are available. Shirley (1990) reported from the subtidal (25–55 m) in Auke Bay, Alaska that larvae of *Priapulus caudatus* are the third most abundant meiofaunal taxon, with densities reaching up to 58,000 m^{-2}. Adults are much fewer (85 m^{-2}). The smallest larvae were found during the winter months and the greatest density is in early spring. Todaro & Shirley (2003) found high densities of *Tubiluchus troglodytes* in a submarine cave in Italy. They estimated 83.3 ± 40.1 individuals/10cm^2 to be present.

Acknowledgements

I thank my former students Birgen Holger Rothe, Maria Freese and Katharina Buschmann, from whose projects data and figures were used for this manuscript. Many thanks to Martin Sørensen for allowing the publication of his *Meiopriapulus* images. Renate Walter kindly helped in making SEM photos. The cooperation in a project of Rick Hochberg allowed the collection of *Tubiluchus corallicola* in the Bahamas. Alessandro Garcia Martinéz helped with collection in Tenerife.

Literature

Aarnio, K., Bonsdorff, E. & Norkko, A. (1998): Role of *Halicryptus spinulosus* (Priapulida) in structuring meiofauna and settling macrofauna. *Mar. Ecol. Prog. Ser.* 163: 145–153.

Adrianov, A. V. & Malakhov, V. V. (1991): First findings of dwarf priapulids of the genus *Tubiluchus* (Priapulida, Tubiluchidae) in Oceania (New-Hebrides). *Zool. Zhurn.* 70: 23–32.

Adrianov, A. V. & Malakhov, V. V. (1996a): *Priapulida: Structure, Development, Phylogeny, and Classification*. KMK Scientific Press, Moscow.

Adrianov, A. V. & Malakhov, V. V. (1996b): The phylogeny, classification and zoogeography of the class Priapulida. I. Phylogeny and classification. *Zoosyst. Rossica* 4: 219–238.

Adrianov, A. V. & Malakhov, V. V. (1996c): The phylogeny, classification and zoogeography of the class Priapulida. II. Revision of the family Priapulidae and zoogeography of priapulids. *Zoosyst. Rossica* 5: 1–6.

Adrianov, A. V. & Malakhov, V. V. (1996d): Comparative Morphological Analysis of the Organization of Cephalorhynch Worms, the Phylogeny and Systematics of the Phylum Cephalorhyncha. 5. Phylogeny and Systematics. *Hydrobiol. J.* 33: 257–266.

Adrianov, A. V. & Malakhov, V. V. (1999): *Cephalorhyncha of the World Ocean*. KMK Scientific Press, Moscow.

Adrianov, A. V. & Malakhov, V. V. (2001a): Symmetry of priapulids (Priapulida). 1. Symmetry of adults. *J. Morphol.* 247: 99–110.

Adrianov, A. V. & Malakhov, V. V. (2001b): Symmetry of priapulids (Priapulida). 2. Symmetry of larvae. *J. Morphol.* 247: 111–121.

Adrianov, A. V., Malakhov, V. V., Chesunov, A. V. & Tsetlin, A. B. (1989): A dwarf priapulid *Tubiluchus arcticus* sp. n. from the white sea (Priapulomorpha, Tubiluchidae). *Zool. Zhurn.* 68. 126–131.

Adrianov, A. V., Reunov, A. A. & Malakhov, V. V. (1992): Fine structure of the testes and spermatogenesis characteristics of the priapulid *Halicryptus spinulosus* (Cephalorhyncha, Priapulida) from the White Sea. *Zool. Zhurn.* 71. 31–39.

Afzelius, B. A. & Ferraguti, M. (1978): The spermatozoon of *Priapulus caudatus* Lamarck. *J. Submicrosc. Cytol.* 10: 71–80.

Alberti, G. & Storch, V. (1983): Fine structure of developing and mature spermatozoa in *Tubiluchus* (Priapulida, Tubiluchidae). *Zoomorphology* 103: 219–227.

Alberti, G. & Storch, V. (1986): Zur Ultrastruktur der Protonephridien von *Tubiluchus philippinensis* (Tubiluchidae, Priapulida). *Zool. Anz.* 217: 259–271.

Alberti, G. & Storch, V. (1988): Internal fertilization in a meiobenthic priapulid worm: *Tubiluchus philippinensis* (Tubiluchidae, Priapulida). *Protoplasma* 143: 193–196.

Alberti, G. & Storch, V. (1989): Zur Feinstruktur des weiblichen Geschlechtstraktes von *Tubiluchus philippinensis* (Tubiluchidae, Priapulida). *Zool. Anz.* 222: 12–26.

Amor, A. (1975): Algunas estructuras cuticulares desconoscidas de *Acanthopriapulus horridus* (Théel 1911). *Physis Sect. A, Buenos Aires* 34: 441–444.

Ankar, S. & Sigvaldadottir, E. (1981): On the food composition of *Halicryptus spinulosus* von Siebold. *Ophelia* 20: 45–51.

Apel, W. (1885): Beitrag zur Anatomie und Histologie des *Priapulus caudatus* (Lam.) und des *Halicryptus spinulosus* (v. Siebold). *Z. Wiss. Zool.* 42: 459–529.

Baird, W. (1868): Monograph of the species of the worms belonging to the subclass Gephyrea; with a notice of such species as are contained in the collection of the British Museum. *Proc. Zool. Soc. Lond.* 1868: 76–114.

Baltzer, F. (1934): Priapulida. In: Krumbach, T. (ed.) *Handbuch der Zoologie Vol. 2, Vermes Amera,* pp. 403–448. De Gruyter, Berlin.

Budd, G. E. & Jensen, S. (2000): A critical reappraisal of the fossil record of bilaterian phyla. *Biol. Rev.* 74: 253–295.

Busch, A., Schumacher, U. & Storch, V. (2001): Histochemistry of lectin-binding sites in *Halicryptus spinulosus* (Priapulida). *Acta Histochem.* 103: 9–19.

Calloway, C. B. (1975): Morphology of the introvert and associated structures of the priapulid *Tubiluchus corallicola* from Bermuda. *Mar. Biol.* 31: 161–174.

Candia-Carnevali, M. D. & Ferraguti, M. (1979): Structure and ultrastructure of the muscles in the priapulid *Halicryptus spinulosus*. Functional and phylogenetic remarks. *J. Mar. Biol. Ass. U.K.* 59: 737–744.

Carlisle, D. B. (1959): On the exuvia of *Priapulus caudatus* Lamarck. *Ark. Zool.* 12: 79–81.

Carter, J. C. H. (1966): A relict priapulid from northern Labrador. *Nature* 211: 438–439.

Conway Morris, S. (1977): Fossil priapulid worms. *Spec. Pap. Palaeont.* 20: 1–95.

Coull, B.C. (1970): Shallow water meiobenthos of the Bermuda platform. *Oecologia* 4: 325–357.

De Lamarck, J. B. (1816): Histoire naturelle des animaux sans vertebrès presentant les caracteres generaux et particuliers des ces animaux, leur distribution, leurs classes, leurs familles, leurs genres, et la citation des principales especes qui s'y rapportent; precedee d'une introduction offrant la determination des caracteres essentiels de l'animal, sa distinction du vegetal et des autres corps naturels, enfin, l'exposition des principes fondamentaux de la zoologie. 3, pp. 1–586. Verdiere, Paris.

De Quatrefages, M. A. (1847): Mémoire sur l´Echiure de Gaertner (*Echiurus gaertnerii* Nob.). *Ann. Sci. Nat. Ser. 3, Zool.* 7: 307–343.

Dong, X. -P., Donoghue, P. C. J., Cheng, H. & Liu, J. -B. (2004): Fossil embryos from the middle and late Cambrian period of Hunan, south China. *Nature* 427: 437–240.

Dunn, C. W. et al. (2008): Broad phylogenomic sampling improves resolution of the animal tree of life. *Nature* 452: 754–749

Ehlers, E. (1862): Ueber die Gattung *Priapulus* (Lam.). Ein Beitrag zur Kenntnis der Gephyreen. *Z. Wiss. Zool.* 11: 205–252.

Elder, H. Y. & Hunter, R. D. (1980): Burrowing of *Priapulus caudatus* (Vermes) and the significance of the direct peristaltic wave. *J. Zool. Lond.* 191: 333–351.

Fänge, R. (1950): Haemerythrin in *Priapulus caudatus* Lam. *Nature* 165: 613–614.

Fänge, R. & Mattisson, A. (1961): Funktion of the caudal appendage of *Priapulus caudatus. Nature* 190: 1216–1217.

Ferraguti, M. & Garbelli, C. (2006): The spermatozoon of a "living fossil": *Tubiluchus troglodytes* (Priapulida). *Tiss. Cell* 38: 1–6.

Franzén, Å. (1983): Priapulida. In: Adiyodi, K. G. & Adiyodi, R. G. (eds) *Reproductive biology of invertebrates Vol. 2: Spermatogenesis and sperm function*, pp. 269–274. Wiley and Sons, New York.

Friedrich, H. & Langeloh, H. P. (1936): Untersuchungen zur Physiologie der Bewegung und des Hautmuskelschlauches bei *Halicryptus spinulosus* und *Priapulus caudatus. Biol. Zentralbl.* 56: 249–260.

Hammarsten, O. D. (1913): Beiträge zur Entwicklung von *Halicryptus spinulosus* (von Siebold). *Zool. Anz.* 41: 501–505.

Hammarsten, O. D. (1915): Zur Entwicklungsgeschichte von *Halicryptus spinulosus* (von Siebold). *Z. Wiss. Zool.* 112: 527–571.

Hammarsten, O. D. & Runnström, J. (1920): Cyto-physiologische Beobachtungen an den Hinterleibdrüsen und den Wanderzellen von *Priapulus caudatus* (Lam.). *Bergens Mus. Aarb., (Naturvid. Raekke)* 1917/1918 13, 1–15.

Hammond, R. A. (1970a): The burrowing of *Priapulus caudatus. J. Zool. Lond.* 162: 469–480.

Hammond, R. A. (1970b): The surface of *Priapulus caudatus* (Lamarck). *Z. Morph. Tiere* 68: 225–268.

Hammond, R. A. (1980): The hydraulic system in a burrowing worm, *Priapulus caudatus. J. Zool. Lond.* 192: 489–499.

Han, J., Shu D., Zhang Z. & Liu, J. (2004): The earliest-known ancestors of recent Priapulomorpha from the Early Cambrian Chengjiang Lagerstätte. *Chin. Sci. Bull.* 49: 1860–1868.

Han, J. & Hu, S. (2006): Response to the "Discussion on the systematic position of the early Cambrian priapulomorph worms". *Chin. Sci. Bull.* 51: 250–256.

Harvey, T. H. P., Dong, X. & Donoghue, P. C. J. (2010): Are palaeoscolecids ancestral ecdysozoans? *Evol. Dev.* 12: 177–200.

Higgins, R. P. & Storch, V. (1989): Ultrastructural observations of the larva of *Tubiluchus corallicola* (Priapulida). *Helgoländer Meeresunters.* 43: 1–11.

Higgins, R. P. & Storch, V. (1991): Evidence for the direct development in *Meiopriapulus fijiensis*. *Trans. Am. Microsc. Soc.* 110: 37–46.

Higgins, R. P., Storch, V. & Shirley, T. C. (1993): Scanning and transmission electron microscopic observations on the larva of *Priapulus caudatus* (Priapulida). *Acta Zool.* 74: 301–319.

Horst, R. (1881): Die Gephyrea gesammelt während der zwei ersten Fahrten des "Willem Barents". II. Priapulidae. Niederländisches *Arch. Zool.* 1 (Suppl.): 13–38.

Hou, X. -G., Bergström, J. Wang, H. -F., Feng, X. -H. & Chen, A. -L. (1999): *The Chengjiang Fauna – Exceptionally Well-preserved Animals from 530 Million Years Ago*. Yunnan Science and Technology Press, Kunming. (in Chinese with English abstract)

Hou, X. -G., Aldridge, R., Bergström, J., Siveter, D. J., Siveter, D. J. & Feng, X. -H. (2004): *The Cambrian fossils of Chengjiang, China: the Flowering of Early Animal Life*. Blackwell Publishers, Oxford.

Huang, D., Vannier, J. & Chen, J. (2004a): Anatomy and lifestyles of Early Cambrian priapulid worms exemplified by *Corynetis* and *Anningvermis* from the Maotianshan Shale (SW China). *Lethaia* 37: 21–33.

Huang, D., Vannier, J. & Chen, J. (2004b): Recent Priapulidae and their Early Cambrian ancestors: comparison and evolutionary significance. *Geobios* 37: 217–228.

Huang, D., Vannier, J. & Chen, J. (2006): Discussion on the systematic position of the Early Cambrian priapulomorph worms. *Chin. Sci. Bull.* 51: 243–249.

Hunter, R. D., Moss, V. A. & Elder, H. Y. (1983): Image analysis of the burrowing mechanisms of *Polyphysia crassa* (Annelida: Polychaeta) and *Priapulus caudatus* (Priapulida). *J. Zool. Lond.* 199: 305–323.

Janssen, H. H. (1994): Pharyngeal glands in *Halicryptus spinulosus* von Siebold, 1849 (Priapulida). *Arch. Fish. Mar. Res.* 42: 89–98.

Janssen, H. H. & Oeschger, R. (1992): The body wall of *Halicryptus spinulosus* (Priapulida) – Ultrastructure and changes induced by hydrogen sulfide. *Hydrobiologia* 230: 219–230.

Janssen, R., Wennberg, S. & Budd, G.E. (2009): The hatching larva of the priapulid worm *Halicryptus spinulosus*. *Frontiers Zool.* 6:8 (doi: 10.1186/1742-9994-6-8).

Joffe, B. I. & Kotikova, E. A. (1988): Nervous system of *Priapulus caudatus* and *Halicryptus spinulosus* (Priapulida). *Proc. Zool. Inst. USSR Acad. Sci.* 183: 52–77.

Juterzenka, K. von & Reichardt, W. (1991): Epizoic bacterial growth on a benthic macro-invertebrate *Halicryptus spinulosus* Priapulida. *Kieler Meeresforsch.* 8: 107–111.

Kirsteuer, E. (1976): Notes on adult morphology and larval development of *Tubiluchus corallicola* (Priapulida), based on in vivo and scanning electron microscopic examinations of specimens from Bermuda. *Zool. Scripta* 5: 239–255.

Kirsteuer, E. & van der Land, J. (1970): Some notes on *Tubiluchus corallicola* (Priapulida) from Barbados, West Indies. *Mar. Biol.* 7: 230–238.

Kirsteuer, E. & Rützler, K. (1973): Additional notes on *Tubiluchus corallicola* (Priapulida), based on scanning electron microscope observations. *Mar. Biol.* 20: 78–87.

Koren, J. & Danielssen, D. C. (1875): Bidrag til de norske Gephyreers Naturhistorie. *Nyt Magazin Naturv.* 21: 108–138.

Kümmel, G. (1964): Die Feinstruktur der Terminalzellen (Cyrtocyten) an den Protonephridien der Priapuliden. *Z. Zellforsch. Mikrosk. Anat.* 62: 468–484.

Lang, K. (1939): Über die Entwicklung von *Priapulus caudatus* Lam. *Kungl. Fysiogr. Sällsk. Lund. Förhandl.* 9: 80–87.

Lang, K. (1948): On the morphology of the larva of *Priapulus caudatus Lam. Arkiv. Zool.* 41 A: 1–8.

Lang, K. (1951): *Priapulus caudatus* Lam. and *Priapulus caudatus* forma *tuberculato-spinosus* Baird represent two different species. *Arkiv. Zool.* 2: 565–568.

Lang, K. (1953): Die Entwicklung des Eies von *Priapulus caudatus* Lamarck und die systematische Stellung der Priapuliden. *Arkiv. Zool.* 5: 321–348.

Langeloh, H. P. (1936): Sinnesphysiologische Untersuchungen an Priapuliden (*Priapulus caudatus* und *Halicryptus spinulosus). Biol. Zentralbl.* 56: 260–267.

Lebsky, V. K. (1974): Fine structure of integumentary epithelium of *Priapulus caudatus* in moulting. *Tsitologiya* 16: 903–907.

Lemburg, C. (1995a): Ultrastructure of the introvert and associated structures of the larvae of *Halicryptus spinulosus* (Priapulida). Zoomorphology 115: 11–29.

Lemburg, C. (1995b): Ultrastructure of sense organs and receptor cells of the neck and lorica of the *Halicryptus spinulosus* larva (Priapulida). *Microfauna Marina* 10: 7–30.

Lemburg, C. (1998): Electron microscopical localization of chitin in the cuticle of *Halicryptus spinulosus* and *Priapulus caudatus* (Priapulida) using gold-labelled wheat germ agglutinin: phylogenetic implications for the evolution of the cuticle within the Nemathelminthes. *Zoomorphology* 118: 137–158.

Lemburg, C. (1999): *Ultrastrukturelle Untersuchungen an den Larven von* Halicryptus spinulosus *und* Priapulus caudatus. *Hypothesen zur Phylogenie der Priapulida und deren Bedeutung für die Evolution der Nemathelminthes*. Cuvillier, Göttingen.

Lemburg, C. & Schmidt-Rhaesa, A. (1999): Priapulida. In: Knobil, E. & Neill, J. D. (eds) *Encyclopedia of Reproduction Vol. 3*, pp. 1053–1058. Academic Press, San Diego.

Linné, C. (1758): *Systema Naturae, Tenth Edition*. Holmiae Salvius.

Linné, C. (1767): *Systema Naturae, Twelfth Edition*. Holmiae Salvius.

Lüling, K. H. (1940): Über die Entwicklung des Urogenitalsystems der Priapuliden. Ein Beitrag zur Anatomie und Histologie dieser Tiere. *Z. Wiss. Zool.* 153: 136–180.

Lüling, K. H. (1943): Über eine neue Coccidie im Darm der Priapuliden: *Priapulus caudatus* (Lam.) und *Halicryptus spinulosus* (V. Sieb.). *Arch. Protistenkd.* 96: 39–74.

Malakhov, V. V. (1978): A new representative of sedentary priapulids: *Chaetostephanus cirratus* sp. n. *Zool. Zhurn.* 58: 1410–1412.

Mattisson, A. & Fänge, R. (1973): Ultrastructure of erythrocytes and leucocytes of *Priapulus caudatus* (de Lamarck) (Priapulida). *J. Morphol.* 140: 367–380.

Mattisson, A., Nilsson, S. & Fänge, R. (1974): Light microscopical and ultrastuctural organization of muscles of *Priapulus caudatus* (Priapulida) and their responses to drugs, with phylogenetic remarks. *Zool. Scripta* 3: 209–218.

McLean, N. (1984a): Ultrastructure of a coccidium (Apicomplexa: Sporozoea: Coccidia) in *Priapulus caudatus* (Priapulida). *J. Protozool.* 31: 241–247.

McLean, N. (1984b): Amoebocytes in the lining of the body cavity and mesenteries of *Priapulus caudatus* (Priapulida). *Acta Zool.* 65: 75–78.

McLean, N. (1985a): Intestinal musculature of *Priapulus caudatus* (Priapulida). *Ophelia* 24: 217–221.

McLean, N. (1985b): A coccidium, a bacterium, and two viruses in *Priapulus caudatus* (Priapulida). *J. Invertebr. Pathol.* 46: 343–345.

Menzies, R. J. (1959): *Priapulus abyssorum*, new species, the first abyssal priapulid. *Nature* 184: 1585–1586.

Merriman, J. A. (1981): Cuticular structures of the priapulid *Halicryptus spinulosus*: A scanning electron microscopical study. *Zoomorphology* 97: 285–295.

Michaelsen, W. (1889): Die Gephyreen von Süd-Georgien nach der Ausbeute der Deutschen Station von 1882–1883. *Jahrbuch Hamburg. Wiss. Anst.* 6: 41–51.

Moritz, K. (1972): Zur Feinstruktur integumentaler Bildungen bei Priapuliden (*Halicryptus spinulosus* und *Priapulus caudatus*). *Z. Morph. Tiere* 72: 203–230.

Moritz, K. & Storch, V. (1970): Über den Aufbau des Integumentes der Priapuliden und Sipunculiden (*Priapulus caudatus* Lamarck, *Pascolion strombi* (Montagu)). *Z. Zellforsch. Mikrosk. Anat.* 105: 55–64.

Moritz, K. & Storch, V. (1971): Elektronenmikroskopische Untersuchungen eines Mechanorezeptors von Evertebraten (Priapulida, Oligochaeta). *Z. Zellforsch. Mikrosk. Anat.* 117: 226–234.

Morse, M. P. (1980): A new interstitial priapulid from Fiji. *Am. Zool.* 20: 893.

Morse, M. P. (1981): *Meiopriapulus fijiensis* n. gen., n. sp.: An interstitial priapulid from coarse sand in Fiji. *Trans. Am. Microsc. Soc.* 100: 239–252.

Murina, G. -V. V. (1984): The first information on quantitive distribution and trophic macrozoobenthos groupings on the Alboran Sea Banks. *Ecol. Morya* 17: 65–70.

Nielsen, C. (2012): *Animal Evolution: Interrelationships of the Living Phyla, Third Edition.* Oxford University Press, Oxford.

Nilsson, A. & Fänge, R. (1967): The digestive fluid of *Priapulus caudatus* (Lam.). *Comp. Biochem. Physiol.* A 22: 927–931.

Nørrevang, A. (1963): Fine structure of the solenocyte tree in *Priapulus caudatus* Lamarck. *Nature* 198: 700–701.

Nørrevang, A. (1965): Oogenesis in *Priapulus caudatus* Lamarck. An electron microscopical study correlated with light microscopical and histochemical findings. *Vidensk. Medd. Dan. Naturhist. Foren.* 128: 1–84.

Oeschger, R. (1990): Long-term anaerobiosis in sublittoral marine invertebrates from Western Baltic Sea: *Halicryptus spinulosus* (Priapulida), *Astarte borealis* and *Arctica islandica* (Bivalvia). *Mar. Ecol. Prog. Ser.* 59: 133–143.

Oeschger, R. & Janssen, H. H. (1991): Histochemical studies on *Halicryptus spinulosus* (Priapulida) with regard to environmental hydrogen sulfide resistance. *Hydrobiologia* 222: 1–12.

Oeschger, R. & Schmaljohann, R. (1988): Association of various types of epibacteria with *Halicryptus spinulosus* (Priapulida). *Mar. Ecol. Prog. Ser.* 48: 285–293.

Oeschger, R. & Storey, K. B. (1990): Regulation of glycolytic enzymes in the marine invertebrate *Halicryptus spinulosus* (Priapulida) during environmental anoxia and exposure to hydrogen sulfide. *Mar. Biol.* 106: 261–266.

Oeschger, R. & Vetter, R. D. (1992): Sulfide detoxification and tolerance in *Halicryptus spinulosus*: A multiple strategy. *Mar. Ecol. Prog. Ser.* 86: 167–179.

Oeschger, R., Peper, H., Graf, G. & Theede, H. (1992): Metabolic responses of *Halicryptus spinulosus* (Priapulida) to reduced oxygen levels and anoxia. *J. Exp. Mar. Biol. Ecol.* 162: 229–241.

Park, J. -K., Rho, H. S., Kristensen, R. M., Kim, W. & Giribet, G. (2006): First molecular data on the phylum Loricifera – an investigation into the phylogeny of Ecdysozoa with emphasis on the positions of Loricifera and Priapulida. *Zool. Sci.* 23: 943–954.

Parra, S. & Benito, J. (1994): Presence of the genus *Tubiluchus* Van der Land, 1968 (Priapulida) on the continental shelf of the Cantabrian Sea. *Bol. Inst. Esp. Oceanogr.* 10: 211–216.

Paulay, G. & Holthuis, B. V. (1994): Biology and morphology of living *Meiopriapulus fijiensis* Morse (Priapulida). In: Wilson, W. H., Stricker, S. A. & Shinn, G. L. (eds) *Reproduction and Development of Marine Invertebrates*, pp. 158–165. John Hopkins Press, London.

Por, F. D. (1973): Priapulida from deep bottoms near Cyprus. *Israel J. Zool.* 21: 525–528.

Por, F. D. (1983): Class Seticoronaria and phylogeny of the phylum Priapulida. *Zool. Scr.* 12: 267–272.

Por, F. D. & Bromley, H. J. (1974): Morphology and anatomy of *Maccabeus tentaculatus* (Priapulida: Seticoronaria). *J. Zool. Lond.* 173: 173–197.

Punin, M. Y. & Zaitseva, O. V. (2002): The first data on the intestine nervous system organization in priapulids. *Dokl. Biol. Sci.* 387: 559–561.

Purasjoki, K. J. (1944): Beiträge zur Kenntniss der Entwicklung und Ökologie der *Halicryptus spinulosus*-Larve (Priapulida). *Ann. Zool. Soc. Zool. Bot. Fenn.* 9: 1–14.

Rehkämper, G., Storch, V., Alberti, G. & Welsch, U. (1989): On the fine structure of the nervous system of *Tubiluchus philippinensis* (Tubiluchidae, Priapulida). *Acta Zool.* 70: 111–120.

Remane, A. & Schulz, E. (1964): Die Strandzonen des Roten Meeres und ihre Tierwelt. *Kieler Meeresforsch.* 20: 5–17.

Richter, S. et al. (2010): Invertebrate neurophylogeny: suggested terms and definitions for a neuroanatomical glossary. *Frontiers Zool.* 7: 29.

Rothe, B. H., Schmidt–Rhaesa, A. & Todaro, M. A. (2006): The general muscular architecture in *Tubiluchus troglodytes* (Priapulida). *Meiofauna Marina* 15: 79–86.

Rothe, B. H. & Schmidt-Rhaesa, A. (2010): Structure of the nervous system in *Tubiluchus troglodytes* (Priapulida). *Invertebr. Biol.* 129: 39–58.

Saldarriaga, J. F. & Storch, V. (1997): Coccidian intestinal parasites in the Priapulidae (Priapulida). *Parasitol. Res.* 83: 233–240.

Saldarriaga, J. F., Voss-Foucart, M. -F., Compère, P., Goffinet, G., Storch, V. & Jeuniaux, C. (1995): Quantitative estimation of chitin and proteins in the cuticle of five species of Priapulida. *Sarsia* 80: 67–71.

Sanders, H. L. & Hessler, R. R. (1962): *Priapulus atlantisi* and *Priapulus profundus*. Two new species of Priapulids from bathyal and abyssal depth of the North Atlantic. *Deep-Sea Res.* 9: 125–130.

Scharff, R. (1885): On the skin and nervous system of *Priapulus* and *Halicryptus*. *Quart. J. Microsc. Sci.* 25: 193–213.

Schauinsland, H. (1887): Zur Anatomie der Priapuliden. *Zool. Anz.* 10: 171–173.

Schram, F. R. (1973): Pseudocoelomates and a nemertine from the Illinois pennsylvanian. *J. Paleontol.* 47: 985–989.

Schreiber, A. & Storch, V. (1992): Free cells and blood proteins of *Priapulus caudatus* Lamarck (Priapulida). *Sarsia* 76: 261–266.

Schreiber, A., Storch, V., Powilleit, M. & Higgins, R. P. (1991): The blood of *Halicryptus spinulosus* (Priapulida). *Can. J. Zool.* 69: 201–207.

Schreiber, A., Eisinger, M. & Storch, V. (1996): Allozyme characterize sibling species of bipolar Priapulida (*Priapulus, Priapulopsis). Polar Biol.* 16: 521–526.

Schulz, E. (1931): Kurze Notiz zur Biologie von *Priapulus caudatus* Lam. *Zool. Anz.* 96: 61–63.

Shapeero, W. L. (1961): Phylogeny of the Priapulida. *Science* 133: 879–880.

Shapeero, W. L. (1962): The epidermis and cuticle of *Priapulus caudatus* Lamarck. *Trans. Am. Microsc. Soc.* 81: 352–355.

Shirley, T. C. (1990): Ecology of *Priapulus caudatus* Lamarck, 1816 (Priapulida) in an Alaskan subarctic ecosystem. *Bull. Mar. Science* 47: 149–158.

Shirley, T. C. & Storch, V. (1999): *Halicryptus higginsi* n.sp. (Priapulida) – a giant new species from Barrow, Alaska. *Invert. Biol.* 118: 404–413.

Solovey, I. V. (1983): Histochemical and autoradiographic study of the cuticle epithelium in priapulids *Priapulus caudatus* and *Halicryptus spinulosus. Tsitologiya* 25: 24–32.

Solovey, I. V. (1984): A study of defence reactions of the body fluid cells in the priapulids. 1. Morphology, histochemistry and physiological regeneration of body fluid cells. *Tsitologiya* 26: 275–284.

Solovey, I. V. (1987): A study of defence reactions of the body fluid cells in the priapulids. 3. Phagocytosis and encapsulation of alien bodies. *Tsitologiya* 29: 788–793.

Solovey, I. V. & Chaga, O. Y. (1982): Some defence reactions of blood cells in the priapulids. *Tsitologiya* 24: 475–479.

Sørensen, M. V., Rho, H. S., Min, W. -G. & Kim, D. (2012): A new recording of the rare priapulid *Meiopriapulus fijiensis*, with comparative notes on juvenile and adult morphology. *Zool. Anz.* (in press).

Storch, V. (1991): Priapulida. In: Harrison, F. W. & Ruppert, E. E. (eds). *Microscopic Anatomy of Invertebrates*, pp. 333–350. Wiley-Liss, New York.

Storch, V. & Alberti, G. (1985a): Ultrastructural investigation of the integument of *Tubiluchus philippinensis* (Priapulida, Tubiluchidae). *Zool. Scripta* 14: 265–272.

Storch, V. & Alberti, G. (1985b): Zur Ultrastruktur des Darmtraktes von *Tubiluchus philippinensis. Zool. Anz.* 214: 262–272.

Storch, V. & Higgins, R. P. (1989): The ultrastructure of developing and mature spermatozoa of *Tubiluchus corallicola* (Priapulida). *Trans. Am. Microsc. Soc.* 108: 45–50.

Storch, V. & Higgins, R. P. (1991): Scanning and transmission electron microscopic observations on the larva of *Halicryptus spinulosus* (Priapulida). *J. Morphol.* 210: 175–194.

Storch, V., Alberti, G., Rosito, R. M. & Sotto, F. B. (1985): Some ultrastructural observations on *Tubiluchus philippinensis* (Priapulida), a new faunal element of Philippine coastal waters. *Philippine Sci.* 22: 144–156.

Storch, V., Higgins, R. P. & Morse, M. P. (1989a): The ultrastructure of the body wall of *Meiopriapulus fijiensis* (Priapulida). *Trans. Am. Microsc. Soc.* 108: 319–331.

Storch, V., Higgins, R. P. & Morse, M. P. (1989b): Internal Anatomy of *Meiopriapulus fijiensis* (Priapulida). *Trans. Am. Microsc. Soc.* 108: 245–261.

Storch, V., Higgins, R. P. & Rumohr, H. (1990): Ultrastructure of introvert and pharynx of *Halicryptus spinulosus* (Priapulida). *J. Morphol.* 203: 163–171.

Storch, V., Higgins, R. P., Malakhov, V. V. & Adrianov, A. V. (1994): Microscopic anatomy and ultrasructure of the introvert of *Priapulus caudatus* and *P. tuberculatospinosus* (Priapulida). *J. Morphol.* 220: 281–293.

Storch, V., Higgins, R. P., Anderson, P. & Svavarsson, J. (1995): Scanning and transmission electron microscopic analysis of the introvert of *Priapulopsis australis* and *Priapulopsis bicaudatus* (Priapulida). *Invert. Biol.* 114: 64–72.

Théel, H. (1911): Priapulids and Sipunculids dredged by the Swedish Antarctic Expedition 1901–1903 and the phenomen of bipolarity. *Kungl. Svenska Vet. Akad. Handl.* 47: 3–36.

Todaro, M. A. & Shirley, T. C. (2003): A new meiobenthic priapulid (Priapulida, Tubiluchidae) from a Mediterranean submarine cave. *Ital. J. Zool.* 70: 79–87.

Van der Land, J. (1968): A new aschelminth, probably related to the Priapulida. *Zool. Meded. Leiden* 42: 237–250.

Van der Land, J. (1970): Systematics, zoogeography, and ecology of the Priapulida. *Zool. Verh.* 112: 1–118.

Van der Land, J. (1982): A new species of *Tubiluchus* (Priapulida) from Red Sea. *Neth. J. Zool.* 32: 324–335.

Van der Land, J. (1985): Two new species of *Tubiluchus* (Priapulida) from the Pacific Ocean. *Proc. K. Ned. Akad. Wet., Sect. C* 88: 371–377.

Van der Land, J. (2010): Phylum Priapulida – penis worms. In: Gordon, D. P. (ed.). *New Zealand Inventory of Biodiversity, Vol. 2*, pp. 476–479. Canterbury University Press, Christchurch.

Vannier, J., Calandra, I., Gaillard, C. & Żylińska, A. (2010): Priapulid worms: pioneer horizontal burrowers at the Precambrian-Cambrian boundary. *Geology* 38: 711–714.

Von Salvini-Plawen, L. (1973): Ein neuer Priapulide mit Kleptocniden aus dem Adriatischen Meer. *Mar. Biol.* 20: 165–169.

Von Salvini-Plawen, L. (1974): Zur Morphologie und Systematik der Priapulida: *Chaetostephanus praeposteriens*, der Vertreter einer neuen Ordnung Seticoronaria. *Z. Zool. Syst. Evolut.-forsch.* 12: 31–54.

Von Salvini-Plawen, L. (1977): Caudofoveata (Mollusca), Priapulida und apode Holothurien (*Labidoplax, Myriotrochus*) bei Banyuls und im Mittelmeer allgemein. *Vie Milieu* 27: 55–81.

Von Siebold, C.T. (1849): Beiträge zur Fauna Preussens. *Neue Preuss. Provincialbl.* 7: 184–185.

Von Siebold, C.T. (1862): Über *Halicryptus spinulosus* (v. Sieb.). *Zusatz. Z. Wiss. Zool.* 11: 413–415.

Weber, R. E. & Fänge, R. (1980): Oxygen equilibrium of *Priapulus* hemerythrin. *Experientia* 36: 427–428.

Wennberg, S. A., Janssen, R. & Budd, G. E. (2008): Early embryonic development of the priapulid worm *Priapulus caudatus. Evol. Dev.* 10: 326–338.

Wennberg, S. A., Janssen, R. & Budd, G. E. (2009): Hatching and earliest larval stages of the priapulid worm *Priapulus caudatus. Invertebr. Biol.* 128: 157–171.

Wesenberg-Lund, E. (1929): Some remarks on the biology and anatomy of the genus *Priapulus* and its distribution in Danish and adjacent waters. *Vidensk. Medd. Dan. Naturhist. Foren.* 88: 165–202.

Westheide, W. (1990): *Meiopriapulus fijiensis* Morse (Priapulida) from South Andaman, another example of large-scale geographic distribution of an interstitial marine meiofauna taxon. *Proc. Biol. Soc. Wash.* 103: 784–788.

Wills, M. A. (1998): Cambrian and recent disparity: the picture form priapulids. *Paleobiology* 24: 177–199.

Zhinkin, L. (1949): Rannie stadii razvitija *Priapulus caudatus. Proc. Acad. Sci. USSR* 65: 409–412.

Zhinkin, L. & Korsakova, G. (1953): Rannie stadii razvitija *Halicryptus spinulosus. Proc. Acad. Sci. USSR* 88: 571–573.

Birger Neuhaus

5. Kinorhyncha (= Echinodera)

Introduction

Diagnosis. Meiofaunal; body composed of eversible introvert (or head) including protrusible mouth cone, neck, and 11 trunk segments. Mouth cone with nine outer oral styles and up to three rings of 5 or 10 inner oral styles. Remaining head with up to seven rings of pentaradial symmetrically arranged, elongated scalids, and one ring of posterior trichoscalids. Neck with 6–16 placids, which may fuse with the first trunk segment. Trunk with one to four segmental cuticular plates, one to three of them ventral (= sternal or episternal and midsternal plates) and one dorsal (= tergal plate) or one ring-like plate; species-specific distribution of sensory spots on all segments; most species with one pair of lateral terminal spines, often also with additional lateral and middorsal spines on various segments. Ring-like brain with anterior perikarya – central neuropil – posterior perikarya, head and mouth cone retractor muscles, pharynx protractor muscles, segmental ventral and dorsal longitudinal muscles, one pair of dorsoventral muscles per segment, and one pair of protonephridium in segments 8 to 9. Primary body cavity with narrow spaces between inner organs. Development via six juvenile stages with a thinner cuticle and less clearly defined cuticular plates.

General. Kinorhyncha comprises exclusively marine, holobenthic, free-living, meiofaunal species of 0.13–1.04 mm body length. The smallest species are *Cephalorhyncha liticola* (127–210 µm), *Echinoderes parrai* (160 µm), *Dracoderes gallaicus* (182–245 µm), *C. nybakkeni* (185 µm), and *Echinoderes wallaceae* (188–260 µm), the largest species are *Pycnophyes greenlandicus* (855–1,035 µm), *P. borealis* (770–990 µm), and *P. spitsbergensis* (915–925 µm) (Higgins 1983, 1986c; Higgins & Kristensen 1988; Higgins & Korczynski 1990; Adrianov 1995; GaOrdóñez et al. 2008; Sørensen 2008b; Sørensen et al. 2012a). Species are dioecious and reproduce sexually; sexes are usually dimorphic in selected characters. The animals inhabit muddy to sandy sediments or occur on algae and invertebrates from the intertidal zone to the deep sea. Kinorhyncha feeds on bacteria, algae, diatoms, and/or detritus. Locomotion of the specimens is facilitated by protruding and retracting the introvert with its up to 89 spinose scalids and by sliding possibly with the help of mucus. The number of species amounts to 191 (plus two species in a paper submitted by Sánchez et al.) based on descriptions from adult stages plus 50 species based on descriptions from juvenile stages. Kinorhynch species are grouped into 21 genera using as distinguishing characters (1) habitus, (2) arrangement, shape, and surface structures of cuticular plates, (3) size, shape, and arrangement of inner and outer oral styles as well as scalids, and (4) size, shape, and distribution of spines, sensory spots, gland cell outlets and papillae. The phylogenetic relationships within Kinorhyncha and of this taxon to its closest relatives Loricifera and Priapulida are unknown. Kinorhyncha has usually been grouped in Aschelminthes (= Nemathelminthes) or more recently in Cycloneuralia.

History of research. Dujardin (1851) was the first to describe a kinorhynch, which he discovered on July 1 in 1841 at Saint Malo and regarded phylogenetically somewhere between crustaceans and worms. The early scholars of Kinorhyncha studied exclusively species of *Echinoderes* and intensively debated their results and those of their colleagues with quite conflicting views; knowledge of the morphology of the recently discovered taxon was developing and misinterpretations occurred (Dujardin 1851; Claparède 1863; Greeff 1869; Schepotieff 1907a,b, 1908; Metschnikoff 1865, 1869; Pagenstecher 1875; Panceri 1878). Reinhard (1881, 1885, 1887) was the first to investigate species of *Pycnophyes*, which he still assigned to *Echinoderes*. Subsequently, Zelinka published a series of articles (Zelinka 1894, 1896, 1907, 1908, 1912) in order to correct reports and systematic assignments of Greeff, Metschnikoff, Reinhard, and Schepotieff. In one study, he cited the species names *Echinoderella capitata, Pycnophyes carinata*, and *P. flaveolatus*, which at this time were nomina nuda, because no identifying characters were given (Zelinka 1912). Zelinka also cited and provided characters of *P. communis* Zelinka, 1908 and *Kinorhynchus giganteus* (Zelinka, 1908), which would not be formally described as new to science until his monograph in 1928. However, Zelinka mentioned descriptive characters for the latter two species, making these species names valid. During the annual meeting of the German zoological society, Zelinka (1896) demonstrated plates of his future monograph. He also reported about the spermiogenesis of Kinorhyncha during the VIIIth international congress of zoology at Graz 1910 (Zelinka 1912). Several authors must have gotten deeper insight into Zelinka's manuscript of his monograph, whose publication was postponed because of events following World War I (Higgins 1971). Among these scientists were A. Remane and R. Southern; the latter mentioned several species including identifying characters from that manuscript, and inadvertently became author of *Echinoderes worthingi* and *P. calmani* (see Southern 1914). Zelinka (1928) completely revised the systematic system of Reinhard (1885, 1887), described 12 species from adult stages (in addition to four species described earlier) and 34 species from juvenile stages, and for the first time provided information on all aspects of the morphology and biology of this taxon. He is still one of the very few scientists who has ever observed live Kinorhyncha for many hours at a time (Zelinka 1928). His contribution "still remains the most singularly important work on Kinorhyncha" to quote the words of Higgins (1971: p. 26) who provided an overview of kinorhynch research (comp. also Remane 1936).

Subsequently, individual contributions were published by several authors until the 1960s (Abe 1930; Blake 1930; Lou 1934; Johnston 1938; Zanefeld 1938; Lang 1936, 1949, 1953; Nyholm 1947a-c; Chitwood 1951; Bacescu & Bacescu 1956; Gerlach 1956; Omer-Cooper 1957a,b; Timm 1958; McIntyre 1962; Reimer 1963; Kirsteuer 1964; Bacescu 1968). Remane's (1928, 1929, 1936) comprehensive compilations refer mainly to Zelinka's monograph from 1928 with some new original information.

A major step forward in kinorhynch taxonomy was accomplished by the studies of Robert P. Higgins and co-workers after 1960. They described 60 new species, redescribed 14 species, and established six new genera (Higgins 1960, 1961, 1964a,b, 1965b, 1966a,b, 1967, 1968, 1969a,b, 1977a,b, 1978b, 1982, 1983, 1985, 1986c, d, 1990, 1991; Brown & Higgins 1983; Higgins & Korczynski 1989; Higgins & Rao 1979; Higgins & Kristensen 1988; Higgins & Shirayama 1990; Higgins & Adrianov 1991; Adrianov & Higgins 1996; Pardos et al. 1998; Martorelli & Higgins 2004; Neuhaus et al. 2013). In addition, R. P. Higgins modified the bubbling method for recovery of kinorhynchs from the sediment and significantly improved it to the bubble & blot method (Higgins 1964c, 1971, 1988; Neuhaus 2003). Also, he introduced the Cobb aluminium slide into kinorhynch methodology invented originally for nematology; this slide allows observation of specimens also from the reverse side (Higgins 1971, 1988). Last but not least, he assembled over decades a worldwide collection of Kinorhyncha at the National Museum of Natural History, Smithsonian Institution, Washington, D. C.; much of this collection still awaits scientific study.

TEM investigations started in the 1970s and provided more information about inner organs (Moritz & Storch 1972a,b; Merriman & Corwin 1973; Nyholm 1976, 1977; Nyholm & Nyholm 1976a,b, 1982b, 1983; Brown 1983, 1985, 1989). Adrianov, Malakhov, and co-workers published a series of TEM investigations about most organ systems of Kinorhyncha with the focus on *Pycnophyes kielensis* and additional data on *Cephalorhyncha asiatica, Echinoderes filispinosus*, and *Kinorhynchus yushini* (Adrianov et al. 1989, 1990a,b, 1993; Yushin et al. 1989, 1990; Adrianov & Malakhov 1990, 1991a–d, 1994, 1999b). More detailed reconstructions of various organs have allowed deeper insight into the functional morphology of Kinorhyncha (Neuhaus 1988: protonephridium of *P. kielensis*, Neuhaus 1991, 1994: alimentary canal and body cavity of *Zelinkaderes floridensis, P. dentatus* and *P. kielensis*, Neuhaus 1997: head sensory organ of *P. dentatus* and *P. kielensis*; Kristensen & Hay-Schmidt 1989: protonephridium of *Echinoderes aquilonius*; Nebelsick 1992a: sensory spots of *E. capitatus*, Nebelsick 1992b: gland cell and spines of *E. capitatus*, Nebelsick 1993: introvert, mouth cone, and nervous system of *E. capitatus*). Confocal laser scanning studies focused on the muscle system of *Antygomonas* sp. and juvenile and adult *P. kielensis* as well as on serotonergic-like immunoreaction in the nervous system of *Antygomonas paulae, E. spinifurca*, and *Zelinkaderes brightae* (see Müller & Schmid-Rhaesa 2003; Rothe & Schmidt-Rhaesa 2004; Schmidt-Rhaesa & Rothe 2006; Herranz et al. submitted).

In addition to articles by individual authors (d'Hondt 1973; Sheremetevskij 1974; Huys & Coomans 1989; Nebelsick 1990; Bauer-Nebelsick 1995, 1996; Song & Chang 2001; Lemburg 2002a,b; Yamasaki & Kajihara 2012), contributions to taxonomy derived from the more extensive studies of Sørensen (27 new species, four new genera), Adrianov (26 new species, one new genus), Pardos (10 new species and two in an submitted article), Neuhaus (six new species, two new genera), and co-workers (Adrianov 1989, 1995; GaOrdóñez et al. 2008; Pardos et al. 1998; Adrianov & Malakhov 1999a; Sørensen et al. 2000, 2005, 2007, 2010a–c, 2012a,b,c; Adrianov et al. 2002a–c; Neuhaus 2004; Neuhaus & Blasche 2006; Sørensen 2006, 2008a,b; Sørensen & Rho 2009; Sørensen & Thormar 2010; Thormar & Sørensen 2010; Sánchez et al. 2011, submitted; Herranz et al. 2012; Neuhaus & Sørensen 2012; Ostmann et al. 2012; Neuhaus et al. 2013).

More recent reviews of Kinorhyncha focus on selected topics such as reproduction (Nyholm & Nyholm 1983; Needham 1989; Adrianov & Malakhov 1999b), ultrastructure (Kristensen & Higgins 1991; Adrianov & Malakhov 1994), phylogenetic relationships (Adrianov & Malakhov 1995a,b, 1996), taxonomy, systematics, and biogeography (Adrianov & Malakhov 1999a; Sørensen & Pardos 2008), and ultrastructure, biology, and systematics (Neuhaus &

Higgins 2002). The current book chapter of this volume summarizes as many aspects as possible of kinorhynch morphology, biology, development, systematics, and biogeography since Zelinka's monograph in 1928.

Current studies of Kinorhyncha still suffer from problematic procedures in the past and nowadays, such as: (1) varying quality of illustrations and textual descriptions; (2) common lack of photographic documentation; (3) lack of type material; (4) deterioration of type material because of embedding in insufficient media, such as Hoyer's mounting medium; (5) few specimens of a given species studied; (6) few localities sampled; (7) identified specimens collected with public funding but partly stored in private collections not available and at risk of loss, if the respective scientist leaves the institution; (8) single-attempt publications by scientists with little or no experience in the taxon Kinorhyncha; and (9) ecological studies without discrimination of kinorhynch species. However, during the last decade more standardized description protocols have emerged, and photographic documentation both by light microscopy and SEM has improved significantly, mainly because of the joint efforts of Robert P. Higgins, Fernando Pardos, Martin V. Sørensen, this author, and their co-workers. Light microscopical and SEM examinations now complement each other; light microscopy offering the advantages of studying both the dorsal and ventral side of specimens mounted on Cobb aluminium slides and of subcuticular structures and SEM offering the advantage of the higher optical resolution of surface structures.

In this book chapter, the number of described species refers to species still valid today. Results will be assigned to individual species if ever feasible. This procedure allows us to realize at a glance the large gaps in our knowledge of virtually all aspects of kinorhynch morphology, development, biogeography, biology, and systematics. Also, future students of Kinorhyncha will be able to follow taxonomic changes more easily, which are quite sure to come, because the monophyly of most genera and higher taxa has not been proven, and the relationships within Kinorhyncha are unresolved. Functional scenarios are presented explicitly in this review in order to stimulate debate about the evolution of Kinorhyncha.

5.1. Morphology

General and external morphology

The body of Kinorhyncha is divided into the introvert, neck, and 11-segmented trunk (Figs 5.1.1–5.1.3). The term "segment" has been used in the early studies of Kinorhyncha with a varying number of 11–13, including introvert and neck in some studies (Claparède 1863; Greeff 1869; Schepotieff 1907a, b; Metschnikoff 1865, 1869; Pagenstecher 1875; Panceri 1878; Reinhard 1881, 1887; Zelinka 1896, 1907, 1908, 1913). In his monograph, Zelinka coins the term "Zonit" (= zonite), probably to distinguish the segmentation of Kinorhyncha ("locomotory segmentation in the sense of Hatschek") from the real metamery of Articulata (Zelinka 1928: p. 387). The term "zonite" does not seem to have convinced most subsequent scientists and has rarely been applied (e.g., Higgins 1960, 1961; Nebelsick 1990; Bauer-Nebelsick 1995, 1996). In this book chapter, the definition of a segment by Scholtz (2002: p. 200) is followed: "A segment is an antero-posteriorly repeated body unit, which can be defined by a set of sub-structures or characters in a specific spatio-temporal correlation." No a priori homology of the segments of Kinorhyncha, Arthropoda, and Annelida is implied by the use of the term segment. Since Zelinka's (1928) monograph, the number of kinorhynch segments including head, neck, and trunk has stabilized at 13, with the exception of *Campyloderes vanhoeffeni* for which Zelinka (1913, 1928) claims 14 segments. Johnston (1938) describing *C. macquariae* corrects this misinterpretation and counts 13 segments.

The trunk is divided into 11 segments, both externally in cuticular plates (ring-like or up to four plates: Figs 5.1.3, 5.1.30, 5.1.31 B, 5.1.32 A, 5.1.36 A, B), sensory spots, papillae, and spines, as well as internally in ventral and dorsal longitudinal muscles, dorsoventral muscles, gland cells, sensory cells, and the central nervous system. These segmental characters of the trunk do not occur in either the introvert or neck (Zelinka 1928; Kristensen & Higgins 1991; Adrianov & Malakhov 1994; Neuhaus 1994; Neuhaus & Higgins 2002). Therefore, the term "segment" should not be applied to the introvert and neck and, consequently, counting of segments should be limited to the 11 trunk segments (Neuhaus & Higgins 2002: p. 621). This view had been taken earlier by Bauer-Nebelsick (1995, 1996) without giving any comments. Also, more recent examinations have adopted this position. Consequently, any scientist studying Kinorhyncha ought to be extremely cautious about the location of characters on which segments, when comparing articles of different authors.

Terminology. Zelinka's (1928) study represents a major breakthrough in kinorhynch terminology. Many of his Latin or German terms have subsequently been translated into English (Tab. 1). More recently, several new descriptive terms, suggestions for measurements, length relations of different characters, and reference positions of

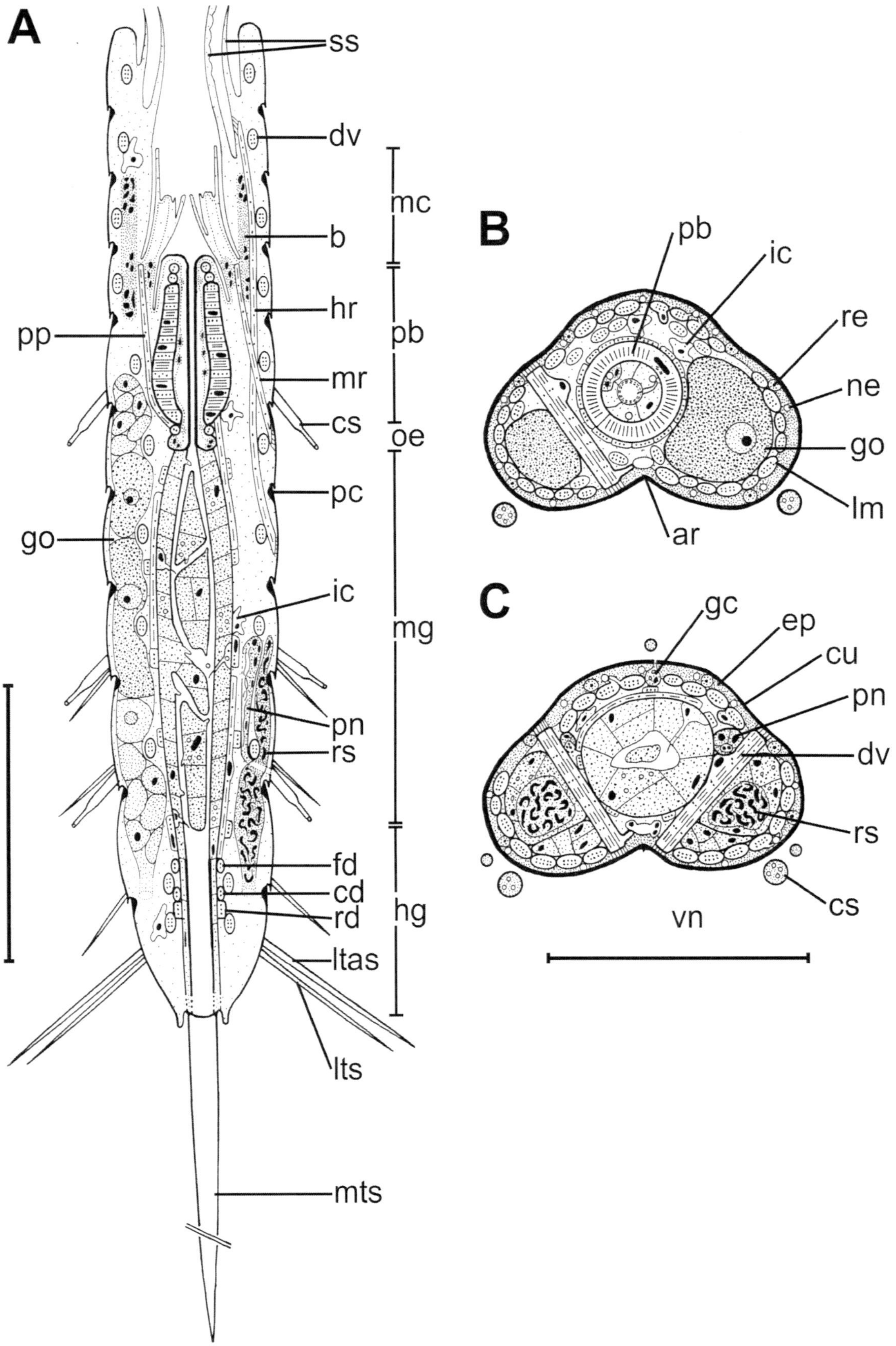

Fig. 5.1.1. Schematic reconstruction of cyclorhagid morphology based on female *Zelinkaderes floridensis*, epidermis not illustrated cellularly in A–C, tanycytes omitted in A, (after Neuhaus 1994; © Verlag Dr. Friedrich Pfeil), lines in A and C = 100 µ, line in C valid for B and C. A, longitudinal section through entire specimen with withdrawn introvert; B, C, cross-sections through trunk in segment 5 (B) and segment 9 (C), respectively.

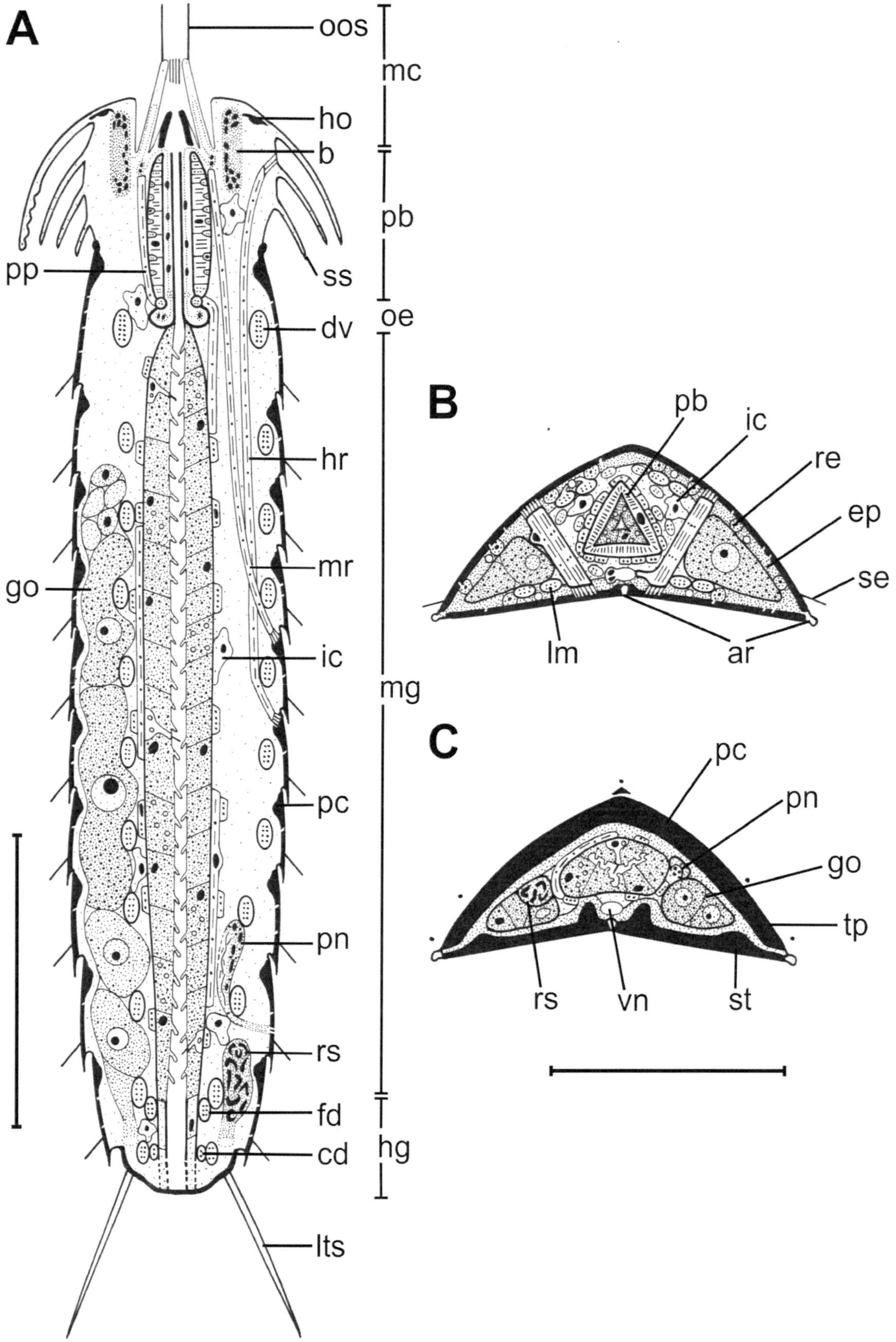

Fig. 5.1.2. Schematic reconstruction of homalorhagid morphology based on female *Pycnophyes dentatus*, epidermis not illustrated cellularly in A–C, tanycytes and inner oral styles omitted in A, (after Neuhaus 1994; © Verlag Dr. Friedrich Pfeil). A, longitudinal section through entire specimen with protruded introvert, line = 200 μ; B, C, cross-sections through trunk in segment 4 (B) and as combined cross-section through pachycyclus of segment 9 (C, right side) and of segment 10 (C, left side), introvert withdrawn, line in C = 100 μ, line valid for B and C.

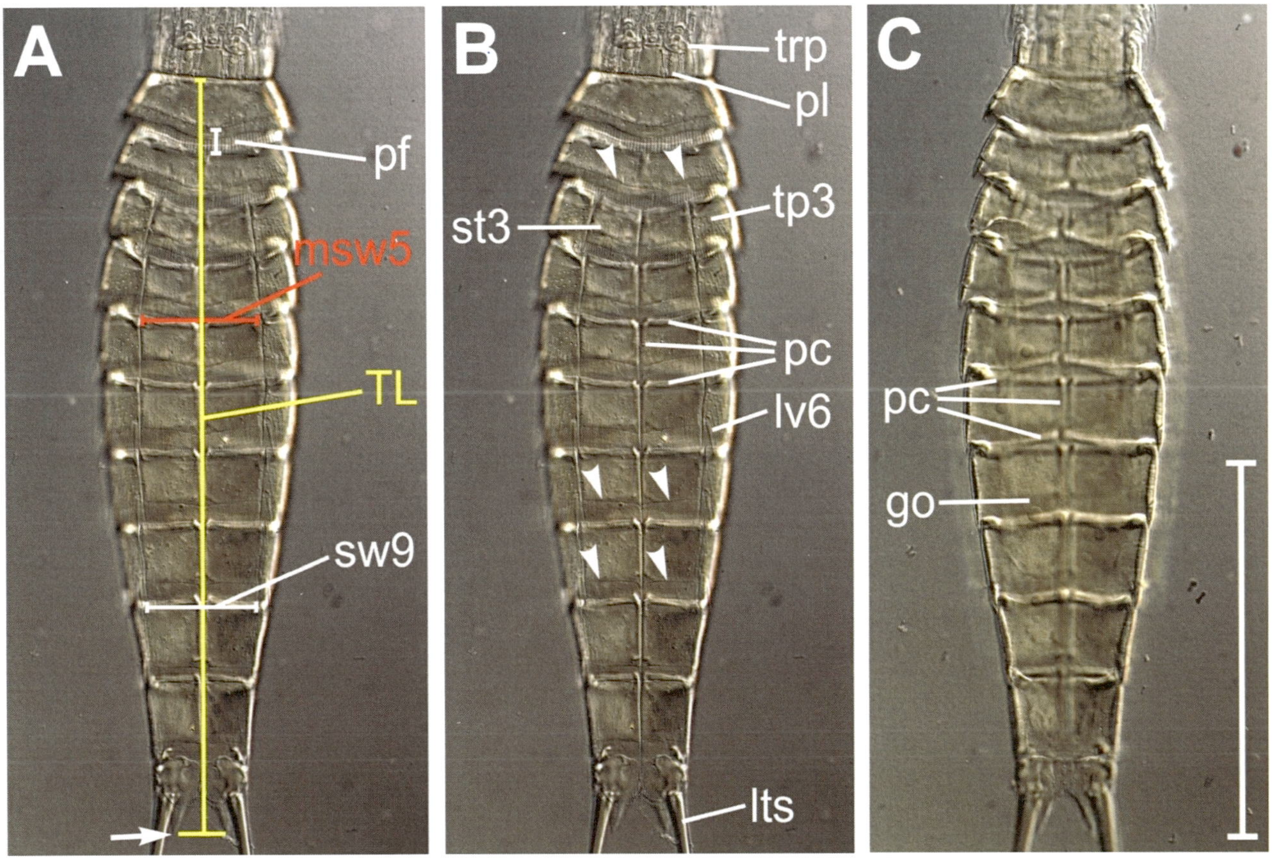

Fig. 5.1.3. LM micrographs of female *Echinoderes setiger*, habitus. A, measurement of trunk length and (maximum) standard width, ventral view, arrow points to tip of tergal extension in segment 11; B, C, morphological characters, ventral view (B) and optical section (C) revealing ventral, lateral, and dorsal pachycycli, arrowheads mark intersegmentary joint line; line in C = 100 μ, line valid for A–C.

trunk characters have been introduced by Higgins (1983) (Tab. 1), but reference positions are illustrated only in 1998 (Pardos et al. 1998; Sørensen & Pardos 2008). Reference positions have been initially understood by R. P. Higgins as longitudinal lines over the entire trunk on which characters appear and have been shown to the author of this review by Higgins already in 1989. Higgins's illustrations has been published much later with his permit by Sørensen and Pardos (2008: Fig. 5.1.5). There are several problems related to assigning a trunk character to a general reference position: (1) characters may be slightly out of line of a reference position; (2) characters may be doubled in individual segments and positions (especially in Homalorhagida); (3) segments possess different diameters, so it may become difficult to identify the same reference position on different segments of the same specimen; and (4) specimens are mounted in different views between exactly dorsoventral and lateral, obscuring certain positions to some degree (Neuhaus & Sørensen 2012). In order to avoid creating many intermediate positions, one approach is to define a reference position not as a narrow line but as a slightly broader band and assign a character to the nearest general reference position (Fig. 5.1.4; Sørensen et al. 2010a; Sánchez et al. 2011; Neuhaus & Sørensen 2012). However, recent suggestions of a modified terminology for Homalorhagida but not Cyclorhagida (Sørensen et al. 2010a; Sánchez et al. 2011) do not seem to be very appropriate and are not followed here. The main drawback of the new terminology is that the old term midlateral position is redefined and marks now a totally different position, namely much more to the lateroventral position, which seems to agree with the older term lateral accessory position. Probably, the new subdorsal position contains the old subdorsal + laterodorsal positions, the new laterodorsal position probably the old laterodorsal + midlateral + sublateral positions, the new term midlateral position replaces the old term lateral accessory position (Sørensen et al. 2010a; Sánchez et al. 2011). This proposal does not seem to facilitate more easy recognition of the position of a trunk character. It will inevitably lead to confusion with previously published species descriptions and in future phylogenetic analysis. For species of the cyclorhygid genus *Echinoderes*, Thormar and Sørensen (2010) define the reference positions in relation to the position of the neck placids, and it remains to see how useful this suggestion is.

Tab. 1. Morphological terms of introvert and trunk characters including remarks on synonomy and conflicting usage (comp. also text).

term (introduced by)	description	taxon	remarks
introvert			
scalid (Zelinka 1907)	general term for introvert appendages, includes primary scalids, spinoscalids, and trichoscalids	all genera	Zelinka 1913: Skalide, Haken; comp. also spinoscalid, inner oral style, and helioscalid
primary (spino)scalid (Higgins 1990)	recurvated, scythe-like scalid of ring 01 with socket + distal element and blunt tip, defines border of the 10 introvert sectors (Figs 21 C, 38 A)	all genera	Higgins 1990: primary scalid; Kristensen & Higgins 1991: clavoscalid; emendet term suggested here: primary spinoscalid
spinoscalid (Boykin 1965)	recurvated, scythe-like scalid of rings 01-06(07) with socket + distal element and spinose tip (Figs 21 C, 34 A, 35 C, 38 A)	all genera	Remane 1936, Moritz & Storch 1972b: Stachelskalide (literal translation: spinoscalid); Lang 1949: spinescalid; Brown 1985, 1989: type 2 scalid = spinoscalid in ring 01 + type 1 scalid = smooth scalid in rings 02–04 and fringed scalid in rings 05–06; Nebelsick 1990, 1993: type 1 spinoscalid in ring 01 + type 2 spinoscalids in rings 02–07; Sørensen et al. 2007 and later publications: spinoscalid in ring 01 + scalid in rings 02–06
leaf-like scalid (Sørensen et al. 2012b)	scalid with broad basis and fringed lateral edges	*Echinoderes cernunnos, E. microaperturus, E. tchefouensis*	
trichoscalid (Zelinka 1928)	short scalid from posteriormost row, covered by regularly arranged bristles, not arranged pentaradial symmetrically (Figs 29 A, 34 A, 38 A)	all genera	Zelinka 1928: Trichoskalide
prescalid (Sørensen et al. 2010d)	in juvenile stages area where a protoscalid will develop in a subfrontal growing zone, usually identified by dense or long cuticular hairs or by an elongate or triangular structure	juvenile *Antygomonas incomitata*	
protoscalid (Neuhaus 1995)	spinose anlage in a subfrontal growing zone where a fully developed scalid will appear in the next juvenile stage (Fig. 35 C)	juvenile stages: *Antygomonas incomitata, Paracentrophyes praedictus, Kinorhynchus phyllotropis*	Brown 1985: proscalid; Sørensen et al. 2010d
eyelid-like structure (Sørensen et al. 2007)	wing-like structure between scalids, possibly corresponding to basal plate	*Zelinkaderes brightae*	
mouth cone			
outer oral style (Neuhaus 1994)	style originating on top of mouth cone, 9 styles in single ring (Figs 21 C, 27 B, 34 B, C, 38, 42)	all genera	Zelinka 1928: Munddolch, Stylus; Higgins 1983: primary oral style; Nebelsick 1993: oral style
inner oral style (Neuhaus 1994)	style extending from mouth cone into inner lumen, usually 3 rings of styles (Fig. 34 B, C)	all genera	Zelinka 1928: akzessorischer Stylus; Higgins 1983: buccal tooth = helioscalid; Brown 1985, 1989: internal scalid = scleroscalid 1 + 2 + helioscalid; Kristensen & Higgins 1991, Nebelsick 1993: pharyngeal style

(Continued)

Tab. 1. (*Continued*)

term (introduced by)	description	taxon	remarks
helioscalid (Brown 1985, 1989)	inner oral style of innermost ring, tubular with basal cuticular fringe resembling sunrays (name!)	all genera	not a scalid but an inner oral style (see text)
cuticular surface of neck and trunk			
condyle (Higgins 1969a)	knobby, conical projection on placid (Fig. 14 C)	*Condyloderes*	Adrianov et al. 2002c; Martorelli & Higgins 2004; Sørensen et al. 2010b
primary fringe (GaOrdóñez et al. 2000)	cuticular, comb-like processes at posterior margin of each segment (Figs 14 D, 17 D, 18 C, 20 C, D, 21 B, 22 B, 23 B-D, 24 A, B)	most genera	Higgins 1983: pectinate fringe
secondary fringe (GaOrdóñez et al. 2000)	line(s) of small cuticular processes (usually triangular scales) at anterior margin of segmental plates (Figs 7 A, 8 B, 9 C, 12 D, 22 B, 24 B, 37 B, D)	*Antygomonas, Campyloderes, Centroderes, Dracoderes, Echinoderes, Meristoderes, Semnoderes, Sphenoderes, Tubulideres, Kinorhynchus, Pycnophyes*	Higgins 1983; Sørensen et al. 2010c; Sørensen et al. 2012a; Neuhaus et al. 2013
intersegmentary joint line (Herranz et al. 2012)	marks the point of attachment of the flexible intersegmental cuticle to the inner side of the stiff sternal and tergal plate cuticle, anterior of the primary pectinate fringe, visible in LM (Fig. 3 B)		
free flap (Herranz et al. 2012)	posterior part of sternal and tergal cuticle overlapping subsequent segment partly, includes primary fringe (Fig. 40 D)		
scale (Neuhaus & Sørensen 2012)	cuticular, short, triangular to shingle-like process or projection (Figs 11 B, C, 12, 28, 29, 34 D, 37 C, D)	*Antygomonas, Campyloderes, Centroderes, Condyloderes, Semnoderes, Sphenoderes, Tubulideres* (only on posterior segments), *Paracentrophyes, Kinorhynchus, Pycnophyes*	Moore 1973; Higgins 1983; Song & Chang 2001; Sørensen 2007; Sørensen et al. 2007, 2009, 2010 a-c; Sørensen 2008c; Neuhaus et al. 2013
cuticular hair (this book chapter)	any cuticular process with a pointed tip not originating from a depression	all genera	Blake 1930: not setae but cuticular hairs; Thormar & Sørensen 2010: cuticular filiform extension or projection
bristle-like cuticular hair (Sørensen & Thormar 2010)	cuticular, pointed, elongate process not originating from depression (Fig. 11 B)	*Antygomonas, Centroderes, Wollunquaderes*	Sørensen et al. 2010d: acicular hairs; Neuhaus et al. 2013
bracteate cuticular hair (Sørensen 2006)	cuticular, pointed hair originating from a circular depression with an anterior, marginal, short thorn (SEM: Figs 21 B, 22 B–D, 23, 24 A, B, 25 B, D), depression appears as perforation site in LM (Fig. 20 C)	*Cephalorhyncha, Echinoderes, E. rex* (thorn missing)	Higgins 1983: perforation site = depression of hair; Huys & Coomans 1989; Nebelsick 1992a; Adrianov & Malakhov 1999a; Sørensen et al. 2000, 2005; Adrianov et al. 2002b; GaOrdóñez et al. 2008; Lundbye et al. 2011
leaf-like hair (Sørensen 2007)	cuticular, broadened hair with a longitudinal cleft, originating with a small pedicle from a circular depression	*Antygomonas, Semnoderes, Sphenoderes, Wollunquaderes, Zelinkaderes* (miniaturized, no pedicle), *Triodontoderes* (no pedicle)	Higgins 1990; Bauer-Nebelsick 1995: bifid cuticular bristle; Sørensen et al. 2007: bristle, 2009, 2010c; Sørensen 2008c; Sørensen & Rho 2009; Sørensen & Thormar 2010

postmarginal spiculae (Sørensen et al. 2010d)	cuticular, elongate, spinose processes at the posterior margin of a segment, represent the primary fringe in juvenile Cyclorhagida (Fig. 7 D)	juvenile stages: *Antygomonas, Campyloderes, Centroderes, Condyloderes, Echinoderes, Zelinkaderes*	Neuhaus & Sørensen 2012; Neuhaus et al. 2013; Neuhaus own obs.
punctuate sculpturing (Higgins 1983)	small, cone-like elevation of cuticle, arranged as multiples in a patch	*Pycnophyes iniorhaptus*	
cuticular ridge (this book chapter)	slightly elevated, ridge- or ripple-like cuticle (Fig. 40 B)	*Pycnophyes dentatus*	Higgins 1983: cuticular striation
reticulate ridges (this book chapter)	cuticular, slightly elevated ridges or ripples connected to a net (Fig. 40 A)	*Pycnophyes australensis, P. canadensis, P. dentatus, P. faveolus, P. furugelmi, P. galtsovae, P. newguiniensis, P. parasanjuanensis, P. sanjuanensis, P. tubuliferus, Pycnophyes* sp. 2	Higgins 1961; Higgins & Korczynski 1989; Neuhaus 1993: net-like sculpturing; Adrianov & Higgins 1996; Adrianov & Malakhov 1999a: cuticular netting-like mosaic (sculpturing), reticulate cuticle, areolate microrelief; Lemburg 2002a, b: reticulate sculpturing; Sánchez et al. submitted
hexagonal sculpturing (this book chapter)	hexagonal pattern of cuticle elevating slightly above surface like cobbled pavement (Fig. 9 C)	*Cateria*	
attachment area of dorsoventral muscles (Neuhaus 1995)	circular, oval to elongate oval, or two oval area(s) devoid of hairs, often wrinkled surface (Figs 11 B, C, 12 D, 14 E, F, 27 D, 31 B, 33 C, 34 D, 37 B, 39 C)	all genera	Higgins 1983: muscle scar
sensory spot type-1 to 5 (Nebelsick 1992a: type-1, -2, -3; Sørensen et al. 2009: type-4; this book chapter: type-5)	**type-1:** circular, oval, or droplet-shaped patch of numerous micropapillae + 1 central pore (+ 1 lateral pore) (Fig. 23 C); **type-2:** circular, oval, or droplet-shaped patch of numerous micropapillae + 1 central pore + 1 lateral very short tube with terminal pore (Figs 11 B, 12, 37 C, D); **type-3:** conus-like or stalked base with few micropapillae + >1 pores (Figs 14 D–F, 33 C, 34 D); **type-4:** circular patch of numerous micropapillae + 1 slightly elongate tube with terminal pore; **type-5:** one ring of few micropapillae around central pore (+ outside ring 1–2 very short tube(s) with terminal pore)(Figs. 21 D, 22 B, C)	**type-1:** *Antygomonas paulae, Echinoderes capitatus, E. gizoensis, E. kanni, E. kozloffi, Paracentrophyes praedictus*; **type-2:** *A. paulae, Centroderes spinosus, Kinorhynchus distentus, K. phyllotropis, K. trisetosus, Pycnophyes greenlandicus*; **type-3:** *Antygomonas, Tubulideres seminoli, Zelinkaderes, Paracentrophyes praedictus* s13, *Pycnophyes cryopygus, P. greenlandicus*, juvenile *K. phyllotropis*, juvenile *P. dentatus*, juvenile *P. kielensis*; **type-4:** *Antygomonas incomitata*; **type-5:** *Echinoderes intermedius, Kinorhynchus yushini, Paracentrophyes anurus, Pycnophyes* sp. 2	Boykin 1965: sensory pit; Higgins 1983: = cuticular scar in *Echinoderes*; Nebelsick 1990; Neuhaus 1993, 1995; Adrianov & Malakhov 1994; Bauer-Nebelsick 1995, 1996; Sørensen 2006: *Nanaloricus*-flosculum = sensory spot type-5, 2007; Sørensen et al. 2007, 2010a; Thormar & Sørensen 2010; Neuhaus et al. 2013; Sánchez et al. submitted
papilla (Sørensen et al. 2009)	short conical to slightly elongate elevation, covered with cuticular micropapillae which may be elongate, terminal pore (Figs 10 B, 15 B)	*Campyloderes, Condyloderes, Wollunquaderes majkenae* (?)	Neuhaus 1993: sensory papilla = sensory spot type-3 of juvenile *P. dentatus* and *P. kielensis*; Sørensen & Thormar 2010: tubule (?); Sørensen et al. 2010b: appendage; Neuhaus & Sørensen 2012
intra- and subcuticular trunk characters			
gland cell outlet type-1, -2 (Neuhaus & Blasche 2006)	**type-1:** small central cone or plate in a circular depression, connected with subcuticular gland	**type-1:** *Fissuroderes* (♀: some outlets with cuticular, funnel-shaped structure),	Thormar & Sørensen 2010: pore field = gland cell outlet type-1, single opening gland outlet = gland cell outlet

(*Continued*)

Tab. 1. *(Continued)*

term (introduced by)	description	taxon	remarks
	cell directly below (LM), pore field (SEM)(Figs 17 A, 18 A, 19 A); **type-2:** larger, with pear-shaped sclerotized walls, wide circular opening (LM)(Figs 9 C , 20 A, B, 22 D, 23 B)	*Campyloderes*; **type-2:** *Fissuroderes, Echinoderes eximus, E. rex* (modified: with fine, cuticular hairs around opening and anterior cuticular furrows)	type-2; Lundbye et al. 2011; Neuhaus & Sørensen 2012; Yamasaki & Kajihara 2012: not gland cell outlet type-2 but fringed tubule
gland cell type-1, -2 (GaOrdóñez et al. 2000)	**type-1:** vesicles electron-lucent with electron-dense, tubular substructure (TEM); **type-2:** numerous microvilli, piles of membranes, oepens as pore field (TEM)	**type-1:** *Echinoderes cantabricus, E. hispanicus, K. giganteus, P. greenlandicus*; **type-2:** *E. aquilonius, E. cantabricus, E. capitatus, E. hispanicus, K. phyllotropis*	Moritz & Storch 1972a; Brown 1985; Kristensen & Higgins 1991; numbering of types of gland cell is independent of numbering of types of gland cell outlets
pachycyclus (Zelinka 1928)	cuticular thickening (= apodeme) at anterior and paraventral margin of segments 2–11, attachment site of dorsal and ventral longitudinal as well as diagonal muscles, head and mouth cone retractors, develop stronger with age (Figs 15, 17–20, 26, 36)	all genera but especially developed in *Echinoderes, Kinorhynchus, Pycnophyes*	Zelinka 1928: Mittelleiste = cuticular thickening at paraventral margin of segments 2–11; may develop to cone-like anteromesial thickening in more posterior segments of *Kinorhynchus* and *Pycnophyes*
anteromesial (or midventral) thickening (of pachycyclus) (Higgins 1983)	cone- to kidney- or bean-like cuticular thickening (= apodeme) at paraventral margin of more posterior segments, develops stronger with age (Figs 39 B, 40 B)	*Kinorhynchus, Pycnophyes*	Zelinka 1928: Mittelwulst
appendages of trunk			
seta (this book chapter)	thin, short, straight, tubular appendage with pore and blunt tip, articulated (Figs 33 C, 37 B, C)	Homalorhagida, *Dracoderes* (s5, 8, 10; ?), *Echinoderes ohtsukai* (s8, 10; ?)	Zelinka 1928: Sinneshaar; Blake 1930: not setae but cuticular hairs (seta as term first mentioned); Remane 1936: Sinnesborste; Lang 1949, 1953: sensory seta; Higgins 1961: sensory hair; Sørensen et al. 2012a, Yamasaki & Kajihara 2012: not tubule but seta (?)
tube (this book chapter)	long, flexible, tubular, thin-walled appendage with blunt tip and pore, articulated	many species of *Echinoderes, Kinorhynchus,* and *Pycnophyes* in s2	Zelinka 1928: Klebröhre of Homalorhagida = adhesive tube; Higgins 1983, GaOrdóñez et al. 2008: adhesive tube
tubule (GaOrdóñez et al. 2000)	long, thin-walled, cylindrical, hollow appendage with a (slightly) broader base and wing-like lateral extensions (preaparation artefact?, see text), articulated (Fig. 28 A, B)	*Dracoderes abei* (s2, 5, 8, 10), *D. gallaicus*(s2, 5, 8, 10), *Echinoderes capitatus* (s2, 5–10), *E. cantabricus* (s1, 2, 5, 8, 10), *E. collinae* (s2, 5, 10), *E. hispanicus* (s1, 2, 5, 8,), *E. intermedius* (s2, 5, 8, 10), *E. isabelae* (s2, 5, 7, 8), *E. neospinosus* (s5), *E. ohtsukai* (s5,10), *E. parrai* (s2, 5, 8), *E. spinifurca* (s2, 10), *Meristoderes* ssp. (s2, 5, (8), 10), *Tubulideres seminoli* (s2, 7)	Pardos et al. 1998: some appendages not tubules but spines (?); GaOrdóñez et al. 2000, 2008; Sørensen et al. 2000: not genital seta but tubule; Sørensen 2005; Sørensen et al. 2007: tubules larger, no lateral extensions; Sørensen 2008b: spines of *E. capitatus*; Sørensen & Thormar 2010: not tubule but papilla (?); Herranz et al. 2012; Sørensen et al. 2012a, c; Yamasaki & Kajihara 2012: lv8 not tubule but spine (?)
fringed tubule (Sørensen 2006)	tubular, very short thin-walled appendage with blunt tip which extends into thin cuticular hairs, articulated	*Echinoderes truncatus* (Panama)	Yamasaki & Kajihara 2012: not gland cell outlet type-2 but fringed tubule

acicular spine (Higgins 1990)	elongate spine with pointed tip, articulated	all genera except *Kinorhynchus* and *Neocentrophyes*	
crenulated spine (Sørensen & Rho 2009)	flexible, distally annulated or segmented spine, articulated, in ♂ only (Figs 12 B, 13 A, 30 D)	*Centroderes, Dracoderes, Echinoderes neospinosus, Triodontoderes anulap, Tubulideres, Wollunquaderes, Zelinkaderes*	Higgins 1990: segmented or moniliform; Higgins & Shirayama 1990; Bauer-Nebelsick 1995; Adrianov & Malakhov 1999a; GaOrdóñez et al. 2008; Sørensen & Thormar 2010; Sørensen et al. 2012a; Neuhaus et al. 2013
cuspidate spine (Higgins 1990)	bottle-like shape: basal half broad and narrowly oval in cross-section, hirsute, distal portion tapering abruptly and round in cross-section, with pore, articulated (Figs 7 A, B, D, 25 B, 27 A, C, D, 29 A, B, 30 B)	*Antygomonas, Cateria, Condyloderes, Semnoderes, Sphenoderes, Triodontoderes, Wollunquaderes, Zelinkaderes*	Zelinka 1928; Gerlach 1956; Higgins 1967, 1968, 1969a; Nebelsick 1990; Bauer-Nebelsick 1996; Sørensen 2007; Sørensen et al. 2007, 2009, 2010b, c; Sørensen & Rho 2009; Sørensen & Thormar 2010
spinose process (Neuhaus 1993)	spine-like process without articulation, pointed tip (Fig. 33)	juvenile Homalorhagida; adult *Kinorhynchus spinosus, Neocentrophyes, Paracentrophyes, Pycnophyes arctous, P. chukchiensis, P. furugelmi*	Lang 1949; Higgins 1969b, 1983, 1991; Adrianov & Malakhov 1999a
taxonomic measurements			
segment (Dujardin 1851)	a segment is an antero-posteriorly repeated body unit which can be defined by a set of sub-structures or characters in a specific spatio-temporal correlation (Scholtz 2002); segment length measured from anterior pachycyclus to primary fringe	all genera	Claparède 1863; Greeff 1869; Schepotieff 1907a, b; Metschnikoff 1865, 1869; Pagenstecher 1875; Panceri 1878; Reinhard 1881, 1887; Zelinka 1896, 1907, 1908, 1913; Zelinka 1928: "Zonit" = segment, also: Higgins 1960, 1961; Nebelsick 1990; Bauer-Nebelsick 1995, 1996
trunk length (= TL) (Higgins 1983)	measured on the midline from the anterior margin of segment 1 to the posterior margin of segment 11, excluding spines but including sternal and tergal extensions of segment 11 (Fig. 3 A)	all genera	
sternal (or standard) width (= sw) (Higgins 1983)	measured at the anteroventral margin of the sternal plates of a given segment (Fig. 3 A)	all genera	
maximum sternal width (= msw) (Higgins 1983)	measured at the anteroventral margin of the widest pair of sternal plates of a specimen (Fig. 3 A)	all genera	
lateral terminal spine (= lts) (Higgins 1983)	acicular lateroventral spine in segment 11 (= terminal segment of trunk)(Fig. 1 A)	all Cyclorhagida, *Paracentrophyes, Pycnophyes*	
lateral terminal accessory spine (= ltas) (Higgins 1983)	acicular spine in segment 11 in lateral accessory position (Fig. 1 A)	most Cyclorhagida (comp. Tab. 8)	
midterminal spine (= mts) (Higgins 1983)	acicular middorsal spine in segment 11 at the posteriormost tip just above the anus (Fig. 1 A)	adult Cyclorhagida except *Dracoderes* and Echinoderidae, all juvenile Cyclorhagida	segment 11 may possess a second middorsal spine more anteriorly

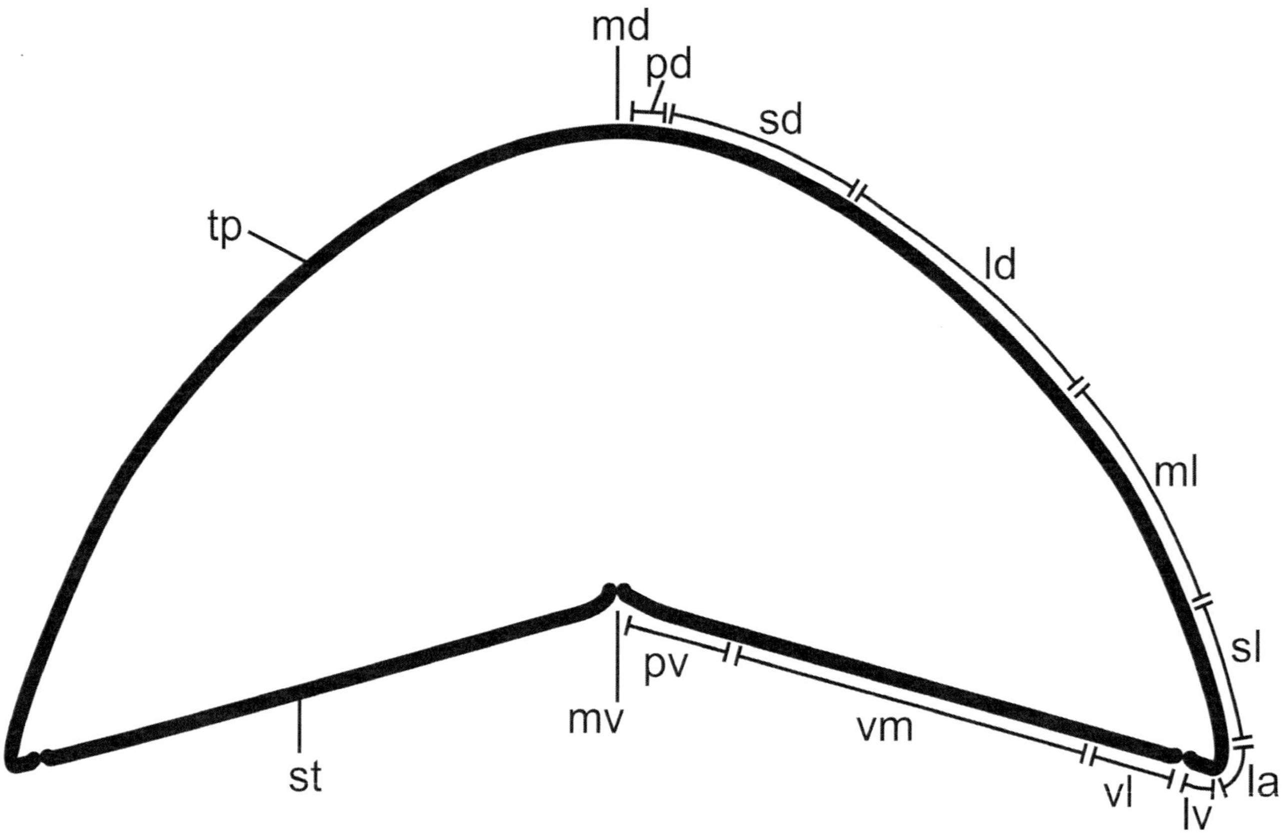

Fig. 5.1.4. Reference position terminology after Pardos et al. (1998), partly emendet by Sørensen et al. (2010a) and Sánchez et al. (2011). Original.

Zelinka (1928: pp. 24–26, text-Figs 3, 4) is the first to recognize and illustrate the regular arrangement of the scalids for *Pycnophyes communis* in a polar-coordinate diagram and in a Mercator-projection diagram divided into 10 sectors and 7 rings. Inspired by Zelinka and based on own reflections, Brown (1985, 1989) provides several types of diagrams in order to illustrate the arrangement of head appendages; she also includes now outer and inner oral styles in one of her diagrams. The polar-coordinate diagram has been formalized and emended by Higgins (1990), and has been adopted or only slightly modified by other authors ever since (Fig. 5.1.5; Neuhaus 1995; Sørensen 2007, 2008b; Sørensen et al. 2007, 2009, 2010a–d, 2012a,c; Sørensen & Rho 2009; Herranz et al. 2012; Sánchez et al. submitted). However, counting of introvert sectors and of inner oral styles varies between different authors (e.g., comp. Figs 5.1.5 and 5.1.6 with Figs 5.1.34 A and 5.2.2 and with Zelinka 1928: text-Fig. 3). More recent emendations are the inclusion of neck placids in the diagram (Sørensen 2008b; Sørensen & Pardos 2008) and indication of the size of the trichoscalid plates (Fig. 5.1.5; Ostmann et al. 2012).

A summary of currently used terminology of external morphological characters and a synonymy of previously used names is provided in Tab. 1. Here, new names are also introduced for several characters with the intention of suggesting as neutral as possible morphological terms, which do not refer to an unproven function. It must be expected that terminology will undergo further changes with increasing knowledge of more kinorhynch taxa.

Introvert – scalids. The introvert carries a series of up to seven rings of recurvated appendages. Zelinka (1907) coins the term "Skalide" (= scalid) for the anterior six rings distinguishing between ring 01 and rings 02–06 scalids and in 1928 the term "Trichoskalide" (= trichoscalid) for the seventh ring (Zelinka 1928). Subsequently, Boykin (1965) replaces scalid by spinoscalid and Brown (1985, 1989) confuses the situation even more by applying spinoscalid to ring 01 and scalid to rings 02–06 (Tab. 1). Sørensen and co-workers take up this latter suggestion (Sørensen 2007 and thereafter) whereas Higgins (1990), Nebelsick (1990, 1993; Bauer-Nebelsick 1995, 1996), and Neuhaus (1995) do not. Higgins (1990) names the scalids of ring 01 primary scalids, because they border the 10 sectors of the introvert (Tab. 1). In summary, it makes sense to have a common term for all introvert appendages, namely scalid, excluding mouth cone appendages and to differentiate further between primary spinoscalids in ring 01, spinoscalids in rings 02–07, and trichoscalids in ring 07. This view is followed here (Tab. 1).

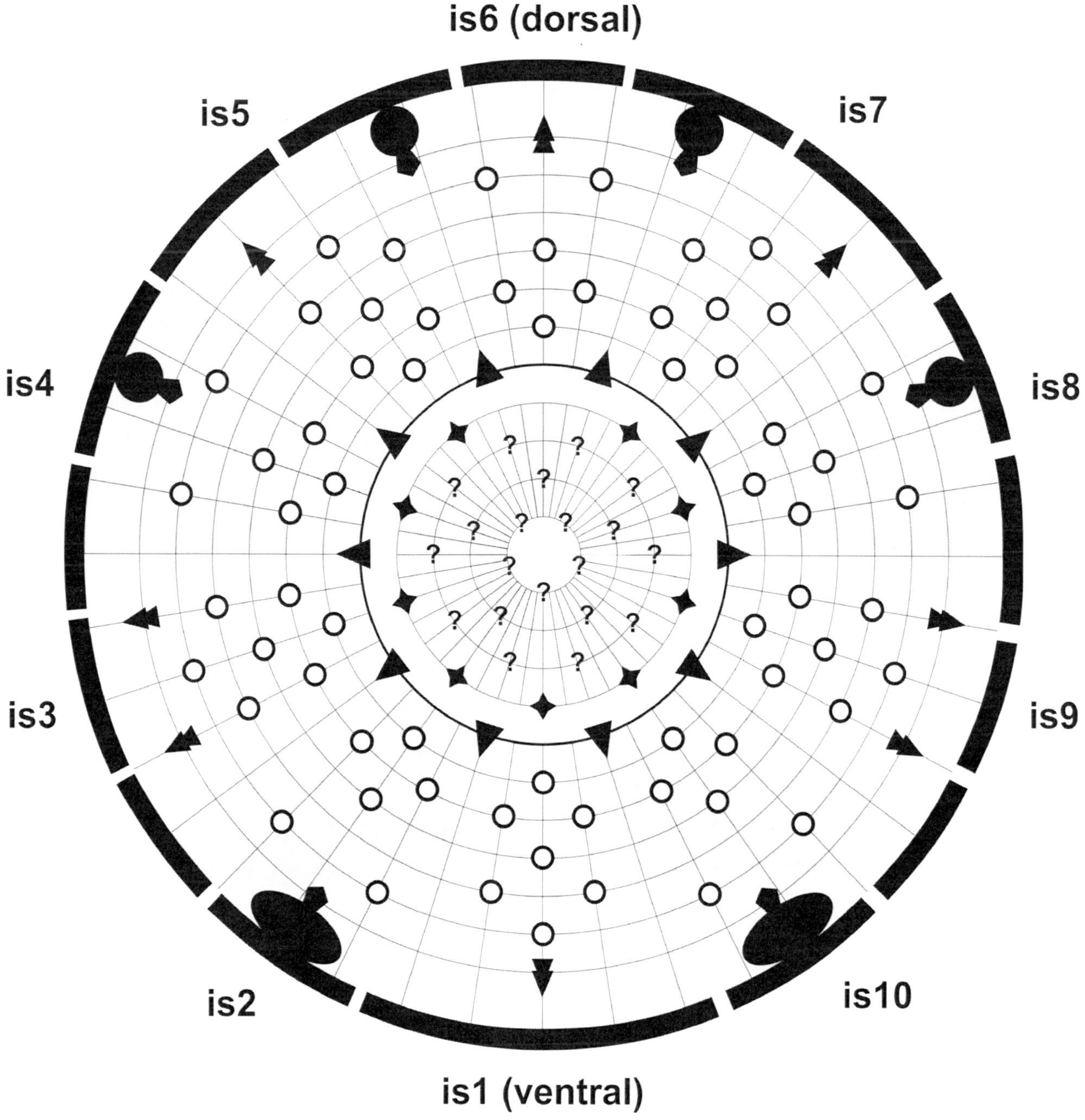

Fig. 5.1.5. Polar-coordinate diagram of the outer (stars) and inner (?) oral styles, primary spinoscalids (triangles), spinoscalids (circles), leaf-like scalids (two triangles), placids (black bent bars around diagram), trichoscalids (pentacons), and trichoscalid basal plates (large black circular and oval symbols) in even-numbered (unshaded) and odd-numbered (shaded) introvert sectors of *Echinoderes cernunnos, E. microaperturus*, and *E. tchefouensis*; staggered arrangement of scalids indicated by radial and circular lines; (after Sørensen et al. 2012b; © Creative Commons License).

The elongated spinoscalids are arranged pentaradial symmetrically and staggered (Figs 5.1.5, 5.1.6, 5.2.2). The introvert can be divided into 10 sectors with all even-numbered sectors showing the same arrangement of spinoscalids and all odd-numbered sectors as well. However, some even- or odd-numbered sectors may differ, e.g., in *Cephalorhyncha liticola, Dracoderes abei, D. gallaicus, Echinoderes spinifurca, Meristoderes herranzae*, and *Pycnophyes* sp. 2 (Fig. 5.1.6), or the presence of a trichoscalid plate may sometimes cause the loss of a lateral scalid so that in *E. applicitus, M. herranzae*, and *M. macracanthus* even the distribution of scalids within the sector becomes asymmetrical (Fig. 5.1.6; Sørensen 2008b; Sørensen & Pardos 2008; Herranz et al. 2012; Ostmann et al. 2012; Sørensen et al. 2012a, c; Sánchez et al. submitted). Leaf-like scalids and ring 07

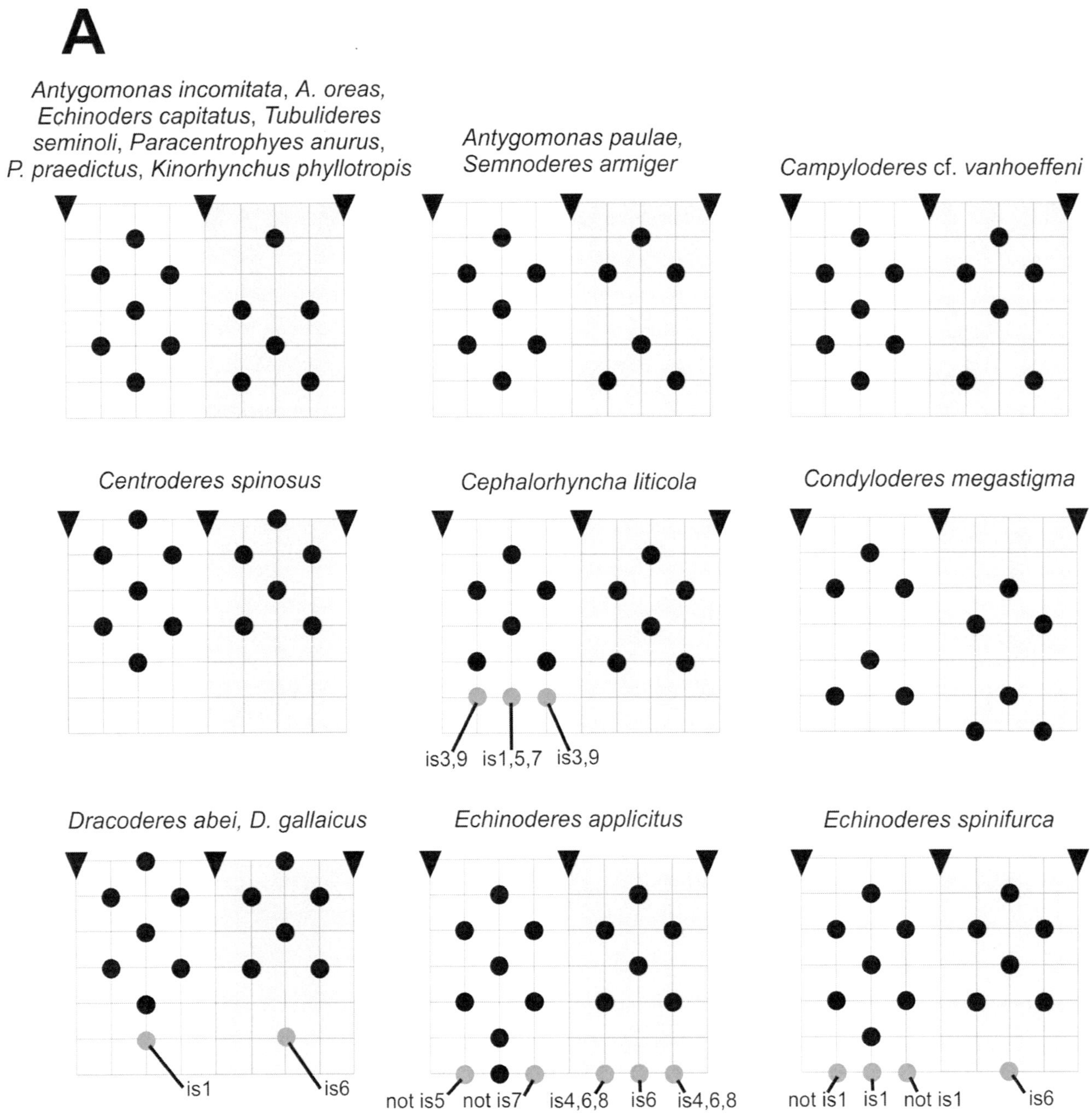

Fig. 5.1.6. *(Continued)*

spinoscalids may also show an asymmetrical distribution in *E. cernunnos*, *E. microaperturus*, and *E. tchefouensis* (Fig. 5.1.5; Sørensen et al. 2012b).

Although only light microscopy has been available at his time, Zelinka (1928: text-Figs 3, 4) recognizes in adult *Pycnophyes communis* that the second ring of spinoscalids in even sectors (todays counting ring 03) originate at a different level than the second ring spinoscalids of the odd sectors (ring 02). In his Fig. 3, the ventral scalids are illustrated at the top and the dorsal scalids at the bottom of the drawing; however, the midventral and middorsal position are indicated slightly laterally to the correct morphological situation. Inter alia, this unconventional orientation probably causes Neuhaus (1995: pp. 189–190, Tab. IV) to state that the scalid arrangement of the adult specimens differs from that of the last juvenile stage of this species (Zelinka 1928: Pl. 17: Fig. 5). This is not correct, neither are the scalid arrangements suggested for the J-6 and adult of this species in Tab. IV by Neuhaus (1995); my current interpretation of the scalid arrangement of *P. communis* is given in Fig. 5.1.6 B.

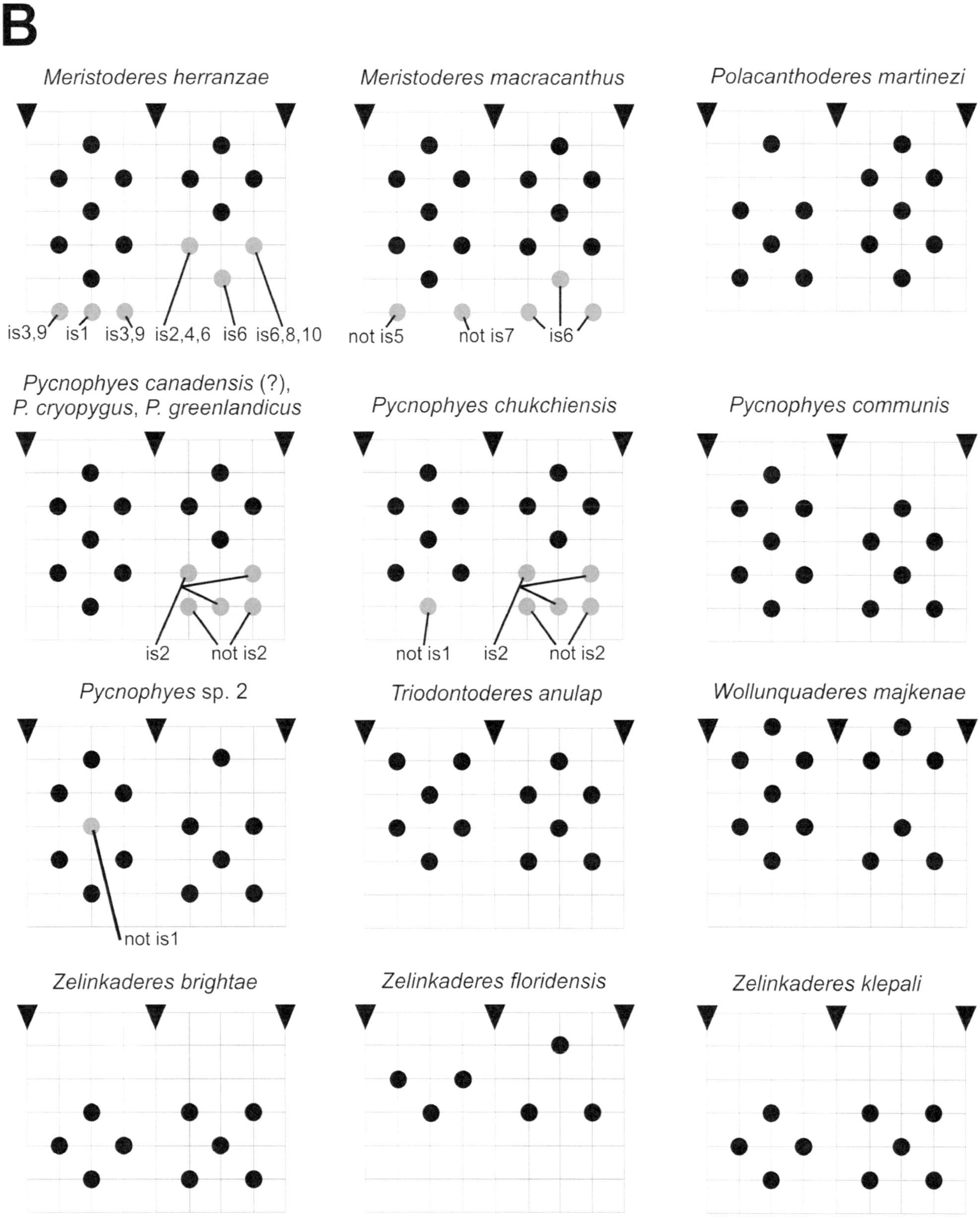

Fig. 5.1.6. A, B, simplified, diagrammatic arrangement of the primary spinoscalids (triangles) and spinoscalids (black dots) in even-numbered (unshaded) and odd-numbered (shaded) introvert sectors of different species; staggered arrangement indicated by vertical and horizontal lines (= rings); position of grey-shaded scalids varies among sectors and does not follow pentaradial symmetry. Original, information from Zelinka (1928), Higgins (1983, 1990), Brown (1985, 1989), Nebelsick (1993), Bauer-Nebelsick (1995, 1996), Neuhaus (1995), Ziemer (2001), Sørensen (2007, 2008b), Sørensen et al. (2007, 2009, 2010a, b, d, 2012a), Sørensen & Pardos (2008), Sørensen & Rho (2009), Sørensen & Thormar (2010), Herranz et al. (2012), Neuhaus & Sørensen (2012), Ostmann et al. (2012), Neuhaus et al. (2013), Sánchez et al. (submitted).

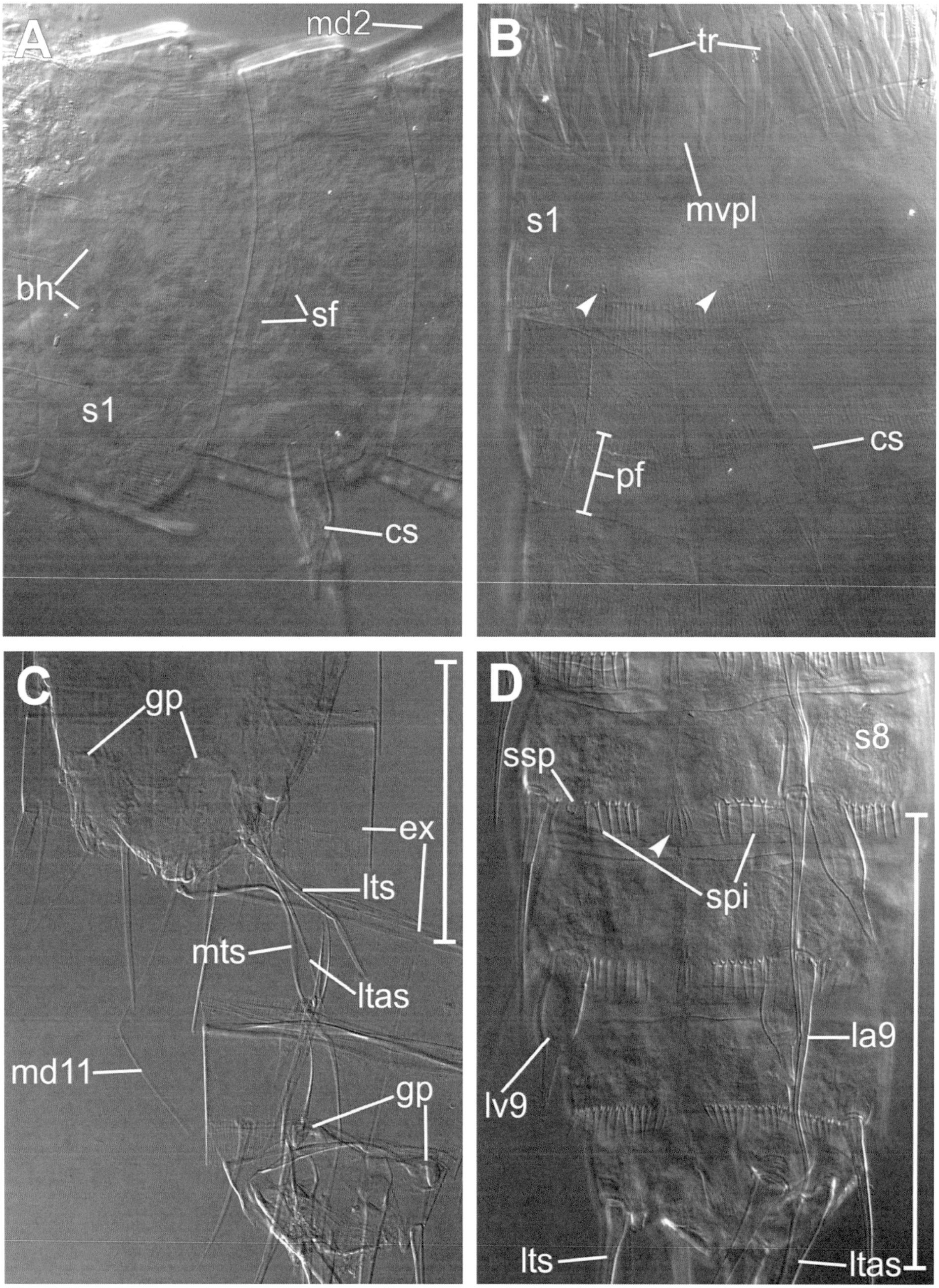

Fig. 5.1.7. LM micrographs of *Antygomonas paulae* (A, D) and *A. oreas* (B, C). A, segments 1 and 2 of adult specimen, right side (image mirrored); B, segments 1–4 of female, arrrowheads mark sensory spots; C, posterior segments of female stage-1 moulting to female stage-2, notice gonopores both in the exuvia and in the moulting specimen; D, segments 8–11 of juvenile stage, ventral view, midventral patch of spiculae (arrowhead) separated from remaining ventral spiculae; lines in C and D = 100 μ, line in D valid for A, B, and D. Original.

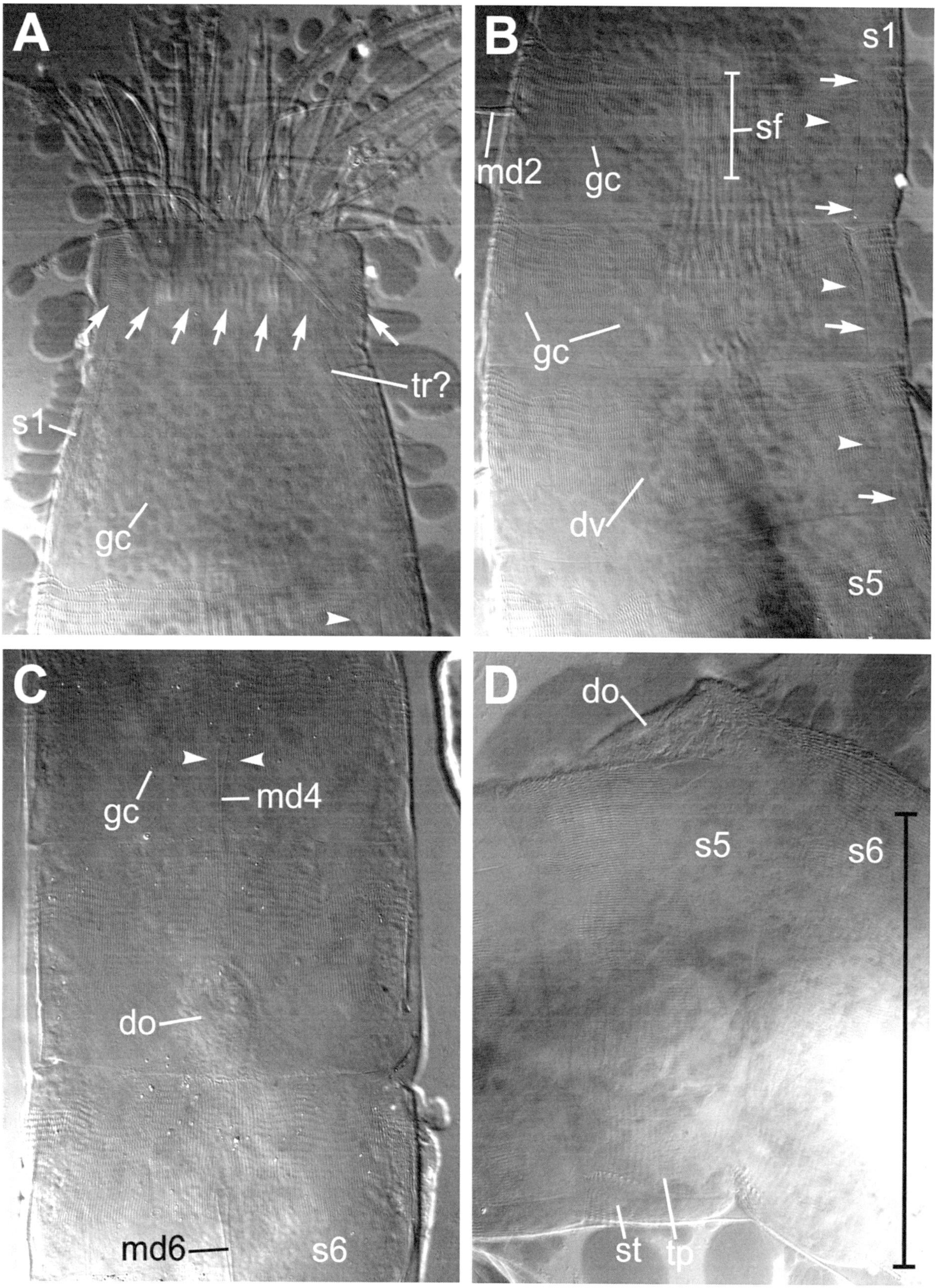

Fig. 5.1.8. LM micrographs of adult *Cateria styx*. A, B, segments 1 and 2, right side, arrows mark cuticular sculpturing anterior of trichoscalids (?), notice small triangular scales in segment 1; B, segments 2–4, right side, arrows mark junction between tergal and sternal plates; C, segments 4–6, dorsal view, dorsal organ hidden below surface; D, segments 5–6, right side (image mirrored), dorsal organ protruded; arrowheads point to sensory spots; line in D = 100 µ, line valid for A–D. Original.

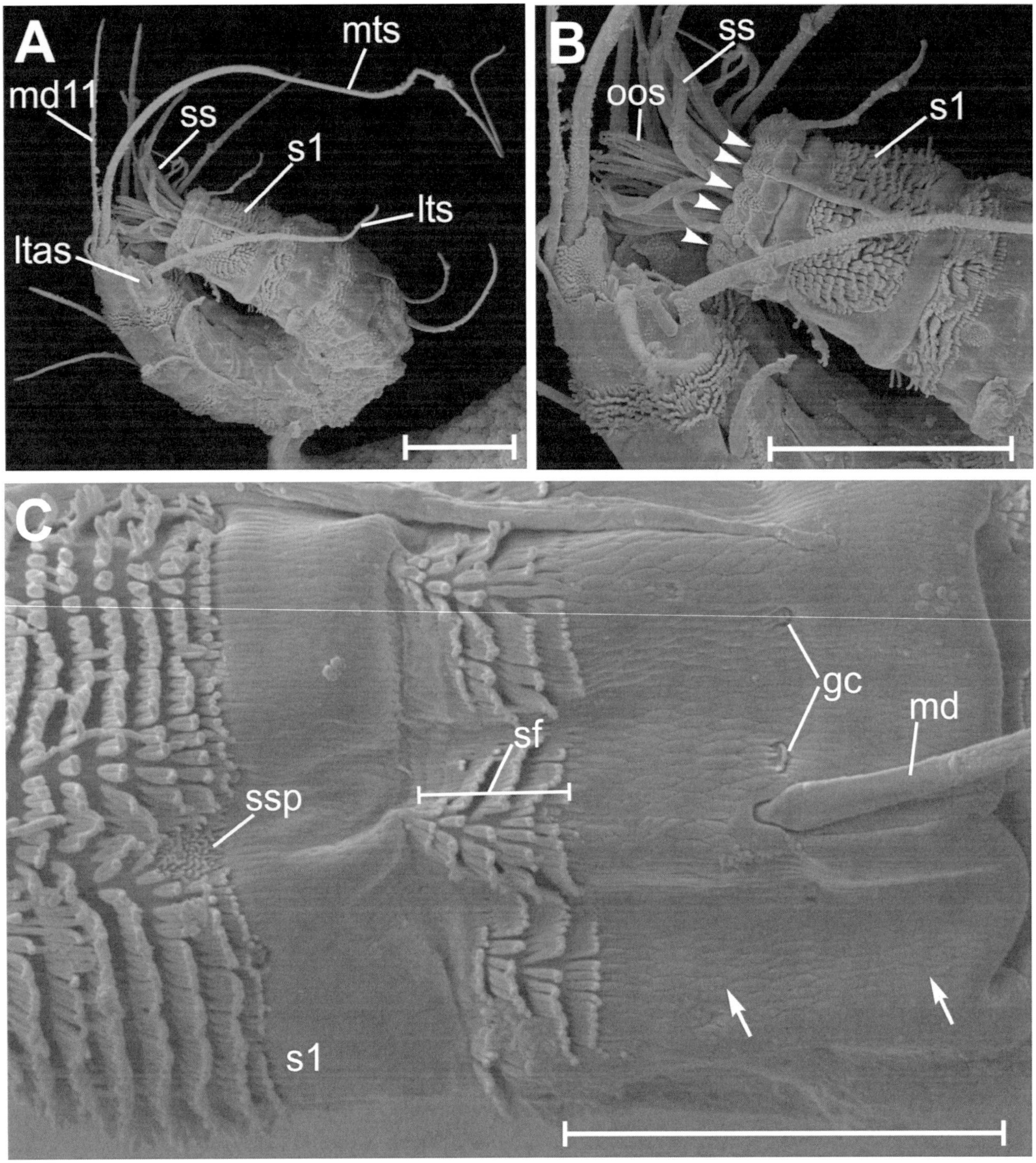

Fig. 5.1.9. SEM micrographs of *Cateria* sp. A, habitus, left side, line = 40 µ; B, anterior and posterior end, left side, arrowheads mark cuticular sculpturing anterior of trichoscalids, line = 40 µ; C, segments 1–2, dorsal view, arrows point to elongate honeycomb-like, hexagonal sculpturing pattern of cuticular surface, line = 20 µ. Original by A. Kegel.

Each spinoscalid consists of a sheath-like basis or socket articulating with a spinose distal element, which ends bluntly in ring 01 and is pointed in rings 02–06(07). The socket reveals an anterior fringe of elongate cuticular bristles in ring 01 and shorter bristles in the remaining spinoscalids. The spinose distal element may show fringes of small cuticular processes or may be smooth (Figs 5.1.21 C, 5.1.34 A, 5.1.35 C, 5.1.38 A; Zelinka 1928; Brown 1989; Higgins 1990; Neuhaus 1995; Sørensen 2007, 2008b; Sørensen et al. 2007, 2009, 2010a, 2012a, b; Sørensen & Rho 2009; Sørensen & Thormar 2010; Herranz et al. 2012; Neuhaus & Sørensen 2012; Neuhaus et al. 2013).

Spinoscalids appear in 4–6 rings. Whereas the primary spinoscalids are expressed in all species studied,

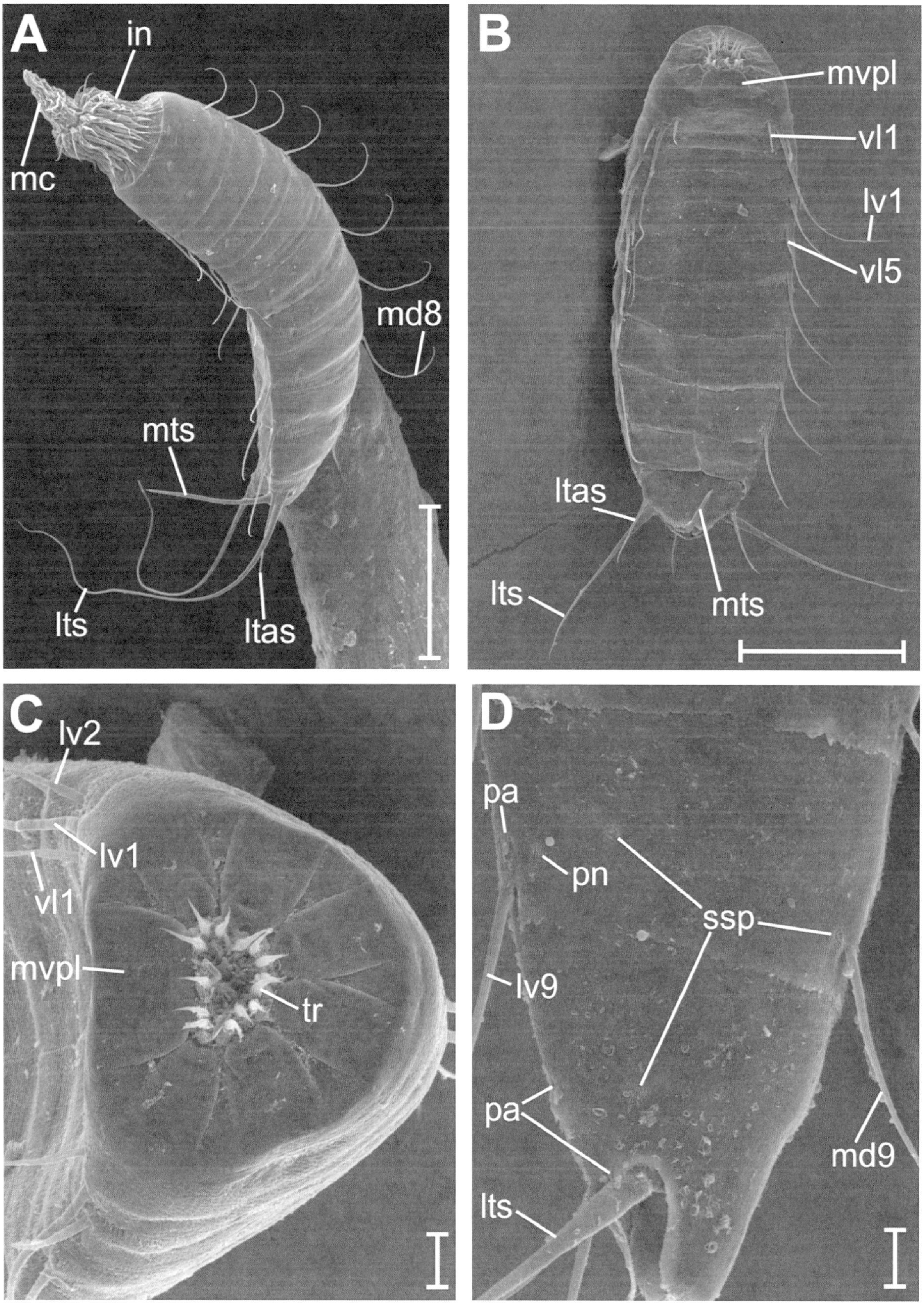

Fig. 5.1.10. SEM micrographs of adult *Campyloderes* cf. *vanhoeffeni*. A, B, habitus, left side (A) and ventral view (B), notice elongated lateroventral spine in segment 1 and length of lateral terminal spine in comparison to midterminal spine, lines in A and B = 100 μ; C, placids and segment 1, frontal view, line = 10 μ; D, segments 10 and 11, left side, notice protonephridium opening, papillae, and sensory spot, line = 10 μ. A and D Original, B and C after Neuhaus & Sørensen 2012; © Elsevier Limited).

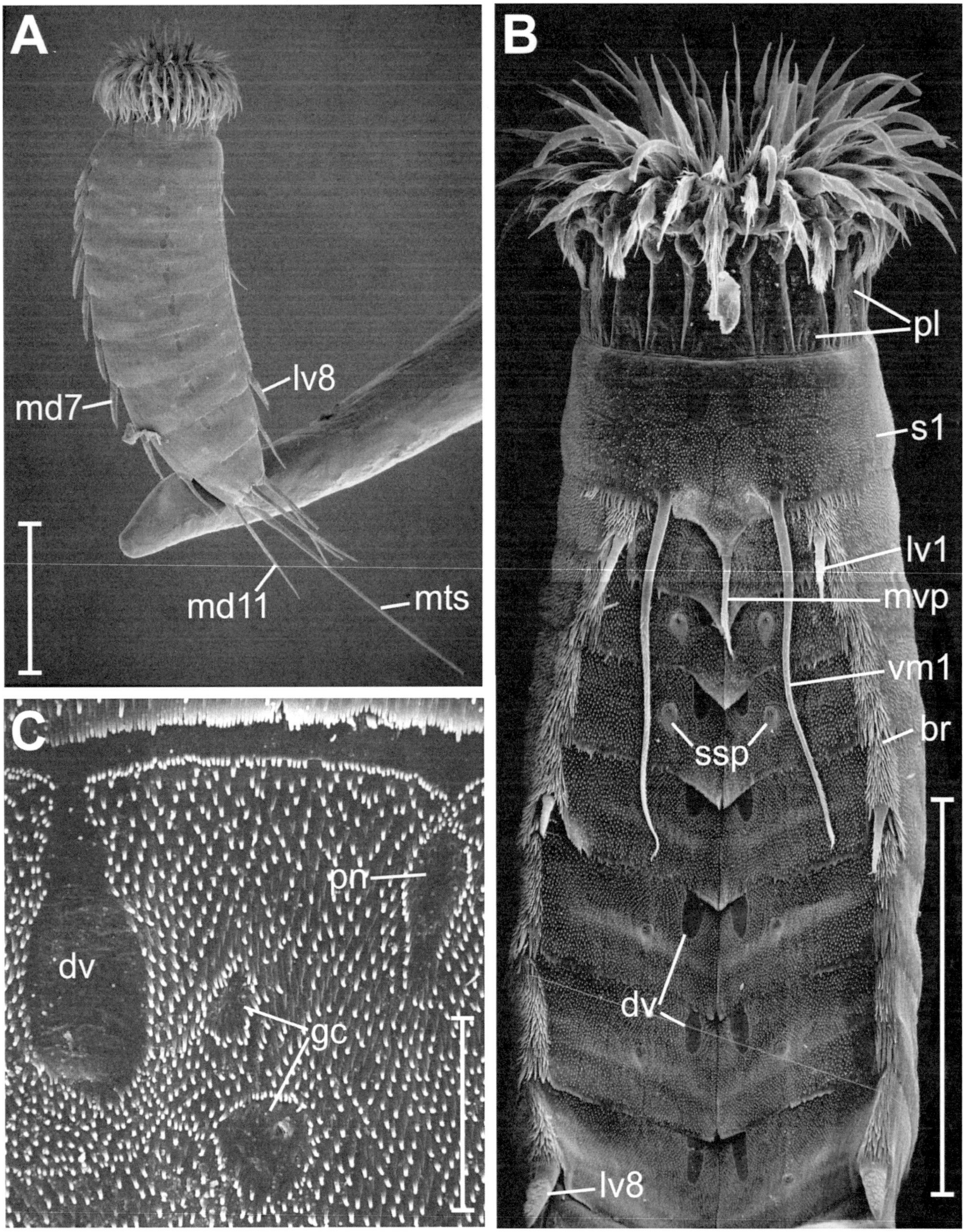

Fig. 5.1.11. SEM micrographs of adult *Centroderes spinosus*. A, habitus of male, right side, Original, line = 100 µ; B, segments 1–8, ventral view, notice scales covering cuticle, Original, line = 100 µ; C, segment 9, right side, notice protonephridium and gland cell openings and attachment area of dorsoventral muscle, (after Neuhaus et al. 2013; © Station de Biologique Roscoff), line = 10 µ.

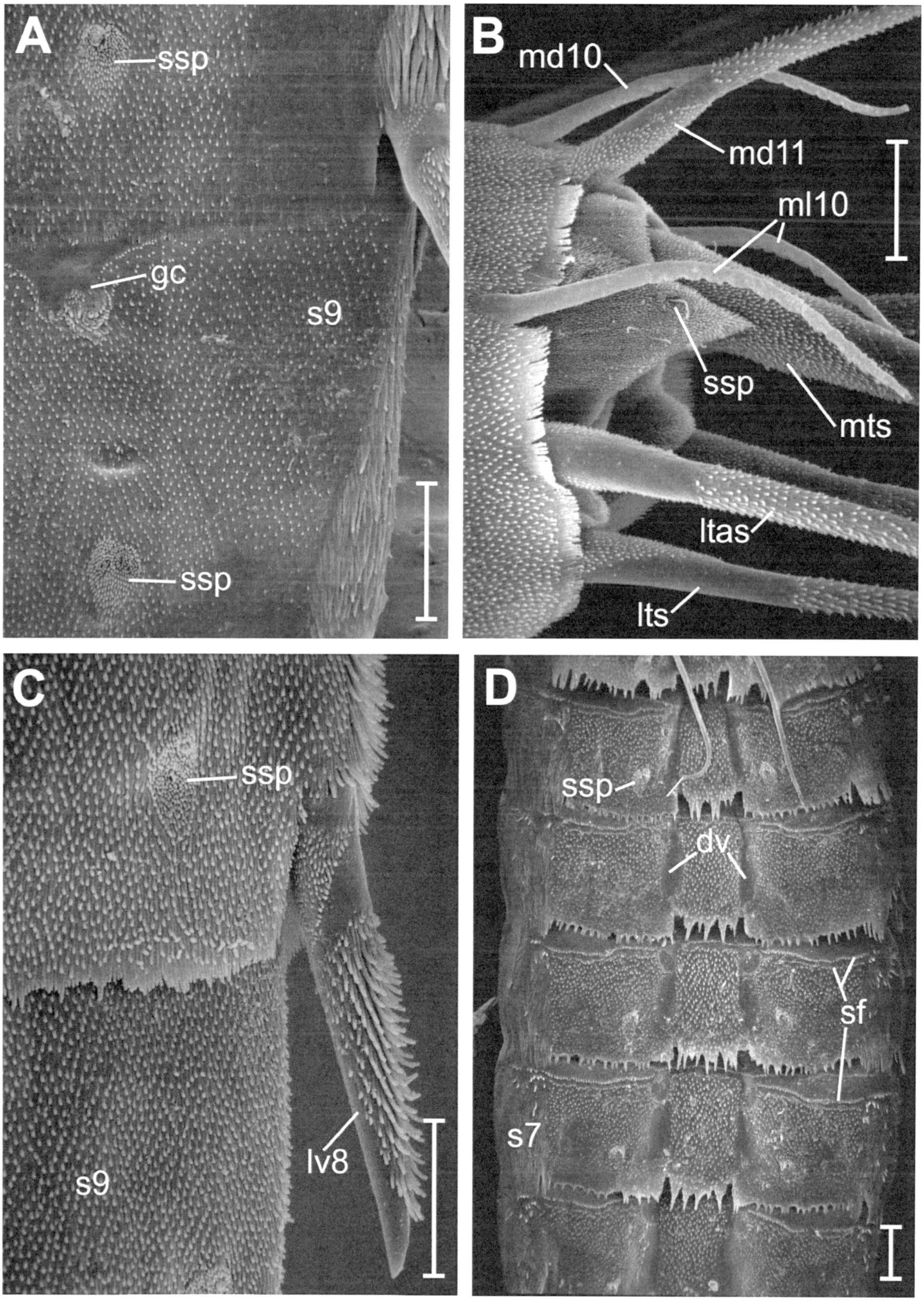

Fig. 5.1.12. SEM micrographs of *Centroderes spinosus*. A, segments 8–9, ventral view, notice sensory spots and female-specific gland cell opening, (after Neuhaus et al. 2013; © Station de Biologique Roscoff), line = 10 µ; B, segments 10–11 of male, left side, notice crenulated middorsal and subdorsal spine in segment 10, line = 10 µ, Original; C, segment 8 of male, ventral view, line = 10 µ, Original; D, segments 4–8 of juvenile stage, ventral view, notice attachment areas of dorsoventral muscles, (after Neuhaus et al. 2013; © Station de Biologique Roscoff), line = 10 µ.

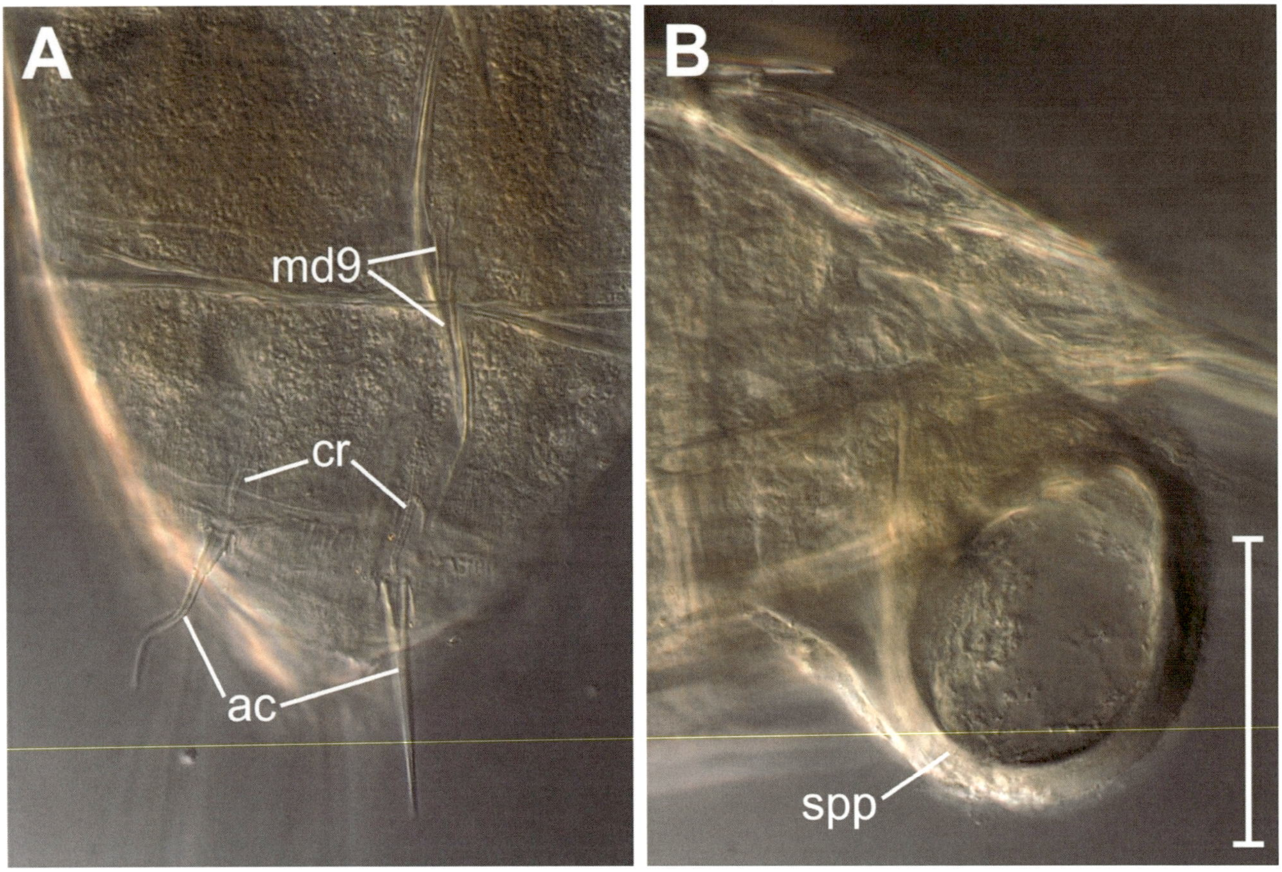

Fig. 5.1.13. LM micrographs of adult *Centroderes* sp. A, segments 9–11 of male type-1 moulting to male type-2, dorsal view, notice crenulated spines in exuvial acicular spines; B, segments 10–11 of male with remnants of spermatophore, left side; line in B = 50 µ, line valid for A and B. Original.

spinoscalids of other rings may be missing or may be replaced by shorter cuticular spines measuring about 1/3 of the length of a regular spinoscalid in *Campyloderes* cf. *vanhoeffeni* or by eyelid-like structures and by spines and fringes in *Zelinkaderes brightae* (e.g., Sørensen et al. 2007; Neuhaus & Sørensen 2012). Primary spinoscalids often reveal distally an incomplete outer annulation or pseudosegmentation at that site facing the introvert. The primary spinoscalids of *Campyloderes* cf. *vanhoeffeni* show 8–31 internal septae; this charatcer is unique among Kinorhyncha (Neuhaus & Sørensen 2012).

Trichoscalids are considerably shorter than spinoscalids and consist of a series of small rings of cuticle with numerous cuticular bristles (name!) (Figs 5.1.7 B, 5.1.14 B, C, 5.1.17 A, 5.1.29 A, 5.1.34 A, 5.1.38 A; Zelinka 1928; Moritz & Storch 1972b). They may originate above a smaller or larger, oval, circular, or triangular, sclerotized trichoscalid basal plate (Figs 5.1.3 B, 5.1.17 A, 5.1.20 A–C, 5.1.21 B, 5.1.22 A; Zelinka 1928) and appear in numbers of 6, 9, or 14 (Tab. 8); consequently, these scalids do not follow the pentaradial arrangement of the spinoscalids.

Introvert – oral styles. The protrusible mouth cone represents the beginning of the alimentary canal (Fig. 5.1.42). Nine outer oral styles and 20 inner oral styles in three rings become visible in the protruded introvert (Figs 5.1.1, 5.1.2, 5.1.21 C, 5.1.27 B, 5.1.34 B, C, 5.1.38, 5.1.42); a fourth ring of inner oral styles may only be recognized at the TEM level, e.g., in *Zelinkaderes floridensis* (Fig. 5.1.42); the middorsal outer oral style is missing, but a basal extension indicates its anlage (Zelinka 1928; Remane 1936; Brown 1989; Nebelsick 1993; Neuhaus 1991, 1994). Adrianov and Malakhov (1990) report 10 outer oral styles for *K. yushini*, but this is doubted by Nebelsick (1993) and has not been noted for any other kinorhynch. Outer oral styles are fused over their entire length (*Campyloderes*) and consist of one element (*Condyloderes*, *Triodontoderes*: Fig. 5.1.27 B, *Kinorhynchus*: Fig. 5.1.38, *Pycnophyes*); alternating styles reveal one or three elements (*Dracoderes*, *Sphenoderes*, *Neocentrophyes*, *Paracentrophyes*: Fig. 5.1.34 B, C); styles are composed of two elements (*Antygomonas*, *Centroderes*, *Cephalorhyncha*, *Echinoderes*: Fig. 5.1.21 C, *Meristoderes*, *Polacanthoderes*,

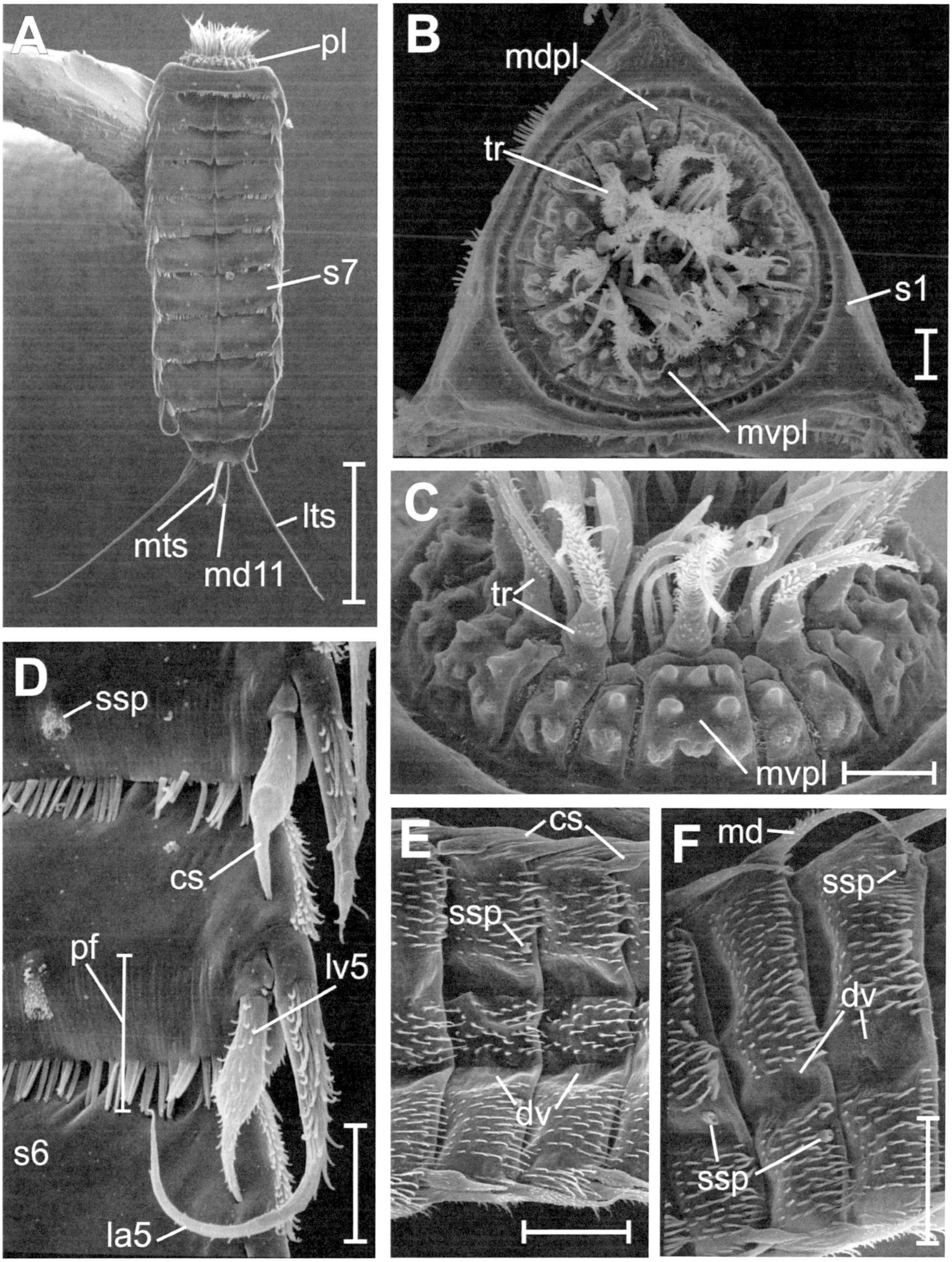

Fig. 5.1.14. SEM micrographs of adult *Condyloderes* sp. A, habitus, ventral view, line = 100 µ; B, anterior end, introvert withdrawn, frontal view, notice triangular cross-section of trunk, line = 10 µ; C, placids with knobby projections, frontoventral view, line = 10 µ; D, segments 4–5, ventral view, notice cuspidate spines, line = 10 µ; E, segments 6–7 of J-5, ventral view, line = 20 µ; F, segments 2–4 of J-5, right side (image mirrored), line = 20 µ; notice attachment area of dorsoventral muscles in E and F. Original.

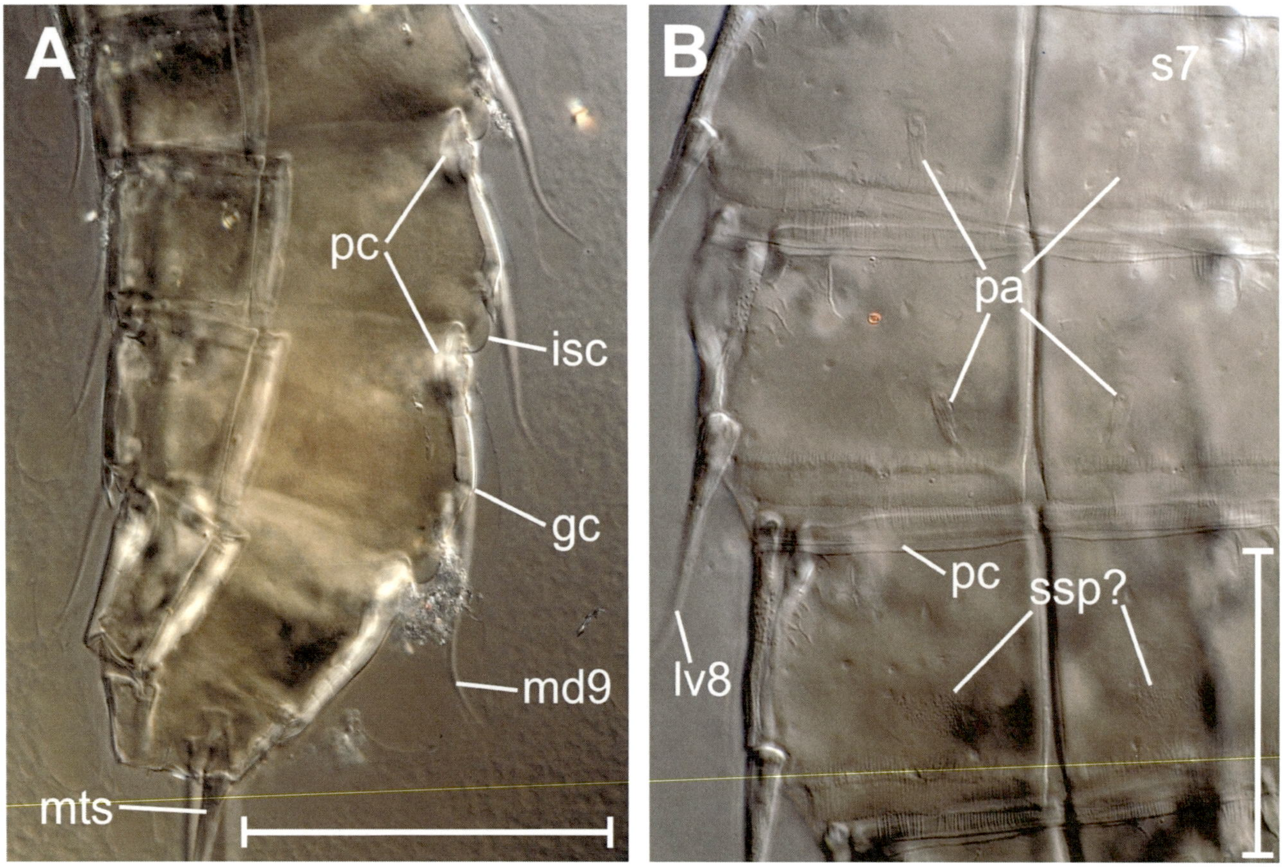

Fig. 5.1.15. LM micrographs of adult *Condyloderes* sp. A, optical section through segments 8–11 of female, lateral view, line = 100 µ; B, segments 7–9 with female-specific large papillae and large sensory spot (?), ventral view, line = 50 µ. Original.

Semnoderes, Tubulideres, Wollunquaderes, Zelinkaderes), or styles are composed of three elements (*Cateria*) (Tabs 8, 9; Zelinka 1912, 1928; Gerlach 1956; Higgins 1967, 1968, 1969a,b, 1983b, 1990; Brown 1989; Higgins & Shirayama 1990; Nebelsick 1990, 1993; Sørensen 2007, 2008b; Sørensen et al. 2007, 2009, 2010a, 2012a,b; Sørensen & Rho 2009; Sørensen & Thormar 2010; Herranz et al. 2012; Neuhaus & Sørensen 2012; Sánchez et al. submitted). In *Triodontoderes anulap, Triodontoderes* sp. (Fig. 27 B), and species of *Zelinkaderes*, outer oral styles may also reveal lateral, wing-like extensions (Higgins 1990; Bauer-Nebelsick 1995; Sørensen & Rho 2009; Sørensen et al. 2007). Similar extensions show an inner fish-bone-like pattern in *Wollunquaderes majkenae* (see Sørensen & Thormar 2010).

Inner oral styles of the anterior rings appear tubular to more or less oval in cross-section, whereas styles of the posteriormost ring articular oblique-diagonal and show basally fringes of thin, cuticular spinose processes (Fig. 5.1.34 B, C; Tab. 1; Brown 1985, 1989; Nebelsick 1993; Neuhaus 1995; Sørensen 2007; Sørensen et al. 2007, 2009, 2012a; Sørensen & Rho 2009; Neuhaus et al. 2013). The latter have been termed helioscalids by Brown (1985, 1989), because the basal fringes are arranged like the rays of the sun. The term is a bit unfortunate, because it implies a relationship to the spinoscalids and trichoscalids of the introvert which may or may not exist. More neutral terms are preferable, and the terms inner and outer oral styles are suggested (Tab. 1).

Neck. In Cyclorhagida, a series of small, trapezoidal to triangular cuticular plates with rounded edges, so-called placids, partly close the trunk except for a central area, if the introvert is withdrawn (see also below; Figs 5.1.3 B, 5.1.7 B, 5.1.10 C, 5.1.11 B, 5.1.14 B, C, 5.1.17 A, B, 5.1.20, 5.1.21 B, D, 5.1.22 A, 5.1.23 A, 5.1.25 B, C, 5.1.26, 5.1.29 A, C, 5.1.30 A, 5.1.33 A, 5.1.40 A, 5.2.5 D). The cuticle folds between the more sclerotized placids and may give the impression of one pair of "interstitial placids" between neighbouring placids, e.g. in species of *Antygomonas, Centroderes, Echinoderes*, and *Fissuroderes* (Figs 5.1.17 A–5.1.20 A, D; Zelinka 1928; Nebelsick 1990; Bauer-Nebelsick 1996; Sørensen 2007; Neuhaus own obs.). Kristensen and Higgins (1991: Fig. 5.1.27) demonstrate for *E. aquilonius* that electron-dense cuticle occurs both in the the placids and in the interstitial placids;

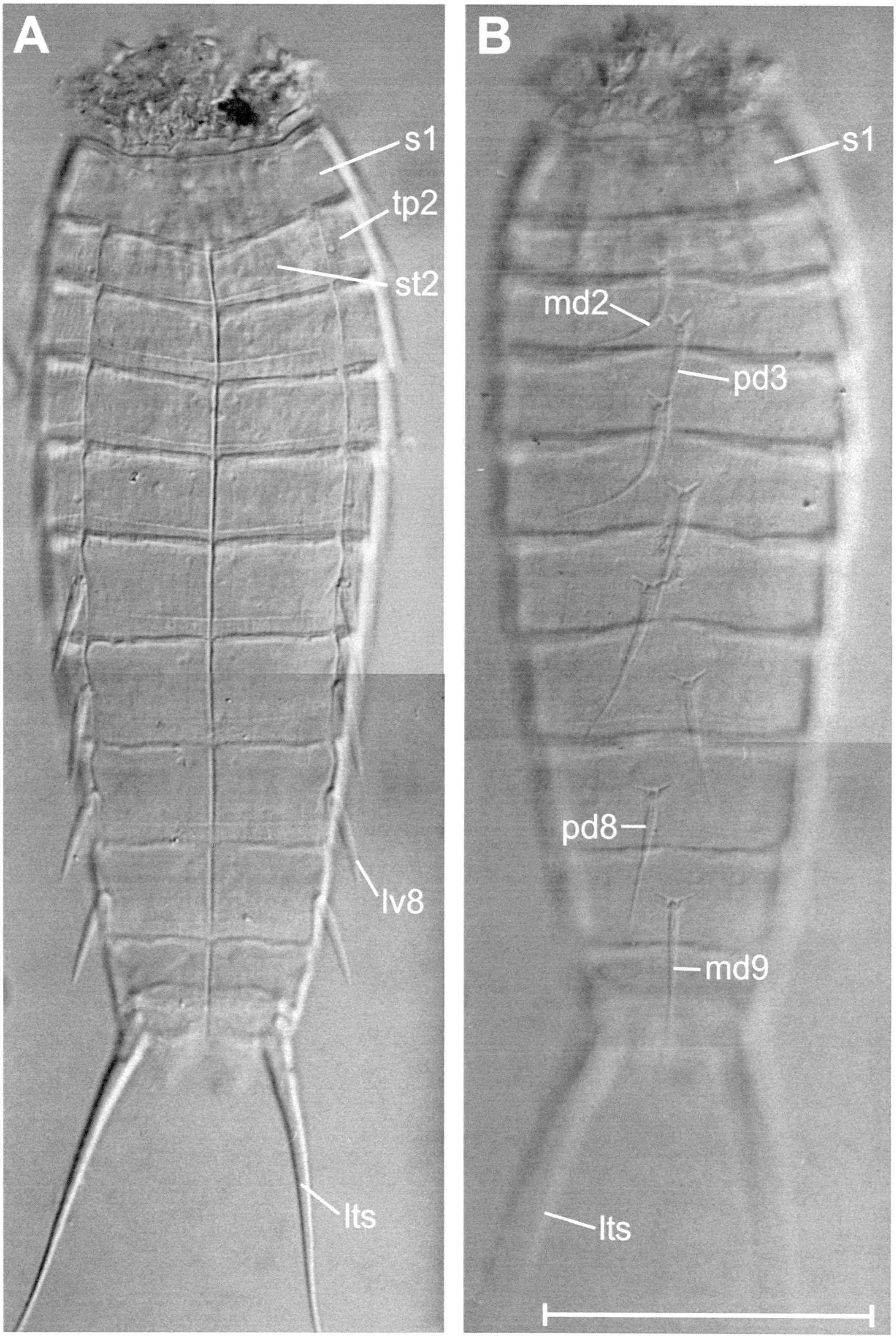

Fig. 5.1.16. LM micrographs of female *Dracoderes abei*, holotype. A, B, habitus, ventral (A) and dorsal (B) view, line in B = 100 μ, line valid for A and B. Original.

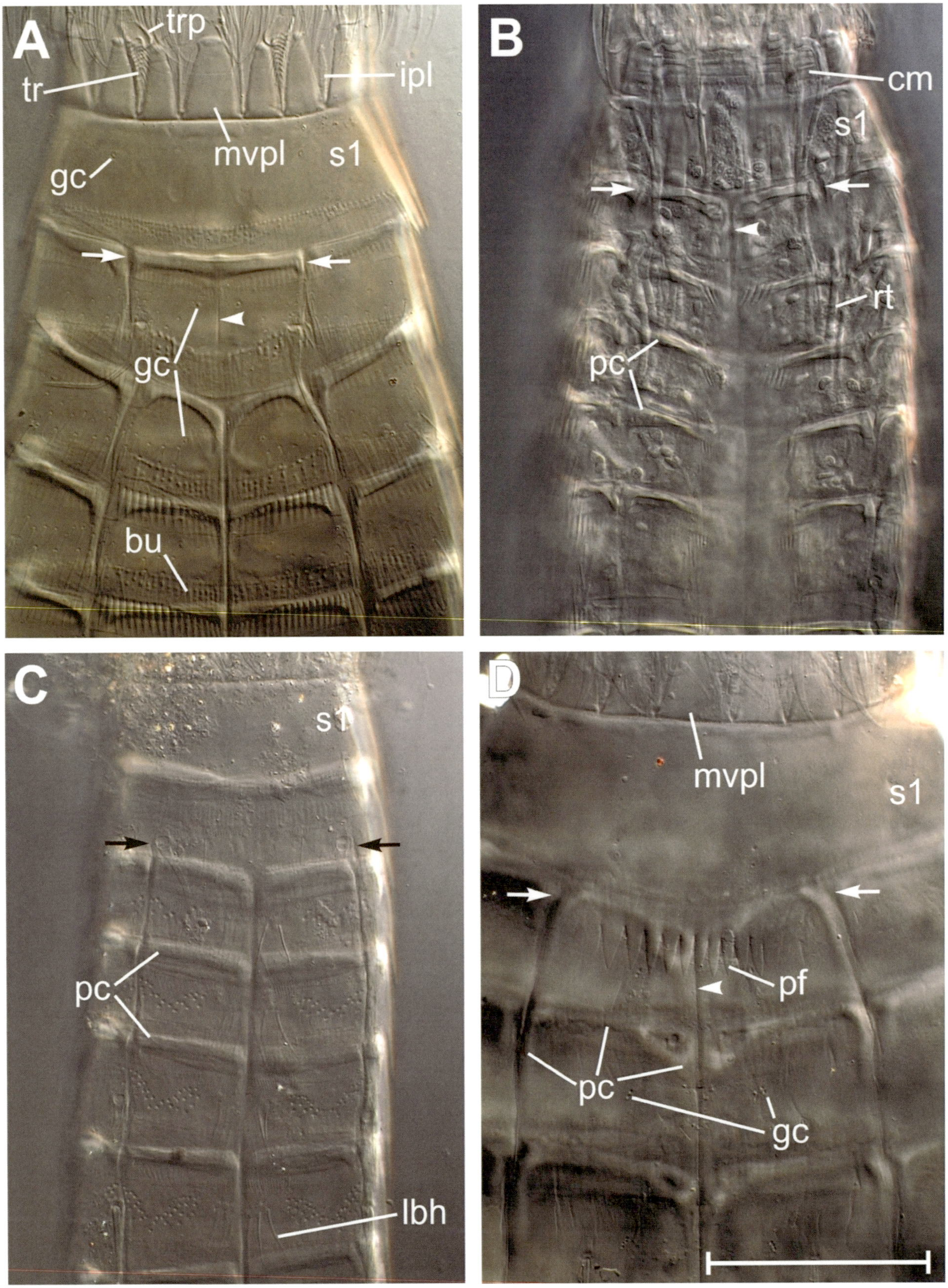

Fig. 5.1.17. LM micrographs of adult Echinoderidae, ventral view. A, segments 1–4 of male *Cephalorhyncha asiatica*; B, segments 1–6 of female *Fissuroderes thermoi*; C, segments 1–6 of male *Meristoderes macracanthus*; D, segments 1–4 of female *Polacanthoderes martinezi*; arrows mark tergosternal junction, arrowheads point to midventral junction; line in D = 50 μ, line valid for A–D. Original.

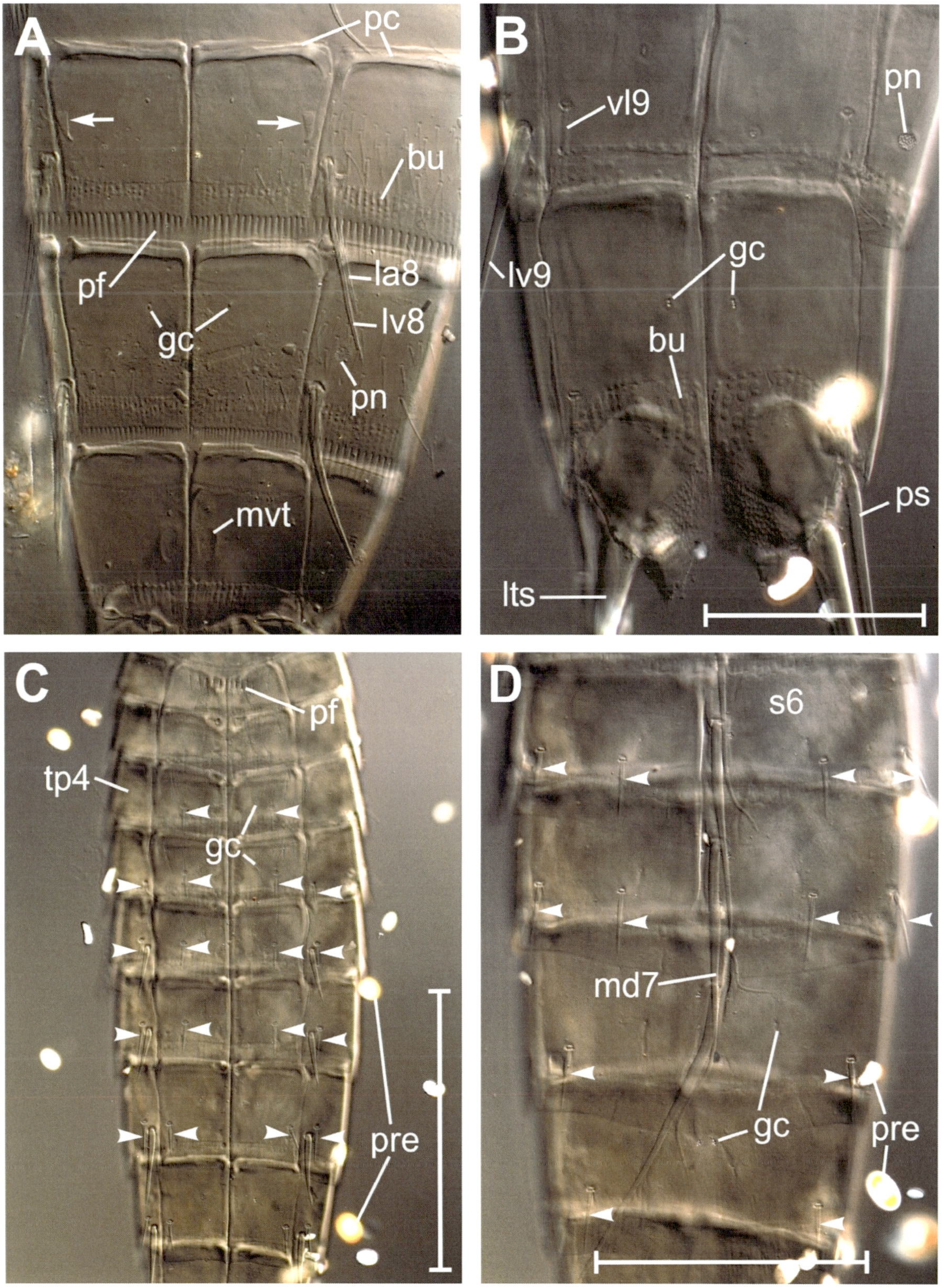

Fig. 5.1.18. LM micrographs of adult Echinoderidae. A, segments 8–10 of female *Cephalorhyncha asiatica* with subcuticular funnel-like structures (arrows) associated with gland cell outlet, ventral view, stacked image; B, segments 9–11 of male *Polacanthoderes martinezi* with protonephridium sieve plate, notice widely fused button-like structures and lack of lateroventral spine on the left side of segment 9, ventral view; C, D, segments 2–9 (C) and segments 6–9 (D) of female *Polacanthoderes* with multiple name-giving spines (arrowheads), ventral (C) and dorsal (D) view; line in B = 50 µ, line valid for A and B; lines in C and D = 100 µ. Original.

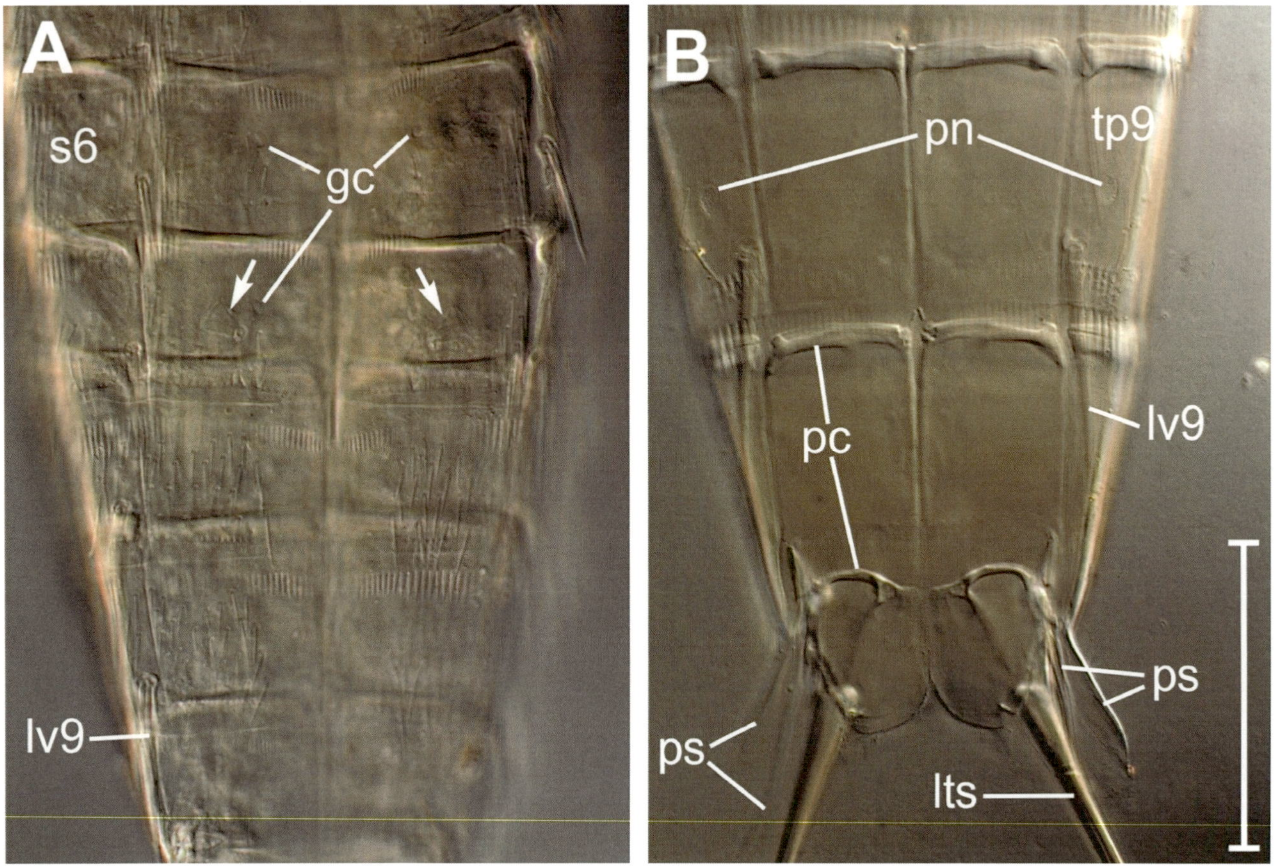

Fig. 5.1.19. LM micrographs of adult Echinoderidae, ventral view. A, segments 6–10 of female *Fissuroderes thermoi* with subcuticular funnel-like structures (arrows) associated with gland cell outlet; B, segments 9–11 of male *Cephalorhyncha asiatica* with spinose and broad penile spines; line in B = 50 μ, line valid for A and B. Original.

here, the electron-dense cuticle is much thinner than in the placids, but probably serves as reinforcement structure of the otherwise more flexible cuticle between the placids. No information at TEM level is available for species of *Antygomonas*, *Centroderes*, and *Fissuroderes*. Whereas 16 placids appear in most Cyclorhagida, 14 are reported for *Campyloderes* cf. *vanhoeffeni* and *Triodontoderes anulap*, nine for species of *Dracoderes* (Tab. 8). Possibly, the very broad midventral placid of *C.* cf. *vanhoeffeni* (Fig. 5.1.10 C) results from the fusion of the midventral and left and right paraventral placids, because species of *Centroderes* and *Condyloderes* from the same family Centroderidae reveal two narrow placids neighbouring the midventral placid, whereas *Campyloderes* possesses just one narrow placid in this position (Figs 5.1.11 B, 5.1.14 B, C; Tabs 8, 9). The size of the remaining placids alternates between broader and narrower in all three genera (Higgins 1969a; Sørensen et al. 2010c; Neuhaus & Sørensen 2012; Neuhaus et al. 2013). It remains open whether or not species of *Cateria* possess separated placids at all (comp. Figs 5.1.8 A, 5.1.9 B).

The surface of the placids is usually smooth, but species of *Condyloderes* show conspicuous, species-specific knobbly projections (Fig. 5.1.14 B, C; Higgins 1969a; Martorelli & Higgins 2004; Sørensen et al. 2010b). Placids may exhibit the same width, may be broader midventrally, or may have different patterns of placids of alternating width (Tabs 8, 9). Placids may also fuse basally with trunk segment 1 in species of *Cateria*, *Triodontoderes*, and *Zelinkaderes* (Figs 5.1.8 A, 5.1.9 B, 5.1.30 A). In the latter two genera, placids are separated incompletely anteriorly and appear tripartite (Tabs 8, 9; Higgins 1990; Bauer-Nebelsick 1995; Sørensen et al. 2007; Sørensen & Rho 2009).

In Homalorhagida, trapezoidal placids with rounded edges (Figs 5.1.40 A, 5.1.54 D) shield the trunk completely, if the introvert is withdrawn (Fig. 5.1.33 D). This closing mechanism is made possible by the contraction of the dorsoventral muscles of segment 1, which pulls the ventral towards the dorsal cuticle. In species of *Neocentrophyes* and *Paracentrophyes* with a weakly sclerotized cuticle, the ventral flexible cuticle appears as a single sternal plate (Figs 5.1.32 A, 5.1.33 B, 5.1.34 A). In species of *Kinorhynchus* and *Pycnophyes* with a strongly sclerotized cuticle, the thicker and more stiff ventral cuticle is divided into three sternal plates (Fig. 5.1.36 A, B; Zelinka 1928; Higgins 1969b, 1983; Adrianov & Malakhov 1999a;

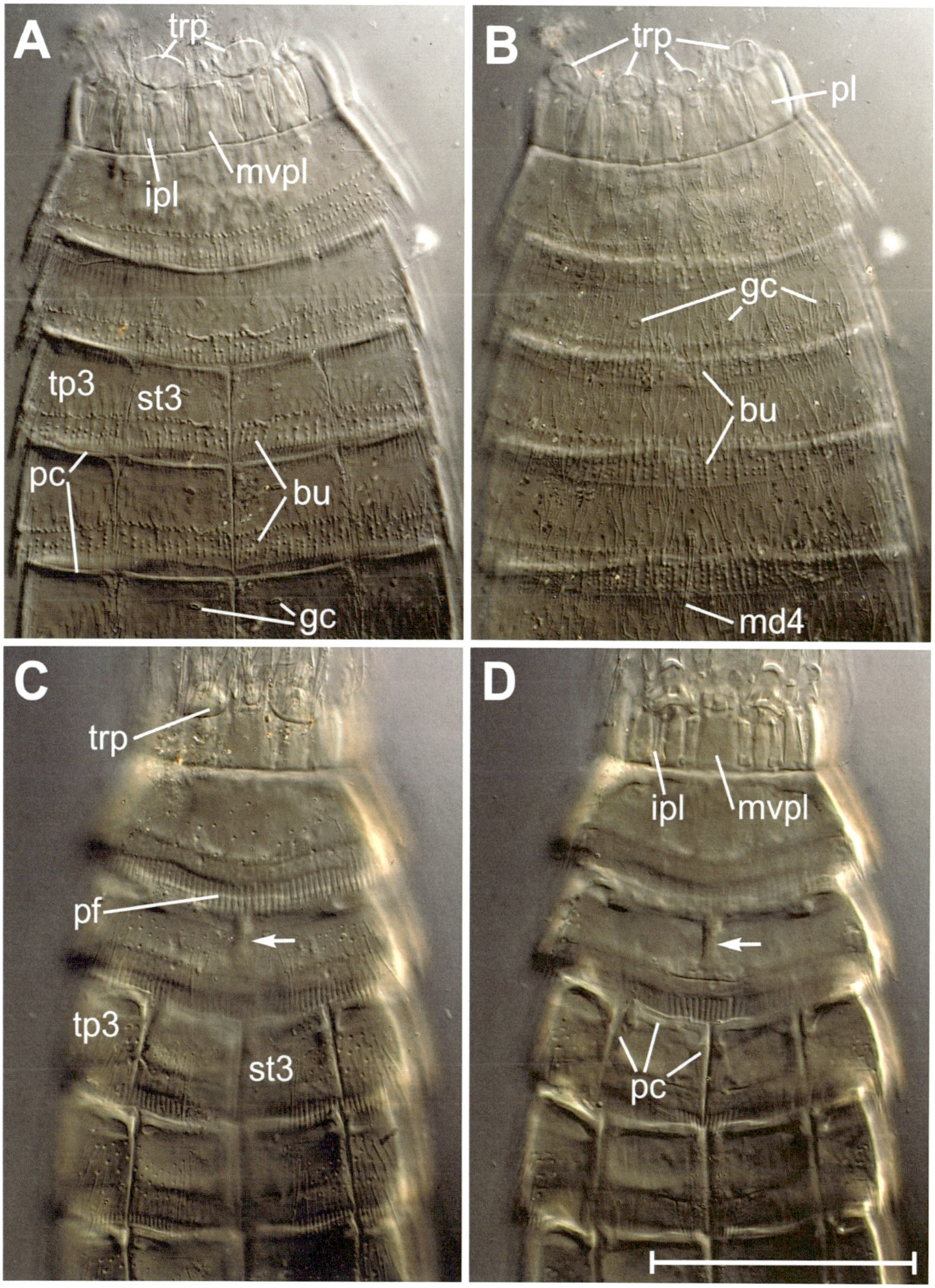

Fig. 5.1.20. LM micrographs of female *Echinoderes setiger*. A, B, segments 1–4 of young adult with poorly developed pachycycli and button-like structure cuticular structures, ventral (A) and dorsal (B) view; C, D, segments 1–4 of mature adult with strong pachycycli and numerous cuticular "pores" (C) but without button-like structures, ventral view at two different focal levels, arrows mark midventral thickening cuticular thickening in segment 2; notice oval ventral and circular dorsal trichoscalid plates; line in D = 50 μ, line valid for A–D. Original, specimens donated by M. Bright.

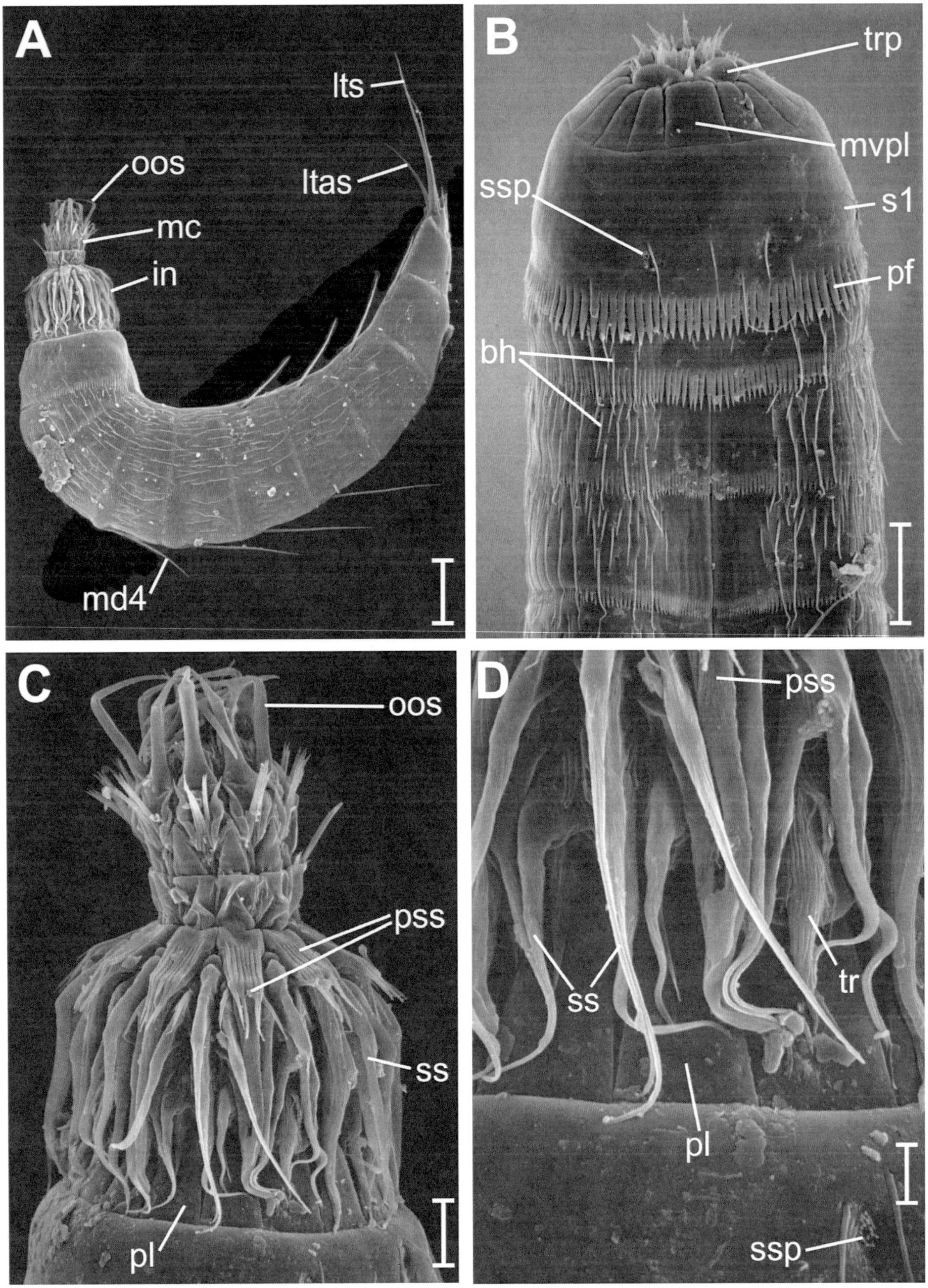

Fig. 5.1.21. phs of adult *Echinoderes remanei*. A, habitus, right side, line = 40 μ; B, segments 1–4, ventral view, line = 20 μ; C, everted introvert, left side, line = 10 μ; D, trichoscalids, left side, line = 4 μ. Original.

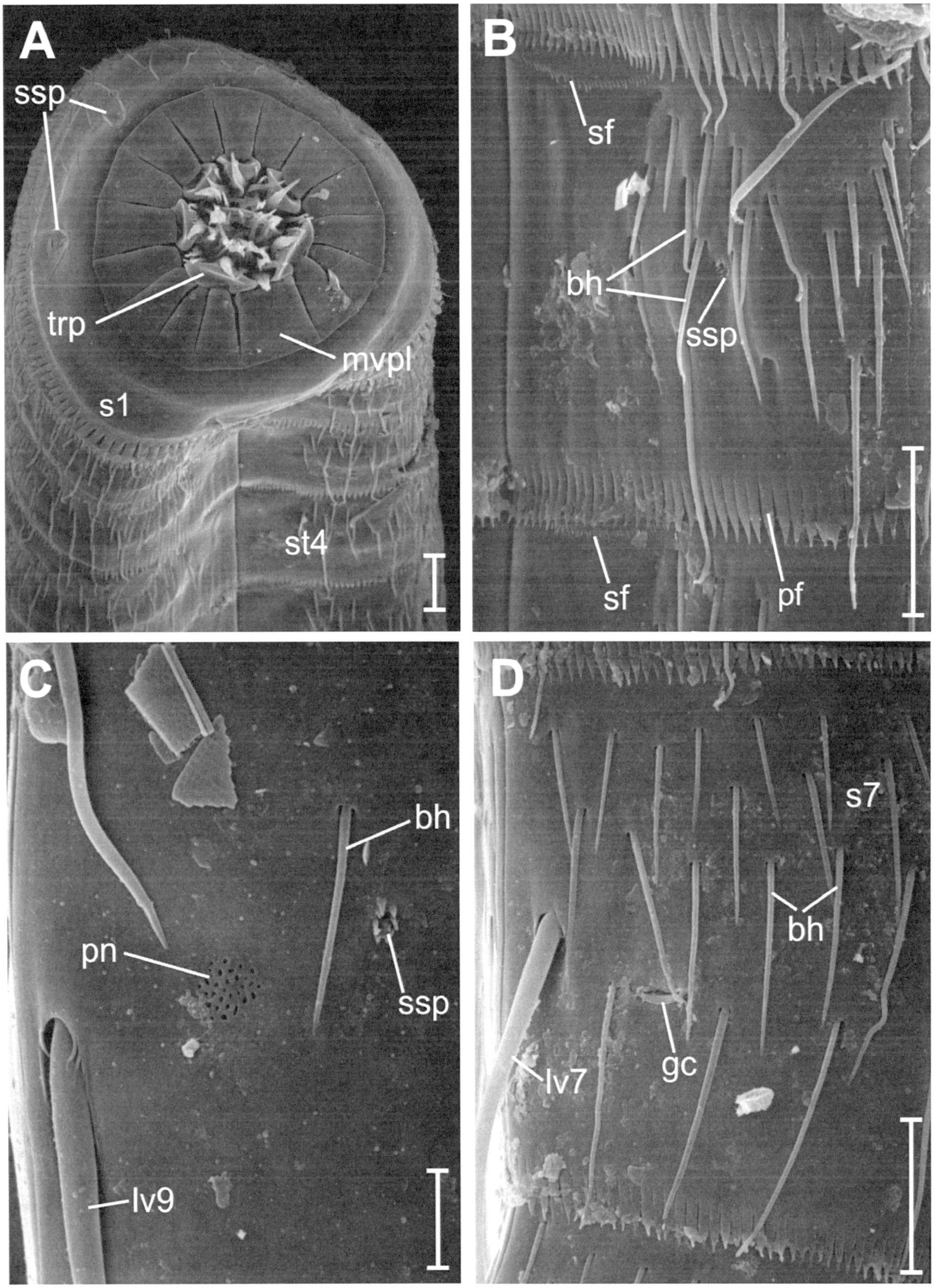

Fig. 5.1.22. SEM micrographs of adult *Echinoderes remanei*. A, anterior end, frontoventral view, line = 10 µ; B, segment 5, ventral view, notice secondary fringe and sensory spot, line = 10 µ; C, segment 9, left side, notice protonephridium sieve plate, sensory spot, bracteate hair, and spine, line = 4 µ; D, segment 7, left side, notice opening of gland cell type-2, line = 4 µ. Original.

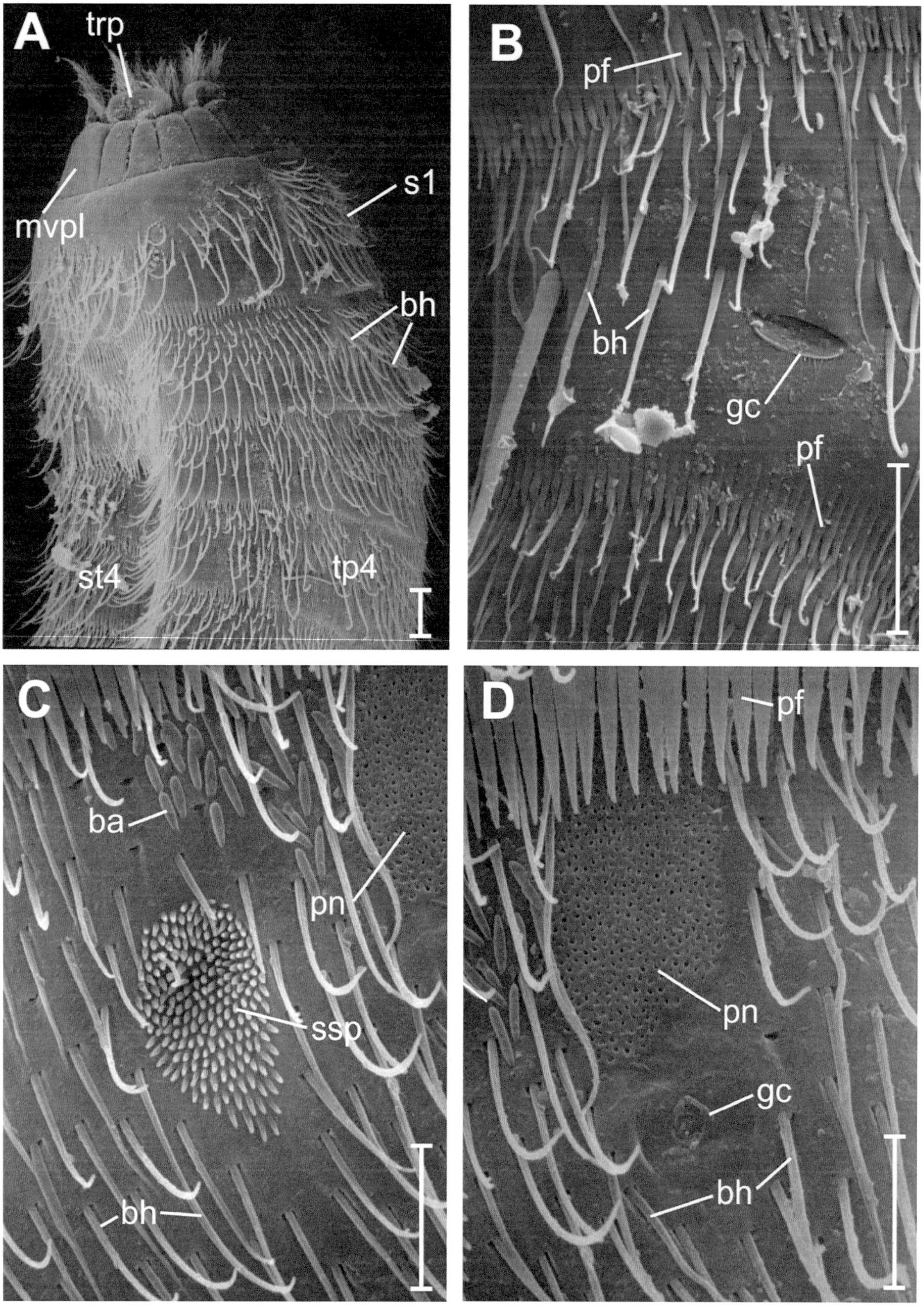

Fig. 5.1.23. SEM micrographs of adult *Echinoderes bookhouti* (A, B) and *E. caribiensis* (C, D). A, segments 1–4, left side, notice numerous bracteate hairs, line = 10 μ; B, segment 8, left side, notice opening of gland cell type-2; C, D, segment 9, right side, sensory spot (C) and protonephridium sieve plate (D), line = 4 μ. Original.

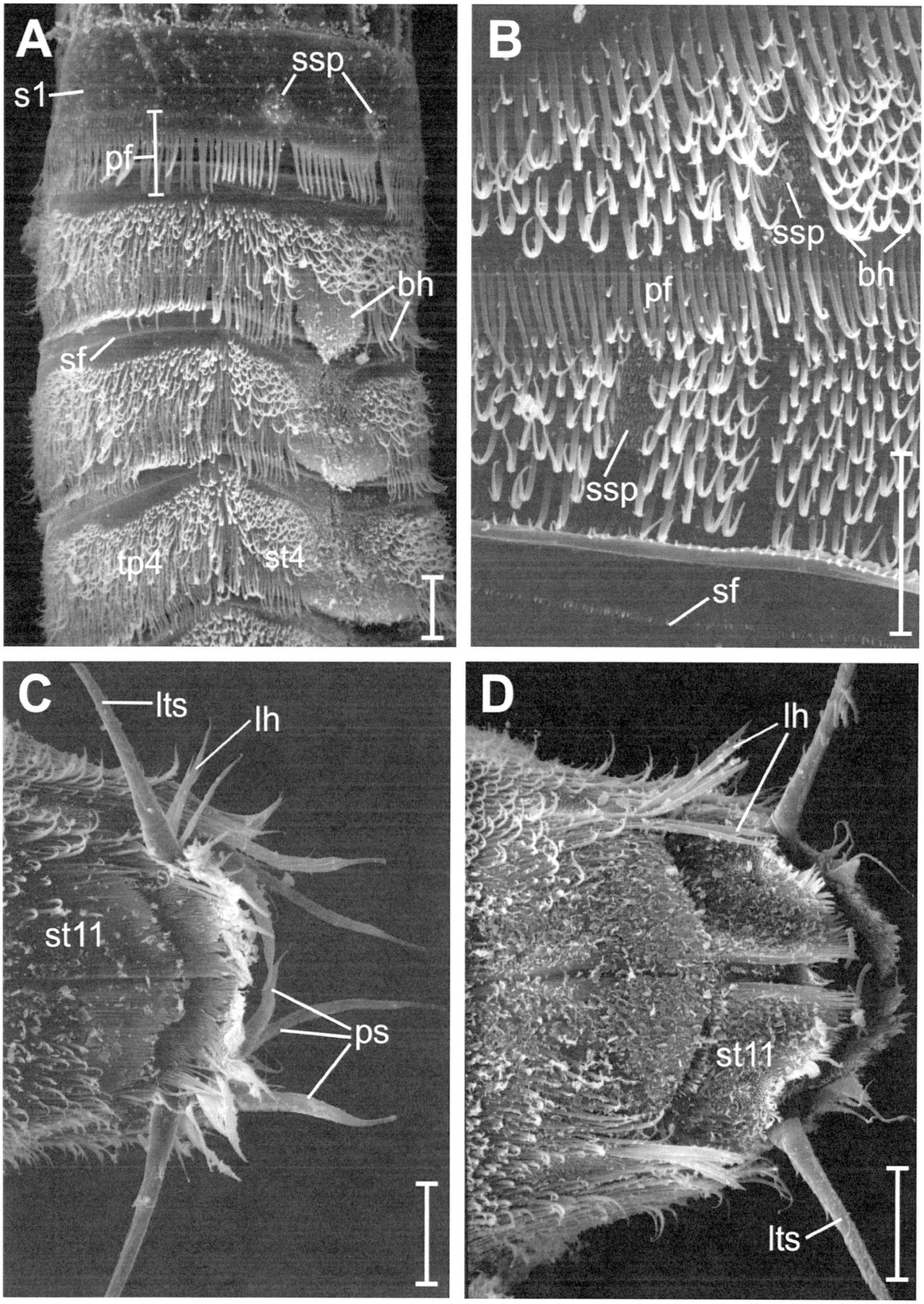

Fig. 5.1.24. SEM micrographs of adult *Echinoderes coulli*. A, segments 1–4, lateroventral view (right), line = 10 μ; B, segments 2–3, left side, line = 10 μ; C, D, segments 10–11, ventral view, male with penile spines (C), female without penile spines (D), line = 10 μ. Original.

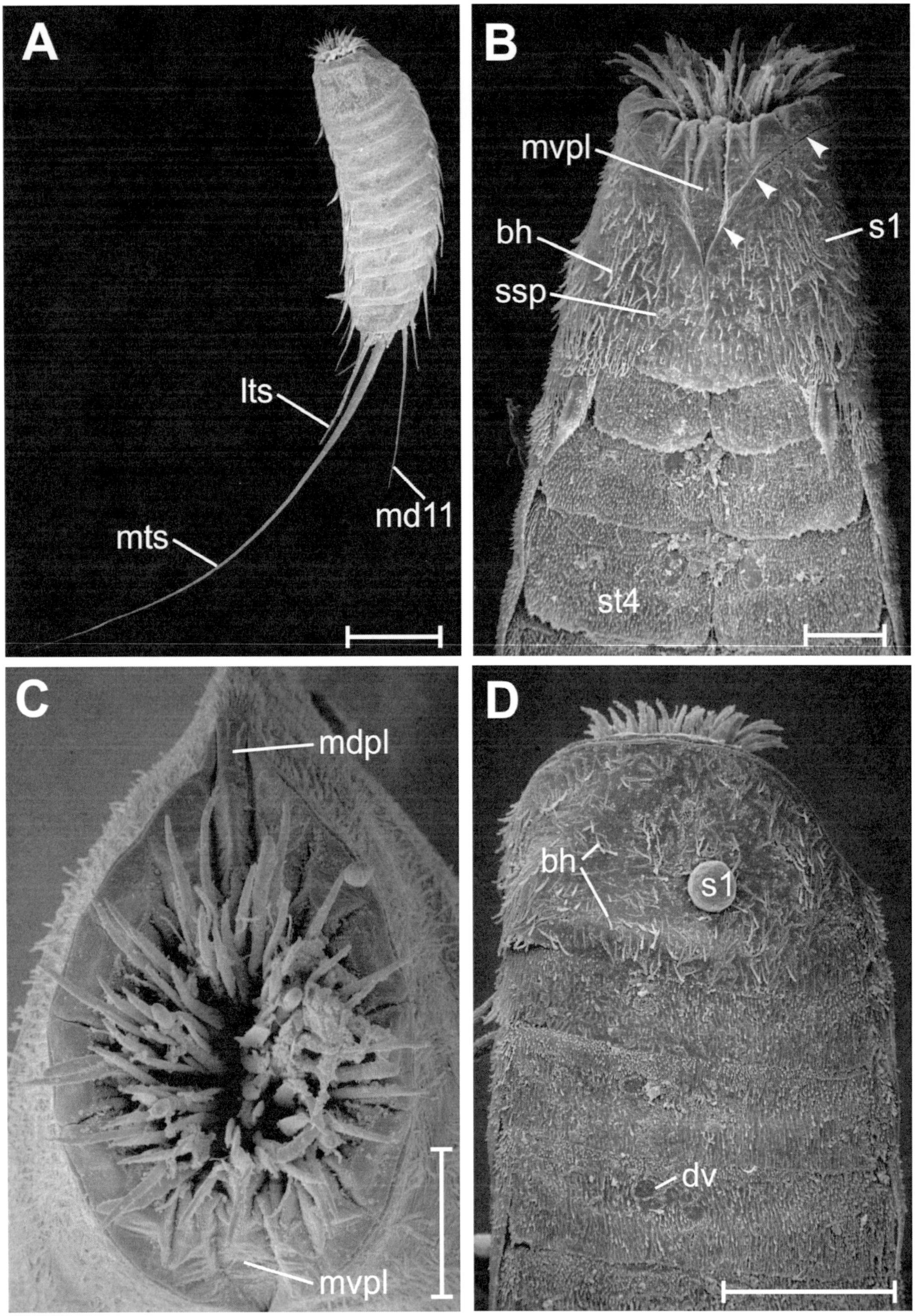

Fig. 5.1.25. SEM micrographs of adult *Semnoderes* sp. A, habitus, left side, notice extremely elongated midterminal spine, line = 100 μ; B, segments 1–4, ventral view, arrowheads mark anterior margin of segment 1, line = 20 μ; C, frontal view, line = 20 μ; D, segments 1–4, left side, line = 40 μ; notice anterior clamshell-like closing apparatus (B–D). Original.

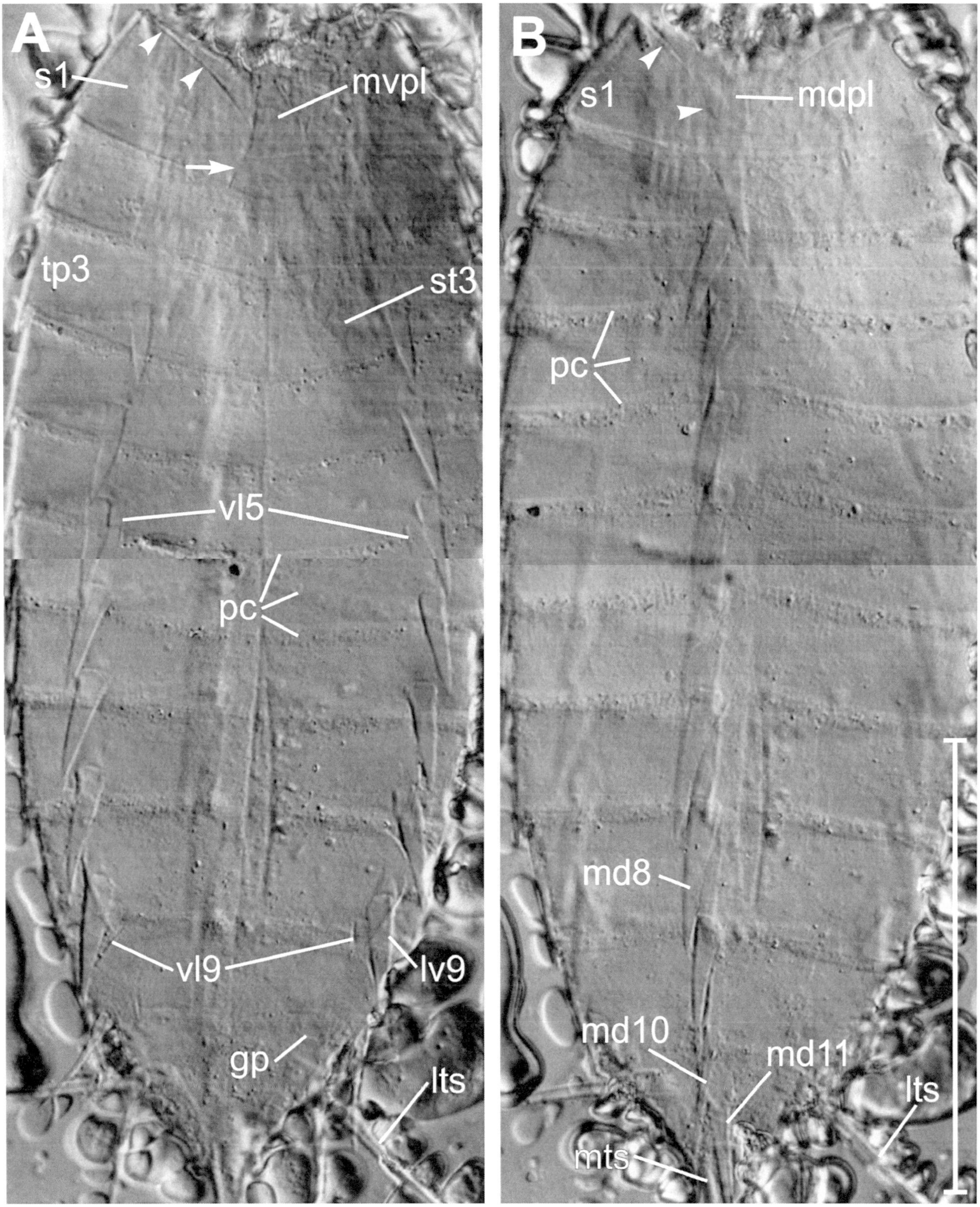

Fig. 5.1.26. LM micrographs of adult *Sphenoderes indicus*, holotypic female. A, B, habitus, ventral (A) and dorsal (B) view, arrowheads mark anterior border of segment 1, arrow points to midventral plate in segment 1, line in B = 100 µ, line valid for A and B; note anterior clamshell-like closing apparatus; notice cavities in Hoyer's embedding medium close to specimen. Original.

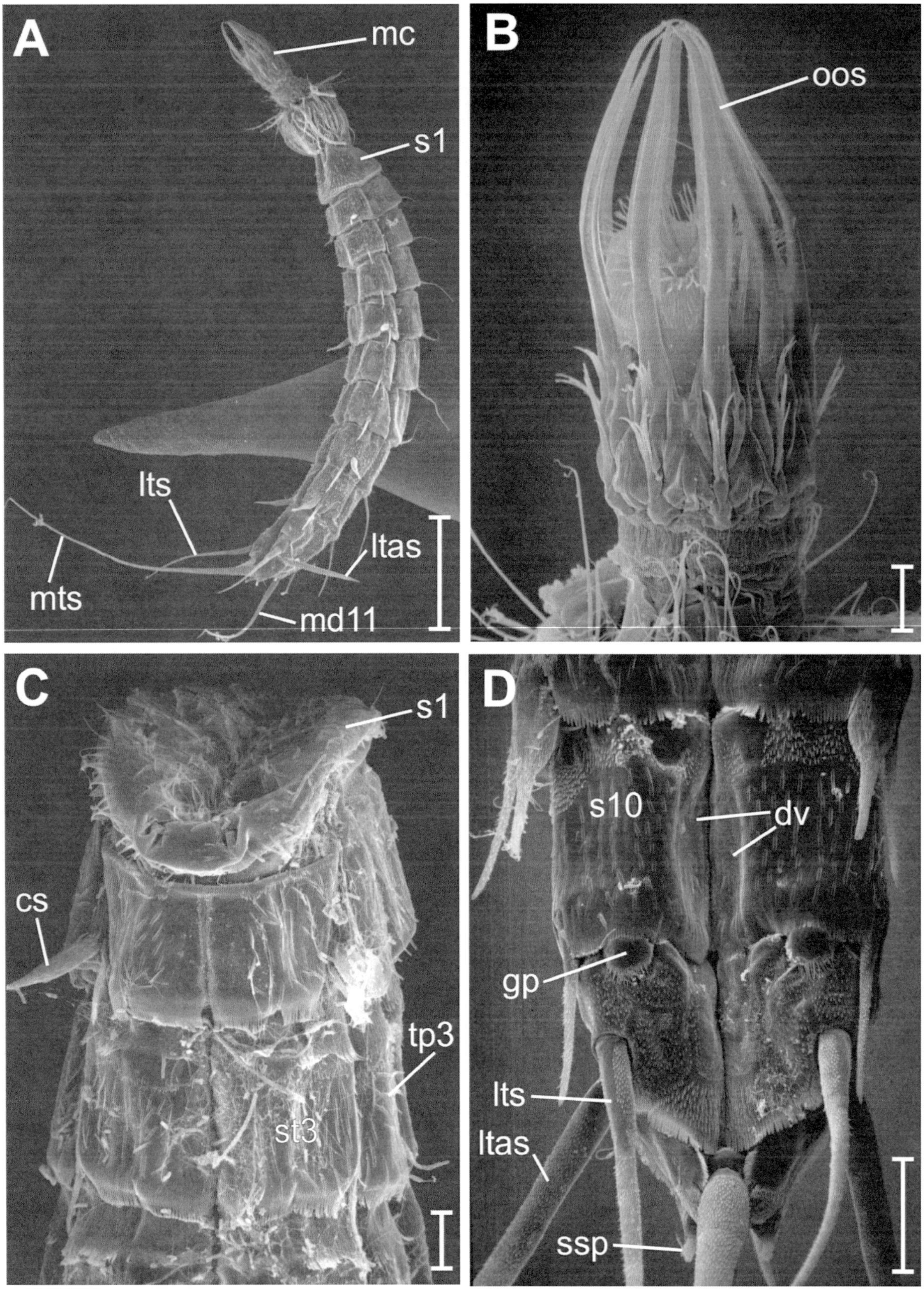

Fig. 5.1.27. SEM micrographs of adult *Triodontoderes* sp. A, habitus of male, left side, line = 100 µ; B, outer oral styles with lateral cuticular flap, ventral view, line = 100 µ; C, segments 1–3 ventral view, notice totally withdrawn segment 1, line = 10 µ; D, segments 10–11 of female with gonopores, line = 20 µ. Original.

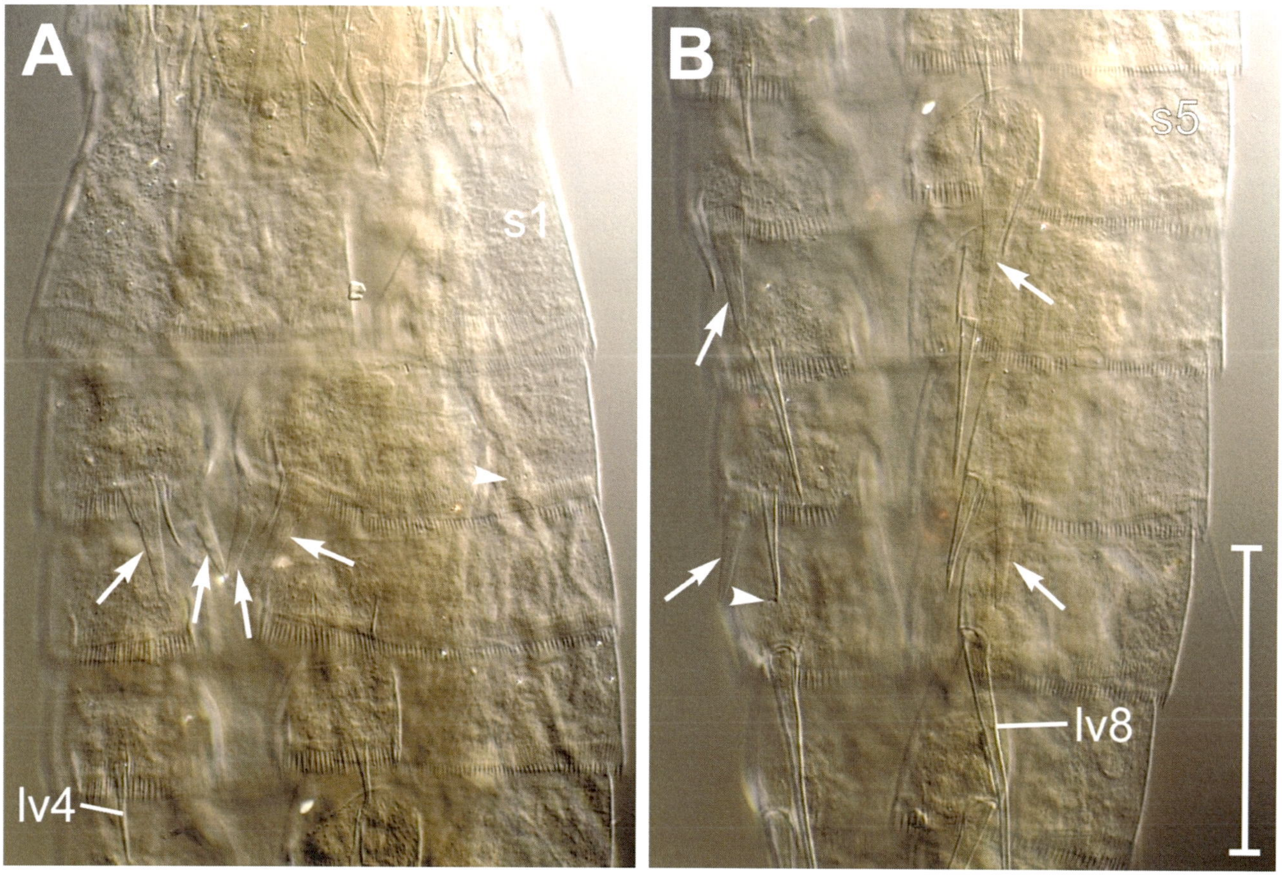

Fig. 5.1.28. LM micrographs of adult *Tubulideres seminoli*. A, B, segments 1–4 (A) and segments 5–9 (B), ventral view; arrows mark large tubuli, arrowheads point to sensory spots, line in B = 50 µ, line valid for A and B. Original.

Sørensen et al. 2010a). Whereas the number of dorsal placids is always four in Homalorhagida (Figs 5.1.33 D, 5.1.40 A, 5.2.5 D), the number of ventral placids varies even within genera between two and four (two placids: e.g., *Kinorhynchus belizensis*, *K. deirophorus*, *K. distentus*, *K. erismatus*, *K. phyllotropis*, *K. stenopygus*, *K. trisetosus*, *Pycnophyes communis*, *P. corrugatus*, *P. dolichurus*, *Pycnophyes* sp. 2; three placids: *Neocentrophyes*, *Paracentrophyes*: Fig. 5.1.33 D; four placids: e.g., *K. apotomus*, *K. giganteus*, *Pycnophyes ecphantor*, *P. emarginatus*, *P. iniorhaptus*, *P. longicornis*) (Zelinka 1928; Higgins 1969b, 1983; Brown & Higgins 1983; Sørensen et al. 2010a; Sánchez et al. 2011, submitted).

Trunk – habitus. If seen from the ventral side, Kinorhyncha reveals a spindle- or cigar-like shape in Cyclorhagida (Figs 5.1.3 B, C, 5.1.9 A, 5.1.10 B, 5.1.11 B, 5.1.14 A, 5.1.16, 5.1.26) and a more rectangular or box-like shape in Homalorhagida (Figs 5.1.33 A, B, 5.1.36 A, 5.1.37 A, 5.1.39 A). The central trunk may show a triangular (*Antygomonas*, *Condyloderes*: Fig. 5.1.14 B, *Neocentrophyes*, *Paracentrophyes*: Fig. 5.1.33 D, *Kinorhynchus*, *Pycnophyes*), more or less circular (*Zelinkaderes*), or more or less oval cross-section with the longest diameter vertically (*Centroderes*, *Semnoderes*: Fig. 5.1.25 C) or horizontally (*Echinoderes*: Fig. 5.1.22 A, *Dracoderes*). In any case, the dorsoventral muscles pull the sternal cuticular plates to a varying degree towards the interior of the trunk, resulting in a slight to deep midventral indentation. In the latter case, sternal plates may be hidden and cannot be studied morphologically. In species with a thin trunk cuticle such as *Triodontoderes* sp., the tergal plate may also be deformed and a deep cleft occurs (Fig. 5.1.27 A; Neuhaus own obs.).

The anterior margin of segment 1 is midventrally and middorsally indented (*Antygomonas*: Fig. 5.1.7 A, B) or deeply incised (*Semnoderes*: Fig. 5.1.25 B, D, *Sphenoderes*: Fig. 5.1.26), but even in the remaining Cyclorhagida (Figs 5.1.10 C, 5.1.11 B, 5.1.14 A, 5.1.16 A, 5.1.17, 5.1.20, 5.1.21, 5.1.22 A, 5.1.23 A, 5.1.25 B, D, 5.1.26, 5.1.29 A, C; Zelinka 1928; Higgins 1969a, 1983; Nebelsick 1990; Bauer-Nebelsick 1996; Pardos et al. 1998; Neuhaus & Blasche 2006; Sørensen et al. 2007, 2009, 2010c; Sørensen & Rho 2009; Sørensen & Thormar 2010; Herranz et al. 2012; Neuhaus & Sørensen 2012; Neuhaus et al. 2013). In species of *Cateria*, *Triodontoderes*, and *Zelinkaderes*, the placids

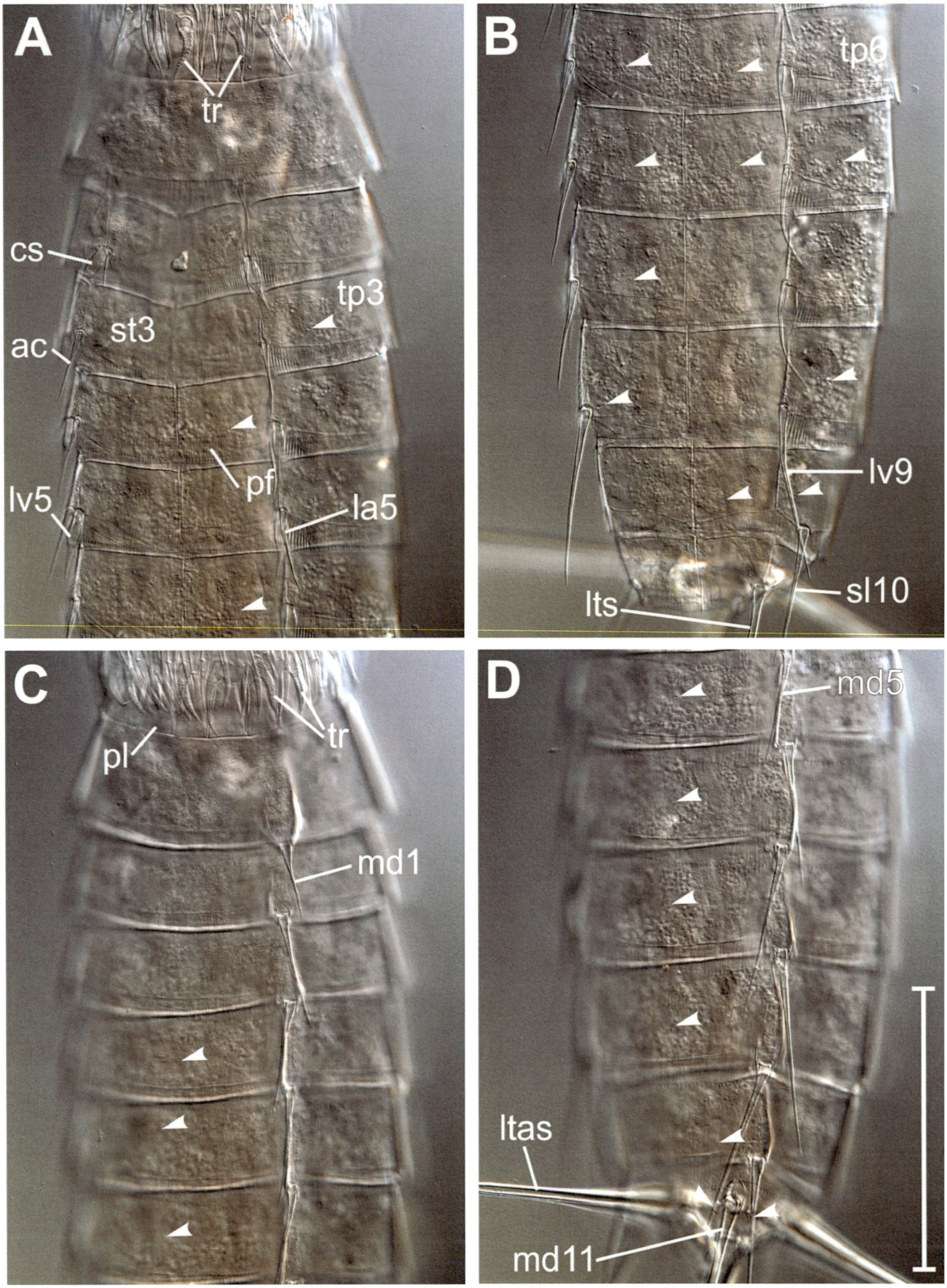

Fig. 5.1.29. LM micrographs of female *Wollunquaderes majkenae*. A, B, segments 1–6 (A) and segments 6–11 (B), ventral view; C, D, segments 1–6 (C) and segments 6–11 (D), dorsal view; arrowheads mark sensory spots; notice small scales on cuticular surface; line in D = 100 μ, line valid for A–D. Original.

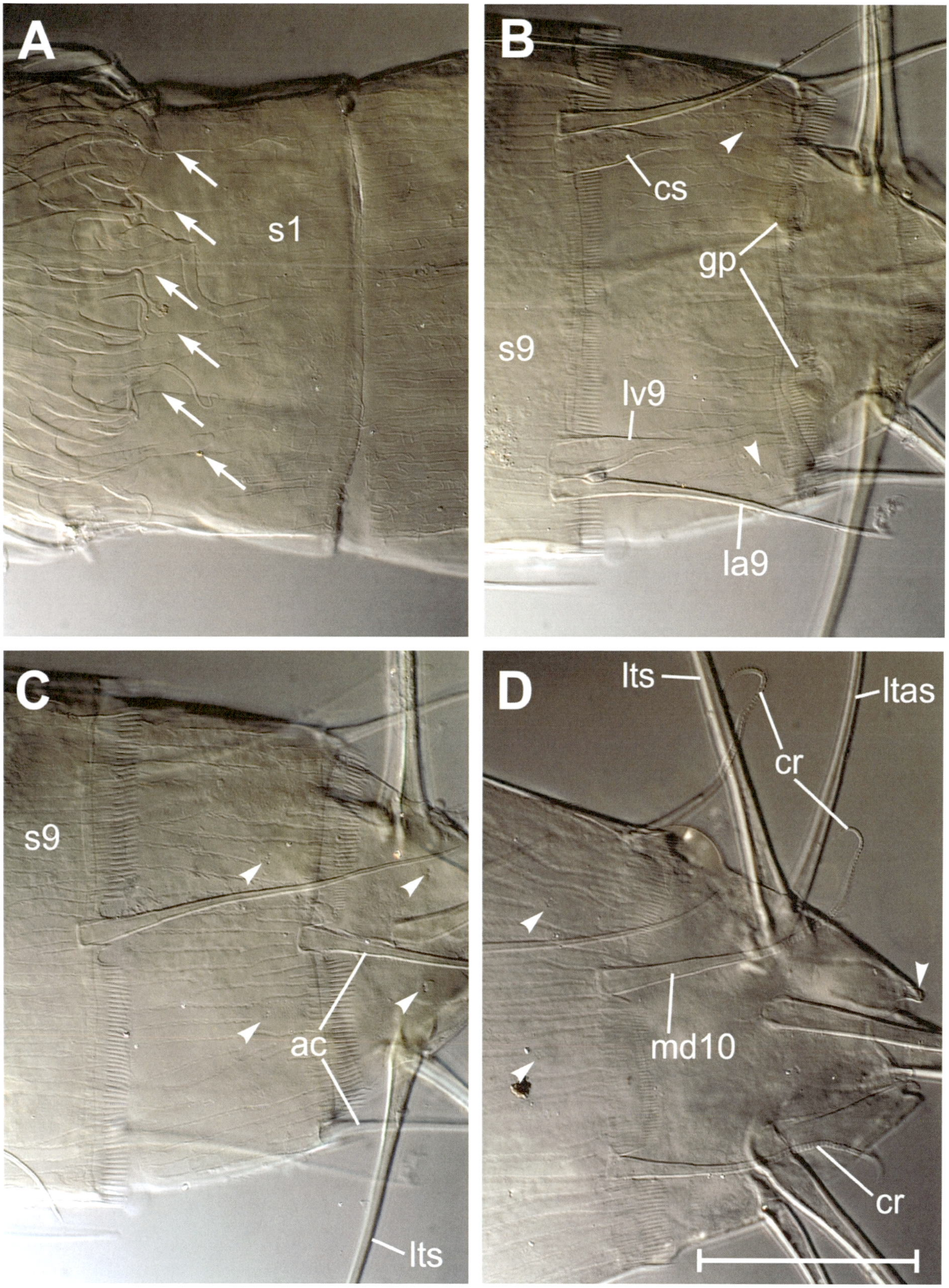

Fig. 5.1.30. LM micrographs of adult *Zelinkaderes klepali*. A, segment 1 of female, dorsal view, arrows point to lateral margin of placids fused with segment 1; B, segments 10–11 of female, ventral view; C, D, segments 10–11 of female (C) and of male (B), dorsal view; arrowheads mark sensory spots, notice regular striation pattern of trunk cuticle and female acicular (C) versus male crenulated spines (D) in segment 10; line in D = 50 µ, line valid for A–D. Original.

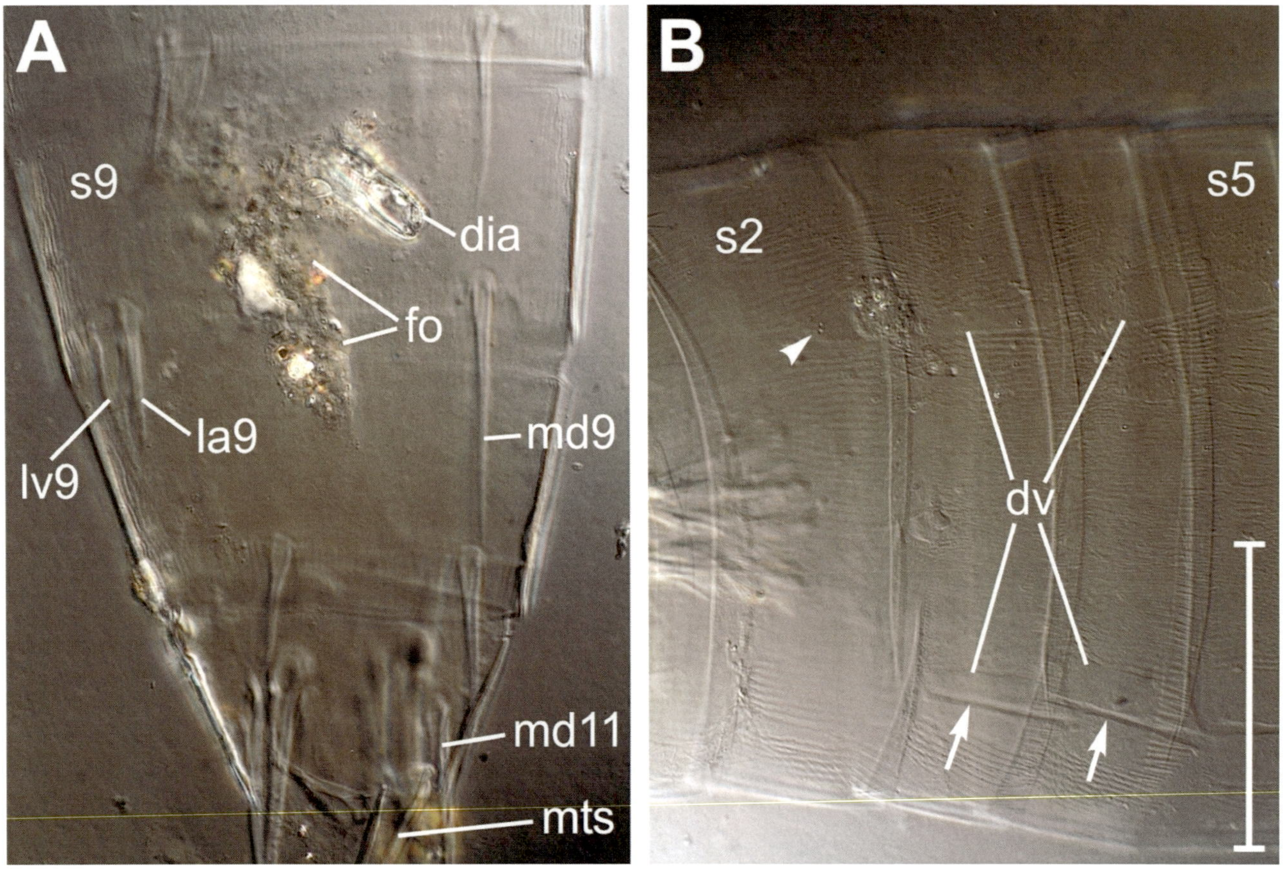

Fig. 5.1.31. LM micrographs of female *Zelinkaderes floridensis*. A, segments 9–11 with diatom shell and other remnants of food, optical section; B, segments 5–7, lateroventral view, notice ring-like cuticle in segment 2, arrows mark midventral junction of cuticular plates in segments 3 and 4; arrowheads point to sensory spots; line in B = 50 μ, line in B valid for A and B. Original.

fuse basally with the first trunk segment (see also above; Figs 5.1.8 A, 5.1.9 B, 5.1.30 A; Gerlach 1956; Higgins 1968, 1990; Bauer-Nebelsick 1995; Sørensen et al. 2007; Sørensen & Rho 2009). In species of *Triodontoderes* and *Zelinkaderes*, the placids and a large part of segment 1 may be withdrawn together with the introvert (Fig. 5.1.27 C). In Homalorhagida, the anterior margin of the tergal plate of the first trunk segment may be even, scalloped (Fig. 5.2.5 D), dentate, denticulate (Fig. 5.1.40 A), coronate, serrate, or possess a reticulate sculpturing (Fig. 5.1.40 A; Higgins 1983; Adrianov & Malakhov 1999a; Sánchez et al. submitted).

Trunk – cuticular plates. The cuticle of each segment appears either as a ring or separated into one to up to four ventral (= sternal) and dorsal (= tergal) plates (Figs 5.1.3, 5.1.30, 5.1.31 B, 5.1.32 A, 5.1.36 A, B; Tabs 8, 9). Trunk segments 1 and 2 vary considerably in their shape (segment 1) and composition of cuticular plates, whereas segments (2)3–10 reveal one tergal and two sternal plates in most genera. In *Cateria*, segments 1–6 (*C. styx*) or 1–7 (*C. gerlachi*) show one tergal and one midsternal plate, segments 7(8)–10 a single tergal plate joining midventrally, and segment 11 is ring-like; deep notches occur laterally where spines originate and ventromedially in segments 7–10 (Fig. 5.1.8; Gerlach 1956; Higgins 1968). In species of *Triodontoderes*, segment 1 seems to be ring-like with weak tergosternal junctions, segments 2–4 exhibit two sternal and one tergal plate, but there is only a single tergal plate in segments 5–11 (Sørensen & Rho 2009; Neuhaus own obs.). In segments 3–10 of *Zelinkaderes* and in segment 2 of *Tubulideres*, a single tergal plate is found articulating midventrally (Figs 5.1.30 B, 5.1.31 B; Higgins 1990; Bauer-Nebelsick 1995; Sørensen et al. 2007).

The cuticle is ring-like in segments 1 and 2 of species of *Echinoderes* and *Zelinkaderes* (Figs 5.1.3 B, 5.1.20 A, C, 5.1.21 B, 5.1.24 A, 5.1.30 A, 5.1.31 B; Zelinka 1928; Higgins 1990; Bauer-Nebelsick 1995; Sørensen et al. 2007). Mature specimens of *E. angustus*, *E. eximus*, *E. setiger*, and *E. tubilak* reveal a strong, midventral, cuticular reinforcement in segment 2 (Fig. 5.1.20 C, D; Neuhaus & Blasche 2006), which is missing in the young adult stage (Fig. 5.1.20 A). Species of several genera possess a habitus very similar

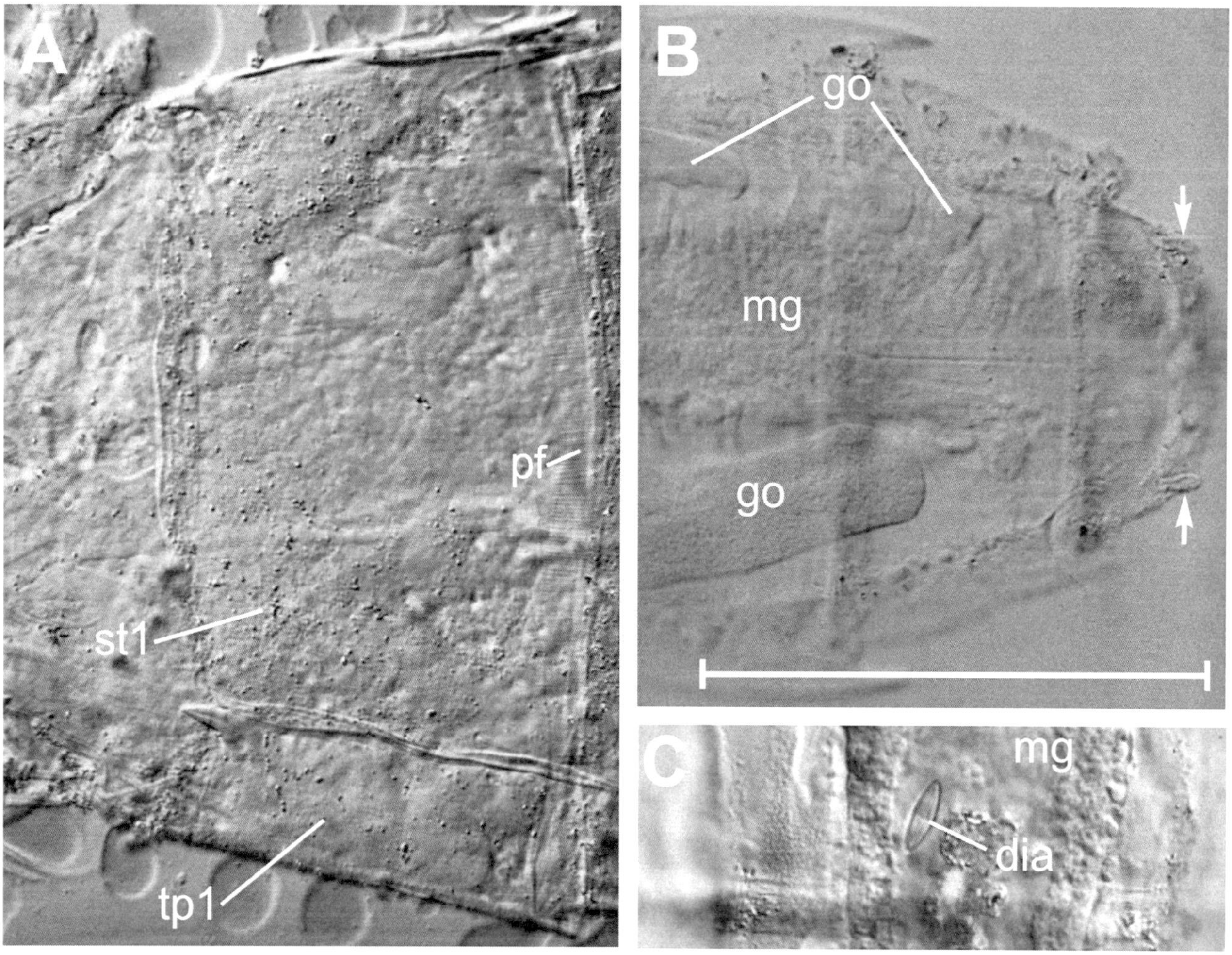

Fig. 5.1.32. LM micrographs of adult *Neocentrophyes*. A, segment 1 of male *N. intermedius*, ventral view; B, C, female *N. satyai*, segments 9–11 (B) and segment 8 with diatom in midgut (C), ventral view (B), optical section (C), arrows mark anlagen of lateral terminal spines in B, line in C = 100 µ, line valid for A–C. Original.

to *Echinoderes*, and only careful observation will allow differentiation of the genera. Although species of *Meristoderes* reveal a ring-like 2nd segment, a partial or even complete lateral tergosternal articulation and a vaguely visible midsternal junction can be found (Herranz et al. 2012; Sørensen et al. 2012c). This character is only recognizable with LM or even only with SEM in favourite specimens (Sørensen et al. 2012c). More thorough investigations will probably result in the finding of all degrees of expression of the tergosternal junctions and of species erroneously placed in *Echinoderes* instead of *Meristoderes*. Species of *Cephalorhyncha* appear to have a tergal and a single sternal plate with a barely recognizable midventral junction in segment 2 (Fig. 5.1.17 A; Adrianov & Malakhov 1999a; Neuhaus & Blasche 2006; Sørensen 2008b). Last but not least, species of *Dracoderes*, *Fissuroderes*, and *Polacanthoderes* show a tergal and two sternal plates in segment 2 and in *Dracoderes* yield alternating paradorsal instead of middorsal spines (Figs 5.1.16 A, B, 5.1.17 B, D; Neuhaus & Blasche 2006; Sørensen 2008b; Sørensen et al. 2012a).

The most peculiar situation occurs in species of *Semnoderes* and *Sphenoderes*, which express a clamshell-like closing apparatus. In *Semnoderes*, the first segment is deeply incised, and a narrow, elongate midventral and middorsal placid is located in the slit (Fig. 5.1.25 B, C; Zelinka 1928; Sørensen et al. 2009). In *Sphenoderes*, the indentation is considerably broader midventrally and filled by a large triangular placid (Fig. 5.1.26 A, B). Conflicting evidence is available as to whether a midventral and middorsal cuticular plate occurs (*S. indicus*, two *Sphenoderes* sp.) or not (*S. poseidon*)(see also below; Higgins 1969a; Sørensen et al. 2010c). Weakly developed tergosternal junctions have been reported for segment 1 of *Antygomonas paulae*, *Triodontoderes anulap*, and *Sphenoderes indicus* (here also tergotergal junctions; Fig. 5.1.26 A) but have not been confirmed for *A. incomitata*, *A. oreas*, and *S. poseidon* (see Higgins 1969a; Nebelsick 1990;

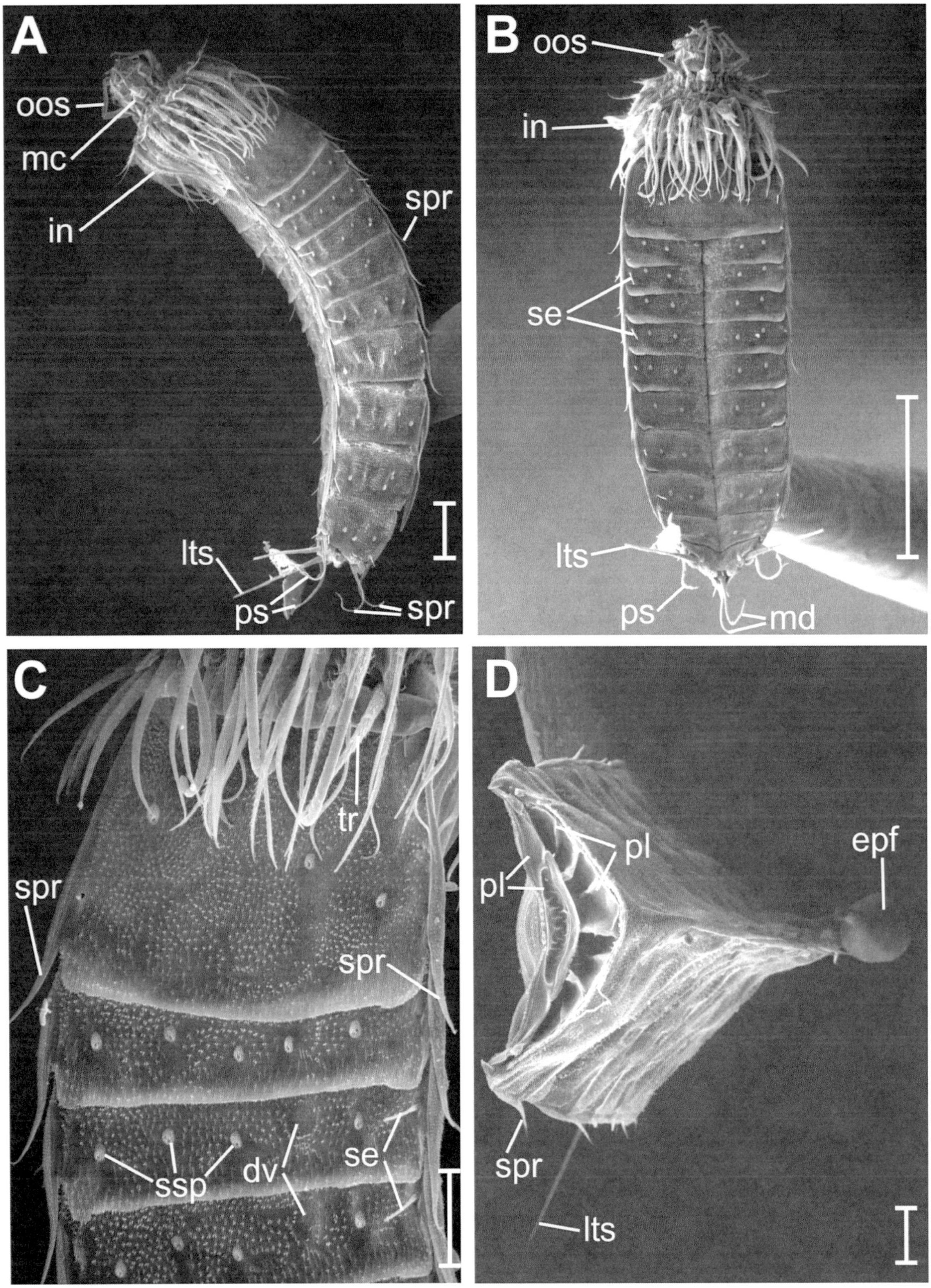

Fig. 5.1.33. SEM micrographs of adult *Paracentrophyes praedictus*. A, B, habitus of male, left side (A) and ventral view (B), line in A = 40 μ, line in B = 100 μ; C, segments 1–4, right side, notice sensory spots elevated above the trunk surface and middorsal and lateroventral spinose processes, line = 20 μ; D, habitus, frontal view, ventral side to the left, notice triangular cross-section and epizoic protozoan, line = 20 μ. Original.

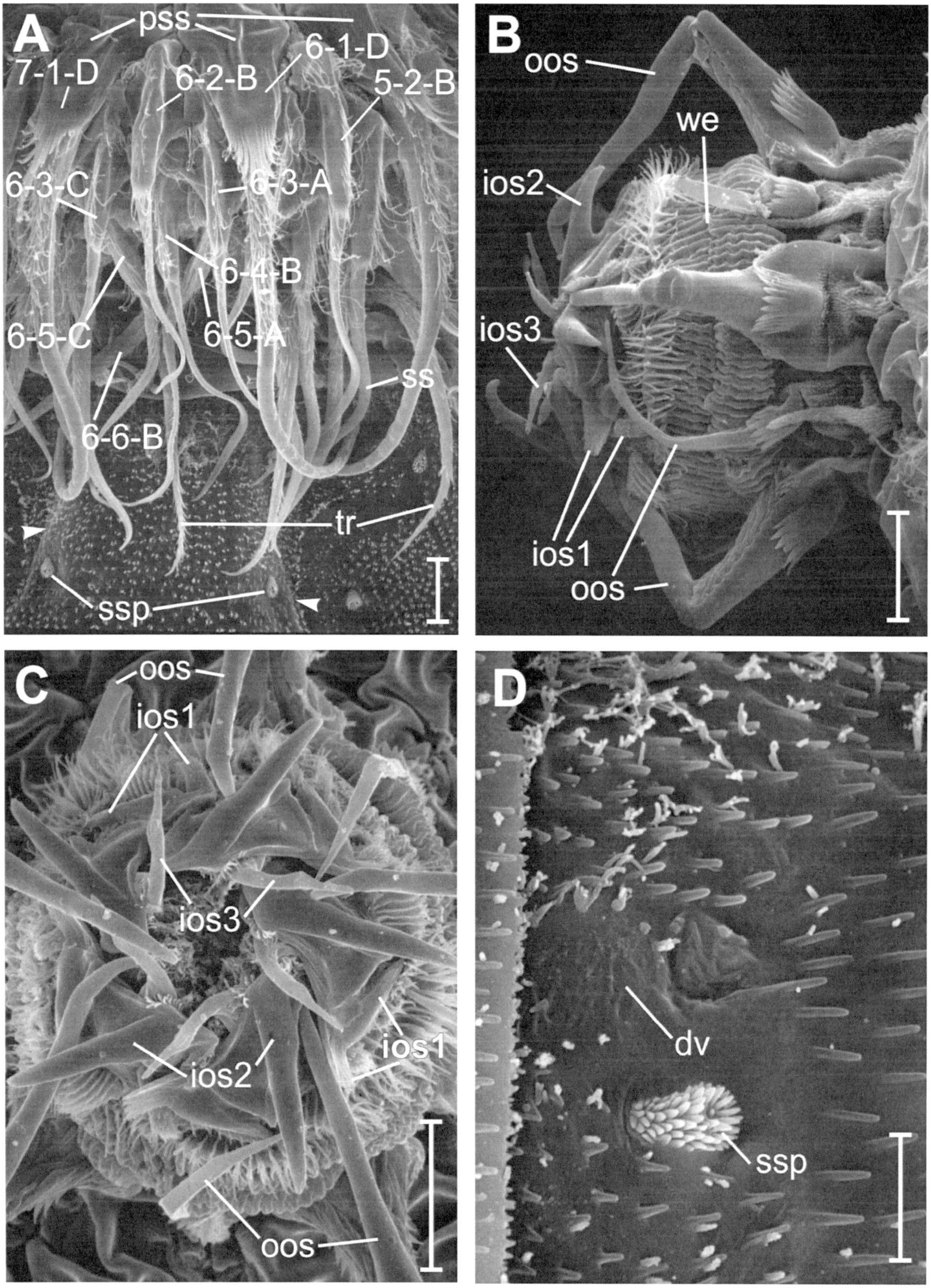

Fig. 5.1.34. SEM micrographs of adult *Paracentrophyes praedictus*. A, introvert and segment 1 of male, ventral view, notice indication of midsternal cuticular plate in segment 1 (arrowheads), labelling of scalids refers to introvert sector-ring-subsector according to Fig. 5.2.2, line = 20 µ; B, protruded mouth cone, right side, notice outer oral styles of alternating length, line = 10 µ; C, inner oral styles in protruded mouth cone, frontal view, line = 10 µ (after Neuhaus 1995; © John Wiley and Sons); D, sensory spot elevated above surface with sensory cilium jutting out of opening in right sternal plate of segment 2, ventral view, line = 4 µ. Original.

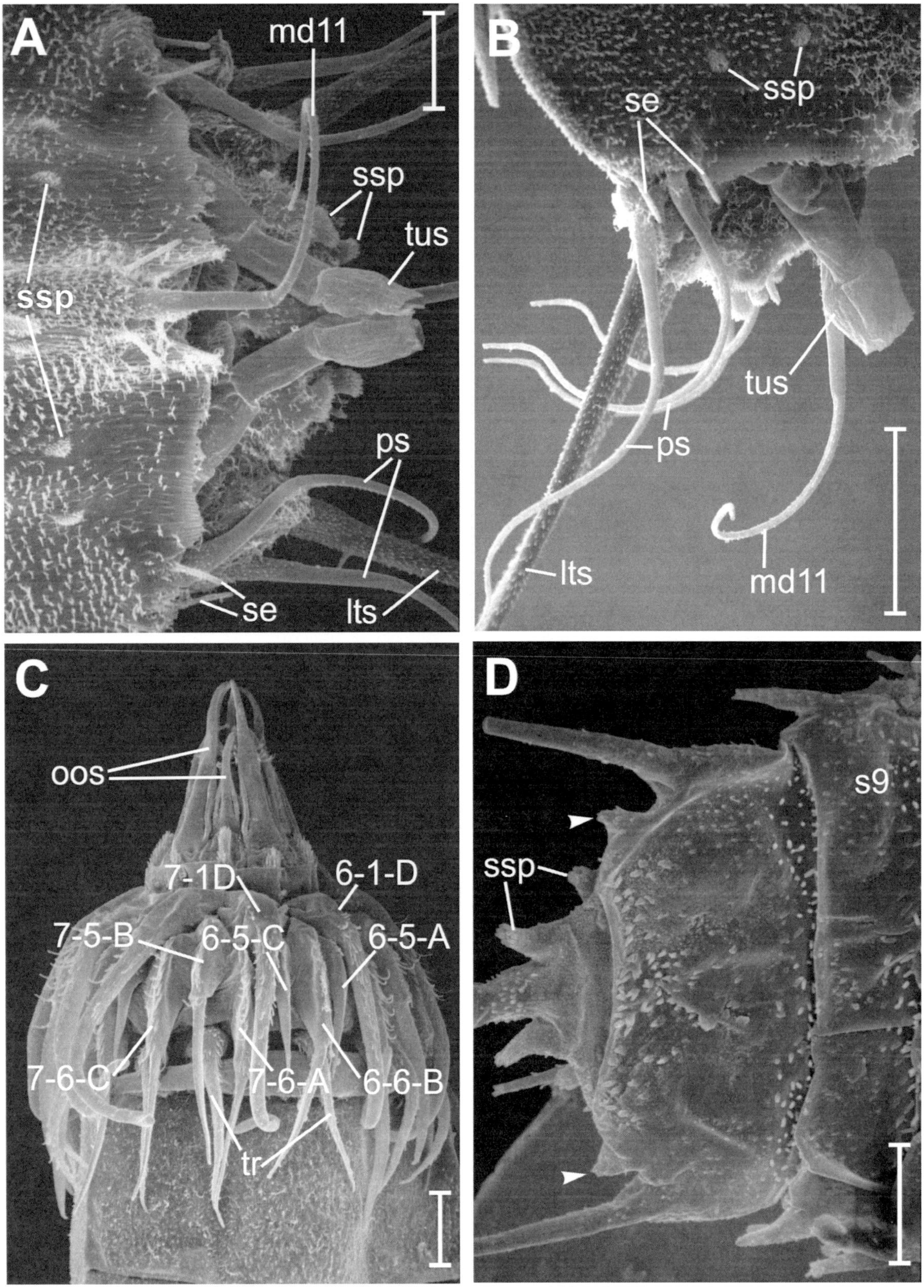

Fig. 5.1.35. SEM micrographs of *Paracentrophyes praedictus*. A, B, segment 11 of male with tubular structure possibly related to copulation, dorsal view (A) and left side (B), line in A = 10 µ, line in B = 20 µ; C, head and segment 1 of J-2 with spinose anlagen of scalids 6-5-A and 6-5-C, ventral view, notice outer oral styles of alternating length, labelling of scalids refers to sector-ring-subsector according to Fig. 5.2.2, line = 10 µ; D, anlagen of lateral terminal spines as spinose processes (arrowheads) of J-2, ventral view, line = 10 µ. Original.

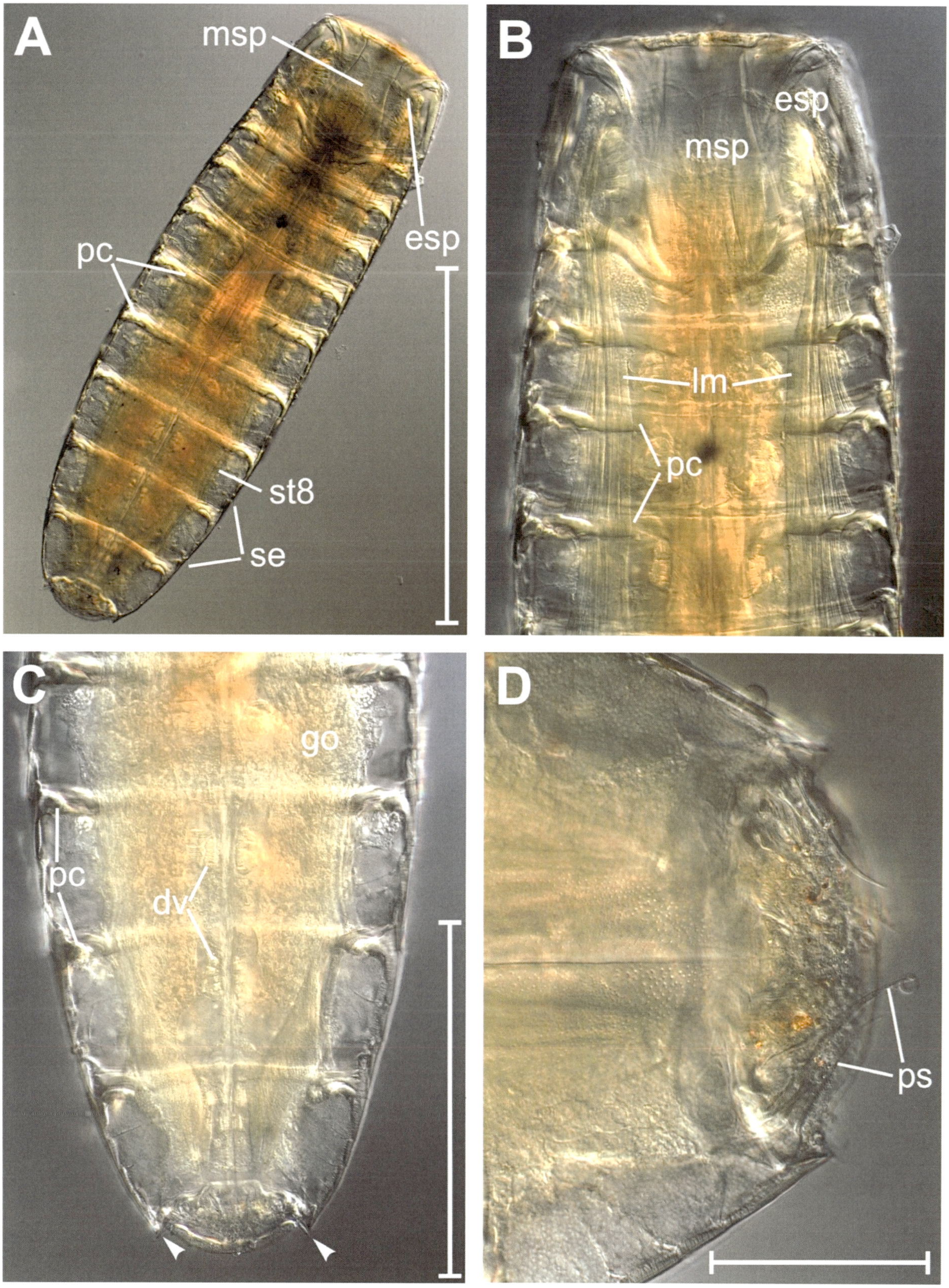

Fig. 5.1.36. LM micrographs of male *Kinorhynchus phyllotropis*. A, habitus, ventral view, line = 400 µ; B, C, segments 1–5 (B) and segments 7–11 (C) with ventral longitudinal muscles (B) and gonads (C), optical section, arrowheads mark anlagen of lateral terminal spines in C, line in C = 200 µ, line valid for B and C; D, segment 11 with penile spines, ventral view, line = 50 µ. Original.

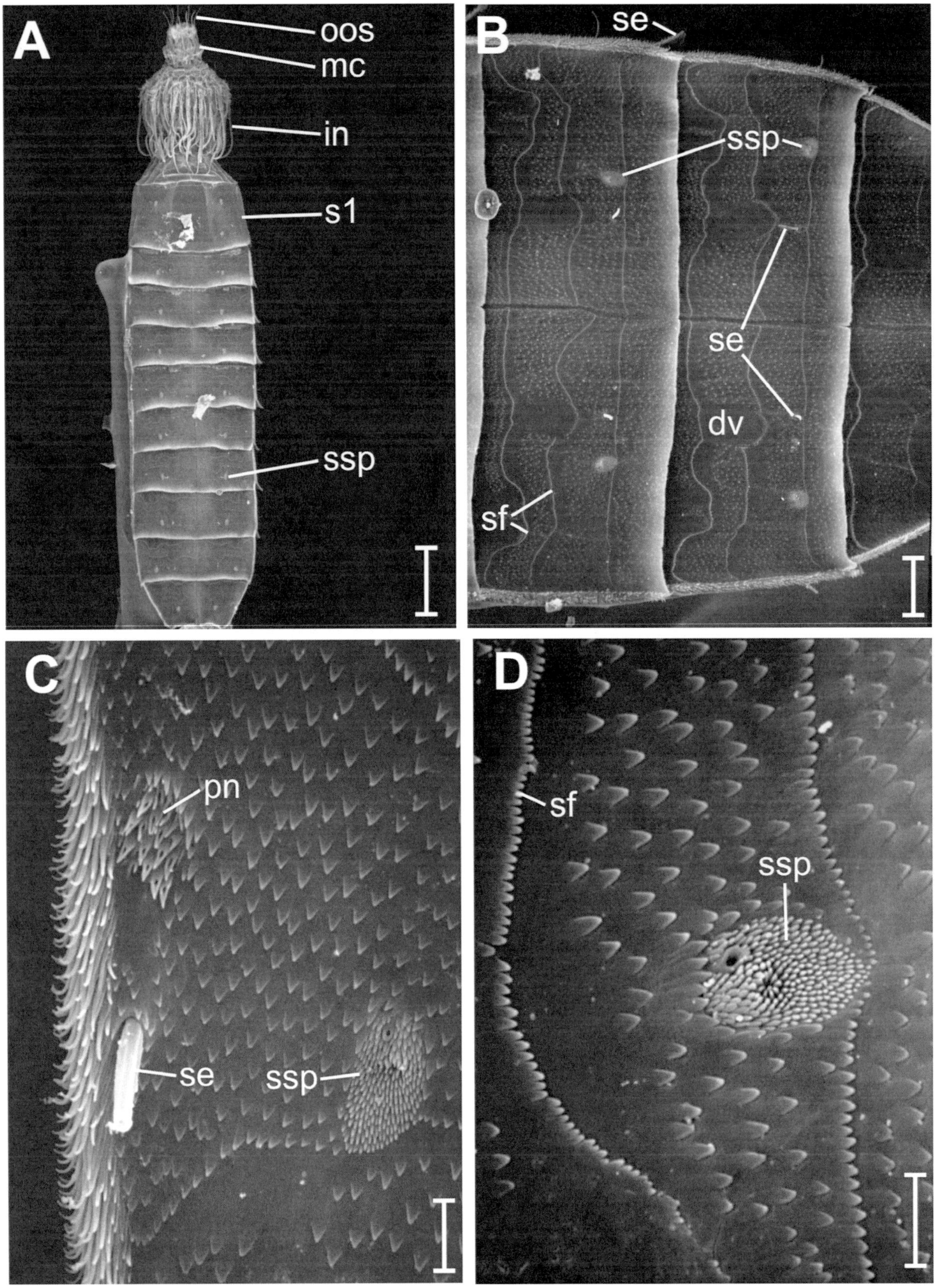

Fig. 5.1.37. SEM micrographs of adult *Kinorhynchus* cf. *erismatus*. A, habitus, ventral view, line = 100 µ; B, segments 8–10, ventral view, line = 20 µ; C, segment 9, left side, notice protonephridium openings at tip of small tubes, seta, and sensory spot, line = 4 µ; D, left sternal plate of segment 4, ventral view, notice sensory spot, line = 4 µ. Original.

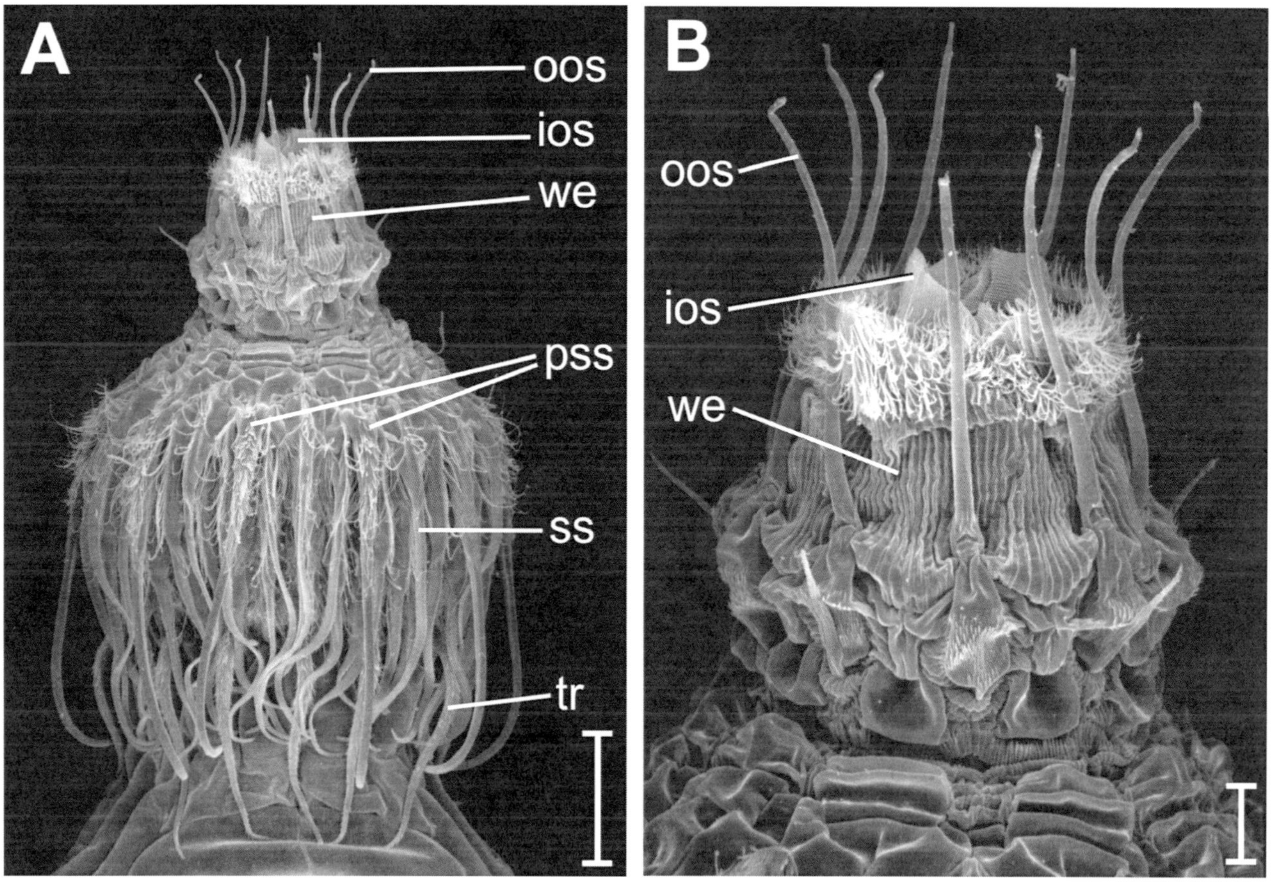

Fig. 5.1.38. SEM micrographs of adult *Kinorhynchus* cf. *erismatus*, ventral view. A, introvert everted, line = 40 μ; B, mouth cone protruded, line = 10 μ. Original.

Bauer-Nebelsick 1996; Sørensen 2007; Sørensen & Rho 2009; Sørensen et al. 2010c). Possibly, contraction of the dorsoventral muscles in segment 1 may be responsible for the impression of junctions between cuticular plates. This phenomenon is well known for juvenile stages (see section 5.2 Reproduction and development, Development).

Only in *Cephalorhyncha asiatica, C. nybakkeni, Echinoderes cernunnos, E. eximus, Fissuroderes papai, Meristoderes elleae, M. glaber, M. imugi,* and *Meristoderes* sp. does the terminal segment consist of two sternal plates and two instead of one tergal plates (Neuhaus & Blasche 2006; Sørensen et al. 2012b, c). This unusual combination of cuticular plates in segment 11 tempts Sørensen et al. (2012b) to vaguely consider a closer potential relationship of the species mentioned before, although the authors are aware that this condition is in conflict with the organization of the 2nd trunk segment defining the genera *Cephalorhyncha, Echinoderes, Fissuroderes,* and *Meristoderes* nowadays.

In Homalorhagida, segment 1 reveals a tergal plate and a single sternal plate in *Neocentrophyes* (Fig. 5.1.32 A) and in *Paracentrophyes*, where it is indistinctly separated anteriorly into three plates (Fig. 5.1.34 A; see also previous discussion about such junctions in Cyclorhagida); a tergal, a midsternal and two episternal plates are found in *Kinorhynchus* and *Pycnophyes* (see also above; Fig. 5.1.36 A, B; Zelinka 1928; Higgins 1969b, 1983; Sørensen et al. 2010a).

Trunk – cuticular surface. The cuticle of segments 1–10 overlaps the subsequent segment partly as a cuticular flap (Figs 5.1.15 A, 5.1.40 D). The flap begins at the intersegmentary joint line (Fig. 5.1.3 B; Thormar & Sørensen 2010; Herranz et al. 2012) and ends posteriorly in the primary fringe which may be strongly pectinate (Figs 5.1.14 D, 5.1.17 D, 5.1.18 C, 5.1.20 C, D, 5.1.21 B, 5.1.22 B, 5.1.23 B-D, 5.1.24 A, B), reveal individual spinose cuticular processes (e.g., *Campyloderes*), or possess postmarginal spiculae in juvenile Cyclorhagida (Fig. 5.1.7 D; Tab. 1). The cuticle of the terminal segment may end in sternal and tergal spinose, blade-like, or truncate extensions. At the anterior margin of a segment, a secondary fringe may exist as one to few linear secondary fringes of very small scales, which may also be arranged in one to several wavy lines (Figs 5.1.7 A, 5.1.8 B, 5.1.9 C, 5.1.12 D, 5.1.22 B, 5.1.24 B, 5.1.37 B, D). Other, often species-specific cuticular sculpturing may occur such as bracteate hairs (Figs 5.1.21 B,

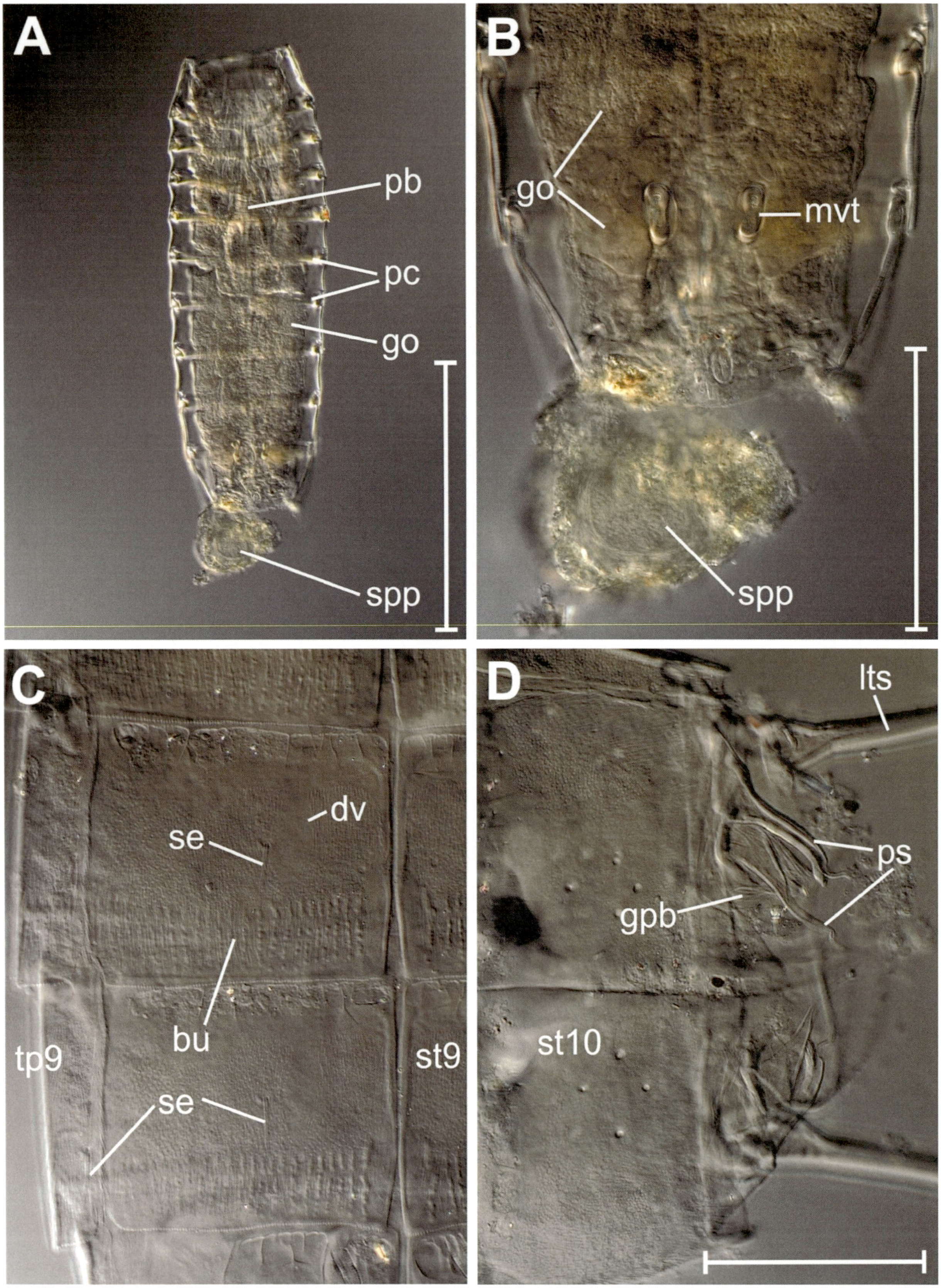

Fig. 5.1.39. LM micrographs of adult *Pycnophyes kielensis*. A, habitus of female with spermatophore, optical section, line = 300 μ; B, segments 10–11 with partly detached spermatophore, line = 100 μ; C, right sternal plates of segments 8–9 with regularly arranged button-like structures, ventral view; D, segments 10–11 with penile spines, ventral view; line in D = 50 μ, line valid for C and D. Original.

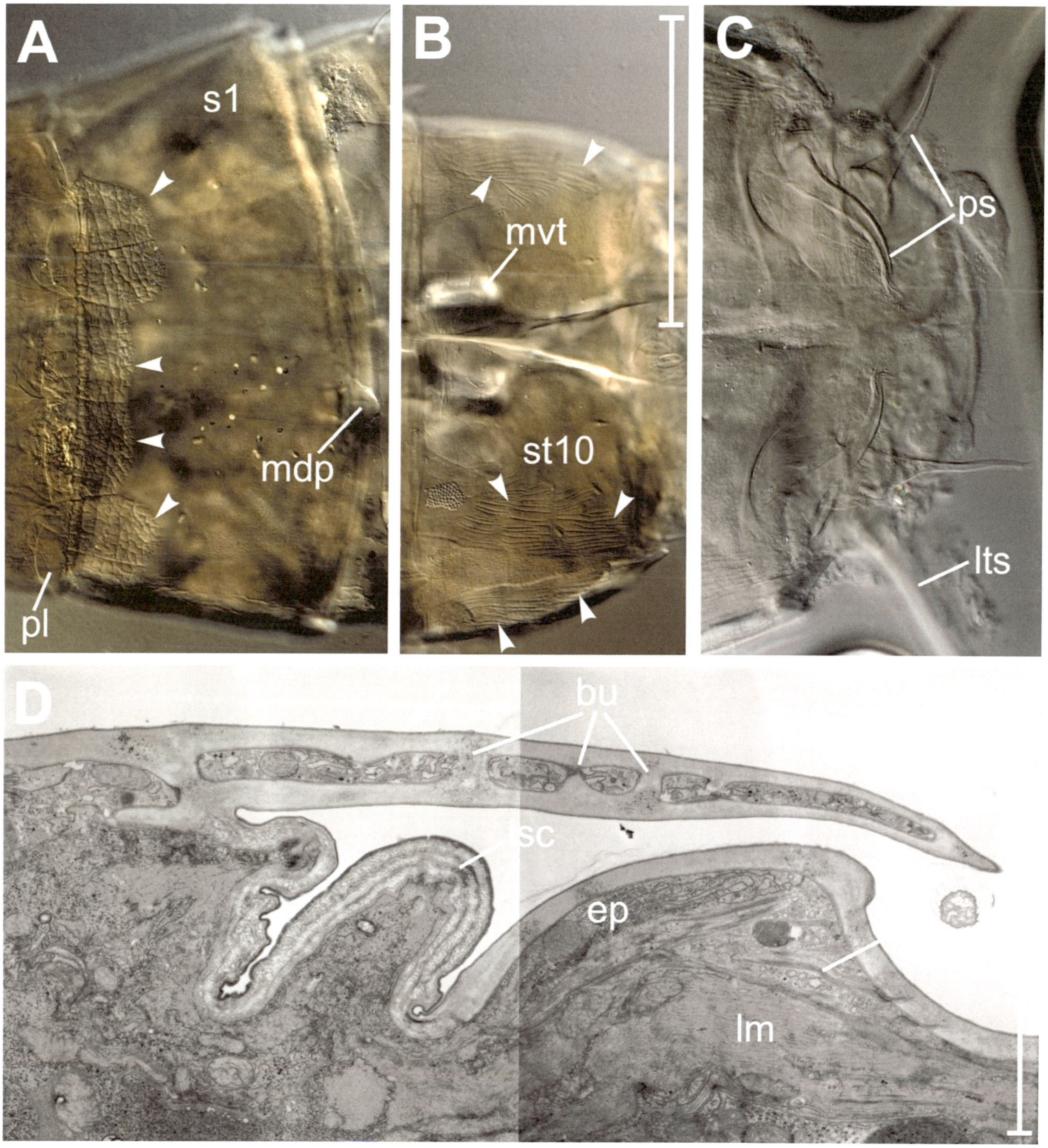

Fig. 5.1.40. LM (A-C) and TEM (D) micrographs of adult *Pycnophyes dentatus*. A, segment 1 with reticulate ridges (arrowheads), dorsal view; B, segment 10 with ventral and lateral cuticular ridges (arrowheads), ventral view; C, segments 10–11 with penile spines, ventral view; D, longitudinal section through dorsal cuticular flap (arrows) of segment 6 of young male with button-like structure cuticular structures; line in B = 100 μ, line valid for A and B; line in C = 50 μ; line in D = 5 μ. Original.

5.1.22 B, D, 5.1.23, 5.1.24 A, B, 5.1.25 B, D), long bracteate hairs (Fig. 5.1.17 C), regularly arranged bifid cuticular bristles giving the appearance of a longitudinal cuticular striation in species of *Zelinkaderes* (Fig. 5.1.30), bristles (Fig. 5.1.11 B), scales (Figs 5.1.11 B, C, 5.1.12, 5.1.28, 5.1.29, 5.1.34 D, 5.1.37 C, D), punctuate sculpturing, cuticular ridges (Fig. 5.1.40 B), reticulate ridges (Fig. 5.1.40 A), and hexagonal sculpturing (Fig. 5.1.9 C) (Tab. 1).

Older specimens of species of *Echinoderes* and of *Pycnophyes* reveal a species-specific pattern of "pores" in light microscopy (Figs 5.1.20 C, 5.2.5 D, 5.8.2 B; e.g., Higgins 1983; Neuhaus 1993; Pardos et al. 1998; Sørensen

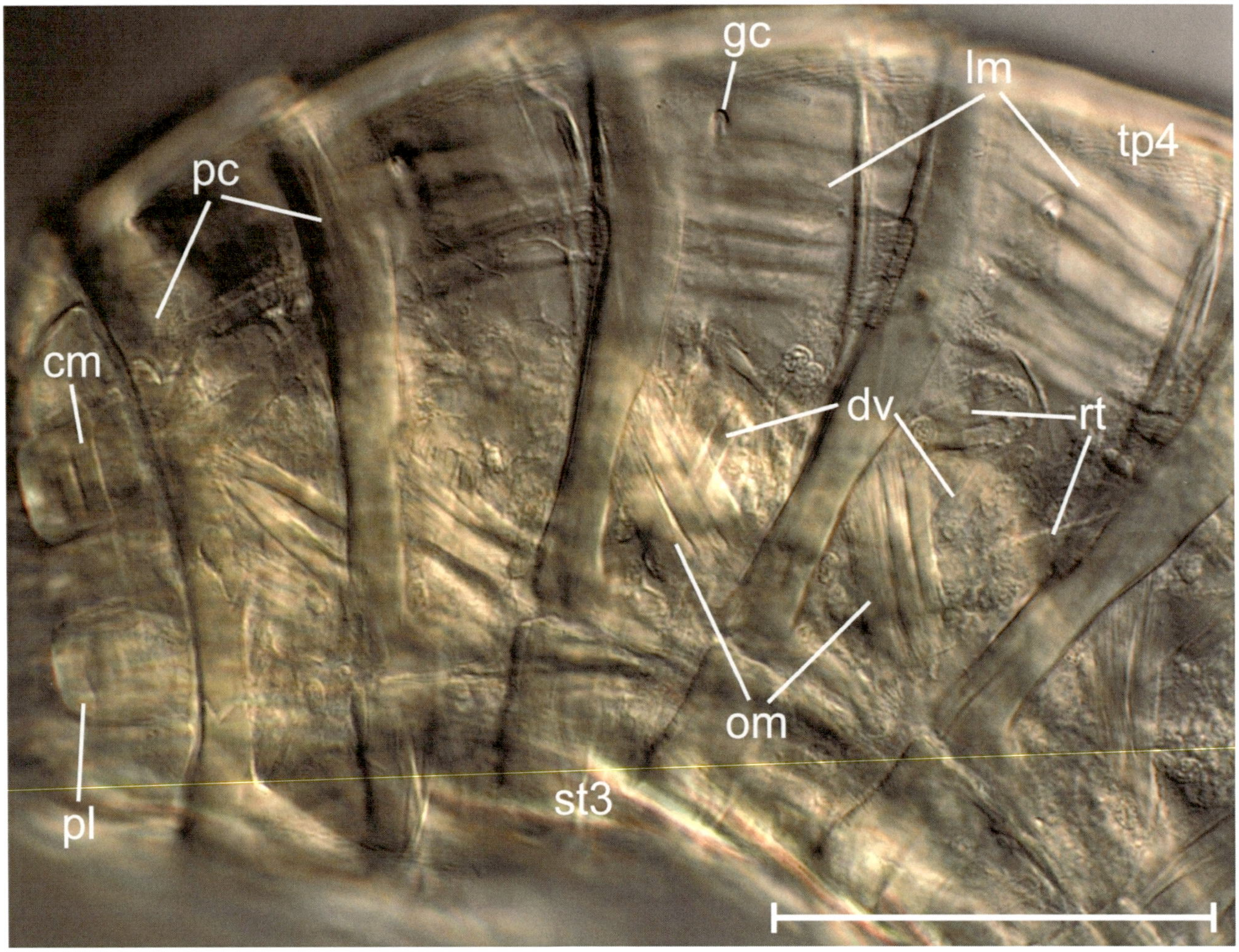

Fig. 5.1.41. LM micrographs of male *Campyloderes* cf. *vanhoeffeni*. Segments 1–4 with muscle system, right side, image mirrored; line = 50 µ. Original.

2006; Lundbye et al. 2011), which are interpreted as "perforation sites with cuticular hairs" by Higgins (1983). Some of these hairs at the perforation sites are bracteate cuticular hairs (Sørensen 2006). It seems that small epidermal processes penetrate the trunk cuticle at these "pores" almost completely (GaOrdóñez et al. 2000), but TEM data is scarce (Kristensen & Hay-Schmidt 1989; Kristensen & Higgins 1991).

Trunk – sensory spots. Kinorhyncha possess a species-specific pattern of bilateral symmetrically arranged sensory spots on the trunk plates. However, asymmetrical distribution and variation occur and have been published for *Campyloderes* cf. *vanhoeffeni*, *Centroderes spinosus*, *Paracentrophyes praedictus*, *Pycnophyes dentatus*, and *P. kielensis* (Neuhaus 1993, 1995; Neuhaus & Sørensen 2012; Neuhaus et al. 2013).

Sensory spots were not recognized as such by early scholars of Kinorhyncha, but noted as paradorsal, thinned, cuticular areas in *Campyloderes vanhoeffeni* by Zelinka (1913, pp. 433, 436, Pl. 39: Fig. 1: "verdünnte Stellen im Hautskelett") or as paradorsal cavities of the dorsal spinose process in *Kinorhynchus giganteus*, *Pycnophyes communis*, *P. ponticus*, and *P. robustus* (1928: "Hohlraum der Rückenzacken"). In *P. aulacodes*, the atria of the paradorsal sensory spots have been described as "butterfly-like structures" and "subcuticular atria" (Sánchez et al. 2011: p. 330, Fig. 8 A, C). The images shown do not favour a subcuticular but an intracuticular though sub-surface position of the atria. Generally, a sensory spot appears in SEM as a round to oval or droplet-like area with one to two and occasionally three pores and numerous cuticular micropapillae, which may be reduced in species of the Echinoderidae. A cilium may or may not jut out of a pore. An intracuticular conical to flask-shaped atrium or a subcuticular cuticular tube with or without a basal, funnel-like extension are usually found and relatively easy to observe in light microscopy.

In type-2 but not in type-1 sensory spots (Fig. 5.1.23 C), one of the pores is located at the tip of a very

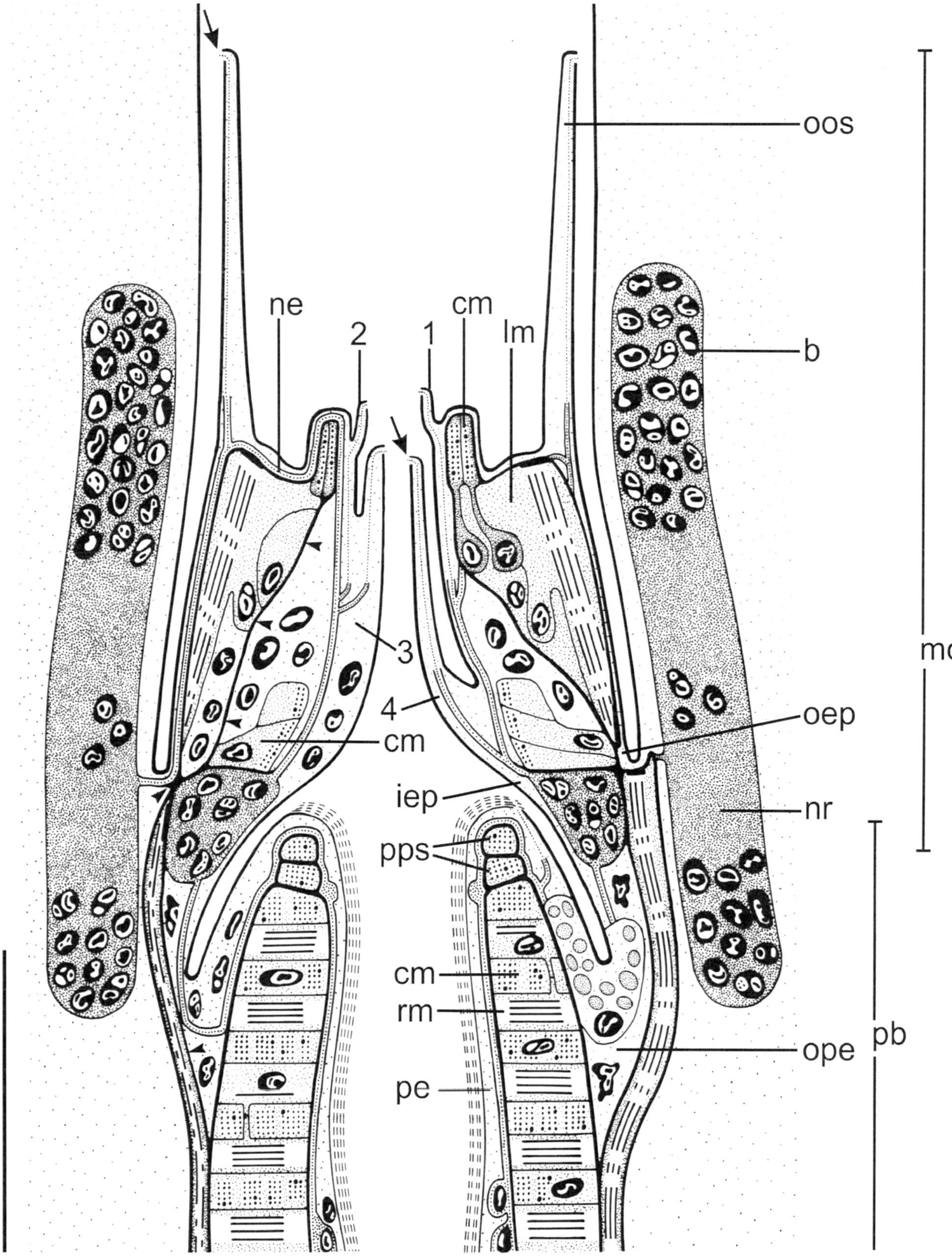

Fig. 5.1.42. Reconstruction of the brain, mouth cone, and anterior pharynx of *Zelinkaderes floridensis*, longitudinal section; notice innervation of the pharynx through the mouth cone; arrowheads point to extracellular matrix separating outer and inner mouth cone epithelium; number refer to inner oral styles of rings 1–4; arrows mark subapical opening of oral styles, (after Neuhaus 1994; © Verlag Dr. Friedrich Pfeil), line = 20 μ.

short tube whereas the other pore is encircled by a ring of micropapillae (Figs 5.1.11 B, 5.1.12, 5.1.36 C, D; Tab. 1). Type-3 sensory spots appear at the tip of a cone-like basis especially on the terminal trunk segment or in juvenile *Antygomonas incomitata, Condyloderes* sp., *Paracentrophyes praedictus, Pycnophyes dentatus*, and *P. kielensis* (Figs 5.1.14 D-F, 5.1.33 C, 5.1.34 D; Tab. 1; Adrianov et al. 1989, 1990a; Adrianov & Malakhov 1994; Kristensen & Higgins 1991; Nebelsick 1992a; Neuhaus 1993, 1995; Sørensen et al. 2009, 2010a-d, 2012a; Neuhaus & Sørensen 2012; Neuhaus et al. 2013). This type of sensory spots has also been named flosculum, because it resembles the sensory organs of Priapulida and Loricifera (Adrianov et al. 1989, 1990a). Type-4 sensory spots consist of a circular patch of numerous micropapillae with one slightly elongate tube with a pore (Sørensen et al. 2009). In type-5 sensory spots, the number of micropapillae are reduced to one ring of few micropapillae around a central pore (Figs. 5.1.21 D, 5.1.22 B, C; this book chapter). This organ has been mentioned for *E. intermedius* and named *Nanaloricus*-flosculum, because a similar receptor organ is known from the loriciferan genus with the same name (Sørensen 2006). It also appears in several undescribed species of *Pycnophyes* associated with two gland cell outlets outside the ring of micropapillae (Sánchez, pers. obs.). From *Condyloderes megastigma* and *Condyloderes* sp., a female-specific, huge, potential sensory spot is known (Fig. 5.1.15 B; Sørensen et al. 2010b).

Trunk – gland cell outlets. Kinorhyncha reveals a species-specific pattern of gland cells; at least in *Echinoderes cantabricus* and *E. hispanicus* two types of gland cells exist (Nebelsick 1992b; GaOrdóñez et al. 2000; Neuhaus & Sørensen 2012; Neuhaus et al. 2013). Gland cells may not be recognizable by light microscopy, because (1) secretions have been released during fixation, (2) preservation may be insufficient, or (3) specimens have been mounted in Hoyer's mounting medium, which clears all inner organs over time. Gland cell outlets have often not been described as such, but may have been identified as cuticular scars (e.g., Higgins 1983).

In *Echinoderes capitatus, E. cantabricus, E. gizoensis*, and *E. hispanicus*, gland cell type-1 openings appear as a pore field anteriorly in a segment (Tab. 1; Nebelsick 1992b; GaOrdóñez et al. 2000; Thormar & Sørensen 2010). In the light microscope, *Cephalorhyncha asiatica*, species of *Fissuroderes* and *Campyloderes* cf. *vanhoeffeni* yield a gland cell outlet in the same part of a segment, but as a circular depression of the surface with a central, conical "plug" (Figs 5.1.17 A, 5.1.18 A, 5.1.19 A; Tab. 1; Neuhaus & Blasche 2006; Neuhaus & Sørensen 2012). In the same position, *Polacanthoderes martinezi* reveals two to few fine pores (Figs 5.1.17 D, 5.1.18 B, D). *Campyloderes* cf. *vanhoeffeni* also shows an egg-shaped, sub- and intra-cuticular atrium. The outlet is feebly visible or not recognizable at all in SEM. On the ventral plates, ventromedial outlets are at least in the posterior segments recognized as inconspicuous notches or interruptions in the posteriormost secondary fringe. The outlet is covered by several small hair-like cuticular processes. However, openings such as a porous area or an actual hole are apparently not present (Neuhaus & Sørensen 2012). In *Centroderes spinosus*, gland cells seem to open through funnel-like cuticular tubes of sensory spots but also in a pore field (Fig. 5.1.11 C; Neuhaus et al. 2013).

The type 2-gland cell outlet of species of *Fissuroderes* and *Echinoderes eximus* is significantly larger than the outlet of type 1 and shows pear-shaped sclerotized walls, a wide circular to oval opening, and an underlying gland cell (Neuhaus & Blasche 2006). Similar gland cell outlets also appear in *Cateria* sp. (Fig. 5.1.9 C), *E. bookhouti* (Fig. 5.1.23 B), *E. remanei* (Fig. 5.1.22 D), and *E. setiger* (Fig. 5.1.20 A, B). These outlets are modified in *E. rex* by a row of numerous fine cuticular hairs around the opening and anterior cuticular furrows (Lundbye et al. 2011). A more recent study suggests that type-2 gland cell outlets are more widely distributed among the Echinoderidae than previously assumed (Sørensen pers. obs.).

Female-specific gland cells and outlets occur in species of *Cephalorhyncha* (Fig. 5.1.18 A), *Echinoderes*,

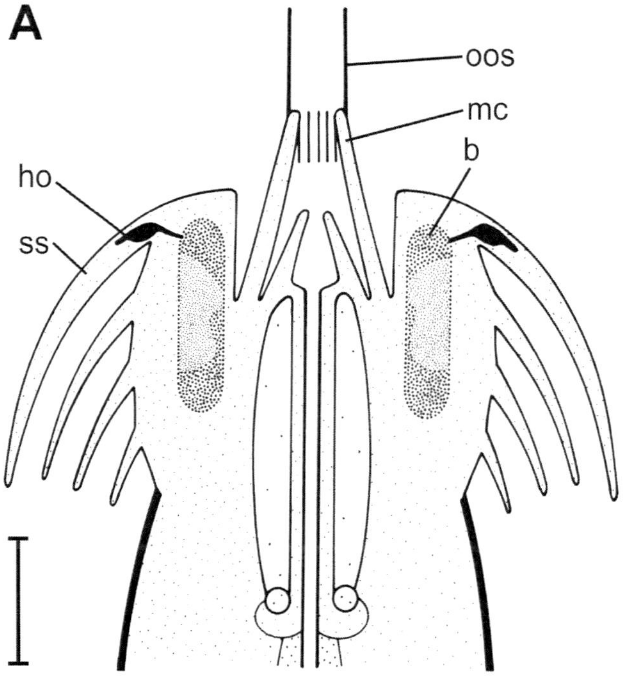

Fig. 5.1.43. *(Continued)*

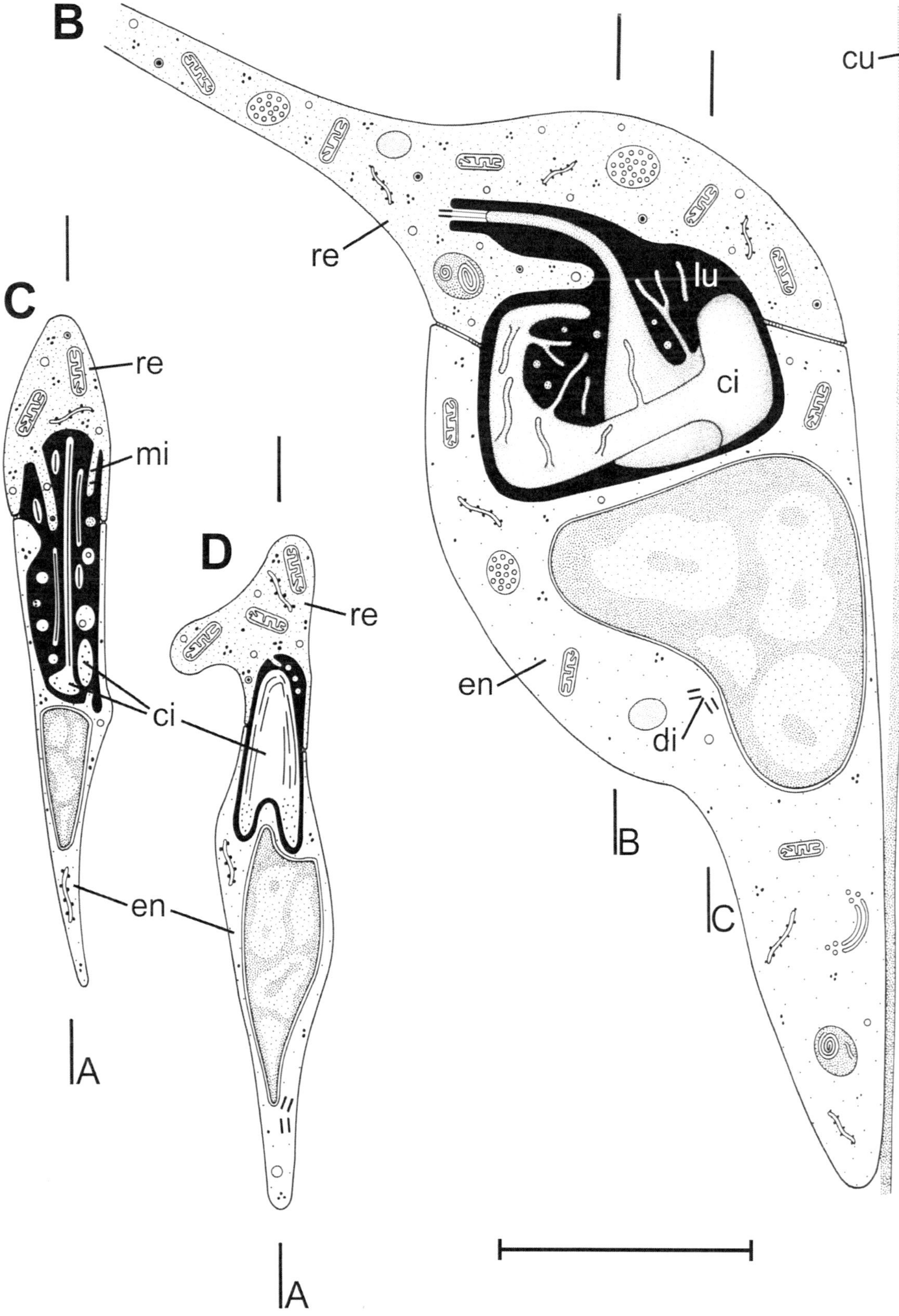

Fig. 5.1.43. Reconstruction of the head sensory organ of adult *Pycnophyes dentatus* (after Neuhaus 1997; © Springer), line in A = 100 μ, line in B = 5 μ, line in B valid for B–D. A, Position of organ in everted introvert; B–D, right organ in longitudinal (B) and parasagittal sections (C, D), microvilli in receptor lumen not illustrated in B, bars indicate level of section.

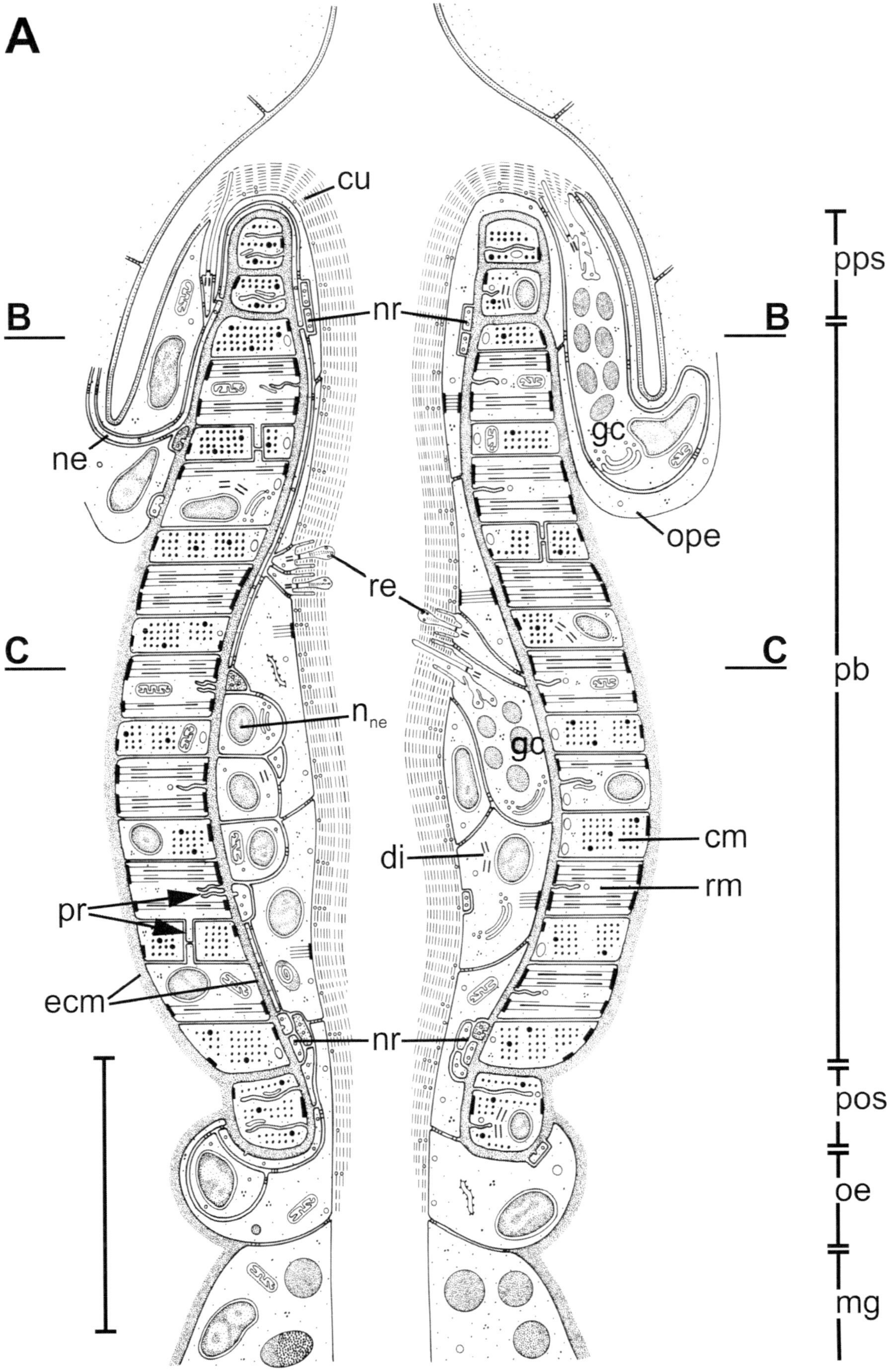

Fig. 5.1.44. *(Continued)*

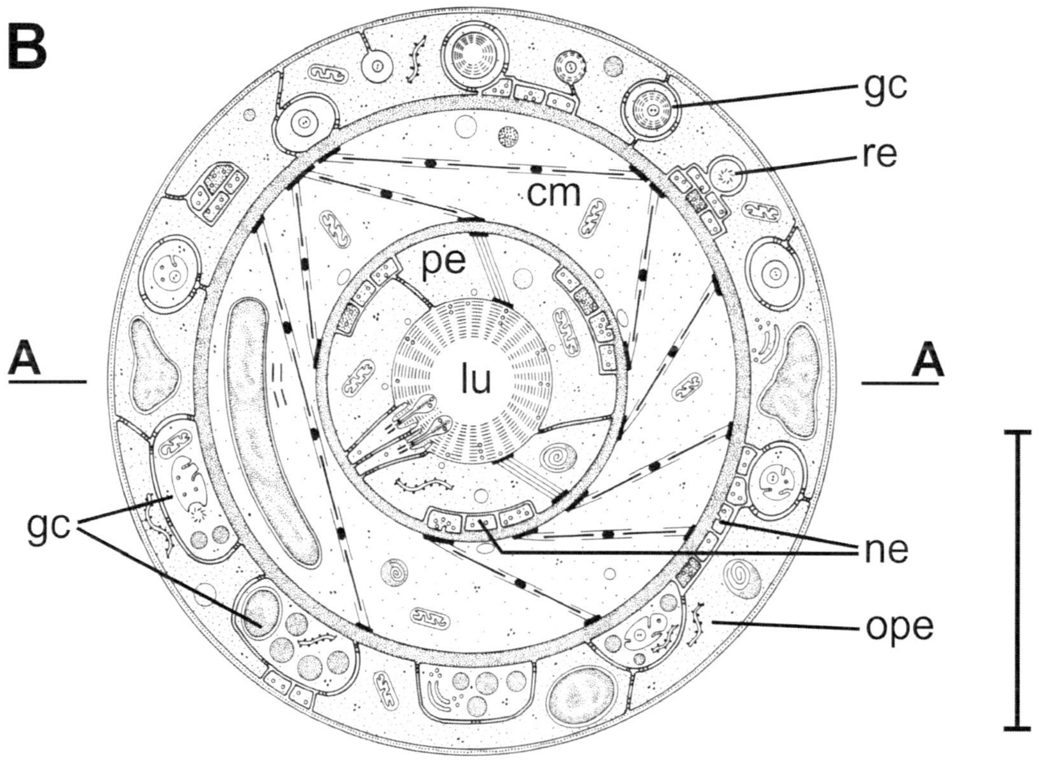

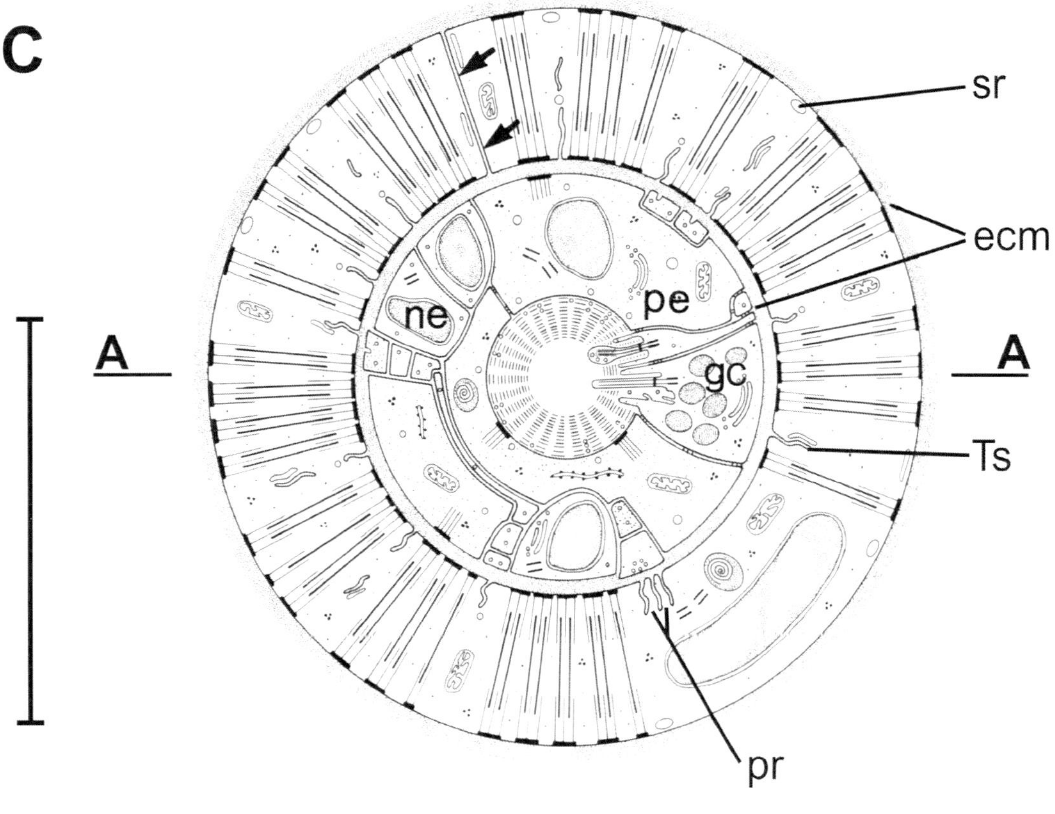

Fig. 5.1.44. Reconstruction of the pharynx and esophagus of *Zelinkaderes floridensis*, (after Neuhaus 1994; © Verlag Dr. Friedrich Pfeil), lines = 15 μ. A, longitudinal section, inner oral styles not shown; B, cross-section through ring muscle cell and outer pharyngeal epithelium, gland and receptor cells drawn at different levels; C, cross-section through radial muscle cell, arrows mark cell membrane of muscle cell joining itself; bars in A–C indicate level of section.

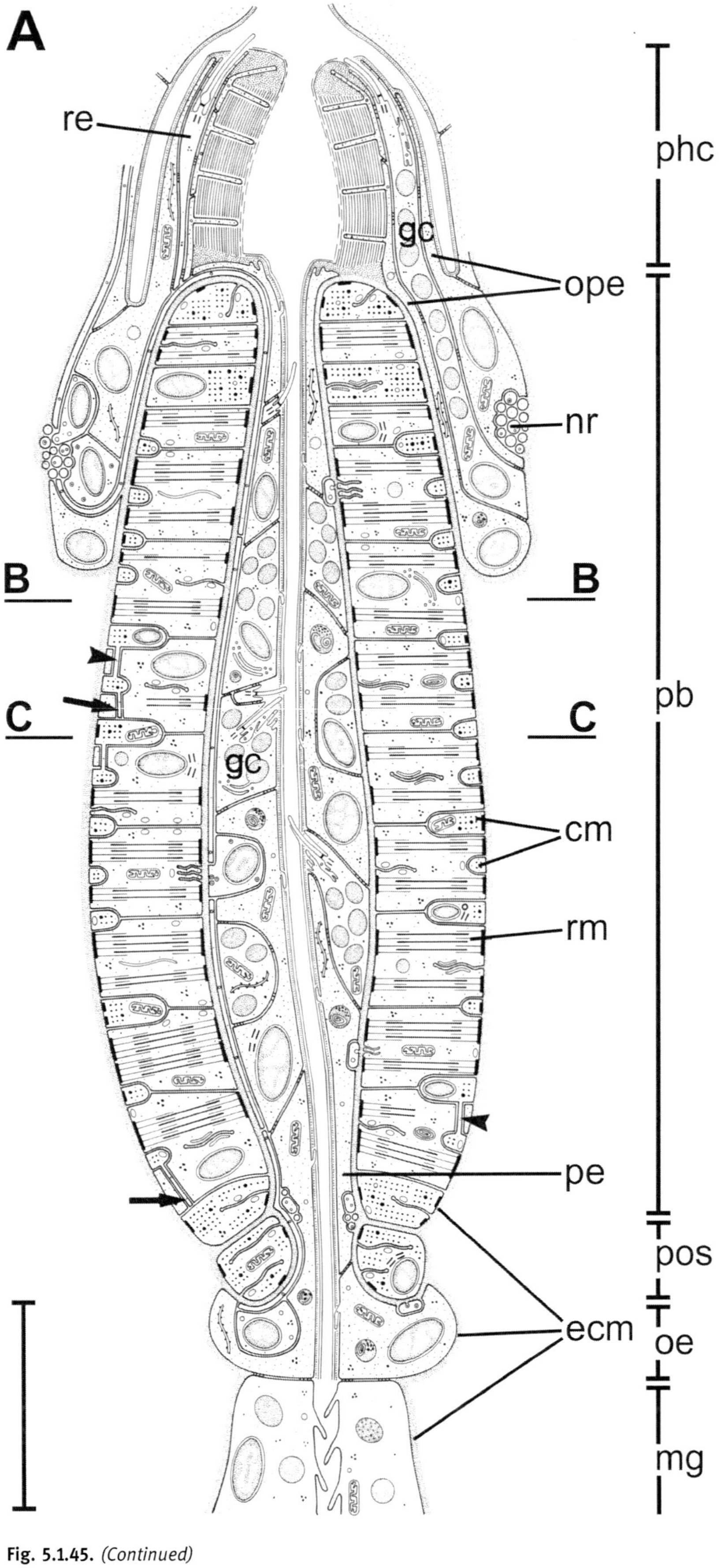

Fig. 5.1.45. *(Continued)*

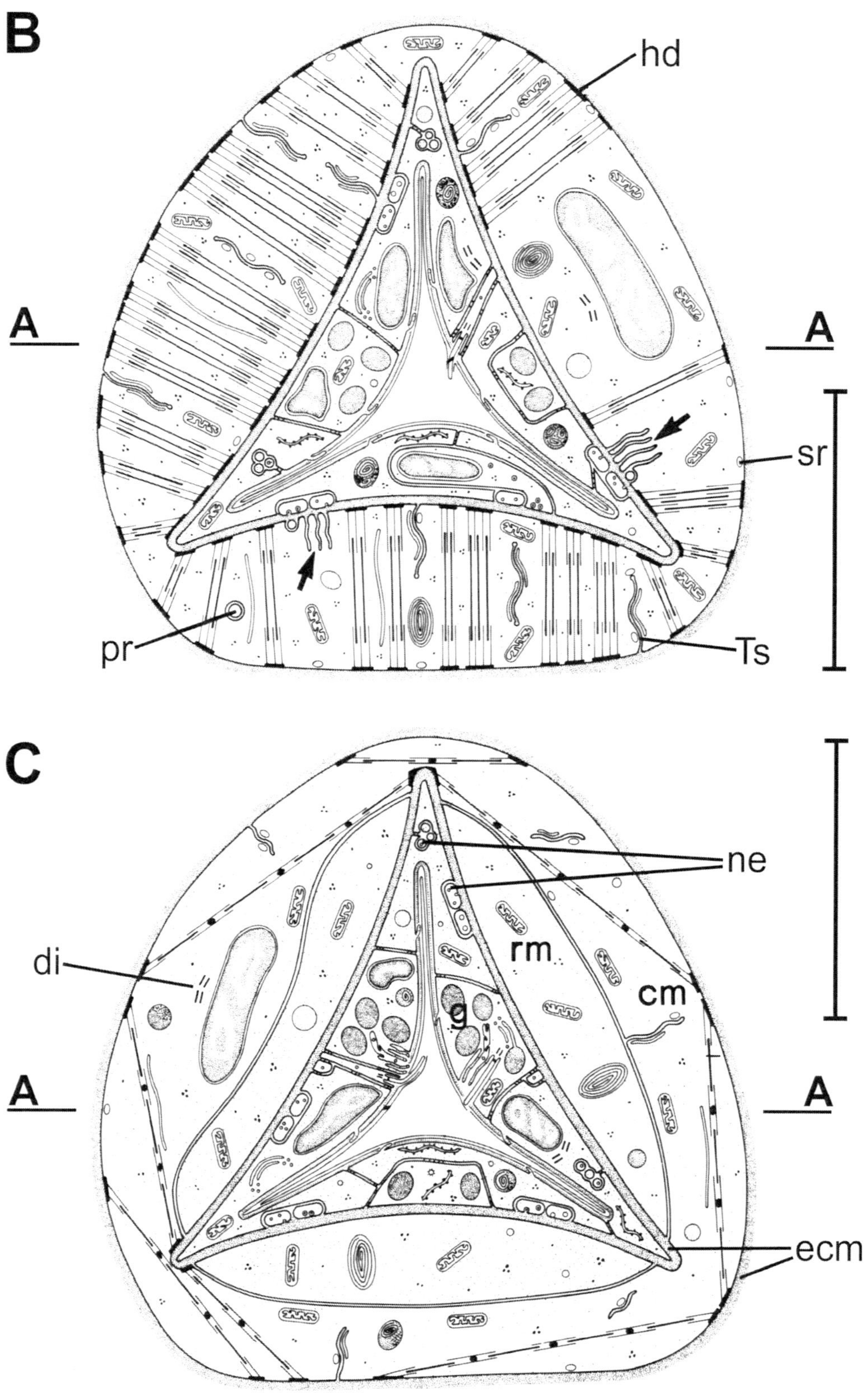

Fig. 5.1.45. Reconstruction of the pharynx and esophagus of *Pycnophyes dentatus*, (after Neuhaus 1994; © Verlag Dr. Friedrich Pfeil), lines = 20 µ. A, longitudinal section, inner oral styles not shown, arrow heads point to connecting cell process between the two circular parts of each ring muscle cell, arrows hint to cell processes towards neighbouring ring muscle cells; B, C, cross-section through radial (B) and ring muscle cell (C), respectively, arrows mark cell processes of muscle cells to nerve cells, process of a ring muscle cell in B; bars in A–C indicate level of section.

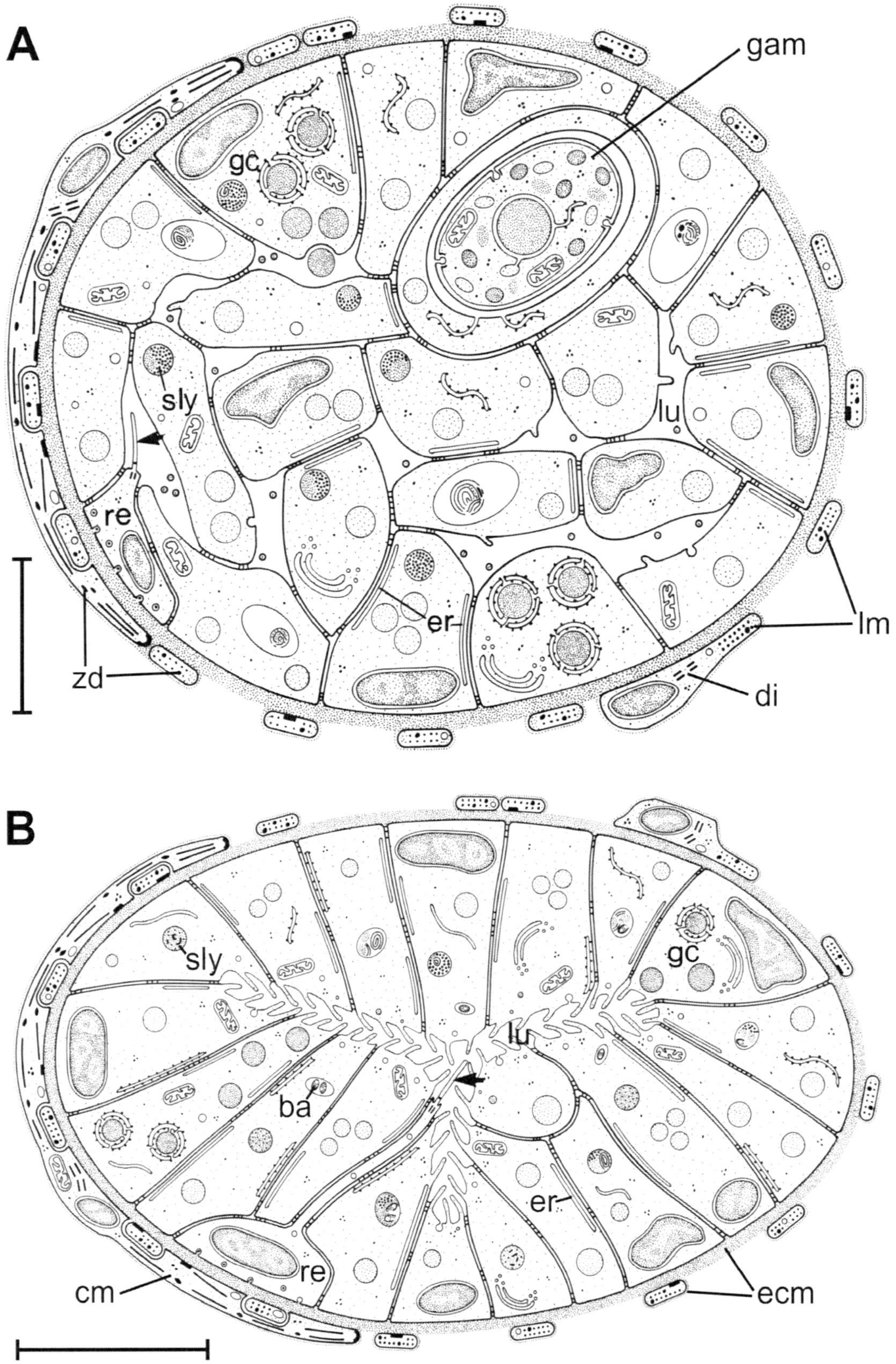

Fig. 5.1.46. Reconstruction of the midgut of *Zelinkaderes floridensis* (A) and *Pycnophyes* (B), cross-section; arrows point to sensory cilia; notice branched (A) and triradiate lumen (B); (after Neuhaus 1994; © Verlag Dr. Friedrich Pfeil), lines = 20 µ.

Fissuroderes (Fig. 5.1.19 A), and *Centroderes* (Fig. 5.1.12 A; see also below).

Trunk – papillae. Neuhaus (1993) refers to the sensory spots type-3 of juvenile *Pycnophyes dentatus* and *P. kielensis* as sensory papillae, because these sensory organs elevate cone-like above the body surface opposite to the "flat" sensory spots of adult specimens of these species. Meanwhile, a more elaborated definition of different types of sensory spots is available (Nebelsick 1992a; Sørensen et al. 2009), and it is recommended to follow these suggestions (Tab. 1). A papilla is now understood as a very short tube covered densely by micropapillae which may be slightly elongate (Figs 5.1.10 B, 5.1.15 B; Sørensen et al. 2009). Such papillae may be female-specific in posterior segments of species of *Condyloderes* but may have been overlooked occasionally (Fig. 5.1.15 B; Higgins 1969a; Sørensen et al. 2010b; Neuhaus own obs.).

The length and size of the papillae of both different specimens and within a specimen of *Campyloderes* cf. *vanhoeffeni* vary considerably (Neuhaus & Sørensen 2012). In the light of the latter study, the sublateral tubules in segment 9 of *Wollunquaderes majkenae* may have to be regarded as papillae (Tab. 1) although they are more slender than the female ventral papillae in segments 8 and 9 of this species (Sørensen & Thormar 2010). In any case, application of the term tubule to *W. majkenae* is unfortunate, because this term relates originally to flexible, thin-walled, elongate processes with a slightly broader base in species of *Echinoderes* (see GaOrdóñez et al. 2000).

Trunk – appendages. Quite different types of appendages appear on the trunk and have been named confusingly and inconsistently (see below). Current usage and definitions are summarized in Tab. 1. The arrangement of appendages is species-specific, but variation has been demonstrated for *Campyloderes* cf. *vanhoeffeni, Centroderes spinosus, Paracentrophyes praedictus, Pycnophyes dentatus*, and *P. kielensis* (see Neuhaus 1993, 1995; Neuhaus & Sørensen 2012; Neuhaus et al. 2013). Appendages appear usually near the posterior margin of a segment, but deep notches may be found in some taxa such as species of *Antygomonas, Cateria, Tubulideres*, and *Wollunquaderes* (see Gerlach 1956; Higgins 1968; Bauer-Nebelsick 1996; Sørensen 2007; Sørensen et al. 2007; Sørensen & Thormar 2010).

Acicular spines are found in Cyclorhagida and in a lateral terminal position in some Homalorhagida (Figs 5.1.1, 5.1.2, 5.1.12 C). Among the latter, adult *Neocentrophyes* and *Paracentrophyes* as well as juvenile *Kinorhynchus* and *Pycnophyes* reveal spinose processes lateroventrally and middorsally, instead (Fig. 5.1.33; Higgins 1969b, 1983; Sørensen et al. 2010a). Lateral terminal spines do not occur in species of *Neocentrophyes* and *Kinorhynchus*, but anlagen of such spines can be traced (Fig. 5.1.32 B). Middorsal spinose processes especially in posterior segments also appear in adult *Pycnophyes arctous, P. chukchiensis, P. furugelmi*, and lesser developed in *P. carinatus, P. greenlandicus, P. odhneri*, and *Kinorhynchus spinosus* (see Zelinka 1928; Lang 1949; Kristensen & Higgins 1988; Higgins 1991; Adrianov & Malakhov 1999a). Possibly, the expression of middorsal and laterodorsal spinose processes represents a plesiomorphic character for Homalorhagida, because the spinose processes of the Homalorhagida seem to be homologous to the acicular spines of the Cyclorhagida (see also section 5.2 Reproduction and development, Development).

Middorsal spines occur in almost all Cyclorhagida, but some of these spines are offset to a paraventral position in *Dracoderes* and *Tubulideres* (Fig. 5.1.16 B; Higgins & Shirayama 1990; Sørensen et al. 2007, 2012a). Lateral or ventral acicular spines may reach segment length or may be extremely elongated in species of *Campyloderes* and *Centroderes* (Figs 5.1.10 A, B, 5.1.11 B; Zelinka 1913; Johnston 1938; Higgins 1986b; Song & Chang 2001; Neuhaus & Sørensen 2012; Neuhaus et al. 2013). The midterminal spine may reach a considerable length up to the length of the trunk or even longer in species of *Antygomonas, Cateria* (Fig. 5.1.9 A), *Centroderes* (Fig. 5.1.11 A), *Semnoderes* (Fig. 5.1.25 A), *Triodontoderes* (Fig. 5.1.27 A), *Tubulideres, Wollunquaderes*, and *Zelinkaderes*.

Cuspidate spines are limited to certain genera of Cyclorhagida (Figs 5.1.7 A, B, D, 5.1.25 B, 5.1.27 A, C, D, 5.1.29 A, B, 5.1.30 B; Tab. 1). The cuspidate spines of *Semnoderes armiger* and *S. armatus* comb. nov. have been suggested to represent a mechanoreceptor (Zelinka 1928, p. 99: "Tastorgan"), but such spines possess a terminal pore in other species (Higgins 1990), which favours more a secretive function. Crenulated spines appear only in the male of several Cyclorhagida on the 10th or 11th segment (Figs 5.1.12 B, 5.1.13 A, 5.1.30 D; Tabs 1, 2). The occurrence of such spines in Kinorhyncha suggests that the spines probably evolved at least twice, once on segment 10 and once on segment 11. A midventral spinose process at the posterior margin of the first trunk segment is known only from *Centroderes spinosus* and may be hidden by detritus or the collapsed sternal plates (Neuhaus et al. 2013).

Several definitions of terms of the trunk appendages are a bit misleading. A secretive function of the so-called adhesive tubes ventromedially in segment 2 of species of

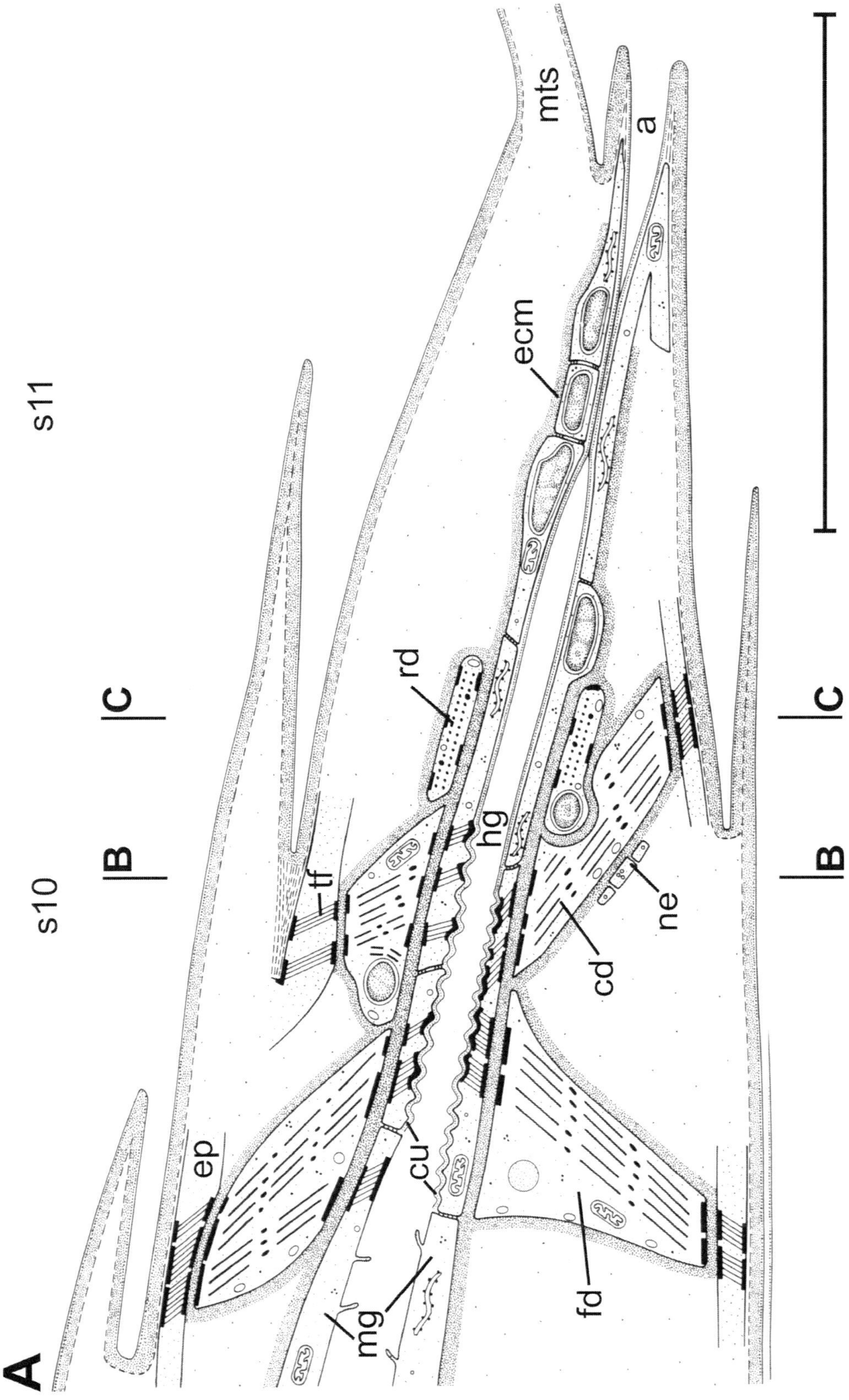

Fig. 5.1.47. *(Continued)*

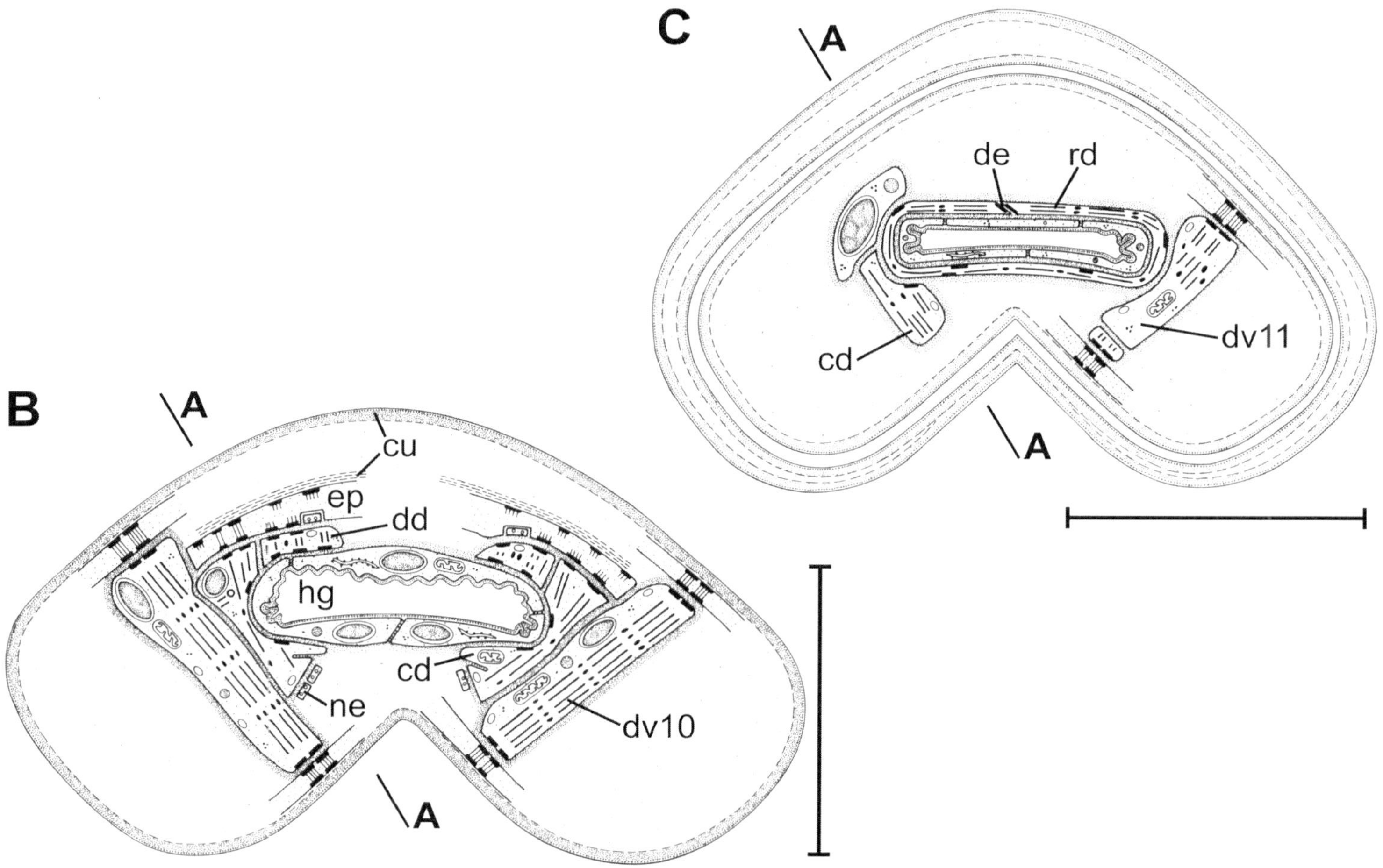

Fig. 5.1.47. Reconstruction of the hindgut of *Zelinkaderes floridensis*, (after Neuhaus 1994; © Verlag Dr. Friedrich Pfeil), lines = 20 µ. A, longitudinal section, notice epicuticle of caudal hindgut sticking together; B, C, cross-sections; bars in A–C indicate level of section.

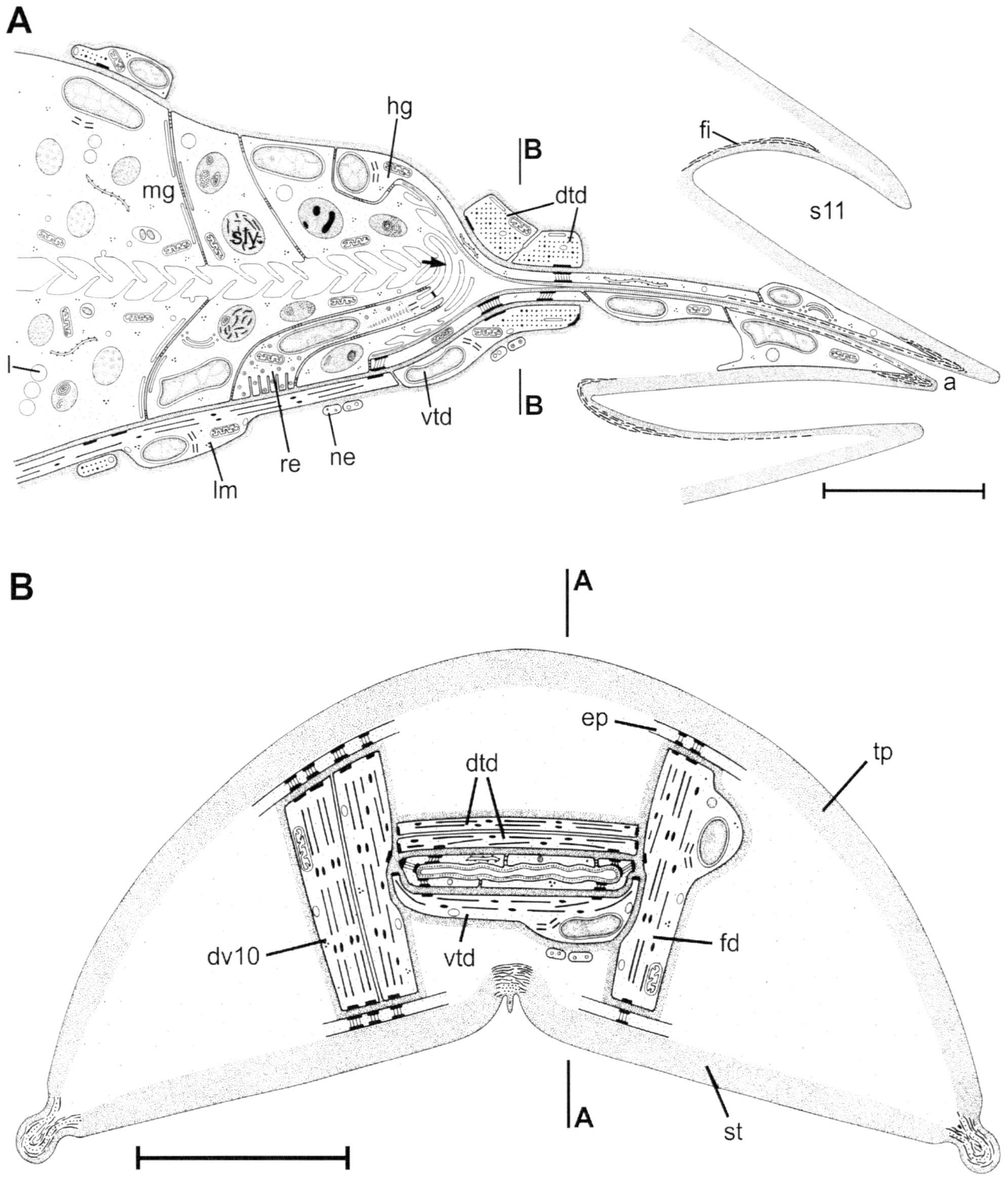

Fig. 5.1.48. Reconstruction of the hindgut of *Pycnophyes*, (after Neuhaus 1994; © Verlag Dr. Friedrich Pfeil), lines = 20 µ. A, longitudinal section, notice epicuticle of caudal hindgut sticking together, arrow points to sensory cilium; B, cross-section; bars in A–B indicate level of section.

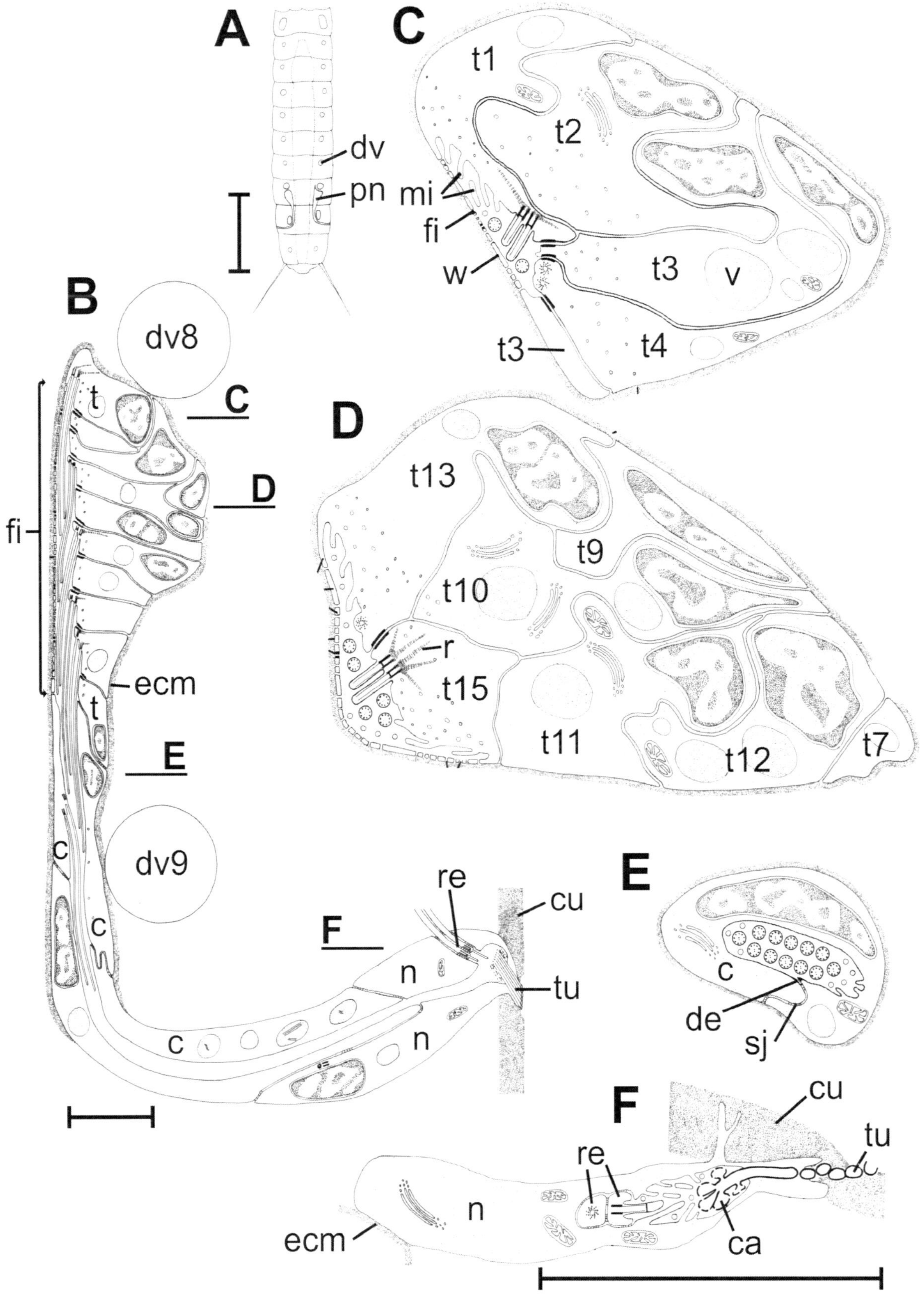

Fig. 5.1.49. Reconstruction of the right protonephridium of female *Pycnophyes kielensis*, (after Neuhaus 1988; © Springer), line in A = 200 µ, lines in B and F = 10 µ, line in F valid for C–F. A, position of protonephridium; B, oblique horizontal section; C–F, cross-sections; bars in A indicate level of cross-section.

Echinoderes, *Kinorhynchus*, and *Pycnophyes* has been assumed (e.g., Zelinka 1928; Higgins 1983; GaOrdóñez et al. 2008), but detailed morphological studies are lacking. Therefore, it is suggested to use the more neutral term "tube" instead (Tab. 1). A "tube" refers to a flexible, tubular, thin-walled, articulated appendage with a blunt tip and pore (Tab. 1). A "tubule" is literally a small tube – but actually about the same size as a tube and is defined as a long, thin-walled, cylindrical, hollow appendage with a slightly broader base and wing-like lateral extensions (Tab. 1; GaOrdóñez et al. 2000: p. 168). It cannot be totally excluded that the lateral extensions represent in fact preparation artefacts originating from folding of the thin-walled cuticle of tubules. Tubules are named "setae" by Nebelsick (1992b) in *E. capitatus*. Seta is also used for species of *Kinorhynchus*, *Neocentrophyes*, *Paracentrophyes*, and *Pycnophyes*, but the lateral, wing-like, cuticular extensions seem to be missing in these species (Figs 5.1.33 C, 5.1.37 B, C; Higgins 1983; Higgins & Kristensen 1988; Sørensen et al. 2010a; Neuhaus own. obs.). However, a "fringed tubule" (Sørensen 2006) means a straight tubular, very short, thin-walled appendage with a blunt tip, which extends into thin cuticular hairs (Tab. 1). To complicate matters even further, "tubule" has been more recently applied to different appendages (Fig. 5.1.28 A, B; Tab. 1) that do not fully fit the original description provided by GaOrdóñez et al. (2000). Partly, this situation arises from more detailed, recent SEM studies resulting in a bunch of new cuticular appendages whose homology may only be verified or falsified by both TEM investigations and a thorough, comparative revision of a larger amount of taxa. In this book chapter, conflicting or inconsistent use of terms has been mentioned in Tab. 1 and in the text.

Setae have been recognized by Zelinka (1928) and assumed to be receptive ("Sinneshaare") in species of *Echinoderes*, *Kinorhynchus*, and *Pycnophyes*. The tubules of *E. capitatus* are connected with a subcuticular gland cell (Nebelsick 1992b). However, the tubules of *E. cantabricus* are hollow, but do not seem to agree in their position with the position of the gland cells (GaOrdóñez et al. 2000); so their function remains open.

Integument

Epidermis and cuticle. The epithelium has been assumed to be syncytial (Zelinka 1928), but TEM studies confirm its cellular nature (Moritz & Storch 1972a; Kristensen & Higgins 1991; Neuhaus 1991, 1994; Adrianov & Malakhov 1994). Pigments and zooxanthellae were found in the epidermis by Zelinka (1928), but have not been confirmed in the TEM (comp. section 5.8 Ecology). *Echinoderes ferrugineus* and *E. subfuscus* are reported to show a rusty brownish colour of the cuticle, *E. citrinus* a bright yellow, *Centroderes spinosus* and *Pycnophyes rugosus* a yellow-green body coloration (Zelinka 1928). The cuticle is secreted by the epidermis and covers the introvert, neck, trunk, foregut, hindgut, and protonephridium, gland cell, and sensory spot tubules (Adrianov et al. 1990b; Neuhaus 1993; Adrianov & Malakhov 1994; GaOrdóñez et al. 2000). The trunk cuticle consists of a chitinous basal layer, which is fine-granular in TEM sections, and a monolamellar, membrane-like epicuticle. Fibrillar material of differing diameters appears in the flexible areas connecting the sternal and tergal plates laterally, anteriorly, and posteriorly in species with a thick cuticle such as *Kinorhynchus* and *Pycnophyes* (Fig. 5.1.48 A, B). In these species, epidermal microvilli and, at muscle attachment sites, also epidermal tonofilaments extend into the cuticle (Jeuniaux 1975; Brown 1985; Adrianov et al. 1990b; Kristensen & Higgins 1991; Adrianov & Malakhov 1994; Neuhaus et al. 1996; GaOrdóñez et al. 2000; Neuhaus & Higgins 2002). The microvilli appear as "pores" in light microscopy (see above) and have been suggested to serve in osmotic and gas exchange with the environment as well as in construction of the trunk cuticle (Zelinka 1928: p. 43). Experimental data about the former hypothesis is lacking, but the latter proposal seems reasonable.

In *Echinoderes aquilonius*, electron-dense amorphous cuticle connects sternal and tergal plates, whereas in *E. cantabricus* and *E. kozloffi* amorphous cuticle with fine fibrils seems to exist, but thicker fibrills occur in *Pycnophyes kielensis* (Merriman & Corwin 1973; Adrianov et al. 1990b; Kristensen & Higgins 1991; Adrianov & Malakhov 1994; GaOrdóñez et al. 2000). The cuticle of species of *Zelinkaderes* seems to be unique by virtue of its longitudinal striation pattern of the trunk cuticle recognizable with light microscopy (Figs 5.1.30, 5.1.31 B; Higgins 1990; Bauer-Nebelsick 1995; Sørensen et al. 2007). This striation is reflected at the TEM level by a more electron-dense type of cuticle alternating with a more electron-lucent type of cuticle (Bauer-Nebelsick 1995: Fig. 5.1.42). Brown (1985) suggests that in *Kinorhynchus phyllotropis* secretions are delivered through epidermal microvilli to the surface of the cuticle. However, the microvilli do not seem to penetrate the cuticle completely in this species, nor in *Pycnophyes dentatus*, *P. greenlandicus*, and *P. kielensis* (see Kristensen & Higgins 1991; Neuhaus own obs.), and the droplets on the cuticle of *K. phyllotropis* may represent preparation artefacts.

The existence of a chitinous cuticle has been claimed for a long time (Claparède 1863; Reinhard 1887; Zelinka 1907, 1928) and denied by others, because the cuticle dissolves in 5% potassium or sodium hydoxide (Remane 1936: p. 276). Boykin (1965) was the first to demonstrate its presence in *Kinorhynchus ilyocryptus* by the chitosan-iodine colour test. Later, Jeuniaux (1975) treated specimens of *Antygomonas* sp. with hot bicarbonate and subsequent staining with Congo Red in comparison to cuticle digestion with chitinase. Studies on *Centroderes* sp., *Echinoderes* sp., *Zelinkaderes floridensis*, *Paracentrophyes praedictus*, *Pycnophyes dentatus*, and *Pycnophyes* sp. with the lectin wheat-germ agglutinin (WGA) bound to gold-labelled antibodies reveal that chitin is actually deposited in the basal fine-granular layer of the cuticle and in the matrix embedding the fibres of the flexible cuticle. Control incubations with chitotriose (competitive inhibition of binding to chitin) and with chitinase (degradation of chitin) corroborate that WGA indeed binds to chitin in the cuticle of Kinorhyncha (Neuhaus et al. 1996; Neuhaus & Higgins 2002; Neuhaus own obs.).

The trunk cuticle possesses thickenings towards the interior, so-called apodemes like the segmental pachycycli and midventral thickenings (Figs 5.1.15, 5.1.17, 5.1.20, 5.1.26, 5.1.36, 5.1.39 A, 5.1.40 B; Zelinka 1928: "Mittelwülste") to which muscles attach. The pachycycli occur at the anterior margin of segments 2–11 and paraventrally in many species, but are most strongly expressed in species with a thick cuticle belonging to *Echinoderes, Kinorhynchus*, and *Pycnophyes*. In species of the latter two genera, a ball-and-socket articulation may be expressed in the pachycyclus of segments 2–10 consisting of a sternal cuticular ring and a tergal cuticular process (Zelinka 1928). Especially in species of *Echinoderes, Kinorhynchus*, and *Pycnophyes*, the adult trunk cuticle increases significantly in thickness, pachycycli and, in *Kinorhynchus* and *Pycnophyes* also midventral thickenings, develop (stronger), ball-and-socket articulations of sternal and tergal plates become more prominent, and the cuticle turns yellowish-brownish (tanning?) with increasing age of the adult individual (Zelinka 1928: p. 309; Lang 1936; Kristensen & Higgins 1991; Adrianov & Malakhov 1994; GaOrdóñez et al. 2000; Neuhaus & Higgins 2002). Species of *Kinorhynchus* and *Pycnophyes* have been distinguished inter alia on the basis of presence or absence of midventral thickenings, although at least Zelinka (1928: p. 323) is aware of the problem (Zelinka 1928; Higgins 1983; Adrianov 1995; Adrianov & Malakhov 1999a; Sørensen & Pardos 2008). Considering that these apodemes seem to develop considerably with age and are weakly expressed if at all in young adults (Zelinka 1928: p. 323; Sánchez et al. submitted; Neuhaus own obs.), this character becomes obsolete in species discriminations (Lang 1936).

After the final moult from the last juvenile stage, young adult specimens of species of *Cephalorhyncha, Echinoderes, Polacanthoderes, Kinorhynchus*, and *Pycnophyes* develop a series of regularly arranged button-like structures ("Knöpfchenreihen" of Zelinka 1928) at the posterior margin of the trunk plates overlapping the following segment (Figs 5.1.17 A, 5.1.18 A, B, 5.1.20 A, 5.1.39 C, 5.1.40 D; Zelinka 1928: *E. dujardinii, E. worthingii, K. giganteus, P. communis, P. flaveolatus*; Lang 1949: *K. spinosus, P. odhneri, P. sculptus*; Higgins 1961, 1964b, 1966a: *K. cataphractus, K. ilyocryptus, P. beaufortensis, P. egyptensis, P. sanjuanensis*; Neuhaus 1993: *P. dentatus, P. kielensis*; Adrianov & Malakhov 1999a: *P. newguinensis*; Lemburg 2002a, b: *P. australensis*; Sánchez et al. 2011: *P. dolichurus*). TEM studies reveal that the structures actually represent cuticular pillars, which seem to fix the young adult, supposedly still soft cuticle (Fig. 5.1.40 D; Neuhaus 1993). With increasing age, the button-like structures increase in size and finally fuse (Fig. 5.1.18 B; Zelinka 1928), so probably the cuticle will fill the entire posterior overlapping part of the sternal and tergal plate, the cuticular free flap.

Gland cells. Gland cells do not seem to be related to tubules, spines, or sensory spots in *Echinoderes capitatus, E. cantabricus*, and *E. hispanicus* (see Nebelsick 1992b; GaOrdóñez et al. 2000). In *Centroderes spinosus*, light microscopical observations suggest that gland cells may open through funnel-like cuticular tubes of sensory spots (Neuhaus et al. 2013). Also in *Pycnophyes greenlandicus*, gland cells are associated with sensory spots (Kristensen & Higgins 1991).

Each unicellular, merocrine gland cell of *Echinoderes capitatus* delivers its secretions into a lumen filled with numerous microvilli and possibly stacks of membranes. The cavity opens to the outside via several branching, convoluted canals in a small pore field (Nebelsick 1992b). In *E. aquilonius, E. cantabricus, E. hispanicus*, and *Kinorhynchus phyllotropis*, gland cells (= type-2 of GaOrdóñez et al. 2000) seem to reveal a similar morphology, but possess only a single outleading canal in the latter two species; a few photographs have been published, but more detailed reconstructions are lacking for these four species (Brown 1985; Kristensen & Higgins 1991; GaOrdóñez et al. 2000). GaOrdóñez et al. (2000) postulated a series of different stages in the development of the outleading pore field with up to 25 pores; more probable, a variable amount of mucous secretions covers the pores in different gland cell outlets. Also, cuticular structures including, e.g., subcuticular tubes of sensory spots and protonephridium are moulted together with the trunk cuticle; post-moulting changes in the cuticle, except for an increase in thickness and tanning processes, are not very likely (Neuhaus 1993; Neuhaus & Higgins 2002).

In light microscopy, female-specific gland cell outlets in *Cephalorhyncha asiatica, C. nybakkeni, Echinoderes collinae, E. gizoensis, E.* cf. *kozloffi, E. spinifurca*, possibly in other species of *Echinoderes*, and in species of *Fissuroderes* show a funnel-like structure (Figs 5.1.18 A, 5.1.19 A) and in species of *Centroderes* a bulbous structure; ultrastructural data is missing (Neuhaus & Blasche 2006; Thormar & Sørensen 2010; Neuhaus et al. 2013). Vesicles in type-1 gland cells of *E. cantabricus, E. hispanicus, Kinorhynchus giganteus*, and *Pycnophyes greenlandicus* are electron-lucent with an electron-dense tubular inner substructure; in the latter two species, each gland cell also possesses a cilium with a large rootlet (Moritz & Storch 1972a; Kristensen & Higgins 1991; GaOrdóñez et al. 2000). Type-1 gland cells of *E. cantabricus* and *E. hispanicus* do not seem to be related in their position to the hollow tubules with their terminal pore; the release sites of these gland cells are unknown (GaOrdóñez et al. 2000).

Epidermal or gland cell secretions have been suggested to protect the animal from abrasion during locomotion through the sediment (Brown 1985) or to "have a lubricating function for movements between segments" (GaOrdóñez et al. 2000: p. 177). Gland cells associated with tubules in species of *Echinoderes, Kinorhynchus*, and *Pycnophyes* are supposed to be adhesive during copulation (Zelinka 1928; Higgins 1983; Kristensen & Higgins 1991), but from the single case of copulation observed this does not seem to be very likely (see section 5.2 Reproduction and development, Reproductive biology).

Scalids. The primary spinoscalids of ring 01 of *Echinoderes capitatus* exhibit several ciliary sensory cells. Subsequent spinoscalids of rings 02–06 and trichoscalids possess a gland cell, which opens through a pore field at the base of each scalid (Nebelsick 1993). Spinoscalids of rings 02–06 contain basally a cell with numerous dictyosomes and stacks of membranes, which may represent microvilli; in *Kinorhynchus giganteus* and *K. phyllotropis*, the cell opens to the outside dorsally at the base of the spinoscalid (Moritz & Storch 1972b; Brown 1989). This cell is supposed to be secretive (Kristensen & Higgins 1991). Spinoscalids of *K. giganteus* and *K. phyllotropis* yield a regular pattern of electron-dense cuticle appearing as layers of bands oriented cross-wise on top of each other (type-1) or in a more complex system of rods and lamellae forming cellular elements (type-2) in a more electron-lucent matrix (Moritz & Storch 1972b; Brown 1989). These electron-dense, cuticular modifications are interpreted here as cuticular reinforcements in the cuticle to add stability to the relatively fragile spinoscalids.

Trichoscalids of *Kinorhynchus giganteus* do not reveal a terminal pore and seem to possess a sensory cell with a single cilium (Moritz & Storch 1972b).

Musculature

The body musculature consists of: (1) one pair of bundles of ventral and dorsal longitudinal muscles extending between the pachycycli of subsequent segments (Figs 5.1.1 B, C, 5.1.2 B, 5.1.36 B, 5.1.41); (2) one pair of segmental dorsoventral muscles between the sternal and tergal trunk plate (Figs 5.1.1 A, B, 5.1.2 A–C, 5.1.41); (3) lateral oblique muscles (Cyclorhagida only) between two subsequent segments (segments 1, 1–2, or 1–3) or between the tergal plates of subsequent segments (segments 2–8, 3–8, or 4–8) (Fig. 5.1.41); (4) circular muscles at the base of the scalids of rings 06 and 07; (5) 10 (Homalorhagida) or 16 (Cyclorhagida) head retractors (= outer retractors) between the base of the head scalids of rings 05 and 06 and posterior trunk segments (Figs 5.1.1 A, B, 5.1.2 A, B, 5.1.41); (6) mouth cone retractors (= inner retractors) between the base of the mouth cone and posterior segments (Figs 5.1.1 A, 5.1.2 A, 5.1.41); (7) pharynx protractors between the posterior end of the pharynx and the base of the mouth cone (Figs 5.1.1 A, 5.1.2 A); (8) circular muscles connecting the placids in the neck region (Figs 5.1.17 B, 5.1.41); and (9) a muscle net around at least the posterior part of the male gonads (Zelinka 1894, 1908, 1928; Kristensen & Higgins 1991; Neuhaus 1991, 1994; Nebelsick 1993; Adrianov & Malakhov 1994; Müller & Schmidt-Rhaesa 2003; Rothe & Schmidt-Rhaesa 2004; Schmidt-Rhaesa & Rothe 2006; Neuhaus & Blasche 2006). The latter has not been confirmed by confocal laser scanning studies of *Pycnophyes kielensis* (Rothe & Schmidt-Rhaesa 2004; Rothe pers. comm.).

Ventral and dorsal longitudinal muscles generally extend in adult specimens between neighbouring pachycycli. However, at least some muscles of subsequent segments may meet at a pachycyclus or may even continue over a pachycyclus in adult *Pycnophyes kielensis* (comp. also section 5.2 Reproduction and development, Development; Schmidt-Rhaesa & Rothe 2006) and in *Zelinkaderes brightae* (see Herranz et al. submitted; M. Herranz, pers. comm.). One pair of paradorsal triangular muscles seems to be attached to each middorsal spine of *Antygomonas* sp. (see Müller & Schmidt-Rhaesa 2003: Fig. 5.1.2 G). Most certainly, such muscles do not exist and labelling activity originates from binding to actin filaments in the circumciliary microvilli of the paradorsal sensory spots (Müller and Schmidt-Rhaesa pers. comm.). Additional muscles are connected with the alimentary canal, namely circular and longitudinal muscles in the mouth cone (Fig. 5.1.42), pharyngeal ring and radial muscles Figs 5.1.44, 5.1.45), pre- and postpharyngeal sphincters (Figs 5.1.44, 5.1.45), a net of longitudinal and circular muscles around the midgut (Fig. 5.1.46), and dilalator muscles at the hindgut (Figs 5.1.47, 5.1.48; Zelinka 1928; Neuhaus 1991, 1994; Adrianov & Malakhov 1994; Müller & Schmidt-Rhaesa 2003; Rothe & Schmidt-Rhaesa 2004; comp. section 5.1 Morphology, Intestinal system).

TEM observations are limited to *Echinoderes aquilonius, Zelinkaderes floridensis, Pycnophyes dentatus, P. flaveolatus, P. greenlandicus*, and *P. kielensis* (see Nyholm & Nyholm 1976b; Kristensen & Higgins 1991; Neuhaus 1991, 1994; Adrianov & Malakhov 1994). Confocal laser scanning studies of the phalloidin-labelled muscle system are available for *Antygomonas* sp. and *Pycnophyes kielensis* (see Müller & Schmidt-Rhaesa 2003; Rothe & Schmidt-Rhaesa 2004; Schmidt-Rhaesa & Rothe 2006).

Ultrastructure. All muscle cells that contact the cuticle attach via desmosomes, epidermal intermediate filaments, hemidesmosomes, and thin filaments into the body cuticle of species of *Echinoderes* and *Pycnophyes* with a strongly sclerotized cuticle; thin filaments do not extend into the body cuticle of *Zelinkaderes floridensis* with its thin cuticle (Nyholm & Nyholm 1976b; Kristensen & Higgins 1991; Neuhaus 1991, 1994; Adrianov & Malakhov 1994). On the outside of the trunk cuticle, the attachment sites of the dorsoventral muscles appear as oval or elongate-oval areas devoid of cuticular hairs or scales (Figs 5.1.11 B, C, 5.1.12 D, 5.1.14 E, F, 5.1.25 D, 5.1.27 D, 5.1.31 B, 5.1.33 C, 5.1.34 D, 5.1.37 B, 5.1.39 C; Higgins & Kristensen 1988; Higgins 1990; Neuhaus 1993; Sørensen 2007; Sørensen et al. 2007; Sørensen & Rho 2009; Sørensen & Thormar 2010; Herranz et al. 2012; Neuhaus & Sørensen 2012; Neuhaus et al. 2013). In *Echinoderes aquilonius, Z. floridensis, Pycnophyes dentatus, P. greenlandicus*, and *P. kielensis* but not in *E. capitatus*, head and mouth cone retractor as well as pharynx protractor muscles attach to the cuticle of the introvert via extremely elongated epidermal cells with a thick bundle of intermediate filaments called tanycytes (Kristensen & Higgins 1991; Neuhaus 1991, 1994; Nebelsick 1993). Dorsoventral muscles, longitudinal muscles, oblique muscles, retractors, and protractors are multisarcomeral, cross-striated, and reveal isolated, spindle-shaped Z-elements (Zelinka 1928; Nyholm & Nyholm 1976b; Kristensen & Higgins 1991; Adrianov & Malakhov 1994; Neuhaus & Higgins 2002). Such a fine structural arrangement facilitates supercontraction of the muscles, so the thick filaments of successive sarcomers pass the isolated Z-elements and overlap partly (Nyholm & Nyholm 1976b). Longitudinal, dorsoventral, and at least some circular and dilatator muscles of the hind gut send cell processes to the nerve cells and not vice versa as in most other invertebrates (Neuhaus 1994; Neuhaus & Higgins 2002).

Function of body muscles. Ventral, dorsal, and lateral bending of the trunk is facilitated by contraction of the segmental ventral and dorsal longitudinal muscles of both sides and of one side, respectively. Species of *Echinoderes* in particular are able to bend their trunk dorsad (Zelinka 1928). Bending is quite limited in species of *Kinorhynchus* and *Pycnophyes* with a thick cuticle, whereas species of *Antygomonas, Cateria*, and *Zelinkaderes* with a thin cuticle can actually reach the posterior segments with their heads (Fig. 5.1.9 A). Cyclorhagida also possess oblique muscles, which support lateral bending of the animals (Zelinka 1928; Remane 1928; Adrianov & Malakhov 1994; Neuhaus & Blasche 2006; Neuhaus & Sørensen 2012; Neuhaus et al. 2013). Circular muscles below the neck placids serve to close the trunk anteriorly when the head is retracted. The pharynx protractors enable the pharynx to be pulled forward in the everted introvert in order to suck in food.

Locomotion requires the alternating action of the dorsoventral muscles, and of the head and mouth cone retractors. The introvert is everted as a consequence of the contraction of the dorsoventral muscles, which diminishes the space between the sternal and tergal trunk plates and thereby increases the body pressure in the small, liquid-filled body cavity. The stiff posterior cuticle overlapping the thin intersegmental cuticle between subsequent segments and possibly the slight contraction of the dorsal and ventral longitudinal muscles may prevent the flexible intersegmental cuticle from bulging outwards as a result of the increased body pressure. When the head everts, scalids first move forward, then plow backward through the water and sediment particles, pulling the animal through the sediment. The minute size of the kinorhynch and the quick movement of its introvert means that the water has to be imagined more as a viscous fluid than as a thin fluid. Mouth cone and head retractors finally withdraw the introvert into the trunk, and the cycle can start again. Forward movement of the specimen is supported by stretching of the trunk; the dorsal and ventral longitudinal muscles shorten the trunk length before the introvert is withdrawn (Zelinka 1894, 1928; Neuhaus & Higgins 2002: p. 625). *Pycnophyes communis* may move 4–5 mm per minute on a slide (Zelinka 1928: p. 174). Species usually inhabiting the algal layer of the sediment may even climb the overgrown walls of an aquarium (Dujardin 1851).

Alternatingly, kinorhynchs may also move by sliding gently forward without using the introvert. It seems the animals leave a "mucus tunnel" after them, and the motion is created by secretion of mucus from their gland cells (Nyholm 1947c: pp. 33–34; Sørensen pers. obs.).

Nervous system

The intraepithelial nervous system is composed of the circumenteric brain, 10 longitudinal nerve strands fusing to five cords, usually two commissures per trunk segment, and the foregut nervous system in *Echinoderes capitatus*. All 10 longitudinal nerve strands originate from the forebrain, extend first frontally, turn caudally, and fuse from segment 2 on to two subdorsal nerve cords, two ventrolateral nerve cords, and the midventral nerve cord in *E. capitatus* (see Nebelsick 1993). Whereas 12 longitudinal nerve strands are reported for *Zelinkaderes floridensis* (Fig. 5.1.1 B, C), eight seem to occur in *E. aquilonius*, *Pycnophyes greenlandicus*, and *P. kielensis*, and seven in *P. dentatus* (Fig. 5.1.7 B). These longitudinal nerves innervate the introvert, neck, and trunk, and only occasionally show perikarya in posterior segments (Zelinka 1894, 1908, 1928; Adrianov & Malakhov 1990, 1994; Kristensen & Higgins 1991; Neuhaus 1991, 1994; Nebelsick 1993).

Central nervous system. The central nervous system comprises the ring-like brain which seems to be interrupted midventrally in the anterior forebrain of *Antygomonas paulae*, *E. capitatus*, *E. spinifurca*, and *Z. brightae* and the ventral longitudinal nerve cord, which originates from the forebrain as two nerve strands. These fuse to one cord in the above mentioned species. One ventral cord is also reported for *Zelinkaderes floridensis*, *Pycnophyes dentatus*, and *P. kielensis* (North Sea), but not for *E. aquilonius*, *P. greenlandicus*, and *P. kielensis* (White Sea). The brain consists of a 10-lobed forebrain with numerous perikarya, a midbrain with neuropil and few perikarya at the inner periphery, and a hindbrain with numerous perikarya (Fig. 5.1.42; Zelinka 1894, 1908, 1928; Boykin 1965; Adrianov & Malakhov 1990, 1994; Kristensen & Higgins 1991; Neuhaus 1991, 1994; Nebelsick 1993; Neuhaus & Higgins 2002; Herranz et al. submitted). The brain is fixed to the epidermis in its anterior half and moves forward and backward together with the everted and retracted introvert. Extracellular material covers the posterior part of the brain. Glia cells and four types of nerve cells occur in the brain of *E. capitatus* (see Nebelsick 1993). Serotonergic-like immunoreactivity appears as two anterior, incomplete rings and two posterior complete rings in the forebrain of *Antygomonas paulae*, *E. spinifurca*, and *Z. brightae* (see Herranz et al. submitted).

The ventral nerve cord reveals one pair of ganglia per segment at least in *E. capitatus*, *P. greenlandicus*, and *P. kielensis* (White Sea); in the posterior trunk segments of *E. capitatus* and *Z. floridensis*, the nuclei appear to be distributed over the entire length of the segment lateral of the central neuropil and to form one large caudal ganglion (Zelinka 1894, 1908, 1928; Adrianov & Malakhov 1990, 1994; Kristensen & Higgins 1991; Neuhaus 1991, 1994; Nebelsick 1993; Neuhaus & Higgins 2002). In *Antygomonas paulae*, *E. spinifurca*, and *Z. brightae*, serotonergic-like immunoreactivity originates as two strands from the second of the two anterior incomplete rings; each strand separates again into two strands; two of these strands join shortly after in *E. spinifurca*, so three strands occur in the ventral nerve cord in this species and four strands in *A. paulae* and *Z. brightae*. The central nerve strand terminates in segment 9 and the lateral strands make a closed loop in segment 10 (*A. paulae*, *Z. brightae*) or terminate more laterally with large somata in *E. spinifurca*. Herranz et al. (submitted) suggest that the loop may be related to the existence of a midterminal spine in *A. paulae* and *Z. brightae*, whereas the lack of this spine in *E. spinifurca* supposedly leads to the lateral termination. In *E. spinifurca*, somata of the lateral nerve strands with serotonergic-like immunoreactivity are concentrated in segmental ganglia, but somata are distributed more regularly along the strands in *A. paulae* and *Z. brightae* (see Herranz et al. submitted).

Foregut nervous system. The foregut is innervated through the mouth cone from the posterior midbrain (*Zelinkaderes floridensis*, *Pycnophyes dentatus*, *P. kielensis*) or anterior hindbrain (*Echinoderes capitatus*); the nine outer mouth cone nerves run first frontally, pass the anterior circular muscles, partly fuse to five strands, and continue than caudally to a basal circular nerve ring in the mouth cone (Fig. 5.1.42; Neuhaus 1991, 1994; Nebelsick 1993; Adrianov & Malakhov 1994). An anterior nerve ring in the mouth cone as reported for *E. capitatus* and *P. greenlandicus* (see Kristensen & Higgins 1991; Nebelsick 1993) has not been found in *Z. floridensis, P. dentatus*, and *P. kielensis* (Fig. 5.1.42; Neuhaus 1991, 1994). Three nerve cords extend from the basal circular nerve ring with its perikarya in the mouth cone of *Z. floridensis* (Fig. 5.1.44 B, C), six such nerves are found in *P. dentatus, P. greenlandicus*, and *P. kielensis* (Fig. 5.1.45 B, C). At least three ring-like commissures connect the pharyngeal nerve strands (Fig. 5.1.44 C); a fourth ring occurs in the short esophagus. Pharyngeal muscle cells send small cell processes to the nerve cells (Figs 5.1.44, 5.1.45; Nyholm & Nyholm 1976a; Kristensen & Higgins 1991; Neuhaus 1991, 1994). Serotonergic-like immunoreactivity occurs as one ring in the mouth cone of *Antygomonas paulae, E. spinifurca*, and *Z. brightae*; from the anterior, incomplete nerve ring in the forebrain, several neurites connect to the mouth cone ring (Herranz et al. submitted).

Supposedly, subpharyngeal ganglia innervate the muscles of the alimentary canal (Kristensen & Higgins 1991). Such an innervation has not been traced in *E. capitatus, Z. floridensis, P. dentatus*, and *P. kielensis* and is highly unlikely, because the pharyngeal muscles are already innervated from the midbrain through the mouth cone (Neuhaus 1991, 1994; Nebelsick 1993). The midgut muscle net is supposed to represent an autonomous system controlled by sensorimotoric cells in the midgut epithelium (Figs 5.1.46 B, 5.1.48 A; Neuhaus 1991, 1994; see also section 5.1 Morphology, Sensory structures).

Sensory structures

Receptors of the head. Detailed TEM investigations of introvert sensory organs are limited to the outer and inner oral styles of *Echinoderes capitatus* as well as to the cephalic sensory organs of *Pycnophyes dentatus* and *P. kielensis* (see Nebelsick 1993; Neuhaus 1997). Spinoscalids and trichoscalids of *E. aquilonius, E. capitatus, Kinorhynchus giganteus, K. phyllotropis*, and *P. kielensis* exhibit one to several ciliary sensory cells that communicate with the outside through a terminal pore, suggesting a chemoreceptive function (Moritz & Storch 1972a, b; Brown 1989; Adrianov & Malakhov 1990; Kristensen & Higgins 1991; Nebelsick 1993). Trichoscalids of *E. aquilonius* and *K. giganteus* do not seem to possess a terminal pore (Moritz & Storch 1972b), so mechanoreception is supposed by Kristensen and Higgins (1991). Spinoscalids of rings 02–06 or alternatively of ring 01 contain basally a cell with numerous dictyosomes and piles of membranes, which may represent microvilli; cilia have not been traced, the cell opens to the outside dorsally at the articulation of the scalid (Moritz & Storch 1972a, b; Brown 1989). This cell is supposed to be secretive (Kristensen & Higgins 1991), whereas Moritz and Storch (1972b) do not exclude a mechanoreceptive function.

The outer oral styles of *Echinoderes capitatus* and *Zelinkaderes floridensis* reveal three monociliary sensory cells with up to 23 μm long receptive cilia and elongated microvilli with microtubuli or filaments. There seems to be no terminal pore in *E. capitatus*, but it is present in *Z. floridensis*. In the latter species, two sensory cells yield basally an electron-dense, regularly perforated structure (Nebelsick 1993; Neuhaus 1994). Ciliary sensory cells also occur in *K. phyllotropis, P. dentatus*, and *P. greenlandicus* (see Brown 1989; Neuhaus 1994). Possibly, outer oral styles are chemosensory in function, because they connect to the environment through a pore.

Each inner oral style of the anteriormost ring of *Echinoderes capitatus* shows one monociliary sensory cell enveloped by a glia sheath cell. In the second ring, a multiciliary sensory cell with three cilia is found in each style. Each ring-03 style possesses two glia sheath cells as well as one multiciliary sensory cell with four cilia and a basal electron-lucent vesicle (Nebelsick 1993). Few to five sensory cilia with a length of 9–13 μm are traced in *Zelinkaderes floridensis, Kinorhynchus phyllotropis*, and *Pycnophyes dentatus* (see Brown 1989; Neuhaus 1994). Inner oral styles exhibit a subterminal pore, which indicates a chemoreceptive function of the sensory cells.

Red pigmented, dorsolateral eyes with a lens have been reported for *Echinoderes canariensis, E. citrinus, E. collinae, E. dujardinii, E. ehlersi, E. ferrugineus, E. subfuscus*, and *E. worthingii*; pigmentation disappears after fixation of the specimen (Zelinka 1928; Sørensen 2006). These eyes have not yet been studied by TEM. In the Pycnophyidae, one pair of ventrolateral, unpigmented cephalic sensory organs exists in *Kinorhynchus phyllotropis, Pycnophyes dentatus, P. greenlandicus*, and *P. kielensis* at the basis of the ring-01 spinoscalids (Fig. 5.1.43; Brown 1985; Kristensen & Higgins 1991; Neuhaus 1997). In the everted introvert, the organs appear at the anterior end of the animals with the cilia facing laterally (Fig. 5.1.43 A). Each organ is innervated from the forebrain and consists of a sensory cell with a single modified cilium and an enveloping cell, which commonly border the lumen of the organ. The receptory cilium increases significantly in size behind the basal body and branches into numerous ciliary processes (Fig. 5.1.43 B-D; Neuhaus 1997). Such processes have also been traced in *K.*

phyllotropis and *P. greenlandicus* but interpreted as microvilli (Brown 1985; Kristensen & Higgins 1991). However, these "microvilli" do contain microtubuli as do the ciliary processes of *P. dentatus* and *P. kielensis* and are consequently suggested as branching cilia by Neuhaus (1997). A cephalic sensory organ has not been traced in *Z. floridensis*; otherwise, no ultrastructural data is available for the Cyclorhagida. The cephalic sensory organ of the Pycnophyidae does not show any open connection to the environment but possesses a highly enlarged surface as the result of the numerous branches of the cilium. Therefore, a photosensory function is suggested for the organ (Neuhaus 1997). A cephalic organ as described before is postulated for the ground pattern of at least the Pycnophyidae (Neuhaus 1997).

Receptors of the integument. The only reconstruction of a kinorhynch sensory spot at TEM level is the one available for *Echinoderes capitatus*. Here, the organ consists of two monociliary sensory cells and a sheath cell. The cilia penetrate into or through the cuticle via two subcuticular tubes, suggesting a chemoreceptive function. One sensory cell shows nine conspicuous circumciliary microvilli with filaments and is therefore similar to the collar receptors of other invertebrates; the other cell forms an ensheathing basal cell process. Gland cells may occur close to the sensory spot, but do not reveal any connection with the sense organ (Nebelsick 1992a). Light microscopical studies indicate a close relationship of gland cells and sensory cells in *Centroderes spinosus*, and in *Pycnophyes greenlandicus* and *P. kielensis* these cells appear in close proximity at the TEM level (Adrianov & Malakhov 1990; Adrianov et al. 1989, 1990a; Kristensen & Higgins 1991; Neuhaus et al. 2013). TEM images of collar receptors are also known for *Kinorhynchus phyllotropis* (see Brown 1989). These sensory organs are supposed to be mechanoreceptive and to measure movements of the cuticular trunk plates (Adrianov & Malakhov 1990), but this appears to contradict the fact that the cilia of sensory spots are in contact with the environment via pores.

Gland cells of *Kinorhynchus giganteus* and *K. phyllotropis* possess at least one cilium with a large ciliary rootlet, whereas no cilia are reported for *Echinoderes cantabricus*, *E. capitatus*, and *E. hispanicus* (see Moritz & Storch 1972a; Brown 1985; Nebelsick 1992b; GaOrdóñez et al. 2000). Gland cells with piles of membranes and a canal through the trunk cuticle in *E. aquilonius* and *K. phyllotropis* are assumed to be chemoreceptive (Brown 1985; Kristensen & Hay-Schmidt 1989; Kristensen & Higgins 1991).

A single record (Brown 1985) mentions a seta associated with a ciliated gland cell in *Kinorhynchus phyllotropis*.

Nothing is known about the ultrastructure of female-specific (1) gland cell outlets with a funnel-like structure in species of *Fissuroderes*, (2) gland cell outlets with a bulbous structure in species of *Centroderes*, and (3) papillae and a patch of micropapillae in species of *Campyloderes* and *Condyloderes* (see Neuhaus & Blasche 2006; Sørensen et al. 2010b; Neuhaus & Sørensen 2012; Neuhaus et al. 2013).

The mid-dorsal spine of *Echinoderes capitatus* reveals a single basal sensory cell with three sensory cilia and no direct connection to the outside, indicating a mechanoreceptive function (Nebelsick 1992b). No enveloping cell occurs as reported by Kristensen and Higgins (1991). In *E. cantabricus* and *E. hispanicus*, lateral spines yield probably at least one sensory cilum (GaOrdóñez et al. 2000). In *E. capitatus*, shorter spines appear, which differ from the middorsal spine in that they show a gland cell and a sensory cell with a circle of microvilli around the cuticular tube but no cilium; the organ's lumen opens via a canal through the spine at its tip. Microvilli of the gland cell encircle the lumen inside the cuticular tube (Nebelsick 1992b).

All 15 specimens of *Cateria styx* studied possess a protrusible, cone-like dorsal organ with a strongly plicated surface at the border of the 5th and 6th segment (Fig. 5.1.8 C, D; Gerlach 1956; Higgins 1968). Such an organ has not been found in any other kinorhynch; its function is unknown.

Receptors of inner organs. In the inner epithelium of the pharynx, several (*Pycnophyes*) or 14 (*Zelinkaderes floridensis*) sensory cells occur, usually associated in pairs with one gland cell (Figs 5.1.44, 5.1.45). The single cilium is swollen anteriorly (*Z. floridensis*) or not (*Pycnophyes*), surrounded by circumciliary microvilli, and reaches the pharyngeal lumen. The receptor cell contacts nerve cells basally. In addition, processes of nerve cells terminate below the pharyngeal cuticle. In the outer pharyngeal epithelium, at least eight receptor cells appear. The electron-dense cilium either does not reach the mouth cone lumen (*Z. floridensis*) or it does (*Pycnophyes*) (Neuhaus 1991, 1994).

The muscle net of the midgut does not require innervation from the central nervous system, because the two or five supposedly monociliary sensorimotor cells in the midgut of *Z. floridensis* and *P. kielensis* are assumed to measure the fullness of the midgut and directly activate the peristalsis of the muscle net via their dense-core vesicles (Figs 5.1.46 B, 5.1.48 A;

Neuhaus 1991, 1994; Neuhaus & Higgins 2002; comp. also section 5.1 Morphology, Intestinal system). These presumptive receptors may also serve to control defecation, because some sensory cells are located near the hind gut and the longitudinal muscle cells connect to the hind gut dilatators. The muscle cells of the midgut are connected via gap junctions, which certainly facilitate communication of these cells (Neuhaus 1991, 1994).

The nephroporous cell of the protonephridium of *Echinoderes aquilonius* is penetrated by one to three processes of sensory cells and in *Pycnophyes kielensis* by two to three sensory cells with a modified cilium and numerous, small electron-dense vesicles (Fig. 5.1.49 A, F; Neuhaus 1988; Kristensen & Hay-Schmidt 1989). These cells are suggested to measure the osmolarity of the filtered nephridial fluid (Neuhaus 1988) and by their electron-dense vesicles possibly influence filtering activity. After having traced the sensorimotor cell-muscle association in the midgut, the protonephridium-sensory cell association may represent the second autonomous system in Kinorhyncha without involvement of the central nervous system. Unfortunately, the cell bodies of the sensory cells have not been traced in any of the species studied, and potential connections to nerve cells can currently not be excluded.

Intestinal system

The straight intestinal system with a terminal mouth and anus consists of a cuticle-lined foregut including the mouth cone, pharyngeal crown (in many species), pharynx, and very short esophagus, the non-ciliated midgut, and the cuticle-lined hindgut (Figs 5.1.1 A, 5.1.2 A). Opposite to light microscopical studies (Zelinka 1928), the alimentary canal is cellular throughout and does not reveal any syncytia; the foregut is innervated directly from the midbrain (Fig. 5.1.42; Neuhaus 1991, 1994; Nebelsick 1993; Adrianov & Malakhov 1994). TEM observations are available for *Cephalorhyncha asiatica*, *Echinoderes aquilonius*, *E. capitatus*, *Zelinkaderes floridensis*, *Kinorhynchus yushini*, *K. phyllotropis*, *Pycnophyes dentatus*, *P. greenlandicus*, *P. flaveolatus*, and *P. kielensis* (Merriman & Corwin 1973; Nyholm & Nyholm 1976a; Brown 1989; Neuhaus 1991, 1994; Nebelsick 1993; Adrianov & Malakhov 1990, 1994; Kristensen & Higgins 1991). The net of muscles around the midgut of *Antygomonas* sp. and *P. kielensis* has been investigated by phalloidin-labeling in connection with confocal laser scanning microscopy (Müller & Schmidt-Rhaesa 2003; Rothe & Schmidt-Rhaesa 2004; Schmidt-Rhaesa & Rothe 2006). A functional scenario for the uptake of food, digestion, and defecation is given in section 5.8 Ecology, Food and food uptake.

Mouth cone. The mouth cone is a protrusible, ring-like, epidermal fold around the anterior pharynx with nine outer orals styles and 20 inner oral styles arranged in three rings; inner oral styles of the first ring appear at two slightly offset levels in *Zelinkaderes floridensis* and *Kinorhynchus phyllotropis*, but are regarded as styles of a single ring (Fig. 5.1.42; Zelinka 1928; Remane 1936; Brown 1989; Nebelsick 1993; Neuhaus 1991, 1994). All oral styles possess ciliary receptor cells and a terminal or subapical opening. Inner oral styles nearly fill the mouth cone lumen. Each outer oral style can be moved by two to four longitudinal muscle cells in *Antygomonas* sp., *Z. floridensis*, and *K. phyllotropis* (see Brown 1989; Adrianov & Malakhov 1990; Nebelsick 1993; Neuhaus 1991, 1994; Müller & Schmidt-Rhaesa 2003). One or two circular muscle cells serve to constrict the anterior mouth cone lumen in *Echinoderes capitatus*, *Z. floridensis*, and *P. communis* and may assist to reject unwanted particles (Zelinka 1928; Nebelsick 1993; Neuhaus 1994). One circular muscle cell attaches to each basal inner oral style of *E. capitatus*, *Z. floridensis*, and *P. communis* and allows some movement of the styles (Fig. 5.1.42). A third circular muscle seems to exist at least in *Antygomonas* sp. and *E. capitatus* (see Zelinka 1928; Nebelsick 1993; Neuhaus 1991, 1994; Müller & Schmidt-Rhaesa 2003). At its anterior end, the mouth cone of *Paracentrophyes praedictus*, *K. giganteus*, *K. phyllotropis*, *Pycnophyes dentatus*, *P. greenlandicus*, and *P. kielensis* reveals longitudinal cuticular rods, which are supposed to act as filters for food particles (Figs 5.1.2 A, 5.1.34 B, C, 5.1.38); such a weir is missing in *E. capitatus* and *Z. floridensis* (Moritz & Storch 1972a; Higgins 1983; Brown 1989; Adrianov & Malakhov 1990; Neuhaus 1991, 1994; Nebelsick 1993; Kristensen & Higgins 1991).

Pharynx. The pharynx is pear-shaped in species of *Antygomonas*, *Centroderes*, and *Zelinkaderes*, but cylindrical in species of *Echinoderes*, *Kinorhynchus*, and *Pycnophyes*. Nuclei concentrate in the expanded part of the pharynx or are distributed more evenly in the epithelium of the cyclindrical organ (Figs 5.1.44 A, 5.1.45 A; Zelinka 1928; Nebelsick 1990; Neuhaus 1991, 1994, own obs.). The pharynx consists of an inner epithelium with monociliary receptor cells, monociliary gland cells, and three or six longitudinal nerve strands and an outer muscle bulb with radial and circular muscle cells. The outer muscle bulb contains 13–15 monosarcomeral radial and 13–15 oligosarcomeral circular muscle cells. Almost all muscle cells envelop the pharyngeal epithelium and lumen as a solid, continuous ring except the radial muscle cells of *Zelinkaderes floridensis*, where two ends meet (Figs 5.1.44, 5.1.45; Neuhaus 1991, 1994; Müller & Schmidt-Rhaesa 2003). In *Campyloderes* cf. *vanhoeffeni*, 20 circular and 20 radial muscle cells seem to exist (Neuhaus & Sørensen 2012). The muscle bulb is completely ensheathed by extracellular matrix. Anteriorly, a cuticular pharyngeal crown exists at least in species of *Centroderes, Dracoderes, Echinoderes, Kinorhynchus*, and *Pycnophyes*, but not in *Zelinkaderes*. Ten monociliary gland cells of the outer pharyngeal epithelium open at the anterior end of the pharynx and pharyngeal crown, respectively, into the mouth cone lumen (Figs 5.1.44 A, B, 5.1.45 A). One or two prepharyngeal sphincters occur at least in species of *Antygomonas, Echinoderes*, and *Zelinkaderes*, but not in Homalorhagida. A postpharyngeal sphincter is reported for species of *Antygomonas, Zelinkaderes, Pycnophyes*, and possibly *Kinorhynchus*. The lumen of the pharynx is triradiate with one apex directed dorsally, round, nine-lobed, or oval (Figs 5.1.44 B, C, 5.1.45 B, C; Zelinka 1928; Moritz & Storch 1972a; Merriman & Corwin 1973; Nyholm & Nyholm 1976a; Brown 1989; Nebelsick 1990; Adrianov & Malakhov 1990, 1994; Kristensen & Higgins 1991; Neuhaus 1991, 1994; Müller & Schmidt-Rhaesa 2003; Sørensen et al. 2012a).

The pharyngeal cuticle consists of a fine-granular basal layer, which is missing in *Zelinkaderes floridensis* and a multilamellar instead of a monolamellar epicuticle as in the body cuticle. This design may be related to food uptake, because the pharyngeal cuticle must follow the widening pharyngeal lumen. In some species, lobes (*Echinoderes aquilonius*) or epithelial folds (*Pycnophyes dentatus, P. kielensis*) may compensate the dilating pharyngeal lumen during the ingestion of food particles. However, no epithelial folds are traced in *Z. floridensis*. Therefore, the tufts of the "epi"cuticle (Fig. 5.1.44 B, C) are suggested to separate and stretch, and consequently the supposedly less elastic, basal, fine-granular cuticle layer is missing (Neuhaus 1991, 1994).

Circular and radial muscle cells alternate in *Zelinkaderes floridensis*, but circular muscle cells appear only in the periphery of the bulb in *Pycnophyes dentatus* and *P. kielensis* (Figs 5.1.44, 5.1.45). Here, each circular muscle cell usually splits into two cell bodies (*P. kielensis*), one between two consecutive radial muscle cells and one in the periphery of a radial muscle cell. The latter cell body is often (*P. dentatus*) or entirely (*P. flaveolatus*) missing in other species. Each two cell bodies of a circular muscle cell are connected by cell processes and contact neighboring circular cells by cell processes. In *Z. floridensis*, cell processes of radial muscle cells penetrate the neighboring circular muscle cell and vice versa. In this way, all circular muscle cells are connected with one another and all radial muscle cells contact each other, thereby allowing coordinated action of muscle cells (Figs 5.1.44, 5.1.45; Nyhom & Nyholm 1976a; Neuhaus 1991, 1994). Circular muscle cells have been taken as glycogen storage and fold artefacts by Merriman and Corwin (1973) or termed chorda muscle fibers by Nyholm and Nyholm (1976a).

The homalorhagid pharyngeal muscle cell arrangement is evaluated as superior in function over the cyclorhagid situation. In the cyclorhagid *Zelinkaderes floridensis*, radial and circular muscles alternate, facilitating dilatation of the pharyngeal lumen only in the area of the radial muscles. In *Pycnophyes*, the radial muscle cells attach to the epithelium over the entire length of the pharynx; the circular muscles are more evenly distributed in the periphery of the bulb, and allow a more homogeneous antagonism to the radial muscles (Neuhaus 1994) or a peristaltic wave, moving the food particles in the pharynx from anterior to posterior (Nyholm & Nyholm 1976a). The radial muscles of *Z. floridensis* possess three longitudinal nerve strands as opposed to six such strands and an enlarged T-system in *Pycnophyes*, supposedly offering a more sophisticated control of muscle cell activity. For the morphological and functional reasons mentioned above, the alternating sequence of circular and radial muscle cells of Cyclorhagida is regarded as plesiomorphic for Kinorhyncha (Neuhaus 1994).

Esophagus. The epithelium of the short esophagus bulges between the longitudinal muscles of the midgut muscle net and the pharynx protractors, which both attach to the posterior pharynx (Figs 5.1.44 A, 5.1.45 A). The epithelium does not differ from the pharyngeal epithelium and does not contain any gland cells (Neuhaus 1991, 1994) that have been suggested previously by some authors (Zelinka 1928; Gerlach 1956; Kristensen & Higgins 1991) but doubted by others (Remane 1936; Brown 1985).

Midgut. In cross-section, the lumen of the midgut appears round (*Echinoderes*), stretched (*Kinorhynchus giganteus, Pycnophyes communis*), triradiate and possibly slightly branched (*P. greenlandicus, P. kielensis*: Fig. 5.1.46 B), or consists of narrow intercellular gaps (*Zelinkaderes floridensis*: Fig. 5.1.46 A). The midgut of *Z. floridensis, P. dentatus*, and *P. kielensis* contains more than 95%

epithelial cells, a small amount of gland cells, and two or five sensorimotor receptor cells. A loose network of longitudinal and circular muscle cells surrounds the midgut in these species, *Antygomonas* sp., *P. carinatus*, and *P. communis* (Fig. 5.1.46; Zelinka 1928; Adrianov & Malakhov 1991d, 1994; Neuhaus 1991, 1994; Müller & Schmidt-Rhaesa 2003; Rothe & Schmidt-Rhaesa 2004). The monociliary sensorimotor cells contain plenty of dense-core vesicles and always face the muscle cells of the midgut; nerve cells do not exist near the presumptive sensorimotor cells (Neuhaus 1991, 1994). Midgut epithelial cells of *P. kielensis* from the White Sea are reported to contain vesicles with undigested and dividing bacteria, which are assumed to represent endosymbionts (Yushin et al. 1989, 1990; Adrianov et al. 1993; Adrianov & Malakhov 1994). This has not been corroborated for the same species from the North Sea (Neuhaus 1991, 1994; Neuhaus & Higgins 2002; comp. also section 5.8 Ecology, Food and food uptake).

Hindgut. In cross-section, the lumen of the cuticle-lined hindgut is triradiate and folded in *Echinoderes aquilonius* but a transverse broad slit in *Zelinkaderes floridensis*, *Kinorhynchus yushini*, *Pycnopyhes communis*, *P. dentatus*, *P. greenlandicus* (?), and *P. kielensis* (Figs 5.1.47 B, C, 5.1.48 B; Zelinka 1928; Kristensen & Higgins 1991; Neuhaus 1991, 1994; Adrianov & Malakhov 1994). Most certainly because of the body pressure of the animals, midgut epithelial cells reach far into the hindgut lumen and may give the impression of a "closed" gut (Kristensen & Higgins 1991). More caudally, the ventral and dorsal hindgut epicuticles stick together (Figs 5.1.47 A, 5.1.48 A; Neuhaus 1991, 1994). The "peritrophic membrane" of the frontal hindgut of *Cephalorhyncha asiatica* (Adrianov & Malakhov 1994: Fig. 86 A) is interpreted by Neuhaus and Higgins (2002) as the hindgut cuticle, which lacks the basal fine-granular layer found more caudally. The hindgut cells of *E. aquilonius* are supposed to serve in osmoregulation (Kristensen & Higgins 1991), but specific structures for this purpose have not been traced in *Z. floridensis*, *P. dentatus*, and *P. kielensis* (Neuhaus 1991, 1994).

Seven muscle cells next to the dorsoventral muscles of the terminal two segments are associated with the hindgut and function as dilatators of the rectum (Figs 5.1.47, 5.1.48). In segment 10 of *Zelinkaderes floridensis*, the paired frontal, dorsal, and caudal dilatators attach dorso- and ventrolaterally to the hindgut epithelium, caudally followed by a circular dilatator (Fig. 5.1.47). In *Pycnophyes dentatus* and *P. kielensis*, the dorsal dilatators appear as two dorsal transversal dilatators, and the circular dilatator occurs only as a ventral transverse dilatator. Frontal and caudal dilatators send a cell process to the ventral longitudinal nerve cord and the frontal dilatator also to a dorsal longitudinal nerve cord in order to get innervated (Fig. 5.1.48; Neuhaus 1991, 1994). Transverse or circular muscle cells have been reported for *Cateria styx*, *Echinoderes aquilonius*, *P. communis*, and *P. greenlandicus* and interpreted as sphincter; the remaining dilatator muscles have been attributed to the dorsoventral muscles as in *P. communis* or may have been overlooked because of their small size (Zelinka 1894, 1928; Remane 1936; Gerlach 1956; Kristensen & Higgins 1991). Zelinka (1928: p. 127) already recognizes that the frontal muscle must have a dilatatory function.

Body cavities and amoebocytes

The body cavity of *Cephalorhyncha asiatica*, *Echinoderes aquilonius*, *Pycnophyes dentatus*, *P. greenlandicus*, and *P. kielensis* is not as spacious as stated by Zelinka (1928) except in the introvert and around the pharynx, but otherwise limited to interstices between epidermis and inner organs. Extracellular matrix covers all inner organs completely but not always the epidermis; the body cavity itself appears as a granular matrix of different density in TEM (Kristensen & Hay-Schmidt 1989; Adrianov & Malakhov 1991d; Kristensen & Higgins 1991; Neuhaus 1991, 1994).

Numerous polymorphous amoebocytes with small, electron-dense vesicles, without cilia or myofilaments, and usually surrounded by extracellular matrix are distributed in the body cavity of *Echinoderes aquilonius*, *Pycnophyes dentatus*, *P. greenlandicus*, and *P. kielensis* (Figs 5.1.1 A, B, 5.1.2 A, B; Kristensen & Hay-Schmidt 1989; Adrianov & Malakhov 1991d; Kristensen & Higgins 1991; Neuhaus 1991, 1994). Amoebocytes appear singly or in groups of two to four; extracellular matrix does not always separate neigboring amoebocytes, indicating potential recent cell division (Neuhaus 1991, 1994). The latter studies corroborate Zelinka's (1928) finding that amoebocytes (= "Wanderzellen" of Zelinka) may develop from cell divisions in the body cavity. Neuhaus (1991, 1994) does not observe any precursor cells of amoebocytes or separation from midgut cells as suggested by Zelinka (1928).

The function of the amoebocytes is unknown. Zelinka (1928) observes migration in the body cavity and supposes phagocytosis of zooxanthellae, but the latter have not been found in species of *Pycnophyes*

Tab. 2. Primary and secondary sexual organs in different genera and indication of a preadult life history stage in different species (see remarks). – = stage not found yet.

genus	characters in ♀ (stage-1)	characters in ♀ (stage-2)	characters in ♂ (stage-1)	characters in ♂ (stage-2)	remarks	references
Cyclorhagida						
Antygomonas	acicular md + ml spine in segment 10; gonopores	acicular md + ml spine in segment 10; gonopores	–	acicular md + ml spine in segment 10	*A. oreas*: 1 stage-1 ♀ moulting	Bauer-Nebelsick 1996: Fig. 25; Sørensen 2007; Sørensen et al. 2009
Campyloderes	acicular md spine in segments 10–11; papillae in segments 6–7; gonopores	no md + ml spine in segments 10 + 11; ventral papillae in segments (5) 6–7; gonopores	–	acicular md + sl spine in segment 10 + md in segment 11	*C.* cf. *vanhoeffeni*: 7 stage-1 ♀ and 11 J-6 with acicular spine md + sl in segment 10 + md in segment 11	Neuhaus 2004; Neuhaus & Sørensen 2012
Centroderes	md + ml spine in segment 10; no gonopores; developing gonad	no md + ml spine in segment 10; gland cell outlet in segments 7–9; gonopores	crenulated [sic] md + ml spine in segment 10	crenulated md + ml spine in segment 10	*C. spinosus*: 1 stage-1 ♂ moulting, 8 stage-1 ♀	Neuhaus et al. 2013; Neuhaus own obs.
Centroderes ssp.	md + ml spine in segment 10; no gonopores	no md + ml spine in segment 10; gonopores	acicular [sic] md + ml spine in segment 10	crenulated md + ml spine in segment 10	2 stage-1 ♀ moulting, 1 stage-1 ♂ moulting (Fig. 5.1.13 A), 12 stage-1 ♀, 1 stage-1 ♂ & 26 juveniles with acicular spine md + ml in segment 10	Neuhaus, Sørensen, Pardos, and Higins own obs.
Condyloderes	–	ventral papilla in segments 7, 8, and/or 9	–			Sørensen et al. 2010b; Neuhaus own obs.
Cateria	–		–	–		Gerlach 1956; Higgins 1968
Dracoderes	–	gonopores	–	1 crenulated + 2 acicular lv penile spines in segment 11	*D. orientalis*: 3 acicular (?) lv penile spines in segment 11	Higgins & Shirayama 1990; Adrianov & Malakhov 1999a; Sørensen et al. 2012a
Cephalorhyncha	–	ltas in segment 11	–	3 lateral penile spines in s11; no ltas in segment 11		Adrianov & Malakhov 1999a; Neuhaus & Blasche 2006; Sørensen 2008b
Echinoderes	–	ltas in segment 11	–	3 lateral penile spines in segment 11; no ltas in segment 11	♂ *E. neospinosus*: 1 acicular + 2 crenulated penile spines; ♂ *E. andamanensis*, ?*E. bengalensis*, *E. kristenseni*, *E. lanceolatus*: sl seta or spine in segment 10, 2 lateral penile spines in segment 11, no ltas in segment 11; ♂ *E. newcaledoniensis*, *E. remanei*: 1 lateral penile spine in segment 11, ltas in segment 11; ♂ *E. maxwelli*, *E. rex*: 2 lateral penile spines; ♂ *E. abbreviatus*, *E. applicitus*, *E. cantabricus*,	Zelinka 1928; Karling 1954; Higgins 1964a, 1967, 1974, 1979b, 1983, 1985; Higgins & Rao 1979; Higgins & Kristensen 1988; Huys & Coomans 1989; Pardos et al. 1998; Adrianov & Malakhov 1999a; Sørensen 2000, 2006; Adrianov et al. 2002a, b; Chang & Song 2002; Sørensen et al. 2005, 2012b;

					E. eximus, E. higginsi, E. hispanicus, E. levanderi, E. maxwelli, E. newcaledoniensis, E. ohtsukai, E. peterseni, E. riedli: sl seta or spine in segment 10, 3 lateral penile spines in s11, no ltas in segment 11; ♀ *E. applicitus, E. bengalensis* (?), *E. capitatus, E. isabelae, E. maxwelli, E. ohtsukai, E. rex, E. teretis*: no ltas; ♀ *E. collinae, E. gizoensis, E.* cf. *kozloffi, E. spinifurca*: gland cell outlet with funnel-shaped subcuticular structure, ltas in segment 11	GaOrdóñez et al. 2008; Thormar & Sørensen 2010; Lundbye et al. 2011; Ostmann et al. 2012; Yamasaki & Kajihara 2012
Echinoderes coulli	morphotype 1: regular length of lts; no ltas	morphotype 2: very short lts; no ltas	–	3 lateral penile spines; regular length of lts; no ltas	2 ♀ morphotypes, have not been observed to moult into each other	Higgins 1977b; Higgins & Fleeger 1980
Fissuroderes	–	ltas in segment 11; gland cell with funnel-shaped subcuticular structure vm in segment 7 or vl in segment 8; gonopores	–	3 lateral penile spines in segment 11; no ltas in segment 11		Neuhaus & Blasche 2006
Meristoderes	–	ltas in segment 11	–	3 lateral penile spines in segment 11; no ltas in segment 11		Herranz et al. 2012; Sørensen et al. 2012c
Polacanthoderes	–	ltas in segment 11; gonopores	–	3 lateral penile spines in segment 11; no ltas in segment 11		Sørensen 2008b
Semnoderes	–	ventral papillae in segments 8–9; gonopores	–			Zelinka 1928; Sørensen 2006; Sørensen et al. 2009
Semnoderes sp.	s. remark	ventral papillae in segments 8–9; acicular sl spine in segment 10; gonopores	–	s. remark	3 specimens of preadult but not juvenile stage without clear ♂ or ♀ characters probably moult into adult	Neuhaus own obs.
Sphenoderes	–	gonopores	–			Higgins 1969a; Sørensen et al. 2010c
Triodontoderes	–	acicular md + lv spine in segment 10; gonopores	–	crenulated md + ld spine in segment 10		Sørensen & Rho 2009
Tubulideres	s. remark	acicular md + sd spine in segment 10; gonopores	–	crenulated md + sd spine in segment 10	3 specimens with acicular md + sd spine in s10 but no gonopores and not juvenile	Sørensen et al. 2007; Neuhaus own obs.

(*Continued*)

Tab. 2. (*Continued*)

genus	characters in ♀ (stage-1)	characters in ♀ (stage-2)	characters in ♂ (stage-1)	characters in ♂ (stage-2)	remarks	references
Wollunquaderes	acicular sl + md spine in segment 10; no papillae in s8-9; no gonopores	acicular sl + md spine in segment 10; papillae in s8-9; gonopores	–	crenulated ml + md spine in segment 10	*W. majkenae*: 3 stage-1 females (Fig. 5.1.29 B)	Sørensen & Thormar 2010; Neuhaus own obs.
Zelinkaderes	spines: n. a.; gonopores	acicular md + ld spine in segment 10; gonopores	–	crenulated md + ld spine in segment 10	*Z. floridensis*: 2 stage-1 ♀ moulting, 1 stage-1 ♂ moulting	Higgins 1990: Fig. 65; Bauer-Nebelsick 1995; Sørensen et al. 2007
Homalorhagida						
Neocentrophyes	–		–	2 lateral flexible penile spines in segment 11; md flexible spine in segment 11		Higgins 1969b
Paracentrophyes	–	md short spinose process in segment 10	–	md flexible spine in segment 10; flexible penile spine ld in segment 10 + ld in segment 11		Zelinka 1928; Higgins 1983; Neuhaus 1995; Sørensen et al. 2010a
Kinorhynchus	–	gonopores	–	vm adhesive tubule in segment 2; 2 vl penile spines in segment 10	*K. apotomus*, *K. belizensis*, *K. deirophrous*, *K. distentus* (?), *K. erismatus*, *K. stenopygus*, *K. trisetosus*: 3 vl penile spines; *K. langi*: 4 vl penile spines	Lang 1953; Higgins 1964b, 1965, 1974, 1982, 1983; Brown & Higgins 1983; Brown 1983; Adrianov & Malakhov 1999a
Pycnophyes	–	gonopores	–	vm adhesive tubule in segment 2; 2 vl penile spines in segment 10; penile bristles around gonopore	♂ *P. furugelmi*, *P. neuhausi*: 3 vl penile spines in segment 10; ♂ *P. chilensis*, *P. dolichurus*, *P. ecphantor*, *P. egyptensis*, *P. longicornis*: no vm adhesive tubule in segment 2; ♂ *P. arctous*: 2 or 3 vl penile spines in segment 10 (?)	Lang 1953; Higgins 1961, 1965, 1966a, 1974, 1983; Higgins & Kristensen 1988; Neuhaus 1993, 1999; Adrianov 1995; Adrianov & Higgins 1996; Adrianov & Malakhov 1999a; Murakami et al. 2001; Lemburg 2002a, b; Martorelli & Higgins 2004; Sánchez et al. 2011

Gametes

Female and male gametes as well as epithelial cells of the gonads are supposed to originate from the apical cell at the anterior end of the gonad (Zelinka 1928), but a later study leaves this hypothesis open (Kristensen & Higgins 1991). The spermatozoa may be 50–140 μm long. The size varies from 29% of the body length of *Echinoderes capitatus*, over 25% of *E. dujardinii*, and 18% of *Kinorhynchus giganteus* and *Pycnophyes carinatus*, to 8% of *P. communis* (see Zelinka 1912, 1928). The enormous size and vermiform shape of the spermatozoa prompted an early investigator to mistake the sperm for embryos (Greeff 1869). TEM data on female and male gametes is limited to *Echinoderes aquilonius, Zelinkaderes floridensis, Pycnophyes communis, P. flaveolatus, P. greenlandicus, P. kielensis*, and *Kinorhynchus phyllotropis* (see Brown 1983; Nyholm 1976, 1977; Nyholm & Nyholm 1982, 1983; Adrianov & Malakhov 1991b, 1994, 1999b; Kristensen & Higgins 1991; Neuhaus & Higgins 2002).

Female gametes. In the centre of each gonad, one giant oocyte with a large nucleus and nucleolus as well as numerous yolk vesicles develops. Consequently, oocytes become mature in each gonad one by one, but usually only in either the left or the right gonad at a time (Fig. 5.1.3 C). The smaller cells anterior and posterior of the oocyte are assumed to represent both immature germ cells and nutritive cells; the latter seem to be reabsorbed during oogenesis in *Echinoderes aquilonius, Pycnophyes communis, P. flaveolatus*, and *P. greenlandicus* (see Zelinka 1894, 1928; Nyholm & Nyholm 1983; Kristensen & Higgins 1991). Cells in the posterior part of the gonad are regarded as primary oocytes (Kristensen & Higgins 1991).

Male gametes in testis. In mature male gonads, spermatogonia, spermatocytes, spermatids, and spermatozoa develop from anterior to posterior (Higgins 1974; Nyholm & Nyholm 1982; Adrianov & Malakhov 1991b, 1994, 1999b). A spermatocyte contains 10 chromosomes in *Pycnophyes carinatus* and 14 chromosomes in *P. communis* according to Zelinka (1928: p. 150). The cigar-shaped mature spermatozoa contain a central rod-shaped nucleus, which is partly surrounded by three different vesicles that bulge above the surface (Zelinka 1928; Brown 1983; Nyholm & Nyholm 1982; Adrianov & Malakhov 1999b). Tubes with a central electron-dense filament face the vesicles. An acrosomal structure and a middle piece without a nucleus are missing (Nyholm 1976; Nyholm & Nyholm 1982; Adrianov & Malakhov 1994, 1999b). One vesicle type develops from mitochondria and is supposed to serve as a storage organelle for either spermatozoan movement, the female, or for the zygote (Nyholm 1976). The cilium of *P. communis* and *P. flaveolatus* seems to originate in the posterior region of the mature cell (Nyholm & Nyholm 1982), but at the anterior end of *P. kielensis*; here, the cilium runs along the entire length of the spermatozoan cell (Adrianov & Malakhov 1994, 1999b). The free end of the cilium measures just 10–15 μm in all three species. However, locomotion of *Echinoderes dujardinii* and of *P. communis* sperm seems to be limited to the anterior two thirds of the cell (Zelinka 1928; Nyholm & Nyholm 1982).

Spermatozoa of Cyclorhagida have not been studied by TEM.

Sperm in spermatophore. Occasionally, female *Kinorhynchus langi, K. ilyocryptus, K. phyllotropis, Pycnophyes dentatus, P. flaveolatus, P. frequens, P. greenlandicus*, and *P. kielensis* (Fig. 5.1.39 A, B) reveal a brownish mucous mass of up to 190 mm diameter at their posterior ends, covered by detritus and diatoms; the mass may or may not possess one or two spherical structures inside with about 140 intertwined spermatozoa and a similar number of spermatids in *K. phyllotropis* (Nyholm 1947c; Higgins 1965, 1974; Brown 1983; Kristensen & Higgins 1991; Adrianov & Malakhov 1999b; Neuhaus own obs.). The morphology of these cells agrees with the same cell type in the testis. Consequently, the brownish mass has been interpreted as either eggs (Nyholm 1947c; Higgins 1965, 1974) or a spermatophore (Brown 1983; Kristensen & Higgins 1991; Adrianov & Malakhov 1999b). The mucous mass may be found also on more anterior segments of *K. phyllotropis* but not in *P. dentatus* and *P. kielensis* (see Brown 1983; Neuhaus own obs.). Reports for Cyclorhagida are only available for a single specimen of *Centroderes spinosus* (Fig. 5.1.13 B).

Sperm in receptaculum seminis. Whereas Zelinka (1928) reports sperm in the receptaculum, Remane (1936) and Boykin (1965) assume a purely glandular function. However, TEM studies confirm the observations by Zelinka. Each receptaculum seminis of *Echinoderes aquilonius* contains some 20–30 spermatozoa, which differ from those in the testis (Kristensen & Higgins 1991); this is very much in accordance with a report by Zelinka (1928) for *Pycnophyes communis*. Spermatozoa of *E. aquilonius* seem to lack cilia and locomotion is supposed to be facilitated by microvilli. Some spermatozoa seem to be reabsorbed by the epithelial cells; in other spermatozoa, microvilli appear to be embedded in the epithelium (Kristensen & Higgins 1991). In *Kinorhynchus phyllotropis*, more than 20 spermatozoa lose their elongate shape, becoming polyhedral and larger in diameter anteriorly with a now polymorphous nucleus; sperm heads face the epithelial lining of the receptaculum, the cilia remain in the organ's lumen (Brown 1983). Alterations of spermatozoa of *P. flaveolatus* include irregularly shaped cell bodies with condensed and uncondensed, polymorphous nuclei, myelination of spermatozoa, lack of cell membranes, groups of four to six flagella surrounded by a single cell membrane, and cells with a regular nucleus supposed to divide amitoticly into 8–12 cells (Nyholm 1976, 1977). These findings prompt Nyholm (1977) to hypothesize "morphological hermaphroditism" in the receptaculum seminis of *P. flaveolatus*. However, Kristensen and Higgins (1991) suggest that the cells with the regular nucleus represent epithelial cells of the receptaculum which reabsorb spermatozoa. This position is also taken by Adrianov and Malakhov (1994, 1999b) and Neuhaus and Higgins (2002).

The incidence of spermatophores in *Kinorhynchus phyllotropis* is about one in 88 females (Brown 1983). Spermatophores occur in females of different ages in *Pycnophyes dentatus* and *P. kielensis* probably at a similar rate (Neuhaus own obs.). Females may reveal spermatozoa in their receptaculum seminales shortly after hatching from the last juvenile stage at a time when the gonads are far from being mature (Zelinka 1928); therefore, the latter author assumes that sperm has to be stored in the receptaculum for up to six months until the female gonads have developed far enough for fertilization. Nyholm and Nyholm (1983) doubt that the dramatically transformed spermatozoa in the receptaculum may be able to fertilize the oocytes. However, a different hypothesis is suggested here. Considering the low incidence rate of spermatophores and that the spermatophore of *K. phyllotropis* contains both mature spermatozoa and immature spermatids (Brown 1983), it seems reasonable that the mature spermatozoa serve to immediately fertilize mature oocytes while they pass the receptaculum, whereas immature spermatids may be stored in the receptaculum until additional oocytes become ready for fertilization. Consequently, surplus spermatozoa may be reabsorbed in the receptaculum and spermatids would have to finish their development in the receptaculum of the female. If this idea is correct, a female needs to be inseminated just once or a few times to receive sperm for her lifetime.

5.2. Reproduction and development

Reproductive biology

Data on the ratio of the sexes in a given population is scarce. *Echinoderes coulli*, *Pycnophyes carinatus*, and *P. communis* reveal the same percentage of male and female specimens in field samples, whereas females dominate males in *E. dujardinii* in the ratio 3:1; no data is available for any other species (Zelinka 1928; Higgins & Fleeger 1980).

Copulation. Copulation has been reported once for Kinorhyncha, namely for *Pycnophyes kielensis*. The specimens are attached to each other with the ventral side of their posterior ends. The animals face more or less in opposite directions. A brownish mucous mass envelops the posterior ends. This finding suggests that the brownish mass of the spermatophore originates from copulation and should not be regarded as eggs (Neuhaus 1999). Ventromedial adhesive tubes in segment 2 of *Pycnophyes* and *Kinorhynchus* have been assumed to aid in copulation (Zelinka 1928), but this hypothesis is not corroborated by Neuhaus (1999, own obs.). However, mating behaviour has not been studied by this author, so the adhesive tubes may be involved at an earlier stage of mating.

For the thin and flexible spines of *Neocentrophyes* and *Paracentrophyes* (Fig. 5.1.33 A, B) a sensory function is assumed by Neuhaus and Higgins (2002). One male *P. praedictus* yields one pair of everted tubular structures dorsally at the border of the 10th and 11th segment (Fig. 5.1.35 A, B; Neuhaus own obs.). The tubes appear tripartite from the outside with the swollen posteriormost part covered by numerous, small knobbly projections. Terminally, the tubes show a smooth end and do not seem to be ruptured (Fig. 5.1.35 A, B). A probable explanation is that the tubes are everted artificially in the process of preparation of the specimen for SEM studies. However, the tubes may also reflect a natural situation insofar as they may possibly represent the cuticle-lined part of the gonoducts, which is everted in order to aid in the transfer of sperm to the female. Amazingly, the tubes appear dorsally displaced and not ventrally as expected from studies of other Homalorhagida for gonopores (Brown 1983; Adrianov & Malakhov 1991c, 1994).

The more rigid penile spines of *Kinorhynchus phyllotropis* (Fig. 5.1.36 D) contain epithelial cells but no muscles and no internal duct, but *Echinoderes filispinosus* seems to reveal ciliary sensory cells and an internal gland cell duct in its spines (Brown 1983; Adrianov & Malakhov 1994, 1999b). It remains also open how penile spines contribute to copulation. Brown (1983: p. 264) suggests that the spines may "prevent loss of the spermatophore after its extrusion" whereas Neuhaus and Higgins (2002) suspect that contraction of the dorsoventral muscles increases the body pressure and helps to insert these spines into the female gonopores in order to keep the gonopore open and to anchor female and male.

The observation of copulation, spermatophores, and of sperm-filled receptaculum seminales in many species (see above) suggests internal fertilization in Kinorhyncha.

Deposition of eggs, parental care. Nyholm (1947c) reports that the last juvenile stage of *Pycnophyes communis* or *P. flaveolatus* deposits eggs in the just shed exuvia. He also notices that these species lay eggs "in the canals excavated by the animals in the uppermost detritus layer" (Nyholm 1947c: pp. 33–34). Lang (1963: p. 260) finds two exuvia of *P. flaveolatus* with five and six eggs, respectively, and raises the question whether the animals die after deposition of eggs or whether the specimens moult as adults.

One female *Echinoderes kozloffi* has been observed to actively cover her egg with a thick layer of detritus, bacteria, and diatoms until the egg reaches 260–300 μm in diameter. The female secretes a sticky substance and uses "the scalids to consolidate and smooth the mass" of glue and particles (Kozloff 1972, 2007: p. 202). The thick layer obscures observation of the embryo and makes searching for eggs difficult in the sediment (Kozloff 1972, 2007). The egg is spherical (diameter 54 μm) or ovoid (60 μm × 72 μm) in shape with an egg shell about 1 μm thick and surrounded by a second, dense, about 30–40 μm thick layer; eggs are oligolecithal (Kozloff 2007). It seems that a female lays one egg at a time.

Cleavage

Cleavage has been studied in a single species, namely *Echinoderes kozloffi*; embryonic development appears to take about 10 days until hatching (Kozloff 1972, 2007). Polar bodies have not been traced. The kinorhynch seems to show a holoblastic cleavage (Kozloff 2007: Figs 2, 3) leading to an embryo with no conspicuous cavity between blastomeres. Later, the embryo elongates, bends inside the egg, and organs such as the alimentary canal develop (Kozloff 2007).

Development

Postembryonic development is direct, there are no larval organs. The first juvenile stage (J-1) of six stages hatches with seven to eight separated segments and a posterior area consisting of the anlagen of three to four segments. The new trunk segments appear in a subcaudal differentiation zone in stages following J-1 (Fig. 5.2.1; Zelinka 1928; Brown 1985; Neuhaus 1993, 1995; Sørensen et al. 2010d); in different species, additional segments become recognizable at different life history stages (Tab. 3). The subcaudal differentiation zone of Kinorhyncha with its anlagen of the segments should not be confused with the subcaudal growing zone of Articulata where segments develop de novo (Scholtz 2002). Hatching from an egg has been observed in a single species, namely *Echinoderes kozloffi*. Here, the J-1 (115–120 μm long) hatches from the egg (about 70 μm long) by stretching its body and protruding its head until the egg shell ruptures. It possesses an alimentary canal, a head with several rings of scalids, and nine trunk segments including the posterior area of anlagen of segments (Kozloff 1972). This observation lays to rest previous studies by Nyholm (1947a, c) who suggests development of *Echinoderes elongata* via a series of seven juvenile stages from a three-segmented stage without scalids, gradually increasing in segment number from stage to stage. Subsequent studies also confirm a first juvenile stage with at least eight trunk segments (seven separated plus four segments as anlagen) and fully functional scalids, as well as a series of not more than six juvenile stages (Tab. 3).

The complete series of six juvenile stages is known for *Antygomonas incomitata*, *Echinoderes bookhouti*, *Zelinkaderes floridensis*, *Paracentrophyes praedictus* (Fig. 5.2.1), *Pycnophyes australensis*, *P. beaufortensis*, *P. dentatus*, *P. kielensis*, and *Kinorhynchus phyllotropis* (Tab. 3; Higgins 1974, 1990; Brown 1985; Neuhaus 1993, 1995; Lemburg 2002a, b; Sørensen et al. 2010d). Individual stages have been noted for *Cateria gerlachi* (J-4, J-5), *C. styx* (J-1–J-5), *Echinoderes capitatus* (J-6), *E. cavernus* (J-1, J-4, J-5), *E. coulli* (J-3–J-6), *E. dujardinii* (J-3, J-4, J-6), *E. gerardi* (J-6), *E. riedli* (J-4–J-6), *E. subfuscus* (J-6), *E. sublicarum* (J-3–J-6), *Kinorhynchus giganteus* (J-6), *Pycnophyes calmani* (J-6), *P. carinatus* (J-6), *P. communis* (J-6), *P. cryopygus* (J-1–J-5), *P. echinoderoides* (J-6), *P. flaveolatus* (J-6), and *P. kielensis* (J-1, J-3, J-4, J-6) (Zelinka 1928; Higgins 1968, 1977a, b, 1978b; Higgins & Kristensen 1988; Adrianov et al. 1990c; Adrianov & Malakhov 1994; Sørensen et al. 2000). Investigations of the postembryonic development have to be regarded with some reservation, because all studies are exclusively based on a

Tab. 3. Characters of juvenile stages of different species.

characters species	juvenile stage: segments separated	plates separated mv in stage	stage: lateral terminal spine appears as	stage: lateral terminal accessory spine appears as	remarks	reference
Cyclorhagida						
Antygomonas incomitata	J-1: s1-8 + anlagen of s9-11 J-2–J-4: s1-9 + anlagen of s10-11 J-5, J-6: s1-11	J-6: s2-10 adult: s2-11	J-2: spine	J-2: spine	J-1–J-6: midterminal spine; md spine: s4 + 6 + 8 + 9 in J-1, s1-10 in J-2, s1-11 from J-3 on	Sørensen et al. 2010d
Echinoderes bookhouti	J-1: s1-8 + anlagen of s9-11 J-2, J-3: s1-9 + anlagen of s10-11 J-4–J-6: s1-11	n. a.	J-1: spine	J-1: spine	J-1–J-3: midterminal spine; md spine: s4-8 in J-1, s4-9 in J-2 & J-3, s4-10 in J-4, s4-9 in J-5, s4-8 from J-6 on	Higgins 1974
Zelinkaderes floridensis	J-1–J-2: s1-8 + anlagen of s9-11 (J-3 not found, see text) J-4, J-5: s1-9 + anlagen of s10-11 J-6: s1-11	J-6: s3-11	J-5: spinose anlage J-6: spine	J-4: spinose anlage J-5: spine	J-1–J-6: midterminal spine; md spine: s4 + 6 + 8 + 9 in J-1 & J-2, s4 + 6 + 8-10 in J-4 & J-5, s4 + 6 + 8-11 from J-6 on	Higgins 1990; Neuhaus own obs.
Homalorhagida						
Paracentrophyes praedictus	J-1: s1-8 + anlagen of s9-11 with 2 lv spinose processes J-2–J-6: s1-11	J-6: s2-10 adult: s1-11	J-2: spinose anlage J-3: articulated anlage J-4–J-6: spine	J-3: spinose anlage	J-1–J-6: midterminal process; md spinose process: s2-9 in J-1, s1-10 from J-2 on	Neuhaus 1995
Pycnophyes australensis	J-1: s1-7 + anlagen of s8-11 with lv spinose process J-2: s1-7 + anlagen of s8-11 with 2 lv spinose processes J-3, J-4: s1-8 + anlagen of s9-11 with 2 lv spinose processes J-5, J-6: s1-9 + anlagen of s10-11 with lv spinose process	J-2: s2-7 J-3, J-4: s2-8 J-5, J-6: s2-9 adult: s1-11	J-3: spinose anlage J-4: articulated anlage J-5: articulated anlage elongated J-6: articulated anlage adult: spine	J-2: spinose anlage	J-1–J-3: midterminal process; md spinose process: s4 + 6 + 8 in J-1, s4 + 6 + 8 + 9 in J-2, s2-9 in J-3 and J-4, s2-10 from J-5 on	Lemburg 2002a, b
P. beaufortensis	J-1, J-2: s1-7 + anlagen of s8-11 J-3: s1-8 + anlagen of s9-11 J-4, J-5: s1-9 + anlagen of s10-11 J-6: s1-11	n. a.	J-1: spinose anlage J-4: spine	J-2: spinose anlage	J-1–J-3: midterminal process; md spinose process: s1-8 in J-1, s1-9 in J-2, s1-10 from J-3 on	Higgins 1974
P. cryopygus	J-1: s1-7 + anlagen of s8-11 J-2: s1-8 + anlagen of s9-11 J-4: s1-9 + anlagen of s10-11 J-5, J-6: s1-11	probably J-6	J-4: spine	n. a.	J-1–J-3: midterminal process; md spinose process: s4 + 6 + 8 in J-1, s2-10 in J-2, s1-10 from J-3 on	Higgins & Kristensen 1988

P. dentatus	J-1: s1-7 + anlagen of s8-11 with lv spinose process J-2: s1-7 + anlagen of s8-11 with 2 lv spinose processes J-3: s1-7 + anlagen of s8-11 with 3 lv spinose processes J-4: s1-8 + anlagen of s9-11 with 2 lv spinose processes J-5: s1-9 + anlagen of s10-11 with spinose process J-6: s1-11 with spinose process	J-3: s2-7 J-4: s2-8 J-5: s2-9 J-6: s2-10 adult: s1-11	J-4: spinose anlage J-5: articulated anlage J-6: articulated anlage elongated adult: spine	J-3: spinose anlage	no midterminal process; md spinose process: s4 + 6 + 8 + 9 in J-1, s4 + 6-9 in J-3, s2-4 + 6-9 in J-4, s1-9 from J-5 on	Neuhaus 1993
P. kielensis	J-1: s1-7 + anlagen of s8-11 with lv spinose process J-2: s1-7 + anlagen of s8-11 with 2 lv spinose processes J-3: s1-7 + anlagen of s8-11with 3 lv spinose processes J-4: s1-8 + anlagen of s9-11 with 2 lv spinose processes J-5: s1-8 + anlagen of s9-11 with 2 lv spinose processes J-6: s1-11with lv spinose process	J-3: s2-7 J-4: s2-8 J-5: s2-8 J-6: s2-10 adult: s1-11	J-4: spinose anlage J-5: articulated anlage J-6: spine	J-3: spinose anlage	no midterminal process; no md spinose process	
Kinorhynchus phyllotropis	J-1–J-5: s1-7 + anlagen of s8-11 J-6: s1-11		J-4: spinose anlage	J-5: spinose anlage	no midterminal process; md spinose process: s2-10 from J-1 on	Brown 1985

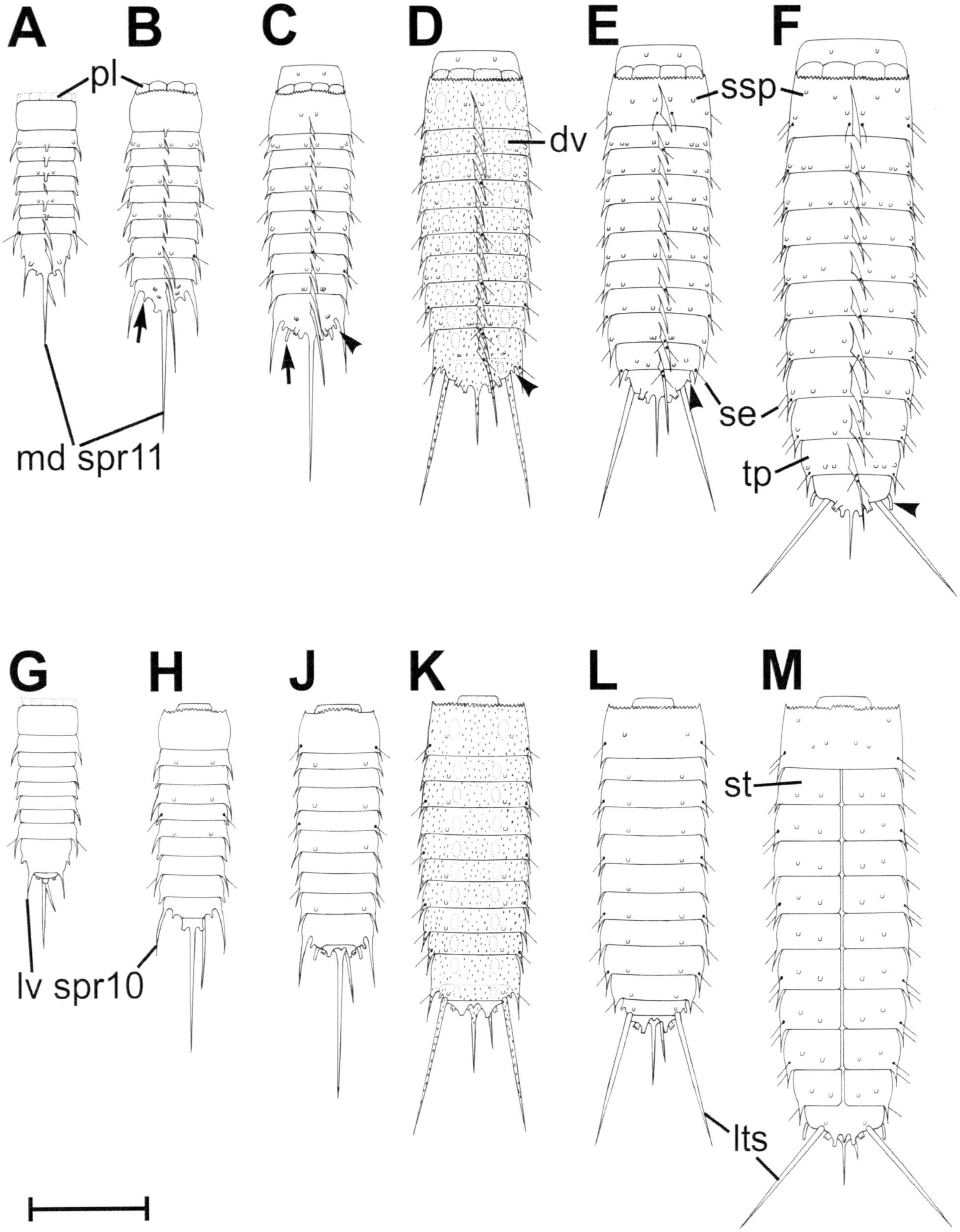

Fig. 5.2.1. Juvenile stages of *Paracentrophyes praedictus* in dorsal view (A–F) and in ventral view (G–M), (after Neuhaus 1995; © John Wiley and Sons), line in H = 100 µ, line valid for A-M. A, G, J-1; B, H, J-2; C, J, J-3; D, K, J-4; E, L, J-5; F, M, J-6; distribution of cuticular hairs and attachment site of dorsoventral muscles only indicated in J-4 (D, K); arrows point to anlagen of lateral terminal spines, arrowheads mark anlagen of lateral terminal accessory spines.

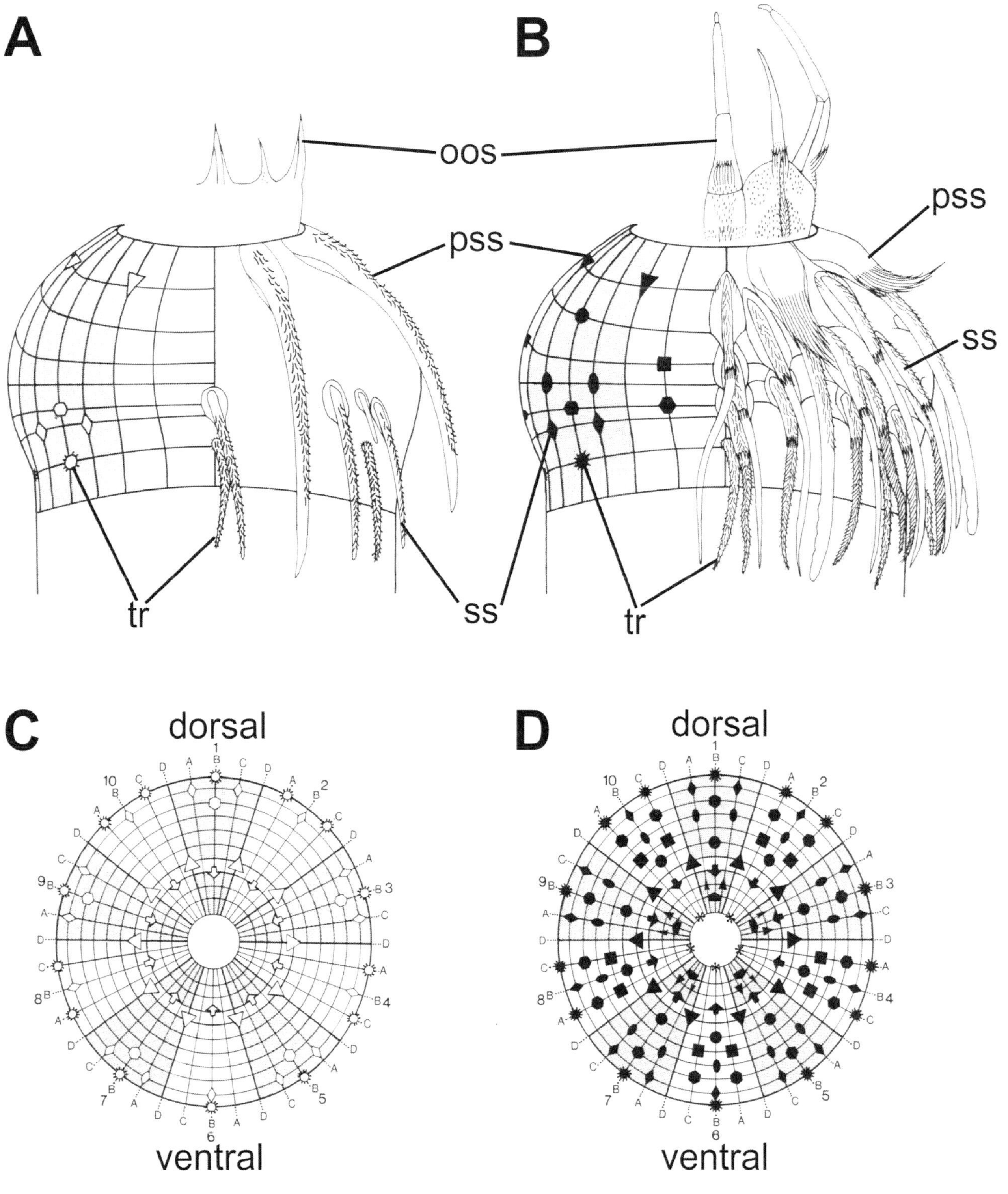

Fig. 5.2.2. Schematic arrangement of scalids and oral styles of the 1st (A, C) and the 6th juvenile stage (B, D) of *Paracentrophyes praedictus*, (after Neuhaus 1995; © John Wiley and Sons). A, B, ventral view of introvert; C, D, polar-coordinate diagram of the distribution of scalids and oral styles. Open symbols represent anlagen of scalids or oral styles, black symbols differentiated scalids or oral styles.

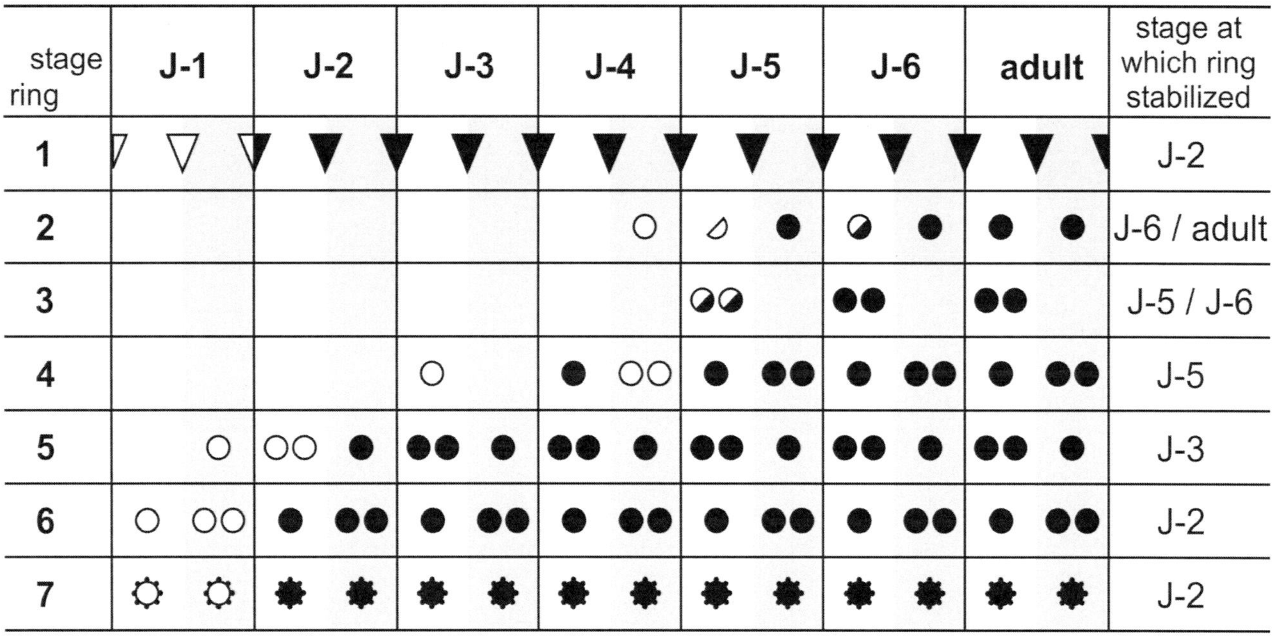

Fig. 5.2.3. Diagrammatic summary of the development of primary spinoscalids (triangles), spinoscalids (dots) and trichoscalids (asterisks) in juvenile stages J-1 through J-6 and adult specimens of *Paracentrophyes praedictus*; only an even-numbered (unshaded) and odd-numbered (shaded) sector of introvert shown for each stage, (after Neuhaus 1995; © John Wiley and Sons); for exact position of trichoscalids refer to Fig. 5.2.2; an open symbol represents the anlage of a scalid, a black symbol a differentiated scalid; a half circle refers to a spinoscalid present in some individuals as anlage, but missing in others; a half black circle symbolizes a spinoscalid expressed as anlage in some specimens, but as a differentiated scalid in other individuals.

limited number of preserved specimens, except observations on *P. kielensis* and *P. dentatus*, which derive from more than 1,200 cultivated specimens including both hatched individuals and the respective exuvia prooving directly specific juvenile stages moulting from each other (Neuhaus 1993).

In *Z. floridensis*, e.g., Higgins's J-2 is distinguished from his J-1 by four new spines and one additonal segment; just two spines and one segment are added in stage J-3, two spines develop in J-4, but no morphological changes except different sizes are described for stages J-5 and J-6 (Higgins 1990). By chance, Neuhaus (own obs.) finds one stage with characters intermediate between Higgins's J-1 and J-2 revealing one additional lateroventral spine in segment 11. This new stage (now J-2?) reduces the amount of changes between Higgins's J-1 and J-2 slightly, but one additional stage (J-3?) not traced yet is expected to exist. Consequently, Higgins's stages J-4 through J-6 are now combined as a single stage J-6, because no addition of morphological characters seems to occur between these stages. In summary, *Z. floridensis* probably still shows 6 juvenile stages, but in a different composition than originally described.

The Homalorhagida *Paracentrophyes praedictus* (Fig. 5.2.1 A-C, G-J), *Pycnophyes australensis*, *P. beaufortensis*, and *P. cryopygus* possess a midterminal process at least in the early three juvenile stages, whereas species of *Kinorhynchus*, *P. dentatus* and *P. kielensis* do not (Tab. 3; comp. also section 5.4 Phylogeny; Zelinka 1928; Higgins 1961, 1962, 1974; Brown 1985; Higgins & Kristensen 1988; Neuhaus 1993, 1995; Lemburg 2002a, b).

Changes of morphological characters in juvenile stages. Introvert scalids of juvenile stages are pentaradial symmetrically arranged as in the adult stage. In the three species where head development has been studied, the number of scalids increases during the ontogeny, new scalids are introduced as prescalids and subsequent protoscalids in a subfrontal differentiation zone in *Antygomonas incomitata* or as protoscalids in *Paracentrophyes praedictus* (Figs 5.1.35 C, 5.2.2, 5.2.3) and *Kinorhynchus phyllotropis*; in the next developmental stage, each protoscalid differentiates into a regular scalid (Brown 1985; Neuhaus 1995; Sørensen et al. 2010d). Prescalids are defined as the area where a protoscalid will develop, usually identified by dense or long cuticular hairs or by an elongate or triangular structure (Fig. 5.1.35 C; Tab. 1; Sørensen et al. 2010d). Protoscalids are spinose anlagen where a scalid will appear in the next stage (Tab. 1; Brown 1985: proscalid; Neuhaus 1995). Stages J-5 and J-6 of *P. praedictus* express scalids in two different patterns depending on whether a protoscalid in rings 02 and 03 appears in J-5 or

J-6, respectively (Fig. 5.2.3); these patterns do not seem to be related to the gender (Neuhaus 1995). In early ontogenetic stages, at least *A. incomitata* and *P. praedictus* also reveal outer oral styles differing in size and length in every second position; the smaller oral styles represent spinose anlagen (Neuhaus 1995; Sørensen et al. 2010d).

The trunk cuticle of juvenile Kinorhyncha is always much thinner and more flexible than in adult specimens and lacks pachycycli (Zelinka 1928). Based on phalloidin labelling in confocal laser scanning studies, the ventral and dorsal longitudinal muscles of *Pycnophyes kielensis* extend from segments 1–7 or 1–8 in stages J-1 through J-5, from segments 1–9 in J-6, and from segments 1–10 in the adult stage. The longitudinal muscles appear to be continuous and not separated into segmental units in any juvenile stage and may continue over the pachycyclus even in adult specimens in some cases (Schmidt-Rhaesa & Rothe 2006). However, TEM studies of the J-1 of *Pycnophyes kielensis* reveal at least some segmental cell membranes between subsequent longitudinal muscles suggesting that an individual muscle cell may not continue over all segments (Neuhaus own obs.). The lack of dorsal longitudinal muscles in segment 11 of the last juvenile stage has been reported previously for *P. communis* and *P. kielensis* (see Zelinka 1928; Rothe & Schmidt-Rhaesa 2004). Whereas the first two juvenile stages of *P. kielensis* possess dorsoventral muscles in segments 1–10, all remaining life history stages reveal these muscles also in segment 11; consequently, the development of the dorsoventral muscles is faster than the development of the externally recognizable segments (Schmidt-Rhaesa & Rothe 2006).

At least in early stages of the postembryonic development, no sternal plates can be identified (Tab. 3; Zelinka 1928). However, juvenile stages of at least *Antygomonas incomitata*, *A. paulae* (Fig. 5.1.7 D), *Centroderes spinosus* (Fig. 5.1.12 D), *Condyloderes* sp. (Fig. 5.1.14 E), *Echinoderes cavernus*, *Paracentrophyes praedictus*, *Pycnophyes dentatus*, and *P. kielensis* seem to possess a midventral sternal plate, which may show cuticular scales or hairs more robust than laterally. This impression originates from a contraction of the dorsoventral muscles, which leads to longitudinal indentation of the weak juvenile cuticle (comp. also Fig. 5.1.14 F; Neuhaus 1993, 1995; Sørensen et al. 2000, 2010d; Neuhaus et al. 2013). Light microscopical investigations do not corroborate a midventral sternal plate in the species mentioned previously, and several illustrations are somewhat misleading in this respect (Neuhaus 1993: Figs 1, 8; Sørensen et al. 2000: Fig. 8 B, 2010d: Fig. 1). At the posterior margin of most segmental cuticular plates, cyclorhagid juvenile stages but not the adult possess conspicuous postmarginal spiculae or spinose processes (Fig. 5.1.7 D; Zelinka 1928; Sørensen et al. 2010d; Neuhaus & Sørensen 2012; Neuhaus et al. 2013).

During postembryonic development, the number of sensory spots, gland cell outlets, setae, spines, and to some degree muscles increases from stage to stage (Fig. 5.2.1; Zelinka 1928; Brown 1985; Higgins & Kristensen 1988; Higgins 1990; Neuhaus 1993, 1995; Lemburg 2002a,b; Rothe & Schmidt-Rhaesa 2004; Schmidt-Rhaesa & Rothe 2006; Sørensen et al. 2010d). Sensory spots may change their morphology from being elevated above the cuticle surface (type-3) to integrated into the cuticle surface (type-2) both within juvenile ontogeny and from juvenile to adult stage (Neuhaus 1993; Sørensen et al. 2010d). Cyclorhagid spines are relatively longer in early juvenile stages than in later stages and in adults, especially middorsal spines on posterior segments (Zelinka 1928; Higgins 1968, 1977a, b; Higgins & Kristensen 1988; Adrianov et al. 1990b; Adrianov & Malakhov 1994; Sørensen et al. 2000). Homalorhagida possess articulated spines only as a lateral terminal spine; in addition, middorsal, midterminal, and lateroventral spinose processes of variable length may occur (Fig. 5.2.1; Zelinka 1928; Higgins 1969b, 1983, 1991; Sørensen et al. 2010a). In the cyclorhagid *Zelinkaderes floridensis* and in species of the homalorhagid *Paracentrophyes* and *Pycnophyes*, the lateral terminal spine develops in different juvenile stages via a spinose anlage (Figs 5.1.35 D, 5.2.1 B, D, E) and an articulated anlage (Fig. 5.2.1 C) to the final articulated spine (Tab. 3; Zelinka 1928; Higgins 1983, 1990; Brown 1985; Neuhaus 1993, 1995; Lemburg 2002a, b). These ontogenetic findings suggest that the middorsal, midterminal, and lateroventral spinose processes of the Homalorhagida may represent the homologous structures of the middorsal, midterminal, and lateroventral spines, respectively, of the Cyclorhagida (Neuhaus 1993).

During hatching, the specimen vigorously everts and withdraws its introvert until the old cuticle ruptures at the anterior margin of the first trunk segment, and the specimen leaves the exuvia here. All cuticular elements are moulted, including body cuticle, foregut, hindgut, sensory spot tubules, setae, gland cell ducts, and protonephridium tubules (Zelinka 1928; Neuhaus 1993, 1995; Neuhaus & Sørensen 2012).

Number of life history stages, neoteny? Developing gonads become visible in the last juvenile stage of *Campyloderes* cf. *vanhoeffeni*, but also at least in species of *Pycnophyes* (Zelinka 1928; Neuhaus & Sørensen 2012). This observation prompts Zelinka (1896) and possibly Nyholm (1947c) to assume paedogenetic development in

Kinorhyncha. Later, Zelinka (1928) corrects his previous statement, because juvenile stages do not possess any gonopore.

During the postembryonic development of *Antygomonas incomitata*, paradorsal sensory spots are transformed from type-3 to type-2 gradually in successive juvenile stages from anterior to posterior segments except segment 11; this finding is interpreted as "some paedomorphic structure may be conserved, even in adult stages, on the most posterior segments", which may indicate a "progenetic evolution"(Sørensen et al. 2010d: p. 880). Such a statement appears somewhat misleading, because neotenic development requires inevitably that a specimen becomes mature at an earlier developmental stage – and six juvenile stages are documented by Sørensen et al. (2010d) as in all other Kinorhyncha studied so far (Tab. 3). Retarded development of individual characters does not necessarily mean neotenic evolution of species in Kinorhyncha.

Adult specimens of the long-known genera *Echinoderes*, *Pycnophyes*, and *Kinorhynchus* reveal a strongly sclerotized body cuticle with clearly separated sternal and tergal trunk plates, pachycycli, and numerous pores in the body cuticle; in contrast, juvenile stages exhibit a translucent, thin cuticle without notable cuticular pores, no clearly recognizable borders between sternal and tergal cuticular plates, no pachycycli, but in Cyclorhagida with spiculae or spinose processes at the posterior border of most segments (Zelinka 1928; Higgins 1974, 1983, 1990; Brown 1985; Neuhaus 1993; Lemburg 2002a, b; Sørensen et al. 2010d; Neuhaus & Sørensen 2012; Neuhaus et al. 2013). Species of, e.g., *Antygomonas*, *Cateria*, *Condyloderes*, *Zelinkaderes*, *Paracentrophyes*, and *Neocentrophyes* yield at least some of the "juvenile characters" mentioned before and have, therefore, been repeatedly termed "neotenic" or "aberrant" taxa (Gerlach 1956; Higgins 1968, 1969a,b, 1971, 1974, 1983; Kristensen & Higgins 1991; Adrianov & Malakhov 1996; Sørensen et al. 2010d). However, life history studies of *A. incomitata*, *Z. floridensis*, and *Paracentrophyes praedictus* yield that these species do possess the same number of juvenile stages as species with a strong cuticle with "adult characters" such as *Echinoderes bookhouti*, *Kinorhynchus phyllotropis*, *Pycnophyes australensis*, *P. dentatus*, and *P. kielensis* (see Higgins 1974; Brown 1985; Neuhaus 1993, 1995; Lemburg 2002a,b; Sørensen et al. 2010d). So species with "juvenile cuticular characters" in the adult stage do not become mature at an earlier life history stage than species with a thick cuticle. Consequently, neoteny cannot be demonstrated for Kinorhyncha. Whereas a homogeneous cuticular mass connects the sternal and tergal plates of *Echinoderes aquilonius* and *E. kozloffi*, distinct fibrils appear in the same area in *Pycnophyes greenlandicus*, *P. dentatus*, and *Kinorhynchus giganteus* (Moritz & Storch 1972a; Merriman & Corwin 1973; Kristensen & Higgins 1991; Neuhaus 1993). From these observations and from an outgroup comparison with the closest relatives of Kinorhyncha, Loricifera and Priapulida, Neuhaus (1995) concludes that a weakly sclerotized cuticle without clear separation into cuticular plates represents probably the plesiomorphic condition for Kinorhyncha and that a thick cuticle evolved probably at least twice independently in both Cyclorhagida and Homalorhagida.

For *Cateria styx*, one male out of 15 adult specimens is illustrated in a sagittal section (Gerlach 1956: Fig. 14), but the 14 adult specimens collected by Higgins (1968) from a nearby site are all of undetermined sex; Higgins also finds 140 juvenile specimens at the same locality. The second species of this genus, *Cateria gerlachi*, is known from 17 adult specimens of undetermined sex (Higgins 1968). A potential third species includes two females and many specimens of undetermined sex (Kegel and Neuhaus own obs.). At this stage, it is not fully understood why sampling of species of *Cateria* reveals so few clearly adult specimens and so many specimens of undetermined sex as well as juvenile stages.

Recent observations of the Cyclorhagida *Antygomonas oreas* (Fig. 5.1.7 C), *Campyloderes* cf. *vanhoeffeni*, *Centroderes spinosus*, *Centroderes* ssp., *Semnoderes* sp., *Wollunquaderes majkenae*, and *Zelinkaderes floridensis* suppose that a preadult life history stage may exist in these species (Tab. 2) and possibly in other species less thoroughly searched for this developmental stage so far. *Tubulideres seminoli* represents another good candidate in this series, but more thorough examinations are necessary (Neuhaus own obs.). The stage-1 female may possess gonopores (*A. oreas*: Fig. 5.1.7 C, *Z. floridensis*) or not (*Campyloderes* cf. *vanhoeffeni*, *Centroderes spinosus*, *Centroderes* sp., *W. majkenae*: Fig. 5.1.29 B, possibly *T. seminoli*). *Echinoderes coulli* shows two female morphotypes, which may or may not represent consecutive life history stages (Tab. 2; Higgins 1977b; Higgins & Fleeger 1980). Lang (1963: p. 260) reports two exuvia of *P. flaveolatus* with eggs and wonders whether the animals die after deposition of eggs or whether the specimens moult as adults. The finding of a preadult life history stage in at least cyclorhagid Kinorhyncha is well in accordance with observations of a similar stage in the Priapulida and of the postlarva in Loricifera, the closest relatives of Kinorhyncha (Higgins & Kristensen 1986; Higgins & Storch 1991; Kristensen 1991; Lemburg & Schmidt-Rhaesa 1999; Kristensen & Gad 2004; Heiner & Neuhaus 2007; Kristensen et al. 2007; Heiner 2008; Gad 2009a,b).

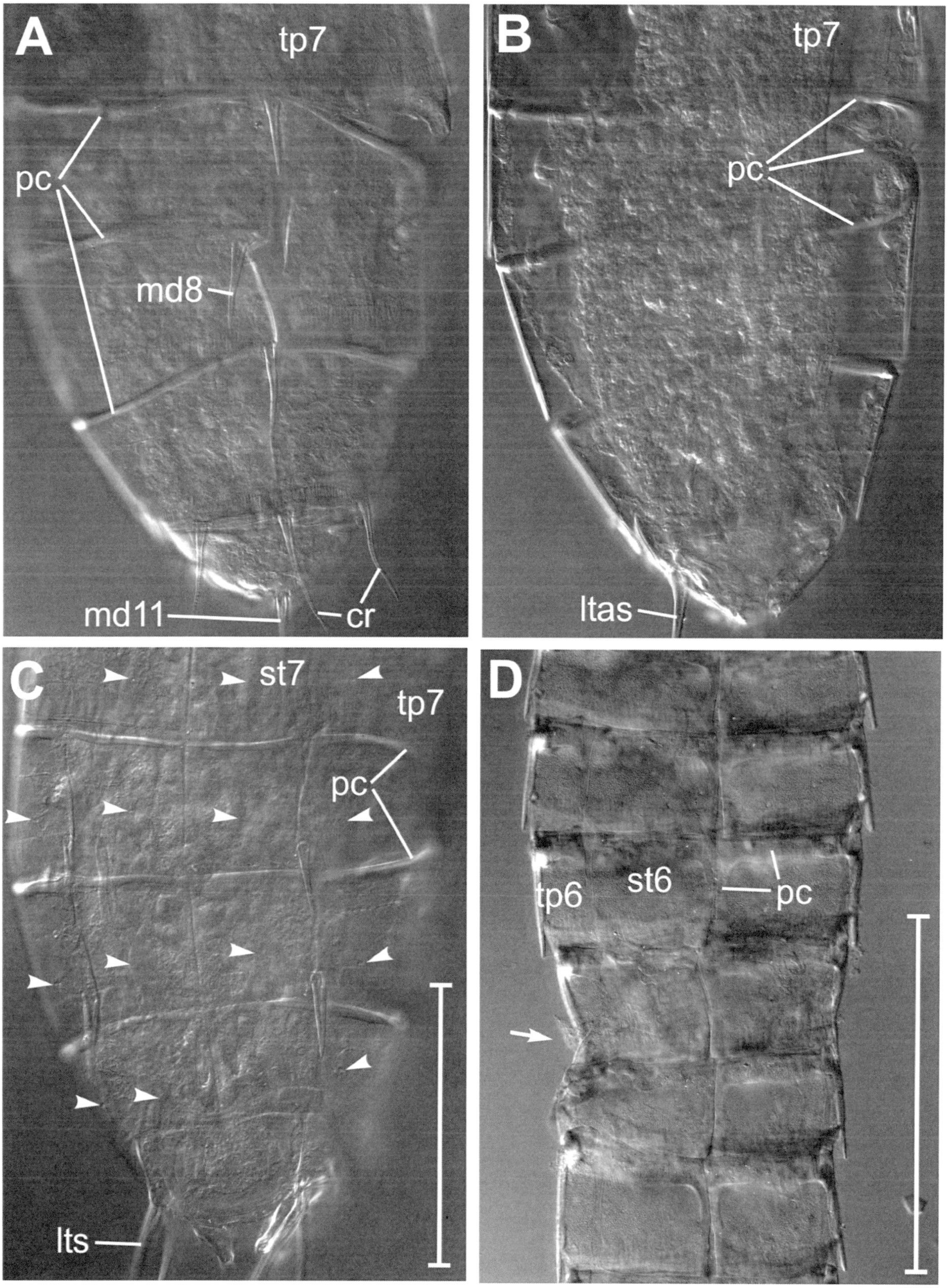

Fig. 5.2.4. LM micrographs of developmental anomalies. A–C, segments 8–11of *Centroderes* sp. in dorsal view (A), optical section (B), and ventral view (C, image mirrored), notice missing pachycyclus on right side of segment 9 (A), line in C = 100 μ, line valid for A–C; D, segments 4–9 of *Pycnophyes kielensis* in ventral view with distorted segments 7 and 8 (arrow), line in D = 200 μ; arrowheads mark sensory spots. Original.

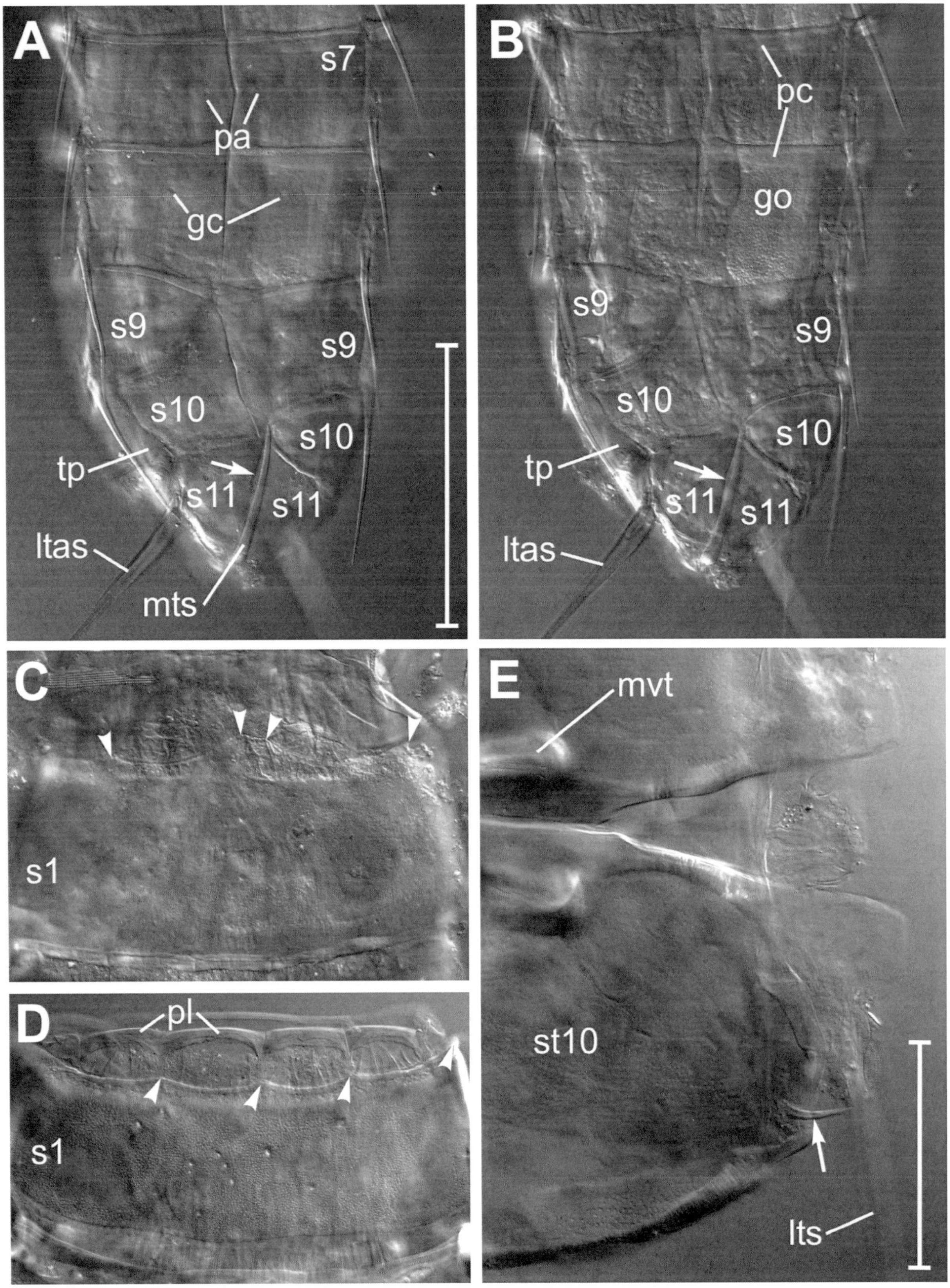

Fig. 5.2.5. LM micrographs of developmental anomalies. A, B, segments 7–11 of female *Campyloderes* cf. *vanhoeffeni* at two different focal levels, arrows point to midventral articulation between sternal plates of segment 11; C, D, segment 1 of female *Pycnophyes kielensis* with distorted (C) and regular (D) anterior margin of segment 1, arrowheads mark anterior margin of tergal plate, dorsal view; E, segments 10 and of female *Pycnophyes dentatus* with anlage of right lateral terminal accessory spine (arrow), ventral view; line in A = 100 µ, line valid for A–D; line in E = 50 µ. Original.

Developmental anomalies. Spines, sensory spots, papillae, gland cell outlets, and setae are generally arranged bilateral symmetrically. However, these structures may be missing, doubled, replaced by another structure, occur in an intermediate position between reference positions, or appear as anlage in one to several segments on one or both sides in both juvenile and adult stages (Nebelsick 1990; Neuhaus 1993, 1995; Neuhaus & Sørensen 2012; Neuhaus et al. 2013). Malformation of the midventral spinose process of *Centroderes spinosus* or of spines of *Campyloderes* cf. *vanhoeffeni* and *Pycnophyes communis* may result in bifurcation or even trifurcation of the structure (Zelinka 1928; Neuhaus & Sørensen 2012; Neuhaus et al. 2013). Spines may not develop to their full length on one or both sides in *Campyloderes* cf. *vanhoeffeni*, *Echinoderes dujardinii*, and *Centroderes* sp. (Zelinka 1928; Neuhaus & Sørensen 2012; Neuhaus, Pardos, Sørensen, and Higgins own obs.). A lateral terminal accessory [sic] spine may be recognizable as a short anlage in *Pycnophyes dentatus* (Fig. 5.2.5 E) and *P. kielensis* although species of the Homalorhagida never express such a spine (Neuhaus 1993: Fig. 6 H, own obs.). Also, an anlage of a terminal lateral spine may exist in species of *Kinorhynchus*, which never possess a fully developed spine, because specimens with this character would be placed in the genus *Pycnophyes* (Zelinka 1928; Higgins 1983; Higgins & Adrianov 1991). Last but not least, segments may deviate from their regular appearance (Fig. 5.2.4 D, comp. Fig. 5.2.5 C with Fig. 5.2.5 D).

More dramatic deformaties concern the sternal or ventral plate of more posterior segments. In *Pycnophyes communis*, the sternal plate of one side of segment 9 may be missing entirely (Zelinka 1928). Sternal plates of a female *Campyloderes* cf. *vanhoeffeni* are significantly distorted in segments 9–11, whereas tergal plates of these segments are not effected (Fig. 5.2.4 A-C; Neuhaus & Sørensen 2012). The tergal plates of segments 7 and 8 of *P. communis* and of segments 8 and 9 of *Centroderes* sp. fuse partly on one side, leading to oblique segmental borders in the first but not in the second case (Fig. 5.2.5 A, B; Zelinka 1928; Neuhaus, Pardos, Sørensen, and Higgins own obs.).

5.3. Physiology

Information on physiological parameters of Kinorhyncha at species level is limited to a single study. Hummon (1975) reports the dry weight of fresh *Echinoderes bookhouti* from Beaufort (USA: North Carolina) as 2.0 ± 0.1 $^{\mu g}/_{individual}$ and its oxygen consumption increasing from 7 $^{nl}/_{animal \cdot h}$ at 26°C to 23 $^{nl}/_{animal \cdot h}$ at 34°C, both at 32% salinity. Alexandrow (2001) calculates the energy content of an unidentified kinorhynch from the Black Sea with 387 µm body length as 2,313 Joule per individual and the specific carbon content as 16.5% of the dry weight. Yingst (1978) reports 42.80 $^{ng}/_{individual}$ of ATP in unidentified kinorhynchs from New Haven (USA: Connecticut). Two unidentified species of Kinorhyncha inhabit the Lynher estuary (U. K.) with 25.3 $^{individuals}/_{10\ cm^2}$, which equals a biomass of 8 $^{mg\ C}/_{10\ cm^2}$, respirate 50 $^{mg\ C}/_{m^2 \cdot year}$, and consume 133 $^{mg\ C}/_{m^2 \cdot year}$ (Warwick et al. 1979).

5.4. Phylogeny

The phylogeny of Kinorhyncha is widely unresolved. Early phylogenetic arguments have been provided by Lang (1949) resulting in a unification of Cyclorhagata and Conchorhagata in what is nowadays called Cyclorhagida. Later, Adrianov and Malakhov (1996: p. 26, 1999a: p. 240) propose characters considered to belong to the "morphological prototype" or "Ur-Kinorhyncha". Adrianov and Malakhov (1994) also outline an evolutionary scenario and "suggest evolutionary lines of primitive (plesiomorphic) and advanced (apomorphic) states of the main morphological characteristics" (Adrianov & Malakhov 1996: p. 26, 1999a: p. 240). However, Cyclorhagida, Protorhagae, and Cyclorhagae represent clearly paraphyletic taxa in their scheme of the phylogenetic relationships (Adrianov & Malakhov 1994: Fig. 118, 1996: Fig. 5.1.7). In addition, it does not always become obvious which character has to be regarded as plesiomorphic or apomorphic and for which reason.

Individual characters have been compared within Kinorhyncha and evaluated as plesiomorphic or apomorphic in the frame of the description of new taxa, life history stages, or organ systems (Neuhaus 1993, 1994, 1995; et al. 2007, 2009, 2010b,c, 2012a; Sørensen & Rho 2009; Sørensen & Thormar 2010; Herranz et al. 2012; Neuhaus & Sørensen 2012). More comprehensive phylogenetic hypotheses are available for species of *Echinoderes* and for the Echinoderidae (GaOrdóñez et al. 2008; Sørensen 2008c). GaOrdóñez et al. (2008) propose that middorsal and lateroventral spines of 55 species of *Echinoderes* may have been reduced in two evolutionary lines. The hypothesis is based on two morphotypes, A and B, which differ in a more fusiform versus a more bulbous general trunk appearance and small sensory spots with short cuticular micropapillae versus large sensory spots with long and numerous cuticular micropapillae. Middorsal spines are supposed to be lost in alternate segments in morphotype A, but in contiguous segments in morphotype B. However, the analysis focuses only on different character states of middorsal and lateral spines. The results are nowadays in conflict with more recent ideas about one to several subgroups within *Echinoderes* (see Lundbye et al. 2011; Ostmann et al. 2012; Yamasaki & Kajihara 2012), because the species of the subgroups are distributed over the two morphotypes of GaOrdóñez et al. (2008).

Sørensen (2008c) analyzes 36 characters in three species of *Cephalorhyncha*, two species of *Echinoderes*, five species of *Fissuroderes*, one species of *Polacanthoderes*, and six other kinorhynchs as outgroups of the Echinoderidae. Here, *Polacanthoderes* appears as the most basal taxon of the Echinoderidae, followed by an unresolved group of non-monophyletic *Fissuroderes* and monophyletic *Cephalorhyncha* and *Echinoderes*. However, the study is influenced by choosing other Kinorhyncha as outgroups for trunk segment characters and by including just two species of *Echinoderes*, a taxon that seems to be much more diverse than previously assumed (see below; Tab. 2). The number of sternal and tergal plates per given segment differs within Kinorhyncha (Tab. 8). Therefore, the hypothesis about the specific taxon that has preserved the plesiomorphic condition for Kinorhyncha inevitably depends on the organisation of the trunk that is assumed for the ground pattern of the Kinorhyncha in comparison with its closest relatives, Loricifera and Priapulida.

The outgroup comparison with Loricifera and Priapulida does not yield a concrete condition for the ground pattern of Kinorhyncha, because the three taxa differ too much in this respect. Consequently, any hypothesis about the evolution of cuticular trunk plates within Kinorhyncha should be based on the postulated ground pattern of the Kinorhyncha and not on a comparison with selected kinorhynch taxa, which even differ from each other in this character set. The hypothesis favoured in this book chapter suggests that in comparison with Loricifera and Priapulida Kinorhyncha retains characters of the ground pattern of Bilateria (Ax 1987), namely small body dimension and body almost round in cross-section, bilateral symmetrically, and with a tube-like cuticle (Neuhaus 1991, 1994). Possibly, species of *Zelinkaderes* preserve a state close to this hypothesis in their ring-like segments 1 and 2, in their single tergal plate joining midventrally in segments 3–10, and in their placids (still) in the process of differentiation from the first trunk segment. Further indication comes from the postembryonic development: in particular early juvenile stages of all taxa studied in more detail reveal a more or less ring-like segmental cuticle indistinctly divided into a sternal and a tergal plate (Brown 1985; Higgins & Kristensen 1988; Higgins 1990; Neuhaus 1993, 1995; Lemburg 2002a,b; Sørensen et al. 2010d). In addition, according to phalloidin-labelling findings at least some ventral and dorsal longitudinal muscles of *Zelinkaderes brightae* (see Herranz et al. submitted; M. Herranz, pers. comm.) and of young adult *Pycnophyes kielensis* (also corroborated by TEM studies of Schmidt-Rhaesa & Rothe 2006) may extend over segmental borders; in all juvenile *P. kielensis*, the longitudinal muscles seem to continue over almost all segments (Schmidt-Rhaesa & Rothe 2006), but this has not been confirmed by TEM investigations. These observations may support the hypothesis that Kinorhyncha originates from an unsegmented ancestor or an ancestor with fewer than 11 segments, and the longitudinal muscles still reflect this situation in the ontogeny (*P. kielensis*) or even as an adult (*Z. brightae*). Further studies are necessary to corroborate the above-mentioned findings.

Alternatively, morphological characters of Kinorhyncha could be mapped on a not yet existing tree derived from genetic characters of at least Kinorhyncha, Priapulida, and Loricifera. This approach may be of limited value, if few genes and few species of the three taxa are included in the analysis and if any combination of Kinorhyncha, Loricifera, and Priapulida shares only few synapomorphies in the characters analysed. A previous comparison of morphological characters of the three major taxa suggests that the latter problem occurs here, arguments for any of the three possible relationsships between Kinorhyncha, Loricifera, and Priapulida can be found but in a low number (Neuhaus 1991, 1994).

Species of *Echinoderes* reveal considerable variation inter alia in sexual dimorphic characters (Tab. 2). Recent studies also favor that *E. applicitus, E. capitatus, E. coulli, E. isabelae, E. maxwelli, E. rex*, and *E. teretis* may be more closely related to each other than to other species of *Echinoderes*, because they lack the lateral terminal accessory spine in both sexes and have no or just one middorsal spine (Lundbye et al. 2011; Ostmann et al. 2012; Yamasaki & Kajihara 2012). Higgins (1977b: p. 347) interprets the “anal cirri” of *E. bengalensis* as “lateral spinose extensions of the terminal tergal or sternal plates”. Thus, this species may also lack lateral terminal accessory spines (as well as middorsal spines) in both sexes and would, consequently, belong to this group of seven species. Among these seven or eight species, *E. applicitus*,

E. coulli, *E. ohtsukai*, *E. rex*, and *E. teretis* may group closer together, because these species share a considerably elongated protonephridium sieve plate followed posteriorly by an additional single larger pore (Brown 1985; Horn 1978; Sørensen 2008b; Lundbye et al. 2011; Ostmann et al. 2012; Yamasaki & Kajihara 2012). Considering the diversity of characters in species of *Echinoderes* (Tabs 2, 8), it would not be too surprising if this genus of the Cyclorhagida turned out to be paraphyletic.

Kinorhynchus and *Pycnophyes* may represent paraphyletic taxa among the Homalorhagida, because species of these genera differ only slightly in their morphology (Remane 1936) and exhibit some variation of sexual dimorphism (Tab. 2). In addition, *Paracentrophyes praedictus*, *Pycnophyes australensis*, *P. beaufortensis*, and *P. cryopygus* reveal a midterminal spinose process at least in the early three juvenile stages, whereas species of *Kinorhynchus*, and at least *Pycnophyes dentatus*, and *P. kielensis* do not (Zelinka 1928; Higgins 1961, 1962, 1974; Brown 1985; Higgins & Kristensen 1988; Neuhaus 1993, 1995; Lemburg 2002a,b). An articulated midterminal spine occurs in the Cyclorhagida *Antygomonas incomitata*, *Echinoderes bookhouti*, and *Zelinkaderes floridensis* (Tab. 3) and is regarded as a plesiomorphic character. The articulated lateral terminal spines of *Paracentrophyes praedictus*, *Pycnophyes dentatus*, and *P. kielensis* develop via spinose anlagen (Neuhaus 1993, 1995). Therefore, spinose processes are interpreted as homologous to articulated spines. The existence of a midterminal spinose process in early juvenile stages of the Homalorhagida *Paracentrophyes praedictus*, *Pycnophyes australensis*, *P. beaufortensis*, and *P. cryopygus* is evaluated as plesiomorphic condition of the Homalorhagida, whereas the lack of a midterminal spinose process in species of *Kinorhynchus*, *Pycnophyes dentatus*, and *P. kielensis* is postulated as apomorphic. Consequently, Neuhaus (1993: p. 188) has suggested that the genus *Pycnophyes* may be paraphyletic.

In summary, an analysis of the phylogenetic relationships within Kinorhyncha would at present suffer significantly from the insufficient knowledge of most species for which redescriptions are badly necessary. At least, many characters in the ground pattern of Kinorhyncha including the autapomorphic characters of this taxon can be currently postulated (Tab. 4); arguments are given in earlier studies of Neuhaus (1991, 1993, 1994, 1995). These characters resemble partly the characters assumed by Adrianov and Malakhov (1994, 1996, 1999a) for the "Ur-Kinoryncha", but in the latter studies character state polarity is not always clearly articulated and argumented (Neuhaus & Higgins 2002).

5.5. Systematics

Classification

Some confusion appears in the literature (Remane 1928, 1929, 1936; Zelinka 1928; Higgins 1971; Adrianov & Malakhov 1999a) concerning the names of and within the taxon Kinorhyncha and the time these names have been introduced (corrected in Tabs 6, 7, 9). Whereas Dujardin (1851) calls his new species simply "l'Echinodère", Claparède (1863) formally assigns it to a genus and names it *Echinoderes dujardinii*. No unifying scientific name is provided for species of *Echinoderes*, probably also because it takes some time to realize that the taxon represents a group of its own. The name "Kinorhyncha (*Echinoderes*)" is first used in the seldom-cited contribution of Reinhard (1885) in Russian. Subsequently, this very same article is published again in German, almost without any major changes (Reinhard 1887), and this time it receives much more attention, especially in the German-speaking scientific community (Remane 1928, 1929, 1936; Zelinka 1928). Reinhard (1885: p. 296, 1887: p. 463) chooses the name "Kinorhyncha" (kinein = Greek for move, rhynchos = Greek for snout), because he wants to avoid confusion with, e.g., Echinodermata, and because he misidentifies the posterior extensions of the esophagus bulging between the pharynx protractors as brain and consequently the protrusible introvert as protrusible snout of the alimentary canal (Zelinka 1928: p. 38). However, Zelinka (1928: p. 224) latinizes the German name "die Echinoderen" and refers to the Kinorhyncha as "Echinodera" (echinos = Greek for hedgehog, deres = Greek for neck; terms refer to spiny scalids of the introvert, which Dujardin in 1851 obviously interprets as neck, *Echinoderes* represents the first described genus of Kinorhyncha). Other authors prefer the name "Echinoderida" derived from the first family erected within Kinorhyncha, Echinoderidae (e.g., Johnston 1938; Nyholm 1947c; Lang 1949, 1953; McIntyre 1962). In summary, the first introduced scientific group name is Kinorhyncha, but the morphologically correct name would be Echinodera.

Higher categories. Bütschli (1876) is cited (e.g., Higgins 1971) to have established the first family name of Kinorhyncha, Echinoderidae, but this is probably based on a misunderstanding of the German language. Bütschli (1876: p. 390 ff.) mentions species of *Echinoderes* as "die Echinoderen", which is simply the Germanized plural of the genus's name. The correct Germanized family name would have been "die Echinoder**id**en" – which is

Tab. 4. Characters in the ground pattern of Kinorhyncha (modified from Neuhaus 1991, 1993, 1994, 1995). Autapomorphic characters are marked in bold and italics.

Habitus:

- small body dimension (less than 2 mm),
- body almost round in cross-section, ventrally slightly flattened,
- ***adult with introvert, neck, and 11 trunk segments (nervous system, epidermal glands, and musculature follow outer segmentation of trunk)***,
- scalids of introvert organized in rings as several rings of anterior spinoscalids and one posterior ring of 14 trichoscalids; spinoscalids staggered, elongated, arranged pentaradial symmetrically, with articulation between basal and distal part; ring-01 spinoscalids with blunt tip, remaining spinoscalids pointed,
- neck with 16 placids, basally fused with trunk segment 1,
- trunk bilateral symmetrically.

Integument:

- cellular epidermis with diplosome but without locomotory cilia,
- cuticle ensheathing trunk in even thickness, without differentiation into segmental plates or articulations between the latter, ***segments 3-11 with midventral longitudinal suture***, neck with at least 14 placids,
- body cuticle with fine-granular basal layer and monolamellar, membrane-like epicuticle (fibrillar material below the cuticle in some areas),
- ***movable, paired lateral and unpaired middorsal spine in segments 1-10; midterminal, paired lateral terminal, and lateral terminal accessory spine in segment 11***;
- cuspidate lateral spine at least in some segments, one pair in the first juvenile stage already.

Mouth cone:

- ***mouth cone as ring-like epidermal fold around the pharynx and protrusible,***
- ***9 outer oral styles,***
- 10 + 5 + 5 ***inner oral styles*** with subapical opening,
- oral styles with monociliary receptors,
- ***frontal circular muscle*** composed of few cells, multicellular (10 cells?) basal circular muscle,
- ***basal nerve ring***,
- innervation of the foregut by the brain only.

Pharynx and esophagus:

- two-layered pharynx with inner epithelium and outer muscle bulb, with diplosome,
- pharyngeal lumen presumably circular in cross-section,
- ring and radial muscle cells alternate in bulb, all muscle cells ring-shaped,
- all muscle cells with T-system and cell processes to [sic] nerve cells,
- radial muscle cells monosarcomeral, ring muscle cells multisarcomeral,
- postpharyngeal sphincter consisting of a single multisarcomeral ring muscle cell,
- ***pharyngeal cuticle with*** fine-granular basal layer and ***multilamellar epicuticle***,
- at least 1 gland cell in the inner pharyngeal epithelium and 10 cells in the outer epithelium,
- ***gland cells*** monociliary, ***with extremely thin microvilli in the atrium***, secretion apocrin,
- 1-2 monociliary receptors near each gland cell, isolated sensory cells in pharynx and esophagus,
- 3 longitudinal nerve cords (ventrally and laterally) and several commissures at least in front of and behind the postpharyngeal sphincter,
- esophagus without any digestive gland cell.

Midgut:

- aciliary epithelial and gland cells with apocrin secretion,
- at least 2 *sensorimotor* ***cells with*** cilium at tip of cell process, ***dense-core vesicles***, no contact to nerve cells,
- network of inner longitudinal and outer circular musculature around esophagus and midgut, T-system weakly developed, multisarcomeral, and with diplosome.

(Continued)

Tab. 4. *(Continued)*

Hindgut:

- ***hindgut lumen transverse slit-like***,
- thin epithelium with tonofilaments,
- cuticle with fine-granular basal layer, thin osmiophilic zone, and monolamellar epicuticle,
- hindgut bound by sticking dorsal and ventral epicuticle,
- ***one pair of frontal, caudal, and dorsal dilatators and one unpaired circular dilatator***,
- muscle cells multisarcomeral, weakly developed T-system, and nuclei in cell extensions.

Body cavity:

- primary body cavity, not bounded by (myo-)epithelial cells,
- polymorphous amoebocytes with diplosome but without cilia or myofilaments.

Nervous system:

- nervous system intraepidermally,
- circumenteric brain, differentiated into 3 regions from frontal to caudal: perikarya – neuropile – perikarya,
- large, unpaired, ventral longitudinal nerve cord with lateral and ventral perikarya, additional longitudinal nerves,
- sensory spots of trunk with cuticular micropapillae arranged in a ring around a central pore and additional micropapillae more peripheral; each sensory spot with 1-3 monociliary receptors, each receptor with 9 circumciliary microvilli,
- scalids with ciliary receptors.

Musculature:

- 10 pharyngeal protractors,
- both introvert and mouth cone with retractors (= outer and inner retractors),
- protractors and retractors attaching to intorvert cuticle via extremely elongated epidermal cells with intermediate filaments (= tanycytes),
- paired bundle of segmental dorsal and ventral longitudinal muscle,
- one pair of segmental dorsoventral muscles,
- body musculature multisarcomeral with spindle-shaped Z-elements,
- muscles attach to the cuticle via desmosomes, epidermal tonofilaments, and hemidesmosomes; no extracellular filaments in the cuticle.

Excretory system:

- one pair of protonephridium in segments 8 and 9 with several terminal cells and at least one canal and one nephropore cell,
- compound filter formed by several terminal cells, the latter biciliary.

Reproduction:

- separate sexes,
- female with paired receptaculum seminales, presumably internal fertilization.

Postembryonic development:

- direct development through 6 juvenile stages + possibly one (pre)adult stage (noticed only in cyclorhagid species so far, Tab. 3),
- juvenile stages with thin, weakly sclerotized cuticle,
- J-1 with at least 7 separated segments + anlagen of 4 segments recognizable by light microscopy, segments not divided into sternal and tergal plates,
- 2-3 new trunk segments expressed in subcaudal differentiation zone in stages following J-1,
- number of scalids increasing during ontogeny, new scalids introduced as indicative anlagen of protoscalids (= prescalids) and/or spinose anlagen (= protoscalids) in a subfrontal growing zone,
- fully differentiated scalids organized in rings (see Habitus).

actually the name used first by Zelinka (1894). Therefore, the latter publication becomes the birth place of the family Echinoderidae.

In a first attempt to group the species of Kinorhyncha, Reinhard (1885, 1887) introduces without a formal rank the names Acerca, Monocerca, and Bicerca refferring to the spine pattern at the terminal segment. Zelinka (1896) recognizes this artificial grouping and, based on the closing apparatus and other morphological characters, divides the Kinorhyncha into the orders Cyclorhagae and Homalorhagae. Subsequently, he proposes a system (Tab. 5) which is followed in many aspects still today (Zelinka 1907). Current writing of these names is Cyclorhagida and Homalorhagida introduced by Higgins (1964b, includes synonomy) (Tab. 9); Chitwood (1959) uses both terms with different ranks as subtaxa in his complex classification system.

Remane (1929) changes the family names Pentacontidae to Semnoderidae and Mesitoderidae to Campyloderidae, probably because the family names have not been derived from the names of the genera. Not much later, he transfers "*Trachydemus*" (since 1974 *Kinorhynchus*) to the family Pycnophyidae and cancels the Trachydemidae, because species of *Kinorhynchus* and *Pycnophyes* agree widely in their morphology and differ mainly in the lack or presence of the lateral terminal spine (Remane 1936). He also does not regard *Echinoderella* as a "natural" genus, but does not draw any formal taxonomic conclusions from this. Lang (1949) discusses the phylogeny of Kinorhyncha extensively and suggests two alternative systems, one closely related to Remane's classification and one uniting Cyclorhagata and Conchorhagata in a taxon Heterorhagae (Tab. 5), which is preferred by him and followed by later authors. Sheremetevskij (1974: p. 985) substitutes the name *Trachydemus* with *Kinorhynchus*, because the former name was preoccupied by Chevrolat who described a genus of Coleoptera and by Reuter who named 1900 a genus of Hemiptera. A couple of years later, Higgins (1981, 1988) summarizes kinorhynch classification. Finally, Higgins (1983) formally synonomizes the names of species described from juvenile stages and lays to rest the "larval" taxa names (Tab. 7).

There is a range of publications modifying the category or names of taxa or introducing new ranks without offering new insights into the phylogeny of the Kinorhyncha (e.g., Pearse 1936; Gerlach 1956; Chitwood 1959; Higgins 1964b). Gradually, new species and genera are described and new families erected to accomodate the new characters of these new taxa (Tab. 5; Gerlach 1956; Higgins 1968, 1969a,b, 1990; Higgins & Shirayama 1990; Adrianov & Malakhov 1994, 1996, 1999a). Few authors laudably refrain from creating new categories of higher ranks, if a new genus does not fit in any of the known taxa, and include the new species as incertae sedis, because the phylogeny of Kinorhyncha is widely unresolved (Nebelsick 1990; Sørensen et al. 2007; Sørensen & Thormar 2010).

The classification of Kinorhyncha suffers still today from superordinated ranks like suborders and families including only one to two genera, but these categories do not reflect the evolution of the Kinorhyncha (Tabs 5, 9). It may, e.g., very well turn out that the deep incision of the clamshell-like closing apparatus of the Conchorhagae *Semnoderes* and *Spenoderes* may represent the final evolutionary stage in a line from a *Zelinkaderes*-like stage, revealing placids fused with trunk segment 1 via an *Antygomonas*-like stage with a slight midventral and middorsal indentation of the first trunk segment. Such a scenario would immediately lead to cancellation of a suborder Protorhagae uniting the Antygomonidae and Zelinkaderidae (Adrianov & Malakhov 1994, 1996, 1999a). Also, a taxon Protorhagae is paraphyletic (Bauer-Nebelsick 1996) based on the argumentation and scheme of phylogenetic relationships provided by Adrianov and Malakhov (1994, 1996). Therefore, in this book chapter all suborders are rejected and alphabetical lists of families and taxa incertae sedis in the two orders Cyclorhagida and Homalorhagida are proposed (Tab. 9) until the phylogeny of Kinorhyncha is better understood. It is forseeable that one of these two taxa will turn out to be paraphyletic, because the Cyclorhagida seem to contain several taxa like *Zelinkaderes* and *Triodontoderes* with presumed more original characters and because Homalorhagida may be closer related to a subtaxon of Cyclorhagida; currently, Sørensen et al. (2012a) suggest the Dracoderidae as this subtaxon (see also Sánchez et al. submitted). It is also predictable that the extensive on-going species descriptions and revisions will generate a wealth of new characters and insights into the evolution of Kinorhyncha.

Species. Today, 191 species of Kinorhyncha (plus two species in a paper submitted by Sánchez et al.) are known from descriptions of adult stages (Tab. 6) and 50 species based on descriptions of juvenile stages (Tab. 7). The early scholars of Kinorhyncha start an unfortunate tradition of naming not only adult life history stages as species but also juvenile stages (Claparède 1863; Greeff 1869; Panceri 1878; Reinhard 1881, 1885). This procedure continues in the 20th century (Steiner 1919; Zelinka 1928; Timm 1958) . Even worth, although Zelinka has been fully aware about the postembryonic development of successive juvenile stages (Zelinka 1928: pp. 213–214), he creates specific genus names for juvenile stages, because these stages "live in their typical form for some extended

Tab. 5. Different classification systems of Kinorhyncha.

Zelinka 1907 (1928):					
class (order)	**order (suborder)**	**suborder (tribus)**	**family**	**genus**	**juvenile "genera"**
Echinodera	Homalorhagae		Pycnophyidae	*Pycnophyes*	*Hyalophyes, Centrophyes, Leptodemus*
			Trachydemidae	*Trachydemus* [1]	
	Conchorhagae		Pentacontidae	*Semnoderes*	
	Cyclorhagae	Notosomata	Echinoderidae	*Echinoderes*	*Habroderes, Centropsis*
				Echinoderella	*Habroderella, Hapaloderes*
			Centroderidae	*Centroderes*	
		Xenosomata	Mesitoderidae	*Campyloderes*	
Lang 1949:					
order	**suborder**	**tribus**	**family**	**genus**	
Echinodera	Homalorhagae		Pycnophyidae	*Pycnophyes*	
			Trachydemidae	*Trachydemus*	
	Heterorhagae	Conchorhagata	Semnoderidae	*Semnoderes*	
		Cyclorhagata	Echinoderidae	*Echinoderes*	
				Echinoderella [2]	
			Centroderidae	*Campyloderes* [3]	
				Centroderes	
Higgins 1990, Higgins & Shirayama 1990:					
phylum	**order**	**suborder**	**family**	**genus**	
Kinorhyncha	Cyclorhagida	Cyclorhagae	Echinoderidae	*Echinoderes*	
			Dracoderidae	*Dracoderes*	
			Centroderidae	*Centroderes*	
				Campyloderes	
				Condyloderes	
			Zekinkaderidae	*Zelinkaderes*	
		Conchorhagae	Semnoderidae	*Semnoderes*	
				Sphenoderes	
		Cryptorhagae	Cateriidae	*Cateria*	
	Homalorhagida	Homalorhagae	Neocentrophyidae	*Neocentrophyes*	
				Paracentrophyes	
			Pycnophyidae	*Pycnophyes*	
				Kinorhynchus	
Adrianov & Malakhov 1994, 1999a:					
phylum	**order**	**suborder**	**family**	**genus**	
Kinorhyncha	Cyclorhagida	Cyclorhagae	Echinoderidae	*Echinoderes*	
			Dracoderidae	*Dracoderes*	
			Cephalorhynchidae	*Cephalorhyncha*	
			Centroderidae	*Centroderes*	
				Campyloderes	
				Condyloderes	
		Protorhagae	Antygomonidae	*Antygomonas*	
			Zekinkaderidae	*Zelinkaderes*	
		Conchorhagae	Semnoderidae	*Semnoderes*	
				Sphenoderes	

(Continued)

Tab. 5. (*Continued*)

Adrianov & Malakhov 1994, 1999a:				
phylum	**order**	**suborder**	**family**	**genus**
		Cryptorhagae	Cateriidae	*Cateria*
	Homalorhagida	Homalorhagae	Neocentrophyidae	*Neocentrophyes*
				Paracentrophyes
			Pycnophyidae	*Pycnophyes*
				Kinorhynchus

[1] since 1974 = *Kinorhynchus*. [2] Comment by Lang: all species of *Echinoderella* differ only in the lack of eyes but not in other morphological characters from *Echinoderes* and should, therefore, be included in the latter genus, except *Echinoderella capitata* which should keep its genus name. [3] Johnston (1938) has shown that *Campyloderes* possesses 11 trunk segments like all other Kinorhyncha, so Lang sees no need for a tribus of its own for this genus.

time before they moult into another form" (author's translation from Zelinka 1928: p. 218). The species known only from juvenile stages are generally regarded as nomina dubia (Tab. 7), but names are available and valid according to the International Code of Zoological Nomenclature, because a new species may be described from any life history stage (ICZN 1999). It may be possible to assign most "species" described from juvenile stages to species based on adult stages, but there is a long way to go to unravel the postembryonic development of the relevant species.

Type species of the genera *Campyloderes, Semnoderes, Pycnophyes*, and *Trachydemus* as well as *Echinoderes* and *Fissuroderes* have not been formally designated in the original publication. For the latter two taxa, this is corrected in Tab. 6. The former four genera have been monotypic at the time of description, so there is no need for subsequent designation according to the International Code of Zoological Nomenclature (ICZN 1999).

Common names. In most languages, the name Kinorhyncha is applied almost unchanged like "the kinorhynchs" (English) and "die Kinorhynchen" (German), but the name may be modified to Kinorrincos, Quinorrinc(h) os, and Cinorrincos in Spanish publications (Pardos 2004) and to Chinorinchi in Italian studies (Mari & Morselli 1987). As Kinorhyncha is unknown to most laymen, few common names are in use. "Mud dragon" or "spiny-crown worms" make it into the Encyclopedia of Life and Wikipedia (http://eol.org; http://en.wikipedia.org) and "stekelwormen" to the Netherlands Wikipedia site (http://nl.wikipedia.org). In China, the literal translation of "mud dragon" has been introduced by Higgins (pers. comm.). German webpages list "Hakenrüssler" as a translation from Kinorhyncha (http://www.dahmstierleben.de; http://de.wikipedia.org), but "Igelnackentiere" would refer more to the introvert scalids and represents a proper translation of Echinodera. Common names for kinorhynch species are generally not available in any language except in Japanese for the seven species known from Japanese waters introduced by Murakami (2005). This recent tradition is continued by Yamasaki and Kajihara (2012) for the eighth Japanese species.

Study of specimens. Among the technical impediments impacting current studies of Kinorhyncha are the lack of type material deposited in museum collections, insufficient embedding media, and choice of slide. More comprehensive collections of Kinorhyncha are found at the National Museum of Natural History, Smithsonian Institution, Washington D.C., at the Natural History Museum of Denmark, Copenhagen, at the Museum für Naturkunde Berlin, and at the Zoological Museum of the Moscow State University. Additional, identified specimens have been and still are partly kept in private collections available to a limited degree.

Embedding media requiring complete dehydration via organic solvents such as neutral Canada balsam and Euparal tend to harden specimens so much that they break into pieces (Sørensen & Pardos 2008; Neuhaus & Sørensen 2012). Media containing chloral hydrate such as Berlese's medium, Hoyer's mounting medium, and Faure's medium clear specimens to a degree that they are no longer traceable on the slide (Higgins 1983; Neuhaus own obs.). Also after several years, slides with Hoyer's and CMCP-10 (= carboxy-methylcellulose-phenol) and polyvinyllactophenol will reveal different types of crystals, often close to the specimens (Figs 5.5.1 C, D, 5.5.2 A, B). Frequently, Hoyer's medium, Aquatex, Berlese's medium, and polyvinyllactophenol will after few years of storage exhibit cavities and cracks near the specimen (Figs 5.1.26, 5.5.1 A–C, 5.5.2 A, B, D). Hoyer's medium seems also to disintegrate into its components both near the specimen (Fig. 5.5.2 C) and anywhere under the coverslip. Fluoromount-G (Sørensen & Pardos 2008)

Tab. 6. Alphabetical list of species described from adult stage, synonymy, distribution, habitat, depth, and references with biogeographical information.

species	original description; synonymy; notes	type locality, distribution, habitat, depth	references
Echinoderes abbreviatus Higgins, 1983	*E. abbreviatus* Higgins, 1983: pp. 10–15, Figs 8–19, Tab. 1	NW Atlantic Ocean: Belize: Twin Cays: gray calcareous sediment with detritus and sand 1–2 m (type locality)	
Dracoderes abei Higgins & Shirayama, 1990	*D. abei* Higgins & Shirayama, 1990: pp. 940–946, Figs 1–5, type species by original designation; Sørensen et al. 2012a: pp. 214–222, Figs 2–6, Tabs 2–3, redescription	NW Pacific Ocean: Japan: central Seto Inland Sea: Mukaishima: sandy mud 3 m (type locality) or mud 17–44 m, SE Japan: Tanabe: sandy mud 14 m, S South Korea: Gamak Bay: mud 5–16 m, E China Sea: mud 57–104 m	Sørensen et al. 2012b
Echinoderes agigens Bacescu, 1968	*E. agigens* Bacescu, 1968: p. 239, Pl. 1: Fig. 3, Pl. 6: Fig. 1A–C; *E.* aff. *dujardini* of Bacescu et al. (1963): p. 138, Figs 1A-C; Bacescu 1968: *E. dujardini* of Marinov (1964)(?) = *E. agigens*	Black Sea: Bulgaria (type locality)	
Echinoderes andamanensis Higgins & Rao, 1979	*E. andamanensis* Higgins & Rao, 1979: pp. 77–79, Fig. 1	N Indian Ocean: Bay of Bengal: South Andaman Islands: Havelock Island: intertidal algal mud (type locality), Little Andaman Island: intertidal mud	Rao 1993
Echinoderes angustus Higgins & Kristensen, 1988	*E. angustus* Higgins & Kristensen, 1988: pp. 7–11, Figs 1–16, Tab. 1	N Atlantic Ocean: W Greenland: Disko Island: brown mud 9 m (type locality), sandy mud or mud 7–300 m	
Kinorhynchus anomalus Lang, 1953	*Trachydemus anomalus* Lang, 1953: pp. 6–8, Figs 5–6; *Kinorhynchus anomalus* comb. nov. Sheremetevskij, 1974: p. 985, *Kinorhynchus* substitute name for *Trachydemus* (preoccupied); Brown & Higgins 1983: pp. 165–168, Figs 4–5, redescription, designation of lectotype; Martorelli & Higgins 2004: pp. 95–97, Figs 18–25, emended description	SE Pacific Ocean: Chile: E of Punta Dirección: Bahía Ralún: Estero Reloncaví: sand with mud and detritus 35–40 m (type locality), Peñon Blanco: Canal Moraleda: intertidal rock pools; SW Atlantic Ocean: Argentina: Chubut: San Jorge Gulf: stomach of shrimp *Pleoticus muelleri*	
Triodontoderes anulap Sørensen & Rho, 2009	*T. anulap* Sørensen & Rho, 2009: pp. 1270–1278, Figs 2–7, Tabs 1–2, type species by original designation	Pacific Ocean: Micronesia: Chuuk Islands: coralline sand with detritus 6–27 m (type locality), Saipan: Managaha Island: coral sand 0 m	Sørensen et al. 2012b
Paracentrophyes anurus Sørensen, Pardos, Herranz & Rho, 2010	*P. anurus* Sørensen et al., 2010a: pp. 47–53, 58–59, Figs 2–6, 7A, B, 8A, B, Tabs 1–3, 5	NW Pacific Ocean: S South Korea: Korea Strait: mud with shell fragments 103 m (type locality)	
Kinorhynchus apotomus Higgins, 1983	*K. apotomus* Higgins, 1983: pp. 82–85, Figs 192–203	NW Atlantic Ocean: Belize: Twin Cays: coralline mud with detritus and sand 3 m (type locality)	
Echinoderes applicitus Ostmann, Nordhaus & Sørensen, 2012	*E. applicitus* Ostmann et al., 2012: pp. 81–90, Figs 2–6, Tabs 2–3	E Indian Ocean: Java: Segara Anakan lagoon: intertidal mangrove mud and sand (type locality)	
Echinoderes aquilonius Higgins & Kristensen, 1988	*E. aquilonius* Higgins & Kristensen, 1988: pp. 11–17, Figs 17–32, Tab. 2	N Atlantic Ocean: W Greenland: Disko Island: sandy mud 200 m (type locality) or mud 60–300 m	Winkelmann & Ziemer 1999

(*Continued*)

Tab. 6. (*Continued*)

species	original description; synonomy; notes	type locality, distribution, habitat, depth	references
Pycnophyes arctous Adrianov, 1999 in Adrianov & Malakhov, 1999	*P. arctous* Adrianov, 1999 in Adrianov & Malakhov, 1999a: pp. 168–170, 280–281, Figs 5.62–5.65	Arctic Ocean: Russia: Fram Strait: (Y)Ermak Plateau: mud 345–441 m (type locality)	
Pycnophyes argentinensis Martorelli & Higgins, 2004	*P. argentinensis* Martorelli & Higgins, 2004: pp. 92–95, Figs 12–17	SW Atlantic Ocean: Argentina: Chubut: San Jorge Gulf: stomach of shrimp *Pleoticus muelleri* (type locality)	
Echinoderes arlis Higgins, 1966	*E. arlis* Higgins, 1966: pp. 518–519, Figs 1–2; Adrianov & Malakhov 1999a: pp. 121–124, 259–260, Figs 5.19–5.22, redescription	Arctic Ocean: USA: Alaska: N of Point Barrow: 419–747 m (type locality), Russia: Fram Strait: (Y)Ermak Plateau: mud 345 m	
Semnoderes armiger Zelinka, 1928	*S. armiger* Zelinka, 1928: pp. 289–293, Pl. 27: Figs 1–5, 10, type species by monotypy	Mediterranian Sea: N Adriatic Sea: Italy: Gulf of Trieste: 22 m (type locality), Gulf of Venice: Chioggia: from sponge *Hymeniacidon sanguinea*, NE Spain: Blanes: coarse *Dentalium*-sand 55 m, E Spain: Denia: mud 15 m; Black Sea: Bulgaria, Caucasus and W Crimea: fossiline mud 50–100 m, Romania; NE Atlantic Ocean: Bay of Biscay: NW Spain: Asturias to Basque Country: muddy sand 49–444 m, France: Roscoff: sand, E North Sea: W Sweden: Tjärnö: brown mud 30–53 m, Sweden: Gullmar Fjord: sandy mud 20–70 m, off NE Scotland: Fladen Ground: greenish gray mud 146 m, W Scotland: Loch Nevis: brown mud 101 m	Nyholm 1947c; McIntyre 1961, 1962, 1964; Bacescu & Bacescu 1956; Marinov 1964; Bacescu 1968; d'Hondt 1970; Sheremeteskij 1974; Mari & Morselli 1987; Sørensen et al. 2009; Sánchez et al. 2012
Cephalorhyncha asiatica (Adrianov, 1989)	*Echinoderes asiaticus* Adrianov, 1989: pp. 18–19, Figs 1a, b, 2; *Cephalorhyncha asiatica* comb. nov. Adrianov, 1999 in Adrianov & Malakhov, 1999a: pp. 161–164, 276–277, Figs 5.58–5.61, redescription, type species by original designation; Yamasaki et al. 2012: pp. 25-30, Figs 2-4, Tabs 2-3, redescription	NW Pacific Ocean: Russia: Sea of Japan: Peter the Great Bay: Gulf of Vostok: mud 5 m (type locality), Sea of Japan: Japan: Hokkaido Island: Akkeshi: 13 m muddy sand	
Echinoderes aspinosus Sørensen, Rho, Min, Kim & Chang, 2012	*E. aspinosus* Sørensen et al., 2012b: 5 pp., Figs 2–3, Tab. 2	NW Pacific Ocean: W South Korea: Yeonpyeong Island: subtidal sand (type locality)	
Pycnophyes aulacodes Sánchez, Pardos, Herranz & Benito, 2011	*P. aulacodes* Sánchez et al., 2011: pp. 327–333, Figs 6–8, Tabs 6–7	NE Atlantic Ocean: NW Spain: Galicia: Ría de Ares: mud to coarse sand 19–45 m (type locality), Galicia: mud intertidal to 28 m, Asturias: muddy sand 76–168 m; Mediterranean Sea: S Spain: Almuñecar: mud to coarse sand 15–20 m, E Spain: Blanes: midfine sand, Denia: fine sand, Garrucha: mud to coarse sand; NE Atlantic Ocean: Faroe Islands: Kaldbak Fjord, E North Sea: Kattegat: W Sweden: Tjärnö, Denmark: Hirsholmene	Sørensen et al. 2012a; Sánchez et al. 2012, submitted
Echinoderes aureus Adrianov, Murakami & Shirayama, 2002	*E. aureus* Adrianov et. al., 2002a: pp. 51–59, Figs 2–5, Tab. 1 Sørensen et al. 2012b: *E. lanceolatus* possibly synonym with *E. aureus*	NW Pacific Ocean: Japan: Honshu: Tanabe Bay: from brown algae *Padina arborescens* in intertidal pools (type locality)	

Pycnophyes australensis Lemburg, 2002	*P. australensis* Lemburg, 2002a: pp. 174–189, Figs 27, Tabs 1–2; Lemburg 2002b: Figs 2, 6, erratum	SW Pacific Ocean: N Australia: Magnetic Island: intertidal sand (type locality)	
Pycnophyes barentsi Adrianov, 1999 in Adrianov & Malakhov, 1999	*P. barentsi* Adrianov, 1999 in Adrianov & Malakhov, 1999a: pp. 170–173, 281–282, Figs 5.66–5.69	Arctic Ocean: Barents Sea: Stockman (= Shtokman) oil-field: mud (type locality)	
Pycnophyes beaufortensis Higgins, 1964	*P. beaufortensis* Higgins, 1964b: pp. 481–485, Figs 1–5;	NW Atlantic Ocean: USA: North Carolina: Beaufort: muddy sand 1 m (type locality), South Carolina: Bear's Bluff: muddy sand, Florida: Marineland; W Atlantic Ocean: N Panama: Bocas del Torro: gray mud with shell and spicule fragments 20 m	Higgins 1978a; Sørensen 2006
Kinorhynchus belizensis Higgins, 1983	*K. belizensis* Higgins, 1983: pp. 85–89, Figs 204–219, Tab. 16	NW Atlantic Ocean: Belize: Twin Cays: gray calcareous sediment with detritus and sand 1–2 m (type locality) or coralline mud with detritus and sand 3 m	
Echinoderes bengalensis (Timm, 1958)	*Echinoderella bengalensis* Timm, 1958: pp. 107–108, Figs 1A, 2A–F; *Echinoderes bengalensis* comb. nov. Higgins, 1960: p. 89; Higgins 1977b: p. 346, *E. sonadiae* = juvenile stage of *E. bengalensis*; *Echinoderella sonadiae* Timm, 1958 (pp. 107–108, Fig. 1B) = *Echinoderes sonadiae* comb. nov. Higgins, 1960: p. 89	N Indian Ocean: Bay of Bengal: Bangladesh: Sonadia Island: off Cox's Bazar: intertidal mud (type locality), E India: Waltair Beach: intertidal medium-grained sand 15 cm below surface	Rao & Ganapati 1968
Echinoderes bermudensis Higgins, 1982	*E. bermudensis* Higgins, 1982: pp. 310–312, Figs 3–5	NW Atlantic Ocean: Bermuda: Castle Harbor (type locality), Trunk Island & mouth of Flatts Island: medium coral sand with or without detritus 3 m	Sørensen 2001
Echinoderes bispinosus Higgins, 1982	*E. bispinosus* Higgins, 1982: pp. 307–310, Figs 1–2	NW Atlantic Ocean: Bermuda: Castle Harbor (type locality), mixed fine and coarse sediment until 14 m	Higgins 1986b
Echinoderes bookhouti Higgins, 1964	*E. bookhouti* Higgins, 1964b: pp. 489–492, Figs 17–18	NW Atlantic Ocean: USA: North Carolina: Beaufort: muddy sand 1 m (type locality), Florida: Marineland, in algae *Enteromorpha intestinalis*	Higgins 1977b
Pycnophyes borealis Higgins & Korczynski, 1989	*P. borealis* Higgins & Korczynski, 1989: pp. 2061–2063, Figs 17–32	Arctic Ocean: Beaufort Sea: Canada: Northwest Territories: Tuktoyaktuk Harbour 15 m (type locality)	
Echinoderes brevicaudatus Higgins, 1977	*E. brevispinosus* Higgins, 1966a: pp. 118–121, Figs 1–2; *E. brevicaudatus* Higgins, 1977a: p. 21, = substitute name for *E. brevispinosus*	Red Sea: Egypt: Al-Ghardaqa (= Hurghada): sandy coral mud 30 m (type locality)	
Zelinkaderes brightae Sørensen, Heiner, Ziemer & Neuhaus, 2007	*Z. brightae* Sørensen et al., 2007: pp. 257–261, Figs 7–12, Tabs 4–6	NW Atlantic Ocean: USA: Florida: 6 miles off Fort Pierce: *Dentalium* sand 15 m (type locality)	
Pycnophyes calmani Southern, 1914	*P. Calmani* Southern, 1914: pp. 4, 69–70, inadvertantly becomes original description;	N Atlantic Ocean: Ireland: Clew Bay: 26 m (type locality), North Sea: S of Clay Deep: mud 50 m, Scotland:	Lang 1936; Remane 1936; Huys & Coomans 1989

(*Continued*)

Tab. 6. (*Continued*)

species	original description; synonymy; notes	type locality, distribution, habitat, depth	references
	P. calmani of Zelinka 1928: pp. 321–323, Pl. 13: Figs 8–9, Pl. 15: Figs 6–7, emended description; *Hyalophyes calmani* Zelinka, 1928: pp. 337–338, Pl. 9: Fig. 9, = juvenile stage of *P. calmani*; Lang 1949: p. 14, *P. calmani* = junior synonym of *P. communis*	St. Andrews Bay: 8–22 m; W Baltic Sea: Germany: Bay of Kiel, Øresund: Gilleleje Flak: muddy sand 22 m	
Pycnophyes canadensis Higgins & Korczynski, 1989	*P. canadensis* Higgins & Korczynski, 1989: pp. 2057–2061, Figs 1–16	Arctic Ocean: Beaufort Sea: Canada: Northwest Territories: Tuktoyaktuk Harbour 15 m (type locality); N Atlantic Ocean: NE Greenland: 25–150 m	Ziemer 2001
Echinoderes canariensis Greeff, 1869	*E. canariensis* Greeff, 1869: p. 20, Pl. 2: Figs 6–7; Pardos et al. 1998: p. 196, species inquirenda because juvenile stage; this book chapter: description based on adult with head + neck (= Greeff's body segment 1) + 11 trunk segments (comp. Greeff 1869: pp. 5–7)	NE Atlantic Ocean: Spain: Canary Islands: Lanzarote (type locality)	
Echinoderes cantabricus Pardos, Higgins & Benito, 1998	*E. cantabricus* Pardos et al., 1998: pp. 197–200, Figs 3–4, Tabs 1, 3, 4	NE Atlantic Ocean: Bay of Biscay: NW Spain: Cantabria: Bay of Santoña: mud to sandy mud 3–15 m (type locality), Galicia: mud to fine sand and shell gravel intertidal to 28 m, S Spain: Andalusia to Murcia: mud or fine sand and shell gravel or coarse sand 8–30 m	Sørensen et al. 2012a; Sánchez et al. 2012
Echinoderes capitatus (Zelinka, 1928)	*Echinoderella capitata* Zelinka, 1928: pp. 246–248, Pl. 3: Figs 3–7; *Habroderella capitata* Zelinka, 1928: p. 260, Pl. 27: Figs 8–9, = juvenile stage of *E. capitata*; *Echinoderella capitata* Zelinka, 1912: pp. 520–521, Figs 2, 8, nomen nudum; Remane 1936: p. 349, Fig. 274a, *Habroderella trispinosa* Zelinka, 1928 = juvenile stage of *Echinoderella capitata* (?); Nyholm 1947c: p. 23, *Hapaloderes minimus* Zelinka, 1928 = juvenile stage of *Echinoderella capitata* (?); *Echinoderes capitata* comb. nov. Karling, 1954: p. 191; Higgins 1983: p. 37, emended name *Echinoderes capitatus*; Nebelsick 1992a: emended description	Mediterranean Sea: Italy: Gulf of Trieste: 12–22 m, Bay of Naples: mud 35 m (type locality)	
Echinoderes caribiensis Kirsteuer, 1964	*E. caribiensis* Kirsteuer, 1964: pp. 388–389, Figs 1a–d; Higgins 1968: p. 30, *Habroderes manglaris* Kristeuer, 1964 (pp. 389–392, Fig. 2a) and *Habroderella orientalis*	NW Atlantic Ocean: Venezuela: Bay of Mochima: sandy mud between mangrove *Rhizophora mangle* 10–20 cm (type locality)	

	Kirsteuer, 1964 (pp. 392–393, Figs 3b, c) = juvenile stages of *E. caribiensis*		
Pycnophyes carinatus Zelinka, 1928	*P. carinatus* Zelinka, 1928: pp. 328–330, Pl. 11: Figs 1, 2, 9, 10; *Hyalophyes carinatus* Zelinka, 1928: pp. 342–344, Pl. 6: Figs 3, 11, Fig. 66, = juvenile stage of *P. carinatus*; *P. carinata* Zelinka, 1912: pp. 520–527, Figs 5, 10, nomen nudum	Mediterranean Sea: Italy: Gulf of Trieste: 12–23 m, Bay of Naples: mud 35 m (type locality), S Spain: Almuñecar: mud to coarse sand 18–20 m	Sánchez et al. 2012
Kinorhynchus cataphractus (Higgins, 1961)	*Trachydemus cataphractus* Higgins, 1961: pp. 86–87, Figs 11–14; *Kinorhynchus cataphractus* comb. nov. Sheremetevskij, 1974: p. 985, substitute name for *Trachydemus* (preoccupied)	NE Pacific Ocean: USA: Washington: Sucia Islands: Echo Bay: mud 30 m (type locality), Orcas Island: East Sound: mud 32 m, Stuart Island: Reid Harbor: mud 10 m; NE Pacific Ocean: Canada: SW Vancouver Island: very fine silt 109 m	Brinkhurst 1987; Adrianov & Malakhov 1999a
Echinoderes cavernus Sørensen, Jørgensen & Boesgaard, 2000	*E. cavernus* Sørensen et al., 2000: pp. 169–178, Figs 1–8, Tabs 1–3	SW Pacific Ocean: Australia: New South Wales: near Tuncurry: coarse coralligenous sand with detritus 32–34 m (type locality)	
Echinoderes cernunnos Sørensen, Rho, Min, Kim & Chang, 2012	*E. cernunnos* Sørensenet al., 2012b: 7 pp., Figs 4–6, Tabs 3–4	NW Pacific Ocean: Korea Strait: between Tsushima Island and mainland: mud 96 m (type locality), mud 113 m	
Pycnophyes chilensis Lang, 1953	*P. chilensis* Lang, 1953: pp. 3–6, Figs 2–3	SW Pacific Ocean: Chile: Gulf of Ancud: Quellin Island: 300 m (type locality)	
Pycnophyes chukchiensis Higgins, 1991	*P. chukchiensis* Higgins, 1991: pp. 185–188, Figs 1–2	Arctic Ocean: Chukchi Sea: USA: Alaska: soft sediment 210 m (type locality); N Atlantic Ocean: NE Greenland: 20–520 m	Ziemer 2001
Echinoderes citrinus Zelinka, 1928	*E. citrinus* Zelinka, 1928: pp. 242–243, Pl. 4: Figs 3–4	Mediterranean Sea: Italy: Bay of Naples: 35 m (type locality)	
Echinoderes collinae Sørensen, 2006	*E. collinae* Sørensen, 2006: pp. 61–67, Figs 5–8, Tabs 7–9	NW Atlantic Ocean: N Panama: Bocas del Torro: gray mud with shell and spicule fragments 20 m (type locality)	
Pycnophyes communis Zelinka, 1908	*P. communis* Zelinka, 1908: pp. 632–641, Figs 2–4, 11, internal morphology, type species by monotypy; Zelinka 1912: pp. 520–524, Figs 3, 9, spermiogenesis; Zelinka 1928: pp. 325–328, Pl. 9: Figs 1–3, 5–7, Pl. 14: Figs 6, 7, Pl. 16: Fig. 1, description; *Hyalophyes communis* Zelinka, 1928: pp. 338–339, Pl. 9: Figs 4–8, Pl. 17: Figs 1–2, Pl. 18: Fig. 5, = juvenile stage of *P. communis*; Lang 1949: p. 14, *P. calmani* = junior synonym of *P. communis*	Mediterranean Sea: Italy: Gulf of Trieste: mud 12–23 m, Bay of Naples: 25–35 m (type locality); NE Atlantic Ocean: E North Sea: Sweden: Gullmar Fjord: 60–120 m; Baltic Sea: Kadetrinne: 25–30 m	Nyholm 1947c; Reimer 1963

(Continued)

Tab. 6. (*Continued*)

species	original description; synonomy; notes	type locality, distribution, habitat, depth	references
Pycnophyes corrugatus Higgins, 1983	*P. corrugatus* Higgins, 1983: pp. 56–60, Figs 111–126, Tab. 9	NW Atlantic Ocean: Belize: Twin Cays: corraline mud with detritus and sand 3 m (type locality)	
Echinoderes coulli Higgins, 1977	*E. coulli* Higgins, 1977b: pp. 342–348, Figs 1–16, Tab. 1	NW Atlantic Ocean: USA: South Carolina: North Inlet Estuary: intertidal mud with smooth cordgrass *Spartina alterniflora* (type locality), North Carolina: Otway: mud with *Juncus roemerianus*, Georgia: Sapelo Island; SW Pacific Ocean: New Zealand: North Island (?)	Higgins 1978b; Higgins & Fleeger 1980; Coull & Wells 1981
Pycnophyes cryopygus Higgins & Kristensen, 1988	*P. cryopygus* Higgins & Kristensen, 1988: pp. 36–47, Figs 81–133, Tab. 6	N Atlantic Ocean: W Greenland: Disko Island: sandy mud 200 m (type locality), mud 9–300 m, S Greenland: Ikka Fjord: 20–32 m	Winkelmann & Ziemer 1999; Ziemer 2001
Kinorhynchus deirophorus Higgins, 1983	*K. deirophorus* Higgins, 1983: pp. 89–93, Figs 220–235, Tab. 17	NW Atlantic Ocean: Belize: Twin Cays: gray calcareous sediment with detritus and sand 1–2 m (type locality)	
Pycnophyes dentatus (Reinhard, 1881)	*Echinoderes dentatus* Reinhard, 1881: p. 589; Reinhard 1885 (pp. 263–265, Pl. 8: Figs 1–18), 1887 (pp. 438–440, Pl. 20: Figs 1–18): emendet description; *Pycnophyes dentatus* comb. nov. Zelinka, 1928: pp. 314–316, Pl. 20: Figs 3, 3a, 4, 4a, redescription; Sheremetevskij 1974: pp. 980–982, Fig. 2.1–4, *P. dentatus* = *E. dentatus* of Reinhard (1881), = *E. dubius* of Reinhard (1887), and = *P. ponticus* partim of Zelinka (1928), *Echinoderes pellucidus* = juvenile stage of *P. dentatus* (?)	Black Sea: Ukraine: Odessa (type locality), Caucasus and W Crimea: 10–150 m, Romania; E Baltic Sea: Germany: Bay of Kiel; NE Atlantic Ocean: North Sea: Ireland: Clew Bay: 4-26 m, England: SW Isle of Man: muddy sand or mud 58-110 m, Scotland: St. Andrews Bay: 8 m, NW Germany: Island of Sylt: intertidal mud, The Netherlands: Scheveningen: intertidal sand, Bay of Biscay: NW Spain: Asturias: 49–95 m, NW Spain: Galicia: mud to coarse *Amphioxus*-sand intertidal to 53 m; Mediterranean Sea: S Spain: Andalusia: mud to coarse sand 8–30 m, E Spain: Garrucha: mud to coarse sand 4–15 m, Denia: mud to fine sand 12–15 m, Blanes: muddy sand 28–38 m	Southern 1914; Zelinka 1928; Zaneveld 1938; Bacescu & Bacescu 1956; Bruce et al. 1963; Bacescu 1968; Sheremetevskij 1974; Neuhaus 1993; Adrianov & Malakhov 1999a; Sørensen et al. 2012a; Sánchez et al. 2012
Kinorhynchus distentus Higgins, 1983	*K. distentus* Higgins, 1983: pp. 93–98, Figs 236–257, Tab. 18	NW Atlantic Ocean: Belize: Twin Cays: coralline mud with detritus and sand 3 m (type locality)	
Pycnophyes dolichurus Sánchez, Pardos, Herranz & Benito, 2011	*P. dolichurus* Sánchez et al., 2011: pp. 323–327, 331–333, Figs 3–5, Tabs 4, 5	NE Atlantic Ocean: NW Spain: Galicia: La Coruña: Ría de Ares: fine or coarse sand 45 m (type locality)	
Echinoderes druxi d'Hondt, 1973	*E. druxi* d'Hondt, 1973: pp. 235–38, Fig. 2; Higgins 1978b: p. 179, possibly juvenile stage of *E. riedli*; Higgins 1983 : p. 41, species inquirenda	Mediterranean Sea: Algeria: E of Ghazaouet: Sydna-Youcha beach: intertidal sand 20 cm below surface (type locality)	
Echinoderes dujardinii Claparède, 1863	*E. Dujardinii* Claparède, 1863: pp. 90–92, Pl. 16: Figs 7–13; "l'*Echinodère*" of Dujardin 1851: p. 158–160, Pl. 3: Figs 1–5; Higgins 1964a: listed as type species; this book chapter: herewith designated as type species; probably not *E. dujardinii* of Tokioka (1949) and of Sudzuki (1976);	NE Atlantic Ocean: France: St. Vaast la Hougue (type locality), St. Malo, Roscoff: sandy mud, Ireland: Clew Bay: from weeds 4–7 m, Blacksod Bay, The Netherlands: Scheveningen: intertidal sand; NE Atlantic Ocean: W Baltic Sea: Bay of Kiel, North Sea: Helgoland, W coast of Sweden: Gullmar Fjord: <20 m, Norway: Bergen, Bay of	Metschnikoff 1865, 1869; Pagenstecher 1875; Panceri 1878; Schepotieff 1907a, b; Southern 1914; Remane 1936; Zaneveld 1938; Nyholm 1947a;

	for details of highly complex synonomy refer to Remane (1928: pp. 76, footnote 81–82; 1936: pp. 332–333, Fig. 273), Zelinka (1928: pp. 228–235, Pl. 3: Figs 1–2, 14–17, Pl. 10: Figs 1–5, 7–23, redescription; pp. 248–250. Pl. 4: Figs 1–2, juvenile stage), Higgins (1977a: pp. 4–13, Figs 1–13, Tab. 1, redescription; 1985: pp. 797–799, Fig. 4), and Adrianov & Malakhov (1999a: pp. 129–130); Sánchez-Tocino et al. 2011: pp. 180–183, Figs 1–4, emendet description	Biscay: NW Spain: shell gravel 2 m; Mediterranean Sea: S Spain: Andalusia to Murcia: mud to coarse sand or on algae *Stypocaulon scoparium* 5–30 m, SE Spain: Denia: mud 15 m, Mallorca: Porto-Pi, Italy: Salerno, Bay of Naples: 1–35 m or *Caulerpa*-mud 4–5 m or from algae 1 m, Adriatic Sea: Croatia: Rovinj, Slovenia: Portoroz: from algae 3 m, Italy: Gulf of Trieste: 3–22 m, Brindisi, Gulf of Venice: Chioggia: from sponge *Hymeniacidon sanguinea*	Mari & Morselli 1987; Sánchez et al. 2012
Pycnophyes echinoderoides Zelinka, 1928	*P. echinoderoides* Zelinka, 1928: p. 313, Pl. 11: Fig. 3; *Hyalophyes echinoderoides* Zelinka, 1928: pp. 336–337, Pl. 11: Figs 3, 4, 6, = juvenile stage of *P. echinoderoides*; Higgins 1983: pp. 77, 80, species indeterminata	Mediterranean Sea: Italy: Bay of Naples: mud 35 m (type locality)	
Pycnophyes ecphantor Higgins, 1983	*P. ecphantor* Higgins, 1983: pp. 60–64, Figs 127–142, Tab. 10	NW Atlantic Ocean: Belize: Twin Cays: gray calcareous sediment with detritus and sand 1–2 m (type locality)	
Pycnophyes egyptensis Higgins, 1966	*P. egyptensis* Higgins, 1966a: pp. 123–125, Figs 5–9	Red Sea: Egypt: Al-Ghardaqa (= Hurghada)(type locality)	
Echinoderes ehlersi Zelinka, 1913	*E. ehlersi* Zelinka, 1913: pp. 419–424, Pl. 39: Figs 6–9; Higgins & Rao 1979: pp. 78–83, Fig. 2, redescription	SW Indian Ocean: Zanzibar (type locality), E Indian Ocean: Bay of Bengal: South Andaman Islands: Havelock Island: intertidal algal mud	
Meristoderes elleae Sørensen, Rho, Min, Kim & Chang, 2012	*M. elleae* Sørensen et al., 2012c: 5 pp., Figs 9–11, Tab. 6	NW Pacific Ocean: Strait of Korea: S South Korea: mud with tiny shells 109 m (type locality), 70 km S of Jeju Island: mud with tiny shells 118 m, W of Jeju Island: mud with tiny shells 126 m	
Echinoderes elongatus (Nyholm, 1947)	*Echinoderella elongata* Nyholm, 1947c: pp. 5, 23–24, Pl. 1: Fig. 6, Fig. 12; *Echinoderes elongata* comb. nov. Karling, 1955: p. 191; Higgins 1985: p. 787, emended name *Echinoderes elongatus*	NE Atlantic Ocean: E North Sea: W coast of Sweden (type locality), W Scotland: Loch Hourn: 110 m, Loch Nevis: brown mud 101 m	McIntyre 1961, 1962
Pycnophyes emarginatus Higgins, 1983	*P. emarginatus* Higgins, 1983: pp. 64–68, Figs 143–158, Tab. 11	NW Atlantic Ocean: Belize: Twin Cays: coralline mud with detritus and sand 3 m (type locality), gray calcareous sediment with detritus and sand 1–2 m	
Kinorhynchus erismatus Higgins, 1983	*K. erismatus* Higgins, 1983: pp. 98–102, Figs 258–273, Tab. 19	NW Atlantic Ocean: Belize: Twin Cays: coralline mud with detritus and sand 3 m (type locality)	
Echinoderes eximus Higgins & Kristensen, 1988	*E. eximus* Higgins & Kristensen, 1988: pp. 17–22, Figs 33–48, Tab. 3	N Atlantic Ocean: W Greenland: Disko Island: sandy mud 300 m (type locality) or brown mud 60–200 m	Winkelmann & Ziemer 1999
Echinoderes ferrugineus Zelinka, 1928	*E. ferrugineus* Zelinka, 1928: pp. 241–242, Pl. 3: Figs 8–11	Mediterranean Sea: Italy: Bay of Naples: from algae 1-35 m, Slovenia: Portoroz: mud or from algae 3 m (type locality); W Baltic Sea: Germany: Bay of Kiel: coarse sand with detritus	Remane 1954

(Continued)

Tab. 6. (*Continued*)

species	original description; synonomy; notes	type locality, distribution, habitat, depth	references
Pycnophyes faveolus Brown, 1999 in Adrianov & Malakhov, 1999	*P. faveolus* Brown, 1999 in Adrianov & Malakhov, 1999a: pp. 181–182, 285, Fig. 5.80A-F, erroneously cited as valid species from unpublished PhD Thesis of R. Brown (1985), inadvertantly becomes valid species	SW Pacific Ocean: Australia: Sydney Harbour: Port Jackson: Hunter Bay: sandy mud 4–6 m (type locality)	Brown 1985
Echinoderes filispinosus Adrianov, 1989	*E. filispinosus* Adrianov, 1989: pp. 19–21, Fig. 1c	NW Pacific Ocean: Sea of Japan: Russia: Gulf of Vostok: mud 5 m (type locality)	Adrianov & Malakhov 1999a
Kinorhynchus fimbriatus Higgins, 1982	*K. fimbriatus* Higgins, 1982: pp. 312–316, Figs 6–9	NW Atlantic Ocean: Bermuda: Castle Harbor (type locality), mixed fine and coarse sediment subtidally	Higgins 1986b
Pycnophyes flaveolatus Zelinka, 1928	*P. flaveolatus* Zelinka, 1928: pp. 323–325, Pl. 13: Figs 1–4; *P. flaveolatus* Zelinka, 1908: Fig. 10, nomen nudum; *Hyalophyes flaveolatus* Zelinka, 1928: pp. 340–341, Pl. 7: Figs 1–2, = juvenile stage of *P. flaveolatus*; Lang 1936: p. 6, Figs 5–7	Mediterranean Sea: Italy: Gulf of Trieste: mud 18–22 m, Bay of Naples: mud 35 m (type locality); NE Atlantic Ocean: E North Sea: W coast of Sweden: Gullmar Fjord: ca. 40 m; W Baltic Sea: Germany: Bay of Kiel, Øresund: Gilleleje Flak: muddy sand 22 m, E of Vedbæk: *Laminaria*, red algae, sand, and stones 13–18 m, Sweden: Stockholm area	Lang 1936; Remane 1936; Nyholm 1947c; Lang 1963
Zelinkaderes floridensis Higgins, 1990	*Z. floridensis* Higgins, 1990: pp. 5–11, Figs 1–69, type species by original designation	NW Atlantic Ocean: USA: Florida: 33 km off Fort Pierce: muddy sand 140 m (type locality)	
Pycnophyes frequens Blake, 1930	*P. frequens* Blake, 1930: p. 5, Figs 1–3; Higgins 1965: pp. 66–69, Figs 1–5, redescription	NW Atlantic Ocean: USA: Maine: Mount Desert Island (type locality), Mount Desert Island: gray mud 15 m, Massachusetts: Cape Cod: Buzzards Bay, Cape Cod: Vineyard Sound: sand to fine silt with or without eel grass <1–5 m, Canada: Nova Scotia: Bay of Fundy: Minas Basin: estuarine saltmarsh pool: sand 20 cm (?)	Wieser 1960; Higgins 1965; Parker 1975; Bromley & Bleakney 1979
Pycnophyes furugelmi Adrianov, 1999 in Adrianov & Malakhov, 1999	*P. furugelmi* Adrianov, 1999 in Adrianov & Malakhov, 1999a: pp. 183–186, 285–286, Figs 5.81–5.85	NW Pacific Ocean: Sea of Japan: Peter the Great Bay: near Furugelmi Island: muddy sand 80 m (type locality)	
Meristoderes galatheae Herranz, Thormar, Benito, Sánchez & Pardos, 2012	*M. galatheae* Herranz et al., 2012: 6 pp., Figs 8–10, Tabs 4–5	SE Pacific Ocean: Melanesia: Salomon Islands: Ghizo Island: coral sand 28 m (type locality)	
Dracoderes gallaicus Sørensen, Herranz, Rho, Min, Yamasaki, Sánchez & Pardos, 2012	*D. gallaicus* Sørensen et al., 2012: pp. 222–228, Figs 7–10, Tabs 2–3	NE Atlantic Ocean: NW Spain: Galicia: Ría de Vigo: mud 21 m (type locality); Mediterranean Sea: SE Spain: Gibraltar: Algeciras Bay: mud 16 m	
Pycnophyes galtsovae Adrianov, 1999 in Adrianov & Malakhov, 1999	*P. galtsovae* Adrianov, 1999 in Adrianov & Malakhov, 1999a: pp. 186–188, 286–2876, Figs 5.86–5.88	Arctic Ocean: Barents Sea: Stockman (= Shtokman) oil-field: mud (type locality)	
Echinoderes gerardi Higgins, 1978	*E. gerardi* Higgins, 1978b: pp. 172–176, Figs 1–8; Higgins 1985: pp. 797–799, Fig. 4: probably not *E. dujardinii* of Metschnikoff (1865), Pagenstecher (1875), and Zelinka (1928) from Mediterranean Sea, and not *E. dujardinii* of Marinov (1964) from Black Sea but *E. gerardi*	Mediterranean Sea: Gulf of Tunis: Korbous: debris of sponge *Tethya aurantium* in shallow water (type locality)	

Cateria gerlachi Higgins, 1968	*C. gerlachi* Higgins, 1968: pp. 32–37, Figs 10–13, 20–25, Tabs 3–4; Gerlach 1969: p. 162, not *C. styx* of Rao & Ganapati (1966, 1968); not *C. styx* of Nagabhushanan (1972) and of Rao (1972) from India but *C. gerlachi* (pers. obs. R. P. Higgins)	N Indian Ocean: Bay of Bengal: India: Waltair: intertidal sand 20–60 cm below surface (type locality), Bay of Bengal: Middle Andaman Islands: Long Island: intertidal medium and coarse sand with little detritus 30 cm below surface, South Andaman Islands: Rutland: intertidal sand	Rao & Ganapati 1966, 1968; Higgins & Rao 1979; Rao 1987b
Kinorhynchus giganteus (Zelinka, 1908)	*Trachydemus giganteus* Zelinka, 1908: p. 639, Fig. 7, type species by monotypy; Zelinka 1912: pp. 520–521, Fig. 4, spermium; Zelinka 1928: pp. 294–296, Pl. 15: Figs 1–5, 5, description; *Leptodemus giganteus* Zelinka, 1928: p. 303, Pl. 8: Fig. 1, = juvenile stage of *T. giganteus*; *Kinorhynchus giganteus* comb. nov. Sheremetevskij, 1974: p. 984, substitute name for *Trachydemus* (preoccupied) by inference, type species by original designation	Mediterranean Sea: Italy: Gulf of Trieste: 17–19 m, Bay of Naples: mud 35 m; Baltic Sea: Germany: Bay of Kiel (type locality)	
Echinoderes gizoensis Thormar & Sørensen, 2010	*E. gizoensis* Thormar & Sørensen, 2010: pp. 71–83, Figs 2–6, Tabs 1–2	SE Pacific Ocean: Melanesia: Salomon Islands: Ghizo Island: coral sand from small ledges on steep rocky walls 14 m (type locality)	
Meristoderes glaber Sørensen, Rho, Min, Kim & Chang, 2012	*M. glaber* Sørensen et al., 2012c: 5 pp., Figs 12–13, Tab. 7	NW Pacific Ocean: NW South Korea: near Deogjeok Island: SW of Incheon: sand 30 m (type locality)	
Pycnophyes greenlandicus Higgins & Kristensen, 1988	*P. greenlandicus* Higgins & Kristensen, 1988: pp. 47–54, Figs 134–167, Tab. 7	N Atlantic Ocean: W Greenland: Disko Island: brown mud 200 m (type locality), mud 60–300 m, NE Greenland: 20–420 m, S Greenland: Ikka Fjord: 20–32 m	Winkelmann & Ziemer 1999; Ziemer 2001
Meristoderes herranzae Sørensen, Rho, Min, Kim & Chang, 2012	*M. herranzae* Sørensen et al., 2012c: 11 pp., Figs 2–5, Tab. 2	NW Pacific Ocean: E South Korea: Ulleung Basin: sand 200 m (type locality), S South Korea: Gamak Bay: mud 4 m, N of Jeju Island: mud with tiny shells 103 m	
Echinoderes higginsi Huys & Coomans, 1989	*E. higginsi* Huys & Coomans, 1989: pp. 213–218, Figs 2–4, Tab. 1	NE Atlantic Ocean: North Sea: The Netherlands: N of Westersheld mouth: fine sand 8 m (type locality)	
Fissuroderes higginsi Neuhaus, 2006 in Neuhaus & Blasche, 2006	*F. higginsi* Neuhaus, 2006 in Neuhaus & Blasche, 2006: pp. 23–27, Figs 2–3, Tabs 1, 6, no type designation; this book chapter: herewith designated as type species	SW Pacific Ocean: New Zealand: S of Chatham Rise: 630-885 m (original locality), North Island: Firth of Thames: 40 m	
Echinoderes hispanicus Pardos, Higgins & Benito, 1998	*E. hispanicus* Pardos et al., 1998: pp. 200–204, Figs 5–6, Tabs 2–4	NE Atlantic Ocean: Bay of Biscay: NW Spain: Cantabria: Bay of Santoña: mud to sandy mud 3–15 m (type locality), NW Spain: Galicia: mud 3 m, S Spain: Andalusia to Murcia: mud or fine sand and shell gravel or coarse sand 8–30 m, E Spain: Garrucha: 8 m, NE Spain: Blanes: muddy to fine sand 28–38 m	Sánchez et al. 2012

(Continued)

Tab. 6. (*Continued*)

species	original description; synonomy; notes	type locality, distribution, habitat, depth	references
Echinoderes horni Higgins, 1983	*E. horni* Higgins, 1983: pp. 15–20, Figs 20–31, Tab. 2	NW Atlantic Ocean: Belize: Twin Cays: gray calcareous sediment with detritus and sand 1–2 m (type locality)	
Meristoderes imugi Sørensen, Rho, Min, Kim & Chang, 2012	*M. imugi* Sørensen et al., 2012c: 3 pp., Figs 6–8, Tabs 3–4	NW Pacific Ocean: S South Korea: 100 km S of Jeju Island: mud 113 m (type locality)	
Kinorhynchus ilyocryptus (Higgins, 1961)	*Trachydemus ilyocryptus* Higgins, 1961: pp. 84–86, Figs 7–10; *Kinorhynchus ilyocryptus* comb. nov. Sheremetevskij, 1974: p. 985, substitute name for *Trachydemus* (preoccupied)	NE Pacific Ocean: USA: Washington: Orcas Island: East Sound: mud 32 m (type locality), Sucia Islands: Echo Bay: mud 30 m, Stuart Island: Reid Harbor: mud 10 m, California: Tomales Bay, Canada: SW Vancouver Island: very fine silt to fine sand 109–142 m	Higgins 1964b; Brinkhurst 1987; Adrianov & Malakhov 1999a
Echinoderes imperforatus Higgins, 1983	*E. imperforatus* Higgins, 1983: pp. 20–26, Figs 32–43, Tab. 3	NW Atlantic Ocean: Belize: Twin Cays: gray calcareous sediment with detritus and sand 1–2 m (type locality)	
Antygomonas incomitata Nebelsick, 1990	*A. incomitata* Nebelsick, 1990: pp. 144–152, Figs 2–16, Tab. 1, type species by original designation	Mediterranean Sea: Croatia: Bay of Vestar: medium-grained sand 4 m (type locality), Italy: Taranto Bay: Porto Cesareo: coarse coralline debris with mud 14 m, NE Spain: Blanes: fine sand 11–30 m	Sørensen et. al. 2009, 2010d; Sánchez et al. 2012
Sphenoderes indicus Higgins, 1969	*S. indicus* Higgins, 1969a: pp. 10–12, Figs 21–23, Tab. 3, type species by original designation	N Indian Ocean: Bay of Bengal: India: Visakhapatnam: brown sandy mud 40 m (type locality), 5 km SE Kakinada Bay: brown mud 40 m, 12 km off shore from largest tributary of Coloroon River near Porto Novo (= Parangipettai): brown sandy mud 22 m, Gulf of Kutch (= Kachchh): near "Rozi" W Jamnagar: gray-brown mangrove mud 6 m	
Pycnophyes iniorhaptus Higgins, 1983	*P. iniorhaptus* Higgins, 1983: pp. 68–72, Figs 159–174, Tab. 12	NW Atlantic Ocean: Belize: Twin Cays: gray calcareous sediment with detritus and sand 1–2 m (type locality)	
Neocentrophyes intermedius Higgins, 1969	*N. intermedius* Higgins, 1969b: pp. 117–121, Figs 1–3, only ♂♂ known, type species by original designation; Higgins 1983: p. 47, *N. satyai* is possibly junior synonym (= ♀ of this species)	W Indian Ocean: NW Madagascar: Nosy Bé: gray-brown sandy mud 1–2 m (type locality)	
Echinoderes intermedius Sørensen, 2006	*E. intermedius* Sørensen, 2006: 67–76, Figs 9–12, Tabs 10–11	NW Atlantic Ocean: N Panama: Bocas del Torro: sand (type locality)	
Echinoderes isabelae GaOrdóñez, Pardos & Benito, 2008	*E. isabelae* GaOrdóñez et al., 2008: pp. 96–100, Figs 2–4, Tabs 1–2	NE Atlantic Ocean: Bay of Biscay: NW Spain: Cantabria: subtidal until 485 m (type locality); W Mediterranean Sea: NE Spain: Blanes: muddy to fine sand 28–38 m	Sánchez et al. 2012
Echinoderes kanni Thormar & Sørensen, 2010	*E. kanni* Thormar & Sørensen, 2010: pp. 83–94, Figs 7–11, Tabs 3–4	SE Pacific Ocean: Melanesia: Salomon Islands: Ghizo Island: coral sand at the foot of a steep sandy slope 30 m (type locality), coral sand from small ledges on steep rocky walls 14 m	

Pycnophyes kielensis Zelinka, 1928	*P. kielensis* Zelinka, 1928: pp. 332–334, Pl. 20: Figs 10–12; Lang 1936: p. 6, Figs 1–4; Reimer 1963: pp. 443–444, Fig. 2, emended description, *Hyalophyes pellucidus* = juvenile stage of *P. kielensis*; Bacescu 1968 (p. 242, Pl. 7: Fig. 2A-D), Sheremetevskij 1974 (p. 982, Fig. 2.5–6): = *P. ponticus* of Bacescu & Bacescu (1956), partim, and of Marinov (1964); Remane 1936: p. 358, *Hyalophyes longisetosus* = juvenile stage of *P. kielensis* (?); *Echinoderes pellucidus* Reinhard, 1881 (p. 590), 1885 (pp. 267–268, Pl. 10: Figs 59–60, Pl. 11: Figs 61–72), 1887 (pp. 442, Pl. 21: Figs 59–72), redescription; *Hyalophyes pellucidus* comb. nov. Zelinka, 1928: pp. 339-340, Fig. 65; Remane 1929: p. 247, = juvenile stage of *Pycnophyes* sp.; *E. pellucidus* = *Pycnophyes pellucidus* comb. nov. Bacescu, 1968: p. 243; Sheremetevskij 1974: p. 982, *E. pellucidus* = juvenile stage of *Pycnophyes dentatus* (?)	W Baltic Sea: Germany: Bay of Kiel: harbor (type locality), Greifswalder Bodden: gray sand or sandy mud 3–6 m, W Baltic Sea: Øresund: E of Vedbæk: *Laminaria*, red algae, sand, and stones 13–18 m, Hollænder Dybet: mud 16 m; E North Sea: NW Germany: Island of Sylt: intertidal mud; Arctic Ocean: White Sea; Black Sea: Caucasus and Crimea: 10–150 m, Romania	Lang 1936; Bacescu & Bacescu 1956; Reimer 1963; Bacescu 1968; Sheremetevskij 1974; Arlt 1977; Neuhaus 1993; Adrianov & Malakhov 1999a; Tchesunov et al. 2008
Zelinkaderes klepali Bauer-Nebelsick, 1995	*Z. klepali* Bauer-Nebelsick, 1995: pp. 60–73, Figs 2–43, Tab. 1	Red Sea: N Bay of Safaga: carbonate sand 0.5–1.5 m (type locality)	
Echinoderes koreanus Adrianov, 1999 in Adrianov & Malakhov, 1999	*E. koreanus* Adrianov, 1999 in Adrianov & Malakhov, 1999a: pp. 135–137, 263–264, Fig. 5.35	NW Pacific Ocean: Sea of Japan: E South Korea: Ulsan Bay: gray mud 22 m (type locality)	
Echinoderes kozloffi Higgins, 1977	*E. kozloffi* Higgins, 1977a: pp. 13–18, Figs 14–21, Tab. 2; Higgins 1977a: not *Echinoderes* sp. of Kozloff (1972) and not *E. dujardini* of Merriman & Corwin (1973) but *E. kozloffi*	NE Pacific Ocean: USA: Washington: San Juan Island: North Bay: intertidal (type locality)	Adrianov & Malakhov 1999a; Kozloff 2007
Echinoderes krishnaswamyi Higgins, 1985	*E. krishnaswamyi* Higgins, 1985: pp. 787–791, Fig. 1	N Atlantik Ocean: North Sea: England: Plymouth: N of Eddystone Rock: 50 m (type locality)	
Echinoderes kristenseni Higgins, 1985	*E. kristenseni* Higgins, 1985: pp. 791–795, Fig. 2	NE Atlantik Ocean: N France: Roscoff: *Dentalium* sand (type locality), Bay of Biscay: NW Spain: Galicia: maërl 8 m	Sánchez et al. 2012
Echinoderes lanceolatus Chang & Song, 2002	*E. lanceolatus* Chang & Song, 2002: pp. 204–210, Figs 1–2, Tab. 1; Sørensen et al. 2012b: *E. lanceolatus* possibly junior synonym of *E. aureus*	NW Pacific Ocean: S South Korea: Beomseom (type locality), S and E coast	
Kinorhynchus langi (Higgins, 1964)	*Trachydemus langi* Higgins, 1964b: pp. 486–488, Figs 12–16; *Kinorhynchus langi* comb. nov. Sheremetevskij, 1974: p. 985, substitute name for *Trachydemus* (preoccupied)	NW Atlantic Ocean: USA: North Carolina: Beaufort: muddy sand 1 m (type locality), Florida: Marineland, Texas: off Galveston Island: Buccaneer Oil Field: muddy sand with shell hash 19–20 m	Harper et al. 1981

(*Continued*)

Tab. 6. *(Continued)*

species	original description; synonomy; notes	type locality, distribution, habitat, depth	references
Echinoderes levanderi Karling, 1955	*E. levanderi* Karling, 1954: pp. 189–192, Figs 1–6; Karling 1954: p. 189, not *Echinoderes* cf. *setigera* of Levander (1900), and not *E. subfuscus* of Zelinka (1928), partim, and not *Echinoderella* sp. of Purasjoki (1945) but *E. levanderi* (?); Huys & Coomans 1989: p. 218, emended description	E Baltic Sea: SW Finland: Tvärminne: muddy sand 9–12 m (type locality), coast from Åland to Helsinki: sand and mud 9–55 m, Sweden: Stockholm Archipel: Sandhamn: muddy sand 10–15 m, Askö-Landsort: mud to coarse sand 10–50 m, S Baltic Sea: Poland: 10–40 m	Purasjoki 1945; Drzycimski 1975; Ankar & Elmgren 1976; Pekkarinen 1985
Cephalorhyncha liticola Sørensen, 2008	*C. liticola* Sørensen, 2008a: p. 232, Fig. 1C, D, original publication on paper according to ICZN (1999); *C. liticola* Sørensen, 2008b: pp. 230e11–230e17, Figs 6–11, Tabs 3, 4, emendet description, electronic publication	Pacific Ocean: USA: Hawai: Oahu: intertidal sand 70 cm below surface (type locality)	
Pycnophyes longicornis Higgins, 1983	*P. longicornis* Higgins, 1983: pp. 72–76, Figs 175–190, Tab. 13	NW Atlantic Ocean: Belize: Twin Cays: gray calcareous sediment with detritus and sand 1–2 m (type locality) or coralline mud with detritus and sand 3 m	
Campyloderes macquariae Johnston, 1938	*C. macquariae* Johnston, 1938: pp. 6–12, Figs 1–7, identical with *C. kerguelensis* comb. nov. (?) described by Zelinka (1913) as *C. vanhöffeni* var. *kerguelensis*; Higgins 1967: pp. 77–79, Figs 1–2, redescription; Moore 1973: pp. 344–353, Pl. 1, Figs 2–4, Tabs. 1–4, redescription; Song & Chang 2001: pp. 209–214, Figs 1–2, Tab. 1, emended description; Neuhaus 2004 (pp. 13, 17), Neuhaus & Sørensen 2012: “almost indistinguishable species complex of *Campyloderes vanhoeffeni-C. vanhoeffeni* var. *kerguelensis-C. macquariae*”; Neuhaus & Sørensen 2012: not *C. macquariae* of Song & Chang (2001) but *C.* cf. *vanhoeffeni*	SW Pacific Ocean: Australia: Macquarie Island: just below low tide mark (type locality); SW Pacific Ocean: New Caledonia: Baie St. Vincent: coral detritus 18 m; NW Atlantic Ocean: USA: New Hampshire: Seabrook; NE Atlantic Ocean: North Sea: NE coast of Britain: Northumberland: Beadnell, Craster, and St. Mary’s Island: holdfasts of kelp *Laminaria hyperborea* 1–5 m, Yorkshire: Robin Hood’s Bay, Burniston, and Flamborough: holdfasts of kelp *Laminaria hyperborea* 1–5 m, Berwickshire: Marsden, Co. Durham, and St. Abb’s: holdfasts of kelp *Laminaria hyperborea* 1–5 m	Higgins 1967, 1977b; Moore 1973
Meristoderes macracanthus Herranz, Thormar, Benito, Sánchez & Pardos, 2012	*M. macracanthus* Herranz et al., 2012: 9 pp., Figs 1–7, Tabs 2–3, type species by original designation	Mediterranean Sea: E Spain: Blanes: muddy sand 11–38 m (type locality), N Sardinia: Maddalena Archipelago: Cala Ferrigno: algae fom jetty 1.5 m	
Kinorhynchus mainensis (Blake, 1930)	*Trachydemus mainensis* Blake, 1930: p. 7, Figs 4–6; Higgins 1965: pp. 69–71, Figs 6–10, redescription; *Kinorhynchus mainensis* comb. nov. Sheremetevskij, 1974: p. 985, substitute name for *Trachydemus* (occupied)	NW Atlantic Ocean: USA: Maine: Mount Desert Island (type locality), Mount Desert Island: gray mud 15 m, Massachusetts: Cape Cod: Buzzards Bay, Cape Cod: Vineyard Sound: sand to fine silt with or without eel grass <1–5 m, Delaware: Lewes	Wieser 1960; Higgins 1965; Parker 1975
Wollunquaderes majkenae Sørensen & Thormar, 2010	*W. majkenae* Sørensen & Thormar, 2010: pp. 262–274, Figs 1–6, Tabs 1–2, type species by original designation	SW Pacific Ocean: Coral Sea: fine coral sand with shell gravel 58 m (type locality)	

Echinoderes malakhovi Adrianov, 1999 in Adrianov & Malakhov, 1999	*E. malakhovi* Adrianov, 1999 in Adrianov & Malakhov, 1999a: pp. 139–141, 265–266, Figs 5.39–5.40	SW Pacific Ocean: New Zealand: North Island: Bay of Plenty: near Whale Island: sand 60 m (type locality)	
Polacanthoderes martinezi Sørensen, 2008	*P. martinezi* Sørensen, 2008a: pp. 231–232, Fig. 1A, B, original publication on paper according to ICZN (1999) , type species by original designation; *P. martinezi* Sørensen, 2008b: pp. 230e3–230e11, Figs 1–5, Tabs 1–2, emendet description, electronic publication	S Atlantic Ocean: N of South Shetland Islands: 2,274–2,290 m (type locality)	
Echinoderes masudai Abe, 1930	*E. Masudai* Abe, 1930: pp. 39–43, Pl. 1: Figs 1–2; Higgins 1960 (p. 89), 1966a (p. 123): nomen dubium; Higgins 1983: p. 37, species indeterminata	NW Pacific Ocean: Japan Setonaikai Sea: near Hiroshima: Gogoshima Island (type locality)	
Pycnophyes maximus Reimer, 1963	*P. maximus* Reimer, 1963: p. 445–446, Fig. 3	W Baltic Sea: Germany: Kadetrinne: 25–30 m (type locality)	
Echinoderes maxwelli (Omer-Cooper, 1957)	*Echinoderella maxwelli* Omer-Cooper, 1957b: pp. 214–216, Figs 1, 2; Omer-Cooper 1957a: p. 486, *Echinoderella* n. sp.; *Echinoderes maxwelli* comb. nov. Higgins, 1960: p. 88	SE Atlantic Ocean: South Africa: Greater Kleinemond estuary (type locality)	
Condyloderes megastigma Sørensen, Rho & Kim, 2010	*C. megastigma* Sørensen et al., 2010b: pp. 236–241, Figs 2–6, Tabs 1–2	NW Pacific Ocean: S South Korea: Korea Strait: coarse mud with shell fragments 90–109 m (type locality)	
Echinoderes microaperturus Sørensen, Rho, Min, Kim & Chang, 2012	*E. microaperturus* Sørensen et al., 2012b: 9 pp., Figs 7–10, Tabs 5–6	NW Pacific Ocean: S South Korea: Korea Strait: 100 km S of Jeju Island: mud 113 m (type locatlity), mud 96–137 m, E China Sea: mud 74–79 m	
Pycnophyes mokievskii Adrianov, 1995	*P. mokievskii* Adrianov, 1995: pp. 1559–1564, Figs 11–16	Arctic Ocean: West Spitsbergen: Isfjord: mud 6 m (type locality)	
Echinoderes multisetosus Adrianov, 1989	*E. multisetosus* Adrianov, 1989: p. 21, Fig. 1d	NW Pacific Ocean: Sea of Japan: Russia: Peter the Great Bay: Bay of Vostok: mud 5 m (type locality)	Adrianov & Malakhov 1999a
Condyloderes multispinosus (McIntyre, 1962)	*Centroderes multispinosus* McIntyre, 1962: p. 505, Fig. 1; *Condyloderes multispinsous* comb. nov. Higgins, 1969a: pp. 3–6, Figs 1–9, Tab. 1, type species by original designation	N Atlantic Ocean: North Sea: W Scotland: Loch Nevis: brown mud 101 m (type locality), Loch Torridon: 137 m, off NE Scotland: Fladen Ground: greenish gray mud 146 m, The Netherlands: N of Westerscheldt Mouth: fine sand 8 m	McIntyre 1961, 1964; Huys & Coomans 1989
Echinoderes neospinosus GaOrdóñez, Pardos & Benito, 2008	*E. neospinosus* GaOrdóñez et al., 2008: pp. 100–104, Figs 5–7, Tabs 3–4	NE Atlantic Ocean: Bay of Biscay: NW Spain: Cantabria: subtidal until 485 m (type locality)	
Pycnophyes neuhausi Higgins, 2004, in Martorelli & Higgins, 2004	*P. neuhausi* Higgins, 2004 in Martorelli & Higgins, 2004: pp. 88–92, Figs 6–11	SW Atlantic Ocean: Argentina: Chubut: San Jorge Gulf: stomach of shrimp *Pleoticus muelleri* (type locality)	
Echinoderes newcaledoniensis Higgins, 1967	*E. newcaledoniensis* Higgins, 1967: pp. 79–84, Figs 3–10, Tab. 1	SW Pacific Ocean: New Caledonia: Baie St. Vincent: coral detritus and sandy mud subtidally (type locality)	
Pycnophyes newguiniensis Adrianov, 1999 in Adrianov & Malakhov, 1999	*P. newguiniensis* Adrianov, 1999 in Adrianov & Malakhov, 1999a: pp. 193–195, 289–290, Figs 5.95–5.97	SW Pacific Ocean: Bismarck Sea: New Guinea: E New Britain Island: sand 10 m (type locality)	

(Continued)

Tab. 6. (*Continued*)

species	original description; synonomy; notes	type locality, distribution, habitat, depth	references
Pycnophyes newzealandiensis Adrianov, 1999 in Adrianov & Malakhov, 1999	*P. newzealandiensis* Adrianov, 1999 in Adrianov & Malakhov, 1999a: pp. 195–198, 290–292, Figs 5.98–5.101	SW Pacific Ocean: New Zealand: North Island: near Whale Island: sand 60 m (type locality)	
Fissuroderes novaezealandia Neuhaus, 2006 in Neuhaus & Blasche, 2006	*F. novaezealandia* Neuhaus, 2006 in Neuhaus & Blasche, 2006: pp. 27–29, Figs 4–5, Tabs 2, 6	SW Pacific Ocean: New Zealand: Chatham Rise: 527 m (type locality), S of Chatham Rise: 1,254–1,468 m	
Cephalorhyncha nybakkeni (Higgins, 1986)	*Echinoderes nybakkeni* Higgins, 1986c: pp. 268–273, Figs 1–2; *Cephalorhyncha nybakkeni* comb. nov. Neuhaus & Blasche, 2006: pp. 42, 43, 45, Figs 11F–M, emended description	NE Pacific Ocean: USA: California: Carmel Beach: intertidal sand 10–70 cm below surface but above groundwater level (type locality)	
Echinoderes obtuspinosus Sørensen, Rho, Min, Kim & Chang, 2012	*E. obtuspinosus* Sørensen et al., 2012b: 6 pp., Figs 11–12, Tabs 7–8	NW Pacific Ocean: S South Korea: Korea Strait: Jeju Island: Munseum Islet: intertidal algae (type locality), E South Korea: harbour of Geumjin (18 km SE of Gangneung): from hermit crab	
Pycnophyes odhneri Lang, 1949	*P. odhneri* Lang, 1949: pp. 18–19, Fig. 5; Lang 1953: p. 3, Fig. 4, *Pocnyphyes odhneri,* lapsus calami, emended description	S Atlantic Ocean: South Georgia Island: Cumberland Bay: clay 250–310 m (type locality)	
Echinoderes ohtsukaii Yamasaki & Kajihara, 2012	*E. ohtsukai* Yamasaki & Kajihara, 2012: pp. 110–117, Fig. 2–8, Tabs 1–2	NW Pacific Ocean: Japan: Honshu: Seto Inland Sea: intertidal muddy sand (type locality)	
Antygomonas oreas Bauer-Nebelsick, 1996	*A. oreas* Bauer-Nebelsick, 1996: pp. 8–21, Figs 3–29, Tab. 1	NE Pacific Ocean: Fieberling Guyot: calcareous or basaltic sand 580–640 m with strong tidal bottom currents (type locality); Atlantic Ocean: Bermuda: Trunk Island & mouth of Flatts Island: medium coral sand with or without detritus 3 m	Levin et al. 1994; Sørensen 2001
Dracoderes orientalis Adrianov, 1999 in Adrianov & Malakhov, 1999	*D. orientalis* Adrianov, 1999 in Adrianov & Malakhov, 1999a: pp. 159–161, 274–275, Figs 5.56–5.57	NW Pacific Ocean: Sea of Japan: E South Korea: Ulsan Bay: gray mud or gray sandy mud 22–34 m (type locality)	
Pycnophyes oshoroensis Yamasaki, Kajihara & Mawatari 2012	*P. oshoroensis* Yamasaki et al. 2012: pp. 30–35, Figs 5–7, Tabs 4, 5	NW Pacific Ocean: Sea of Japan: Japan: Hokkaido Island: Oshoro: 8 m muddy sand (type locality)	
Echinoderes pacificus Schmidt, 1974	*E. pacificus* Schmidt, 1974: pp. 182–187, Figs 1–2; Higgins 1977a: p. 21, Figs 26–31, redescription	E Pacific Ocean: Galapagos Islands: Santa Cruz Island: 0.5–1 m (type locality)	
Semnoderes pacificus Higgins, 1967	*S. pacificus* Higgins, 1967: pp. 85–88, Figs 11–12, Tab. 2	SW Pacific Ocean: New Caledonia: Baie St. Vincent: coral detritus and sandy mud 18 m (type locality), USA: California: off Redondo Beach; NW Atlantic Ocean: N Panama: Bocas del Torro: sand (?)	Sørensen 2006
Fissuroderes papai Neuhaus, 2006 in Neuhaus & Blasche, 2006	*F. papai* Neuhaus, 2006 in Neuhaus & Blasche, 2006: pp. 29–33, Figs 5H–O, 6A-C, 7, Tabs 3, 6	SW Pacific Ocean: New Zealand: N of Chatham Rise: 1,849–1,957 m (type locality)	
Condyloderes paradoxus Higgins, 1969	*C. paradoxus* Higgins, 1969a: pp. 6–9, Figs 10–20, Tab. 2	N Indian Ocean: Bay of Bengal: India: Visakhapatnam: brown sandy mud 40 m (type locality); NW Pacific Ocean: S South Korea: Korea Strait: mud 132 m (*C.* cf. *paradoxus*)	Sørensen et al. 2012b

Kinorhynchus paraneapolitanus Sheremetevskij, 1974	*K. paraneapolitanus* Sheremetevskij, 1974: pp. 984–985, Fig. 2.15–18; Higgins & Adrianov 1991: pp. 330–333, Figs 1–8, redescription	Black Sea: Russia: Caucasus: Lazarevskoye: mud with mollusc shells 90 m (type locality)	Adrianov & Malakhov 1999a
Pycnophyes parasanjuanensis Adrianov & Higgins, 1996	*P. parasanjuanensis* Adrianov & Higgins, 1996: pp. 237–244, Figs 1–24	NE Pacific Ocean: USA: Washington: near Friday Harbor Laboratory: mud 20 m (type locality); N Atlantic Ocean: NE Greenland: 60–478 m	Adrianov & Malakhov 1999a; Ziemer 2001
Echinoderes parrai GaOrdóñez, Pardos & Benito, 2008	*E. parrai* GaOrdóñez et al., 2008: pp. 104–107, Figs 8–10, Tabs 5–6	NE Atlantic Ocean: Bay of Biscay: NW Spain: Cantabria: subtidal until 485 m (type locality)	
Antygomonas paulae Sørensen, 2007	*A. paulae* Sørensen, 2007: pp. 156–166, Figs 1–5, Tabs 1–2	NW Atlantic Ocean: USA: Florida: 6 miles off Fort Pierce: *Dentalium* sand 15 m (type locality)	
Echinoderes pennaki Higgins, 1960	*E. pennaki* Higgins, 1960: pp. 86–89, Fig. 1; Higgins 1977a: pp. 18–21, Figs 22–25, redescription	NE Pacific Ocean: USA: Washington: Orcas Island: East Sound: mud 32 m (type locality), Deception Pass: Rosario Beach: intertidal on sea grass rhizomes of *Phyllospadix scouleri*	Chitwood 1964
Echinoderes peterseni Higgins & Kristensen, 1988	*E. peterseni* Higgins & Kristensen, 1988: pp. 22–28, Figs 49–64, Tab. 4	N Atlantic Ocean: W Greenland: Disko Island: brown mud 9 m (type locality), Faroe Bank: in nodules of red algae *Lithothamni glaciale*: 160–270 m	Kristensen 2005
Kinorhynchus phyllotropis Brown & Higgins, 1983	*K. phyllotropis* Brown & Higgins, 1983: pp. 162–165, Figs 1–3, Tab. 1	SW Pacific Ocean: Australia: Sydney Harbor: Port Jackson: Hunter Bay: sandy mud 6 m (type locality)	Brown 1985
Echinoderes pilosus Lang, 1949	*E. pilosus* Lang, 1949: p. 17, Fig. 4; Pallares 1966: pp. 103–106, Pls 1, 2, emended description; Higgins 1986d: pp. 400–405, Figs 1–10, redescription	S Atlantik Ocean: South Georgia Island: Grytviken: from old kelp cast up on shore (type locality), Argentina: Puerto Deseado: from intertidal and subtidal algae	Pallares 1966
Pycnophyes ponticus (Reinhard, 1881)	*Echinoderes ponticus* Reinhard, 1881: pp. 589–590; Reinhard 1885 (pp. 265–267, Pl. 8: Figs 19–24, Pl. 9: Figs 25–42), 1887 (pp. 440–442, Pl. 20: Figs 19–42): emended description; *Pycnophyes ponticus* (Reinhard, 1887[sic]) comb. nov. Zelinka, 1928: pp. 330–332, Pl. 11: Fig. 11, Pl. 13: Figs 5–6, emended description; Bacescu 1968: p. 242, Pl. 7: Fig 1A–C, *E. ponticus* of Reinhard (1887) and *P. ponticus* of Zelinka (1928), both from Odessa; Sheremetevskij 1974: p. 982, Fig. 2.5–6, *E. ponticus* of Reinhard (1887) but not *P. ponticus* of Zelinka (1928) and of Bacescu & Bacescu (1956);	Black Sea: Ukraine: Odessa (type locality), Caucasus and Crimea: 10–150 m, Romania; Mediterranean Sea: Bay of Naples: mud 35 m; W Baltic Sea: Germany: Bay of Kiel	Reinhard 1887; Bacescu & Bacescu 1956; Remane 1936; Partaly 2003

(*Continued*)

Tab. 6. *(Continued)*

species	original description; synonomy; notes	type locality, distribution, habitat, depth	references
	Bacescu 1968 (p. 243, Pl. 6: Fig. 4), Sheremetevskij 1974 (pp. 983–984, Fig. 2.8–12): not *Pycnophyes ponticus* of Zelinka (1928) from Naples and Kiel but *P. neapolitanus* of Bacescu (1968); Sheremetevskij 1974: pp. 980–982, Fig. 2.1–4, not *P. ponticus* of Zelinka (1928), partim, and of Marinov (1964) but *P. dentatus* (Reinhard, 1881); Higgins 1983: pp. 77, 79–80, *P. neapolitanus* of Bacescu (1968) and Sheremetevskij (1974) = *P. ponticus* of Zelinka (1928)		
Semnoderes ponticus Bacescu & Bacescu, 1956	*S. ponticus* Bacescu & Bacescu, 1956: pp. 545–546, Fig. 6	Black Sea: Romania (type locality)	Bacescu 1968
Sphenoderes poseidon Sørensen, Rho & Kim, 2010	*S. poseidon* Sørensen et al., 2010c: pp. 473–483, Figs 2–6, Tabs 1–2	NW Pacific Ocean: S South Korea: Korea Strait: 79–126 m (type locality)	
Paracentrophyes praedictus Higgins, 1983	*P. praedictus* Higgins, 1983: pp. 44–54, Figs 69–110, Tab. 8, type species by original designation	NW Atlantic Ocean: Belize: Twin Cays: coralline mud with detritus and sand 3 m (type locality)	Sørensen et al. 2010a
Paracentrophyes quadridentatus (Zelinka, 1928)	*Pycnophyes quadridentatus* Zelinka, 1928: pp. 310–311, Pl. 12: Figs 3–4; *Pycnophyes quadridentatus* Zelinka, 1928 and *Pycnophyes flagellatus* Zelinka, 1928 (pp. 311–313, Pl. 14: Figs 4, 5) combined as *Paracentrophyes quadridentatus* comb. nov. Higgins, 1983: p. 45	Mediterrenean Sea: Italy: Bay of Naples: 25 m (type locality); NE Atlantic Ocean: Gulf of Biscay: NW Spain: Asturias to Basque Country: muddy sand 127–272 m, NE England: off Blyth, W Scotland: Loch Torridon: 137 m, Loch Nevis: brown mud 101 m, Faroe Islands: Streymoy: sandy mud 42 m (?), Sweden: Gullmar Fjord (?)	Nyholm 1947c; McIntyre 1961, 1962; Bamber 1989; Sørensen et al. 2010a; Sánchez et al. 2012
Kinorhynchus rabaulensis Adrianov, 1999 in Adrianov & Malakhov, 1999	*K. rabaulensis* Adrianov, 1999 in Adrianov & Malakhov, 1999a: pp. 139–141, 301–302, Fig. 5.131	NW Pacific Ocean: Bismarck Sea: New Guinea: E New Britain Island: sand 10 m (type locality)	
Fissuroderes rangi Neuhaus, 2006 in Neuhaus & Blasche, 2006	*F. rangi* Neuhaus, 2006 in Neuhaus & Blasche, 2006: pp. 33–39, Figs 6D–H, 8, Tabs 4, 6	SW Pacific Ocean: New Zealand: N of Chatham Rise: 2,378–3,202 m (type locality)	
Echinoderes remanei (Blake, 1930)	*Echinoderella remanei* Blake, 1930: p. 9, Figs 7–8; *Echinoderes remanei* (Blake, 1930) comb. nov. Karling, 1954: p. 191; Higgins 1964a: pp. 244–247, Figs 1–3, redescription	NW Atlantic Ocean: USA: Maine: Mount Desert Island (type locality), Mount Desert Island: gray mud 15 m, Massachusetts: Buzzards Bay, Texas: off Galveston Island: Buccaneer Oil Field: muddy sand with shell hash 19–20 m	Wieser 1960; Higgins 1964a; Harper et al. 1981
Echinoderes rex Lundbye, Rho & Sørensen, 2011	*E. rex* Lundbye et al., 2011: pp. 42–51, Figs 1–5, Tabs 1–2;	NW Pacific Ocean; S South Korea: Korea Strait: mud with shell particles 13 m (type locality), Gamak Bay: mud 8–16 m	
Echinoderes riedli Higgins, 1966	*E. riedli* Higgins, 1966a: pp. 121–123, Figs 3–4; Higgins 1978b: pp. 176–179, Figs 9–20, redescription	Red Sea: Egypt: Al-Ghardaqa: intertidal sandy mud (type locality); Mediterranean Sea: Gulf of Tunis: Carthage: sediment from stones covered with *Enteromorpha* and diatoms 10–50 cm	

Pycnophyes robustus Zelinka, 1928	*P. robustus* Zelinka, 1928: pp. 319–321, Pl. 12: Figs 1, 2, 5, 6; *P. robustus* Southern, 1914: p. 71, nomen nudum	Mediterranean Sea: Italy: Gulf of Trieste: 17–22 m, Bay of Naples: mud 35 m (type locality)	
Pycnophyes rugosus Zelinka, 1928	*P. rugosus* Zelinka, 1928: pp. 316–319, Pl. 14: Figs 1–3	Mediterranean Sea: Italy: Bay of Naples: mud or from algae 1–35 m (type locality)	
Pycnophyes sanjuanensis Higgins, 1961	*P. sanjuanensis* Higgins, 1961: pp. 82–84, Figs 1–6	NE Pacific Ocean: USA: Washington: Orcas Island: East Sound: mud 32 m (type locality), Canada: SW Vancouver Island: very fine silt to fine sand 107–142 m	Brinkhurst 1987
Neocentrophyes satyai Higgins, 1969	*N. satyai* Higgins, 1969b: pp. 121–123, Figs 4–5, only ♀♀ known; Higgins 1983: p. 47, junior synonym of *N. intermedius* (?; = ♀ of this species?)	N Indian Ocean: Bay of Bengal: India: Visakhapatnam: brown sandy mud 40 m (type locality)	
Pycnophyes schornikovi Adrianov, 1999 in Adrianov & Malakhov, 1999	*P. schornikovi* Adrianov, 1999 in Adrianov & Malakhov, 1999a: pp. 201–203, 293–294, Figs 5.107–5.109	NW Pacific Ocean: Sea of Japan: Russia: Peter the Great Bay: near Furugelmi Island: muddy sand 80 m (type locality)	
Pycnophyes sculptus Lang, 1949	*P. sculptus* Lang, 1949: pp. 19–20, Figs 6–7	S Atlantic Ocean: South Georgia Island: off May Bay: clay 75 m (type locality), off Moraine Bay: gray clay 74 m	
Tubulideres seminoli Sørensen, Heiner, Ziemer & Neuhaus, 2007	*T. seminoli* Sørensen et al., 2007: pp. 249–257, Figs 1–6, Tabs 1–3, type species by original designation	NW Atlantic Ocean: USA: Florida: 6 miles off Fort Pierce: *Dentalium* sand 15 m (type locality)	
Echinoderes sensibilis Adrianov, Murakami & Shirayama, 2002	*E. sensibilis* Adrianov et. al., 2002c: pp. 465–472, Figs 2–7, Tab. 1	NW Pacific Ocean: Japan: Honshu: Tanabe Bay: from calcareous red algae *Corallina pilulifera* in intertidal pools (type locality)	
Echinoderes setiger Greeff, 1869	*Echinoderes setigera* Greeff, 1869: p. 19, Pl. 2: Figs 1–5 by inference; Zelinka 1928: pp. 244–246, Pl. 3: Figs 12, 13, redescription, emended name *Echinoderella setigera*; Nyholm 1947c: p. 23, *Hapaloderes gracilis* = juvenile stage of *E. setigera* (?); Karling 1954: p. 189, not *Echinoderes* cf. *setigera* of Levander (1900) but *E. levanderi* Karling, 1954; Higgins 1982: p. 310, emended name *Echinoderes setiger*	NE Atlantic Ocean: Belgium: Ostend (type locality), North Sea: off NE Scotland: Fladen Ground: greenish gray mud 146 m, Norway: Bergen; Mediterranean Sea: Italy: Gulf of Trieste: 22 m	Schepotieff 1907a, b; Zelinka 1928; McIntyre 1961, 1962
Condyloderes setoensis Adrianov, Murakami & Shirayama, 2002	*C. setoensis* Adrianov et. al., 2002b: pp. 208–215, Figs 2–6	NW Pacific Ocean: Japan: Honshu: Tanabe Bay: mud 15–27 m (type locality), E China Sea: 79 m (*C.* cf. *setoensis*)	Sørensen et al. 2012b
Echinoderes spinifurca Sørensen, Heiner & Ziemer, 2005	*E. spinifurca* Sørensen et al., 2005: pp. 500–507, Figs 1–4, Tabs 1–3; Sørensen & Pardos 2008: Fig. 3, scalids and oral styles	NW Atlantic Ocean: USA: Florida: 6 miles off Fort Pierce: *Dentalium* sand 15 m (type locality); W Atlantic Ocean: N Panama: Bocas del Torro: gray mud with shell and spicule fragments 20 m	

(Continued)

Tab. 6. (*Continued*)

species	original description; synonomy; notes	type locality, distribution, habitat, depth	references
Centroderes spinosus (Reinhard, 1881)	*Echinoderes spinosus* Reinhard, 1881: p. 590; Reinhard 1885 (pp. 270–272, Pl. 9: Figs 43–44, Pl. 10: Figs 45–51), 1887 (pp. 444–445, Pl. 20: Figs 43–47: Pl. 21: 48–51): emended description; *Centroderes spinosus* comb. nov. Zelinka, 1928: pp. 261–262, Figs 43–44, erroneus reproduction of illustrations of Reinhard (1887); Adrianov & Malakhov 1999a, Neuhaus et al. 2013: not *Centroderes agigensis* Bacescu, 1963 (Fig. 1D, figure caption "*Centroderes spinosus agigensis* n. s-sp."), not *Zelinkiana pseudospinosa* Sheremetevskij, 1974, not *Zelinkiana adunca* Sheremetevskij, 1974, not *Zelinkiana agigensis* Sheremetevskij, 1974, not *Reinhardella euxinica* Sheremetevskij, 1974, not *Reinhardella spinosa* comb. nov. Sheremetevskij, 1974 but *C. spinosus*; *C. spinosus* (Reinhard, 1881) and *C. eisigii* Zelinka, 1928 (pp. 262–263, Pl. 1: Fig.1) = *C. spinosus* comb. nov. Neuhaus et al., 2013; Neuhaus et al. 2013: not *Campyloderes adherens* Nyholm, 1947b but *C. spinosus* and not *C. spinosus* of Higgins (1977b, 1986b) and Coull (1970) but *Centroderes* sp.; *C. eisigii* = type species by subsequent designation of Sheremetevskij (1974)	Black Sea: Ukraine: Odessa (type locality), Bulgaria, Romania, Russia: fossiline mud or with shells 50–120 m; Mediterranean Sea: N Adriatic Sea: Croatia: off Rovinj, Italy: Bay of Naples: mud 35 m, NE Spain: off Blanes: muddy sand or coarse *Dentalium* sand 55 m, SE Spain: Gibraltar: Algericas Bay: muddy sand 80 m; NE Atlantic Ocean: Bay of Biscay: NW Spain: Asturias to Basque Country: muddy sand 49–444 m, Ireland: Casheen Bay: off Kilkieran Bay: muddy sand 14 m, North Sea: England: off Blyth: sandy mud, E North Sea: Kattegat: W Sweden: Gullmar Fjord: brown mud 45–65 m, W Sweden: Tjärnö: brown mud 30 m	Zelinka 1928; Nyholm 1947c; Bacescu & Bacescu 1956; Bacascu et al. 1963; Bacescu 1968; Marinov 1964; Bamber 1989; Adrianov & Malakhov 1999a; Neuhaus et al. 2013; Sánchez et al. 2012
Kinorhynchus spinosus (Lang, 1949)	*Trachydemus spinosus* Lang, 1949: pp. 21–22, Fig. 8; Lang 1953: p. 3, Fig. 7, illustration; *Kinorhynchus spinosus* comb. nov. Sheremetevskij, 1974: p. 985, substitute name for *Trachydemus* (preoccupied)	S Atlantic Ocean: Falkland Islands: Port Louis: Berkeley Sound: gravel and ooze 2–4 m, ooze with shells 8 m, Port William: sand and gravel 12 m (type locality)	
Pycnophyes spitsbergensis Adrianov, 1995	*P. spitsbergensis* Adrianov, 1995: pp. 1555–1559, Figs 1–10, 17–31	Arctic Ocean: West Spitsbergen: Isfjord: mud 6 m (type locality); N Atlantic Ocean: W Greenland: Disko Island: 60 m	Winkelmann & Ziemer 1999
Echinoderes steineri (Chitwood, 1951)	*Echinoderella steineri* Chitwood, 1951: p. 113, Fig. unnumbered; *Echinoderes steineri* Higgins, 1960: p. 89; Higgins 1966a: p. 123, nomen dubium; Higgins 1983: pp. 37, 41, species indeterminata	NW Atlantic Ocean: USA: Texas: Aransas Bay: 1.2 m (type locality)	
Kinorhynchus stenopygus Higgins, 1983	*K. stenopygus* Higgins, 1983: pp. 102–107, Figs 274–301, Tab. 20	NW Atlantic Ocean: Belize: Twin Cays: coralline mud with detritus and sand 3 m (type locality), gray calcareous sediment with detritus and sand 1–2 m	

Echinoderes stockmani Adrianov, 1999 in Adrianov & Malakhov, 1999	*E. stockmani* Adrianov, 1999 in Adrianov & Malakhov, 1999a: pp. 148–150, 268–269, Figs 5.49–5.50	Arctic Ocean: Barents Sea: Stockman (= Shtokman) oil-field: mud (type locality)	
Condyloderes storchi Higgins, 2004, in Martorelli & Higgins, 2004	*C. storchi* Higgins, 2004, in Martorelli & Higgins, 2004: pp. 86–88, Figs 2–5, Tabs 1–2	SW Atlantic Ocean: Argentina: Chubut: San Jorge Gulf: stomach of shrimp *Pleoticus muelleri* (type locality)	
Cateria styx Gerlach, 1956	*C. styx* Gerlach, 1956: pp. 120–124, Pls 40–42, type species by monotypy; Delamare Deboutteville 1957: pp. 36–37, Figs a-h, emendet description; Higgins 1968: pp. 25–32, Figs 2–9, 14–19; Tabs 1–2, redescription, lectotype designation; Gerlach 1969: p. 162, not *C. styx* of Rao & Ganapati (1966, 1968) but *C. gerlachi* not *C. styx* of Nagabhushanan (1972) and of Rao (1972) from India but *C. gerlachi* (pers. obs. R. P. Higgins)	SW Atlantic Ocean: Brazil: Macaé: intertidal coarse sand 40–90 cm below surface (type locality), Caveleiro beach: intertidal medium sand 20–50 cm below surface, Ubatuba; SE Atlantic Ocean: Angola: Lobito; SE Pacific Ocean: Chile: Cachagua Beach & Renaca Beach N of Valparaíso	Delamare Deboutteville 1957; Higgins 1968; Brown & Higgins 1983
Echinoderes subfuscus Zelinka, 1928	*E. subfuscus* Zelinka, 1928: pp. 235–236, Fig. 39, not *E. dujardinii* of Metschnikoff (1865, 1869) and not of Levander (1900) but *E. subfuscus*; *Habroderes subfuscus* Zelinka, 1928: pp. 250–251, Pl. 27: Figs 6–7, = juvenile stage of *E. subfuscus*	Mediterranean Sea: Italy: Gulf of Trieste, Slovenia: Portoroz (type locality), NE Atlantic Ocean: North Sea: Helgoland, W Baltic Sea: Kadetrinne: 25–30 m, E Baltic Sea: Åland Islands: Kobbaklintar: 211 m (?)	Levander 1900; Reimer 1963
Echinoderes sublicarum Higgins, 1977	*E. sublicarum* Higgins, 1977b: pp. 348–354, Figs 17–30, Tab. 2	NW Atlantic Ocean: USA: South Carolina: North Inlet Estuary: from hydroid *Eudendrum* colonies at pilings (type locality)	Higgins 1978b
Zelinkaderes submersus (Gerlach, 1969)	*Cateria submersa* Gerlach, 1969: pp. 162–167, Figs 2–8, Tab. 1; *Zelinkaderes submersus* comb. nov. Higgins, 1990: pp. 11–12, Figs 70–79, redescription	N Atlantik Ocean: North Sea near Denmark: N of Helgoland: medium coarse sand 30–46 m (type locality)	
Echinoderes svetlanae Adrianov, 1999 in Adrianov & Malakhov, 1999	*E. svetlanae* Adrianov, 1999 in Adrianov & Malakhov, 1999a: pp. 150–152, 270–271, Fig. 5.51	Arctic Ocean: Barents Sea: White Sea: Kandalaksha Bay: sand 15–20 m (type locality)	Tchesunov et al. 2008
Echinoderes tchefouensis Lou, 1934	*E. tchefouensis* Lou, 1934: p. 5–7, Pl. 1, Figs 1–5; Higgins 1960: p. 89, nomen dubium; Higgins 1983: p. 37, species indeterminata Sørensen et al. 2012b: 8 pp., Figs 13–15, Tabs 9–10, redescription, neotype designation	NW Pacific Ocean: China: Shandong Province: Tchefou (= Yantai) (type locality), E China Sea: mud 79 m (neotype locality), mud 57–104 m, S South Korea: Gamak Bay: mud 13 m, Geum-do: mud 9 m, Korea Strait: mud 96–140 m, Malaysia: Borneo: Sipadan: coral sand 13 m, Phillipines: Batangas: coral sand 62 m, Saipan: Managaha Island: coral sand 0 m	Sørensen et al. 2012a, b
Echinoderes teretis Brown, 1999 in Adrianov & Malakhov, 1999	*E. teretis* Brown, 1999 in Adrianov & Malakhov, 1999a: pp. 152–153, 271, Fig. 5.52; erroneously cited as valid species from unpublished PhD Thesis of R. Brown (1985), inadvertantly becomes valid species; Yamasaki & Kajihara 2012: erroneously assume that name is not available	SW Pacific Ocean: Australia: New South Wales: Sydney Harbour: Port Jackson: Lane Cove River: intertidal coarse mud, Broken Bay: Brisbane Water: coarse mud 3–5 m (type locality), Batemans Bay: inter- & subtidal mud between mangroves or with sea-grass *Posidonia australis*	Brown 1985; Nicholas & Sørensen 2009

(Continued)

Tab. 6. (*Continued*)

species	original description; synonomy; notes	type locality, distribution, habitat, depth	references
Fissuroderes thermoi Neuhaus & Blasche, 2006	*F. thermoi* Neuhaus & Blasche, 2006: pp. 39–42, Figs 6J-L, 10, 11A-E, Tabs 5–6	NE Pacific Ocean: off Costa Rica: 85–98 m (type locality)	
Kinorhynchus trisetosus Higgins, 1983	*K. trisetosus* Higgins, 1983: pp. 108–114, Figs 302–335, Tab. 21	NW Atlantic Ocean: Belize: Twin Cays: gray calcareous sediment with detritus and sand 1–2 m (type locality), coralline mud with detritus and sand 3 m	
Echinoderes truncatus Higgins, 1983	*E. truncatus* Higgins, 1983: pp. 26–31, Figs 44–55, Tab. 4	NW Atlantic Ocean: Belize: Twin Cays: coralline mud with detritus and sand 3 m (type locality), N Panama: Bocas del Torro: gray mud with shell and spicule fragments 20 m	Sørensen 2006
Echinoderes tubilak Higgins & Kristensen, 1988	*E. tubilak* Higgins & Kristensen, 1988: pp. 28–34, Figs 65–80, Tab. 5	N Atlantic Ocean: W Greenland: Disko Island: sandy mud 200 m (type locality) or mud 9–300 m	Winkelmann & Ziemer 1999
Pycnophyes tubuliferus Adrianov, 1989	*P. tubuliferus* Adrianov, 1989: pp. 25–26, Fig. 5; Adrianov & Malakhov 1999a: pp. 204–209, 294–296, Figs 5.114–5.118, redescription; Murakami et al. 2001: pp. 116–12-126, Figs 2–7, Tab. 1, redescription	NW Pacific Ocean: Sea of Japan: Russia: Gulf of Vostok: muddy sand 3 m (type locality), Sea of Japan: E South Korea: Ulsan Bay: gray mud 22 m, Japan: Honshu Island: Tanabe Bay: mud 13–25 m, Japan: Seto Inland Sea: mud 17–25 m, Seto Inland Sea: intertidal muddy sand	Sørensen et al. 2012a; Yamasaki & Kajihara 2012
Echinoderes ulsanensis Adrianov, 1999 in Adrianov & Malakhov, 1999	*E. ulsanensis* Adrianov, 1999 in Adrianov & Malakhov, 1999a: pp. 154–155, 272, Fig. 5.54	NW Pacific Ocean: Sea of Japan: E South Korea: Ulsan Bay: gray mud 22 m (type locality)	
Campyloderes vanhoeffeni Zelinka, 1913	*C. vanhöffeni* Zelinka, 1913: pp. 424–435, Pl. 39: Figs 1–5, type species by monotypy; Vanhöffen 1905: p. 19, *Echinoderes* sp.; Zelinka 1928: pp. 281–288, Figs 51–54, emended description; *Campyloderes vanhöffeni* var. *kerguelensis* Zelinka, 1913: p. 435, Pl. 39: Fig. 11, type locality: S Indian Ocean: Kerguelen Island: Observatory Bay; *Campyloderes kerguelenensis* (sic)(Zelinka, 1913) Johnston, 1938: p. 12 = *Campyloderes vanhöffeni* var. *kerguelensis* Zelinka, 1913; Higgins 1967: p. 77, listed as type species; Moore 1973: p. 341, emended name *Campyloderes vanhoeffeni*; Neuhaus 2004 (pp. 13, 17), Neuhaus & Sørensen 2012: “almost indistinguishable species complex of *Campyloderes vanhoeffeni-C. vanhoeffeni* var. *kerguelensis-C. macquariae*”; Neuhaus & Sørensen 2012: not *C. macquariae* of Song & Chang (2001) but *C.* cf. *vanhoeffeni*	S Indian Ocean: Antarctica: Gauss station N of Kaiser Wilhelm II. Land: coarse sand 385 m (type locality), Kerguelen Island: Observatory Bay; E Indian Ocean: Tansania: Zanzibar canal: 463 m; E Indian Ocean: Australia: Rottnest Island: from *Amphibolis antarctica* 1–2 m, Australia: Christmas Island: 1,427–1,891 m; E Pacific Ocean: Cocos Ridge: 1,048 m, near Galápagos spreading centre: 539–2,995 m; SW Pacific Ocean: Hikurangi Plateau: 1,605–1,940 m, Chatham Rise: 420–885 m; central Pacific Ocean: Manihiki Plateau: Danger Islands Troughs: 4,925 m, Clarion-Clipperton fraction zone : 5,050 m; NW Pacific Ocean: S South Korea: Jeju Island: mud with shells or from brown algae *Ecklonia cava* 12–28 m, E South Korea: Namae: from hermit crab *Pagurus ochotensis* and tunicate *Halocynthia* sp. 30–50 m, Solomon Islands: near Ghizo Island: coral sand 14 m; NE Atlantic Ocean: Denmark: Kattegat: Hirsholmene: 10 m, Faroe Islands: Kaldbak Fjord: 0–0.4 m; NE Atlantic Ocean: SE of Canary Islands: 5,055–5,118 m; Mediterranean Sea: SE Spain: Denia: fine sand 12 m, Grosa Island: coarse sand 10 m, Gibraltar: Algeciras Bay:	Song & Chang 2001; Neuhaus 2004; Neuhaus & Sørensen 2012; Sánchez et al. 2012

		muddy sand 80 m; W Atlantic Ocean: USA: New Hampshire: Seabrook: from *Corallina* holdfasts, Gulf of Mexico: 344–1,386 m; S Atlantic Ocean: King George Island: 411–801 m, E of South Shetland Islands: 2,893 m; E Atlantic Ocean: East Guinea Basin: 5,064 m	
Echinoderes wallaceae Higgins, 1983	*E. wallaceae* Higgins, 1983: pp. 31–36, Figs 56–67, Tab. 5	NW Atlantic Ocean: Belize: Twin Cays: coralline mud with detritus and sand 3 m (type locality), fine sand 25 m	
Echinoderes worthingi Southern, 1914	*E. Worthingi* Southern, 1914: pp. 71–72, inadvertantly becomes original description; *E. worthingii* Zelinka, 1928: pp. 240–241, Pl. 20: Figs 8–9, incorrectly considered a new species; Krishnaswamy 1962: p. 62, incorrectly considers *Pycnophyes dentatus* as a junior synonym of *E. worthingi*; Higgins 1985: pp. 795–797, Fig. 3, redescription	N Atlantik Ocean: Ireland: Blacksod Bay: amongst weeds 4 m (type locality), England: Plymouth Sound: muddy sand 8 m, Sussex: Worthing; NE Atlantic Ocean: France: Roscoff: sandy mud, Bay of Biscay: NW Spain: shell gravel to coarse *Amphioxus*-sand, S Spain: W Andalusia: mud	Higgins 1985; Sánchez et al. 2012
Kinorhynchus yushini Adrianov, 1989	*K. yushini* Adrianov, 1989: pp. 22–25, Figs 3–4; Adrianov & Malakhov 1999a: pp. 224–226, 303–304, Figs 5.133–5.136, redescription; Yamasaki et al. 2012: pp. 36–40, Figs 8–10, Tabs 6–7, redescription	NW Pacific Ocean: Sea of Japan: Russia: Gulf of Vostok: mud 5 m (type locality), S South Korea: Gamak Bay: mud 4 m, Sea of Japan: E South Korea: Ulsan Bay: gray mud 22-34 m, Japan: Seto Inland Sea: mud 17–25 m and intertidal muddy sand, SE Japan: Honshu Island:Tanabe Bay: sandy mud 14 m, Japan: Hokkaido Island: Oshoro: 8 m muddy sand	Adrianov & Malakhov 1999a; Sørensen et al. 2012a; Yamasaki & Kajihara 2012; Sørensen et al. 2012c
Pycnophyes zelinkaei Southern, 1914	*Pycnophyes Zelinkaei* Southern, 1914: pp. 70–71, Pl. 12: Figs 33A-G; *Pycnophyes zelinkaei* of Zelinka 1928: pp. 334–336, Fig. 64, redescription; Higgins 1964b: p. 481, erroneously lists *P. zelinkaei* as type species	N Atlantic Ocean: Ireland: Clew Bay: sand 26–31 m (type locality), North Sea: England: off Blyth: sandy mud, off NE Scotland: Fladen Ground: greenish gray mud 146 m, W Scotland: Loch Nevis: brown mud 101 m, S Spain: Gibraltar: fine sand and shell gravel 30 m	McIntyre 1961, 1962; Bamber 1989; Sánchez et al. 2012

Tab. 7. Alphabetical list of species described from juvenile stage (nomina dubia), synonomy, distribution, habitat, depth, and references with biogeographical information.

species	original description; synonomy; notes	type locality, distribution, habitat, depth	notes
Kinorhynchus acercus (Reinhard, 1885)	*Echinoderes acercus* Reinhard, 1885: pp. 274–275, Pl. 12: Figs 96–97; Reinhard 1887: pp. 446–447, Pl. 22: Figs 96–97; Schepotieff 1907a (pp. 151–152, Fig. 16), 1907b (p. 300, Pl. 17: Fig. 14): emended name *Echinoderes acerca*, emendet description; *Leptodemus acercus* comb. nov. Zelinka, 1928: pp. 297–299, Figs 56–57, p. 360, not *E. acerca* of Schepotieff (1907b) whose specimens are not identifiable; *Pycnophyes acercus* comb. nov. Bacescu, 1968: p. 243; *Kinorhynchus acercus* comb. nov. Higgins, 1983: p. 115	Black Sea: Ukraine: Odessa (type locality); Mediterranean Sea: Italy: Bay of Naples: *Caulerpa*-mud 4–5 m; NE Atlantic Ocean: North Sea: Norway: Bergen	Schepotieff 1907a; Zelinka 1928
Echinoderes arcticus (Steiner, 1919)	*Centropsis arcticus* Steiner, 1919: pp. 178–182, Figs 1–4; Zelinka 1928: pp. 274–276, Fig. 49, emended name *Centropsis arctica*; *Echinoderes arcticus* comb. nov. Higgins, 1983: p. 37	Arctic Ocean: Barents Sea: Russia: Murmansk: from algae (type locality)	
Echinoderes arcuatus (Zelinka, 1928)	*Centropsis arcuata* Zelinka, 1928: pp. 266–267, Pl. 2: Figs 7–8; *Echinoderes arcuatus* comb. nov. Higgins, 1983: p. 37	Mediterranean Sea: Italy: Bay of Naples from algae 1–35 m (type locality)	
Semnoderes armatus (Zelinka, 1928)	*Hapaloderes armatus* Zelinka 1928: p. 227, Pl. 2: Figs 5–6; *Echinoderes armatus* comb. nov. Higgins, 1983: p. 37 this book chapter: *Semnoderes armatus* comb. nov., possibly juvenile stage of *S. armiger*	Mediterranean Sea: Italy: Bay of Naples: mud 15 m (type locality)	*Semnoderes*: long mts + cu spine lv 5, 8, 9 + ac spine md1-10
Pycnophyes biserratus (Zelinka, 1928)	*Centrophyes biserratus* Zelinka, 1928: pp. 354–357, Pl. 5: Figs 6–7, Pl. 7: Fig. 7 ("larva secunda"), Pl. 5: Figs 1–2 ("larva tertia"); Remane 1929: p. 247, = juvenile stage of *Pycnophyes* sp.; *Pycnophyes biserratus* comb. nov. Sheremetevskij, 1974: p. 984, Fig. 2.13–14	Mediterranean Sea: Italy: Bay of Naples: (*Caulerpa*) mud 4–25 m, Gulf of Trieste: 19–22 m (type locality); Black Sea: Caucasus: fossiline mud 75 m; NE Atlantic Ocean: E North Sea: W coast of Sweden	Nyholm 1947c
Echinoderes borealis Greeff, 1869	*E. borealis* Greeff, 1869: p. 20, Pl. 2: Fig. 8; Zelinka 1928: p. 242, Fig. 42, emended differental diagnosis; Higgins 1966a: p. 123, nomen dubium; Higgins 1983: pp. 37, 41, species indeterminata	NE Atlantic Ocean: North Sea (type locality)	
Pycnophyes conspicuus (Zelinka, 1928)	*Hyalophyes conspicuus* Zelinka, 1928: pp. 179, 344–345, Pl. 7: Fig. 3, juvenile stage of *Pycnophyes robustus* (?); Remane 1929: p. 247, = juvenile stage of *Pycnophyes* sp.; *Pycnophyes conspicuus* comb. nov. Higgins, 1983: p. 77	Mediterranean Sea: Italy: Bay of Naples: mud 30 m, Adratic Sea: Gulf of Trieste: mud 19–22 m (type locality)	
Pycnophyes curvatus (Zelinka, 1928)	*Centroderes curvatus* Zelinka, 1928: pp. 346–347, Pl. 7: Fig. 5; Remane 1929: p. 247, = juvenile stage of *Pycnophyes* sp.; *Pycnophyes curvatus* comb. nov. Higgins, 1983: p. 77	Mediterranean Sea: Italy: Bay of Naples: mud 20 m, Gulf of Trieste: 19 m (type locality); NE Atlantic Ocean: E North Sea: W Sweden	Nyholm 1947c

Pycnophyes denticulatus (Zelinka, 1928)	*Centrophyes denticulatus* Zelinka, 1928: pp. 349–350, Pl. 6: Figs 6–7; Remane 1929: p. 247, = juvenile stage of *Pycnophyes* sp.; *Pycnophyes denticulatus* comb. nov. Higgins, 1983: p. 77	Mediterranean Sea: Italy: Bay of Naples: detritus 35 m (type locality), NE Atlantic Ocean: North Sea: off NE Scotland: Fladen Ground: greenish gray mud 146 m	McIntyre 1961, 1962
Pycnophyes diffusus (Zelinka, 1928)	*Centrophyes diffusus* Zelinka, 1928: pp. 351–352, Pl. 7: Fig. 4; Remane 1929: p. 247, = juvenile stage of *Pycnophyes* sp.; *Pycnophyes diffusus* comb. nov. Higgins, 1983: p. 77	Mediterranean Sea: Italy: Gulf of Trieste: 19 m (type locality)	
Kinorhynchus dubius (Reinhard, 1885)	*Echinoderes dubius* Reinhard, 1885: pp. 268–269, Pl. 11: Figs 73–81; Reinhard 1887: pp. 442–443, Pl. 21: Figs 73–81, emended description; *Leptodemus dubius* comb. nov. Zelinka, 1928: pp. 307–308, Pl. 8: Figs 3–5; Remane 1929: p. 247, = juvenile stage of *Pycnophyes* sp.; *Pycnophyes dubius* comb. nov. Bacescu, 1968: p. 243; Sheremetevskij 1974: pp. 980–982, Fig. 2.1–4, *Echinoderes dubius* of Reinhard (1887) [sic] = *Pycnophyes dentatus* = *Echinoderes dentatus* of Reinhard (1881); *Kinorhynchus dubius* comb. nov. Higgins, 1983: p. 115	Black Sea: Ukraine: Odessa (type locality); Mediterranean Sea: Italy: Bay of Naples: *Caulerpa*-mud 4–5 m, Gulf of Trieste	
Echinoderes erinaceus (Zelinka, 1928)	*Habroderes erinaceus* Zelinka, 1928: pp. 251–253, Pl. 4: Fig. 6; *Echinoderes erinaceus* comb. nov. Higgins, 1983: p. 37	Mediterranean Sea: Italy: Bay of Naples: from algae 1 m or mud 35 m (type locality)	
Echinoderes erucus Panceri, 1878	*Echinoderes eruca* Panceri, 1878: p. 6, Fig. 11; *Centropsis eruca* comb. nov. Zelinka, 1928: p. 266, Fig. 47; *Echinoderes erucus* comb. nov. Higgins, 1983: p.37	Mediterranean Sea: Italy: Bay of Naples: Ischia (type locality)	
Echinoderes ferox (Zelinka, 1928)	*Habroderella ferox* Zelinka, 1928: pp. 256–258, Pl. 4: Figs 12–13; *Echinoderes ferox* comb. nov. Higgins, 1983: p. 37	Mediterranean Sea: Italy: Gulf of Trieste (type locality)	
Kinorhynchus forceps (Zelinka, 1928)	*Leptodemus forceps* Zelinka, 1928: pp. 301–302, Pl. 5: Fig. 10; *Kinorhynchus forceps* comb. nov. Higgins, 1983: p. 115	Mediterranean Sea: Italy: Bay of Naples: (*Caulerpa*-) mud 4–20 m (type locality)	
Kinorhynchus forficulus (Zelinka, 1928)	*Leptodemus forficula* Zelinka, 1928: p. 302, Pl. 5: Fig. 11; Remane 1929: p. 247, = juvenile stage of *Pycnophyes* sp.; *Kinorhynchus forficulus* comb. nov. Higgins, 1983: p. 115	Mediterranean Sea: Italy: Bay of Naples: mud 4–5 m (type locality)	
Echinoderes gracilis Zelinka, 1928	*Hapaloderes gracilis* Zelinka, 1928: p. 279, Pl. 1: Figs 7–8; Nyholm 1947c: p. 23, *H. gracilis* = juvenile stage of *Echinoderella setigera* (?); *Echinoderes gracilis* comb. nov. Higgins, 1983: p. 37	Mediterranean Sea: Italy: Bay of Naples: mud 25 m (type locality); NE Atlantic Ocean: E North Sea: W coast of Sweden	Nyholm 1947c
Echinoderes greeffi (Zelinka, 1928)	*Centropsis greeffi* Zelinka, 1928: pp. 272–274, Fig. 48, = *Echinoderes monocercus* of Greeff (1869) and of Metschnikoff (1869); Metschnikoff 1865: pp. 460–461, Pl. 35: Fig. 11, description of *Echinoderes monocercus*; *Echinoderes greeffi* comb. nov. Higgins, 1983: p. 37	NE Atlantic Ocean: North Sea: Helgoland (type locality)	
Echinoderes hyalinus (Zelinka, 1928)	*Habroderella hyalina* Zelinka, 1928: pp. 259–260, Pl. 2: Figs 1–2; *Echinoderes hyalinus* comb. nov. Higgins, 1983: p. 37	Mediterranean Sea: Italy: Bay of Naples: *Caulerpa* -mud 4–5 m (type locality)	

(*Continued*)

Tab. 7. (*Continued*)

species	original description; synonomy; notes	type locality, distribution, habitat, depth	notes
Echinoderes incertus Reinhard, 1885	*Echinoderes incertus* Reinhard, 1885 (p. 262) and 1887 (p. 438): incorrectly cites Pagenstecher as author; Pagenstecher 1875: p. 118, unnamed specimen 1; *Habroderes incertus* comb. nov. Zelinka, 1928: pp. 359–360, species indeterminata; Higgins 1983: p. 37, *Echinoderes incertus*	Mediterranean Sea: Spain: Mallorca: Porto Pi (type locality)	
Echinoderes kowalewskii Reinhard, 1885	*Echinoderes Kowalewskii* Reinhard, 1885: p. 272–273, Pl. 10: Figs 52–58; Reinhard 1887: pp. 445–446, Pl. 21: Figs 52–58; *Hapaloderes Kowalevskii* Zelinka, 1928: pp. 279–280, Fig. 50a–c, emended description; Sheremetevskij 1974: p. 975, =*Centroderes eisigii* (?); Higgins 1983: p. 37, *Echinoderes kowalewskii*	Black Sea: Ukraine: Odessa (type locality)	
Echinoderes lanuginosus (Greeff, 1869)	*Echinoderes lanuginosa* Greeff, 1869: p. 22, Pl. 2: Fig. 9; Schepotieff 1907a: p. 151, emended name *E. lanuginosus*; *Centropsis lanuginosa* Zelinka, 1928: pp. 264–265, Figs 45, 46; Higgins 1983: p. 37, *Echinoderes lanuginosus*	NE Atlantic Ocean: NE North Sea (type locality), Norway: Bergen	Schepotieff 1907a, b
Pycnophyes longihastatus (Zelinka, 1928)	*Centrophyes longihastatus* Zelinka, 1928: pp. 348–349, Pl. 6: Figs 4–5; Remane 1929: p. 247, = juvenile stage of *Pycnophyes* sp.; *Pycnophyes longihastatus* comb. nov. Higgins, 1983: p. 77	Mediterranean Sea: Italy: Bay of Naples: *Caulerpa*-mud 4–5 m or mud 25–35 m, Gulf of Trieste: 19 m (type locality)	
Pycnophyes longisetosus (Zelinka, 1928)	*Hyalophyes longisetosus* Zelinka, 1928: pp. 341–342, Pl. 6: Fig. 2, = juvenile stage of *P. kielensis* (?); Remane 1929: p. 247, = juvenile stage of *Pycnophyes* sp.; *Pycnophyes longisetosus* comb. nov. Higgins, 1983: p. 77	Mediterranean Sea: Italy: Bay of Naples: mud 35 m (type locality); Black Sea: Constanta (?)	Bacescu & Bacescu 1956
Echinoderes meridionalis Panceri, 1878	*E. meridionalis* Panceri, 1878: pp. 5–6, Fig. 8; Schepotieff 1907a (p. 149), 1907b (p. 298, Pl. 17: Fig. 10): emended description; *Habroderes meridionalis* Zelinka, 1928: pp. 254–255, Pl. 2: Figs 11–13, not *E. meridionalis* of Panceri (1878) and of Schepotieff (1907a, b) but *H. meridionalis*; Higgins 1983: p. 37, *Echinoderes meridionalis*	Mediterranean Sea: Italy: Bay of Naples: Ischia (type locality), Bay of Naples: from algae or from sediment 1–35 m, Salerno, Adriatic Sea: Gulf of Trieste: 19 m, Croatia: Rovinj	Schepotieff 1907a, b
Kinorhynchus metschnikowii (Reinhard, 1885)	*Echinoderes metschnikowii* Reinhard, 1885: pp. 273–274, Pl. 12: Figs 88–95; Reinhard 1887: p. 446, Pl. 22: Figs 88–95, emended name *E. Metschnikoffii*; *Leptodemus metschnikoffii* comb. nov. Zelinka, 1928: pp. 299–300, Figs 58–60; *Pycnophyes metschnikoffi* (sic) comb. nov. Bacescu, 1968: p. 243; *Kinorhynchus metschnikowii* comb. nov. Higgins 1983: p. 115	Black Sea: Ukraine: Odessa (type locality)	

Echinoderes minax (Zelinka, 1928)	*Habroderes minax* Zelinka, 1928: p. 253, Pl. 4: Fig. 5; *Echinoderes minax* comb. nov. Higgins, 1983: p. 37	Mediterranean Sea: Italy: Bay of Naples: mud or from algae 1–35 m (type locality)	
Echinoderes minimus (Zelinka, 1928)	*Hapalcderes minimus* Zelinka, 1928: pp. 277–278, Pl. 2: Fig. 16; Nyholm 1947c: p. 23, *Hapaloderes minimus* = juvenile stage of *E. capitata* (?); *Echinoderes minimus* comb. nov. Higgins, 1983: p. 37	Mediterranean Sea: Italy: Bay of Trieste: mud 19 m (type locality)	
Echinoderes minutus Panceri, 1878	*E. minutus* Panceri, 1878: p. 6, Figs 9–10; *Hapaloderes minutus* comb. nov. Zelinka, 1928: pp. 278–279, Pl. 2: Figs 9–10, redescription; Higgins 1983: p. 37, *Echinoderes minutus*	Mediterranean Sea: Italy: Bay of Naples: detritus or from algae 1–35 m (type locality); NE Atlantic Ocean: E North Sea: W coast of Sweden	Nyholm 1947c
Pycnophyes moderatus (Zelinka, 1928)	*Centrophyes moderatus* Zelinka, 1928: pp. 347–348, Pl. 6: Fig. 8, pp. 350–351 (*larva secunda*), Pl 5: Fig. 5 figure caption "forma secunda"; Remane 1929: p. 247, = juvenile stage of *Pycnophyes* sp.; *Pycnophyes moderatus* comb. nov. Higgins, 1983: p. 77	Mediterranean Sea: Italy: Bay of Naples: detritus 35 m, Italy: Gulf of Trieste (type locality); NE Atlantic Ocean: E North Sea: W Coast of Sweden	Nyholm 1947c
Echinoderes monocercus Claparède, 1863	*Echinoderes monocercus* Claparède, 1863: p. 91, Pl.16: Figs 14–16; Metschnikoff 1865: pp. 460–461, Pl. 35: Fig. 11, description, not *E. monocercus* of Claparède (1863)(?), = juvenile stage of *E. dujardinii* (?); Greeff 1869: pp. 20–22, Pl. 2: Fig. 10, *E. monocercus* of Claparède (1863) and of Metschnikoff (1865)(?); Schepotieff 1907a (pp. 150–151), 1907b (p. 300, Pl. 17: Fig. 9): emended description; Steiner, 1919: p. 182, emended name *Echinoderes monocercus*; Zelinka 1928: pp. 357–359, Figs 67–69, *Centropsis monocerca* = *E. monocercus* of Claparède (1863), of Metschnikoff (1865, 1869), and of Schepotieff (1907a, b) but descriptions differ from each other, species indeterminata; Remane 1928: pp. 76, 80–81, Fig. 19, not *Centropsis (E.) monocercus* of Greeff (1869)	NE Atlantic Ocean: France: St. Vaast la Hougue (type locality), North Sea: Norway: Bergen; Mediterranean Sea: Italy: Salerno	Metschnikoff 1869; Schepotieff 1907a, b
Pycnophyes naviculus (Zelinka, 1928)	*Leptodemus navicula* Zelinka, 1928: pp. 308–309, Pl. 8: Fig. 2; Remane 1929: p. 247, = juvenile stage of *Pycnophyes* sp.; this book chapter: *Pycnophyes naviculus* comb. nov.	Mediterranean Sea: Italy: Gulf of Trieste: 19 m (type locality)	
Echinoderes orientalis Adrianov, 1989	*E. orientalis* Adrianov, 1989: pp. 21–22, Fig. 1e; Adrianov & Malakhov 1999: pp. 157 and 273, listed as species described from juvenile stage	NW Pacific Ocean: Sea of Japan: Russia: Gulf of Vostok: mud 5 m (type locality)	
Echinoderes pagenstecheri (Reinhard, 1885)	*Echinoderes Pagenstecheri* Reinhard, 1885 (p. 262) and 1887 (p. 438): incorrectly cites Pagenstecher as author; Pagenstecher 1875: p. 118, unnamed specimen 2; *Centropsis pagenstecheri* comb. nov. Zelinka, 1928: p. 357, Fig. 67, species indeterminata; *Echinoderes pagenstecheri* comb. nov. Higgins, 1983: p. 37	Mediterranean Sea: Spain: Mallorca: Porto Pi (type locality)	

(Continued)

Tab. 7. (*Continued*)

species	original description; synonomy; notes	type locality, distribution, habitat, depth	notes
Echinoderes pallidus (Zelinka, 1928)	*Centropsis pallida* Zelinka, 1928: pp. 267–268, Pl. 2: Figs 3–4; *Echinoderes pallidus* comb nov. Higgins, 1983, p. 37	Mediterranean Sea: Italy: Bay of Naples: from algae 1 m (type locality)	
Echinoderes parallelus (Zelinka, 1928)	*Centropsis parallela* Zelinka, 1928: pp. 269–270, Pl. 1: Figs 4, 5, 11, 12; *Echinoderes parallelus* comb. nov. Higgins, 1983: p. 37	Mediterranean Sea: Italy: Bay of Naples: from algae 1 m, Gulf of Trieste: 18–23 m (type locality)	
Kinorhynchus parvulus (Reinhard, 1881)	*Echinoderes parvulus* Reinhard, 1881: p. 590; Reinhard 1885 (pp. 269–270, Pl. 11: Figs 82–83, Pl. 12: Figs 84–87),1887 (p. 443, Pl. 21: Figs 22–27): emended description; *Leptodemus parvulus* comb. nov. Zelinka, 1928: pp. 305–307, Figs 61–63; Remane 1929: p. 247, = juvenile stage of *Pycnophyes* sp.; *Pycnophyes parvulus* comb. nov. Bacescu, 1968: p. 243; Sheremetevskij 1974: p. 982, = juvenile stage of *Pycnophyes dentatus* (?); *Kinorhynchus parvulus* comb. nov. Higgins, 1983: p. 115	Black Sea: Ukraine: Odessa (type locality)	
Kinorhynchus perlatus (Zelinka, 1928)	*Leptodemus perlatus* Zelinka, 1928: p. 304, Pl. 8: Fig. 6; Remane 1929: p. 247, = juvenile stage of *Pycnophyes* sp.; *Kinorhynchus perlatus* comb. nov. Higgins, 1983: p. 115	Mediterranean Sea: Italy: Bay of Naples: detritus 35 m (type locality)	
Echinoderes pulchellus (Zelinka, 1928)	*Centropsis pulchella* Zelinka, 1928: pp. 270–271, Pl. 1: Figs 2–3; Remane 1936: p. 351, = *Centropsis rosacea* (?); *Echinoderes pulchellus* comb. nov. Higgins, 1983: p. 37	Mediterranean Sea: Italy: Bay of Naples: from algae 1 m (type locality)	
Echinoderes pusillus (Zelinka, 1928)	*Centropsis pusilla* Zelinka, 1928: pp. 265–266, Pl. 27: Fig. 15; *Echinoderes pusillus* comb. nov. Higgins, 1983: p. 37	Mediterranean Sea: Italy: Gulf of Trieste (type locality)	
Pycnophyes rectilineatus (Zelinka, 1928)	*Centrophyes rectilineatus* Zelinka, 1928: p. 347, Pl. 7: Fig. 6; Remane 1929: p. 247, = juvenile stage of *Pycnophyes* sp.; *Pycnophyes rectilineatus* comb. nov. Higgins, 1983: p. 77	Mediterranean Sea: Italy: Bay of Naples: mud 30–35 m (type locality); NE Atlantic Ocean: E North Sea: W Coast of Sweden	Nyholm 1947c
Echinoderes rosaceus (Zelinka, 1928)	*Centropsis rosacea* Zelinka, 1928: pp. 268–269, Pl. 1: Figs, 9–10; Remane 1936: p. 351, may include *Centropsis pulchella*; *Echinoderes rosaceus* comb. nov. Higgins, 1983: p. 37	Mediterranean Sea: Italy: Bay of Naples: detritus or from algae 1–35 m (type locality)	
Kinorhynchus serratus (Zelinka, 1928)	*Leptodemus serratus* Zelinka, 1928: pp. 300–301, Pl. 8: Fig. 7; *Kinorhynchus serratus* Higgins, 1983: p. 115	Mediterranean Sea: Italy: Bay of Naples: mud 35 m (type locality), N Atlantic Ocean: North Sea: England: off Blyth: sandy mud	Bamber 1989
Pycnophyes solidus (Zelinka, 1928)	*Hyalophyes solidus* Zelinka, 1928: p. 344, Pl. 6: Fig. 1; Remane 1929: p. 247, = juvenile stage of *Pycnophyes* sp.; *Pycnophyes solidus* comb. nov. Higgins, 1983: p. 77	Mediterranean Sea: Italy: Bay of Naples: detritus 35 m (type locality)	
Echinoderes spinosus Panceri, 1878	*E. spinosus* Panceri, 1878: p. 7, Figs 12–13; Schepotieff 1907a (p.150, Fig. 15), 1907b (p. 299, Pl. 17: Figs 12, 13): emended description;	Mediterranean Sea: Italy: Bay of Naples: Ischia (type locality), detritus 35 m	Zelinka 1928

	Centropsis spinosus comb. nov. Steiner, 1919: p. 182, aware of Zelinka's yet unpublished monograph introduces Zelinka's generic assignment; Zelinka 1928: pp. 271–272, Pl. 1: Figs 13–14, emended description, *Centropsis spinosa* = *E. spinosus* of Panceri (1878) and of Schepotieff (1907a, b); Higgins, 1983: p. 37: *Echinoderes spinosus*	
Echinoderes splendidus (Zelinka, 1928)	*Habroderes splendidus* Zelinka, 1928: pp. 255–256, Pl. 4: Figs 10–11; *Echinoderes splendidus* comb. nov. Higgins, 1983: p. 37	Mediterranean Sea: Italy: Bay of Naples: from algae 1–5 m or *Caulerpa*-mud 4–5 m (type locality), Slovenia: Portoroz: from algae 3 m
Pycnophyes tenuis (Zelinka, 1928)	*Centrophyes tenuis* Zelinka, 1928: pp. 353–354, Pl. 5: Fig. 9; Remane 1929: p. 247, = juvenile stage of *Pycnophyes* sp.; *Pycnophyes tenuis* comb. nov. Higgins, 1983: p. 77	Mediterranean Sea: Italy: Bay of Naples: *Caulerpa*-mud 4–5 m (type locality)
Echinoderes trispinosus (Zelinka, 1928)	*Habroderella trispinosa* Zelinka, 1928: pp. 258–259, Pl. 4: Figs 7–9; Remane 1936: p. 349, Fig. 274a, = juvenile stage of *Echinoderella capitata* (?)	Mediterranean Sea: Italy: Gulf of Trieste: yellow mud 19–22 m (type locality)
Pycnophyes validus (Zelinka, 1928)	*Centrophyes validus* Zelinka, 1928: pp. 352–353, Pl. 5: Figs 3–4; Remane 1929: p. 247, = juvenile stage of *Pycnophyes* sp.; *Pycnophyes validus* comb. nov. Higgins, 1983: p. 77	Mediterranean Sea: Italy: Gulf of Trieste (type locality)
Kinorhynchus vitreus (Zelinka, 1928)	*Leptodemus vitreus* Zelinka, 1928: p. 305, Pl. 8: Figs 8–9; Remane 1929: p. 247, = juvenile stage of *Pycnophyes* sp.; *Kinorhynchus vitreus* comb. nov. Higgins, 1983: p. 115	Mediterranean Sea: Italy: Bay of Naples: detritus 35 m or *Caulerpa*-mud 4–5 m, Gulf of Trieste: 22 m (type locality)

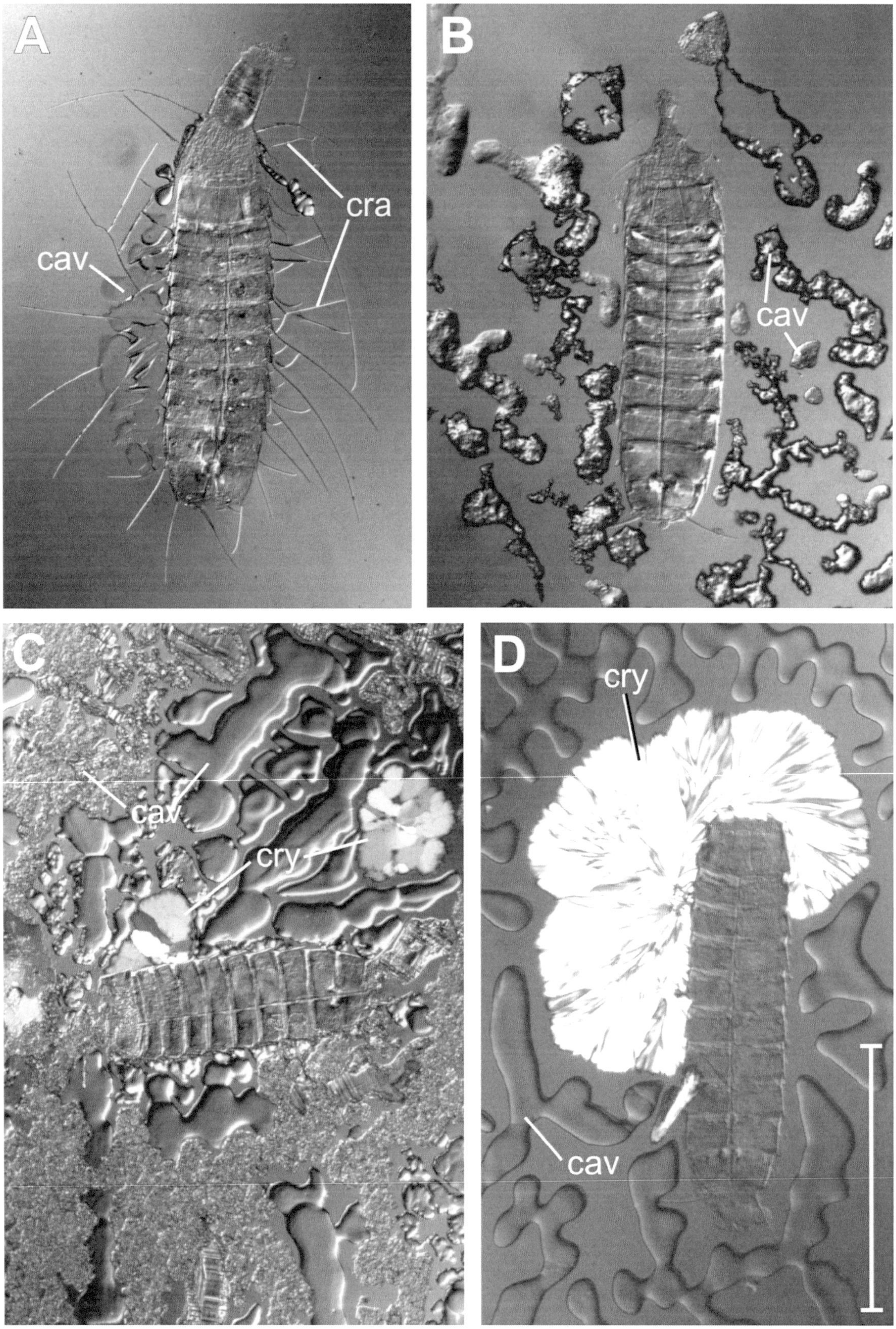

Fig. 5.5.1. LM micrographs of adult *Pycnophyes* ssp. with artefacts of embedding media. A, *Pycnophyes dentatus* in Aquatex; B, *Pycnophyes* sp. in Berlese; C, D, *Pycnophyes kielensis* (C) and *Pycnophyes dentatus* (D) in polyvinyllactophenol; notice different types of crystals; line in D = 500 μ, line valid for A–D. Original.

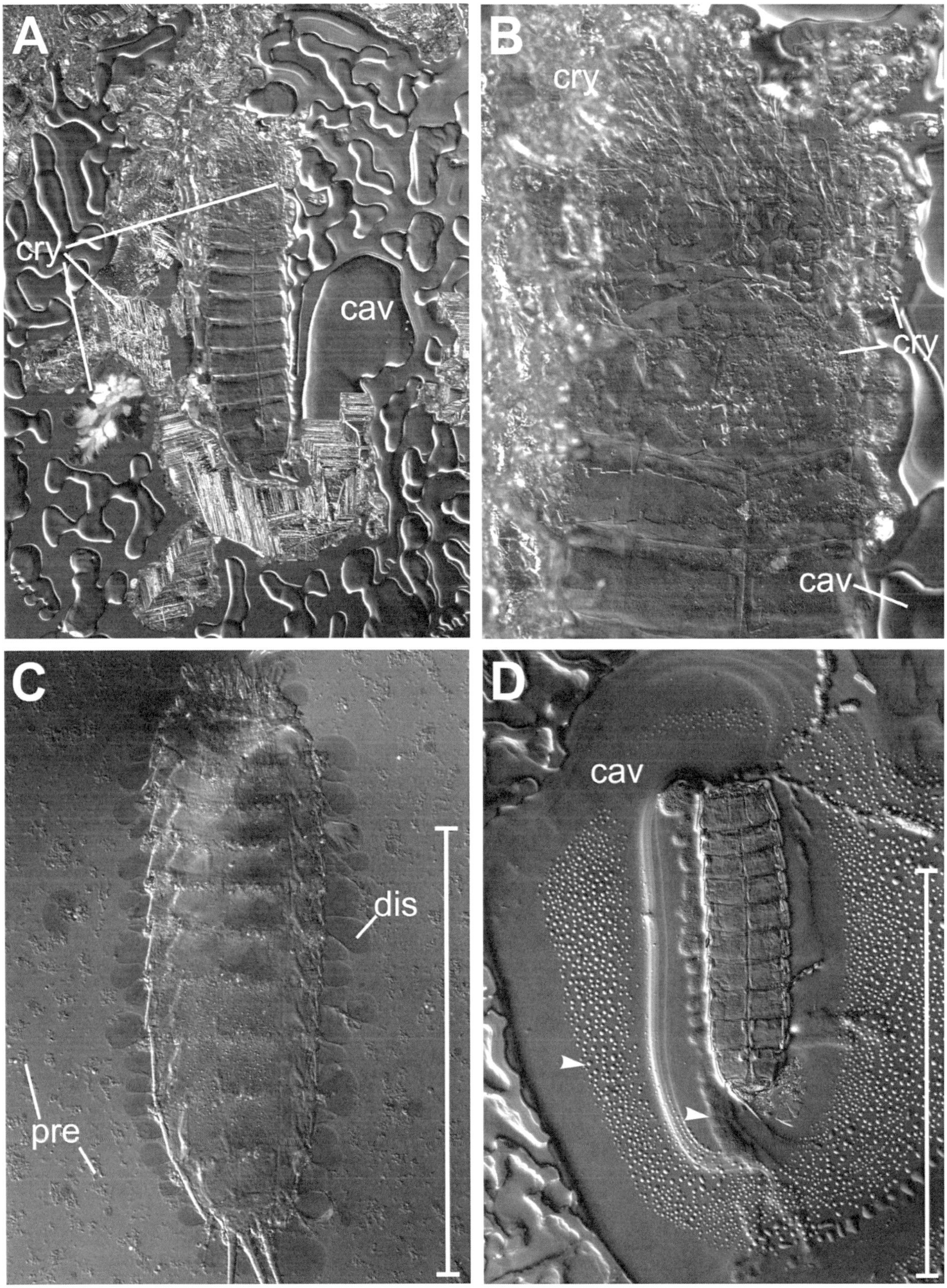

Fig. 5.5.2. LM micrographs of Kinorhyncha with artefacts of embedding media. A, B, *Pycnophyes kielensis* in polyvinyllactophenol at lower (A) and higher magnification (B) with different types of crystals; C, *Condyloderes paradoxus* in disintegrating Hoyer's mounting medium; D, *Pycnophyes dentatus* in unknown medium, notice cavities with different types of precipitations (arrowheads); line in D = 900 μ, line valid for A and D; line in C = 250 μ, line valid for B and C. Original.

has been used for over about 10 years now, but at least some slides reveal relatively small, oval, light-refracting particles both in the embedding medium and close to the specimens (Fig. 5.1.18 B-D; Neuhaus own obs.). Positive experience is currently only available for glycerin-paraffin mounts applied in nematology for decades (Hooper 1970; Neuhaus 2004; Neuhaus & Sørensen 2012). Any slides are recommended to be stored horizontally and checked for leakage or detoriation regularly.

Specimens on regular glass slides can be studied with high-power oil immersion lenses from one side only through the cover slip, because microscope objectives are designed to show their highest resolution in the focal plane immediately below the cover slip. With increasing distance from the cover slip, resolution decreases dramatically, especially in high-power oil immersion objectives. Consequently, a total mount of a kinorhynch on a regular glass slide cannot be observed in all necessary detail on the reverse side opposite to the cover slip. This problem is overcome by the Cobb aluminium slide which allows observation of a specimen also from the reverse side, because the specimen is mounted between two cover slips in an aluminium frame (Higgins 1971, 1988); therefore the Cobb slide is strongly recommended for the light microscopical study of Kinorhyncha and its long-term storage. Care should be taken to use archive-quality labelling paper according to DIN/ISO 9706 and archive-quality scripting. The Higgins-Shirayama slide (H-S slide) assembled from two layers of plastic pieces offers the same advantage as the Cobb slide (Westheide & Purschke 1988), but bends over time significantly upwards in its central area and becomes extremely difficult to use, especially from the reverse side (Neuhaus own obs.). Also, any plastic material will inevitably deteriorate after a couple of decades and is, therefore, not recommended for long-term storage in museum collections.

Different philosophies exist as to whether the holotype should be illustrated as an idealized image of a species with corrections here and there or whether it should be drawn as naturally as possible with the help of a drawing tube including, e.g., omission of missing spines. Unfortunately, drawing tubes are increasingly difficult to obtain. The first approach often results in more aesthetic images, whereas the latter reflects more the real specimen. This author favours the latter procedure, because evidently much type material gets lost or deteriorates considerably over time, and the publication is the only evidence that remains. This also aims to provide as detailed and extensive as possible light microscopical and SEM documentation in the original scientific description.

Key

Of the 21 known genera of Kinorhyncha 18 can be identified with the help of a multiple entry and a dichotomous key by Sørensen and Pardos (2008). The multiple entry key provided in Tab. 8 includes the three genera described since 2008. At species level, keys for a few genera and usually limited to the 20th century are available in individual publications (worldwide: Adrianov & Malakhov 1999a; Arctic: Adrianov 2011; *Echinoderes*: Huys & Coomans 1989, Adrianov et al. 2002a; *Pycnophyes*: Higgins & Kristensen 1988, Adrianov 1995; *Kinorhynchus*: Higgins 1983). The reader is strongly encouraged to refer to the original descriptions to identify Kinorhyncha to species.

Characterization of families and genera

Many characters identifying families and genera are shared with other genera, few features can be regarded as unique (Tab. 9). The latter may partly turn out to represent autapomorphic characters of taxa, but a phylogenetic analysis is missing.

5.6. Biogeography

The biogeography of Kinorhyncha reflects more the sampling strategy of the scientists involved than provides a realistic picture of the distribution of this taxon (Figs 5.6.1–5.6.6). About 77% of the ca. 420 locations of Kinorhyncha identified at species level refer to a depth of less than 100 m, 13% to a depth of 100–500 m, 2% to 500–1,000 m, 3% to >1,000 m, and for 5% of all locations data on the depth cannot be concluded with certainty (Tab. 6).

Continental shelf. Kinorhyncha is best studied at species level in the North Sea, Mediterranean Sea, Black See, along the North American coasts (Figs 5.6.1–5.6.3; Claparède 1863; Reinhard 1881, 1887; Southern 1914; Zelinka 1928; Karling 1955; Higgins 1960, 1961, 1964b, 1966a, 1977a,b, 1978bb, 1986b, 1990; Reimer 1963; Gerlach 1969; Sheremetevskij 1974; Higgins & Kristensen 1988; Higgins & Korczynski 1989; Huys & Coomans 1989; Nebelsick 1990; Neuhaus 1993; Adrianov & Higgins 1996; Pardos et al. 1998; Sørensen et al. 2005, 2007, 2012a; Sørensen 2007; GaOrdóñez et al. 2008; Sânchez et al. 2011, 2012; Herranz et al. 2012). Species lists are available for Belgium (Vandepitte et al. 2010), the British Isles (Bamber 1989, 1997), the European part of

Tab. 8. Diagnostic characters of adult specimens of kinorhynch genera, recognizable in light microscopy (40x objective recommended for many characters).

genus / characters		*Antygomonas*	*Campyloderes*	*Cateria*	*Centroderes*	*Cephalorhyncha*	*Condyloderes*	*Dracoderes*	*Echinoderes*	*Fissuroderes*	*Kinorhynchus*	*Meristoderes*	*Neocentrophyes*	*Paracentrophyes*	*Polacanthoderes*	*Pycnophyes*	*Semnoderes*	*Sphenoderes*	*Triodontoderes*	*Tubulideres*	*Wollunquaderes*	*Zelinkaderes*
habitus	spindle- to cigar-shaped habitus	+	+	+	+	+	+	+	+	+	–	+	–	–	+	–	+	+	+	+	+	+
	box-shaped habitus	–	–	–	–	–	–	–	–	–	+	–	+	+	–	+	–	–	–	–	–	–
scalids	ring 01 scalids chambered	–	8–31	–	–	–	–	–	–	–	–	–	–	–	–	–	–	–	–	–	–	–
	number of trichoscalids	14	14	n. a.	14	6	14	9	6	6	14	6	14	14	6	14	14	14	14	14	14	14
outer oral styles	number of elements	2	fused	3	2	2	1	1 + 3 alternate[1]	2	2	1	2	1 + 3 alternate[1]	1 + 3 alternate[1]	2	1	2	2	1	2	2	2
	with lateral cuticular flap	–	–	–	–	–	–	–	–	–	–	–	–	–	–	–	+	n. a.	+	–	+: fish-bone pattern	+
placids	fused with segment 1	–	–	+	–	–	–	–	–	–	–	–	–	–	–	–	–	–	+	±	–	+
	tripartite anteriorly	–	–	–	–	–	–	–	–	–	–	–	–	–	–	–	–	–	+	–	–	+
	with knobby projections	–	–	–	–	–	+	–	–	–	–	–	–	–	–	–	–	–	–	–	–	–
	number of placids	16	14	n. a.	16	16	16	9	16	16	6 or 8	16	7	7	16	6 or 8	16	16	14	16	16	16
	mv broad + 2x pv narrow + remaining broad[2]	+	–	n. a.	–	–	–	–	–	–	–	–	–	–	–	–	–	–	–	–	+	–
	mv placid very broad + alternating broad and narrow[2]	–	+	n. a.	–	–	–	–	–	–	–	–	–	–	–	–	–	–	–	–	–	–
	mv broad + 2x pv narrow + alternating broad and narrow[2]	–	–	n. a.	+	–	+	–	–	–	–	–	–	–	–	–	–	–	–	–	–	–
	mv broad + pv narrow + broad + 2x narrow[2]	–	–	n. a.	–	–	–	+	–	–	–	–	–	–	–	–	–	–	–	–	–	–
	placids of similar width	–	–	n. a.	–	+	–	–	–	–	–	–	–	–	+	–	–	–	–	–	–	–
	mv broad + remaining narrow[2]	–	–	n. a.	–	–	–	–	+	+	–	+	–	–	–	–	–	–	+	+	–	+
	mv + md placid narrow and elongate + remaining broad[2]	–	–	n. a.	–	–	–	–	–	–	–	–	–	–	–	–	+	–	–	–	–	–
	mv broad and large + pv narrow + >1 broad + >1 narrow[2]	–	–	n. a.	–	–	–	–	–	–	–	–	–	–	–	–	–	+	–	–	–	–

(*Continued*)

Tab. 8. *(Continued)*

genus / characters		*Antygomonas*	*Campyloderes*	*Cateria*	*Centroderes*	*Cephalorhyncha*	*Condyloderes*	*Dracoderes*	*Echinoderes*	*Fissuroderes*	*Kinorhynchus*	*Meristoderes*	*Neocentrophyes*	*Paracentrophyes*	*Polacanthoderes*	*Pycnophyes*	*Semnoderes*	*Sphenoderes*	*Triodontoderes*	*Tubulideres*	*Wollunquaderes*	*Zelinkaderes*
segment 1	ring-like	+	+	–	+	+	+	+	+	+	–	+	–	–	+	–	–	–	+[3]	+	+	+
	1 sternal + 1 tergal plate	–	–	+	–	–	–	–	–	–	–	–	+	+[4]	–	–	–	–	–	–	–	–
	3 sternal + 1 tergal plates	–	–	–	–	–	–	–	–	–	+	–	–	–	–	+	–	–	–	–	–	–
	clamshell-like, anterior, broad incision with large triangular mv placid	–	–	–	–	–	–	–	–	–	–	–	–	–	–	–	–	+	–	–	–	–
	clamshell-like, anterior, narrow incision with narrow, elongate mv placid	–	–	–	–	–	–	–	–	–	–	–	–	–	–	–	+	–	–	–	–	–
	slight mv + md anterior indentation	+	–	–	–	–	–	–	–	–	–	–	–	–	–	–	–	–	–	–	–	–
segment 2	ring-like	–	–	–	–	–	–	–	+	–	–	+[5]	–	–	–	–	–	–	–	–	–	+
	1 tergal plate with mv junction	–	–	–	–	–	–	–	–	–	–	–	–	–	–	–	–	–	–	+	–	–
	1 (mid)sternal + 1 tergal plate	–	–	+	–	+[6]	–	–	–	–	–	–	–	–	–	–	–	–	–	–	–	–
	2 sternal + 1 tergal plates	+	+	–	+	–	+	+	–	+	+	–	+	+	+	+	+	+	+	–	+	–
segments 3 + 4	2 sternal + 1 tergal plates	+	+	–	+	+	+	+	+	+	+	+	+	+	+	+	+	+	+	+	+	–
	1 midsternal + 1 tergal plate	–	–	+	–	–	–	–	–	–	–	–	–	–	–	–	–	–	–	–	–	–
	1 tergal plate with mv junction	–	–	–	–	–	–	–	–	–	–	–	–	–	–	–	–	–	–	–	–	+
segment 5	2 sternal + 1 tergal plates	+	+	–	+	+	+	+	+	+	+	+	+	+	+	+	+	+	–	+	+	–
	1 midsternal + 1 tergal plate	–	–	+	–	–	–	–	–	–	–	–	–	–	–	–	–	–	–	–	–	–
	1 tergal plate with mv junction	–	–	–	–	–	–	–	–	–	–	–	–	–	–	–	–	–	+	–	–	+
segment 6	2 sternal + 1 tergal plates	+	+	–	+	+	+	+	+	+	+	+	+	+	+	+	+	+	–	+	+	–
	1 midsternal + 1 tergal plate	–	–	+	–	–	–	–	–	–	–	–	–	–	–	–	–	–	–	–	–	–
	1 tergal plate with mv junction	–	–	–	–	–	–	–	–	–	–	–	–	–	–	–	–	–	+	–	–	+

segments 8–10	2 sternal + 1 tergal plates	+	+	–	+	+	+	+	+	+	+	+	+	+	+	+	+	+	–	+	+	–
	1 tergal plate with mv junction	–	–	+	–	–	–	–	–	–	–	–	–	–	–	–	–	–	+	–	–	+
trunk appendages	md acicular/ crenulated spines in segments	1–11	1–11; stage-2 ♀: 1–9	2–4, 6, 8–10	1–11; stage-2 ♀: 1–9, 11	4–8[7]	1–9 (10)	2, 9	–; 4; 4, 6; 4, 7; 4–7; 4–8; 4, 6, 8	4–8[7]	–[8]	4 (6, 8)[7]	–[8]	–[8]	4–8[7]	–[8]	1–11	1–11	1–11	1, 2, 9–11	1–11	4, 6, 8–11
	acicular spines alternatingly offset to left and right from md line in segments	–	–	–	–	–	–	3–8	–	–	–	–	–	–	–	–	–	–	–	3–8[9]	–	–
	cuspidate spines	+	–	–	–	–	+	–	–	–	–	–	–	–	–	–	+	+	+	–	+	+
	lv spine in segment 1	–	–	–	–	–	+	–	–	–	–	–	–	–	–	–	–	–	–	–	+	–
	acicular lateral spines at least in some segments of s2–9	+	+	+	+	+	+	+	+	+	–	+	–	–	+	–	+	+	+	+	+	+
	length **vm** spine in s1 >> length segment	–	–	–	+	–	–	–	–	–	–	–	–	–	–	–	–	–	–	–	–	–
	length **lv** spine in s1 >> length segment	–	+	–	–	–	–	–	–	–	–	–	–	–	–	–	–	–	–	–	–	–
	short spine vm, vl, la, sd, and ld in several segments of s4–10	–	–	–	–	–	–	–	–	–	–	–	–	–	+	–	–	–	–	–	–	–
	lateral terminal spine	+	+	+	+	+	+	+	+	+	–	+	–	+	+	+	+	+	+	+	+	+
	lateral terminal accessory spine	+	+	+	+	♀: +, ♂ –[11]	–	–	+ / –[10]	♀: +, ♂ –[11]	–	♀: +, ♂ –[11]	–	–	♀: +, ♂ –[11]	–	+	+	+	+	+	+
	midterminal spine	+	+	+	+	–	+	–	–	–	–	–	–	–	–	–	+	+	+	+	+	+
	length mts >> length lts	+	–	+	+	–	–	–	–	–	–	–	–	–	–	–	+	+	+	+	+	+
	length mts < length lts	–	+	–	–	–	+	–	–	–	–	–	–	–	–	–	–	–	–	–	–	–
	short midterminal spinose process in s11	–	–	–	–	–	–	–	–	–	–	–	+	+	–	–	–	–	–	–	–	–
trunk surface	cuticle with longitudinal striation pattern	–	–	–	–	–	–	–	–	–	–	–	–	–	–	–	–	–	–	–	–	+

[1] single-element outer oral style much shorter than 3-element style.[2] size of placids indicated as neighbouring the midventral placid to the left and right.[3] weak tergosternal junctions possible.[4] anterior, weak, incomplete division into 3 plates.[5] tergosternal junction in posterior but not anterior part of segment difficult to recognize in LM, barely visible midventral junction.[6] barely visible midventral junction.[7] few species are known in these genera, so a diversity of spine patterns like in *Echinoderes* can be expected.[8] may possess md spinose processes without articulation.[9] dorsal spines slightly offset to almost paradorsal position.[10] ♀: +, ♂ – (but with penile spines); except *E. neospinosus*: ♀ and ♂ with ltas; except *E. capitatus, E. coulli, E. isabelae, E. maxwelli, E. rex, E. teretis*: ♀ and ♂ without ltas.[11] no ltas but with penile spines.

Tab. 9. Diagnostic characters of the families and genera of Kinorhyncha Reinhard, 1885; unique features marked in bold. Families according to Adrianov and Malakhov (1999a), modified (Neuhaus & Blasche 2006; Sørensen 2007, 2008b; Sørensen & Rho 2009; Sørensen & Thormar 2010; Sørensen et al. 2010c, 2012a, c; Herranz et al. 2012). Etymology taken from original publication or searched from dictionaries. For sexual dimorphism compare Tab. 2, for higher classification text.

family	identification characters of family	genus (# species adult / # species juvenile)	etymology	identification characters of genus	references
order Cyclorhagida (Zelinka, 1896) Higgins, 1964b					
Antygomonidae Adrianov & Malakhov, 1994	(same as genus)	*Antygomonas* Nebelsick, 1990 (3 / 0)	antyx, antygos = Greek for edge, monas = Greek for unit; refer to cuticle of the trunk consisting of a ring (unit) with 2 edges at the midventral articulation site	spindle- to cigar-shaped habitus; 14 trichoscalids; outer oral styles with 2 elements; 16 placids: mv broad + 2x pv narrow + remaining broad; **segment 1 with mv + md indentation,** ring-like; segments 2–10 with 2 sternal + 1 tergal plate; md acicular spine in segments 1–11 + mts; lateral acicular and cuspidate spines + lts + ltas; length mts >> length lts	Nebelsick 1990; Bauer-Nebelsick 1996; Sørensen 2007; Sørensen et al. 2009
Cateriidae Gerlach, 1956	(same as genus)	*Cateria* Gerlach, 1956 (2 / 0)	kato = Greek for down, refers to the animals living deep below the sediment surface	spindle- to cigar-shaped habitus; trichoscalids # n. a.; outer oral styles with 3 elements; placids # n. a., **placids fused with segment 1; segments 1–6(7) with 1 midsternal + 1 tergal plate, segments 7(8)–10 with 1 tergal plate and mv junction, segment 11 ring-like; cuticle with hook-like denticulation**; md acicular spine in segments 2–4, 6, 8–10 + mts; lateral acicular spines + lts + ltas; length mts >> length lts; no cuspidate spines	Gerlach 1956; Higgins 1968
Centroderidae Zelinka, 1896	spindle- to cigar-shaped habitus; 14 trichoscalids; outer oral styles fused laterally or with 1 or 2 elements; 14 or 16 placids: mv broad + 2x narrow + alternating broad and narrow placids or very broad + alternating broad and narrow placids; segment 1 ring-like; segment 2–11 with 2 sternal + 1 tergal plate; md acicular spine at least on segments 1–9 + mts; lateral spines acicular and sometimes also cuspidate + lts + ltas; length mts >> lts or mts < lts	*Campyloderes* Zelinka, 1907 (2 / 0) (possibly just 1 species, see Tab. 6)	kampylos = Greek for bent or arched, deres = Greek for neck (common suffix of cyclorhagid genera); refers to midterminal spine bent towards ventral side	spindle- to cigar-shaped habitus; **outer oral styles fused over entire length; primary scalids of adult stage with 8–31 internal septae; ring 02 scalids short and acicular**; 14 trichoscalids; placids: **mv placid very broad** + alternating narrow and broad; segment 1 ring-like; segments 2–11 with 2 sternal + 1 tergal plate; md acicular spine in segments 1–11 (stage-2 ♀: segments 1–9) + mts; lateral acicular spines + lts + ltas; **lv acicular spine in segment 1 significantly elongated**; length mts < length lts; **mts with 1 proximal dorsal thin area in the cuticle; ltas with 2 proximal dorsal thin areas in the cuticle**; no cuspidate spines	Zelinka 1913; Moore 1974; Neuhaus 2004; Neuhaus & Sørensen 2012
		Centroderes Zelinka, 1907 (1 / 0)	kentron = Greek for spine, refers to midterminal spine	spindle- to cigar-shaped habitus; 14 trichoscalids; outer oral styles with 2 elements; 16 placids: mv broad + 2x pv narrow + alternating broad and narrow; segment 1 ring-like; segments 2–11 with 2 sternal + 1 tergal plate; **mv spinose process in segment 1**; md acicular spine in segments 1–11 (stage-2 ♀: segments 1–9, 11) + mts; **vm acicular spine in**	Neuhaus et al. 2013

				segment 1 elongated; lateral acicular spines + lts + ltas; length mts >> length lts; no cuspidate spines	
		Condyloderes Higgins, 1969 (5 / 0)	kondylos = Greek for bone joint, refers to knobby thickenings on placids	spindle- to cigar-shaped habitus; 14 trichoscalids; outer oral styles with 1 element; 16 placids: mv broad + 2x pv narrow + alternating broad and narrow, **knobby projections on placids**; segment 1 ring-like; segments 2–11 with 2 sternal + 1 tergal plate; acicular spine md in segments 1–9 (10) + mts; lateral acicular and cuspidate spines including segment 1 + lts + no ltas; length mts < length lts	Higgins 1969a; Sørensen et al. 2010b
Dracoderidae Higgins & Shirayama, 1990	(same as genus)	*Dracoderes* Higgins & Shirayama, 1990 (3 / 0)	drakon = Greek for dragon	spindle- to cigar-shaped habitus; **9 trichoscalids**; outer oral styles with 1 or 3 elements, alternating; **9 placids**: mv broad + pv narrow + broad + 2x narrow; segment 1 ring-like; segments 2–11 with 2 sternal + 1 tergal plate; md acicular spine in segments 2, 9 + **acicular spine in segments 3–8 alternating left and right of dorsal midline**; lateral acicular spines + lts + no ltas; no cuspidate spines	Higgins & Shirayama 1990; Sørensen et al. 2012a
Echinoderidae Zelinka, 1894	spindle- to cigar-shaped habitus; **6 trichoscalids**; outer oral styles with 2 elements; 16 placids of similar width or mv broader + remaining narrow; segment 1 ring-like; segment 2 ring-like or with 1–2 sternal + 1 tergal plate, if single sternal plate than with barely visible mv junction; segments 3–10 with 2 sternal + 1 tergal plate; md acicular spine missing in adult but not juveniles of some species or on at least one segment of segments 4–8; no mts; lateral acicular spines + lts + ltas usually only in ♀ but sometimes missing; no cuspidate spine	*Cephalorhyncha* Adrianov, 1999 in Adrianov & Malakhov, 1999 (3 / 0)	kephalon = Greek for head, rhynchos = Greek for snout	spindle- to cigar-shaped habitus; 6 trichoscalids; outer oral styles with 2 elements; 16 placids of similar width; segment 1 ring-like; segment 2 with 1 sternal + 1 tergal plate, **tergosternal junction in posterior but not anterior part of segment, barely visible midventral junction**; segments 3–10 with 2 sternal + 1 tergal plate; segment 11 with 2 sternal + 1–2 tergal plates; acicular md spine in segments 4–8; lateral acicular spines + lts + ♀: ltas / ♂: no ltas; no cuspidate spines; ♀: gland cell with funnel-shaped structure vl in segment 8 in some species	Adrianov & Malakhov 1999a; Neuhaus & Blasche 2006; Sørensen 2008b; Thormar & Sørensen 2010
		Echinoderes Claparède, 1863 (78 / 26)	echinos = Greek for hedgehog, deres = Greek for neck; refer to spiny scalids of introvert which Dujardin (1851) regarded as neck (Higgins 1971), first described genus of Kinorhyncha	spindle- to cigar-shaped habitus; trunk cuticle well developed; 6 trichoscalids; outer oral styles with 2 elements; 16 placids: mv broad + remaining narrow; segments 1–2 ring-like; segments 3–11 with 2 sternal + 1 tergal plate; md acicular spines missing / in segment 4 / in segments 4, 6 / in segments 4, 7 / in segments 4–7 / in segments 4–8 / or in segments 4, 6, 8; lateral acicular spines + lts + ♀: ltas / ♂ no ltas but with penile spines (except *E. neospinosus*: ♀ and ♂ with ltas; except *E. capitatus, E. coulli, E. isabelae, E. maxwelli, E. rex, E. teretis*: ♀ and ♂ without ltas); no cuspidate spines; ♀ of some species: gland cell with funnel-shaped structure vm in segment 7 or vl in segment 8	Zelinka 1928; Higgins 1977b, 1983; Pardos et al. 1998; Adrianov & Malakhov 1999a; GaOrdóñez et al. 2008; Thormar & Sørensen 2010; Lundbye et al. 2011; Sørensen et al. 2012b
		Fissuroderes Neuhaus &	fissura = Latin for fissure or split, refers to the	spindle- to cigar-shaped habitus; 6 trichoscalids; outer oral styles with 2 elements; 16 placids: mv broad + remaining narrow; segment 1 ring-like; segments 2–10 with 2 sternal +	Neuhaus & Blasche 2006

(*Continued*)

Tab. 9. *(Continued)*

family	identification characters of family	genus (# species adult / # species juvenile)	etymology	identification characters of genus	references
		Blasche, 2006 (5 / 0)	midventral split between the 2 sternal plates of segment 2	1 tergal plate; segment 11 with 2 sternal + 1–2 tergal plates; md acicular spine in segments 4–8; lateral acicular spines + lts + ♀: ltas / ♂ no ltas but with penile spines; no cuspidate spines; ♀: gland cell with funnel-shaped structure vm in segment 7 or vl in segment 8	
		Meristoderes Herranz et al., 2012 (2 / 0)	meristos = Greek for divided, refers to incomplete division of segment 2	spindle- to cigar-shaped habitus; 6 trichoscalids; outer oral styles with 2 elements; 16 placids: mv broad + remaining narrow; segments 1–2 ring-like, **sternal plate of segment 2 with weak tergosternal junction in posterior but not anterior part of segment and barely visible midventral junction**; segments 3–11 with 2 sternal + 1 tergal plate; md acicular spine in segment 4 / or segments 4, 6, 8; lateral acicular spines + lts + ♀: ltas / ♂ no ltas but with penile spines; no cuspidate spines	Herranz et al. 2012
		Polacanthoderes Sørensen, 2008 (1 / 0)	polys = Greek for many, akantha = Greek for spine; refer to the dorsal multiple serial rows of spines	spindle- to cigar-shaped habitus; 6 trichoscalids; outer oral styles with 2 elements; 16 placids of similar width; segment 1 ring-like; segments 2–11 with 2 sternal + 1 tergal plate; no cuticular hairs and perforation sites; md acicular spine in segments 4–8; lateral acicular spines + **short acicular spine vm, vl, la, sd, and ld in several segments of segments 4–10** + lts + ♀: ltas / ♂ no ltas but with penile spines; no cuspidate spines	Sørensen 2008b
Semnoderidae Remane, 1929	spindle- to cigar-shaped habitus; 14 trichoscalids; outer oral styles with 2 elements or 1 and 3 elements which alternate; 16 placids, mv placid large and triangular or narrow and elongate; **segment 1 with deep, clamshell-like, anterior incision**; segments 2–11 with 2 sternal + 1 tergal plate; md acicular spine in segments 1–11 + mts; lateral acicular and cuspidate	*Semnoderes* Zelinka, 1907 (3 / 1)	semnos = Greek for honourable or proud, refers to long clamshell-like segment 1	spindle- to cigar-shaped habitus; 14 trichoscalids; outer oral styles with 2 elements and **lateral cuticular flap**; 16 **placids: mv + md narrow and elongate** + remaining broad; **segment 1 with deep, clamshell-like, anterior, narrow incision**; segments 2–11 with 2 sternal + 1 tergal plate; md acicular spine in segments 1–11 + mts; lateral acicular and cuspidate spines + lts + ltas; length mts >> length lts	Higgins 1967; Sørensen 2006; Sørensen et al. 2009; Neuhaus own obs.
		Sphenoderes Higgins, 1969 (2 / 0)	sphen = Greek for wedge, refers to midventral plate in segment 1	spindle- to cigar-shaped habitus; 14 trichoscalids; outer oral styles with 2 elements; 16 **placids: mv large and triangular** + pv narrow + >1 broad + >1 narrow; **segment 1 with deep, clamshell-like, anterior, broad incision**; segments 2–11 with 2 sternal + 1 tergal plate; md acicular, **broad-triangular** spine in segments 1–11 + mts; lateral acicular and cuspidate spines + lts + ltas, one series of lateral spines in segments 3–9 **broad-triangular**; length mts >> length lts	Higgins 1969a; Sørensen et al. 2010c

	spines + lts + ltas; length mts >> length lts				
Zelinkaderidae Higgins, 1990	spindle- to cigar-shaped habitus; 14 trichoscalids; **outer oral styles with** 1 or 2 elements and **cuticular lateral flap**; 14 or 16 **placids, fused with segment 1, tripartite anteriorly**; segment 1 ring-like, weak tergosternal junctions possible; at least **segments 5–11 with 1 tergal plate and mv junction**; md acicular spine at least in segments 4 + 6 + 8–11+ mts; lateral acicular and cuspidate spines + lts + ltas; length mts >> length lts	*Triodontoderes* Sørensen & Rho, 2009 (1 / 0)	triodus = Greek for trident of Poseidon; refers to placids anteriorly separated incompletely into 3 parts	spindle- to cigar-shaped habitus; 14 trichoscalids; outer oral styles with 1 (?) element and lateral cuticular flap; 14 placids: mv broad + remaining narrow, basally fused with segment 1, tripartite anteriorly; segment 1 ring-like, weak tergosternal junctions possible; segments 2–4 with 2 sternal + 1 tergal plate; segments 5–11 with 1 tergal plate and mv junction; md acicular spine in segments 1–11 + mts; lateral acicular and cuspidate spines + lts + ltas; length mts >> length lts	Sørensen & Rho 2009
		Zelinkaderes Higgins, 1990 (4 / 0)	refers to pioneer of kinorhynch studies Carl Zelinka	spindle- to cigar-shaped habitus; 14 trichoscalids; outer oral styles with 2 elements and lateral cuticular flap; 16 placids: mv broad + remaining narrow, basally fused with segment 1, tripartite anteriorly; segments 1–2 ring-like; **segments 3–4** and 5–11 with 1 tergal plate and mv junction; **cuticle with longitudinal internal and external striation pattern**; md acicular spine in segments 4, 6, 8–11+ mts; lateral acicular and cuspidate spines + lts + ltas; length mts >> length lts	Higgins 1990; Bauer-Nebelsick 1995; Neuhaus 1994; Sørensen et al. 2007; Sørensen & Rho 2009
incertae sedis	–	*Tubulideres* Sørensen et al., 2007 (1 / 0)	tubulus = Latin tube, refers to the ventral tubules in segment 2	spindle- to cigar-shaped habitus; 14 trichoscalids; outer oral styles with 2 elements; 16 placids: mv broad + remaining narrow; segment 1 ring-like; **segment 2 with 1 tergal plate and mv junction**; segments 3–11 with 2 sternal + 1 tergal plate; md acicular spine in segments 1, 2, 9–11 + **acicular spine in segments 3–8 alternating left and right of dorsal midline** + mts; acicular lateral spines + **several ventral tubules in segment 2** + lts + ltas; length mts >> length lts; no cuspidate spines	Sørensen et al. 2007
incertae sedis	–	*Wollunquaderes* Sørensen & Thormar, 2010 (1 / 0)	Wollunqua = god of rain and fertility in Aborigine mythology	spindle- to cigar-shaped habitus; 14 trichoscalids; outer oral styles with 2 elements and **lateral cuticular flaps with fish-bone pattern**; 16 placids: mv broad + 2x pv narrow + remaining broad; segment 1 ring-like; segments 2–11 with 2 sternal + 1 tergal plate; md acicular spine in segments 1–11 + mts; acicular and cuspidate lateral spines including segment 1 + lts + ltas; length mts >> length lts; ♀: short papilla vl in segment 8 and vm in segment 9	Sørensen & Thormar 2010
order Homalorhagida (Zelinka, 1896) Higgins, 1964b					
Neocentrophyidae Higgins, 1969	box-shaped habitus; thin trunk cuticle; 14 trichoscalids; outer oral styles with 1 or 3 elements,	*Neocentrophyes* Higgins, 1969 (2 / 0) (possibly just 1	neos = Greek for new or young, kentron = Greek for spine, phyo =	box-shaped habitus; thin trunk cuticle; 14 trichoscalids; outer oral styles with 1 or 3 elements, alternating; 7 placids: 3 ventral + 4 dorsal; segment 1 with 1 sternal + 1 tergal plate; segments 2–11 with 2 sternal + 1 tergal plate; no md spine	Higgins 1969b

(*Continued*)

Tab. 9. (*Continued*)

family	identification characters of family	genus (# species adult / # species juvenile)	etymology	identification characters of genus	references
	alternating; 7 placids: 3 ventral + 4 dorsal; segment 1 with 1 sternal + 1 tergal plate, sometimes sternal plate with anterior, weak, incomplete division into 3 plates; segments 2–11 with 2 sternal + 1 tergal plate; no md spine but md spinose process + spinose midterminal process; no lateral spines + lts present or not + no ltas; no cuspidate spine	species, see Tab. 6)	Greek for let develop or made by nature (common suffix of homalorhagid genera); refer to "juvenile characters" in adult stage, derived from Zelinka's genus *Centrophyes* described from juvenile stages	but md spinose process + spinose midterminal process; no lateral spines + no lts + no ltas; no cuspidate spine	
		Paracentrophyes Higgins, 1983 (3 / 0)	para = Greek for near or next to, kentron = Greek for spine; refer to close phylogenetic relationship to *Neocentrophyes*	box-shaped habitus; thin trunk cuticle; 14 trichoscalids; outer oral styles with 1 or 3 elements, alternating; 7 placids: 3 ventral + 4 dorsal; segment 1 with 1 sternal plate with anterior, weak, incomplete division into 3 plates + 1 tergal plate; segments 2–11 with 2 sternal + 1 tergal plate; no md spine but md spinose processes + spinose midterminal process; no lateral spines + lts + no ltas; no cuspidate spine	Higgins 1983; Sørensen et al. 2010a
Pycnophyidae Zelinka, 1896	box-shaped habitus; trunk cuticle well developed; 14 trichoscalids; outer oral styles with 1 element; 6 or 8 placids: 2 or 4 ventral + 4 dorsal; segment 1 with 3 sternal + 1 tergal plate; segments 2–11 with 2 sternal + 1 tergal plate; no md spine but sometimes md spinose process; no lateral spines + lts present or not + no ltas; no cuspidate spine	*Kinorhynchus* Sheremetevskij, 1974 (19 / 9)	kinein = Greek for move, rhynchos = Greek for snout; refer to the protrusible head	box-shaped habitus; trunk cuticle well developed; 14 trichoscalids; outer oral styles with 1 element; 6 or 8 placids: 2 or 4 ventral + 4 dorsal; segment 1 with 3 sternal + 1 tergal plate; segments 2–11 with 2 sternal + 1 tergal plate; no md spine but sometimes md spinose process; no lateral spines + no lts + no ltas; no cuspidate spine	Brown & Higgins 1983; Brown 1989; Adrianov & Malakhov 1999a
		Pycnophyes Zelinka, 1907 (47 / 13)	pyknos = Greek for dense or sturdy; refer to box-like habitus (opposite to more slender, cigar-shaped habitus of *Echinoderes* one of the few genera known in 19th century)	box-shaped habitus; trunk cuticle well developed; 14 trichoscalids; outer oral styles with 1 element; 6 or 8 placids: 2 or 4 ventral + 4 dorsal; segment 1 with 3 sternal + 1 tergal plate; segments 2–11 with 2 sternal + 1 tergal plate; no md spine but sometimes md spinose process; no lateral spines + lts + no ltas; no cuspidate spine	Zelinka 1928; Higgins 1983; Adrianov & Malakhov 1999a; Sánchez et al. 2011

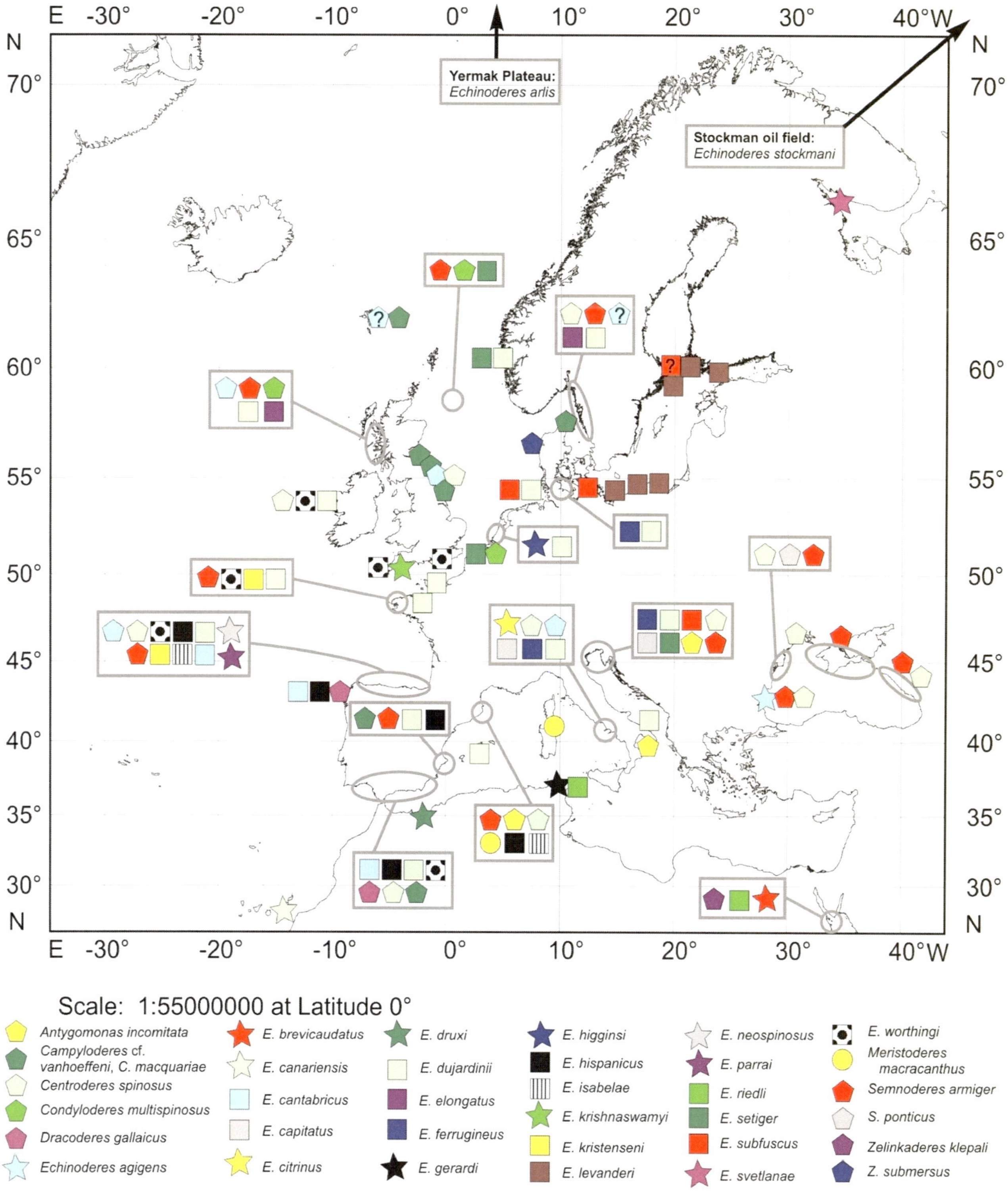

Fig. 5.6.1. Map of the distribution of European Cyclorhagida. A symbol may refer to several neighbouring locations. Symbols in a row indicate species from same site. Created by PanMap Version 0.9.6 (Diepenbroek et al. 2000). Original.

the Atlantic Ocean and for the Mediterranean Sea (Neuhaus & van der Land 2001; Neuhaus 2003; Sandulli & de Zio Grimaldi 2008); the latter study includes also species based on the description from juvenile stages. Intensive more local studies have also been conducted in Belize (Fig. 5.6.3; Higgins 1983), at the Pacific coast of Russia (Fig. 5.6.4; Adrianov 1989; Adrianov & Malkahov 1999a), near New Zealand (Figs 5.6.4, 5.6.6; Neuhaus et al. 2010), and in Japan (Fig. 5.6.4; Abe 1930; Higgins & Shirayama 1990; Adrianov et. al. 2002a-c; Sørensen et al. 2012a;

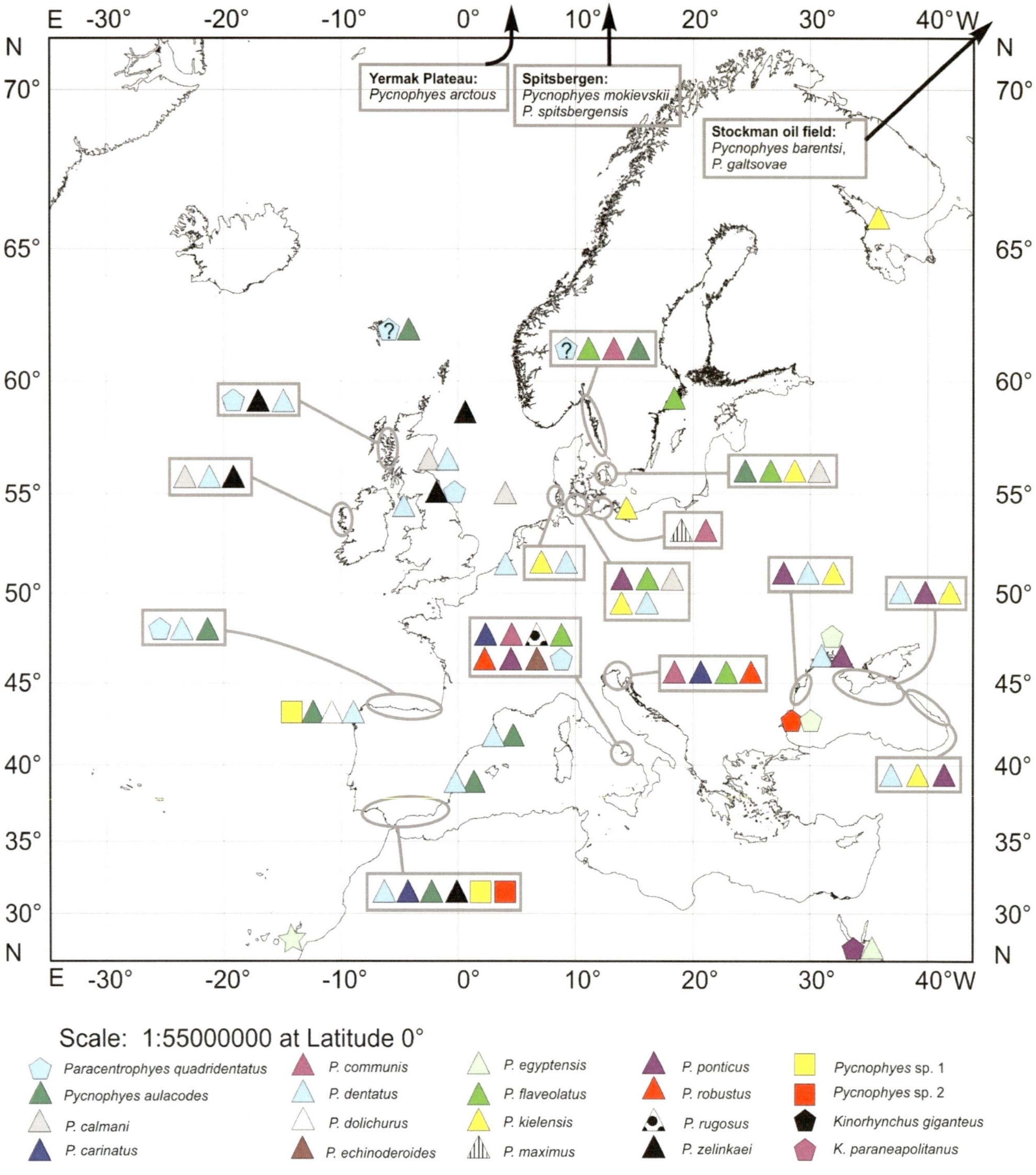

Fig. 5.6.2. Map of the distribution of European Homalorhagida. A symbol may refer to several neighbouring locations. Symbols in a row indicate species from same site. Created by PanMap Version 0.9.6 (Diepenbroek et al. 2000). Original.

Yamasaki et al. 2012), whereas the waters around South Korea have been and are still investigated extensively (Fig. 5.6.4; Adrianov & Malakhov 1999a; Song & Chang 2001; Chang & Song 2002; Sørensen et al. 2010a-c, 2012a, b; Lundbye et al. 2011). Several articles mention ecological parameters of Kinorhyncha from Chinese and Indian waters (Damodaran 1972; Ansari 1978; Fleeger et al. 1983; Fleeger et al. 1983; Rao & Satapathy 1996; Yuqing et al. 2001; Huang et al. 2005; Zhang et al. 2011), but only six species have been formally described (Fig. 5.6.4; Lou 1934; Timm 1958; Higgins 1968, 1969a,b).

From Arctic waters, 11 species of Kinorhyncha have been reported (Figs 5.6.1–5.6.3; Higgins 1966b, 1991; Higgins & Kristensen 1988; Adrianov 1995, 2011;

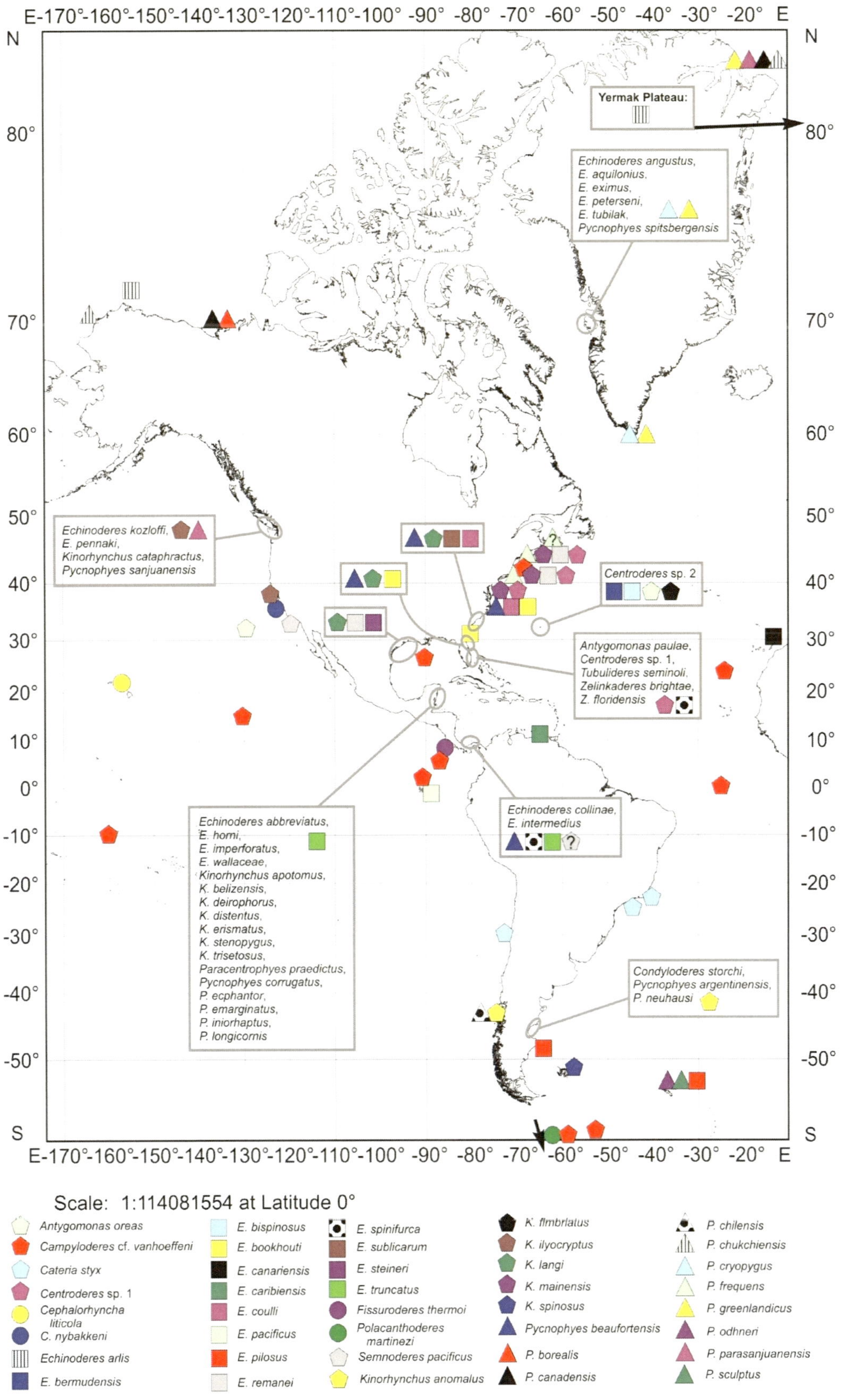

Fig. 5.6.3. Map of the distribution of Kinorhyncha in the East Pacific, Atlantic, and Arctic Ocean. A symbol may refer to several neighbouring locations. Symbols in a row indicate species from same site. Created by PanMap Version 0.9.6 (Diepenbroek et al. 2000). Original.

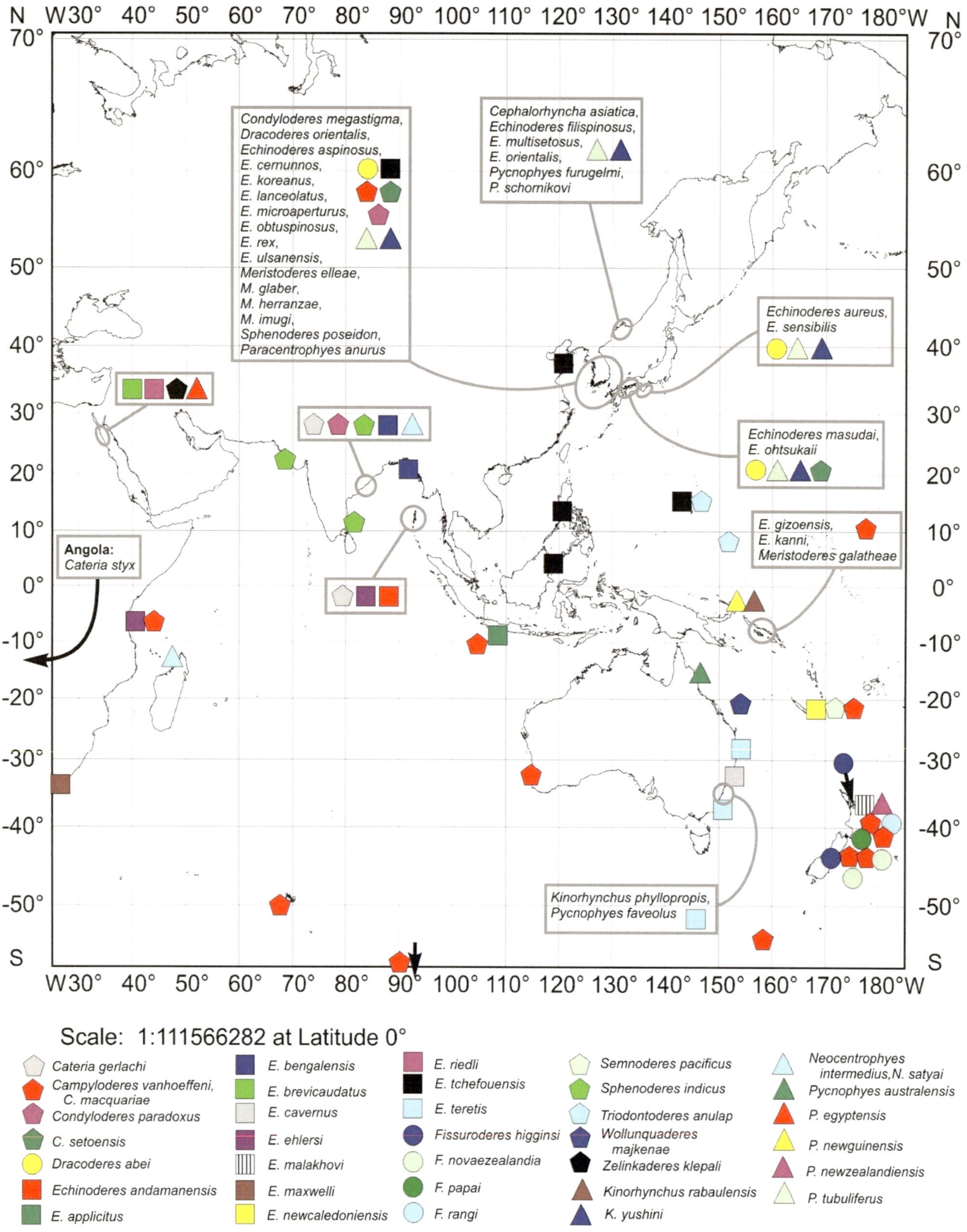

Fig. 5.6.4. Map of the distribution of Kinorhyncha in the Indian and West Pacific Ocean. A symbol may refer to several neighbouring locations. Symbols in a row or line indicate species from same site. Created by PanMap Version 0.9.6 (Diepenbroek et al. 2000). Original.

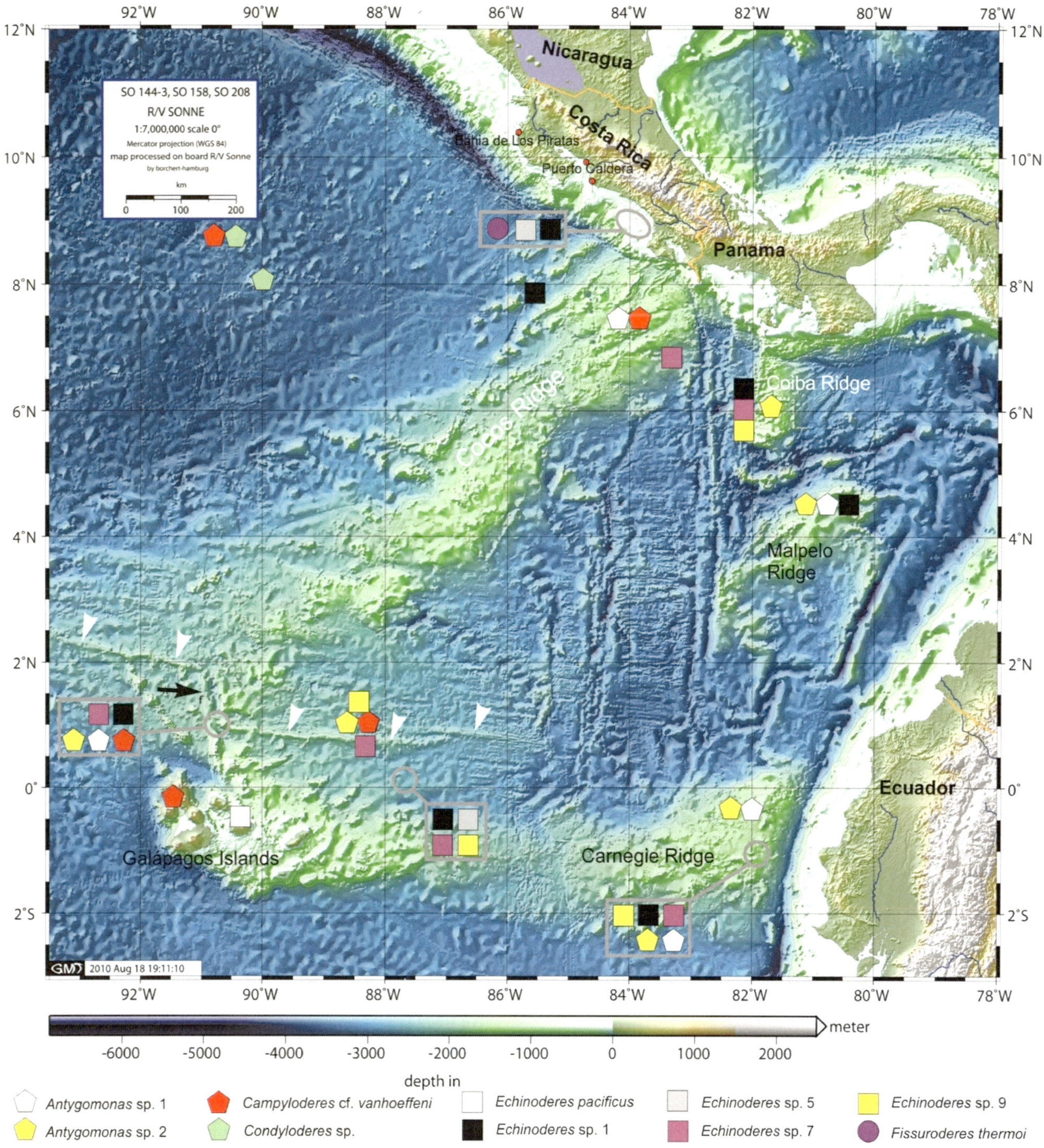

Fig. 5.6.5. Map of the distribution of Kinorhyncha repeatedly found in the Central American East Pacific; arrowheads mark Galápagos spreading centre, arrow points to tranform fault. Map created on board of R/V Sonne by system operators.

Adrianov & Malakhov 1999a). Kinorhyncha also occurs in the Weddell and Ross Sea around Antarctica (Tab. 13; Herman & Dahms 1992; Vanhove et al. 1995; Fabiano & Danovaro 1999), but only *Campyloderes vanhoeffeni* has been identified to species level (Figs 5.6.3, 5.6.4; Zelinka 1913).

Deep sea. Only five species of Kinorhyncha have been described from the deep sea below 1,000 m (Figs 5.6.3–5.6.6; Tab. 6; Neuhaus 2004; Neuhaus & Blasche 2006; Sørensen 2008a,b; Neuhaus & Sørensen 2012). Kinorhyncha are mentioned at group level from the Indian Ocean (Thiel 1966) and from various locations in the

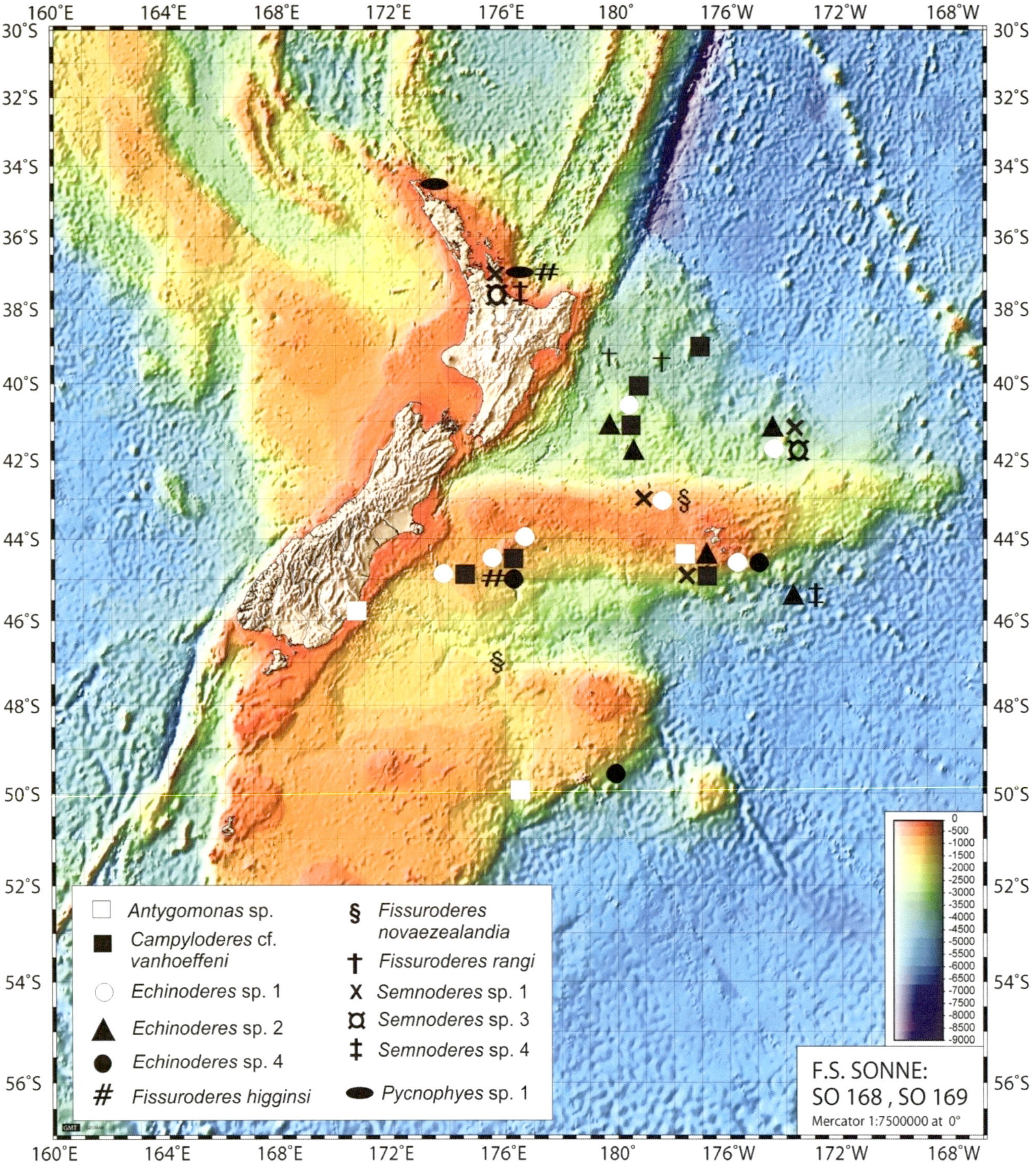

Fig. 5.6.6. Map of the distribution of Kinorhyncha repeatedly found in New Zealand waters. Map created on board of R/V Sonne by system operators.

Pacific Ocean (Eckman & Thistle 1988; Meadows et al. 1994; Danovaro et al. 2002; at species level: Neuhaus & Sørensen 2012), whereas the records of unidentified kinorhynchs from the Atlantic Ocean are much more numerous (Tab. 13).

Three recent expeditions to the Central American East Pacific Ocean around the Galápagos Islands (Neuhaus & Lüter 2003; Neuhaus own obs.) reveal at least 20 species of Kinorhyncha and facilitate the drawing of first conclusions on this taxon from the deep sea such as the extent of minimal distribution areas of individual species (Fig. 5.6.5). Seven species occur on both sides of and even on the Galápagos spreading centre and on Cocos Ridge (Fig. 5.6.5), which rise about 1,000 m above the surrounding sea floor. These observations suggest that submarine ridges do not represent barriers for the dispersal of Kinorhyncha. Possibly, transform faults in the spreading centre may be involved in dissemination. Two species of *Echinoderes* are found both in the deep sea at 1,000–2,400 m depth and on the continental shelf off Costa Rica at 100 m depth (Fig. 5.6.5; Neuhaus & Lüter 2003). Results from two expeditions to New Zealand yield 12 species of Kinorhyncha occurring at several locations and six species appearing both on the continental shelf and in the deep sea (Fig. 5.6.6; Neuhaus 2005). Nothing is known about potential exchange mechanisms between the fauna of the continental shelf and of the deep sea, but turbidity currents, sediment slides, benthic storms (Gross & Williams 1991; Stein 1991; Kontar & Sokov 1994; Masson et al. 1994) and large-scale benthic bottom currents may be involved (Schmitz 1996a, b). Debris flows may continue over several hundred to 1,000 km (Masson et al. 1994). Benthic storms may reach a velocity of 20–40 cm/s and may last for several days up to a month (see review by Vincx et al. 1994).

5.7. Paleontology

No fossil record is known for Kinorhyncha.

5.8. Ecology

Kinorhyncha occupies various habitats mainly in sandy and muddy sediments from intertidal to the deep sea (Tabs 6, 10). Several species also crawl on algae, sea grass, kelp, or on and in invertebrates; others inhabit unusual biota like stromatolid microbial nodules or cold seeps (Tabs 6, 7, 10). Species of *Echinoderes* and *Semnoderes* seem to occur in more sandy sediments in the southern North Sea, whereas species of *Pycnophyes* and *Kinorhynchus* prefer muddy ground and are found in the central North Sea (Huys et al. 1992).

Kinorhyncha represents 1–8% of the total meiofauna at a given locality, occasionally only 0.1% or exceptional 15–33% (de Bovée & Soyer 1974; Parulekar & Ansari 1981; Rao & Satapathy 1996; Dalto & Albuquerque 2000; Mazzola et al. 2000; Cibic et al. 2009; Santos et al. 2009; Grzelak & Kotwicki 2012). Usually, Kinorhyncha does not belong to the most dominant taxa, but in a few cases it ranks third in dominance after Nematoda and Copepoda (de Bovée & Soyer 1974; Jensen 1983; Herman & Dahms 1992; Santos et al. 2009). Kinorhyncha is generally more abundant in sediments of the continental shelf than of the deep sea (comp. Tabs 11 and 13).

Few localities have been sampled extensively or repeatedly, so little information is available for species richness at a given station. The record of 13 species in a single sample is reported from Belize (Higgins 1983), 11 species per location occur off Spain (Sánchez et al. 2012), and 7–9 species in Belize, the Bay of Naples, the Bay of Trieste, and off Spain (Zelinka 1928; Higgins 1983; Sanchez et al. 2012). Data on the density of Kinorhyncha at species level is summarized for several species in Tab. 11.

Inside the sediment, Kinorhyncha occupies the upper 0–1 cm or 0–2 cm and rarely occurs below this depth both on the continental shelf and in the deep sea (Dinet 1973; Ansari 1978; Horn 1978; Yingst 1978; Jensen 1983; Rudnick et al. 1985; Ólafsson & Elmgren 1991; Meadows et al. 1994; Rao & Satapathy 1996; Rysgaard et al. 2000). However, several species seem to have invaded sandy beaches 10–100 cm below the surface independently (Tab. 10). Whereas species of *Cephalorhyncha* and *Echinoderes* occur both in intertidal and sublitoral habitats and in sand well below the surface, *Cateria gerlachi* and *C. styx* have only been reported from sandy sediment well below the surface (comp. Tabs 6 and 10). However, *Cateria* sp. from Sri Lanka seems to occur also in intertidal sediment near the surface (Neuhaus own obs.).

Ecological data on Kinorhyncha often suffers from the lack of identification beyond group level and from sample volumes too small to mirror the density of this more rare taxon. Consequently, stations show great variation in kinorhynch abundance, even if suitable habitats are sampled. Information on autecological parameters of Kinorhyncha at species or at least genus level is limited to few studies (Tab. 11).

Dry weight. Dry weight of fresh *Echinoderes bookhouti* is mentioned as $2.0 \pm 0.1\ ^{\mu g}/_{individual}$, of *Kinorhynchus langi* as $5.5 \pm 0.1\ ^{\mu g}/_{individual}$ (Hummon 1975), of *Pycnophyes frequens* as $3.3\ ^{\mu g}/_{individual}$ (Wieser 1960), and of several species of *Centroderes* as $2.7\ ^{\mu g}/_{individual}$ (Coull 1970). Widbom (1984) calculates the dry weight of formalin-preserved kinorhynchs from Gullmar Fjord (West Sweden) as $2.8\ ^{\mu g}/_{individual}$ (= $2.5\ ^{\mu g}/_{individual}$ ash-free dry weight) if sieved with 200 µm mesh size and as $0.5\ ^{\mu g}/_{individual}$ (= $0.35\ ^{\mu g}/_{individual}$ ash-free dry weight) if sieved with 100 µm mesh size.

Food and food uptake. Based on morphological studies, the following scenario for food uptake, digestion, and

Tab. 10. Habitats of Kinorhyncha. For more details about individual species refer to Tab. 6.

kinorhynch identification	habitat	geographical location	references
Echinoderes applicatus, E. coulli, E. maxwelli, E. ohtsukai, E. teretis	upper few cm of exclusively intertidal mudflats with fluctuating salinities	South Carolina, Java, Australia, South Africa	Omer-Cooper 1957b; Higgins 1978b; Higgins & Fleeger 1980; Brown 1985; Adrianov & Malakhov 1999a; Ostmann et al. 2012; Yamasaki & Kajihara 2012
most species	upper few cm of sandy to muddy bottoms from intertidal to deep sea	worldwide	e.g., Zelinka 1928; Higgins 1983; Adrianov & Malakhov 1999a; Neuhaus & Blasche 2006; Sørensen 2008b; Neuhaus & Sørensen 2012; Sánchez et al. 2012
Cateria gerlachi, C. styx, Cephalorhyncha nybakkeni, C. liticola, Echinoderes bengalensis, E. druxi	10–100 cm below surface of sandy beaches	Bay of Bengal, Brazil, Chile, Angola, Algeria, Hawai	Gerlach 1956; Delamare Deboutteville 1957; Timm 1958; Rao & Ganapati 1966, 1968; Higgins 1968, 1986c; d'Hondt 1973; Higgins & Rao 1979; Brown & Higgins 1983; Sørensen 2008b
Campyloderes macquariae, C. vanhoeffeni, Echinoderes bookhouti, E. dujardinii, E. ferrugineus, E. gerardi, E. obtuspinosus, E. pennaki, E. peterseni, E. pilosus, E. sublicarum, Semnoderes armiger	on algae, kelp, sea grass, or on and in invertebrates (hydroid colonies, sponges, hermit crabs, tunicates); ≤9 specimens may inhabit one frond of *Caulerpa taxifolia* (see Travizi & Zavodnik 2004)	Italy, Spain, Great Britain, South Korea, USA, South Georgia Island, Argentina	Dujardin 1851; Zelinka 1928; Lang 1949; Chitwood 1964; Pallares 1966; Moore 1973; Higgins 1977b, 1978b; Mari & Morselli 1987; Song & Chang 2001; Kristensen 2005; Neuhaus & Sørensen 2012; Sørensen et al. 2012b; Sánchez-Tocino et al. 2011
Kinorhyncha	in laminated layer (1.6 individuals/10 cm^3) of stromatolid microbial nodules made mainly by cyanobacteria, 2.3–3 m depth	Atlantic Ocean: Bermuda	Westphalen 1993
Kinorhyncha	cold seeps: in *Beggiatoa* mats (1.6 individuals/10 cm^3) at hydrocarbon seep, depth 2,215 m; methane gas seeps, depth 70–120 m; periphery of methane seepage populated by *Calyptogena* (4 kinorhynchs/10 cm^2), depth 786 m; inactive pockmark, depth 350 m; mud volcanoe, depth 163 m, CH_4 concentration 2.2 mmol; periphery of mud volcanoe (39 individuals/10 cm^2), depth 5,000 m	Gulf of Mexico; Black Sea; Pacific Ocean: off Oregon; Mediterranean Sea: France, Italy; Caribbean Sea	Olu et al. 1997; Revkov & Sergeeva 2004; Robinson et al. 2004; Sommer et al. 2007; Zeppilli et al. 2011, 2012

Tab. 11. Ecological parameters of Kinorhyncha from the continental shelves in the Atlantic, Indian, and Pacific Ocean.

kinorhynch identification	depth [m]	sediment	temp. [°C]	salinity [‰]	abundance [individuals/10 cm²]	biomass [µg/10 cm²]	geographical location	remarks	reference
Atlantic Ocean									
Kinorhyncha, mainly *Echinoderes* sp.	8–13	mud	15–30	25–35	~10 to ~220	–	USA: Louisiana	seasonal data: population peak in summer, low abundance after hypoxia (0.5–3 mg O_2/l) in summer, peak in spring after recovery; maximum: >9 mg O_2/l	Murrell & Fleeger 1989
Echinoderes coulli	intertidal	mud with *Spartina alterniflora*	–	12–42	19–34	–	USA: South Carolina: Georgetown	species lives in upper 1.5–2.5 cm of sediment; 81% of *E. coulli* survive 1‰ for ≥11 h	Horn 1978
			–	–	12.1–72.2, 0.9–59.1	–		seasonal data for ♂, type-1 ♀, type-2 ♀, and juveniles; ♂ = 48.6% of adults; juveniles most abundant May – September, adults most abundant July – November; population peak in July and September – November	Higgins & Fleeger 1980
Echinoderes bookhouti	1–2	muddy sand	≤35	–	0.2–3.4	–	USA: North Carolina: Beaufort	seasonal data: population peak in June and December	Higgins & Fleeger 1980; Higgins 1983
Pycnophyes beaufortensis					0.4–14			seasonal data: population peak in February/March and August	
Kinorhynchus langi					0.3–3.7			seasonal data: population peak in February and August	
Kinorhyncha	14	silty clay	2–22	–	1.2–22.9 individuals/10 cm²	–	USA: Connecticut: New Haven	kinorhynchs live in upper 0–3 cm of sediment	Yingst 1978
Kinorhyncha	7	silt clay	1–22	27–31	~5 to ~80	–	USA: Rhode Island: Narragansett Bay	seasonal data: population peak in June; kinorhynchs live in upper 0–2 cm of sediment (82%), 2–6 cm (15%), and 6–10 cm (3%)	Rudnick et al. 1985
Pycnophyes frequens, *Kinorhynchus mainensis*	12–30; <1–5	sand to fine silt with or without eel grass	2–22	29.5–32.5	1–39	3–48	Massachusetts: Cape Cod: Buzzards Bay	–	Wieser 1960

(Continued)

Tab. 11. (*Continued*)

kinorhynch identification	depth [m]	sediment	temp. [°C]	salinity [‰]	abundance [individuals/10 cm²]	biomass [µg/10 cm²]	geographical location	remarks	reference
Centroderes ssp.	13	sandy silt with shell fragments	16–27	35–36	1–24	10–60	USA: Bermuda	seasonal data: population peak in September; organic matter: 5–6.24 $^{mg\ C}/_{g\ sediment}$	Coull 1970
Condyloderes multispinosus	101; 146	Loch Nevis (LN): brown mud; Fladen Ground (FG): greenish gray mud	LN: 7–13; FG: 6–8	LN: 33.9–34.3; FG: 35.2	LN: 1.7; FG: 2	–	North Sea: W Scotland: Loch Nevis (LN); off NE Scotland: Fladen Ground (FG)	–	McIntyre 1961, 1964
Semnoderes armiger					LN: 0.6; FG: 1.6				
Echinoderes ssp.					LN: 5.2; FG: 3				
Pycnophyes ssp.					LN: 1.4; FG: 0.4				
Kinorhyncha, mainly *Pycnophyes* sp.	3–16	clayey	0.3–18.1	16–34	0.2–3.3	–	E North Sea: W Sweden: Kungsbackafjorden	seasonal data: population peak in May; reduced density in July because of predation?; 2.6–9.7 mg O_2/l	Nyholm & Olsson 1973
Echinoderes subfuscus (= *E. levanderi?*), *Pycnophyes communis*, *P. maximus*	25–30	–	–	≤18.5	–	–	W Baltic Sea: Kadetrinne	–	Reimer 1936
Pycnophyes kielensis	3–6; 13–18	gray sand or sandy mud	–	6–7; 30	2	–	Baltic Sea: Greifswalder Bodden, Øresund	–	Lang 1936; Reimer 1963; Arlt 1977
Echinoderes levanderi	9–55; 10–50	mud to coarse sand	1.5–3.5; 0–18.4	3.7–6.1; 5.6–9.3	0.22–2.72; 20–325	–; 100–600	Baltic Sea: S Finland: Tvärminne, Sweden: Askö-Landsort	kinorhynchs live in upper 0–1 cm of sediment, density reduced to half in 1–2 cm, almost 0 specimens in 2–6 cm	Purasjoki 1945; Ankar & Elmgren 1976; Ólafsson & Elmgren 1991
Kinorhyncha	33	–	12–20	–	28–285	–	Mediterranean Sea: S France	seasonal data: population peak in June and September	de Bovée & Soyer 1974
Kinorhyncha	25	mud	7–25	35.1–37.7	'92–'95: 1.5–5.5; 1996–'98: 1–47	–	Mediterranean Sea: North Adriatic Sea: Gulf of Trieste	seasonal data: population peak in August and February (1992–1995) or in July (1996–1998); oxygen saturation 95% in December, 55% in September (2–6.7 mg O_2/l)	Vrišer 1997; Vrišer & Vukovic 1999

Echinoderes maxwelli	sublitoral?	–	–	24.3–41.2	–	–	South Africa: Greater Kleinemonde River	in lagoon temporarily isolated from ocean but with freshwater inflow	Omer-Cooper 1957b
Indian Ocean									
Kinorhyncha	6–8	sandy silt and clay	23.9–29.4	8.83–35.35	4–16	13–64	W India: Karwar	kinorhynchs live in upper 0-1 cm of sediment; 2.06–3.76 ml O_2/l	Ansari 1978
Echinoderes sp.	sublitoral	sand with mud or silt or phytal with *Halophyta ovata*	20–33.5	4.8–20.3	sand: 1–68, phytal: 1–36	sand: 60.3–489.1, phytal: 27.3–368.5	Bay of Bengal: E India: Chilka lagoon	seasonal data: population peak of *Echinoderes* sp. in January – March, of *Pycnophyes* sp. in January - April; kinorhynchs live in 0–5 cm of sediment (80–100%); Kinorhyncha = 1.63–3.24% of total meiofauna	Rao 1987a; Rao & Satapathy 1996
Pycnophyes sp.					sand: 1–5, phytal: 1–34				
Echinoderes applicitus	intertidal	medium grain size 10.7-139.2 µm	26.1–33.4	20.4–33.4	0.5-74.5		S Java: Segara Anakan lagoon	10-15% of kinorhynchs infected with possibly sulfur bacteria *Thiothrix* sp., possibly ciliate *Cothurnia* sp., and with unknown rod-shaped specimens (fungi?)	Ostmann et al. 2012
Pacific Ocean									
Echinoderes gizoensis	14	coral sand	29	35	–	–	SE Pacific Ocean: Melanesia: Salomon Islands: Ghizo Island	–	Thormar & Sørensen 2010
Echinoderes kanni	30	coral sand	29	35	–	–			
Echinoderes teretis	3–5	coarse mud	20–23	27.7–30	–	–	Australia: New South Wales: Broken Bay: Brisbane Water: Erina creek	microtidal estuary with salinity range from16–42‰ and temperature of 19.4–25°C in at least some areas	Brown 1985; Paterson et al. 2003
Kinorhyncha	59-80	sand, silt, or clay	9.3–10.1	–	0.3–29.9	7.9–52.5	NE Pacific Ocean: S Yellow Sea: China: SE of Shandong Peninsula	production: 70.94–471.78 µg dry weight/10 cm^2 • year	Yong et al. 2005

defecation is suggested (see also sections 5.1 Morphology, Sensory structures, Intestinal system; Neuhaus 1991, 1994; Neuhaus & Higgins 2002). Food particles are most certainly ingested while the introvert is everted. Ciliary sensory cells in the outer and inner oral styles are probably chemoreceptive and may sense the quality of food. The cuticular weir in the anterior mouth cone of Homalorhagida may prevent unwanted particles from entering the digestive tract. In Cyclorhagida, the anterior circular muscles may serve this same purpose by constricting the mouth cone lumen. In species of *Triodontoderes* and *Zelinkaderes*, the cuticular lateral extensions of neighbouring outer oral styles may overlap distally and create a kind of weir in front of the mouth cone. The pharyngeal protractors will pull the pharynx anteriorly into the mouth cone lumen in order to get the organ as far forward as possible, thereby minimizing the cavity in the mouth cone, and allowing the most efficient usage of the sucking pharynx. The radial muscle cells of the pharyngeal bulb will contract and as a result suck in water and food particles in front of the mouth cone. Pharyngeal receptor cells may sense whether or not food is suitable for uptake. Pharyngeal gland cells may either support the gliding of food within the pharynx or secrete digestive mucus. The postpharyngeal sphincter will prevent back flow of food particles and midgut cells into the pharyngeal lumen driven by the body pressure. Gland cells in the midgut epithelium will provide digestive secretions for lysis of the food particles, which will then be taken up by the epithelial cells. The muscle net around the midgut allows peristalsis of this part of the gut. Peristalsis is probably controlled by several supposedly sensorimotor cells integrated into the midgut epithelium. The same kind of monociliary receptors near the hindgut may also assist in defecation, inter alia via the cells of the muscle net. The latter possess gap junctions, which will certainly serve communication purposes. Also, the longitudinal muscles contact the hind gut dilatators. The frontal, caudal, and dorsal dilatators dilate the hindgut lumen against the body pressure and consequently also suck remnants of food from the midgut. Circular and transversal dilatators aid in rounding the slit-like lumen, and the sticked dorsal and ventral epicuticle will separate. Live observation of *Centroderes* sp. from Florida reveals a very quick defecation event (Neuhaus 1994).

Kinorhyncha seem to feed on detritus, bacteria, algae, and diatoms, but specific data is quite limited. Higgins (1990: Fig. 67) suggests that *Zelinkaderes floridensis* from Florida, USA shows bacteria in its gut, but the author of this book chapter is not totally convinced that the structures mentioned do represent bacteria, because the latter would probably not become recognizable in the light micrograph at the magnification used. Zelinka (1928) states that Kinorhyncha from deeper waters ingest sand and detritus; the sand grains becoming visible in polarized light. Sandy sediment from the Bay of Kiel, Germany reveals $0.7–2.3 \bullet 10^6\ ^{\text{bacteria}}/_{\text{mm}^3\ \text{wet sediment}}$; 49–64% of the bacteria are attached to sand grains, and 36–51% float free in the interstitial water. The upper 0–1 cm of muddy sediment from the same general location contains $46.6–77.7 \bullet 106\ ^{\text{bacteria}}/_{\text{mm}^3\ \text{wet sediment}}$ (Rheinheimer 1985). Nyholm and Nyholm (1976a) calculate the pump volume of the pharynx of *Pycnophyes flaveolatus* as 7,300 µm³. So one specimen may suck in 340–567 bacterial cells from muddy sediments in one pharyngeal pump cycle. Consequently, one should imagine Kinorhyncha as moving through its nutrients and just having to eat its way through its food (Zelinka 1928; Neuhaus 1991).

Cyclorhagid species in the photic zone consume considerable amounts of algae and diatoms like *Navicula* sp., *Pleurosigma* sp., and *Synedra* sp. according to Zelinka (1928). Some diatoms reach about 143 µm in length and may only be ingested by peristaltic movements of the circular muscle cells of the pharynx, which is shorter than the diatom itself (Zelinka 1928). Adult and juvenile *Echinoderes kozloffi* feed on the diatom *Navicula pavillardi* (8 µm × 40 µm); the first juvenile stage ingests its first diatom within minutes after hatching (Kozloff 1972). Specimens of *Zelinkaderes floridensis* yield diatoms in their gut and remnants of food (Fig. 5.1.31 A; Higgins 1990: Fig. 68). Shells of one to five diatoms are found in the midgut of several specimens of *Centroderes spinosus* from Europe (Neuhaus et al. 2013). Occasionally, the midgut of this species also shows remnants of food or erroneously swallowed material appearing as light-refractive or smaller, irregularly shaped, chaotic particles. Also, the homalorhagids *Neocentrophyes satyai* (Fig. 5.1.32 C) and *Pycnophyes kielensis* feed on diatoms (Neuhaus 1993: Fig. 3 E).

Adrianov (1991) and Adrianov and Malakhov (1994) cultivate *Cephalorhyncha asiatica* from Peter the Great Bay, Pacific Ocean together with the diatom *Phaeodactylum tricornutum* and suggest three ways of uptake of diatoms: (1) the kinorhynch collects single diatoms in circular movements of the anterior body with the help of its outer oral styles, which are observed to act as multiple forceps, hook-like inner oral styles break up the diatom shell, the pharynx sucks in the diatom's cellular contents, and the empty shell is discarded by more spinose inner oral styles probably supported by "spitting" the shell out with the pharynx; (2) the kinorhynch collects diatoms between its scalids by circular movements of its anterior body, withdraws the introvert, pushes the mouth cone slightly forward in order to remove the diatoms from the scalids with its inner oral styles, and continues with the diatoms as described before; (3) the kinorhynch bends its head and anterior trunk ventrally, "cleans" the diatoms attached to the lateral terminal spines by pulling its terminal spines through its scalids, and continues as mentioned before. The latter observation has also been reported by Zelinka (1928) for *Echinoderes dujardinii* and cyclorhagid juvenile stages. Adrianov and Malakhov (1994) are aware that the activities of the kinorhynch may be too quick to observe all details sufficiently without the help of a high-speed camera.

Supposedly, kinorhynchs manipulate diatoms with the help of their outer oral styles (Adrianov & Malakhov 1990, 1994; Adrianov 1991). However, the few muscles attaching to the styles (Brown 1989; Nebelsick 1993; Neuhaus 1994; Müller & Schmidt-Rhaesa 2003) will certainly not allow very sophisticated manipulation. Adrianov (1991) and Adrianov and Malakhov (1994) also hypothesize that Kinorhyncha secrete mucus through spines, setae, sensory spots, and pores in order to trap bacteria, diatoms, and detritus on their body and do some kind of "algal gardening". Previous studies on *Kinorhynchus phyllotropis* and *Pycnophyes parasanjuanensis* (see Brown 1985; Adrianov & Higgins 1996) and own observations on *Condyloderes paradoxus*, *P. dentatus* and *P. kielensis* support the observation that older adult specimens are usually covered by diatoms and detrital particles at the anterior margin of segments 2–11 (see below; Neuhaus own obs.; Fig. 5.8.2).

The midgut of *Pycnophyes kielensis* from the White Sea is supposed to reveal modified epithelial cells, so-called reservoir cells or bacteriocytes, osmiophilic vesicles possibly filled with sulfur and to host endosymbiontic bacteria (Yushin et al. 1989, 1990). Based on their TEM investigations, Adrianov et al. (1993) suggest that these bacteria represent chemolithoautotroph sulfur oxidizer. Unfortunately, the bacteria have not been checked to assess whether they actually possess sulfur-oxidizing enzymes, and the osmiophilic vesicles in the epithelial celles have not been proven to contain sulfur at all. Also, the concentration of hydrogen sulphide has not been measured in the exact sediment layer from which the Kinorhyncha have been sampled. Own studies of the gut of *P. kielensis* from intertidal mud in northern Germany (Neuhaus 1994) do not corroborate the bacteriocytes of Adrianov et al. (1993) but do demonstrate undigested individual bacteria surrounded by vesicles. Therefore, it remains open whether the bacteria in the midgut of *P. kielensis* function indeed as (temporary?) endosymbionts or whether the bacteria have been ingested and only partly digested.

In a 3-year study in the Gulf of Trieste, Mediterranean Sea, kinorhynch abundance is in phase with the density of diatoms such as *Navicula* sp. and *Diploneis* sp. and is also correlated with the occurrence of *Nitzschia* sp., *Pinnularia* sp., and *Pleurosigma* sp. (Cibic et al. 2009). However, the gut contents of kinorhynchs have not been checked for the shells of these diatoms. In the Baltic Sea, phytoplankton sedimentates in May after a spring bloom in March and seems positively to influence a peak in kinorhynch abundance in May and June (Ólafsson & Elmgren 1997).

At 1,400–1,460 m depth in Sagami Bay, Japan, Kinorhyncha migrate seasonally about 6 mm upwards in the sediment without reaching the surface; Shimanaga et al. (2000) suggest that this migration is triggered by phytodetritus on the sediment surface. So most certainly the kinorhynchs do not feed directly on the phytodetritus but on the bacteria involved in the decomposition of the phytodetritus or particulate organic matter (see also Nomaki et al. 2008). In the Southern Ocean, kinorhynch abundance at 2,963–2,990 m depth increases after a phytopankton bloom (Veit-Köhler et al. 2011).

Kinorhyncha, both from Sweden and from the Atlantic coast of the USA (Rhode Island), do not take up much fresh carbon from ^{14}C-labelled phytodetritus but seem to prefer carbon from other sources in micro- and mesocosm experiments (Rudnick 1989; Widbom & Frithsen 1995; Nascimento et al. 2008). Within 14 days, Kinorhyncha from the Baltic Sea consume 1.6 ± 0.9 $^{\mu g\ C}/_{\mu g\ dry\ weight\ of\ kinorhynch}$ derived from the cyanobacterium *Aphanizomenon* sp., 5.5 ± 1.3 $^{\mu g\ C}/_{\mu g\ dry\ weight\ of\ kinorhynch}$ originating from the cyanobacterium *Nodularia spumigena*, and 1.7 $^{\mu g\ C}/_{\mu g\ dry\ weight\ of\ kinorhynch}$ descending from the diatom *Skeletonema costatum* (Ólafsson et al. 1999; Nascimento et al. 2008). The Kinorhyncha from the Baltic Sea possibly represent *Echinoderes levanderi*, the only species reported from here by Ankar and Elmgren (1976).

Epifauna. The epifauna of Kinorhyncha consists of diatoms, bacteria, cyanobacteria, fungi, and of the protozoan Ciliophora (Zelinka 1914, 1928; Adrianov & Higgins 1996; Ostmann et al. 2012; Neuhaus own obs.). Diatoms and bacteria mainly occur at the anterior margin of sternal and tergal plates in older adult specimens of *Condyloderes paradoxus* and of *Pycnophyes* such as *P. parasanjuanensis*, *P. dentatus*, and *P. kielensis* (Fig. 5.8.2; Adrianov & Higgins 1996: Figs 7–13, 15–20; Neuhaus own obs.). Juvenile stages and recently hatched adults of the

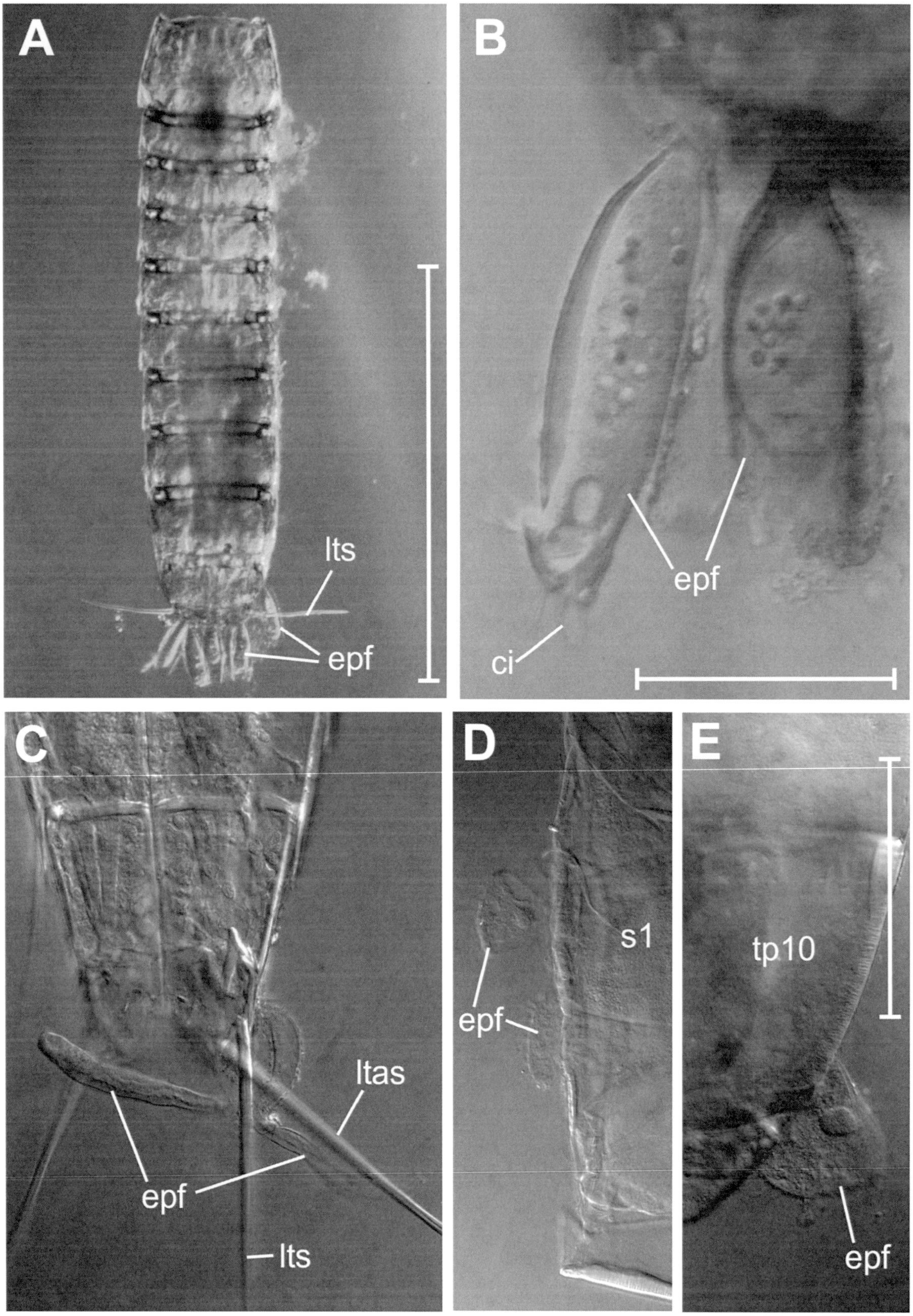

Fig. 5.8.1. Epifauna of Kinorhyncha. A, B, sessile loricate ciliates on *Pycnophyes* sp., Florida; C, unknown epibiont (?) on segment 11 of *Fissuroderes papai*, New Zealand, ventral view; D, E, suctorian ciliate on segment 1 (D) in ventral view and on segment 10 (E) in dorsal view of *Pycnopyhes kielensis*, Germany; line in A = 500 µ, lines in B and E = 50 µ, line in E valid for C–E.

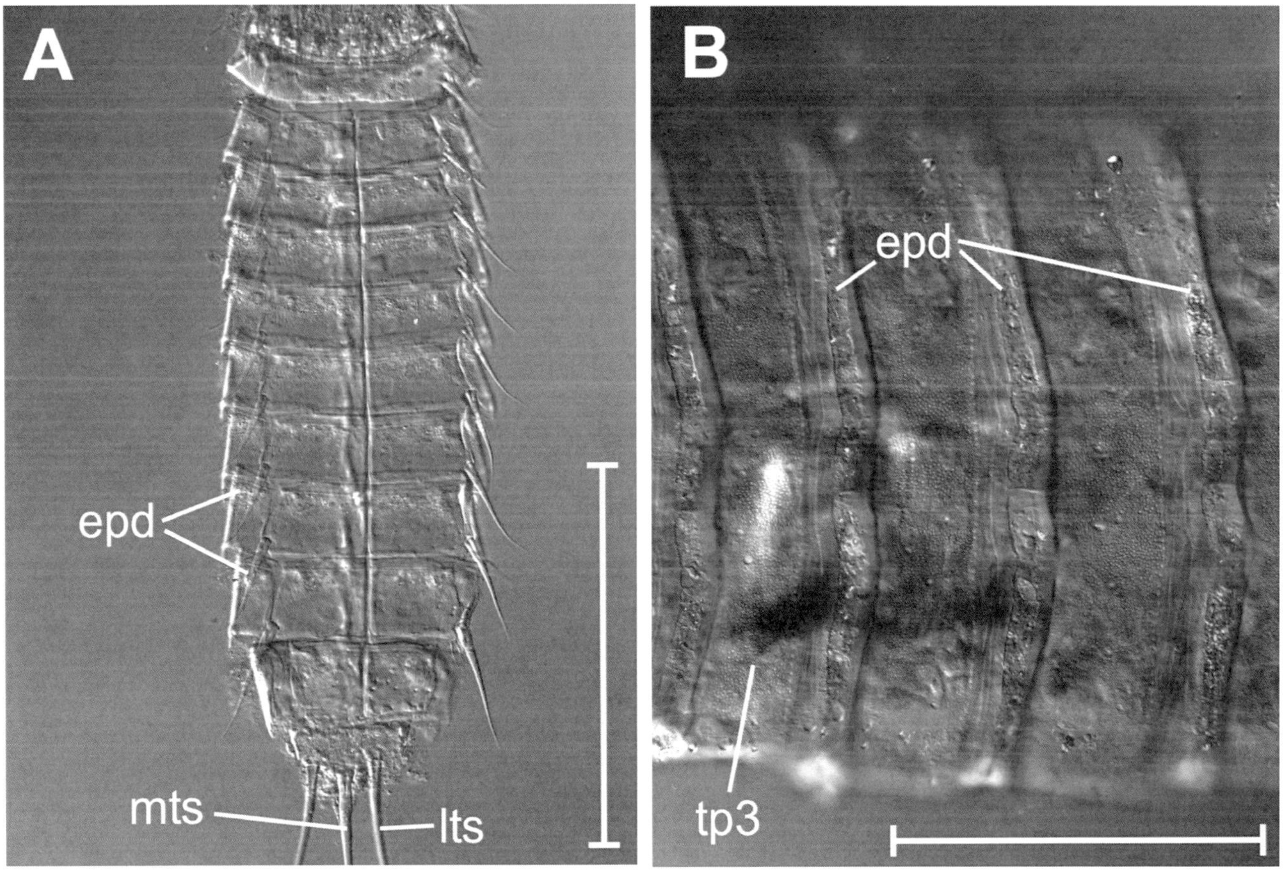

Fig. 5.8.2. Epifauna of Kinorhyncha. A, B, epifaunal bacteria, diatoms, and detritus on *Condyloderes paradoxus* (A), India, ventral view and on *Pycnophyes kielensis* (B), Germany, dorsal view; lines in A and B = 100 μ.

latter two species have not been observed to carry any epifauna (Neuhaus own obs.; but see below). Cyclorhagida seem to be rarely inhabited by epifauna in contrast to Homalohagida (Zelinka 1928).

At its posterior end, 10–15% of *Echinoderes applicitus* from Java carry filamentous bacteria, possibly sulfur bacteria of the genus *Thiothrix*, and rod-shaped organisms (Ostmann et al. 2012). Filamentous bacteria are also reported for *Echinoderes spinifurca* from the tropical Atlantic (Ostmann et al. 2012). Unknown epibionts occur on segment 11 of *Fissuroderes papai*, New Zealand (Fig. 5.8.1 C).

Suctorian ciliate epifauna possesses a funnel-like, wide anterior end with several contractile tubular tentacles and attaches with its narrow posterior end to the trunk cuticle. *Acineta tuba* Zelinka, 1914 has been reported from *Pycnophyes communis* in the Gulf of Trieste (Zelinka 1914, 1928). Suctoria, possibly *Acineta* sp., are also common ectocommensals on *Pycnophyes dentatus* and *P. kielensis* from intertidal mudflats in northern Germany; few to several specimens attach to any free area on the trunk cuticle (Fig. 5.8.1 D, E; Neuhaus own obs.). The suctorian *Limnoricus ponticus* Dovgal & Lozowskiy, 2008 settles on *Pycnophyes* sp. from the west coast of India (Dovgal et al. 2008).

The sessile loricate ciliate *Cothurnia bütschlii* Zelinka, 1914 inhabits the trunk and lateral terminal spine of juvenile and adult *Pycnophyes communis* from the Gulf of Trieste and leaves low elevation marks on the cuticle and particularly well recognizable marks on the lateral terminal spine; species of *Echinoderes* are possibly not infected, because the specimens are able to clean their lateral terminal spine by bending their anterior end and pulling the spine through the scalids of the head (Zelinka 1914, 1928). Sessile ciliates, possibly a species of *Cothurnia*, are also found on *Pycnophyes* sp. from Florida (Fig. 5.8.1 A, B), on *P. parasanjuanensis* from the Pacific coast of Washington, and on *Echinoderes applicitus* from Java (Adrianov & Higgins 1996: Fig. 23; Ostmann et al. 2012; Neuhaus own obs.).

Zooxanthellae. Zelinka (1928) reports individual or clustered yellow cells from the surface of the cuticle and from the epidermis of *Echinoderes dujardinii* and occasionally of *E. capitatus*, specifically appearing ventrally and laterally in *Pycnophyes flaveolatus*, middorsally in *P. carinatus*, and in a middorsal to laterodorsal position in *P. communis*. The zooxanthellae occur regularly but never frequently in the species mentioned before and are supposed to represent endosymbionts. Amoebocytes in the body cavity of *P. communis* seem to hunt the yellow cells occasionally (Zelinka 1928: Pl. 17: Fig. 7). All these findings have not been corroborated, refuted, or even restudied by subsequent authors.

Parasites. Internal protozoal parasites of Kinorhyncha are known as the microsporidian *Kinorhynchospora yushini* Adrianov & Rybakov, 1991 in *Kinorhynchus yushini* from the Russian Pacific coast and an unidentified Apicomplexa in *Zelinkaderes floridensis* from Florida; both parasites appear in midgut cells (Adrianov & Rybakov 1991; Adrianov et al. 1993; Neuhaus 1991: Fig. 17 A, 1994: Fig. 17 A, own obs.). Adrianov and Rybakov (1991) find various life history stages such as meront, pansporoblast, sporont, sporoblast, and mature spores; Neuhaus (1994; own obs.) traces three merozoits and one female gamont in three epithelial cells of a single kinorhynch specimen.

Kinorhyncha as food of macrofauna. Few observations are available for macrofauna feeding on Kinorhyncha (Tab. 12). Most puzzling is the occurrence of *Echinoderes levanderi* mainly enclosed by granulocytes in various tissues of the bivalve *Macoma balthica* (see Pekkarinen 1985, Tab. 12). It remains open how the kinorhynchs reach the bivalve tissues and how closely this is related to food ingestion of *M. balthica*. Also, Ólafsson et al. (1993) do not corroborate feeding of *M. balthica* on Kinorhyncha (Tab. 12).

Tab. 12. Potential macrofauna species feeding on Kinorhyncha.

macrofauna species	kinorhynch identification	observation	geographical location	reference
Polychaeta				
Pholoe minuta	*Pycnophyes* sp.	in gut and faeces, 9 cases	Baltic Sea: W Sweden: Koster: 20–35 m	Pleijel 1983
Opisthosyllis brunnea	*Echinoderes* sp.	1 specimen in gut	Cuba: Punta dell Francés: Island of Pines: inside dead coral: 1 m	San Martín 1991, 2003
Bivalvia				
Macoma balthica	*Echinoderes levanderi*	mainly in gonads but also in gill vessels, kidneys, heart and pericardium, digestive gland, and foot of 28–70% of clams, usually surrounded by granulocytes	Baltic Sea: SW Finland: Tvärminne: mud 20–40 m	Pekkarinen 1985
Macoma balthica	Kinorhyncha (*Echinoderes levanderi*?, see Ankar & Elmgren 1976)	little effect of bivalves on kinorhynch abundance after 5 months of microcosm studies => *M. balthica* probably not feeding on Kinorhyncha	Baltic Sea: E Sweden: Askö-Landsort: mud: 34 m	Ólafsson et al. 1993
Aulacomya ater	*Echinoderes* sp.	in stomach, 6 specimens	SE Pacific Ocean: Chile: Arica	Osorio & Iguain 1987
Crustacea				
Pleoticus muelleri	*Condyloderes storchi*, *Pycnophyes argentinensis*, *P. neuhausi*, *Kinorhynchus anomalus*	in stomachs of 2–4.3% of shrimps, 4 and 6 specimens	SW Atlantic Ocean: Argentina: Chubut: San Jorge Gulf	Albertó et al. 1993; Martorelli 2002; Martorelli & Higgins 2004
Neomysis integer	Cyclorhagida	in stomach of 11.1% of shrimps, kinorhynchs represent 2.9% of stomach contents	NE Atlantic Ocean: SW Spain: Guadalquivir river delta	Vilas et al. 2008
Melicertus kerathurus	*Pycnophyes communis*	in stomach, kinorhynchs represent <1% of stomach contents	Mediterranean Sea: Adriatic Sea: Italy: Lesina: 30 m	Prato et al. 2011

Salinity and temperature. Hummon (1975) reports for *Kinorhynchus langi* from intertidal sediment at Beaufort (USA: North Carolina) the incipient somatic lethal limit after 24 h at 19‰ salinity and 27°C and for *Echinoderes bookhouti* at 19‰ salinity and 14–32°C. *Echinoderes levanderi* in the southern Baltic Sea seems to withstand salinity as low as 3.68‰ (Purasjoki 1945; Karling 1955). Kinorhynch abundance is highest (73.5–93.3 $^{\text{individuals}}/_{10\ \text{cm}^2}$) in the Rio Formoso estuary (Brazil) at 37‰ salinity (Vasconcelos et al. 2004) and even richer (121.4–154.8 $^{\text{individuals}}/_{10\ \text{cm}^2}$) in an Itamaracá mangrove area (Brazil) at 40–44‰ and 28–32° C (Gomes et al. 2002). The lowest temperature noticed for a kinorhynch is –1.85° C at 385 m depth for *Campyloderes vanhoeffeni* from the Antarctic (Zelinka 1913).

Hypoxia. Observations on the effect of hypoxia on Kinorhyncha are somewhat contradictory. Microcosm experiments for 2 months with muddy sediment from Askö-Landort at the Swedish Baltic coast reveal that Kinorhyncha tolerates hypoxic conditions of 0.2 mg dissolved oxygen per liter sea water (Modig & Ólafsson 1998; the species involved is possibly *Echinoderes levanderi*, see Ankar & Elmgren 1976). Kinorhyncha also appear about 1 cm deep in the deoxygenated sediment layer at 300 m depth south of Disko Island, Greenland (Rysgaard et al. 2000). An oxygen saturation as low as 7.2% for two weeks in the northern Adriatic Sea at 25 m depth does not seem to influence kinorhynch abundance (Vrišer 1997). Veit-Köhler et al. (2009) report Kinorhyncha from stations with 0.79 ml O_2/l and 1.52 ml O_2/l at 300 m depth off Chile but not from two stations with lower oxygen content; however, the sampled sediment surface is quite small and may not reflect abundance of more rare taxa sufficiently. Finding of Kinorhyncha in oxygen-minimum zones at 560 m (0.227 ml O_2/l) but not at 300 m (0.013 ml O_2/l) depth off Peru (Neira et al. 2001) may be reasoned in the same way. Near the Crimea Peninsula in the Black Sea, Kinorhyncha occurs at 120 m depth (3.1 ml O_2/l) with 0.32 $^{\text{individuals}}/_{10\ \text{cm}^2}$ and at 140 m depth (2 ml O_2/l) with 0.2 $^{\text{individuals}}/_{10\ \text{cm}^2}$ (Sergeeva et al. 2012). In the Gulf of Mexico, a temporary hypoxic event with 0.5–3 mg O_2/l for about a month in summer seems to account for low abundance data of Kinorhyncha until the population peaks in spring the following year (Murrell & Fleeger 1989; Tab. 11). Probably, kinorhynch live history stages tolerate a low level of dissolved oxygen for some time (Modig & Ólafsson 1998), but reproduction and development may decrease under these conditions (Murrell & Fleeger 1989; Sergeeva et al. 2012).

Deep sea. Kinorhyncha is found regularly in the deep sea (Tab. 13). Deep-sea sediments appear generally well oxygenated "to a depth of at least several decimetres" (see review by Vincx et al. 1994; Meadows et al. 1994). However, Shimanaga et al. (2000) report the oxygenated layer in Sagami Bay, Japan at 1,400–1,460 m to be just 8–30 mm thick with Kinorhyncha occurring at 6–15 mm depth in the sediment; two samples out of 15 reveal kinorhynchs in the deoxygenated layer. The deepest record of Kinorhyncha is 7,800 m in the Atacama Trench off Chile (Danovaro et al. 2002). Here, kinorhynchs reach a density of 46 ± 40 $^{\text{individuals}}/_{10\ \text{cm}^2}$ but seem "to reduce their size for 60%". This "dwarfism" does not seem to be related to reduced food availability, low food quality, or low temperature. Therefore, Danovaro et al. (2002) consider "(a) increased abundance of juveniles at hadal depths; (b) selective predation over large-sized meiofauna; (c) selection of small-sized barophilic/ barotolerant organisms". Several expeditions suggest that kinorhynch juvenile stages are quite common in the material from the deep sea (Neuhaus own obs.).

Reaction to natural events. In the Askö-Landsort area of the Baltic Sea, summer blossoms of the toxin-producing cyanobacterium *Nodularia spumigena* do not affect survival of Kinorhyncha (Nascimento et al. 2008).

Strong tidal bottom currents at two stations 580–640 m deep on Fieberling Guyot (northeast Pacific Ocean) seem to account for a depth distribution of *Antygomonas oreas* up to 10 cm deep in the sediment; amazingly, the kinorhynchs are more numerous at the station exposed to stronger currents (measured 4 m above bottom; Levin et al. 1994; Bauer-Nebelsick 1996; Thistle et al. 1999). However, experimentally increased near-bottom flows lead to significantly decreased abundance of Kinorhyncha (Thistle & Levin 1998). Probably, current measuring 4 m above the ground does not totally reflect the actual situation very close to the bottom.

Experiments in a tidal creek through mudflats in South Carolina (USA) reveal that kinorhynchs are rarely suspended in the water column contrary to other meiofaunal taxa (Palmer & Gust 1985). Most certainly, this observation relates to *Echinoderes coulli*, the only kinorhynch species known from this area despite an intensive search by Higgins (1977b). This species seems to be obligate intertidal, cannot be extracted by air bubbles from the sediment, and is therefore probably not affected by the surface tension of the tidal water currents (Higgins 1977b ; Horn 1978).

Iceberg scouring removes meiofauna from sediment totally but recolonization by kinorhynchs happens after

Tab. 13. Records of Kinorhyncha identified at group level by expeditions with focus on the deep sea. Abundance data has to be regarded with some caution, because samples of different origin have been sieved with different mesh sizes.

geographical location	abundance [individuals/10 cm²]	% of total meiofauna	vertical distribution	depth [m]	reference
Atlantic Ocean					
N Atlantic: Iceland Faroe Ridge	<1,000 m:1.2–4.4; >1,000 m: 0.4–7.6	0.1	0–1 cm (75%), 1–2 cm (17%)	435–2,500	Thiel 1971
NE Atlantic: Norwegian Sea	<1,000 m: 1–8; >1,000 m: 3–6	0.3–1.7	0.3–1.7	283–2,000	Dinet 1974
NE Atlantic: Norwegian Sea	1–11	0.1–1.8	–	2,465–3,608	Dinet 1979
N Atlantic: N of Scotland: Darwin Mounds	0-2–0.4	0.03–0.05	–	954–960	Van Gaever et al. 2004
NE Atlantic: SW of Ireland: Pocupine Seabight & Abyssal Plain	<1,000 m: 2–11; >1,000 m: 1–9	0.1–0.6	–	500–4,500	Pfannkuche 1985
	0.7–4.9	–	–	1,340–4,495	Vincx et al. 1994
NE Atlantic: SW of Ireland: Goban Spur	0.5	0.05–0.08	–	206–1,961	Vanaverbeke et al. 1997
NE Atlantic: Gulf of Biscay	1–5	0.2–0.7	–	1,369–4,725	Dinet & Vivier 1977
NE Atlantic: W of Spain	0.2–1.2	0.02–0.2	–	3,958–4,788	Rutgers van der Loeff & Lavaleye 1986
NE Atlantic: Spain: Madeira	0.8–7.5	–		1,543–4,856	Vincx et al. 1994
NE Atlantic: Spain: Madeira	1	0.5–0.7	upper 0–1 cm	1,238–1,769	Rachor 1975
NE Atlantic: off Mauretania	1.9–7.5	–	–	1,543–4,629	Vincx et al. 1994
Mediterranean Sea: off Corsica	0.5–1.5	0.04–0.3		160–1,220	Soetaert et al. 1991
Mediterranean Sea: off Macedonia	1–2	0.2–1.1	–	130–1,209	Dinet 1976
SE Atlantic: near Walvis Ridge	1–5	0.3–0.6	–	1,440–4,660	Dinet 1973
NW Atlantic: Canada: Nova Scotian Rise	0.5–1	–	upper 0–2 cm	4,626	Thistle et al. 1985
SW Atlantic: French Guyana: Demerara Plain	0.1–0.9	0.1	–	4,440–4,870	Sibuet et al. 1984
Pacific Ocean					
Central Pacific	[0.07–0.13/ml]	0.2–0.4	upper 0–1 cm	5,098–5,298	Meadows et al. 1994
E Pacific: San Diego Trough	0.1–5.5	–	–	1,080	Eckman & Thistle 1988
E Pacific: off Chile: Atacama Trench	0.2–86	–	–	1,050–7,800	Danovaro et al. 2002
Indian Ocean					
W Indian: off Somalia	0.4–2	–	–	2,600–4,690	Thiel 1966
Arctic Ocean					
Barents Sea: Svalbard Shelf & Nansen Basin	<1,000 m: 1–297; >1,000 m: 1–11	0.5–6.1	–	226–3,290	Pfannkuche & Thiel 1987
Antarctic Ocean					
Weddell Sea: off Kapp Norvegia	1–45	–		211–2,080	Vanhove et al. 1995
Weddell Sea: off Halley Bay	0.26–63.5	2.6	–	339–1,958	Herman & Dahms 1992
Ross Sea	1–11	0.1–0.6	–	556–587	Fabiano & Danovaro 1999

50 days at the latest (Lee et al. 2001a, b). A typhoon may rework sediments at 22 m depth and increase abundance of Kinorhyncha by 46% within days (Hua et al. 2010). During recolonization of experimentally defaunated mud in Gullmar Fjord, Kinorhyncha benefit within 9 days from *Amphiura filiformis* (Ophiuroidea) at 1.2 $^{\text{individuals}}/_{10\ \text{cm}^2}$ but not at higher densities (Crowe et al. 1987). Kinorhyncha in microcosm experiments over 2 months from Askö-Landort, Baltic Sea is positively influenced by the amphipod *Monoporeia affinis* at 7.7 $^{\text{individuals}}/_{10\ \text{cm}^2}$ but not at lower densities (Ólafsson & Elmgren 1991). Kinorhyncha from the same location show only half of their original density after 17 days of exposure to decaying *Macoma balthica* (Bivalvia) in microcosm experiments (Ólafsson 1992). Possibly, these observations refer to *Echinoderes levanderi*, which is reported by Ankar and Elmgren (1976) from this locality.

Reaction to human impact. In Hamilton Harbour, Bermuda, pollution by metal traces, tributyltin, and aromatic hydrocarbons significantly reduces kinorhynch abundance (Warwick et al. 1990). Mesocosm experiments at Rhode Island, USA reveal significant decrease of kinorhynch density after exposure to 90 µg petroleum hydrocarbons per liter sea water (Frithsen et al. 1985). Domestic sewage in an estuary in Goa, India seems to influence kinorhynch density in a sandy beach negatively (Ansari et al. 1984). In Pernambuco, Brazil, experimental exposure of sediment in a mangrove estuary with sugar cane fertilizer at a weekly dosage of 188 g/m² leads to reduction of kinorhynch abundance in an asymptotic fashion within a few weeks (Santos et al. 2009). Kinorhyncha also reacts very sensitively to organic pollution by bivalve and fish farming (Mediterranean Sea: ; Grego et al. 2009; Black Sea: Lysykh 2005) and here especially to increasing concentrations of free sulfide (Canada: Sutherland et al. 2007). Whereas meiofaunal abundance may increase under fish farming cages in the long term, kinorhynch density does not recover (Mediterranean Sea:). However, shrimp farming in an estuary in northern Brazil seems to positively affect kinorhynch abundance (de Paula et al. 2006).

Acknowledgements

Specimens of Kinorhyncha have been kindly given on loan by Dr Martin V. Sørensen and Majken T. Tøttrup, Natural History Museum of Denmark, Copenhagen, as well as by Kathryn Ahlfeld, Cheryl Bright, and Dr Jon Norenburg, National Museum of Natural History, Washington, D.C.; Susan Braden and Peter Walt Brown provided SEM services while the author was a postdoctoral fellow at the Smithsonian Institution; all this help is greatly appreciated. Stimulating discussions with Dr Robert P. Higgins, former curator at Washington, Profs Dr Monika Bright, Vienna, Dr. Martin V. Sørensen, Copenhagen, and Dr Fernando Pardos, Madrid, over many years as well as comments on an earlier version of the manuscript by the latter two authors are gratefully acknowledged. Special thanks go to Kristine Kämpf, Karin Meschter, and Antje Schwiering for technical support. Invitation by PD Dr Andreas Schmidt-Rhaesa to contribute to this volume and professional support by Dr Gunda Stöber, Walter de Gruyter Press is greatly appreciated. I wish to express my sincere gratitude to the Smithsonian Institution for granting a postdoctoral fellowship in 1992, to the German Ministry of Education and Research (BMBF) for funding the expeditions SO 144-3, SO 158, SO 168, SO 193, SO 199, and SO 208, and to the SYNTHESYS Programme, financed by the European Community Research Infrastructure Action under the FP6 "Structuring the European Research Area Programme".

Abbreviations

a	anus
ac	acicular spine
ar	articulation between cuticular sternal and tergal plates
b	brain
ba	bacterium
bh	bactreate cuticular hair
br	cuticular bristles
bu	regularly arranged button-like structures
c	canal cell
ca	cuticle-lined cavity
cav	cavity in embedding medium
cd	caudal dilatator
ci	cilium
cm	ring muscle cell
cr	crenulated spine
cra	cracks in embedding medium
cry	crystal in embedding medium
cs	cuspidate spine
cu	cuticle
d	dictyosome
dd	dorsal dilatator
de	desmosome
di	diplosome
dia	diatom
dis	disintegrated embedding medium

do	dorsal organ
dtd	dorsal transversal dilatator
dv12	dorsoventral muscle of segment 12 or its attachment site
E	east
ecm	extracellular matrix
en	enveloping cell
ep	epidermis
epd	epifaunal bacteria, diatoms, and detritus
epf	ciliate epifauna
esp	episternal plate
ex	exuvia
er	endoplasmic reticulum
fd	frontal dilatator
fi	filter of protonephridium
fo	remnants of food in midgut
g	gland cell
gam	female gamont of Apicomplexa
gc	gland cell (outlet)
gy	glycogen
go	gonad
gp	gonopore
gpb	gonopore bristles
gy	glycogen
hd	hemidesmosome
hg	hindgut (epithelium)
ho	head sensory organ
hr	head retractor muscle
ic	interstitial cell
iep	inner epithelium of mouth cone
in	introvert
ios	inner oral style(s)
ipl	"interstitial" placid
is1	introvert sector 1
isc	intersegmental cuticle
J-1	juvenile stage 1
l	lipid vesicle
la	lateral accessory
la9	spine in lateral accessory position of segment 9
lbh	long bactreate cuticular hair
ld	laterodorsal
lh	lanceolate cuticular hairs
lj	lateral junction between sternal and tergal plate
lm	longitudinal muscle
LM	light microscopy
ltas	lateral terminal accessory spine
lts	lateral terminal spine
lu	lumen
lv	lateroventral
lv5	lateroventral spine in segment 5
mc	mouth cone
md	middorsal
md1	middorsal spine in segment 1
mdp	middorsal process
mdpl	middorsal placid
mg	midgut (epithelium)
mi	microvillus
ml	midlateral
ml10	midlateral spine in segment 10
mr	mouth cone retractor muscle
msp	midsternal plate
msw	maximum standard width
mts	midterminal spine
mv	midventral
mvp	midventral process of segment 1
mvpl	midventral placid
mvt	midventral thickening
n	nephroporus cell
N	north
n. a.	data not available
ne	nerve cell(s)
n_{ne}, n_{oe}, n_{pe}	nucleus of nerve, esophageal, pharyngeal cell
nr	nerve ring
oe	esophagus
oep	outer epithelium of mouth cone
om	oblique trunk muscle
oos	outer oral style(s)
ope	outer pharyngeal epithelium
pa	papilla
pb	pharyngeal bulb
pc	pachycyclus
phc	pharyngeal crown
pd	paradorsal
pe	inner pharyngeal epithelium
pf	primary fringe
pl	placid
pn	(outlet of) protonephridium
pos	postpharyngeal sphincter
pp	pharynx protractor
pps	prepharyngeal sphincter
pr	non-contractile process of muscle cell
pre	precipitations
ps	penile spine
pss	primary spinoscalid
pv	paraventral
r	ciliary rootlet
rd	ring-shaped dilatator
re	receptor cell
rm	radial muscle cell

rs	receptaculum seminis
rt	head or mouth cone retractor muscle
S	south
s1	trunk segment 1
sd	subdorsal
se	seta
SEM	scanning electron microscopy
sf	secondary fringe
sj	septate junction
sl	sublateral
sl10	sublateral spine in segment 10
sly	secondary lysosome
sp	sperm
spi	spiculae
spp	spermatophore
spr	spinose process
sr	sarcoplasmic reticulum
ss	spinoscalid
ssp	sensory spot
st	sternal plate
sw10	standard width measured in segment 10
t	terminal cell
tf	tonofilaments
TL	trunk length
tp	tergal plate
tr	trichoscalid
trp	trichoscalid basal plate
Ts	T-system
tu	cuticular tubes of protonephridium
tus	tubular structure possibly related to copulation
v	vesicle
vl	ventrolateral
vl5	ventrolateral spine in segment 5
vm	ventromedial
vm1	ventromedial spine in segment 1
vn	ventral longitudinal nerve cord
vtd	ventral transversal dilatator
w	peripheral cytoplasmic wall of protonephridium
W	west
we	weir of mouth cone
zd	Z-dot

References

Abe, Y. (1930): Das Vorkommen von *Echinoderes* in den japanischen Gewässern. Journal of Science of the Hiroshima University Ser. B Div. 1 Vol.1(3): 39–43, pl. I.

Adrianov, A.V. (1989): The first report on Kinorhyncha of the Sea of Japan. Zoologichesky Zhurnal 68(7): 17–27. [in Russian]

Adrianov, A.V. (1991): Some peculiarities of biology of Cephalorhyncha, Kinorhyncha. Ekologiya Morya 39: 57–61. [in Russian]

Adrianov, A.V. (1995): The first description of kinorhynchs from the Spitsbergen Archipelago (Greenland Sea), with a key to the genus *Pycnophyes* (Homalorhagida, Kinorhyncha). Canadian Journal of Zoology 73: 1554–1566.

Adrianov, A.V. (2011): Phylum Cephalorhyncha, class Kinorhyncha. pp. 41–49 in Buzhinskaja, G.N. (ed.) Illustrated keys to free-living invertebrates of Eurasian Arctic Seas and adjacent deep waters. Vol. 2 Nemertea, Cephalorhyncha, Oligochaeta, Hirudinida, Pogonophora, Echiura, Sipuncula, Phroronida, and Brachiopoda. Alaska Sea Grant, University of Alaska Fairbanks, Fairnbanks.

Adrianov, A.V., Chesunov, A.V. & Spiridonov, S.E. (1990c): Post-embryonic development of *Pycnophyes kielensis* (Kinorhyncha, Homalorhagida). Zoologichesky Zhurnal 69(6): 132–137. [in Russian]

Adrianov, A.V. & Higgins, R.P. (1996): *Pycnophyes parasanjuanensis*, a new kinorhynch (Kinorhyncha: Homalorhagida: Pycnophyidae) from San Juan Island, Washington, U.S.A. Proc Biol Soc Wash 109(2): 236–247.

Adrianov, A.V. & Malakhov, V.V. (1990): Sense organs of Konorhyncha using the example of the order Homalorhagida (Cephalorhyncha, Kinorhyncha). Zoologichesky Zhurnal 69(4): 5–14.

Adrianov, A.V. & Malakhov, V.V. (1991a): The nervous system of Cephalorhyncha, Kinorhyncha. Zoologichesky Zhurnal 70(2): 5–12. [in Russian]

Adrianov, A.V. & Malakhov, V.V. (1991b): The fine structure of spermatozoids and peculiar features of spermatogenesis in *Pycnophyes kielensis* (Homalorhagida, Pycnophyidae). Zoologichesky Zhurnal 70(4): 28–36. [in Russian]

Adrianov, A.V. & Malakhov, V.V. (1991c): Fine structure of the reproductive system in Cephalorhyncha, Kinorhyncha. Zoologichesky Zhurnal 70(5): 28–38. [in Russian]

Adrianov, A.V. & Malakhov, V.V. (1991d): Kinorhynchid body cavity organization. Proceeding of the Academy of Sciences 317: 247–249.

Adrianov, A.V. & Malakhov, V.V. (1994): Kinorhyncha: Structure, development, phylogeny and taxonomy. 262 pp. Nauka Publishing, Moscow. [in Russian]

Adrianov, A.V. & Malakhov, V.V. (1995a): Synopsis of the system of the phylum Cephalorhyncha. Russian Journal of Marine Biology 21(2): 89–96.

Adrianov, A. V. & Malakhov, V.V. (1995b): The phylogeny and classification of the phylum Cephalorhyncha. Zoosystematica Rossica 3(2): 181–201.

Adrianov, A.V. & Malakhov, V.V. (1996): The phylogeny and classification of the class Kinorhyncha. Zoosystematica Rossica 4(1): 23–44.

Adrianov, A.V. & Malakhov, V.V. (1999a): Cephalorhyncha of the world ocean. 328 pp. KMK Scientific Press, Moscow. [in Russian and English]

Adrianov, A.V. & Malakhov, V.V. (1999b): 7. Kinorhyncha. pp. 193–211 in Adiyodi, K.G. & Adiyodi, R.G. (eds.) Reproductive biology of invertebrates, Vol. IX, Part A: Progress in male gamete ultrastructure and phylogeny. John Wiley & Sons, Chichester.

Adrianov, A.V., Malakhov, V.V. & Chesunov, A.V. (1989): Floscules – specific organs of kinorhynchs, priapulids and loriciferans. Dok Akad Nauk SSSR 307: 1277–1279. [in Russian]

Adrianov, A.V., Malakhov, V.V.: Chesunov, A.V. (1990a): Fluscules – specific organs of Kinorhynchia, Priapulida, and Loricifera. Proceedings of the Academy of Sciences of the USSR 307: 545–547.

Adrianov, A.V., Malakhov, V.V. & Yushin, V.V. (1990b): Fine structure of the cuticle of *Pycnophyes kielensis* (Kinorhyncha, Homalorhagida). Zoologichesky Zhurnal 69(1): 5–11. [in Russian]

Adrianov, A.V., Malakhov, V.V. & Yushin, V.V. (1993): Intracellular endosymbionts and parasites in the gut epithelium of kinorhynchs. Russian Journal of Marine Biology 17(5): 271–278.

Adrianov, A.V., Murakami, C. & Shirayama, Y. (2002a): *Echinoderes aureus* n. sp. (Kinorhyncha: Cyclorhagida) from Tanabe Bay (Honshu Island), Japan, with a key to the genus *Echinoderes*. Species Diversity 7: 47–66.

Adrianov, A.V., Murakami, C. & Shirayama, Y. (2002b): Taxonomic study of the Kinorhyncha in Japan. II. *Condyloderes setoensis*, a new species (Kinorhyncha: Cyclorhagida) from Tanabe Bay (Honshu) – first representative of the genus in the Pacific Ocean. Proc Biol Soc Wash 115(1): 205–216.

Adrianov, A.V., Murakami, C. & Shirayama, Y. (2002c): Taxonomic study of the Kinorhyncha in Japan. III. *Echinoderes sensibilis* n. sp. (Kinorhyncha: Cyclorhagida) from Tanabe Bay. Zoological Science 19(4): 463–473.

Adrianov, A.V. & Rybakov, A.V. (1991): *Kinorhynchospora japonica* gen. n., sp. n. (Microsporidia) from the intestine epithelium of *Kinorhynchus yushini* (Homalorhagida, Pychophyidae), from the Sea of Japan. Zoologichesky Zhurnal 70(10): 5–11. [in Russian]

Albertó, E., Scrosati, R.A. & Díaz, G. (1993): Alimentacion del langostino *Pleoticus muelleri* (Crustacea, Decapoda) del Golfo San Jorge, Argentina. Gayana, Zoologia 57(2): 279–284.

Alexandrov, B.G. (2001): Caloric content of the Black Sea invertebrates. I. Zooplankton and meiobenthos. Ekologiya Morya 55: 5–10. [in Russian]

Ankar, S. & Elmgren, R. (1976): The benthic macro- and meiofauna of the Askö-Landsort area (northern Baltic proper). A stratified random sampling survey. Contributions from the Askö Laboratory, University of Stockholm 11: 1–115.

Ansari, Z.A. (1978): Meiobenthos from the Karwar region (central west coast of India). Mahasagar-Bulletin of the National Institute of Oceanography 11(3–4): 163–167.

Ansari, Z. A., Chatterji, A. & Parulekar, A.H. (1984): Effect of domestic sewage on sand beach meiofauna at Goa, India. Hydrobiologia 111: 229–233.

Arlt, G. (1977): Verbreitung und Artenspektrum der Meiofauna im Greifswalder Bodden. Wissenschaftliche Zeitschrift der Wilhelm-Pieck-Universität Rostock, Mathematisch-Naturwissenschaftliche Reihe 26(2): 217–222.

Ax, P. (1987): The phylogenetic system: the systematization of organisms on the basis of their phylogenesis. 340 pp. John Wiley & Sons, Chichester.

Bacescu, M. (1968): Class Kinorhyncha. Determination of fauna of the Black and Azov Seas. Naukova Dumka 1: 237–250. [in Russian]

Bacescu, M. & Bacescu, E. (1956): Kinorhynchii – reprezentanti ai unei clase de animale, nouã pentru fauna Romineascã. Communicãrile Academiei R. P. R. 6(4): 543–549.

Bacescu, M., Dumitrescu, E., Marcus, A., Paladian, G. & Mayer, R. (1963): Données quantitatives sur la faune pétricole de la Mer Noire à Agigea (secteur Roumain), dans les conditions spéciales de l'année 1961. Travaux du Muséum d'Histoire Naturelle "Grigore Antipa" 4: 131–155.

Bamber, R.N. (1989): Kinorhyncha. The marine fauna of the Cullercoats district. Number 25. Kinorhyncha. Report of the Dove Marine Laboratory, 3rd series 38: 1–10.

Bamber, R. (1997): Kinorhyncha. pp. 64–66 in Howson, C.M. & Picton, B. E. (eds.) The species directory of the marine fauna and flora of the British Isles and surrounding seas. Ulster Museum and Marine Conservation Society, Belfast.

Bauer-Nebelsick, M. (1995): *Zelinkaderes klepali* sp. n., from shallow water sands of the Red Sea (Kinorhyncha: Cyclorhagida: Zelinkaderidae). Annalen des Naturhistorischen Museums in Wien, Serie B Botanik und Zoologie 97 B: 57–74.

Bauer-Nebelsick, M. (1996): *Antygomonas oreas* sp. n., a new deep sea kinorhynch from the Pacific Ocean (Kinorhyncha: Cyclorhagida). Annalen des Naturhistorischen Museums in Wien, Serie B Botanik und Zoologie 98 B: 5–22.

Blake, C.H. (1930): Three new species of worms belonging to the order Echinodera. Biological Survey of the Mount Desert Region 4: 3–10.

de Bovée, F. & Soyer, J. (1974): Cycle annual quantitatif du méiobenthos des vases terrigènes côtières distribution verticale. Vie et Milieu 24(1) série B: 141–157.

Boykin, J.C. (1965): The anatomy of *Trachydemus ilyocryptus* (Kinorhyncha). 116 pp. M.S. Thesis, University of Washington.

Brinkhurst, R.O. (ed.) (1987): Distribution and abundance of macroscopic benthic infauna from the continental shelf off southwestern Vancouver Island, British Columbia. Canadian Technical Report of Hydrography and Ocean Sciences 85: 1–86.

Bromley, J.E.C. & Bleakney, S. (1979): Taxonomic survey of benthic fauna in estuarine saltmarsh pools, Minas Basin, Bay of Fundy. Proceedings of the N. S. Institute of Science 29: 411–446.

Brown, R. (1983): Spermatophore transfer and subsequent sperm development in a homalorhagid kinorhynch. Zoologica Scripta 12(4): 257–266.

Brown, R. (1985): Developmental and taxonomic studies of Sydney harbour Kinorhyncha. 193 pp. Ph.D. Thesis, Macquarie University, Sydney.

Brown, R. (1989) Morphology and ultrastructure of the sensory appendages of a kinorhynch introvert. Zool Scr 18(4): 471–482.

Brown, R. & Higgins, R.P. (1983): A new species of *Kinorhynchus* (Homalorhagida, Pycnophyidae) from Australia with a redescription and range extension of other Kinorhyncha from the South Pacific. Zoologica Scripta 12(3): 161–169.

Bruce, J.R., Colman, J.S. & Jones, N.S. (1963): Memoir No. 36. Marine fauna of the Isle of Man and ist surrounding seas. 307 pp. Liverpool University Press, Liverpool.

Bütschli, O. (1876) Untersuchungen über freilebende Nematoden und die Gattung *Chaetonotus*. Zeitschrift für wissenschaftliche Zoologie 26: 363–413, pls. XXIII–XXVII.

Chang, C.Y. & Song, Y.H. (2002): *Echinoderes lanceolatus*, a new kinorhynch from Korea (Kinorhyncha, Cyclorhagida, Echinoderidae). The Korean J Syst Zool 18(2): 203–211.

Chitwood, B.G. (1951): *Echinoderella steineri* new species (Scolecida, Echinodera). The Texas Journal of Science 1: 113–114.

Chitwood, B.G. (1959): The classification of the phylum Kinorhyncha. Proceedings of the XVth International Congress of Zoology, London, 1958: 941–943.

Chitwood, B.G. (1964): European Kinorhyncha from Tomales Bay, California. Forty-fifth Annual Meeting of the Western Society of Naturalists, Abstract of contributed papers, Abstract 6, pp. 2–3.

Cibic, T., Blasutto, O. & Bettoso, N. (2009): Microalgal-meiofaunal interactions in a sublittoral site of the Gulf of Trieste (northern Adriatic Sea, Italy): A three-year study. Journal of Experimental Marine Biology and Ecology 370: 144–154.

Claparède, A.R.E. (1863): Zur Kenntnis der Gattung *Echinoderes* Duj. Beobachtungen über Anatomie und Entwicklungsgeschichte wirbelloser Thiere an der Küste von Normandie angestellt. pp. 90–92, 119, pls. XVI–XVII. Verlag von Wilhelm Engelmann, Leipzig.

Coull, B.C. (1970): Shallow water meiobenthos of the Bermuda platform. Oecologia 4: 325–357.

Coull, B.C., Wells, J.B.J. (1981): Density of mud-dwelling meiobenthos from three sites in the Wellington region. New Zealand Journal of Marine and Freshwater Research 15: 411–415.

Crowe, W.A., Josefson, A.B. & Svane, I. (1987): Influence of adult density on recruitment into soft sediments: a short-term *in situ* sublittoral experiment. Mar Ecol Prog Ser 41: 61–69.

Dalto, A.G. & Albuquerque, E.F. (2000): Meiofauna distribution in a tropical estuary of the south-western Atlantic (Brazil). Vie et Milieu 50(3): 151–162.

Damodaran, R. (1972): Meiobenthos of the mudbanks of Kerala coast. Proceedings of the Indian National Science Academy 38B: 288–297.

Danovaro, R., Gambi, C. & Croce, N.D. (2002): Meiofauna hotspot in the Atacama Trench, eastern South Pacific Ocean. Deep-Sea Research I 49: 843–857.

Delamare Deboutteville, C. (1957): Sur la présence des Echinodères de la famille des Cateriidae Gerlach dans les eaux souterraines littorales de l'Angola. Publiçacoes Culturais, Companhia de Diamantes de Angola (DIAMANG) 34: 33–37.

Diepenbroek, M., Grobe, H. & Sieger, R. (2000): PanMap, http://www.pangaea.de/Software/PanMap.

Dinet, A. (1973): Distribution quantitative du méiobenthos profond dans la région de la dorsale de Walvis (Sud-Ouest Africain). Marine Biology 20: 20–26.

Dinet, A. (1974): Données quantitatives sur le meiobenthos bathyal de la Mer de Norvège. Géochimie organique des Sédiments marins profonds, Orgon 1: Mer de Norvège: 13–14.

Dinet, A. (1976): Etude quantitative du beiobenthos dans le secteur nord de la Mer Egee. Acta Adriatica 18(6): 83–88.

Dinet, A. (1979): A quantitative survey of meiobenthos in the deep Norwegian sea. Ambio Special Report 6: 75–77.

Dinet, A. & Vivier, M.-H. (1977): Le méiobenthos abyssal du Golfe des Gascogne. I. – Considérations sur les données quantitatives. Cahiers de Biologie Marine 18(1): 85–97.

Dovgal, I., Chatterjee, T., Ingole, B. & Nanajkar, M. (2008): First report of *Limnoricus ponticus* Dovgal & Lozowskiy (Ciliophora: Suctorea) as epibionts on *Pycnophyes* (Kinorhyncha) from the Indian Ocean with key to species of the genus *Limnoricus*. Cahiers de Biologie Marine 49: 381–385.

Drzycimski, I. (1975): *Echinoderes levanderi* Karling (Kinorhyncha), a new species in the fauna of Poland from the Southern Baltic Sea. Prezegląd Zoologiczny 19(1): 59–62.

Dujardin, F. (1851): Sur un petit animal marin, l'Echinodère, formant un type intermédiaire entre les Crustacés et les Vers. Annales des Scienes naturelles (3) Zoologie 15: 158–173, pl. 3.

Eckman, J.E. & Thistle, D. (1988): Small-scale spatial pattern in meiobenthos in the San Diego Trough. Deep-Sea Research 35(9): 1565–1578.

Fabiano, M. & Danovaro, R. (1999): Meiofauna distribution and mesoscale variability in two sites of the Ross Sea (Antarctica) with contrasting food supply. Polar Biology 22: 115–123.

Fleeger, J.W., Zhican, T. & Higgins, R.P. (1983): Preliminary study on the ecology of meiobenthic Copepoda and Kinorhyncha off the Changjiang Estuary and adjacent waters. pp. 937–946 in Proceedings of the International Symposium on Sedimentation on the Continental Shelf, with special Reference to the East China Sea, Vol. 2. China Ocean Press, Beijing.

Frithsen, J.B., Elmgren, R. & Rudnick, D.T. (1985): Responses of benthic meiofauna to long-term, low-level additions of No. 2 fuel oil. Marine Ecology Progress Series 23: 1–14.

Gad, G. (2009a): *Culexiregiloricus*, a new genus of Nanaloricidae (Loricifera) from the deep sea of the Guinea Basin (Southeast Atlantic). Zootaxa 2096: 33–49.

Gad, G. (2009b): A clearly identifiable postlarva in the life cycle of a new species of *Pliciloricus* (Loricifera) from the deep sea of the Angola Basin. Zootaxa 2096: 50–81.

GaOrdóñez, D., Pardos, F. & Benito, J. (2000): Cuticular structures and epidermal glands of *Echinoderes cantabricus* and *E. hispanicus* (Kinorhyncha, Cyclorhagida) with special reference to their taxonomic value. The Journal of Morphology 246: 161–178.

GaOrdóñez, D., Pardos, F. & Benito, J. (2008): Three new *Echinoderes* (Kinorhyncha, Cyclorhagida) from North Spain, with new evolutionary aspects in the genus. Zool Anz 247: 95–111.

Gerlach, S.A. (1956): Über einen aberranten Vertreter der Kinorhynchen aus dem Küstengrundwasser. Kieler Meeresforschungen 12(1): 120–124.

Gerlach, S.A. (1969): *Cateria submersa* sp. n., ein cryptorhager Kinorhynch aus dem sublitoralen Mesopsammal der Nordsee. Veröffenlichungen des Instituts für Meeresforschung in Bremerhaven 12: 161–168.

Gomes, C.A.A., dos Santos, P.J.P., Alves, T.N.C., Rosa-Filho, J.S. & Souza-Santos, L.P. (2002): Variação temporal da meiofauna em área de manguezal em Itamaracá – Pernambuco. Atlântica, Rio Grande 24(2): 89–96.

Greeff, R. (1869): Untersuchungen über einige merkwürdige Thiergruppen des Arthropoden- und Wurm-Typus. Archiv für Naturgeschichte 35: 71–121, pls. IV-VII.

Grego, M., de Troch, M., Forte, J. & Malej, A. (2009): Main meiofauna taxa as indicator for assessing the spatial and seasonal impact of fish farming. Marine Pollution Bulletin 58: 1178–1186.

Gross, T.F. & Williams III, A.J. (1991): Characterization of deep-sea storms. Marine Geology 99: 281–301.

Grzelak, K. & Kotwicki, L. (2012): Meiofaunal distribution in Hornsund fjord, Spitsbergen. Polar Biology 35: 269–280.

Harper, D.E., Jr., Potts, D.L., Salzer, R.R., Case, R.J., Jaschek, R.L. & Walker, C.M. (1981): Distribution and abundance of macrobenthic and meiobenthic organisms. pp. 133–177 in Middleditch, B.S. (ed.) Environmental effects of offshore oil production. Marine Science 14. Plenum Press, New York.

Heiner, I. (2008): *Rugiloricus baccatus* sp. nov. (Loricifera – Pliciloricidae) and a ghost-larva with paedogenetic reproduction. Systematics and Biodiversity 6(2): 225–247.

Heiner, I. & Neuhaus, B. (2007): Loricifera from the deep sea at the Galápagos spreading center, with a description of *Spinoloricus turbatio* gen. et sp. nov. (Nanaloricidae). Helgoland Marine Research 61: 167–182.

Herman, R.I. & Dahms, H.U. (1992): Meiofauna communities along a depth transect off Halley Bay (Weddell Sea – Antarctica). Polar Biology 12: 313–320.

Herranz, M., Pardos, F. & Boyle, M.J. (submitted): Comparative morphology of serotonergic-like immunoreactive elements in the central nervous system of kinorhynchs (Kinorhyncha, Cyclorhagida). Journal of Morphology.

Herranz, M., Thormar, J., Benito, J., Sánchez, N. & Pardos, F. (2012): *Meristoderes* gen. nov., a new kinorhynch genus, with the description of two new specie and their implications for echinoderid phylogeny (Kinorhyncha: Cyclorhagida, Echinoderidae). Zoologischer Anzeiger 251: 161–179.

Higgins, R.P. (1960): A new species of *Echinoderes* (Kinorhyncha) from Puget Sound. Trans Am Microsc Soc 79(1): 85–91.

Higgins, R.P. (1961): Three new homalorhage kinorhynchs from the San Juan Archipelago, Washington. Journal of the Elisha Mitchell Scientific Society 77(1): 81–88.

Higgins, R.P. (1962): Postembryonic life history of the homalorhagid Kinorhyncha. American Zoologist 2: 414.

Higgins, R.P. (1964a): Redescription of the kinorhynch *Echinoderes remanei* (Blake, 1930) Karling,1954. Transactions of the American Microscopical Society 83(2): 243–247.

Higgins, R.P. (1964b): Three new kinorhynchs from the North Carolina coast. Bulletin of Marine Sciene of the Gulf and Caribbean 14(3): 479–493.

Higgins, R.P. (1964c): A method for meiobenthic invertebrate collection. American Zoologist 4(3): 291.

Higgins, R.P. (1965): The homalorhagid Kinorhyncha of Northeastern U. S. Coastal Waters. Trans Am Microsc Soc 84(1): 65–72.

Higgins, R.P. (1966a): Faunistic studies in the Red Sea (in winter, 1961–1962), Part II: Kinorhynchs from the area of Al-Ghardaqa. Zoologische Jahrbücher Abteilung für Systematik Ökologie und Geographie der Tiere 93: 118–126.

Higgins, R.P. (1966b): *Echinoderes arlis*, a new kinorhynch from the Arctic Ocean. Pacific Science 20(4): 518–520.

Higgins, R.P. (1967): The Kinorhyncha of New-Caledonia. pp. 75–90 in Expédition Française sur les Recifs coralliens de la Nouvelle-Calédonie 2. Fondation Singer-Polignac, Paris.

Higgins, R.P. (1968): Taxonomy and postembryonic development of the Cryptorhagae, a new suborder for the mesopsammic kinorhynch genus *Cateria*. Trans Am Microsc Soc 87(1): 21–39.

Higgins, R.P. (1969a): Indian Ocean Kinorhyncha: 1. *Condyloderes* and *Sphenoderes*, new cyclorhagid genera. Smithsonian Contributions to Zoology 14: 1–13.

Higgins, R.P. (1969b): Indian Ocean Kinorhyncha: 2. Neocentrophyidae, a new homalorhagid family. Proceedings of the Biological Society of Washington 82: 113–128.

Higgins, R.P. (1971): A historical overview of kinorhynch research. Smithsonian Contributions to Zoology 76: 25–31.

Higgins, R.P. (1974): Chapter 11. Kinorhyncha. pp. 507–518 in Giese, A.C. & Pearce, J.S. (eds.) Reproduction of marine invertebrates. Vol. I Acoelomate and pseudocoelomate metazoans. Academic Press, New York.

Higgins, R.P. (1977a): Redescription of *Echinoderes dujardinii* (Kinorhyncha) with descriptions of closely related species. Smithsonian Contributions to Zoology 248: 1–26.

Higgins, R.P. (1977b): Two new species of *Echinoderes* (Kinorhyncha) from South Carolina. Transactions of the American Microscopical Society 96(3): 340–354.

Higgins, R.P. (1978a): Phylum Kinorhyncha. p. 100 in Zingmark, R.G. (ed.) An annotated checklist of the biota of the coastal zone of South Carolina, University of South Carolina Press, Columbia.

Higgins, R.P. (1978b): *Echinoderes gerardi* n. sp. and *E. riedli* (Kinorhyncha) from the Gulf of Tunis. Transactions of the American Microscopical Society 97(2): 171–180.

Higgins, R.P. (1981): Kinorhyncha. pp. 873–877 in Parker, S.P. (ed.) Synopsis and classification of Living Organisms Volume 1. McGraw-Hill, New York.

Higgins, R.P. (1982): Three new species of Kinorhyncha from Bermuda. Trans Am Microsc Soc 101(4): 305–316.

Higgins, R.P. (1983): The Atlantic barrier reef ecosystem at Carrie Bow Cay, Belize, II: Kinorhyncha. Smithson Contrib Mar Sci 18: 1–131.

Higgins, R.P. (1985): The genus *Echinoderes* (Kinorhyncha: Cyclorhagida) from the English Channel. Journal of the Marine Biological Association of the United Kingdom 65: 785–800.

Higgins, R.P. (1986a): Kinorhyncha. pp. 110–118 in Botosaneanu, L. (ed.) Stygofauna mundi. A faunistic, distributional, and ecological synthesis of the world fauna inhabiting subterranean waters (including the marine interstitial). E.J. Brill & Dr. W. Backhuys, Leiden.

Higgins, R.P. (1986b): Phylum Kinorhyncha. pp. 220–222 in Sterrer, W. & Schöpfer-Sterrer, C. (eds.) Marine fauna and flora of Bermuda: A systematic guide to the identification of marine organisms. Wiley & Sons, New York.

Higgins, R.P. (1986c): A new species of *Echinoderes* (Kinorhyncha: Cyclorhagida) from a coarse-sand California beach. Transactions of the American Microscopical Society 105(3): 266–273.

Higgins, R.P. (1986d): Redescription of *Echinoderes pilosus* (Kinorhyncha: Cyclorhagida). Proc Biol Soc Wash 99: 399–405.

Higgins, R.P. (1988): Kinorhyncha. pp. 328–331 in Higgins, R.P. & Thiel, H. (eds) Introduction to the study of meiofauna. Smithsonian Institution Press, Washington, D.C.

Higgins, R.P. (1990): Zelinkaderidae, a new family of cyclorhagid Kinorhyncha. Smithsonian Contributions to Zoology 500: 1–26.

Higgins, R.P. (1991): *Pycnophyes chukchiensis*, a new homalorhagid kinorhynch from the Arctic Sea. Proceedings of the Biological Society of Washington 104(1): 184–188.

Higgins, R.P. & Adrianov, A.V. (1991): Kinorhyncha from the Black Sea. 1. Redescription of *Kinorhynchus paraneapolitanus*. Transactions of the American Microscopical Society 110(4): 328–336.

Higgins, R.P. & Fleeger, J.W. (1980): Seasonal changes in the population structure of *Echinoderes coulli* (Kinorhyncha). Estuarine and Coastal Marine Science 10: 495–505.

Higgins, R.P. & Korczynski, R.E. (1989): Two new species of *Pycnophyes* (Homalorhagida, Kinorhyncha) from the Canadian coast of the Beaufort Sea. Can J Zool 67: 2056–2064.

Higgins, R.P. & Kristensen, R.M. (1986): New Loricifera from southeastern United States coastal waters. Smithsonian Contributions to Zoology 438: 1–70.

Higgins, R.P. & Kristensen, R.M. (1988): Kinorhyncha from Disko Island, West Greenland. Smithson Contrib Zool 458: 1–56.

Higgins, R.P., Rao, G.C. (1979): Kinorhynchs from the Andaman Islands. Zoological Journal of the Linnean Society 67(1): 75–85.

Higgins, R.P. & Shirayama, Y. (1990): Dracoderidae, a new family of the cyclorhagid Kinorhyncha from the Inland Sea of Japan. Zoological Science 7: 939–946.

Higgins, R.P. & Storch, V. (1991): Evidence for direct development in *Meiopriapulus fijiensis* (Priapulida). Transactions of the American Microscopical Society 110(1): 37–46.

d'Hondt, J.-L. (1970): Inventaire de la fauna marine de Roscoff. Gastrotriches, kinorhynques, rotifères, tardigrades. Éditions de la station biologique de Roscoff Nouvelle série: 1–28.

d'Hondt, J.-L. (1973): Contribution a l'étude de la microfaune interstitielle des plages de l'Ouest Algérien. Vie et Millieu 23(2) série A: 227–241.

Hooper, D.J. (1970): Handling, fixing, staining and mounting nematodes. pp. 39–54 in Southey, J.F. (ed.) Laboratory methods for work with plant and soil nematodes. Ministry of Agriculture, Fisheries and Food, Technical Bulletin 2. Her Majesty's Stationary Office, London.

Horn, T.D. (1978): The distribution of *Echinoderes coulli* (Kinorhyncha) along an interstitial salinity gradient. Transactions of the American Microscopical Society 97(4): 586–589.

Hua, E., Zhang, Z.N., Yu, Z.S. & Zhang, Y. (2010): Preliminary study on the immediate response of the nematode community to typhoon Soudelor. Deep-Sea Research II: 57: 1064–1070.

Huang, Y., Zhang, Z. & Liu, X. (2005): Studies on the community structure of meiofauna and marine nematode at six stations in the southern Yellow Sea, China. Journal of Ocean University of China 4(1): 34–42.

Hummon, W.D. (1975): Respiratory and osmoregulatory physiology of a meiobenthic marine gastrotrich, *Turbanella ocellata* Hummon 1974. Cahiers de Biologie Marine 16: 255–268.

Huys, R. & Coomans, A. (1989): *Echinoderes higginsi* sp. n. (Kinorhyncha, Cyclorhagida) from the southern North Sea with a key to the genus *Echinoderes* Claparède. Zoologica Scripta 18(2): 211–221.

Huys, R., Herman, P.M.J., Heip, C.H.R. & Soetaert, K. (1992): The meiobenthos of the North Sea: density, biomass trends and distribution of copepod communities. ICES Journal of marine Science 49: 23–44.

ICZN, (1999): International Code of Zoological Nomenclature. 4th ed. 306 pp. The Natural History Museum, London.

Jensen, P. (1983): Meiofaunal abundance and vertical zonation in a sublittoral soft bottom, with a test of the Haps corer. Marine Biology 74: 319–326.

Jeuniaux, C. (1975): Principes de systématique biochimique et application a quelques problèmes particuliers concernant les Aschelminthes, les polychètes et les tardigrades. Cahiers de Biologie marine 16: 597–612.

Johnston, T.H. (1938): Report on the Echinoderida. Australasian Antarctic Expedition 1911–1914, Scientific Reports, Series C – Zoology and Botany 10(7): 1–13.

Karling, T.G. (1955): *Echinoderes levanderi* n. sp. (Kinorhyncha) aus der Ostsee. Arkiv för Zoologi 7(10): 189–192.

Kirsteuer, E. (1964): Zur Kenntnis der Kinorhynchen Venezuelas. Zoologischer Anzeiger 173: 388–393.

Kondalarao, B. (1983): Distribution of meiofauna in the Gautami-Godavari estuarine system. Mahasagar-Bulletin of the National Institute of Oceanography 16(4): 453–457.

Kontar, E.A. & Sokov, A.V. (1994): A benthic storm in the northeastern tropical Pacific over the fields of mangeanese nodules. Deep-Sea Research I 41(7): 1069–1089.

Kozloff, E.N. (1972): Some aspects of development in *Echinoderes* (Kinorhyncha). Trans Am Microsc Soc 91(2): 119–130.

Kozloff, E. (2007): Stages of development, from first cleavage to hatching, of an *Echinoderes* (phylum Kinorhyncha: class Cyclorhagida). Cahiers de Biologie Marine 48: 199–206.

Krishnaswamy, S. (1962): Occurence of *Echinoderella* (Echinodereda) off Plymouth. The Annals and Magazine of Natural History 13(5): 61–63.

Kristensen, R.M. (1991): Loricifera – a general and phylogenetic overview. Verhandlungen der Deutschen Zoologischen Gesellschaft 84: 231–246.

Kristensen, R.M. (2005): Fifteen years investigations of the meiofauna of the Faroe Bank (NE Atlantic). Biofar Proceedings 2005: 202–212.

Kristensen, R.M. & Gad, G. (2004): *Armorloricus*, a new genus of Loricifera (Nanaloricidae) from Trezen ar Skoden (Roscoff, France). Cahiers de Biologie Marine 45: 121–156.

Kristensen, R.M. & Hay-Schmidt, A. (1989): The protonephridia of the arctic kinorhynch *Echinoderes aquilonius* (Cyclorhagida, Echinoderidae). Acta Zoologica 70(1): 13–27.

Kristensen, R.M., Heiner, I. & Higgins, R.P. (2007): Morphology and life cycle of a new loriciferan from the Atlantic coast of Florida with an emendet diagnosis and life cycle of Nanaloricidae (Loricifera). Invertebrate Biology 126(2): 120–137.

Kristensen, R.M. & Higgins, R.P. (1991): Chapter 10. Kinorhyncha. pp. 377–404 in Harrison, F.W. & Ruppert, E.E. (eds.) Microscopic anatomy of invertebrates, Vol. 4: Aschelminthes. Wiley-Liss, New York.

Lang, K. (1936): Undersökningar över Öresund. Untersuchungen aus dem Öresund XXI. Einige Kleintiere aus dem Öresund. Lunds Universitets Årsskrift N. F. Avd. 2 31(10): 1–8.

Lang, K. (1949): Echinoderida. Further Zoological Results of the Swedish Antarctic Expedition 1901–1903 4: 1–22.

Lang, K. (1953): Reports of the Lund University Chile Expedition 1948–1949. 9. Echinoderida. Lunds Universitets Årsskrift N. F. Avd. 2 49(4): 3–8.

Lang, K. (1963): 18. The relation between the Kinorhyncha and Priapulida and their connection with the Aschelminthes. pp. 256–262 in Dougherty, E.C. (ed.) The lower Metazoa. Comparative biology and phylogeny. University of California Press, Berkeley.

Lee, H.J., Gerdes, D., Vanhove, S. & Vincx, M. (2001a): Meiofauna response to iceberg disturbance on the Antarctic continental shelf at Kapp Norvegia (Eweddell Sea). Polar Biology 24: 926–933.

Lee, H.J., Vanhove, S., Peck. L.S. & Vincx, M. (2001b): Recolonisation of meiofauna after catastophic iceberg scouring in shallow Antarctic sediments. Polar Biology 24: 918–925).

Lemburg, C. & Schmidt-Rhaesa, A. (1999): Priapulida. pp. 1053–1058 in Knobil, E. & Neill, J.D. (eds.) Encyclopedia of reproduction Vol. 3. Academic Press, New York.

Lemburg, C. (2002a): A new kinorhynch *Pycnophyes australensis* sp. n. (Kinorhyncha: Homalorhagida: Pycnophyidae) from Magnetic Island, Australia. Zoologischer Anzeiger 241: 173–189.

Lemburg, C. (2002b): Erratum. A new kinorhynch *Pycnophyes australensis* sp. n. (Kinorhyncha: Homalorhagida: Pycnophyidae) from Magnetic Island, Australia. Zoologischer Anzeiger 241(3): 1–4.

Levander, K.M. (1900): Über das Herbst- und Winter-Plankton im finnischen Meerbusen und in der Åland-See 1898. Acta Societatis pro Fauna et Flora Fennica 18(5): 1–25.

Levin, L.A., Leithold, E.L., Gross, T.F., Huggett, C.L. & di Bacco, C. (1994): Contrasting effects of substrate mobility on infaunal assemblages inhabiting two high-energy settings on Fieberling Guyot. Journal of Marine Research 52: 489–522.

Lou, T.-H. (1934): Sur la presence d'un nouveau kinorhynque a Tchefou: *Echinoderes tchefouensis* sp. nov. Contribution du Laboratoire de Zoologie Academie Nationale de Peiping 1(4): 1–9, pl. 1.

Lundbye, H., Rho, H.S. & Sorensen, M.V. (2011): *Echinoderes rex* n. sp (Kinorhyncha: Cyclorhagida), the largest *Echinoderes* species found so far. Scientia Marina 75(1): 41–51.

Lysykh, N.M. (2005): The taxonomic composition and quantitative development of the soft bottom meiobenthos under mariculture of *Mytilus galloprovincialis* Lam., 1819 (the Black Sea). Morskoj Ekologiceskij Zhurnal 4(1): 61–68. [in Russian]

Mari, M. & Morselli, I. (1987): Acari (Alacaridi ed Idracnelle) e Chinorinchi nella spugna Hymeniacidon sanguinea (Grant) della laguna veneta. Atti Soc Nat Mat Modena 118: 115–117.

Marinov, T. (1964): On the microzoobenthos fauna of the Black sea (Kinorhyncha and Halacaridae). Bulletin of the Institute of fish culture and fisheries 4: 61–71. [in Russian]

Martorelli, S.R. (2002): Parásitos y epibiontes de langostino *Pleoticus muelleri* (Bate 1888) en el Atlántico Sud-occidental. CIVA 2002: 647–665.

Martorelli, S. & Higgins, R.P. (2004): Kinorhyncha from the stomach of the shrimp *Pleoticus muelleri* (Bate, 1888) from Comodoro Rivadavia, Argentina. Zoologischer Anzeiger 243: 85–98.

Masson, D.G., Kidd, R.B., Gardner, J.V., Huggett, Q.J. & Weaver, P.P.E. (1994): Saharan continental rise: facies distribution and sediment slides. pp. 327–347 in Poag, V.W. & Degrocianski, P.C. (eds.) Geological evolution of Atlantic continental rises. Van Nostrand Reinhold, New York.

Mazzola, A, Mirto, S. & Danovaro, R. (1999): Initial fish-farm impact on meiofaunal assemblages in coastal sediments of the western Mediterranean. Mar Pollut Bull 38(12): 1126–1133.

Mazzola, A., Mirto, S., La Rosa, , T., Fabiano, M. & Danovaro, R. (2000): Fish-farming effects on benthic community structure in coastal sediments: analysis of meiofaunal recovery. ICES Journal of Marine Science 57: 1454–1461.

McIntyre, A.D. (1961): Quantitative differences in the fauna of boreal mud associations. J Mar Biol Assoc U K 41: 599–616.

McIntyre, A.D. (1962): The class Kinorhyncha (Echinoderida) in British waters. J Mar Biol Assoc U K 42: 503–509.

McIntyre, A.D. (1964): Meiobenthos of sub-littoral muds. Journal of the Marine Biological Association of the United Kingdom 44: 665–674.

Meadows, P.S., Reichelt, A.C., Meadows, A. & Waterworth, J.S. (1994): Microbial and meiofaunal abundance, redox potential, pH and shear strength profiles in deep sea Pacific sediments. Journal of the Geological Society, London, 151: 377–390.

Merriman, J.A. & Corwin, H.O. (1973): An electron microscopical examination of *Echinoderes dujardini* Claparède (Kinorrhyncha). Zeitschrift für Morphologie der Tiere 76: 227–242.

Metschnikoff, E.M. (1865): Ueber einige wenig bekannte niedere Thierformen. Zeitschrift für wissenschaftliche Zoologie 15: 450–463, pl. XXXV.

Metschnikoff, E. (1869): Bemerkungen über *Echinoderes*. Bulletin de l'Académie impériale des Sciences de St.-Pétersbourg, Mélanges biologiques 7: 190–194.

Modig, H. & Ólafsson, E. (1998): Responses of Baltic benthic invertebrates to hypoxic events. Journal of Experimental Marine Biology and Ecology 229: 133–148.

Moore, P.G. (1973): *Campyloderes macquariae* Johnston, 1938 (Kinorhyncha: Cyclorhagida) from the northern hemisphere. Journal of Natural History 7: 341–354.

Moritz, K. & Storch, R. (1972a): Zur Feinstruktur des Integumentes von *Trachydemus giganteus* Zelinka (Kinorhyncha). Zeitschrift für Morphologie der Tiere 71: 189–202.

Moritz, K. & Storch, V. (1972b): Über den ultrastrukturellen Bau der Skaliden von *Trachydemus giganteus* (Kinorhyncha). Marine Biology 16(1): 81–89.

Müller, M.C.M. & Schmidt-Rhaesa, A. (2003): Reconstruction of the muscle system in *Antygomonas* sp. (Kinorhyncha, Cyclorhagida) by means of phalloidin labeling and cLSM. The Journal of Morphology 256: 103–110.

Murakami, C. (2005): A brief review of general biology of Kinorhyncha, with proposed new standard Japanese names for the species from Japan. Taxa Proceedings of the Japanese Society of Systematic Zoology 19: 34–41.

Murakami, C., Adrianov, A. & Shirayama, Y. (2001): Taxonomic study of the Kinorhyncha in Japan: I. *Pycnophyes tubuliferus* Adrianov, 1989 (Kinorhyncha: Homalorhagida) from Japan. Publications of the Seto Marine Biological Laboratory 39(2/3): 113–127.

Murrell, M.C. & Fleeger, J.W. (1989): Meiofauna abundance on the Gulf of Mexico continental shelf affected by hypoxia. Continental Shelf Research 9(12): 1049–1062.

Nagabhushanam, A.K. (1972): Studies on the marine intertidal ecology of the Orissa coast. Proceedings of the Indian National Science Academy 38B(3/4): 308–315.

Nascimento, F.J.A., Karlson, A.M.L. & Elmgren, R. (2008): Settling blooms of filamentous cyanobacteria as food for meiofauna assemblages. Linology and Oceanography 53(6): 2636–2643.

Nebelsick, M. (1990): *Antygomonas incomitata* gen. et sp. n. (Cyclorhagida, Kinorhyncha) and its phylogenetic relationships. Zoologica Scripta 19(2): 143–152.

Nebelsick, M. (1992a): Sensory spots of *Echinoderes capitatus* (Zelinka, 1928)(Kinorhyncha, Cyclorhagida). Acta Zoologica 73(3): 185–195.

Nebelsick, M. (1992b): Ultrastructural investigations of three taxonomic characters in the trunk region of *Echinoderes capitatus* (Kinorhyncha, Cyclorhagida). Zool Scr 21(4): 335–345.

Nebelsick, M. (1993): Introvert, mouth cone, and nervous system of *Echinoderes capitatus* (Kinorhyncha, Cyclorhagida) and implications for the phylogenetic relationships of Kinorhyncha. Zoomorphology 113: 211–232.

Needham, A.E. (1989): 10. Kinorhyncha. pp. 207–217 in Adiyodi, K.G. & Adiyodi, R.G. (eds.) Reproductive biology of invertebrates, Vol IV, Part A: Fertilization, development, and parental care. John Wiley & Sons, Chichester.

Neira, C., Sellanes, J., Levin, L.A. & Arntz, W.E. (2001): Meiofaunal distribution on the Peru margin: relationships to oxygen and organic matter availability. Deep-Sea Research I 48: 2453–2472.

Neuhaus, B. (1988): Ultrastructure of the protonephridia in *Pycnophyes kielensis* (Kinorhyncha, Homalorhagida). Zoomorphology 108: 245–253.

Neuhaus, B. (1991): Zur Ultrastruktur, Postembryonalentwicklung und phylogenetischen Verwandtschaft der Kinorhyncha. Ph.D. Thesis, University of Göttingen, pp. 1–190.

Neuhaus, B. (1993): Postembryonic development of *Pycnophyes kielensis* and *P. dentatus* (Kinorhyncha) from the North Sea. Microfauna Marina 8: 163–193.

Neuhaus, B. (1994): Ultrastructure of alimentary canal and body cavity, ground pattern, and phylogenetic relationships of the Kinorhyncha. Microfauna Marina 9: 61–156.

Neuhaus, B. (1995): Postembryonic development of *Paracentrophyes praedictus* (Homalorhagida): neoteny questionable among the Kinorhyncha. Zoologica Scripta 24(3): 179–192.

Neuhaus, B. (1997): Ultrastructure of the cephalic sensory organs of adult *Pycnophyes dentatus* and of the first juvenile stage of *P. kielensis* (Kinorhyncha, Homalorhagida). Zoomorphology 117: 33–40.

Neuhaus, B. (1999): Kinorhyncha. pp. 933–938 in Knobil, E. & Neill, J.D. (eds.) Encyclopedia of reproduction Vol. 2. Academic Press, New York.

Neuhaus, B. (2003): 19. Kinorhyncha. pp. 646–653 in Hofrichter, R. (ed.) Das Mittelmeer – Fauna, Flora, Ökologie, Bd. II/1 Bestimmungsführer Prokaryota, Protista, Fungi, Algae, Plantae, Animalia (bis Nemertea). Spektrum Akademischer Verlag, Heidelberg.

Neuhaus, B. (2004): Description of *Campyloderes* cf. *vanhoeffeni* (Kinorhyncha, Cyclorhagida) from the Central American East Pacific deep sea with a review of a genus. Meiofauna Marina 13: 3–20.

Neuhaus, B. (2005): Die Kinorhyncha der Ausfahrt SO 168: Zoobenthische Untersuchungen in der Tiefsee und auf dem Kontinentalschelf östlich Neuseelands. Statusseminar 2005 Meeresforschung mit FS Sonne, 9.-11 .3 .2005 in Warnemünde, Abstracts, pp. 225–227.

Neuhaus, B. & Blasche, T. (2006): *Fissuroderes*, a new genus of Kinorhyncha (Cyclorhagida) from the deep sea and continental shelf of New Zealand and from the continental shelf of Costa Rica. Zoologischer Anzeiger 245: 19–52.

Neuhaus, B. & Higgins, R.P. (2002): Ultrastructure, biology, and phylogenetic relationships of Kinorhyncha. Integrative and Comparative Biology 42: 619–632.

Neuhaus, B., Higgins, R.P. & Paavo, B.L. (2010): Phylum Kinorhyncha mud dragons. pp. 468–471 in Gordon, D.P. (ed.) New Zealand Inventory of Biodiversity 2, Chaetognatha, Ecdysozoa, Ichnofossils. Canterbury University Press, Christchurch.

Neuhaus, B., Kristensen, R.M. & Lemburg, C. (1996): Ultrastructure of the cuticle of the Nemathelminthes and electron microscopical localization of chitin. Verhandlungen der Deutschen Zoologischen Gesellschaft 89.1: 221.

Neuhaus, B. & van der Land, J. (2001): Kinorhyncha. pp. 159–160 in Costello, M.J., Emblow, C.S. & White, R. (eds.) European register of marine species. A check-list of the marine species in Europe and a bibliography of guides to their identification. Patrimoines naturels 50.

Neuhaus, B. & Lüter, C. (2003): Ausfahrten SO 144–143 und SO 158: Zoobenthische Untersuchungen im zentralamerikanischen Ostpazifik. Statusseminar 2003 Meeresforschung mit FS Sonne, 12.-14. März 2003 in Hamburg, Abstracts, pp. 123–126.

Neuhaus, B. & Sørensen, M.V. (2012): Populations of *Campyloderes* sp. (Kinorhyncha, Cyclorhagida): one species with significant morphological variation? Zoologischer Anzeiger.

Neuhaus, B., Pardos, P., Sørensen, M.V. & Higgins, R.P. (2013): Redescription, morphology, and biogeography of *Centroderes spinosus* (Reinhard, 1881)(Kinorhyncha, Cyclorhagida) from Europe. Cahiers de Biologie Marine 54.

Nicholas, W. & Sørensen, M.V. (2009): Kinorhynchs from sea-grass beds in south eastern Australia – with notes on a new collecting method. Wetlands (Australia) 25(1): 20–24.

Nomaki, H., Ogawa, N.O., Ohkouchi, N., Suga, H., Toyofuku, T., Shimanaga, M., Nakatsuka, T. & Kitazato, H. (2008): Benthic foraminifera as trophic links between phytodetritus and benthic metazoans: carbon and nitrogen isotopic evidence. Marine Ecology Progress Series 357: 153–164.

Nyholm, K.-G. (1947a): Contributions to the knowledge of the postembryonic development in Echinoderida Cyclorhagae. Zoologisk Bidrag fran Uppsalla 25: 423–428.

Nyholm, K.-G. (1947b: *Campyloderes* – an "antarctic" genus of Echinoderida off the West coast of Sweden. Arkiv för Zoologi 39 A(13): 1–6.

Nyholm, K.-G. (1947c): Studies in the Echinoderida. Arkiv för Zoologi 39 A(14): 1–36, pls. I–II.

Nyholm, K.-G. (1976): Ultrastructure of the spermatozoa in Homalorhaga Kinorhyncha. Zoon 4: 11–18.

Nyholm, K.-G. (1977): Receptaculum seminis and morphological hermaphroditism in Homalorhaga Kinorhyncha. Zoon 5: 7–10.

Nyholm, K.-G. & Nyholm, P.-G. (1976a): Ultrastructure of the pharyngeal muscles of Homalorhaga Kinorhyncha. Zoon 4: 121–130.

Nyholm, K.-G. & Nyholm, P.-G. (1976b): Z-bodies and supercontraction in the integumental muscles of Homalorhaga Kinorhyncha. Zoon 4: 131–136.

Nyholm, K.-G. & Nyholm, P.-G. (1982): Spermatozoa and spermatogenesis in Homalorhagha Kinorhyncha. Journal of Ultrastructure Research 78(1): 1–12.

Nyholm, K.-G. & Nyholm, P.-G. (1983): 11. Kinorhyncha. pp. 207–220 in Adiyodi, K.G. & Adiyodi, R.G. (eds.) Reproductive biology of invertebrates, Vol. II: Spermatogenesis and sperm function. John Wiley & Sons, Chichester.

Nyholm, K.-G. & Olsson, I. (1973): Seasonal fluctuations of the meiobenthos in an estuary on the Swedish West coast. Zoon 1: 69–76.

Ólafson, E. (1992): Small-scale spatial distribution of marine meiobenthos: the effects of decaying macrofauna. Oecologia 90: 37–42.

Ólafsson, E. & Elmgren, R. (1991): Effects of biological disturbance by benthic amphipods *Monoporeia affinis* on meiobenthic community structure: a laboratory approach. Marine Ecology Progress Series 74: 99–107.

Ólafsson, E. & Elmgren, R. (1997): Seasonal dynamics of sublittoral meiobenthos in relation to phytoplankton sedimentation in the Baltic Sea. Estuarine, Coastal and Shelf Science 45: 149–164.

Ólafsson, E., Elmgren, R. & Papakosta, O. (1993): Effects of the deposit-feeding benthic bivalve *Macoma balthica* on meiobenthos. Oecologia 93: 457–462.

Ólafsson, E., Modig, H. & van de Bund, W.J. (1999): Species specific uptake of radio-labelled phytodetritus by benthic meiofauna from the Baltic Sea. Mar Ecol Prog Ser 177: 63–72.

Olu, K., Lance, S., Sibuet, M., Henry, P., Fiala-Médioni, A. & Dinet, A. (1997): Cold seep communities of fluid expulsion patterns through mud volcanoes seaward of the Barbados accretionary prism. Deep-Sea Research I 44(5): 811–841.

Omer-Cooper, J. (1957a): *Protohydra* and Kinorhyncha in Africa. Nature 179: 486.

Omer-Cooper, J. (1957b): Deux nouvelles especes de Kinorhyncha en Provenance de l'Afrique du Sud. Bulletin mensuel de la Société Linnéenne de Lyon 26(8): 213–216.

Osorio, C. & Iguain, R. (1987): Kinorhyncha en Chile: Presencia de *Echinoderes* Dujardin, 1851. Boletin de la Sociedad de Biologia de Concepción 58: 119–120.

Ostmann, A., Nordhaus, I. & Sørensen, M.V. (2012): First recording of kinorhynchs from Java, with the description of a new brackish water species from a mangrove-fringed lagoon. Marine Biodiversity 42: 79–91.

Pagenstecher, H.A. (1875): *Echinoderes sieboldii*. Zeitschrift für wissenschaftliche Zoologie 25(supplement): 117–123, pl. VII.

Pallares, R.E. (1966): Nota sobre *Echinoderes pilosus* Lang, 1949 (Aschelminthes, Kinorhyncha). Physis Revista de la Asociacion Argentina de Ciencias naturales 26(71): 101–106.

Palmer, M.A. & Gust, G. (1985): Dispersal of meiofauna in a tidal creek. Journal of Marine Research 43: 179–210.

Panceri, P. (1878): Osservazioni intorno a nuove forme di vermi nematodi marini. Atti della R. Academia delle Scienze Fisiche e Matematiche 7(10): 1–10, pl. 1.

Pardos, F., Higgins, R.P. & Benito, J. (1998): Two new *Echinoderes* (Kinorhyncha, Cyclorhagida) from Spain, including a reevaluation of kinorhynch taxonomic characters. Zoologischer Anzeiger 237: 195–208.

Pardos, F. (2004): La taxonomía biológica: problemas lexicográficos y de traducción. Panace@ 5(17–18): 244–247.

Parker, R.H. (1975): The study of benthic communities. Elsevier Oceanography Series 9: 1–279.

Partaly, Y.M. (2003): Pecularities of successions in biocenoses of fouling in the Sea of Azov. Hydrobiological Journal 39(1): 103–112.

Parulekar, A.H. & Ansari, Z.A. (1981): Bottom fauna of the Malacca Strait. Mahasagar-Bulletin of the National Institutte of Oceanography 14(2): 155–158.

Paterson, K.J., Schreider, M. & Zimmermann, K.D. (2003): Effect of catchment disturbance on seston quality and quantity in Brisbane water, NSW. Estuarine, Coastal and Shelf Science 56: 355–368.

de Paula, J.H.C., Rosa Filho, , J.S., de Souza, A.L.B. & Aviz, D. (2006): A meiofauna como indicadora de impactos de carcinicultura no estuário de Curuça (PA). Boletim do Laboratório de Hidrobiologia 19: 61–72.

Pearse, A.S. (1936): Zoological names. A list of phyla, classes, and orders. 24pp. American Association for the Advancement of Science, Durham.

Pekkarinen, M. (1985): Exoskeletons of kinorhynchs in tissues of the bivalve *Macoma balthica* in the Baltic Sea, southwestern Finland. Annales Zoologici Fennici 22: 407–410.

Pfannkuche, O. (1985): The deep-sea meiofauna of the Porcupine Seabight and abyssal plain (NE Atlantic): population structure, distribution, standing stocks. Oceanologica Acta 8(3): 343–353.

Pfannkuche, O. & Thiel, H. (1987): Meiobenthic stocks and benthic activity on the NE-Svalbard shelf and in the Nansen Basin. Polar Biology 7: 253–266.

Pleijel, F. (1983): On feeding of *Pholoe minuta* (Fabricius, 1780) (Polychaeta: Sigalionidae). Sarsia 68: 21–23.

Prato, E., Pastore, M., Parlapiano, I. & Biandolino, F. (2011): Alimentazione predazione in natura di *Melicertus kerathurus*. pp. 90–99 in Barbieri, S. & Mazzuccato, F. (eds.) La risorsa crostacei nel Medeiterraneo: ricerca, produzione e mercato, 25 e 26 novembre 2010, Convegno. Veneto Agricoltura, Legnaro.

Purasjoki, K.J. (1945): Quantitative Untersuchungen über die Mikrofauna des Meeresbodens in der Umgebung der Zoologischen Station Tvärminne an der Südküste Finnlands. Societas Scientarium Fennica Commentationes Biologicae 9(14): 1–24.

Rachor, E. (1975): Quantitative Untersuchungen über das Meiobenthos der nordostatlantischen Tiefsee. „Meteor" Forschungsergebnisse Reihe D 21: 1–10.

Rao, D.G. (1987a): Ecology of meiobenthos of Rambha Bay in Chilka Lagoon, Bay of Bengal. Journal of the Marine Biological Association of India 29(1–2): 74–85.

Rao, D.G. & Satapathy, S. (1996): Demecology of Kinorhyncha of Chilka lagoon (Bay of Bengal). Journal of the Marine Biological Association of India 38(1–2): 15–24.

Rao, G.C. (1972): On the geographical distribution of interstitial fauna of marine beach sand. Proceedings of the Indian National Science Academy 38B(3/4):164–178.

Rao, G.C. (1987b): Meiofauna of the marine national park, South Andaman. Journal of the Andaman Scientific Association 3(2): 88–97.

Rao, G.C. (1993): Littoral meiofauna of Little Andaman. Records of the Zoological Survey of India, occasional paper 155: 1–120.

Rao, G.C. & Ganapati, P.N. (1966): Occurrence of an aberrant kinorhynch *Cateria styx* Gerlach, in Waltair beach Sands. Current Science 35(8): 212–213.

Rao, G.C. & Ganapati, P.N. (1968): The interstitial fauna inhabiting the beach sands of Waltair Coast. Proceedings of the National Institute of Science of India 34B: 82–125.

Reimer, L. (1963): Zur Verbreitung der Kinorhyncha in der mittleren Ostsee. Zoologischer Anzeiger 171(11/12): 440–447.

Reinhard, W. (1881): Über *Echinoderes* und *Desmoscolex* der Umgegend von Odessa. Zoologischer Anzeiger 4(97): 588–592.

Reinhard, W. (1885): Kinorhyncha (*Echinoderes*), leur structure anatomique et leur place dans le système. Travaux de la Société des naturalistes à l'Université Impériale de Kharkow 19: 205–305, pls. VIII–XII. [in Russian]

Reinhard, W. (1887): Kinorhyncha (*Echinoderes*), ihr anatomischer Bau und ihre Stellung im System. Zeitschrift für wissenschaftliche Zoologie 45: 401–467, pls. XX–XXII.

Remane, A. (1928): Kinorhyncha, VII. d_2. pp. 57–84 in Grimpe, G. & Wagler, E. (ed.), Die Tierwelt der Nord- und Ostsee. Akademische Verlagsgesellschaft, Leipzig.

Remane, A. (1929): Dritte Klasse des Cladus Nemathelminthes, Kinorhyncha = Echinodera. pp. 187–248 in Krumbach, T. (ed.) Handbuch der Zoologie. Eine Naturgeschichte der Stämme des Tierreichse. Sechste Lieferung. Walter de Gruyter & Co, Berlin.

Remane, A. (1936): Kinorhyncha (= Echinodera). pp. 243–382 in Bronn, H.G. (ed.) Klassen und Ordnungen des Tierreichs, Band 4: Vermes, II. Abteilung: Askhelminthes, Trochelminthes, 1. Buch, 2. Teil: Gastrotricha und Kinorhyncha. Akademische Verlagsgesellschaft, Leipzig.

Remane, A. (1954): Zwei „mediterrane" Tierarten in der Kieler Bucht. Faunistische Mitteilungen aus Norddeutschland 1(4): 12.

Revkov, N.K. & Sergeeva, N.G. (2004): Current state of the zoobenthos at the Crimean shores of the Black Sea. pp. 186–214 in Öztürk, B., Mokievsky, V.O. & Topaloğlu, B. (eds.) International workshop on Black Sea benthos, 18.-23.

April 2004. Istanbul. Turkish Marine Research Foundation, TÜDAV Publication No. 20, Istanbul.
Rheinheimer, G. (1985): Mikrobiologie der Gewässer. 262 pp. Gustav Fischer Verlag, Stuttgart.
Robinson, C.A., Bernhard, J.M., Levin, L.A., Mendoza, G.F. & Blanka, J.K. (2004): Surficial hydrocarbon seep infauna from the Blake Ridge (Atlantic Ocean, 2150 m) and the Gulf of Mexico (690–2240 m). P. S. Z. N. Marine Ecology 25(4): 313–336.
Rothe, B.H. & Schmidt-Rhaesa, A. (2004): Probable development from continuous to segmental longitudinal musculature in *Pycnophyes kielensis* (Kinorhyncha, Homalorhagida). Meiofauna Marina 13: 21–28.
Rudnick, D.T. (1989): Time lags between the deposition and meiobenthic assimilation of phytodetritus. Marine Ecology Progress Series 50: 231–240.
Rudnick, D.T., Elmgren, R. & Frithsen, J.B. (1985): Meiofaunal prominence and benthic seasonality in a coastal marine ecosystem. Oecologia 67: 157–168.
Rutgers van der Loeff, M.M. & Lavaleye, M.S.S. (1986): Sediments, fauna and the dispersal of radionuclides at the N.E. Atlantic dumpsite for low-level radioactive waste. Report of the Dutch DORA program. 134 pp. Netherlands Institute for Sea Research, AB Den Burg, Texel.
Rysgaard, S., Christensen, P.B., Sørensen, M.V., Funch, P. & Berg, P. (2000): Marine meiofauna, carbon and nitrogen mineralization in sandy and soft sediments of Disko Bay, Greenland. Aquatic and Microbial Ecology 21: 59–71.
San Martín, G. (1991): Syllinae (Polychaeta: Syllidae) from Cuba and the Gulf of Mexico. Bulletin of Marine Science 48(2): 227–235.
San Martín, G. (2003): Annelida Polychaeta II. Syllidae. Fauna Ibérica 21: 1–554.
Sánchez, N., Pardos, F., Herranz, M., Benito, J. (2011): *Pycnophyes dolichurus* sp. nov. and *P. aulacodes* sp. nov. (Kinorhyncha, Homalorhagida, Pycnophyidae), two new kinorhynchs from Spain with a reevaluation of homalorhagid taxonomic characters. Helgoland Marine Research 65: 319–334.
Sánchez, N., Herranz, M., Benito, J. & Pardos, F. (2012): The phylum Kinorhyncha in Spain: new data from the first intensive sampling campaigns. Zootaxa 3402: 24–44.
Sánchez, N., Herranz, M., Benito, J. & Pardos, F. (submitted): *Pycnophyes almansae* sp. nov. and *Pycnophyes annulatus* sp. nov., two new homalorhagid kinorhynchs (Kinorhyncha, Homalorhagida) from the Iberian Peninsula, with special focus on introvert features. Marine Biology Research.
Sánchez-Tocino, L., Tierno de Figueroa, J.M., López-Rodríguez, M.J. & Liébanas, G. (2011): First record of *Echinoderes dujardinii* Claparède, 1863 (Kinorhyncha, Cyclorhagida) in Iberian Peninsula coast waters. Zoologica Baetica 22: 179–184.
Sandulli, R. & de Zio Grimaldi, S. (2008): Kinorhyncha. Biologia Marina Mediterranea 15(supplement): 220–222.
Santos, P.J.P., Botter-Carvalho, M., do Nascimento, A.B., Jr., Marinho, R.G.C., Carvalho, P.V.V.C. & Valebça, A.P.M.C. (2009): Response of estuarine meiofauna assemblage to effects of fertilizer enrichment in the sugar cane monoculture. Pernambuco, Brazil. Brazilian Journal of Oceanography 57(1): 43–55.
Schepotieff, A. (1907a): 2. Zur Systematik der Nematoideen. Zoologischer Anzeiger 31(5/6): 132–161.
Schepotieff, A. (1907b): Die Echinoderiden. Zeitschrift für wissenschaftliche Zoologie 88(2): 291–326, pls. XVII–XX.
Schepotieff, A. (1908): 3. Zur Kenntnis der Echinoderiden. Zoologischer Anzeiger 32: 585–589.
Schmidt, P. (1974): Interstitielle Fauna von Galapagos X. Kinorhyncha. Mikrofauna des Meeresbodens 43: 1–15.
Schmidt-Rhaesa, A. & Rothe, B.H. (2006): Postembryonic development of dorsoventral and longitudinal musculature in *Pycnophyes kielensis* (Kinorhyncha, Homalorhagida). Integrative and Comparative Biology 46(2): 144–150.
Schmitz, W.J. (1996a): On the world ocean circulation: Volume I. Some global features, North Atlantic circulation. 140 pp. Woods Hole Oceanographic Institution Technical Report WHOI-96-03.
Schmitz, W.J. (1996b): On the world ocean circulation: Volume II. The Pacific and Indian Oceans, a global update. 237 pp. Woods Hole Oceanographic Institution Technical Report WHOI-96-08.
Scholtz, G. (2002): The Articulata hypothesis – or what is a segment? Organisms, Diversity and Evolution 2: 197–215.
Sergeeva, N.G., Gooday, A.J., Mazlumyan, S.A., Kolesnikova, E.A., Lichtschlag, A., Kosheleva, T.N. & Anikeevba, O.V. (2012): Meiobenthos of the oxic/anoxic interface in the southwestern region of the Black Sea: abundance and taxonomic composition. Cellular Origin, Life in Extreme Habitats and Astrobiology 21(5): 369–401.
Sheremetevskij, A.M. (1974): Kinorhyncha of the Black Sea. Zoologichesky Zhurnal 53(7): 974–987.
Shimanaga, M., Kitazato, H. & Shirayama, Y. (2000): Seasonal patterns of vertical distribution between meiofaunal groups in relation to phytodetritus deposition in the bathyal Sagami Bay, central Japan. Journal of Oceanography 56: 379–387.
Sibuet, M., Monniot, C., Desbruyères, D., Dinet, A., Khripounoff, A., Rowe, G. & Segonzac, M. (1984): Peuplements benthiques et caractéristiques trophiques du milieu dans la plaine abyssale de Demerara. Oceanologica Acta 7(3): 345–358.
Soetaert, K., Heip, C. & Vincx, M. (1991): The meiobenthos along a Mediterranean deep-sea transect off Calvi (Corsica) and in an adjacent canyon. P. S. Z. N. I.: Marine Ecology 12(3): 227–242.
Sommer, S., Gutzmann, E. & Pfannkuche, O. (2007): Sediments hosting gas hydrates: oases for metazoan meiofauna. Marine Ecology Progress Series 337: 27–37.
Song, Y.H. & Chang, C.Y. (2001): First record of *Campyloderes macquariae* Johnston (Kinorhyncha, Cyclorhagida, Centroderidae) from the North Pacific. The Korean Journal of Systematic Zoology 17(2): 207–216.
Sørensen, M.V. (2001): On the rotifer fauna of Bermuda, including notes on the associated meiofauna and the description of a new species of *Encentrum* (Rotifera: Ploima: Dicranophoridae). Proceedings of the Biological Society of Washington 114(3): 725–736.
Sørensen, M.V. (2006): New kinorhynchs from Panama, with a discussion of some phylogentically significant cuticular structures. Meiofauna Marina 15: 51–77.
Sørensen, M.V. (2007) A new species of *Antygomonas* (Kinorhyncha: Cyclorhagida) from the Atlantic coast of Florida, USA. Cahiers de Biologie Marine 48: 155–168.
Sørensen, M.V. (2008a): A new kinorhynch genus from the Antarctic deep sea and a new species of *Cephalorhyncha* from Hawaii (Kinorhyncha: Cyclorhagida: Echionoderidae). Organisms, Diversity and Evolution 8: 230–232.
Sørensen, M.V. (2008b): A new kinorhynch genus from the Antarctic deep sea and a new species of *Cephalorhyncha* from Hawaii

(Kinorhyncha: Cyclorhagida: Echionoderidae). Organisms, Diversity and Evolution 8: 230e1–230e18.
Sørensen, M.V. (2008c): Phylogenetic analysis of the Echinoderidae (Kinorhyncha: Cyclorhagida). Organisms, Diversity and Evolution 8: 233–246.
Sørensen, M.V., Accogli, G. & Hansen, J.G. (2010d): Postembryonic development of *Antygomonas incomitata* (Kinorhyncha: Cyclorhagida). The Journal of Morphology 271: 863–882.
Sørensen, M.V., Heiner, I. & Hansen, J.G. (2009): A comparative morphological study of the kinorhynch genera *Antygomonas* and *Semnoderes* (Kinorhyncha: Cyclorhagida). The Journal of Morphology 63: 129–147.
Sørensen, M.V., Heiner, I. & Ziemer, O. (2005): A new species of *Echinoderes* from Florida (Kinorhyncha: Cyclorhagida). Proceedings of the Biological Society of Washington 118(3): 499–508.
Sørensen, M.V., Heiner, I., Ziemer, O. & Neuhaus, B. (2007): *Tubulideres seminoli* gen. et sp. nov. and *Zelinkaderes brightae* sp. nov. (Kinorhyncha, Cyclorhagida) from Florida. Helgoland Marine Research 61: 247–265.
Sørensen, M.V., Herranz, M., Rho, H.S., Min, W.-G., Yamasaki, H., Sánchez, N. & Pardos, F. (2012a): On the genus *Dracoderes* Higgins & Shirayama, 1990 (Kinorhyncha: Cyclorhagida) with a redescription of its type species, *D. abei*, and a description of a new species from Spain. Marine Biology Research 8: 210–232.
Sørensen, M.V., Jørgensen, A. & Boesgard, T.M. (2000): A new *Echinoderes* (Kinorhyncha: Cyclorhagida) from a submarine cave in New South Wales, Australia. Cahier de Biologie Marine 41: 167–179.
Sørensen, M.V. & Pardos, F. (2008): Kinorhynch systematics and biology – an introduction to the study of kinorhynchs, inclusive identification keys to the genera. Meiofauna Marina 16: 21–73.
Sørensen, M.V., Pardos, F., Herranz, M. & Rho, H.S. (2010a): New data on the genus *Paracentrophyes* (Homalorhagida, Kinorhyncha), with the description of a new species from the West Pacific. The Open Zoology Journal 3: 42–59.
Sørensen, M.V. & Rho, H.S. (2009): *Triodontoderes anulap* gen. et sp. nov. – a new cyclorhagid kinorhynch genus and species from Micronesia. J Mar Biol Assoc U K 89(6): 1269–1279.
Sørensen, M.V., Rho, H.S. & Kim, D. (2010b): A New Species of *Condyloderes* (Cyclorhagida, Kinorhyncha) from Korea. Zoological Science 27: 234–242.
Sørensen, M.V., Rho, H.S. & Kim, D. (2010c): A new species of the rare genus *Sphenoderes* (Cyclorhagida, Kinorhyncha), with differential notes on *S. indicus* Higgins, 1969. Marine Biology Research 6: 472–484.
Sørensen, M.V., Rho, H.S. & Min, W.-G., Kim, D. & Chang, C.Y. (2012b): An exploration of *Echinoderes* (Kinorhyncha: Cyclorhagida) in Korean and neighboring waters, with the description of four new species and a redescription of *E. tchefouensis* Lou, 1934. Zootaxa 3368: 161–196.
Sørensen, M.V., Rho, H.S. & Min, W.-G., Kim, D. & Chang, C.Y. (2012c): Occurrence of the newly described kinorhynch genus *Meristoderes* (Cyclorhagida: Echinoderidae) in Korea, with the description of four new species. Helgoland Marine Research.
Sørensen, M.V. & Thormar, J. (2010): *Wollunquaderes majkenae* gen. et sp nov. – a new cyclorhagid kinorhynch genus and species from the Coral Sea, Australia. Marine Biodiversity 40: 261–275.
Southern, R. (1914): Clare Island Survey. Nemathelmia, Kinorhyncha, and Chaetognatha. Proceedings of the Royal Irish Academy 31 Part 54 Sect. 3: 1–80, pls. I-XII.
Stein, R. (1991): Accumulation of organic carbon in marine sediments. Lecture Notes in Earth Sciences 34. 217 pp. Springer, Berlin.
Steiner, G. (1919): Zur Kenntnis der Kinorhyncha, nebst Bemerkungen über ihr Verwandtschaftsverhältnis zu den Nematoden. Zoologischer Anzeiger 50: 177–187.
Sudzuki, M. (1976): Microscopical marine animals scarcely known from Japan. I. Micro- & meio-faunae around Kasado Island in the Seto Inland Sea of Japan. Proceedings of the Japanese Society of Systematic Zoology 12: 5–12, pls. 1–3.
Sutherland, T.F., Levings, C.D., Petersen, S.A., Poon, P. & Piercey, B. (2007): The use of meiofauna as an indicator of benthic organic enrichment associated with salmonid aquaculture. Marine Pollution Bulletin 54: 1249–1261.
Tchesunov, A.V., Kaljakina, N.M. & Bubnova, E.N. (2008): A catalogue of biota of the White Sea Biological Station of the Moscow State University. 384 pp. KMK Scientific Press, Moscow.
Thiel, H. (1966): Quantitative Untersuchungen über die Meiofauna des Tiefseebodens. (Vorläufiges Ergebnis der „Meteor"-Expedition in den Indischen Ozean). Veröffentlichungen des Instituts für Meeresforschung in Bremerhaven, Sonderband 2: 131–147.
Thiel, H. (1971): Häufigkeit und Verteilung der Meiofauna im Bereich des Island-Färöer-Rückens. Berichte der deutschen wissenschaftlichen Kommission für Meeresforschung 22: 99–128.
Thistle, D. & Levin, L.A. (1998): The effect of experimentally increased near-bottom flow on metazoan meiofauna at a deep-sea site, with comparison data on macrofauna. Deep-Sea Research I 45: 625–638.
Thistle, D., Levin, L.A., Gooday, A.J., Pfannkuche, O. & Lambshead, P.J.D. (1999): Physical reworking by near-bottom flow alters the metazoan meiofauna of Fieberling Guyot (northeast Pacific). Deep-Sea Research I 46: 2041–2052.
Thistle, D., Yingst, J.Y. & Fauchald, K. (1985): A deep-sea benthic community exposed to strong near-bottom currents on the Scotian Rise (Western Atlantic). Marine Geology 66: 91–112.
Thormar, J. & Sørensen, M.V. (2010): Two new species of *Echinoderes* (Kinorhyncha: Cyclorhagida) from the Solomon Islands. Meiofauna Marina 18: 67–96.
Timm, R.W. (1958): Two new species of *Echinoderella* (phylum Kinorhyncha) from the Bay of Bengal. Journal of the Bombay Natural History Society 55(1): 107–109.
Tokioka, T. (1949): II. Notes on *Echinoderes* found in Japan. Publ Seto Mar Biol Lab 1(2): 67–69 (= 49–51).
Travizi, A. & Zavodnik, N. (2004): Phenology of *Caulerpa taxifolia* and temporal dynamics of its epibiontic meiofauna in the port of Malinska (Croatia, northern Adriatic Sea): Scientia Marina 68 (supplement 1): 145–154.
Vanaverbeke, J., Soetaert, K., Heip, C. & Vanreusel, A. (1997): The metazoan meiobenthos along the continental slope of the Goban Spur (NE Atlantic). Journal of Sea Research 38: 93–107.
Vandepitte, L., Decock, W. & Mees, J. (2010): Belgian register of marine species, compiled and validated by the VLIZ Belgian Marine Species Consortium. VLIZ Special Publications 46: 1–78.
Van Gaever, S., Vanreusel, A., Hughes, J.A., Bett, B.J. & Kiriakoulakis, K. (2004): The macro- and micro-scale

patchiness of meiobenthos associated with the Darwin Mounds (nort-east Atlantic). J Mar Biol Assoc U K 84: 547–557.
Vanhöffen, E. (1905): Einige zoogeographische Ergebnisse der Deutschen Südpolar-Expedition. Verhandlungen des XV. Deutschen Geographentages zu Danzig 1905, 14–19.
Vanhove, S., Wittoeck, J., Desmet, G., Van den Berghe, B., Herman, R.L, Bak, R.P.M., Nieuwland, J.H., Boldrin, A., Rabitti, S. & Vincx, M. (1995): Deep-sea meiofauna comunities in Antarctica: structural analysis and relation with the environment. Mar Ecol Prog Ser 127: 65–76.
Vasconcelos, D.M., Santos, P.J.P. & Trinidade, R.L.N. (2004): Distribuição espacial da meiofauna no estuário do Rio Formoso, Pernambuco, Brasil. Atlântica, Rio Grande 26(1): 45–54.
Veit-Köhler, G., Gerdes, D., Quiroga, E., Hebbeln, D. & Sellanes, J. (2009): Metazoan meiofauna within the oxygen-minimum zone off Chile: results of the 2001-PUCK expedition. Deep-Sea Research II 56: 1105–1111.
Veit-Köhler, G., Guilini, K., Peeken, I., Sachs, O., Sauter, E.J. & Würzberg, L. (2011): Antarctic deep-sea meiofauna and bacteria react to the deposition of particulate organic matter after a phytoplankton bloom. Deep-Sea Research II 58: 1983–1995.
Vezzulli, L., Marrale, D., Moreno, M. & Fabiano, M. (2003): Sediment organic matter and meiofauna community response to long-term fish-farm impact in the Ligurian Sea (Western Mediterranean). Chemistry and Ecology 19(6): 431–440.
Vilas, C., Drake, P. & Fockedey, N. (2008): Feeding preferences of estuarine mysids *Neomysis integer* and *Rhopalophthalmus tartessicus* in a temperate estuary (Guadalquivir estuary, SW Spain). Estuarine, Coastal and Shelf Science 77: 345–356.
Vincx, M., Bett, B.J., Dinet, A., Ferrero, T., Gooday, A.J., Lambshead, P.J.D., Pfannkuche, O., Soltwedel, T. & Vanreusel, A. (1994): Meiobenthos of the deep Northeast Atlantic. Advances in Marine Biology 30: 1–88.
Vrišer, B. (1997): Seasonal and three-year variability of meiofauna in the Gulf of Trieste (northern Adriatic). Periodicum Biologorum 99(2): 209–212.
Vrišer, B. & Vukovič, B. (1999): Seasonal and long-term variability of meiofauna in the environment frequently affected by hypoxia in the central part of the Gulf of Trieste. Ann Ser Hist Nat 9(2): 203–208.
Warwick, R.M., Joint, I.R. & Radford, P.J. (1979): 27. Secondary production of the benthos in an estuarine environment. pp. 429–450 in Jefferies, R.L. & Davy, A.J. (eds.) Ecological processes in coatsal environments: The 1st European Ecological Symposium and 19th Symposium of the British Ecological Society, Norwich, 12.-16. September 1977. Blackwell Scientific Publications, Oxford.
Warwick, R.M., Platt, H.M., Clarke, K.R., Agard, J. & Gobin, J. (1990): Analysis of macrobenthic and meiobenthic community structure to pollution and disturbance in Hamilton Harbour, Bermuda. J Exp Mar Biol Ecol 138: 119–142.
Westheide, W. & Purschke, G. (1988): 10. Organism processing. pp. 146–160 in Higgins, R.P. & Thiel, H. (eds.) Introduction to the study of meiofauna. Smithsonian Institution Press, Washington, D.C.
Westphalen, D. (1993): Stromatolitoid microbial nodules from Bermuda – a special micro habitat for meiofauna. Marine Biology 117: 145–157.
Widbom, B. (1984): Determination of average individual dry weights and ash-free dry weights in different sieve fractions of marine meiofauna. Marine Biology 84: 101–108.
Widbom, B. & Frithsen, B. (1995): Structuring factors in a marine soft bottom community during eutrophication – an experiment with radio-labelled phytodetritus. Oecologia 101: 156–168.
Wieser, W. (1960): Benthic studies in Buzzards Bay. II: The meiofauna. Limnology and Oceanography 5(2): 121–137.
Winkelmann, J. & Ziemer, O. (1999): Distribution and composition of meiofauna with special attention to the Kinorhyncha in the Mellemfjord, Disko Island (West Greenland). Berichte zur Polarforschung 330: 170–181.
Yamasaki, H. & Kajihara, H. (2012): A new brackish-water species of *Echinoderes* (Kinorhyncha: Cyclorhagida) from the Seto Inland Sea, Japan. Species Diversity 17: 109–118.
Yamasaki, H., Kajihara, H. & Mawatari, S.F. (2012): First report of kinorhynchs from Hokkaido, Japan, including a new species of *Pycnophyes* (Pycnophyidae: Homalorhagida). Zootaxa 3425: 23–41.
Yingst, J.Y. (1978): Patterns of micro- and meiofaunal abundance in marine sediments, measured with the adenosine triphosphate assay. Marine Biology 47: 41–54.
Yong, H., Zhinan, Z. & Xiaoshou, L. (2005): Studies in the community structure of meiofauna and marine nematode at six stations in the southern Yellow Sea, China. Journal of Ocean University of China 4(1): 34–42.
Yuqing, G., Zhinan, Z. & Fanghong, M. (2001): Biomass of meiobenthic in the Bohai Sea, China. Acta Oceanologica Sinica 20(3): 435–442.
Yushin, V.V., Adrianov, A.V. & Malakhov, V.V. (1989): Intracellular Procaryota in enteric epithelium of *Pycnophyes kielensis* (Kinorhyncha, Homalorhagida). Doklady Akademia Nauk SSSR 308(3): 762–763.
Yushin, V.V., Adrianov, A.V. & Malakhov, V.V. (1990): Intracellular procaryotes in gut epithelium of the kinorhynchan *Pycnophyes kielensis* (Kinorhyncha, Homalorhagida). Proceedings of the Academy of Science of the USSR 308: 678–679.
Zaneveld, J.S. (1938) Marine Gastrotricha and Kinorhyncha from Scheveningen. Zoologische Mededeelingen 20: 257–262.
Zelinka, C. (1894): Über die Organisation von *Echinoderes*. Verhandlungen der Deutschen Zoologischen Gesellschaft 4: 46–49.
Zelinka, C. (1896): Demonstration der Tafeln der *Echinoderes*-Monographie. Verhandlungen der Deutschen Zoologischen Gesellschaft 6: 197–199.
Zelinka, C. (1907): Zur Kenntnis der Echinoderen. Zoologischer Anzeiger 32(5): 130–136.
Zelinka, C. (1908): Zur Anatomie der Echinoderen. Zoologischer Anzeiger 33(19/20): 629–647.
Zelinka, C. (1912): Die Spermatozoen der Echinoderen und ihre Genese. pp. 520–527 in Stummer-Traunfels, R.R.v. (ed.) Verhandlungen des VIII. Internationalen Zoologen-Kongresses zu Graz vom 15.-20. August 1910. Gustav Fischer, Jena.
Zelinka, C. (1913): Die Echinoderen der Deutschen Südpolar-Expedition 1901–1903. Deutsche Südpolar Expedition XIV Zoologie VI: 419–437, pl. XXXIX.
Zelinka, C. (1914): Zwei Ektoparasiten der Echinoderen aus der Klasse der Ciliaten. Verhandlungen der Gesellschaft deutscher Naturforscher und Ärzte, 85. Versammlung zu Wien, 22. September 1913, 2. Teil, 1. Hälfte: IV, Nr. 13: 680–683.
Zelinka, K. (1928): Monographie der Echinodera. 396 pp., pls. 1–27. Verlag von Wilhelm Engelmann, Leipzig.

Zeppilli, D., Canals, M. & Danovaro, R. (2012): Pockmarks enhance deep-sea benthic biodiversity: a case study in the western Mediterranean Sea. Diversity and Distribution 18: 832–846.

Zeppilli, D., Mea, M., Corinaldesi, C. & Danovaro, D. (2011) Mud volcanoes in the Mediterranean Sea are hot spots of exclusive meiobenthic species. Progress in Oceanography 91: 260–272.

Zhang, J.-H., Gao, Y. & Fang, H.-D. (2011): Abundance and biomass of meiobenthos in Lingdingyang Bay of Pearl River estuary. Chinese Journal of Applied Ecology 22(10): 2741–2748.

Ziemer, O. (2001): Die Morphologie der *Pycnophyes* (Kinorhyncha, Homalorhagida) von vier Standorten in Grönland unter besonderer Berücksichtigung des Introverts. 109 pp. M.S. Thesis, University of Hamburg.

Iben Heiner Bang-Berthelsen, Andreas Schmidt-Rhaesa
and Reinhardt Møbjerg Kristensen

6. Loricifera

Introduction

Loricifera are very small marine benthic animals. Their discovery was made possible by increasing research on marine meiofauna. Single specimens were seen, but not recognized as such, on a few occasions, but sufficient material was collected from subtidal coarse sediments off Roscoff, France for the first species description by Kristensen (1983). Specimens were collected following an extraction with a "freshwater shock", preferred over the standard anesthetization with magnesium-chloride under time pressure. As loriciferans adhere to the substrate (sand grains) tightly, the freshwater shock appears to be the most suited extraction method.

After the first species, *Nanaloricus mysticus*, was described in 1983 (Kristensen 1983), soon more species were discovered (Higgins and Kristensen 1986), all from comparably shallow waters. The first deep-sea loriciferan, *Pliciloricus hadalis*, was described in 1988 (Kristensen & Shirayama 1988). With the exception of the description of *Nanaloricus khaitatus* (Todaro & Kristensen 1998), no further species were described until 2004; since then the species number has risen to 30. This is, however, still an underestimation of the real species number. New species are now regularly found from different sediments, including the deep sea and seamounts. Experts know of more than 100 undescribed species in the available material, which is not always complete enough to allow a proper species description (see e.g., Gad 2005a, Bang-Berthelsen 2008).

With the growing number of species it has become clear that Loricifera are distributed throughout the marine environment, with the exception of the intertidal zone. What has also become evident is that they exhibit fascinating and complicated life cycles. Loriciferans have a characteristic larval stage, the Higgins-larva, which molts several times, finally into a postlarva and then to the mature adult. In some species further larval stages were discovered, such as the ghost-larva (first described as "simplified parthenogenetic adults" by Gad 2005a,b), making it likely that in some species a sexual and an asexual (parthenogenetic) life cycle can be present. Further specialized larvae were discovered for example in *Titaniloricus inexpectatovus* (Gad 2005d) and *Urnaloricus gadi* (Heiner & Kristensen 2009), making some life cycles very complex or highly specialized. The occurrence of hermaphroditism (see Gad 2005e, Heiner & Neuhaus 2010 mentioning two undescribed species) shows that the reproductive biology of loriciferans never stops to hold new surprises.

Loriciferans are very rarely observed alive; usually they are sorted from fixed sediments. Therefore statements on their ecology are often indirect. Very surprising was, therefore, the report that loriciferans (three yet undescribed species) are among the few inhabitants of permanently anoxic sediments in a hypersaline basin in the Mediterranean Sea (Danovaro et al. 2010). Here, loriciferans were found in much higher numbers than in other sediments.

This series of new discoveries in loriciferan species shows that the discovery of Loricifera is still in its beginning.

6.1. Morphology

Loriciferan specimens are usually collected from sediment samples fixed in formalin. Such fixation allows investigation of the external anatomy by light and scanning electron microscopy, while fixations needed to investigate the internal anatomy in detail or extract DNA cannot be applied. Such fixations have been applied to only a few specimens, therefore ultrastructural information is very restricted and rests on a few specimens that have been fixed in trialdehyde for transmission electron microscopy (see Heiner et al. 2009b). If not otherwise indicated, most information goes back to Kristensen (1991a).

External morphology

The trunk or abdominal region of the body in adult loriciferans is surrounded by a stiff cuticular case, the lorica. The anterior part of the body can be withdrawn into the lorica. It is divided into four regions. These are, from the front, the introvert including the mouth cone, the neck, the thorax and the loricated abdomen. The entire body is less than 300 μm long (Figs. 6.1.1, 6.1.2); however, all adults have more than 200 appendages (scalids) on the head (introvert).

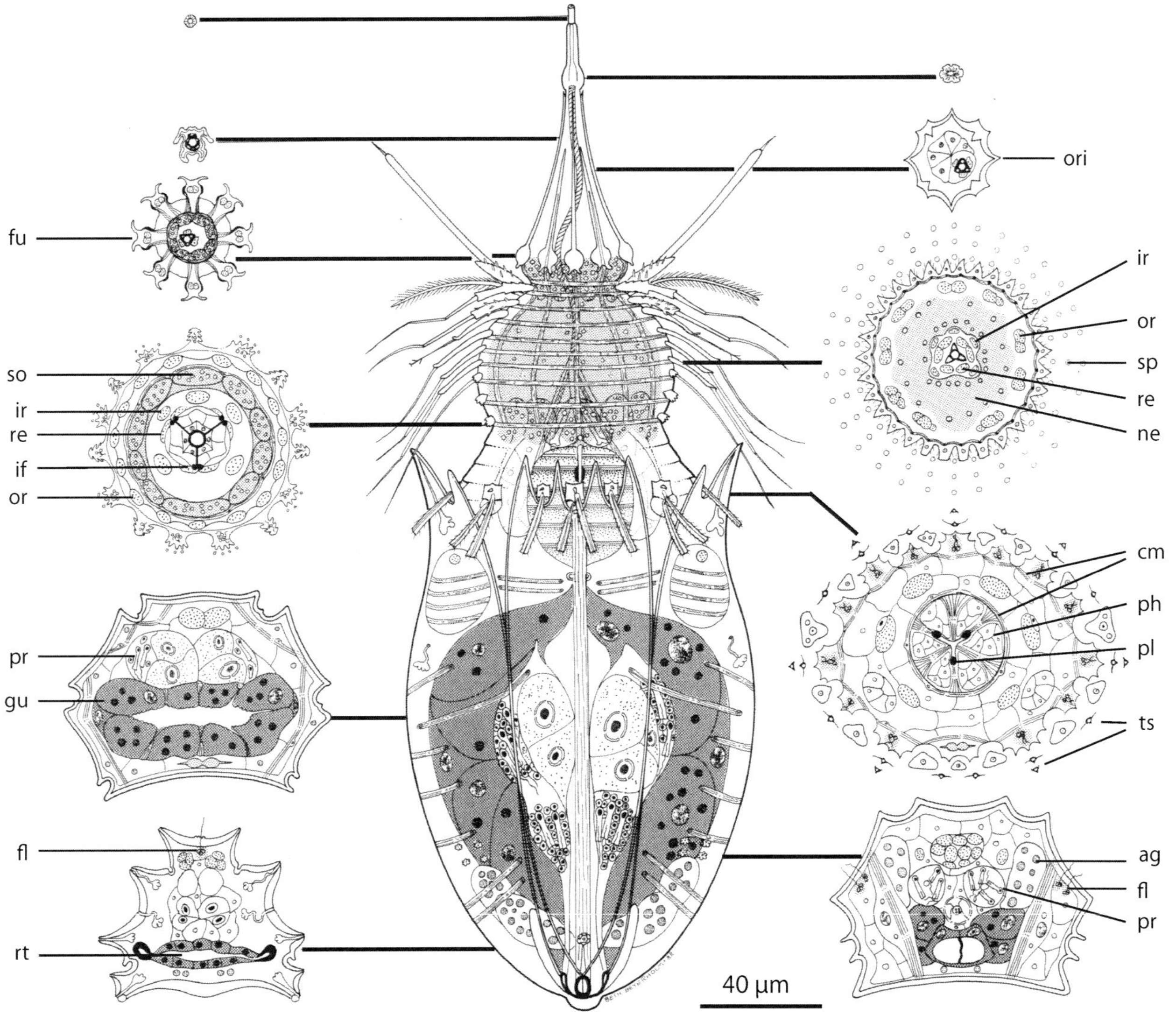

Fig. 6.1.1. Drawing based on SEM and TEM studies. The lateral cross sections are based on semi ultrathin sections (2 μm thick). *Nanaloricus mysticus*, dorsal view of female (modified from Kristensen 1991a). ag, adhesive gland; cm, circular muscle; fl, flosculi; fu, furca; gu, gut; if, internal furca; ir, inner head retractor; ne, neuropil; or, outer head retractor; ori, oral ridges; ph, pharynx; pl, placoids; pr, protonephridia; re, buccal tube retractor; rt, rectum; so, stomata; sp, spinoscalids; ts, trichoscalids.

The mouth cone is a more or less long cone, ending in the terminal mouth opening. Within the mouth opening there are maybe six (Nanaloricidae; Kristensen & Gad 2004) or four (Pliciloricidae) tiny eversible stylets (Gad 2005b,c). However, in the type species for loriciferans, *Nanaloricus mysticus*, the six stylets are lacking. The mouth cone is supported by cuticular ridges. In Nanaloricidae there are eight such oral ridges (called oral stylets in Kristensen 1983) (Figs. 6.1.1, 6.1.3 B, E), in Pliciloricidae there are 6–16 ridges. The oral ridges in Nanaloricidae are divided into two branches, in between is the furca (Fig. 6.1.3 B, E). These serve as attachment for the introvert retractor muscles. In Pliciloricidae, there is a so-called ruff at the base of the mouth cone, which is composed of a varying number of cuticular folds (Fig. 6.1.2).

The introvert carries the majority of appendages, the so-called scalids. Scalids are elongated cuticularized structures with locomotory and sensory function. The anterior scalids are directed anteriorly, with one exception being the posterior scalids, which are directed posteriorly. When the anterior end is inverted into the lorica, the scalids fold over and take up a position next to the mouth cone, pointing anteriorly. The scalids are arranged in eight to nine rings but their shape, number and position varies. Therefore, details will be mentioned below in the systematical section. The eight scalids in the first ring are club- or leaf-shaped and are called clavoscalids. Their orientation is in an anterior direction. Whereas in Pliciloricidae all eight clavoscalids have the same structure, in Nanaloricidae they exhibit sexual dimorphism.

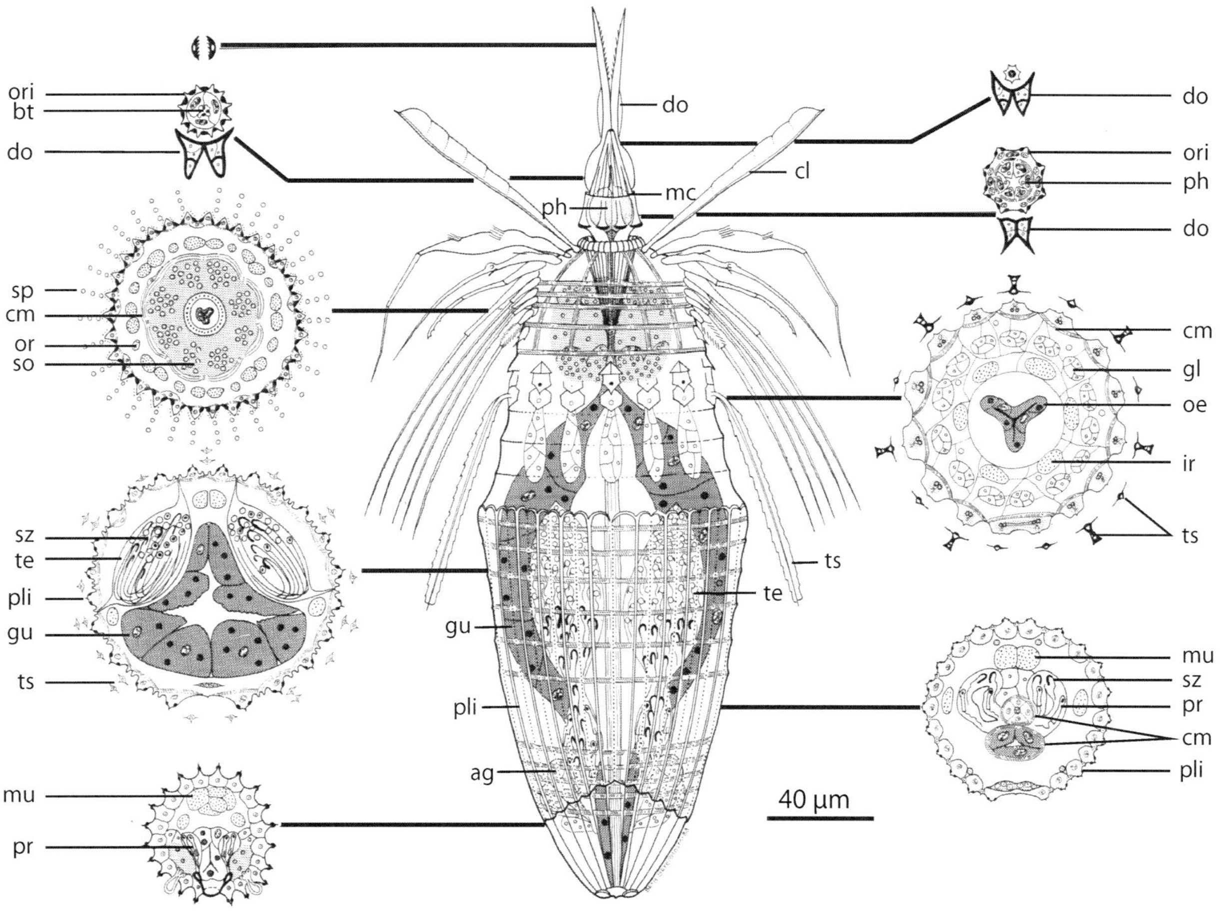

Fig. 6.1.2. Drawing based on SEM and TEM studies. The lateral cross sections are based on semi ultrathin sections (2 µm thick). *Pliciloricus enigmaticus*, dorsal view of male (Modified from Kristensen 1991a). ag, adhesive gland; bt, buccal tube; cm, circular muscle; do, double organ; gl, gland; gu, gut; ir, inner head retractor; mc, mouth cone; mu, muscle; oe, esophagus; or, outer head retractor; ori, oral ridges; ph, pharyngeal bulb; pli, plicae; pr, protonephridia; so, stomata; sp, spinoscalids; sz, spermatozoon; te, testis; ts, trichoscalids.

The two ventral clavoscalids are similar in males and females, but in males the remaining six are apically branched into three (giving in total 20 branches of clavoscalids), whereas they are unbranched in females. The scalids in the following rings are spine-like and are therefore called spinoscalids. In particular in the anterior rings they are usually strong and often segmented; this may hint at a main role in crawling locomotion in the sediment. In all species of Nanaloricidae the spinoscalids of rows 4 to 6 (sr4 to sr6) consist of 30 scalids, each with two small muscles. These scalids are involved in swimming or even in the jumping behavior seen in the two genera *Nanaloricus* and *Armorloricus* (R. M. Kristensen personal observation). The spinoscalids in the posterior rings are finer and may have a mainly sensory function. The scalids in the last ring usually also differ in structure and are short, thorn-like structures. In the genus *Pliciloricus*, the two ventral spinoscalids of the second ring are basally fused and form a strong, anteriorly directed structure, the double organ (Fig. 6.1.2, 6.1.3 F (do)) (Higgins & Kristensen 1986). The function may be to support the everting mouth cone (Kristensen 1991a) or help during molting (Gad 2005c). A secondary double organ is found in *Pliciloricus shukeri* (Heiner & Kristensen 2005) and *P. corvus* (Gad 2005b). The function of the secondary double organ is unknown.

Posterior to the introvert are two regions. The first region is called the neck and the second the thorax. In the neck, three rings of basal plates are present; the plates of the third ring have fine appendages, the trichoscalids. Among these, seven are paired and eight are single in all nanaloricids and the genus *Pliciloricus*. However, the genus *Rugiloricus* has 15 single trichoscalids. Trichoscalids have a finely serrated margin. In males of *Nanaloricus mysticus*, two lateroventral trichoscalids are modified into claw-like claspers, which are thought to function during copulation. New observations (R. M. Kristensen unpublished observations) show that the trichoscalids are involved in the jumping behavior of live adult nanaloricids. The thoracal cuticle is structured by longitudinal and

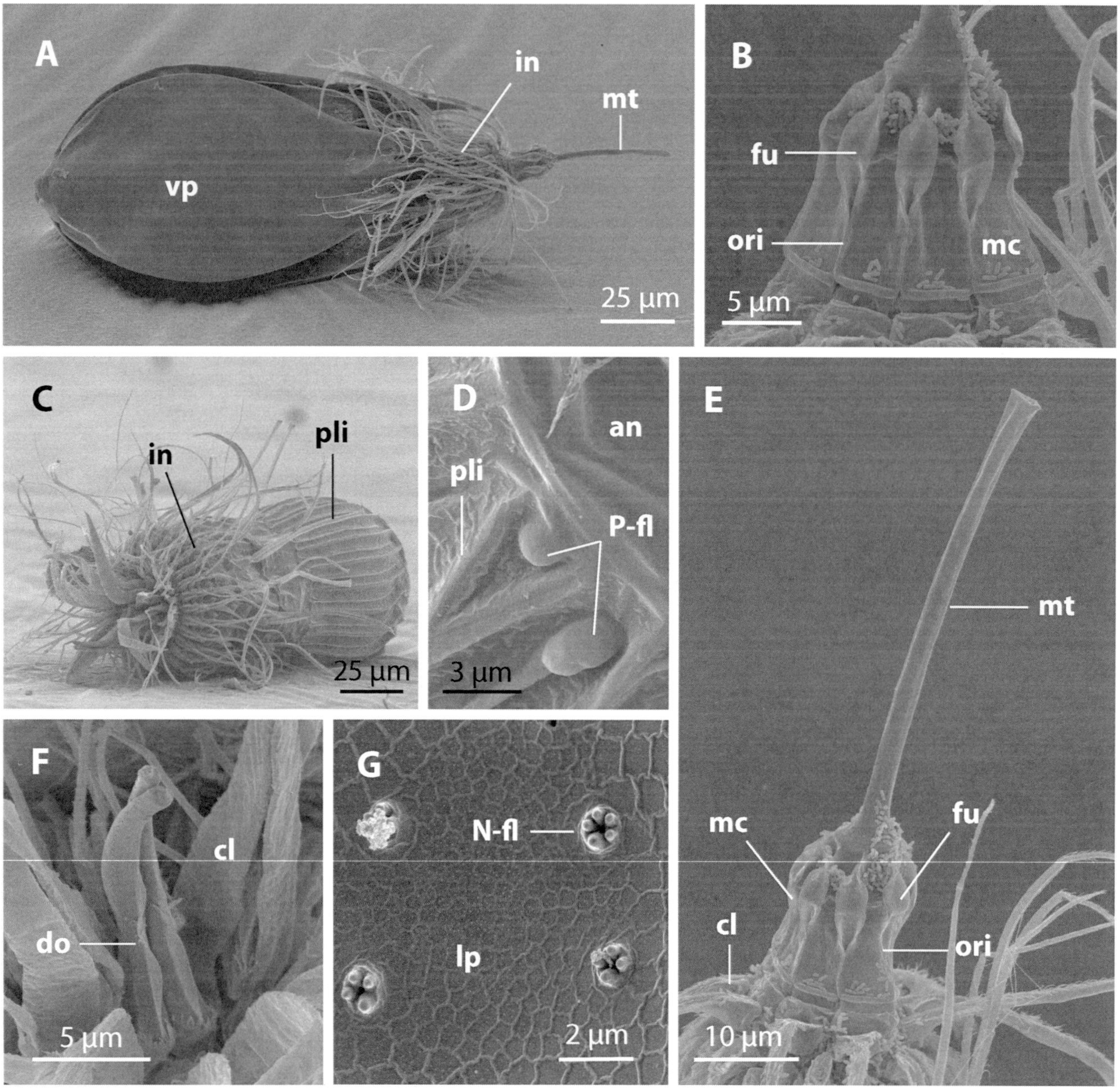

Fig. 6.1.3. SEM micrographs of adult loriciferans. A. *Armorloricus kristenseni*, overview; B, *A. kristenseni*, mouth cone; C, *Pliciloricus leocaudatus*, overview; D, *P. leocaudatus*, anal cone with P-flosculi; E, *A. kristenseni*, mouth tube; F, *P. leocaudatus*, double organ; G, undescribed *Nanaloricus* from Roscoff, N-flosculi. an, anal cone; cl, clavoscalids; do, double organ; fu, furca; in, introvert; lp, lateral plate, mc, mouth cone; mt, mouth tube; N-fl, N-flosculi; ori, oral ridges; P-fl, P-flosculi; pli, plicae; vp, ventral lorica plate.

transversal folds, which obviously are necessary during withdrawing of the introvert.

In Nanaloricidae, the neck region is elongated and longer than the thorax. It is in particular the area between the first row and second row of basal plates that has been elongated with a thin cuticular region. In Nanaloricidae the thorax region is short and almost hidden by the lorica plates. In Pliciloricidae the two regions are of somewhat similar length and here the basal plates are positioned close to each other.

The lorica differs in structure between Nanaloricidae and Pliciloricidae. In the genera *Nanaloricus* and *Armorloricus* it is composed of six large plates, whereas there are eight lorica plates in *Spinoloricus* and ten in *Phoeniciloricus*. On the anterior margins there are 15 spines with gland openings (probably a defense secretion, Kristensen 1991a). Numerous glands open on the lorica plates, the majority in the marginal regions, with tiny pores. The surface structure is a fine honeycomb pattern in *Nanaloricus*, *Spinoloricus* and *Phoeniciloricus* (Fig. 6.1.3 G). *Armorloricus* has a totally smooth cuticular surface (Fig. 6.1.3 A). At the posterior end, on the ventral side, is the anal cone with the cloacal opening at its tip. In Pliciloricidae, the lorica is structured by several longitudinal ridges into a number of plates named plicae; there are 22 plicae in *Pliciloricus* and 30–60 plicae in *Rugiloricus* (Fig. 6.1.2, 6.1.3 C, D). Further fine median ridges can additionally be present on the plicae. In *Pliciloricus* some ventral plates are fused into one larger plate. At the posterior end an anal field composed of several anal plates is separated from the remaining lorica. Terminally it bears the anal cone with the cloacal opening. The surface structure is either a honeycomb pattern or a papillar fine structure. Flosculi, special sensory structures with a single cilium surrounded by four to nine microvilli are present in several genera. Flosculi have a high systematic value in distinguishing species in the different genera.

Loriciferans have a fascinating diversity of larval types. Common to all are the Higgins-larvae. All further larval types are mentioned below under life cycle; for their exact description see the primary literature.

All Higgins-larvae have a broad, conical mouth cone attaching with a broad base to the introvert (Fig. 6.1.4, 6.1.5 A). The introvert and neck together form a continuous region, not well separated from each other. The scalids are shorter as in adults and about half in number. Maximally seven rows are present. The first row of scalids contains clavoscalids as in adults, but these are not as distinct as in adults.

The families Nanaloricidae and Pliciloricidae differ in the way they can close their lorica when the introvert is withdrawn. In contrast to adults, where the long scalids stick out of the animal when the introvert is withdrawn, the entire introvert can be completely withdrawn into the abdomen of the Higgins-larvae. In Nanaloricidae, a pair of ventral plates serves as a closing apparatus when the introvert is withdrawn. In Pliciloricidae and Urnaloricidae, the posterior part of the neck is a ruff or collar (especially well developed in *Pliciloricus*), function as a closing apparatus when the introvert is withdrawn (Fig. 6.1.5 E).

The thorax has five to six rows of small plates and no appendages. The abdominal lorica is structured by a number of longitudinal folds. Three pairs (Nanaloricidae) or two pairs (Pliciloricidae) of spines or setae are present in the anterior, ventral region of lorica (Fig. 6.1.5 C). The setae in Nanaloricidae are positioned close together on a plate. Very conspicuous in the Higgins-larvae is one pair of large toes in the posterior part of the trunk. In Nanaloricidae it has a leaf-like structure made by lateral mucrones (Fig. 6.1.5 A, B). High mobility is very likely, probably a propeller-like movement (Kristensen 1983).

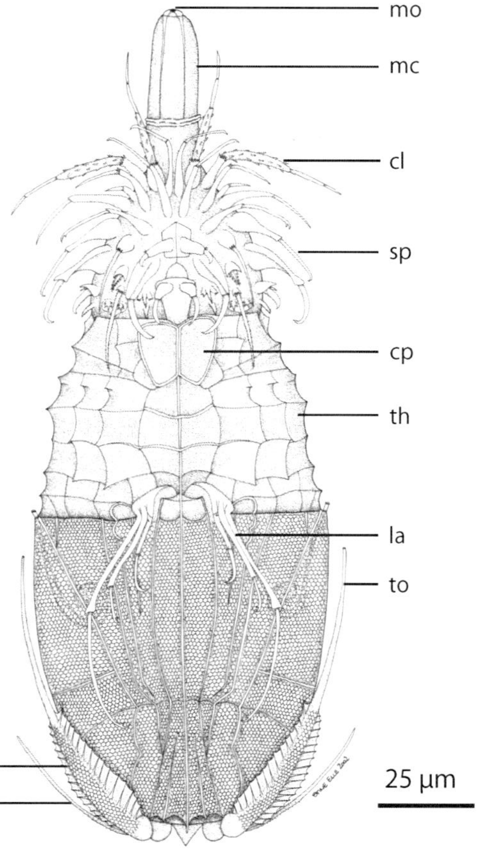

Fig. 6.1.4. Drawing of *Nanaloricus gwenae*, ventral view of Higgins-larva (modified from Kristensen et al. 2007). cl, clavoscalids; cp, closing plate; la, anterior ventral setae; mc, mouth cone; mo, mouth opening; mu, mucrones; se, posterior setae; sp, spinoscalids; th, thorax; to, toes.

However, the toes have also a function in the sediment, seen in crawling movements or when the larva attaches to the sand grains. In Pliciloricidae the toes are either long and slender with a spiny tip as in *Pliciloricus*, or they have a broad base, a smaller midpiece and a pointed tip as in *Rugiloricus*. All toes in loriciferans have toe glands; however, this may not have the dual gland function (Hermans 1983). Two to three pairs of sensory setae or modified spines are found in the terminal part of the lorica as well as the flosculi (Fig 6.1.5 A, B). In nanaloricid larvae two lateral flosculi and a single dorsal flosculum are present.

Integument

The integument of Loricifera consists of a cellular epidermis resting on a basal lamina and a cuticle. The epidermal cells do not interdigitate extensively (Kristensen 1991a). They rest on a homogeneous basal lamina, which is about 10 nm thick (Kristensen 1991a). Cell–cell contacts are insufficiently known. In the pharyngeal bulb,

Fig. 6.1.5. SEM micrographs of larval loriciferans. A, *Armorloricus elegans*, Higgins-larva overview; B, undescribed *Nanaloricus* from Roscoff, Higgins-larva close-up of toes; C, undescribed *Nanaloricus* from Roscoff, Higgins-larva, close-up of anterior ventral setae; D, *Urnaloricus gadi*, mega-larva, overview; E, *U. gadi* mega-larva, close-up of introvert; F, *U. gadi* mega-larva, close-up of gonopore. ab, abdomen; co, collar; go, gonopore; in, introvert; la, anterior ventral setae; mc, mouth cone; mu, mucrones; se, posterior setae; sp, spinoscalids; th, thorax; to, toes.

myoepithelial cells directly contact the cuticle, and these cells also form the five rows of placoids. Where mesodermal muscle cells attach to the cuticle, epidermal cells become very thin and mediate muscular force to the cuticle via tonofilaments (Neuhaus et al. 1997).

The cuticle covers the entire body, including the anterior and the posterior part of the intestinal system. In the trunk region it forms the plates of the lorica; these are thickened regions of the cuticle. The cuticle is composed of three layers, which are named here, from distal to proximal, epi-, exo- and endocuticle. These terms were used in order to compare cuticular layers among ecdysozoan taxa (Schmidt-Rhaesa et al. 1998), thus avoiding a confusing number of different names for comparable layers. This also applies to Loricifera, where different terms have been used for cuticular layers. The three cuticular layers are present in most parts of the cuticle, but show specific or additional features in different regions.

The cuticle in representatives of Nanaloricidae appears to be generally thicker than in representatives of Pliciloricidae. Maximal values of 6.7 µm have been measured in an undescribed nanaloricid species (however, this undescribed species has later been described as *Armorloricus elegans* (see Kristensen & Gad 2004)), and it ranges between 620 nm and this maximal value (Neuhaus et al. 1997). In Pliciloricidae, the cuticle is 170–350 nm thick (Neuhaus et al. 1997). In larvae the cuticle is distinctly thinner compared to adults, measuring less than 1 µm in *Nanaloricus mysticus* (Kristensen 1991a) and 60–90 nm in *Rugiloricus* cf. *carolinensis* (Neuhaus et al. 1997). It nevertheless contains the same three layers.

The distal epicuticle is a thin, trilaminate layer (Neuhaus et al. 1997). The trilaminate character cannot be observed in every specimen (e.g. in *Nanaloricus mysticus*; Neuhaus et al. 1997), but this may be due to unfavorable sectioning; this particular structure can only be revealed in sections exactly perpendicular to the cuticle.

The exocuticle ("intracuticle" in Kristensen 1991a, "amorphous layer" in Neuhaus et al. 1997) appears as one homogeneous layer in *Pliciloricus enigmaticus* (Neuhaus et al. 1997), but is subdivided into two or three layers in Nanaloricidae. In *Nanaloricus mysticus*, there is a thick, homogeneous, moderately electron-dense main layer below the epicuticle and a proximal thinner and more electron-dense layer. This layer is very thin in most regions, but is thickened into ridges in some regions. In *Armorloricus elegans* (the undescribed species of Nanaloricidae in Neuhaus et al. 1997), a distal electron-dense layer, a middle electron-light layer and a proximal electron-dense layer are present, at least in some body regions. Some micrographs have shown this layer to be composed of what appears to be stacks of lamellae (Neuhaus et al. 1997). The exocuticle contains further structures in particular regions. Where lorica plates attach, only the distal layer of the exocuticle remains present; between this layer and the proximal endocuticle is a thin electron-dense plate (Neuhaus et al. 1997). Between the exo- and endocuticle, next to these plates and in lorical spines are accumulations of large fibers (Neuhaus et al. 1997).

The endocuticle includes fine fibers in all species investigated so far. In *Nanaloricus mysticus*, fibers run in the same direction in most parts of the endocuticle, but have a random orientation in the distal part. In *Armorloricus elegans*, fibers appear less ordered than in *N. mysticus*. The endocuticle contains chitin, as shown by staining with gold-labeled wheat germ agglutinin (Neuhaus et al. 1997). For measurements of the cuticular layers see Neuhaus et al. (1997).

Within the pharynx of *Nanaloricus mysticus* only the epi- and the exocuticle could be detected (Neuhaus et al. 1997). The exocuticle is composed of only one moderately electron-dense layer, which is thickened in the edges of the triradiate pharyngeal lumen. Here it contains numerous electron-dense structures interpreted as fibers. Such fibers run parallel to the longitudinal axis of the pharynx at the edges, but extend into the exocuticle between the edges in an orientation perpendicular to the longitudinal axis (Neuhaus et al. 1997).

Musculature

Loriciferans have a complex musculature. Ultrastructural descriptions by TEM are available from *Nanaloricus mysticus* and *Pliciloricus enigmaticus* (Kristensen 1991a, all information below is from this source). Muscular staining with phalloidin has been attempted, but so far has not yielded satisfactory results (Heiner, Rothe & Schmidt-Rhaesa, unpublished results). Stomatogastric muscles are described under Intestinal system.

Below the epidermis, isolated bundles of longitudinal and circular muscles are present. In *N. mysticus*, these muscles are best developed in the introvert, where they form ten rings of circular muscles and several longitudinal bundles (Fig. 6.1.1). The circular musculature is distal of the longitudinal one. The circular muscles correspond in position to the rings of scalids and attach to the base of each scalid. The cuticle is thickened in these attachment sites. It can be assumed that the circular musculature therefore is involved in scalid mobility. The longitudinal muscles can be observed running from the base of the mouth cone to the base of the introvert, but their exact attachment sites are unknown. Circular muscles are also present in the neck, but in the trunk they disintegrate into isolated muscles connecting the lorica plates. In *P. enigmaticus*, a grid of circular and longitudinal muscles occurs throughout the entire body (with the exception of the mouth cone). In the trunk, there are 11 rings of circular musculature and 20–22 longitudinal muscles (Fig. 6.1.2).

Several longitudinally arranged muscles serve as retractors. There are eight longitudinal muscles in the enlarged base of the oral ridges of the oral cone in *N. mysticus*, probably acting as mouth cone retractors. Within the introvert, three groups of retractors are reported in this species. These are, from proximal to distal: six buccal tube retractors, five inner head retractors and 15–24 outer head retractors (Figs. 6.1.6 A, 6.1.7 A, C). These muscles

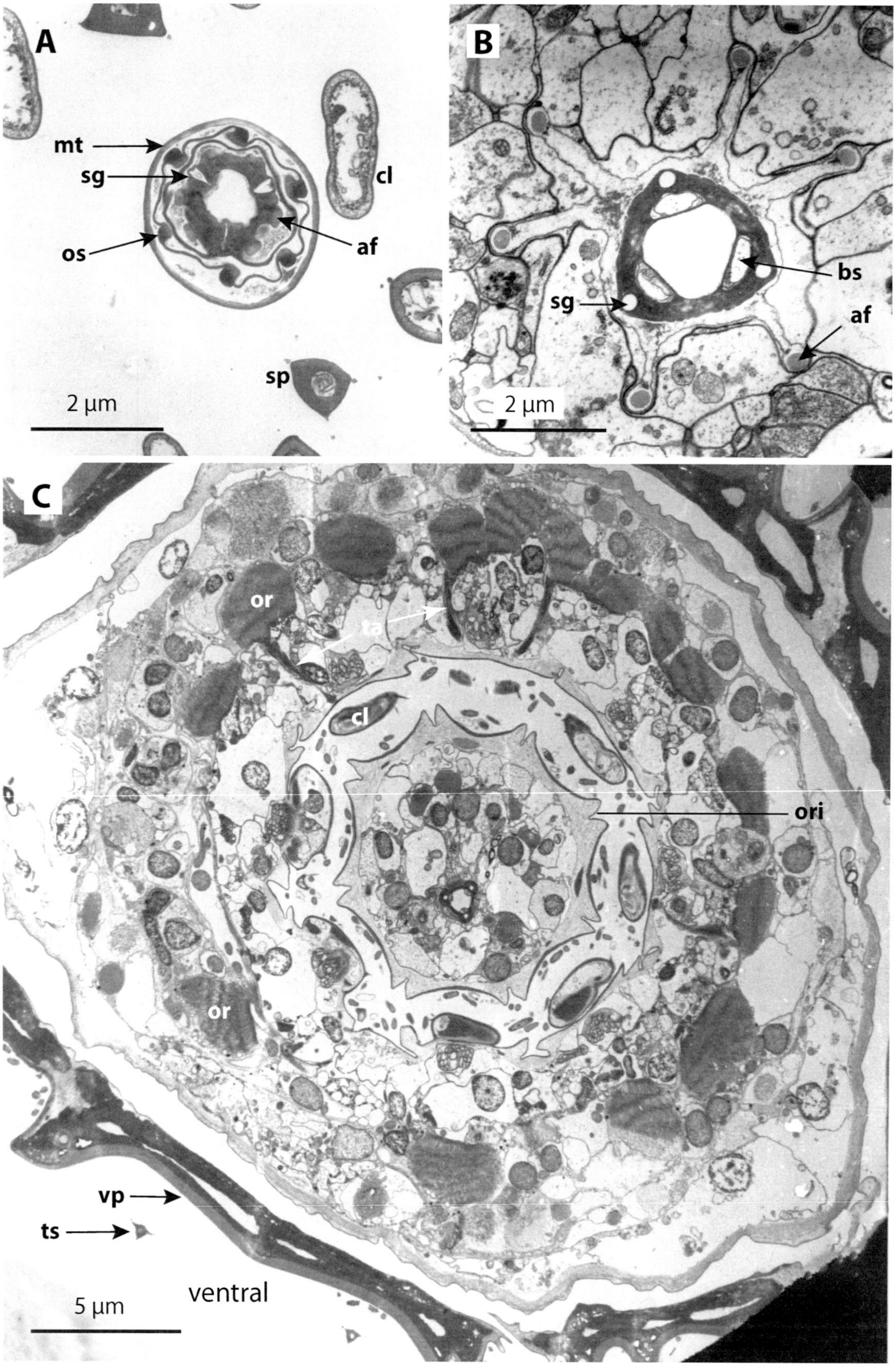

Fig. 6.1.6. TEM micrographs of *Armorloricus elegans*. A, mouth tube with oral stylets; B, buccal tube with attachment fibers; C, tanocyte attachments to the base of the clavoscalids. af, attachment fibers; bs, buccal sensory organ; cl, clavoscalids; mt, mouth tube; or, outer head retractor; ori, oral ridges; os, oral stylets; sg, salivary glands, sp, spinoscalids, ta, tanocytes; ts, trichoscalids; vp, ventral lorica plate.

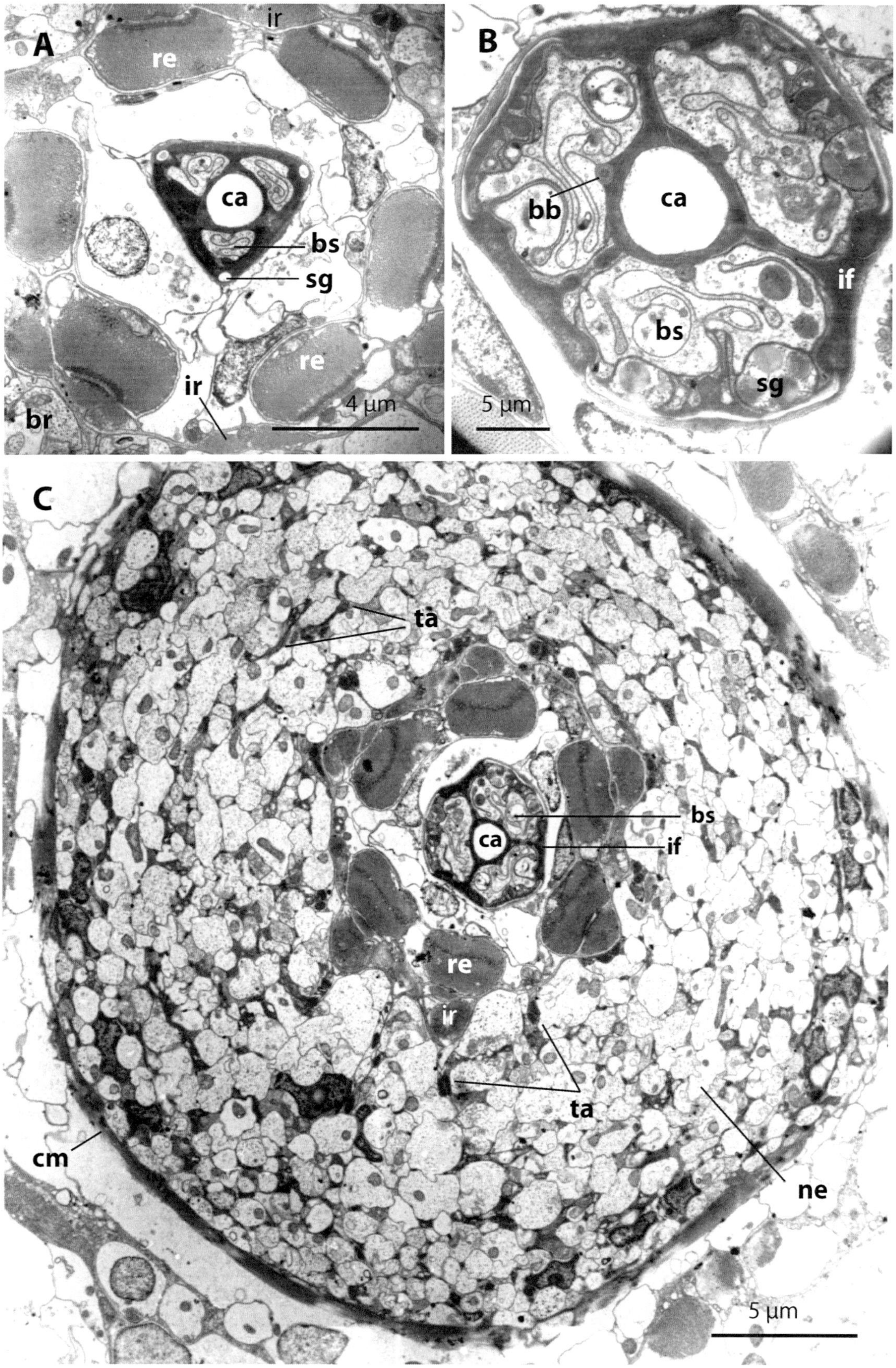

Fig. 6.1.7. TEM micrographs of *Armorloricus elegans*. A, buccal tube and the buccal tube retractors; B, buccal tube with internal furca and basal body (close up of C); C, cross section of the neuropil part of the brain. br, brain; bs, buccal sensory organ; ca, buccal canal; cm, circular muscle; if, internal furca, ir, inner head retractor; ne, neuropil; re, buccal tube retractor; sg, salivary glands; ta, tanocytes.

have been identified from TEM cross sections and their exact attachment sites are not in all cases certain; however, the six buccal tube retractors have very long cuticular attachment fibers running inside a cuticular casing close to the buccal tube. Within the trunk one unpaired dorsal abdominal muscle is present; this appears to split into two portions in its posteriormost part. *Pliciloricus enigmaticus* differs from *N. mysticus*. Longitudinal oral cone muscles and buccal retractors have not been described. The number of figured inner head retractors is six, and that of outer head retractors is 24. In the trunk, the dorsal abdominal muscle is also present, although it is not distinctly split in the posterior part. Additionally, a pair of lateral longitudinal muscles is present in the trunk.

Further muscles occur within the spinoscalids. Usually there is a pair of tiny muscles, but in the enlarged pair of spinoscalids in *P. enigmaticus*, up to five muscles are present. The muscles are present close to the scalid base and it appears plausible that they move the scalid up and down. Furthermore, special muscles have been reported in the mouth cone of *N. mysticus*. Apart from the mouth cone retractors, each of the eight bases of the oral ridges contains a pair of radial muscles running proximally to the anterior tip of the brain. The region of the brain serves as attachment for the inner and outer head retractors as well as, in *P. enigmaticus*, for circular muscles surrounding the brain. Attachment is made possible by two cells (called tanocytes) containing filaments, which are integrated into the brain and connect to the muscle cells by hemidesmosomes (Fig. 6.1.6 C, 6.1.7 C).

As far as current observations show, loriciferan muscles are either cross or obliquely striated. Cross striation has been reported for scalid and pharyngeal muscles, while the longitudinal retractor muscles are obliquely striated.

Nervous system

The nervous system has been described from TEM sections in *Nanaloricus mysticus* and *Pliciloricus enigmaticus* (Kristensen 1991a). Immunohistochemical staining of nervous system components has to date not been performed.

The brain of Loricifera is very large. It is located in the introvert and may take up the majority of space there. The brain surrounds the buccal tube in *N. mysticus* or the esophagus in *P. enigmaticus* (the pharynx is anterior to the brain, in the mouth cone, in Pliciloricidae). It has a tripartite structure and contains, in anteroposterior sequence, a region containing nuclei, a neuropil and again a region of nuclei (Fig. 6.1.7 C). These three regions have been termed fore-, mid- and hindbrain by Kristensen (1991a), but as it is more likely that they reflect a morphological separation of somata and neuropil instead of functionally different regions, it appears more appropriate not to use such terms. In comparison with nematodes, priapulids and kinorhynchs it is clear now that this structure reflects a particular architecture called the cycloneuralian brain. The neuropil forms a closed ring, the somata attach to the entire ring but form distinct clusters (lobes). Anterior of the neuropil, eight lobes of somata are present, posterior of the neuropil there are ten lobes. In cross section of the neuropil more 500 nerve fibers and a few nuclei can be observed (Fig. 6.1.7 C). The tanocytes can also be seen as small dark fibers in the middle of the brain in the same cross section (Fig. 6.1.7 C). The regions of somata also contain the tanocytes (described under 2.3).

Neurites or neurite bundles run from the brain in anterior and posterior direction. The brain innervates all scalids as well as the mouth cone. Scalid innervation may come from different regions of the brain. Most scalids appear to be innervated from anteriorly oriented projections of the brain. Kristensen (1983) reported that innervation of the clavoscalids in *N. mysticus* comes from eight ganglia that are almost fused with the brain (Fig. 6.1.6 C). These eight ganglia appear to be the eight lobes of somata of the brain. The trichoscalids are reported to be innervated by a thorax ganglion (Kristensen 1983). In a posterior direction, ten longitudinal neurite bundles have been observed, of which the two ventral ones are thicker than the remaining ones (Kristensen 1991a,b). Several thickenings called ganglia appear to be present along these ventral longitudinal nerve cords. First, a subpharyngeal ganglion has been reported by Kristensen (1991a,b). It is evident in *P. enigmaticus*, but appears to be fused with the posterior region of the brain in *N. mysticus* (Kristensen 1991a). Then, a thorax ganglion (innervating the trichoscalids) is mentioned by Kristensen (1983, 1991b), but not by Kristensen (1991a). All these descriptions mention a caudal ganglion (= rectal ganglion in Kristensen 1991b). The presence of further pairs of ganglia along the ventral nerve cords is suspected by Kristensen (1991a). The exact connection between all these components as well as the presence of stomatogastric innervation is unknown to date.

Very few data in Kristensen (1983) describe the nervous system of the Higgins-larva. The brain in the introvert is already very large. Two additional ganglia are present, one innervating the locomotory spines and one posterior caudal ganglia innervating structures such as the toes and sensory structures.

An extracellular matrix is not always present around the brain; however, in *Armorloricus elegans,* a thick matrix is found separating the pharyngeal bulb and the brain and a thin matrix separating the brain from the mesodermal derivates (Kristensen 1991a). The position of the brain therefore is intraepithelial.

Sensory structures

The most prominent sensory structures are the scalids on the introvert of all life stages. Further sensory structures are so-called flosculi, larval setae and buccal sensory structures (see under Intestinal system).

All scalids appear to be sensory in function, because at least a cilium as part of a receptor cell has been shown in ultrastructural investigations. The anterior clavoscalids are hollow, up to 20 neurites enter each clavoscalid and end in cilia, which branch several times. One larger or several smaller pores are present at the tip of the clavoscalids. It is likely that they are chemoreceptive. The spinoscalids in the second and third ring possess three to eight receptive cilia, the other spinoscalids a single cilium. The trichoscalids are again hollow and contain one cilium.

Flosculi are elevated sensory structures. In Loricifera, three types are distinguished, the N-flosculi (Fig. 6.1.3 G), the P-flosculi (Fig. 6.1.3 D) and micro-flosculi. The papilate N-flosculi (nanaloricid-flosculi) correspond to flosculi present in kinorhynchs and priapulids. They are composed of one receptor cell with a sensory cilium surrounded by a circle of four to nine microvilli (also known as the collar receptor) resembling a flower (Fig. 6.1.3 G). In adult nanaloricids three to four lateral flosculi are present on each dorsolateral plate of the lorica. The middorsal plate may have one to two flosculi. *Armorloricus davidi* has additionally two pairs of flosculi on the midventral plate of the lorica. N-flosculi are also seen in Higgins-larvae. In *Armorloricus kristenseni* the N-flosculi are reduced to micro-flosculi, which lack the flower-shaped microvilli but have a central cilum as seen in the N-flosculi (Heiner 2004). The P-flosculi (plicilorid-flosculi) are quite different sense organs and found only in adults of Pliciloricidae. The P-flosculum consists of a small, round papilla covered by a very thin cuticle. A depression has been seen in the middle of the P-flosculum, both by DIC-optics and SEM examinations; however, a real pore could not be observed (Fig. 6.1.3 D).

Higgins-larvae have different types of setae with sensory function. In the posterior part, three pairs of setae are present; one pair is called anal setae and two pairs are called flagelliform setae (Fig. 6.1.5 B). The anal setae are stiff, hollow structures with a secretory and a receptor cell. They therefore seem to have both secretory and sensory functions. The flagelliform setae have only a receptor cell and are probably only sensory in function. Further setae are present in the trunk region; these differ between Nanaloricidae and Pliciloricidae. In pliciloricid larvae, two pairs of lateroventral setae including a receptor cell are present. In nanaloricid larvae, three pairs of midventral setae (called spines) are present. Two of these pairs include receptor cells, the third pair seems to lack the receptor cell and this spine is modified to a claw-shaped structure.

Photoreceptors have so far only been found in Higgins-larvae of one single species, *Australoricus oculata* (Heiner et al. 2009a).

Intestinal system

The intestinal system consists of a buccal tube, pharynx, esophagus, midgut and rectum (= hindgut). The fine structure is best investigated for *Nanaloricus mysticus* (see Kristensen 1983, 1991a,b). Differences between Nanaloricidae and Pliciloricidae seem to consist mainly in the structure of the buccal tube, which is simpler in Pliciloricidae (Kristensen 1991a).

The buccal tube consists of a long cuticularized tube connecting the mouth opening with the pharynx. Special buccal tube retractors make the anterior part of the buccal tube able to telescope in and out of the mouth cone. Such movement causes the fine buccal teeth, which are cuticular structures in the distal part of the buccal tube, to protrude from the mouth opening. Three salivary glands open in the distal part of the buccal tube (Fig. 6.1.7 A, B). In the middle and posterior regions of the buccal tube the cuticle is quite thick and includes three gland ducts of the salivary glands as well as three structures including particular fibers; these are interpreted as attachment structures of the buccal tube retractors (Fig. 6.1.6 A, B). Furthermore, three sensory structures consisting of a pair of modified cilia are present in the buccal tube (Fig. 6.1.7 A, B). The lumen of the buccal tube is triangular in the anterior region and hexagonal in the posterior region. The pharynx is composed of monosarcomeral and duosarcomeral muscle cells attaching directly to the pharyngeal cuticle (myoepithelial condition). Gland cells are also present in the pharyngeal bulb (*Armorloricus elegans*, Fig. 6.1.8 C). The cuticle has three rows of thickened elements called placoids. Five placoids are present per row. The pharyngeal lumen is triradiate (Fig. 6.1.8 C), the placoids are present in the distal part of the three radii. Contraction

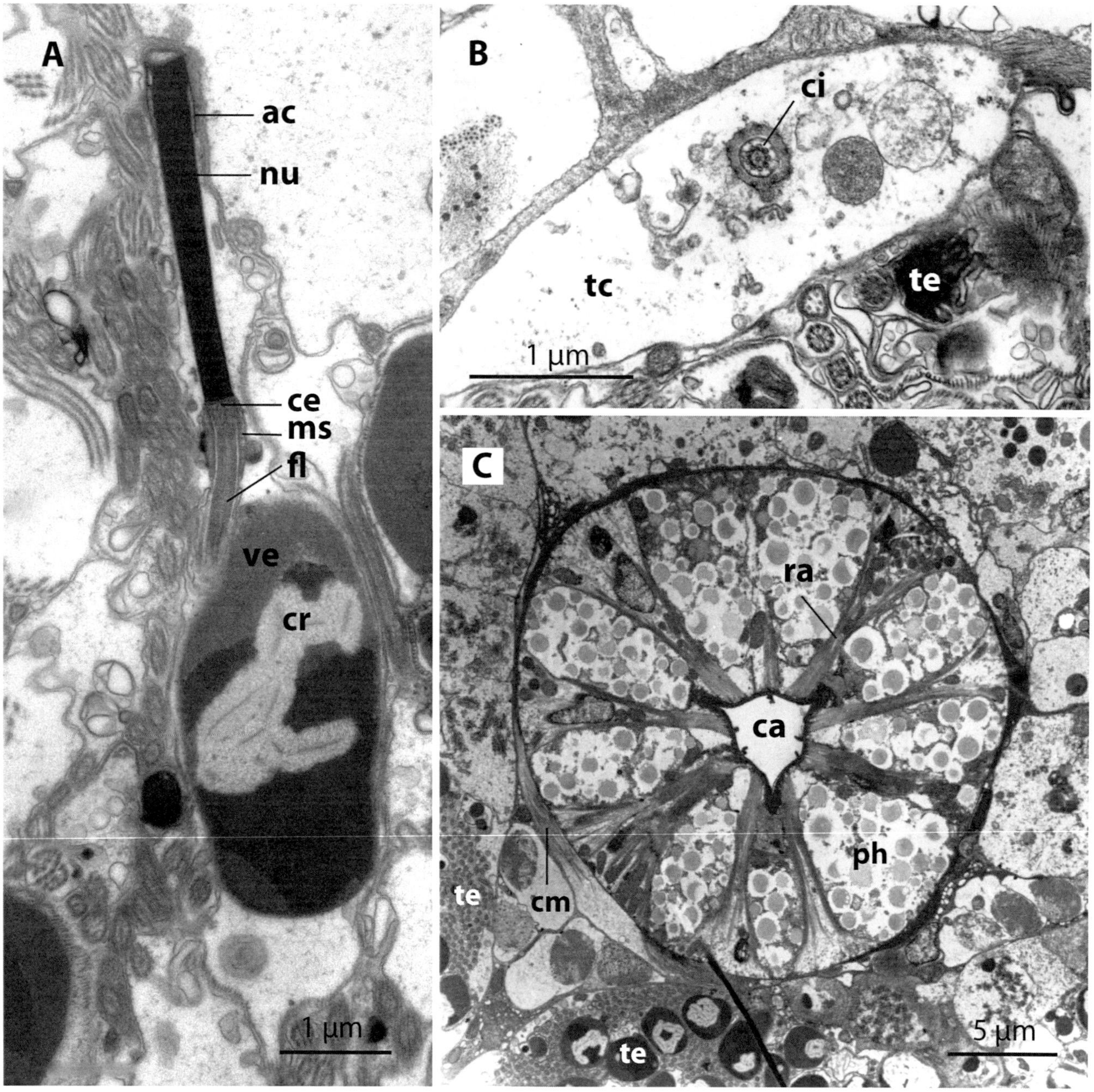

Fig. 6.1.8. TEM micrographs of *Armorloricus elegans*. A, early spermatozoon (modified figure from Heiner et al, 2009b); B, protonephridia; C, cross section of pharyngeal bulb. ac, acrosomal-like cap; ca, buccal canal; ce, centriolar adjunct; ci, cilium; cm, circular muscle; cr, crystal-like structure; fl, flagellum; ms, mitochondrial sheet; nu, nucleus; ph, pharyngeal bulb; ra, radial muscle; tc, terminal cell; te, testis; ve, vesicle.

of the radial pharynx musculature leads to a quick formation of a triangular or even almost circular pharynx lumen. By this action, particles or liquid can be sucked into the pharynx. Such a construction is present in several animals (e.g. nematodes, tardigrades, gastrotrichs and others).

The pharynx is followed by a short cuticularized esophagus leading into the midgut. The midgut consists of large cells with probable absorptive function. Towards the midgut lumen they have microvilli but no cilia. The rectum is a short, cuticularized region leading to the anus.

Body cavities

A body cavity is absent (acoelomate) or only very small in all nanaloricids (Kristensen 2002) both in the larvae and the adults. In pliciloricids the body cavity seems to be larger (Kristensen 1991a). Kristensen (2002) states that in

Pliciloricus and *Rugiloricus* the condition may be acoelomate in the larvae, and pseudocoelomate as adults (Kristensen 1995).

Excretory system

Loriciferans possess a pair of protonephridia, which are located inside the gonads (Fig. 6.1.1, 6.1.2). After some initial ultrastructural data in Kristensen (1991a), Neuhaus & Kristensen (2007) described the protonephridia of three species in more detail. Species investigated were larval *Rugiloricus* cf. *cauliculus*, male adult *Armorloricus elegans* and adult female *Nanaloricus mysticus*, all information is from Neuhaus & Kristensen (2007).

The number of terminal cells in the protonephridium differs among species. In the adult of *A. elegans*, two times two terminal cells are present per gonad, in the larval *R.* cf. *cauliculus* there are four in total and in *N. mysticus* there are more (five to six were counted on one side, eight on the other side). Terminal cells in *A. elegans* each have an individual filter, in *R.* cf. *cauliculus* no filter was observed. *Nanaloricus mysticus* is more complicated in this respect, it was observed that two terminal cells have an individual filter, whereas the other terminal cells form a shared filter. All terminal cells are monociliate; the cilium is surrounded by a circle of nine microvilli, which are connected by extracellular material (Fig. 6.1.8 B). In *A. elegans* the cilium has the "usual" pattern of nine peripheral doublets of microtubuli and two central ones (9×2+2), in *R.* cf. *cauliculus* the central microtubules are lacking (9×2+0). Two canal cells are present in *A. elegans*, four in *R.* cf. *cauliculus*. Their number in *N. mysticus* could not be counted. No canal cells have cilia, but diplosomes may be present. In the male *A. elegans*, there are two sets of protonephridia located inside the gonads (Heiner et al. 2009b). In the first set the protonephridial duct leads with two nephridiopore cells into the lumen of the vasa deferentia and the second set leads into the lumen of the seminal vesicle (Heiner et al. 2009b). In both other species the opening of the protonephridial canal could not be observed.

Reproductive organs

Loriciferans are dioecious. Gonads are paired and dorsal. Ovaries may develop asymmetrically, as a single oocyte matures at a time (Gad 2004). Each gonad has an epithelium, which disintegrates in mature animals. Epidermal cells then become separated by nurse cells (Kirstensen 1991a). Protonephridia in the males of *A. elegans* are found inside the gonads (see above) and together form a recto-urogenital system where sperm, digestive and excretory products are emitted (Heiner et al. 2009b).

In males of *Armorloricus elegans*, paired testes are connected by short seminal ducts (vasa deferentia) to seminal vesicles (Heiner et al. 2009b). The seminal vesicles unite with the hindgut to form a short recto-urogenital canal, which is surrounded by a sphincter. Posterior of the sphincter, a pair of ducts separates and leads to latero-terminal gonopores, whereas the rectum leads to a temporarily closed anus.

In few species, hermaphroditism occurs (see Gad 2005e, Heiner & Neuhaus 2010 where undescribed species are mentioned).

Gametes

Spermiogenesis and early spermatozoa were described from *Armorloricus elegans* in detail by Heiner et al. (2009b), but comparative data from other species is rare. In *A. elegans*, the early spermatozoon consists of an elongate, rod-shaped nucleus, an acrosome, a cilium (=flagellum; with a 9×2+2 pattern of microtubules), a small mitochondrial sheath around the base of the cilium and a large vesicle next to the posterior half of the nucleus and the anterior part of the cilium (Fig. 6.1.8 A). During five spermatid stages it could be observed that the nucleus develops from a round, less electron-dense nucleus, which then elongates and becomes more osmiophilic. The vesicle develops in the second spermatid stage and originates from the Golgi complex. The acrosomal structure seems not to originate from the Golgi complex. Initially, two mitochondria are present, which then fuse into one mitochondrial sheath surrounding the ciliary base. The cilium can be observed from the second spermatid stage onwards.

There is a maturation gradient within the testes from its posterior periphery to the anterior periphery and then in a central direction (Heiner et al. 2009b). Most of the spermiogenesis takes place in the testes; only the maturation from late spermatids to spermatozoa takes place in the vasa deferentia and seminal vesicles.

Light microscopical investigations make it likely that the spermatozoon in *Nanaloricus mysticus* has the same structure as the spermatozoon of *A. elegans* (Heiner et al. 2009b), although spermiogenesis may be different. A very few ultrastructural observations in *Pliciloricus enigmaticus* suggest that the spermatozoon looks different and has a round nucleus and no large vesicle (Heiner et al. 2009b).

6.2. Reproduction and development

Reproductive biology

Most loriciferans are dioecious, but a few hermaphroditic species are known. In an undescribed *Rugiloricus* species the presence of an ovary and a testis is noted (Gad 2005e), and Heiner & Neuhaus (2010) also mention the previous species as well as another undescribed hermaphroditic species.

Although copulation has never been observed, it is likely that loriciferans reproduce by direct sperm transfer and internal fertilization. The claspers in males of Nanaloricidae are most likely structures used during reproduction, presumably to hold on to a female. Kristensen (1983) described "penile spines", but because such spines are present in both sexes (Kristensen 1991a), it is less likely that they are used for sperm transfer. Also the modified type of spermatozoon and the maturation of single oocytes are good indicators for internal fertilization. One undescribed species of the genus *Nanaloricus* from Roscoff clearly has two seminal receptacles filled with mature spermatozoa.

Cleavage and development

In most species, eggs are released into the sediment. Such free eggs are not documented. The few observations on early development can be made in those cases, in which the eggs develop within the shelter of a life cycle stage. This is common in several species of *Rugiloricus*, *Pliciloricus* and the family Urnaloricidae. In *Urnaloricus gadi*, Heiner & Kristensen (2009) observed total radial cleavage followed by the formation of a blastocoel. No further data are known. In *Pliciloricus pedicularis*, in the eggs inside the ghost-larva (named the simplified adult by Gad, 2005c) an unequal holoblastic cleavage is observed with macromeres making the vegetal pole and micromeres the animal pole (Gad 2005c).

Life cycle

The life cycle of loriciferans consists of several larval stages and adults. Each stage moults from the previous one. Over time, a number of different larval types have been described and the life cycle of some species appears to be very complex. In addition, in some species it seems that different pathways are possible. It should be kept in mind that information on the life cycles of loriciferans is derived not from cultures, but from assigning different individuals to one species. A number of specimens are found in moult or may even keep one or two older cuticular cases as a shelter. By this, even strong morphological transitions can be assigned to one particular species.

In Nanaloricidae, the life cycle appears to be more or less simple (compared to Pliciloricidae). From the eggs hatch Higgins-larvae, which moult several times. The final larval stage is called a postlarva; this moults to sexually dimorphic adults. Only sexual reproduction seems to be present in Nanaloricidae (Kristensen et al. 2007). All Higgins-larvae have the characteristic "toes". Postlarvae resemble adults, but lack the seventh row of scalids, are about 80% of the length of adults, have no sexual dimorphism and are of course not sexually mature (Gad 2009a). There has been a discussion on the number of larval stages. Gad (2005b) groups the known larvae into size classes and concludes from this that Nanaloricidae have five larval stages. However, a complete series of larval stages is not known from any species. Size classes may vary between species and size may even be a variable character depending on food supply or other factors (Heiner 2008). Therefore, Heiner (2008) prefers the term "penultimate larva" over "fifth larval stage" for the Higgins-larva that molts into the postlarva.

In Pliciloricidae, life cycles are more complex (see, e.g., Heiner & Kristensen 2009) (Fig. 6.2.1 A). Reductions of some morphological features can occur in some stages and new stages can evolve. Both sexual and asexual modes of reproduction are possible.

In some species, the life cycle roughly follows the one in Nanaloricidae. In these cases, only sexual reproduction appears to be present. In the "*cauliculus*-type", occurring in some *Pliciloricus* and some *Rugiloricus* species, several Higgins-larva stages moult into a free-living postlarva, which molts to the adult. The postlarva has very short scalids and is therefore morphologically more different from the adult than in Nanaloricidae. In the "*carolinensis*-type", occurring in some other *Pliciloricus* and some *Rugiloricus* species, the postlarva is reduced to a thin cuticle remaining inside the last Higgins-larval stage. Hence, in species such as *Pliciloricus leocaudatus*, the adult seems to emerge directly from the last Higgins-larval stage, however a postlarva is present even though it is merely a thin cuticle surrounding the adult (Heiner & Kristensen 2005).

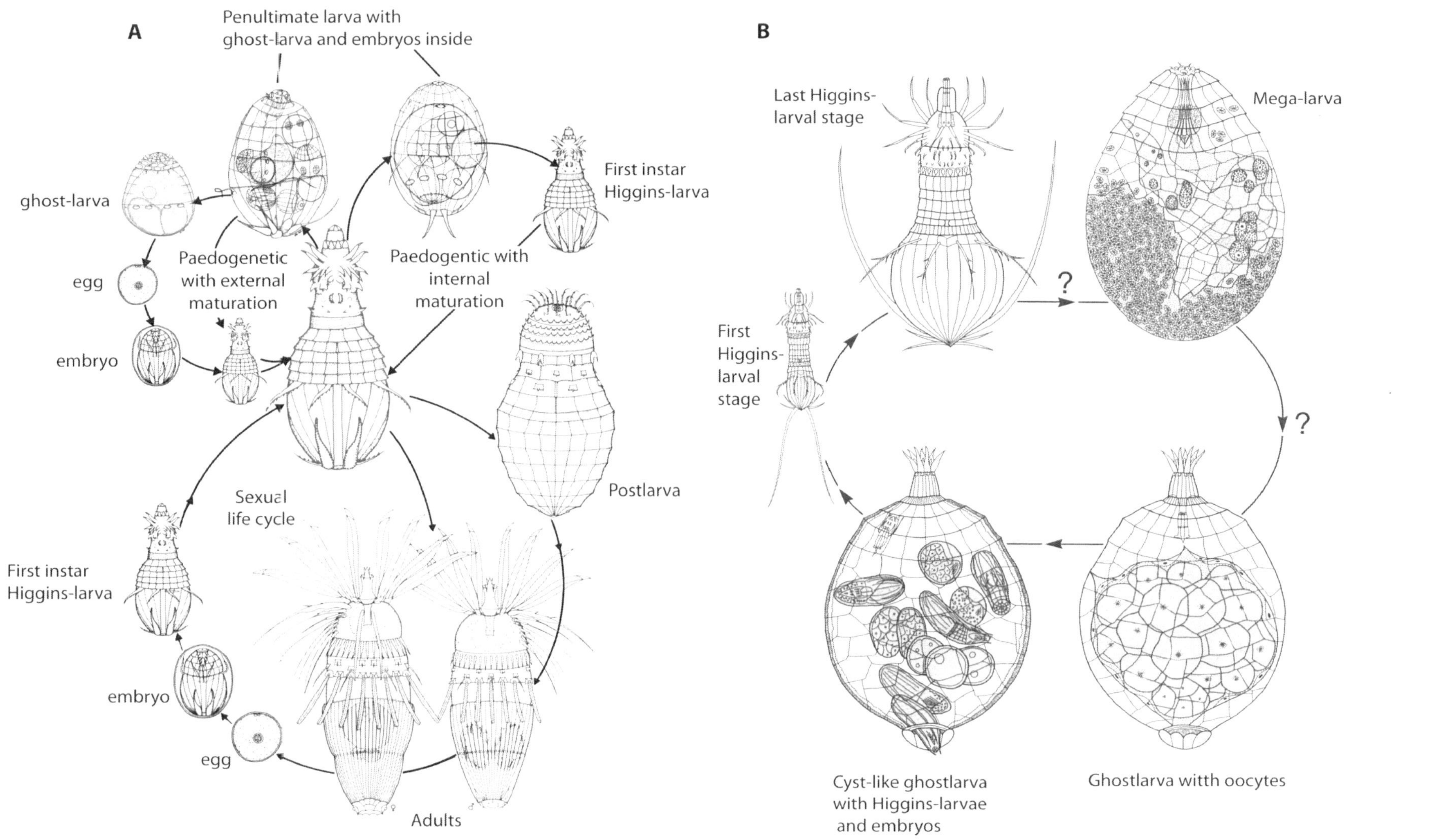

Fig. 6.2.1. Life cycles of Loricifera (modified from Heiner 2008, Heiner & Kristensen 2009). A, The lifecycle of the family Pliciloricidae including two types of sexual life cycles (one including a postlarval stage and one with a very reduced postlarval stage) and two types of asexual life cycles (one paedogenetic with external maturation and one paedogenetic with internal maturation); B, Life cycle of the family Urnaloricidae.

In *Pliciloricus diva*, two types of postlarvae appear to be present in the life cycle (Gad 2009b). The first type has a simplified adult morphology, the introvert contains minute protoscalids. The second type is only present as a very thin cuticle.

In some species of *Rugiloricus*, an asexually reproducing parthenogenetic stage is present (Kristensen 2002; Kristensen & Brooke 2002). This stage molts from the last Higgins-larval stage (Fig. 6.2.1 A). It was later also discovered in *Pliciloricus corvus* (Gad 2005b) and *P. pedicularis* (Gad 2005c) and was then called "simplified parthenogenetic adult". In *Titaniloricus inexpectatovus* Gad (2005d) described a parthenogenetic stage within a Higgins-larva and called this a modified 7th instar Higgins-larva and therefore as a "paedogenetic larva". Heiner (2008) found a parthenogenetic stage in *Rugiloricus bacatus* and called this a ghost-larva (Fig. 6.2.1 A). In all cases mentioned, the parthenogenetic stage is between clearly recognizable Higgins-larvae and adults. Therefore, it is appropriate to interpret it as a larval stage, which is paedogenetic. The term "ghost-larva" is now generally applied to this stage (Heiner 2008, Heiner & Neuhaus 2010). Rather problematic is the interpretation by Gad (2005b,c), who described a very thin cuticle between the cuticle of the last Higgins-larval stage and the ghost-larva; this led to his naming it as a simplified adult stage. However, in such complex, tiny wrinkled animals, a very thin cuticle can easily be misinterpreted and it has been questioned whether such a reduced postlarva really exists (Heiner 2008). Heiner (2008) notes that ghost-larvae were found already in 1986 by Renaud-Mornant, but not recognized as stages in the life cycle of Loricifera.

In most cases, ghost-larvae remain within the last Higgins-larval stage, but in some cases they can also be free-living (*Rugiloricus bacatus*, and two further unnamed species of *Rugiloricus* from the Faroe Bank, which are only known by their ghost-larvae, Heiner 2008). Ghost-larvae produce eggs, from which Higgins-larvae mature. Therefore, different developmental stages are found within the ghost-larvae. In *Pliciloricus corvus* (Gad 2005b), *P. diva* (Gad 2009b) and *Rugiloricus bacatus* (Heiner 2008), only eggs and embryos have been found in the ghost-larvae. In *P. pedicularis* (Gad 2005c), early developmental stages and Higgins-larvae were found. In *Titaniloricus inexpectatovus* (Gad 2005d) one of the last Higgins-larval stages (interpreted as 6th instar by Gad) becomes extremely large in size (about 800 µm) and includes a ghost-larva (interpreted as the reduced 7th instar Higgins-larva by Gad) and a number of different stages ranging from eggs to Higgins-larvae and adults. A possible interpretation is that in the first mentioned species Higgins-larvae are released by the ghost-larvae and that in *T. expectatovus* the last Higgins-larval stage and the ghost-larvae are adapted by their large size to shelter the entire development from eggs to adults.

Finally, a completely different life cycle has been discovered in *Urnaloricus gadi* (Heiner & Kristensen 2009) (Fig. 6.2.1 B). Adults are unknown, but a new stage, the mega-larva (as pre mega-larva and cyst-forming mega-larva), is present, which is not comparable to any larval stage previously known (Fig. 6.1.5 E, F). Most likely, the life cycle works as follows: a pre mega-larva is free-living, oval and about 260 µm long. It has scalids reduced in number, a digestive system and an ovary. It presumably molts into another mega-larval stage, the cyst-forming mega-larva. This is quite large and measures up to 450 µm. In comparison to the previous mega-larval stage it has a large gonopore and the ovary extends and produces a large number of oocytes. The cyst-forming mega-larva moults twice. The cuticles of these two stages are interpreted as a postlarva and a ghost-larva; they remain inside the cyst-forming mega-larva. The eggs develop into embryos and Higgins-larvae, which appear to absorb the tissue of the ghost-larva. During this process, the outermost cuticle (the one of the cyst-forming mega-larva) hardens and becomes a solid cyst. Higgins-larvae finally emerge through the gonopore.

In a few species hermaphroditism occurs (see Gad 2005e, Heiner & Neuhaus 2010 for referances to two undescribed species).

6.3. Phylogeny

Loricifera have almost always been compared primarily to the taxa Priapulida and Kinorhyncha and in most cases, these three are united in a monophyletic taxon called Scalidophora (Lemburg 1995; also Cephaloryncha by Nielsen 2001, but not 2012). Within Scalidophora, internal relationships are not so clear. There are potential synapomorphies supporting a sister-group relationship between Loricifera and Kinorhyncha as well as those supporting a sister-group relationship between Loricifera and Priapulida. Kristensen (1991a), Neuhaus (1993, 1994) and Lemburg (1999) name the following characters as potential synapomorphies for (Loricifera + Kinorhyncha): elongate scalids, scalids in first ring with blunt tip, trichoscalids with basal plates, mouth cone, modified spermatozoa. According to Lemburg (1999; see also Ax 2003) the following characters are potential synapomorphies for (Loricifera + Priapulida): indirect development with larval stage, neck region with rectangular plates in larvae, lorica in trunk of larva, nine short introvert retractors, at least larvae with large dorsocaudal gland, urogenital system (protonephridial ducts lead into gonoducts), urogenital system attached to body wall by mesenterium. Lemburg (1999) discusses both sets of characters in detail and concludes that (Loricifera + Priapulida) is supported by stronger characters; this taxon is named Vinctiplicata. Kristensen (2003) mentioned the similarities between the tardigrades and loriciferans, especially the triradiated myo-epithelial pharyngeal bulb with the cuticular placoids and the telescoping of the mouth cone. The pharyngeal bulb of both the priapulids and kinorhynchs consists of mesodermal muscles and epidermal cells forming the cuticle in the pharyngeal apparatus.

For some considerable time it was not possible to sequence DNA from loriciferan species. The first successful results were published by Park et al. (2006). Loricifera were found to belong to Ecdysozoa, but there was no considerable support for any further relationship. Newer attempts by Sørensen et al. (2008) found support for a closer relationship between Loricifera and Nematomorpha (see also the original description of the loriciferans, Kristensen (1983)). This hypothesis is in contrast to a monophyletic taxon Scalidophora, but Sørensen et al. (2008) developed potential scenarios supporting their hypothesis.

Fossils assigned to Loricifera are very rare. Maas et al. (2009) describe specimens named *Orstenoloricus shergoldii* extracted from Middle Cambrian limestone in Queensland, Australia. These have an anterior region with traverse folds and a posterior, lorica-like region with longitudinal folds. Some extensions in the anterior region of the lorica and in one case also in the posterior region of the lorica are interpreted as the basal part of the sensory setae known only from larval loriciferans and therefore support the interpretation of *Orstenoloricus shergoldii* as fossil loriciferans. Specimens are in the size range between about 250 μm and 500 μm. Another potential fossil loriciferan is larger (about 80 mm) and was found in the Early Cambrian Sirius Passet fauna in North Greenland (Peel 2010). It was named *Sirilorica carlsbergi*. This specimen appears to consist of two circlets of plates, strongly resembling lorica plates. Anterior of this potential lorica are tooth-like structures, which have some resemblance to the last row of scalids on the introvert of loriciferans.

6.4. Diversity

Currently, 30 species are described in ten genera. The existing diversity of Loricifera is certainly much higher, because many analyses of new marine sediments reveal specimens of new, undescribed species (see for example reports on undetermined species from the Galapagos Spreading Center by Heiner & Neuhaus 2007 or from the New Zealand region by Heiner & Neuhaus 2010).

The described species are assigned to three families: Nanaloricidae with ten species in six genera, Pliciloricidae with 19 species in three genera and Urnaloricidae with one single species.

Nanaloricidae

- *Armorloricus davidi* Kristensen & Gad, 2004
- *Armorloricus elegans* Kristensen & Gad, 2004
- *Armorloricus kristenseni* Heiner, 2004
- *Australoricus oculata* Heiner, Boesgaard & Kristensen, 2009
- *Culexiregiloricus trichoscalida* Gad, 2009
- *Nanaloricus gwenae* Kristensen, Heiner & Higgins, 2007
- *Nanaloricus khaitatus* Todaro & Kristensen, 1998
- *Nanaloricus mysticus* Kristensen, 1983
- *Phoeniciloricus simplidigitatus* Gad, 2004
- *Spinoloricus turbatio* Heiner & Neuhaus, 2007

Pliciloricidae

- *Pliciloricus cavernicola* Heiner, Boesgaard & Kristensen, 2009
- *Pliciloricus corvus* Gad, 2005
- *Pliciloricus diva* Gad, 2009
- *Pliciloricus dubius* Higgins & Kristensen, 1986
- *Pliciloricus enigmaticus* Higgins & Kristensen, 1986
- *Pliciloricus gracilis* Higgins & Kristensen, 1986
- *Pliciloricus hadalis* Kristensen & Shirayama, 1988
- *Pliciloricus leocaudatus* Heiner & Kristensen, 2005
- *Pliciloricus orphanus* Higgins & Kristensen, 1986
- *Pliciloricus pedicularis* Gad, 2005
- *Pliciloricus profundus* Higgins & Kristensen, 1986
- *Pliciloricus senicirrus* Gad, 2005
- *Pliciloricus shukeri* Heiner & Kristensen, 2005
- *Rugiloricus bacatus* Heiner, 2008
- *Rugiloricus carolinensis* Higgins & Kristensen, 1986
- *Rugiloricus cauliculus* Higgins & Kristensen, 1986
- *Rugiloricus ornatus* Higgins & Kristensen, 1986
- *Rugiloricus polaris* Gad & Arbizu, 2005
- *Titaniloricus inexpectatovus* Gad, 2005

Urnaloricidae

- *Urnaloricus gadi* Heiner & Kristensen, 2009

In the following, supraspecific taxa are briefly characterized and for the species their occurrence and known material are listed. For exact descriptions of the species see the primary literature.

Nanaloricidae

The diagnosis of Nanaloricidae has been extended several times with the discovery of new species. For an extensive diagnosis see Kristensen et al. (2007), but some of these characters differ in *Culexiregiloricus trichoscalida* (Gad 2009a), even if only a postlarva is described. Some of the characteristic features are:

- mouth cone inserts with broad base on introvert; consists of at least two sections, a basal section with eight oral ridges and oral furcae and a protrusible mouth tube of variable length
- triradiate pharynx with three rows of placoids, each consisting of five macroplacoids
- introvert with nine rows of scalids
- sexual dimorphism in clavoscalids (unknown in *Culexiregiloricus*)
- further rows may contain segmented ("leg-like") scalids, feather-like scalids, trichoscalid-like spinoscalids or slender spinoscalids
- last (ninth) row with 30 beak- or teeth-like scalids on basal plates (unknown in *Culexiregiloricus*)
- neck with three rows of basal plates, third row with eight single and seven double trichoscalids
- sexual dimorphism in trichoscalids, two ventral trichoscalids are hook-like claspers in males (unknown in *Culexiregiloricus*)
- lorica with six to ten plates and hollow anterior spikes of varying number carrying gland openings, lorica smooth or with honeycomb structure
- eight to nine flosculi in the posterior region
- one pair of gonads
- postlarva resembles the adult, but without seventh row of spinoscalids
- Higgins-larvae: mouth cone without armature; introvert with seven rows of scalids; indistinct neck, thorax with five folds or plates (some forming a closing apparatus for the retracted introvert); lorica with longitudinal folds and honeycomb structure; three pairs of locomotory setae on lorica; caudal leaf-shaped toes; three pairs of sensory setae and three flosculi.

Genus *Armorloricus* Kristensen & Gad, 2004. Three described species; some typical characters are (see Kristensen & Gad 2004, Heiner 2004):

- smooth lorica with large midventral and laterodorsal plates
- very long mouth tube
- six oral stylets
- prepharyngeal armature with "manubrium"
- toes of Higgins-larvae are shield-shaped, ending in two diverging pointing tips.

Armorloricus davidi Kristensen & Gad, 2004

- Material: few adult males and females; Higgins-larvae are present from the same locality, but could not be specifically assigned to this species (Kristensen & Gad 2004).
- Occurrence: Trezen ar Skoden, Roscoff, France, 50 m depth, clean shell gravel (Kristensen & Gad 2004).

Armorloricus elegans Kristensen & Gad, 2004

- Material: several hundred adult males, females and postlarvae; Higgins-larvae are present from the same locality, but could not be specifically assigned to this species (Kristensen & Gad 2004).
- Occurrence: Trezen ar Skoden, Roscoff, France, 50–60 m depth, clean shell gravel (Kristensen & Gad 2004).

Armorloricus kristenseni Heiner, 2004

- Material: one adult male, few adult females, postlarvae and Higgins-larvae (Heiner 2004).
- Occurrence: Faroe Bank, North Atlantic, 136–139 m depth, from fine shell gravel and fine shell sand (Heiner 2004).

Genus *Australoricus* Heiner, Boesgaard & Kristensen, 2009. One described species, some typical characters of the Higgins-larva are (see Heiner et al. 2009a):

- broad, cylindrical mouth cone, ending in an almost flat disc with the central mouth opening
- toes with bulbous basal part and spinose terminal part
- a pair of photoreceptors.

Australoricus oculata Heiner, Boesgaard & Kristensen, 2009

- Material: three Higgins-larvae (Heiner et al. 2009a).
- Occurrence: submarine caves (Jim´s Cave and Fish Rock Cave) off the coast of New South Wales, Australia, 34 and 18–20 m, from coarse coralligenous sand (Heiner et al. 2009a).

Genus *Culexiregiloricus* Gad, 2009. One described species, some typical characters are (after Gad 2009a):

- mouth cone with long and slim mouth tube that can be elongated telescopically
- filiform clavoscalids and spinoscalids of the second and third ring, all covered by trichoids
- lorica with eight plates and eight intercalary plicae
- Square-like honeycomb ultrasculture
- 14 lorica spikes
- eight dorsal papillated flosculi.

Culexiregiloricus trichoscalida Gad, 2009

- Material: one postlarva (Gad 2009a).
- Occurrence: Guinea Basin, 4141 m depth, from clayish sediments (Gad 2009a).

Genus *Nanaloricus* Kristensen, 1983. Three described species, some typical characters are:

- mouth cone short and broad with eight oral ridges of two lengths
- six lorica plates with honeycomb structure
- two of the ventral trichoscalids are modified into hook-like claspers in the males
- toes of Higgins-larva with leaf-shaped mucrones.

Nanaloricus mysticus Kristensen, 1983

- Material: few adults and postlarvae; numerous Higgins-larvae (Kristensen 1983).
- Occurrence: Roscoff, France, in 25–30 m depth; *Dentalium*-sand (Kristensen 1983).

Nanaloricus khaitatus Todaro & Kristensen, 1998

- Material: two adult females, three postlarvae and two Higgins-larvae (Todaro & Kristensen 1998).
- Occurrence: Livorno, Italy, 300 m NW of the coast, at 7 m depth; from different sediment types, holotype was obtained from coarse sand (Todaro & Kristensen 1998).

Nanaloricus gwenae Kristensen, Heiner & Higgins, 2007

- Material: one adult female, two postlarvae (one only as exuvium) and five Higgins-larvae (one of them as exuvium) (Kristensen et al. 2007).
- Occurrence: Continental shelf off Fort Pierce, Florida (~10 km from the coast), USA, 15–17 m depth, from coarse sand or shell gravel (Kristensen et al. 2007).

Genus *Phoeniciloricus* Gad, 2004. One described species, some typical characters are (see Gad 2004):

- ten lorica plates with 13 lorica spikes
- spinose toes lacking mucrones but with enlarged bases.

Phoeniciloricus simplidigitatus Gad, 2004

- Material: one postlarva inside a Higgins-larva (Gad 2004).
- Occurrence: Kilinailau Trench Northwest Papua New Guinea, 1813 m depth, from fine sand covered by a layer of volcanic ash (Gad 2004).

Genus *Spinoloricus* Heiner & Neuhaus, 2007. One described species, some typical characters are:

- mouth cone with eight oral ridges of different length
- basal mouth cone pleat (a cuticular reinforcement)
- eighth row with 30 whip-like spinoscalids and 30 alternating plates
- lorica with eight plates with eight intercalary plicae and additional anterior spikes
- Higgins-larva with six rectangular plates in the seventh scalid row and long toes.

Spinoloricus turbatio Heiner & Neuhaus, 2007

- Material: two postlarvae, each with an unsexed adult inside; two Higgins-larvae (Heiner & Neuhaus 2007).
- Occurrence: Galapagos Spreading Center, East Pacific, 2119–2776 m depth, sediment clay (color: black brown gray), containing Foraminifera (mainly *Globigerina*) (Heiner & Neuhaus 2007).

Pliciloricidae

Pliciloricidae was described by Higgins & Kristensen (1986) including ten species from the two genera *Pliciloricus* and *Rugiloricus*. Since then, several new species and a diversity of new life-cycle stages have been discovered. Some typical characters for Pliciloricidae are:

- stalked mouth cone with 6–12 oral ridges
- pharynx in mouth cone (anterior of brain)
- lorica composed of numerous longitudinal elements called plicae
- Higgins-larva: Internal prepharyngeal armature in buccal canal; pharynx with internal armature; neck

region with short, scalid-like structures, probably early trichoscalids; posterior part of neck as ruff (especially well developed in *Pliciloricus*), this is a closing apparatus when introvert is withdrawn; larval lorica with a number of longitudinal furrows; two pairs of sensory setae on anterior, ventral margin of lorica.

Genus *Pliciloricus* Higgins & Kristensen, 1986. 13 described species, some typical characters are (after Gad 2009c):

- large mouth cone
- double organ in second row of spinoscalids
- eight single trichoscalids alternate with seven double trichoscalids
- lorica with about 22 double ridges
- broad transformed midventral plica
- pharynx in mouth cone, brain inside introvert
- voluminous midgut with dorsocaudal cloaca
- Higgins-larva with indistinct neck region and well-developed collar
- lorica of Higgins-larva with 20–22 longitudinal folds, two pairs of short anterior setae and three pairs of posterior setae; toes long, slender and with spiny tip.

Pliciloricus cavernicola Heiner, Boesgaard & Kristensen, 2009

- Material: one adult male (Heiner et al. 2009a).
- Occurrence: submarine cave (Jim´s Cave) off the coast of New South Wales, Australia, ~34 m, from coralligenous sand (Heiner et al. 2009a).

Pliciloricus corvus Gad, 2005

- Material: seven Higgins-larvae (interpreted as 7th instar) containing postlarval exuvia and adult males or females; six Higgins-larvae containing postlarval exuvium and parthenogenetic simplified adults; one younger Higgins larva (interpreted as 5th instar) (Gad 2005b).
- Occurrence: Great Meteor Bank, North Atlantic, 467 m depth, from coarse sand (Gad 2005b).

Pliciloricus diva Gad, 2009

- Material: two adult males; three last-stage Higgins-larvae with internal postlarva containing adult males or females; two last-stage Higgins-larvae with internal reduced postlarva containing parthenogenetic female; two Higgins-larvae (one as exuvia) (Gad 2009b).
- Occurrence: Angola Basin in Southern Atlantic Ocean, near the coast of Namibia, 5423 m depth, from muddy sediments (Gad 2009b).

Pliciloricus dubius Higgins & Kristensen, 1986

- Material: two adult female specimens (Higgins & Kristensen 1986).
- Occurrence: Continental shelf off South Carolina (~200 km east of the coast), USA, 294 m depth, from medium sand (Higgins & Kristensen 1986).

Pliciloricus enigmaticus Higgins & Kristensen, 1986

- Material: several specimens including adult males, females and Higgins-larvae; some molting larvae with adults inside (Higgins & Kristensen 1986).
- Occurrence: Continental shelf off South and North Carolina (~200 km east of the coast), USA, between 294 and 439 m depth, from medium sand (Higgins & Kristensen 1986).

Pliciloricus gracilis Higgins & Kristensen, 1986

- Material: several specimens including one adult male and several Higgins-larvae, none of them with adults inside (Higgins & Kristensen 1986).
- Occurrence: Continental shelf off South and North Carolina (~200 km east of the coast), USA, between 289 and 439 m depth, from medium sand (Higgins & Kristensen 1986).

Pliciloricus hadalis Kristensen & Shirayama, 1988

- Material: one adult male, one adult female, one Higgins-larva (Kristensen & Shirayama 1988).
- Occurrence: Izu-Ogasawara Trench, Western Pacific, 8260 m depth, from upper layers of red clay sediments including layers of silt and sand.

Pliciloricus leocaudatus Heiner & Kristensen, 2005

- Material: few adult males and two Higgins-larvae, one of them with an adult male inside (Heiner & Kristensen 2005).
- Occurrence: Faroe Bank, North Atlantic, 120 and 249 m depth, from carbonate sand and coarse shell gravel (Heiner & Kristensen 2005).

Pliciloricus orphanus Higgins & Kristensen, 1986

- Material: two Higgins-larvae (Higgins & Kristensen 1986).
- Occurrence: Continental shelf off South and North Carolina (~200 km east of the coast), USA, between 289 and 439 m depth, from medium sand (Higgins & Kristensen 1986).

Pliciloricus pedicularis Gad, 2005

- Material: one adult male; four Higgins-larvae (interpreted as 7th instar) containing postlarval exuvia and adult males or females; four Higgins-larvae containing postlarval exuvium and parthenogenetic simplified adults, sometimes together with eggs, embryos and Higgins-larvae; seven Higgins-larvae of different stages (Gad 2005c).
- Occurrence: Angola Basin in Southern Atlantic Ocean, near the coast of Namibia, 5448 m depth, from fine-grained clayish sediments (Gad 2005c).

Pliciloricus profundus Higgins & Kristensen, 1986

- Material: one adult male and several Higgins-larvae (Higgins & Kristensen 1986).
- Occurrence: Continental shelf off South and North Carolina (~200 km east of the coast), USA, between 294 and 439 m depth, from medium sand (Higgins & Kristensen 1986).

Pliciloricus senicirrus Gad, 2005

- Material: two Higgins-larvae (interpreted as 7th instar) containing postlarval exuvia and adults (one male and one unsexed specimen); three Higgins-larvae (two interpreted as 5th instars, one as 3rd instar) (Gad 2005b).
- Occurrence: Great Meteor Bank, North Atlantic, 297 m depth, from coarse sand (Gad 2005b).

Pliciloricus shukeri Heiner & Kristensen, 2005

- Material: one unsexed adult and two Higgins-larvae (Heiner & Kristensen 2005).
- Occurrence: Faroe Bank, North Atlantic, 260 m depth, from fine shelly sand (Heiner & Kristensen 2005).

Genus *Rugiloricus* Higgins & Kristensen, 1986. six described species, some typical characters are:

- lorica with 30–60 plicae
- 15 single trichoscalids
- typically without flosculi or if present, only as a single pair of P-flosculi
- Higgins-larva resembles the one in *Pliciloricus*, but ruff not so clear; toes with broad base, smaller midpiece and pointed tip; two pairs of sensory setae around anal cone.

Rugiloricus bacatus Heiner, 2008

- Material: two adult females; seven postlarvae, four with adults inside; two Higgins-larvae with a postlarva inside; two Higgins-larvae with a ghost-larva inside; one penultimate and one early Higgins-larva; two free ghost-larvae (Heiner 2008).
- Occurrence: Faroe Bank, North Atlantic, 109–260m depth, from fine shell sand to coarse gravel (Heiner 2008).

Rugiloricus carolinensis Higgins & Kristensen, 1986

- Material: one adult (probably a female) within a larva and many Higgins-larvae (Higgins & Kristensen 1986).
- Occurrence: Continental shelf off South and North Carolina (~200 km east of the coast), USA, between 289 and 439 m depth, from medium sand (Higgins & Kristensen 1986).

Rugiloricus cauliculus Higgins & Kristensen, 1986

- Material: one adult male and two adult females; one unsexed specimen and one adult (probably a male) within a postlarval exuvium (Higgins & Kristensen 1986).
- Occurrence: Continental shelf off South and North Carolina (~200 km east of the coast), USA, between 289 and 439 m depth, from medium sand (Higgins & Kristensen 1986).

Rugiloricus ornatus Higgins & Kristensen, 1986

- Material: few male and female adults and few adults within postlarval exuvia (Higgins & Kristensen 1986).
- Occurrence: Continental shelf off North Carolina (~200 km east of the coast), USA, between 430 and 439 m depth, from medium sand (Higgins & Kristensen 1986).

Rugiloricus polaris Gad & Martínez Arbizu, 2005

- Material: one adult male and two Higgins-larvae (Gad & Martínez Arbizu 2005).
- Occurrence: Laptev Sea (Arctic Ocean), 216 m depth, from clayish sediments (Gad & Martínez Arbizu 2005).

Genus *Titaniloricus* Gad, 2005. One described species, some typical characters are:

- giant size of last toed Higgins-larval stage (interpreted as 6th instar, Gad 2005d)

Titaniloricus inexpectatovus Gad, 2005

- Material: one unsexed adult, five Higgins-larvae and five giant Higgins-larvae (interpreted as 6th instar) with several specimens or remnants of adults and younger Higgins-larvae inside (Gad 2005d).
- Occurrence: Angola Basin in Southern Atlantic Ocean, near the coast of Namibia, 5448 m depth, from fine-grained clayish sediments.

Urnaloricidae

Urnaloricidae include a single species, which is known only by larval stages. These larval stages contain two otherwise unknown stages. Some typical characteristics are:

- pre-mega-larval stage
- cyst-forming mega-larval stage with embryos developing inside by consuming the maternal tissue; larval cuticle hardens to solid cyst; Higgins-larvae emerge through broad gonopore.
- Higgins-larva with distinct collar, two ventral setae, two posterior setae and two very long toes.

Genus *Urnaloricus* Heiner & Kristensen, 2009. One described species, but many undescribed, characters as in family.

Urnaloricus gadi Heiner & Kristensen, 2009

- Material: 15 cyst-forming mega-larvae containing ghost-larvae and/or embryos; one pre-mega-larva and 11 Higgins-larvae (Heiner & Kristensen 2009).
- Occurrence: Faroe Bank, North Atlantic, 109–273 m depth, from shelly sand to shell gravel (Heiner & Kristensen 2009).

6.5. Ecology

Loricifera occur only in marine sediments. However they have been found in quite different sediments and different regions. Sediment can be coarse to fine, but coarse sediment appears to be richer in loriciferans compared to finer sediments. For a while it appeared that a good oxygen supply would be a common factor uniting the different sediments where loriciferans have been found, but recently, loriciferans were unexpectedly found to inhabit anoxic sediments. Danovaro et al. (2010) found very high densities of loriciferans from three undescribed species of Loricifera in an anoxic, hypersaline basin in the Mediterranean Sea. By several methods it could be shown that the specimens were really living in this environment. The lack of mitochondria appears to be an adaptation to the anoxic mode of life, as well as a high number of hydrogenosome-like organelles. Probably symbionts support the loriciferan metabolism. Furthermore, symbiotic sulfur bacteria were observed in epidermial cells.

Loriciferans are presumed to eat bacteria and algae, since bacteria have been found in the buccal tube of *Armorloricus elegans*. In Fig. 6.1.6 C these bacteria are seen surrounding the mouth cone (see also fig 1. in Kristensen 1995).

Loriciferans are known from quite shallow depths to the deep sea. Only one record from Chesterfield Islands (Coral Sea) from the intertidal is known (Kristensen 1983).

Literature

Ax, P. (2003): *Multicellular animals*. Springer, Berlin.

Bang-Berthelsen, I. H. (2008): *Phylogeny of Loricifera based on combined morphological and molecular studies*, PhD Thesis.

Danovaro, R., Dell'Anno, A., Pusceddu, A., Gambi, C., Heiner, I. & Kristensen, R. M. (2010): The first metazoa living in permanently anoxic conditions. *BMC Biology* 8:30: 1–10.

Gad, G. (2004): A new genus of Nanaloricidae (Loricifera) from deep-sea sediments of volcanic origin in the Kilinailau Trench north of Papua New Guinea. *Helgoland Mar. Res.* 58: 40–53.

Gad, G. (2005a): *Life history stages of Loricifera: their morphology and position in the life cycle*. Dissertation. Carl von Ossietzky Universität, Oldenburg, Germany, pp. 60 +13 manuscripts.

Gad, G. (2005b): Successive reduction of the last instar larva of Loricifera, as evidenced by two new species of *Pliciloricus* from the Great Meteor Seamount (Atlantic Ocean). *Zool. Anz.* 243: 239–271.

Gad, G. (2005c): A parthenogenetic, simplified adult in the life cycle of *Pliciloricus pedicularis* sp.n. (Loricifera) from the deep sea of the Angola Basin. *Org. Div. Evol.* 5: 77–103.

Gad, G. (2005d): Giant Higgins-larvae with paedogenetic reproduction from the deep sea of the Angola Basin – evidence for a new life cycle and for abyssal gigantism in Loricifera? *Org. Div. Evol.* 5: 59–75.

Gad, G. (2005e). Hermaphroditism in Loricifera – as evidenced by *Rugiloricus doliolius* sp. n. from the Antarctic deep sea. In: Thatje, S., Calcagno, J. A. & Arntz, W. E. (eds) Evolution of antarctic fauna – extended abstracts of the IBMANT/ANDEEP international symposium and workshop in 2003, *Ber. Pol. Meeresforsch.* 507: 140–141.

Gad, G. (2009a): *Culexiregiloricus*, a new genus of Nanaloricidae (Loricifera) from the deep sea of the Guinea Basin (Southeast Atlantic). *Zootaxa* 2096: 33–49.
Gad, G. (2009b): A clearly identifiable postlarva in the life cycle of a new species of *Pliciloricus* (Loricifera) from the deep sea of the Angola Basin. *Zootaxa* 2096: 50–81.
Gad, G. & Martínez Arbizu, P. (2005): First description of an Arctic Loricifera – a new *Rugiloricus*-species from the Laptev Sea. *Mar. Biol. Res.* 1: 313–325.
Heiner, I. (2004): *Armorloricus kristenseni* (Nanaloricidae, Loricifera), a new species from the Faroe Bank (North Atlantic). *Helgoland Mar. Res.* 58: 192–205.
Heiner, I. (2008): *Rugiloricus bacatus* sp. nov. (Loricifera – Pliciloricidae) and a ghost-larva with paedogenetic reproduction. *Syst. Biodiv.* 6: 225–247.
Heiner, I. & Kristensen, R. M. (2005): Two new species of the genus *Pliciloricus* (Loricifera, Pliciloricidae) from the Faroe Bank, North Atlantic. *Zool. Anz.* 243: 121–138.
Heiner, I. & Kristensen, R. M. (2009): *Urnaloricus gadi* nov. gen. et nov. sp. (Loricifera, Urnaloricidae nov. fam.), an aberrant Loricifera with a viviparous pedogenetic life cycle. *J. Morphol.* 270: 129–153.
Heiner, I. & Neuhaus, B. (2007): Loricifera from the deep sea at the Galapagos Spreading Center, with a description of *Spinoloricus turbatio* gen. et sp. nov. (Nanaloricidae). *Helgoland Mar. Res.* 61: 167–182.
Heiner, I. & Neuhaus, B. (2010): Loricifera, loriciferans, corset animals. In : Gordon, D. P. (ed.) *New Zealand Inventory of Biodiversity*, Vol. 2, pp. 472–475, Canterbury University Press.
Heiner, I., Boesgaard, T. M. & Kristensen, R. M. (2009a): First time discovery of Loricifera from marine caves and Australian waters. *Mar. Biol. Res.* 5: 529–546.
Heiner, I., Kürstein Sørensen, K. J. & Kristensen, R. M. (2009b): The spermiogenesis and the early spermatozoa of *Armoloricus elegans* (Loricifera, Nanaloricidae). *Zoomorphology* 128: 285–304.
Hermans, C. I. (1983): The duo-gland adhesive system. *Oceanogr. Mar. Biol. Ann. Rev.* 21: 283–339.
Higgins, R. P. & Kristensen, R. M. (1986): New Loricifera from Southeastern United States coastal waters. *Smiths. Contrib. Zool.* 438: 1–70.
Kristensen, R. M. (1983): Loricifera, a new phylum with Aschelminthes characters from the meiobenthos. *Z. Zool. Syst. Evolutionsforsch.* 21: 163–180.
Kristensen, R. M. (1991a): Loricifera. In: Harrison, F. W. & Ruppert, E. E. (eds) *Microscopic Anatomy of Invertebrates*, Vol. 4, pp. 351–375. Wiley-Liss, New York.
Kristensen, R. M. (1991b): Loricifera – a general biological and phylogenetic overview. *Verh. Dtsch. Zool. Ges.* 84: 231–246.
Kristensen, R. M. (1995): Are Aschelminthes pseudocoelomate or acoelomate? In: Lanzavecchia, G., Valvassori, R. & Candia Carnevali, M. D. (eds) *Body cavities: function and phylogeny*, pp.41–43. Selected Symposia and Monographs U.Z.I., 8, Mucchi, Modena.
Kristensen, R. M. (2002): An introduction to Loricifera, Cycliophora, and Micrognathozoa. *Integr. Comp. Biol.* 42: 641–651.
Kristensen, R. M. (2003): Comparative morphology: do the ultrastructural investigations of Loricifera and Tardigrada support the clade Ecdysozoa? In: Legakis, A., Sfenthourakis, S., Polymeni, R. & Thessalou-Legakis, M. (eds) *The new panorama of animal* evolution, pp. 467–477. Proc. 18th Int. Congr. Zool., Pensoft Publishers, Sofia.
Kristensen, R. M. & Brooke, S. (2002): Phylum Loricifera. In: Young, C. M. (ed.) *Atlas of Marine Invertebrate* Larvae, pp. 179–187. Academic Press, San Diego.
Kristensen, R. M. & Gad, G. (2004): *Armorloricus*, a new genus of Loricifera (Nanaloricidae) from Trezen ar Skoden (Roscoff, France). *Cah. Biol. Mar.* 45: 121–156.
Kristensen, R. M., Heiner, I. & Higgins, R. P. (2007): Morphological and life cycle of a new loriciferan from the Atlantic coast of Florida with an emended diagnosis and life cycle of Nanaloricidae (Loricifera). *Invertebr. Biol.* 126: 120–137.
Kristensen, R. M. & Shirayama, Y. (1988): *Pliciloricus hadalis* (Pliciloricidae), a new loriciferan species collected from the Izu-Ogasawara Trench, Western Pacific. *Zool. Sci.* 5: 875–881.
Lemburg, C. (1995): Ultrastructure of sense organs and receptor cells of the neck and lorica of the *Halicryptus spinulosus* larva (Priapulida). *Microfauna Marina* 10: 7–30.
Lemburg, C. (1999): *Ultrastrukturelle Untersuchungen an den Larven von* Halicryptus spinulosus *und* Priapulus caudatus. Hypothesen zur Phylogenie der Priapulida und deren Bedeutung für die Evolution der Nemathelminthen. Cuvillier, Göttingen.
Maas, A., Waloszek, D., Haug, J. T. & Müller, K. J. (2009): Loricate larvae (Scalidophora) from the Middle Cambrian of Australia. *Mem. Assoc. Austral. Palaeont.* 37: 281–302.
Neuhaus, B. (1993): Postembryonic development of *Pycnophyes kilensis* and *P. dentatus* (Kinorhyncha) from the North Sea. *Microfauna Marina* 8: 163–193.
Neuhaus, B. (1994): Ultrastructure of alimentary canal and body cavity, ground pattern, and phylogenetic relationships of Kinorhyncha. *Microfauna Marina* 9: 61–156.
Neuhaus, B., Kristensen, R. M. & Peters, W. (1997): Ultrastructure of the cuticle of Loricifera and demonstration of chitin using gold-labelled wheat germ agglutinin. *Acta Zool.* 78: 215–225.
Neuhaus, B. & Kristensen, R. M. (2007): Ultrastructure of the protonephridia of larval *Rugiloricus* cf. *cauliculus*, male *Armoloricus elegans*, and female *Nanaloricus mysticus* (Loricifera). *J. Morphol.* 268: 357–370.
Nielsen, C. (2001): *Animal evolution. Interrelationships of the living phyla, Second Edition.* Oxford University Press, Oxford.
Nielsen, C. (2012): *Animal evolution. Interrelationships of the living phyla, Third Edition.* Oxford University Press, Oxford.
Park, J. -K., Rho, H.S., Kristensen, R. M., Kim, W. & Giribet, G. (2006): First molecular data on the phylum Loricifera – an investigation into the phylogeny of Ecdysozoa with emphasis on the positions of Loricifera and Priapulida. *Zool. Sci.* 23: 943–954.
Peel, J. S. (2010): A corset-like fossil from the Cambrian Sirius Passet Lagerstätte of North Greenland and its implications for cycloneuralian evolution. *J. Paleont.* 84: 332–340.
Schmidt-Rhaesa, A., Bartolomaeus, T., Lemburg, C., Ehlers, U. & Garey, J. R. (1998): The position of the Arthropoda in the phylogenetic system. *J. Morphol.* 238: 263–285.
Sørensen, M. V., Hebsgaard, M. B., Heiner, I., Glenner, H., Willerslev, E. & Kristensen, R. M. (2008): New data from an enigmatic phylum: evidence from molecular sequence data supports a sister-group relationship between Loricifera and Nematomorpha. *J. Zool. Syst. Evol. Res.* 46: 231–239.
Todaro, A. & Kristensen, R. M. (1998): A new species and first report of the genus *Nanaloricus* (Loricifera, Nanaloricida, Nanaloricidae) from the Mediterranean Sea. *Ital. J. Zool.* 65: 219–226.

Index

AURC021716240818
301890AU00009B/624
9783110219388